# Observing Global Climate Change

# Observing Global Climate Change

K. YA. KONDRATYEV

*St Petersburg, Russia*

A. P. CRACKNELL

*Dundee, UK*

TAYLOR & FRANCIS
ALERE FLAMMAM
1798 – 1998

**UK** Taylor & Francis Ltd, 1 Gunpowder Square, London EC4A 3DE
**USA** Taylor & Francis Inc., 1900 Frost Road, Suite 101, Bristol, PA 19007



**British Library Cataloguing in Publication Data**

A catalogue record for this book is available from the British Library

ISBN 0 7484 0124 5 (cased)

**Library of Congress Cataloging Publication Data are available**

Cover design by Jim Wilkie

Typeset in Times 10/12pt by Santype International Ltd, Salisbury, UK.

Printed in Great Britain by T. J. International, Padstow, UK.

# Contents

# Preface

The question of anthropogenic effects on the Earth's climate is a topic of intense interest in the worlds of science, politics, and human affairs. For any informed social and political discussion of this topic, it is necessary to draw on the conclusions of appropriate scientific research. It is now clear that one cannot consider anthropogenic effects in isolation; one must simultaneously consider natural changes in the climate. In order to gain an understanding of the nature of climate change, one must thus consider the whole climatic system – the atmosphere, the biosphere, the hydrosphere, the lithosphere and the cryosphere, in all their complexity – and the ways in which natural and anthropogenic activities act and influence that system.

To achieve this understanding, one must develop fully both observations and numerical modelling. There is a fundamental difficulty with the former: a combined system of satellite and conventional (*in situ*) observations is required to obtain global-scale representative data sets for sufficiently long periods of time. In this context, the first meteorological satellite, TIROS-1, was only launched on 1 April 1960, and it has since taken considerable time and effort to improve the quality of data and to broaden the range of parameters retrieved from satellite measurement using the various techniques of remote sensing; this effort is of course still going on today. The World Meteorological Organization has recently decided to develop a Global Climate Observing System (GCOS), which, it is proposed, and if funding permits, will include components such as a Global Ocean Observing System (GOOS) and a Global Terrestrial Observing System (GTOS). It will be some time before a sufficiently complete long-term series of important climatic parameters can be obtained via the combination of conventional and satellite observing systems. We note, in this context, that the first phase of the climate-oriented Earth Observing System (EOS) Programme is likely to begin with the new millennium.

The understanding of climate change can only come from an intensive study of the fundamental physics of the processes involved in the climatic system's behaviour. Problematically, the set of equations that one obtains, and the details of the boundary conditions that these equations must satisfy, are so complicated that an analytical solution of the equations is not feasible; one must resort to numerical techniques. Thus the development of climate predictions or forecasting involves the use of simulation modelling, which despite its recent rapid development is not yet mature. Even the most sophisticated and complex three-dimensional global climate models are still able to simulate only certain basic climate features, because they are frequently incomplete with regard to the consideration – the parametrization – of numerous climatically-significant processes. On the other hand, substantial progress has been made in the assessments of the nature of climate changes, including anthropogenic impacts, during the preparation of reports for the Intergovernmental Panel on Climate Change (IPCC) (Houghton *et al.* 1990, 1992). Equally valuable were the initatives within the World Climate Research Programme (WCRP) and the International Geosphere–Biosphere Programme (IGBP). Finally, we should note the outstanding contribution made by the Second United Nations Conference on the Environment and Development, convened in Rio de Janeiro, Brazil, in June 1992.

It is in this context that we have written this book. We have set out to combine considerations of physical principles and the results of observation and numerical modelling. Whilst we have emphasized recent research, we have not ignored earlier studies of significance. We have used as a foundation K. Ya. Kondratyev's *Global Climate*, published in Russian in St Petersburg, Russia, in 1992. We have completely revised the text, drawing on material from the IPCC reports, particularly the Scientific Assessment produced by the first of the IPCC's three Working Groups, without seeking to reproduce this material *in toto*, as, although these reports provide an overview and summary of knowledge, they deviate from our concern with the fundamental principles of naturally- and anthropogenically-induced climate change, and they are easily accessible to the interested reader. We worked on the preparation for this book through 1993 and 1994; in early 1996 we included a brief allusion to the IPCC report *Climate Change 1995* (Houghton *et al.* 1996).

We are confident that this book will be useful not only for professionals but also for any person wishing to inform their understanding of the problems of climate change.

K. YA. KONDRATYEV AND A. P. CRACKNELL
Athens, St Petersburg and Dundee
1994–1996

# Acknowledgements

We should like to acknowledge the hospitality of those institutions who helped us to consult references and sources, or to consult each other, in particular the Climate Research Unit of the University of East Anglia, the Library of University College London, and the Laboratory of Meteorology within the Department of Applied Physics at the University of Athens.

We should also like to thank Mrs A. P. Kostrova for her translation work on the original Russian text; and Helen Piccou and Lora Sharples of Taylor & Francis Ltd for their creation – and partial rescue! – and revision of the electronic version of the manuscript, from which this book has been typeset.

# Abbreviations

| | |
|---|---|
| ABL | atmospheric boundary layer |
| ACC | antarctic circumpolar current |
| ACP | Applied Climatology Programme |
| ACRIM | a multi-angle spectro-radiometer |
| ADEOS | Advanced Earth Observation Satellite |
| AGC | atmospheric general circulation |
| AGCM | atmospheric general circulation model |
| AIRS | Atmospheric Infra-red Sounder |
| ALT | a radar altimeter |
| AMSU-A | Advanced Microwave Sounder |
| AOT | aerosol optical thickness |
| APE | available potential energy |
| ASR | absorbed solar radiation |
| ASTER | Advanced Radiometer to Measure Thermal Emission/Reflection |
| ATSR | Along-Track Scanning Radiometer |
| AVHRR | Advanced Very High Resolution Radiometer |
| BATS | Biosphere–Atmosphere Transport Scheme |
| CCM | Community Climate Model |
| CDIC | Carbon Dioxide Information Center |
| CDP | Climatic Data Programme |
| CERES | Clouds and the Earth's Radiant Energy System |
| CFCs | chlorofluorocarbons |
| CHAMMP | Computer Hardware Advanced Mathematics & Model Physics |
| CN | condensation nuclei |
| COADS | comprehensive ocean data set |
| COSPAR | Committee on Space Research |
| CRF | cloud-radiation forcing |
| CSMP | Climate System Modelling Programme |
| CTT | cloud top temperature |
| CZCS | Coastal Zone Color Scanner |
| d | day |
| DMS | dimethyl sulphide |
| DMSP | Defense Meteorological Satellite Program |
| DMSPSSM | Defense Meteorological Satellite Program Special Sensor Microwave |
| DTE | downward thermal emission |
| DVI | dust veil index |

| | |
|---|---|
| EAZO | energetically active zones of the ocean |
| EC | extinction coefficient |
| ECMWF | European Centre for Medium Range Weather Forecasts |
| ENSO | El Niño / Southern Oscillation |
| EOS | Earth Observing System |
| EOSP | a scanning spectro-polarimeter |
| ERB | Earth radiation budget |
| ERBE | Earth Radiation Budget Experiment |
| ESA | European Space Agency |
| ESC | extended stratus cloudiness |
| ESMR | Electronic Scanning Multi-channel Radiometer |
| FGGE | First GARP Global Experiment |
| FIFE | First ISLSCP Field Experiment |
| flops | floating-point operations per second |
| GAAREX | Global Atmosphere Aerosol Radiation Experiment |
| GAC | global area coverage |
| GARP | Global Atmospheric Research Programme |
| GAW | Global Atmosphere Watch |
| GCIP | GEWEX Continental Scale International Project |
| GCM | general circulation model |
| GCOS | Global Climate Observing System |
| GEWEX | Global Energy and Water Cycle Experiment |
| GFDL | Geophysical Fluid Dynamics Laboratory |
| GG | greenhouse gases |
| GGI | instruments for coordinate and geodetic information |
| GISS | Goddard Institute for Space Studies |
| GLAS | Goddard Laboratory for Atmospheric Sciences |
| GLRS-A | a laser altimeter |
| GLRS-R | a geodynamic laser range sounder |
| GOES | Geostationary Observational Earth Satellite |
| GOEZS | project to study processes in the euphotic zone |
| GOOS | Global Ocean Observing System |
| GOS | geomagnetic complex instrument |
| GPCP | Global Precipitation Climatology Project |
| Gt | gigatonne |
| GTOS | Global Terrestrial Observing System |
| HAPEX | Hydrological Atmospheric Pilot Experiment |
| HIRDLS | a high-resolution limb sounder |
| HIRIS | a high resolution video-spectrometer |
| HIRS-2 | a high-resolution infra-red spectrometer |
| hPa | hectopascal (100 Pa = 1 mbar) |
| ICSU | International Council of Scientific Unions |
| IGBP | International Geosphere–Biosphere Programme |
| IPCC | Intergovernmental Panel on Climate Change |
| IPEI | an instrument to measure characteristics of ionosphere plasma |
| IRSR | Infra-red Scanning Radiometer |
| ISCCP | International Satellite Cloud Climatology Project |
| ISLSCP | International Satellite Land Surface Climatology Project |
| ITCZ | Inter-tropical Convergence Zone |
| IWC | ice water content |
| IWP | ice water path |
| JERS-1 | Japanese Earth Resource Satellite |
| JGOFS | Joint Global Ocean Flux Study |
| K | kelvin (used for both points and differences on the kelvin temperature scale) |

| | |
|---|---|
| kt | kilotonne |
| LAC | local area coverage |
| LAWS | lidar wind sounder |
| LIS | a lightning video sensor |
| LWC | liquid water content |
| MGACs | minor gaseous and aerosol components |
| MGCs | minor gaseous components |
| MHS | Microwave Humidity Sounder |
| MIMR | Multi-spectral Video Microwave Radiometer |
| MISR | Multi-angle Spectro-radiometer |
| MLS | microwave limb sounder |
| MODIS-N | a mid-resolution nadir-viewing video-spectrometer |
| MODIS-T | a slant-axis, mid-resolution video-spectrometer |
| MOPITT | instrument to measure tropospheric pollution |
| MOS-1 | Japanese oceanographic satellite |
| MSD | mean standard deviation |
| MSR | multi-channel scanning radiometer |
| MSU | Microwave Sounding Unit |
| Mt | megatonne |
| NCAR | National Center for Atmospheric Research |
| NDVI | normalized difference vegetation index |
| NH | northern hemisphere |
| NOAA | National Oceanic and Atmospheric Administration |
| NPP | net primary productivity |
| OGC | ocean general circulation |
| OLR | outgoing longwave radiation |
| OSU | Oregon State University |
| ppb(v) | parts per billion (by volume) |
| ppm(v) | parts per million (by volume) |
| ppt(v) | parts per trillion (by volume) |
| RH | relative humidity |
| SAFIRE | a spectrometer |
| SAGE | Stratospheric Aerosol and Gas Experiment |
| SAR | synthetic aperture radar |
| SAT | surface air temperature |
| SBUV | solar backscattered ultraviolet |
| SC | solar constant |
| SeaWiFS | an ocean colour radiometer |
| SH | southern hemisphere |
| SIRS | Satellite Infra-red Spectrometer |
| SLR | side-looking radar |
| SMM | (1) Stratospheric Measurement Mission<br>(2) Solar-Maximum Mission |
| SMMR | Scanning Multi-channel Microwave Radiometer |
| SOLSTICE | an instrument to measure extra-atmospheric UV solar radiation |
| SPOT | Système pour l'Observation de la Terre |
| SRB | surface radiation budget |
| SSM/I | DMSP Special Sensor Microwave Imager |
| SST | sea surface temperature |
| SSU | Stratospheric Sounding Unit |
| STICKSAT | a scatterometer |
| SWIRLS | an infra-red limb wind sounder |
| TES | an emission spectrometer to sound the troposphere |

| | |
|---|---|
| TIMS | a thermal video-spectrometer |
| TLR | temperature lapse rate |
| TM | Landsat thematic mapper |
| TOA | top of the atmosphere |
| TOC | total ozone content |
| TOGA | Tropical Ocean and Global Atmosphere |
| TOMS | Total Ozone Monitoring Spectrometer |
| TOPEX/ POSEIDON | a French/American satellite |
| TRMM | Tropical Rain Measuring Mission |
| UARS | Upper Atmosphere Research Satellite |
| UCAR | University Corporation on Atmospheric Research |
| UKMO | UK Meteorological Office |
| UNEP | United Nations Environment Programme |
| VISR | Visible Infra-red Scanning Radiometer |
| WC | water content |
| WCAP | World Climate Application Programme |
| WCC-1 | First World Climate Conference |
| WCC-2 | Second World Climate Conference |
| WCP | World Climate Programme |
| WCRP | World Climate Research Programme |
| WMO | World Meteorological Organization |
| WOCE | World Ocean Circulation Experiment |
| WWW | World Weather Watch |
| XIE | an X-ray video-spectrometer |
| yr | year |
| ZAPE | zonal available potential energy |

CHAPTER ONE

# Introduction

## 1.1 Background

Substantial changes in the environment caused by human industrial activities, and their possible climatic impact, have become matters of very great concern, because variations in climate can seriously affect agricultural productivity and many other aspects of human activity. Serious attention has recently been given, for example, to the possible effects of the increase of $CO_2$ and other greenhouse gases in the atmosphere. The influence of nuclear tests on weather and climate has also been widely discussed (Crutzen and Birks 1982, Turco *et al.* 1983, Harwell 1984, London and White 1984, US National Research Council 1985, Bach 1986, 1987, Berger and Labeyrie 1987, Ginzburg 1988, Kondratyev 1988b, Pittock *et al.* 1989).

A study is therefore urgently required of the physical and chemical factors of climate and its changes, including further developments of climate theory, and, in particular, a determination of the relative importance of the contributions of natural causes and of anthropogenic effects to present-day climate changes. Of primary importance are forecasts of future climate that take account of anthropogenic effects (Budyko 1985). However, while the objective of much of modern climatological work is to predict the effects on the climate caused by various human activities, it is not possible to study the effects of these human activities directly. One cannot ignore the natural factors that are constantly causing climate changes irrespective of those human activities. Attempts at modelling the climate system are thus based on the idea of predicting the climate change due solely to natural factors, then of introducing the anthropogenic effects in addition to the natural factors. The effects, in terms of changes in such quantities as mean temperature or mean annual precipitation, duc to human activity, are then determined by the subtraction of the results of two such model calculations.

There are numerous interacting factors causing climate change, which have a complex system of interactions in the climate system:

- atmosphere
- biosphere
- hydrosphere

- lithosphere
- cryosphere.

This complexity presents the principal difficulty that occurs in formulating an adequate theory of climate and in identifying the most important factors responsible for its variations (Hunt 1981, Houghton 1984, 1986, Flöhn 1986, 1987, 1988, Kondratyev 1986a–c, 1988b, 1992b, Washington and Parkinson 1986, Henderson-Sellers 1987, Trenberth 1992, Cracknell 1994a, Vaughan and Cracknell 1994, Kondratyev *et al.* 1996a).

The present level of parametrization of the processes in the climate system is such that climate theory can only be a first approximation to understanding the laws of formation of climate and its changes. For this reason, discussions are still continuing on the nature of climate (Borisenkov and Kondratyev 1988). In this regard, special attention is given to discussion of the problem of climate predictability. Studies of the Earth's palaeoclimates, the climatic features of other planets (Goody and Walker 1972, Berger 1980, Kandel 1980, Kondratyev and Hunt 1982, Henderson-Sellers 1983, Lewis and Prinn 1984, Hunt 1985, Bernhardt 1987, Berger *et al.* 1990), and the specific character of the climate of large industrial cities determined by atmospheric pollution and transformed surface characteristics are all of great importance for understanding present-day climate changes.

Figures 1.1 and 1.2 illustrate a variety of spatial and temporal scales of climate-forming phenomena and processes, while Figure 1.3 characterizes the temporal variability of climate.

Climatic predictability is the key aspect of the problem under discussion. The ability to analyse and forecast climate depends on understanding that the variability of seasonal mean characteristics of climate is caused by short-term internal fluctuations of the climatic system (representing unpredictable climatic noise) and also by long-term fluctuations (the climatic signal).

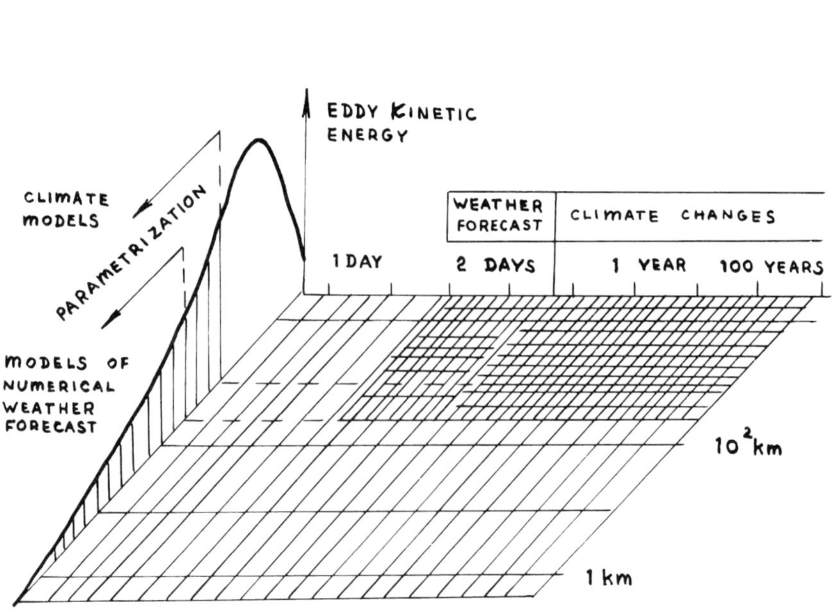


**Figure 1.1** The spatial and temporal scales of the weather- and climate-forming processes.

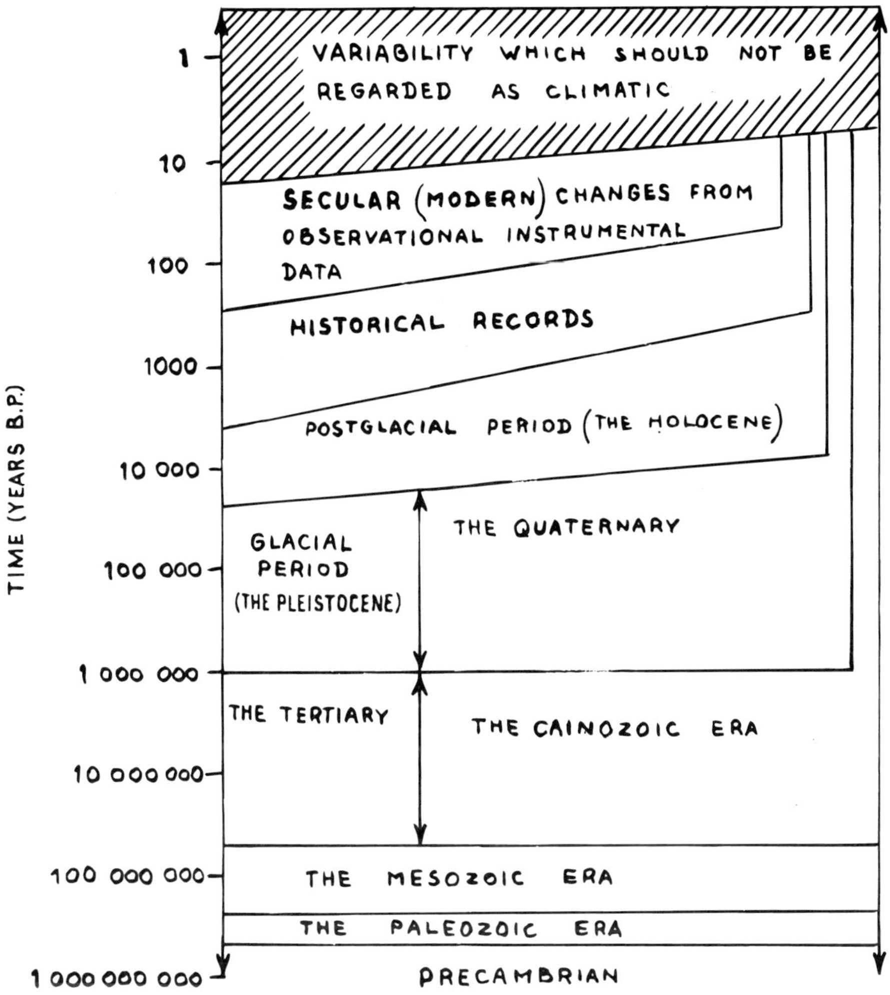


**Figure 1.2** Palaeoclimatic changes.

It is perhaps convenient to distinguish between weather and climate. One finds in practice that weather forecasts are reliable for a period of a few days ahead, that the temporal extent of their reliability is improving, but that the ultimate limit of their reliability may be of the order of 2 to 4 weeks. This can be discussed in terms of determinism versus chaos. One is therefore entitled to ask whether there is any hope of ever being able to make meaningful climate forecast predictions over the scale of a century or two if one has no hope of being able to make reliable weather forecasts for a time such as only 6 or 12 months ahead. This question is particularly pertinent if one takes the numerical modelling approach to climate prediction where one is using general circulation models of the atmospheric ciculation which have been derived from weather forecast models and which are based on the same physical mechanisms and equations as are used in the weather forecast models. General circulation models are based on the physical conservation laws which describe the

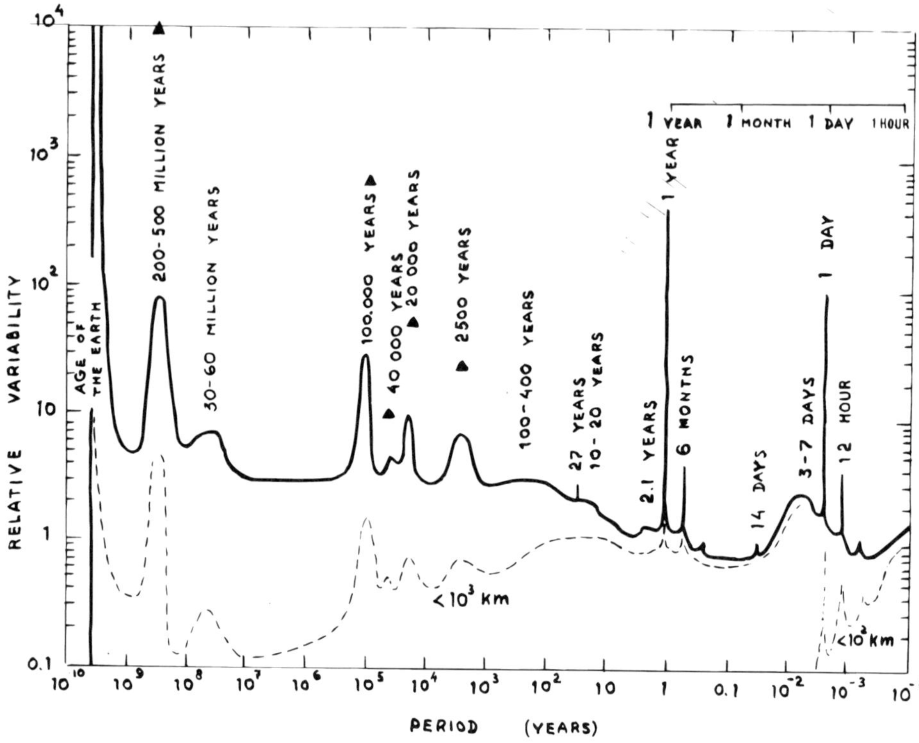


**Figure 1.3** Typical climatic variability.

redistribution of momentum, heat and water vapour by atmospheric motions. All of these processes are formulated in the 'primitive' equations, which describe the behaviour of a fluid (air or water) on a rotating body (the Earth) under the influence of differential heating (the temperature contrast between equator and pole) caused by an external heat source (the Sun). These governing equations are non-linear partial differential equations, whose solution cannot be obtained except by numerical methods. If it is surprising that any success can be achieved with climate models, then it may help to appreciate this if we recall that in looking at climate we are looking at an average of the weather over a reasonably long time, e.g. 1 yr or 10 yr. The chaotic elements, or the statistical fluctuations, in the climate are the weather systems. In looking at climate change we are looking at long-term changes in this average. The long-term fluctuations are potentially predictable, since they are determined either by different external forcing (sea surface temperature anomalies, soil moisture variations, etc.) or by internal atmospheric dynamics. Viewed over a century or a millennium we find that climatic parameters (temperature, rainfall, etc.) are basically stable and vary only slowly.

It is the nature of these slow, long-term variations that are of concern to us in climate studies after the local, short-term fluctuations have been smoothed out. It is the stability of the long-term components that make climate prediction possible. In a climate model we consider the effect of various forcing conditions on a given mean state. Analogies are always dangerous and should not be pushed too far. However, let us draw an analogy with the behaviour of a gas. We can use very successfully an equation of state for a gas to determine the effect of a change in pressure, volume or temperature on the gas. But if one were to seek to study the behaviour of the individual molecules in the gas and try to make predictions of the path of any given molecule, one would be faced with an extremely difficult problem. Moreover, even if one could find the solution, it would probably not be very interesting or useful in terms of enabling one to understand the behaviour of the gas as a whole.

In spite of randomness or of chaos at the small-scale level, one can nevertheless make useful and meaningful calculations at the macroscopic level based on a transition from one mean (or quasi-equilibrium) state to another. Thus we commonly assume in climate prediction that the climate system is in equilibrium with its forcing. That means, as long as its forcing is constant and the slowly varying components alter only slightly in the time scale considered, the mean state of the climate system will be stable and that if there is a change in the forcing, the mean state will change until it is again in balance with the forcing. The time scale of the transition period until an equilibrium state is re-established is determined by the adjustment time of the slowest climate system component, i.e. the ocean. The stable ('quasi-stationary') behaviour of the climate system gives us the opportunity to detect changes by taking time averages. Because the internal variability of the system is so high, the averaging interval has to be long compared with the chaotic fluctuations to detect a statistically significant signal which can be attributed to the external forcing.

Studies of the completed change from one mean state to another are called equilibrium response studies. Studies of the time evolution of the climate change due to an altered forcing, which might also be time dependent, are called transient response experiments.

## 1.2 The factors involved in climate change

We shall now consider the principal factors and processes that determine climate changes. The variables which are commonly used in studying the climate are concerned mainly with the atmosphere. However, we cannot look at the atmosphere alone. This is because processes in the atmosphere are strongly coupled to the oceans, to the land surface and to the parts of the Earth which are covered with ice, i.e. the cryosphere. There is also strong coupling to the biosphere, i.e. to the vegetation and other living systems on the land and in the sea. It is convenient to consider three categories of events that may affect the climate:

1. Events that occur outside the Earth;
2. Natural events on the surface of, or within, the Earth;
3. Human activities.

### 1.2.1 Events that occur outside the Earth

The source of virtually all the energy that drives the atmosphere is the Sun. Thus any changes in the intensity of the solar radiation that is incident on the atmosphere from above will clearly have a significant effect on the weather and on the climate. There is a variety of possible causes of changes in the intensity of this radiation. The amount of energy emitted by the Sun itself may vary, the transmission of the radiation through space from the Sun to the Earth may change since space is not absolutely empty, and the separation and the relative orientation of the Sun and the Earth may change. Let us consider these possibilities in turn.

First there is the variation of the Sun's radiation itself. There are regular periodic variations in the Sun's behaviour; these include (a) pulsation with a period of about 2 hours 40 minutes, (b) rotation on its axis every 27 days and (c) sunspot activity with an 11-yr cycle. Tickell (1977), perhaps seeing things as a historian, has a nice comment on sunspots: 'The connection between sunspots and the weather on Earth is part of universal folklore and cannot easily be dismissed, although the means by which one could affect the other has not been convincingly demonstrated'. He goes on to relate that between 1645 and 1715, the coldest part of the Little Ice Age, was a period of very low sunspot activity, that the estimated intensity of the solar radiation was about 1 per cent lower than its present value, and that this is consistent with a temperature of about 1 °C or so less than at present. It is important to have accurate measurements of the intensity of solar radiation (the solar 'constant') over a period of years but, like all other measurements of parameters related to climatology, it has to be remembered that in climatological terms any period over which we are able to acquire measurements is usually short. Ground-based measurements of the solar constant have been made for a century or so, but it is only within the last 20 years that accurate measurements of the radiation output of the Sun have been made by spacecraft which operate outside the Earth's atmosphere, i.e. by remote sensing. These satellite data do not show any substantial long-term variations in the value of the solar constant. Its short-term variations (on a time scale from seconds to months) do not exceed a few tenths of one per cent. Some studies show correlations between solar activity and climate (Eddy 1977, Labitzke 1987, ICSU 1988, Kondratyev 1989c, Reid 1991, Labitzke and van Loon 1992, Stanhill and Moreshet 1992) but, as yet, no convincing explanations of the physical and chemical mechanisms have been suggested by which solar activity could affect climate. We shall consider the question of the variation of the value of the solar constant in more detail in Section 4.7. The problem of an impact of solar activity on climate has recently been discussed by Kondratyev and Nikolsky (1995a,b).

Secondly, the radiation travelling from the Sun to the Earth may encounter dust clouds and thus undergo some attenuation. Or showers of micro-meteors may enter the Earth's atmosphere causing dust and ice particles in the stratosphere; these then act as a screen and prevent some of the solar radiation from reaching the Earth.

Thirdly, there are the variations in the relative positions and orientations of the Sun and the Earth; this in turn involves three factors. First, the eccentricity of the elliptical orbit of the Earth around the Sun varies with a period of about 100 000 yr. At present the intensity of the solar radiation received at the Earth varies by about 7 per cent during the year, but when the eccentricity of the Earth's orbit is greatest the intensity of the solar radiation received at the Earth varies by up to 30 per cent during the year. A second source of variation arises from the oscillation of the tilt of the Earth's axis of rotation relative to the plane of its orbit; this oscillation has a period of about 40 000 yr. The greater the tilt, the greater will be the difference between summer and winter (the last maximum was 10 000 yr ago). Thirdly, there is the precession of the equinoxes, i.e. the variation in the time of the year when the night and day are of equal length. This arises from the wobble of the Earth around its axis and has a period of 21 000 yr. These three variations clearly have an important effect on the intensity of the solar radiation that reaches the Earth. It was first pointed out by Croll (1867), whose ideas were later developed by Milankovitch (1920), that the major glacial–interglacial cycles over the past few hundred thousand years might be linked with these regular variations in the distribution of solar radiation reaching the Earth. There is a quite impressive correlation between low intensity of solar radiation and large volumes of ice on the Earth's surface, and this provides strong evidence to support the Milankovitch theory.

### 1.2.2 Natural events on the Earth

Most of the natural events on the surface of the Earth that affect the climate will be considered in Sections 1.3.1 to 1.3.5. However, one or two things not covered in those sections will be mentioned here because, for one reason or another, they are not generally included in climate modelling. These include plate tectonic movements and volcanic eruptions.

It is as well to recall that we are, at the moment, not just concerned with short-term climate changes attributable to human activities but, rather, with changes arising from the whole history of the Earth. Thus, although to us the motions of the various tectonic plates (continental drift) on the surface of the Earth may seem small and of little immediate relevance to climate change, they are – over long periods of time – very important. The transfer of energy between the surface of the Earth and the atmosphere depends on whether that surface is ocean, ice or land and, if it is land, on the nature and elevation of the land surface and land cover. Thus the changes in the distribution of the land surfaces and ocean areas of the Earth arising from the movements of the tectonic plates will cause changes in the distribution of the input of energy into the atmosphere. Another aspect of crustal movements is associated with variations in the gravitational effect of the Sun and the Moon on the surface of the Earth. Apart from causing the familiar ocean tides, these effects give rise to stresses on the land surface as well, so that there is a greater chance of volcanic eruption when these stresses are at their maximum. Another factor that needs to be considered in long-term studies is the effect of the variation, and indeed the reversal, of the Earth's magnetic field on incoming radiation.

In discussing extra-terrestrial causes of changes in the intensity of solar radiation that reaches the surface of the Earth we mentioned dust clouds and meteor showers. Volcanic eruptions produce similar effects to those of showers of micro-meteors, though the origin of the particles is terrestrial rather than extra-terrestrial. A volcano blasts large quantities of matter into the sky during an eruption. Some of this reaches the stratosphere as dust particles around which ice can form. Dust is produced in the atmosphere by many causes other than volcanoes, e.g. by strong winds over desert areas. But the quantities arising from major volcanic eruptions are much larger than the quantities arising from other causes. The fate of volcanic dust depends on the latitude of the eruption. A belt of dust particles forms a screen around the Earth at the latitude of the erupting volcano and then widens out in a north–south direction as well. If the erupting volcano is near to the equator the screen spreads out over the whole Earth towards both poles; if the volcano is more than about 20° N or 20° S from the equator the stratospheric screen tends to be confined to one hemisphere. This screen reduces the intensity of solar radiation reaching the surface of the Earth and also reduces the intensity of the terrestrial radiation escaping to outer space; however, the former effect is dominant. Thus the general effect of a volcanic eruption is to lead to a cooling of the Earth. We shall discuss the question of the effect of volcanoes on the climate in Section 6.3.

### 1.2.3 Human activities

#### 1.2.3.1 *Carbon dioxide ($CO_2$)*

The human activity of which we are most aware, in common culture, in terms of its effect on the global environment is the production of $CO_2$ by the burning of fossil fuels. The concentration of $CO_2$ in the atmosphere has risen from about 280 ppm at the time of the Industrial Revolution to over 350 ppm at the present time; however in earlier times the concentration of $CO_2$ was by no means constant and had in fact varied quite widely. At present the concentration of $CO_2$ in the atmosphere is rising fast, not only because of the rapid increase in the consumption of fossil fuels but also because of the destruction of forests, thereby reducing the removal of $CO_2$ from the atmosphere by photosynthesis.

The main effect of increasing the concentration of $CO_2$ in the atmosphere is to change the balance between incoming radiation from the Sun and outgoing radiation from the Earth. The $CO_2$ acts like the glass or transparent plastic in a greenhouse. The bulk of the incoming solar radiation is in the visible wavelength range. After interaction with the Earth's surface much of the energy is converted to longer-wavelength (infra-red) radiation which is blocked by the $CO_2$ (or by the glass or transparent plastic in a greenhouse). The greater the concentration of $CO_2$, the greater the amount of infra-red radiation that is radiated back to the Earth's surface. The $CO_2$ thus warms the lower atmosphere and cools the upper atmosphere. This in turn increases moisture, which traps more infra-red radiation. The effect of rising temperature is to reduce the areas covered by snow and ice, thereby diminishing the amount of heat reflected back into space and increasing the absorption of solar radiation. $CO_2$ is not the only gas contributing to the greenhouse effect in the atmosphere but it is the most important one, after water vapour. We shall discuss the role of $CO_2$ in connection with climate at various later stages in this book.

#### 1.2.3.2 *Ozone ($O_3$)*

Probably the second-best-known effect of human activity on the atmosphere is the depletion of the ozone layer. Basically ozone ($O_3$) is created by the action of sunlight on ordinary oxygen molecules; ozone is destroyed by various natural processes. However, ozone is also being destoyed as a result of human actvities, by oxides of nitrogen and by the decomposition products of chlorofluorocarbons (CFCs, principally $CCl_3F$ and $CCl_2F_2$). The importance of $O_3$ is that in the stratosphere it absorbs ultraviolet from the incoming solar radiation. Without the ozone layer the ultraviolet intensity at ground level would be so high as to render life impossible for many species, including *Homo sapiens.* Even a modest reduction in the concentration of the ozone layer would lead to significant increases in skin cancers and to various other medical problems as well. The other effect of $O_3$ in the stratosphere is that, by absorbing ultraviolet radiation, it increases the temperature of the stratosphere. Thus a reduction in $O_3$ concentration would lead to a lowering of temperature in the stratosphere, with consequential effects on weather and climate.

#### 1.2.3.3 *Development of land areas*

As a result of media coverage, it is the increasing concentration of $CO_2$ in the atmosphere and the depletion of $O_3$ in the stratosphere that spring most immediately to mind when we think about possible effects of human activities on weather and the climate. A few decades ago, at the time of extensive above-ground nuclear testing programmes, we were concerned about the effect of those nuclear explosions on the weather and, possibly, the climate (see Section 1.4) But there are several other factors which should be considered and which, in the long term, may be very important. Tropical deforestation, in a number of developing countries, is often referred to in this context. However, human beings and domestic animals have been affecting the surface of the Earth for a long time.

Tickell (1977) quotes something like 20 per cent of the total area of the continents as having been drastically changed. The cutting down of trees for settlement and agriculture, the slash-and-burn method of cultivation in primitive societies, the overgrazing of land by such animals as goats and cattle, and the overuse, impoverishment and erosion of topsoils have in the past affected the heat and water balance in various areas. The amount of solar radiation absorbed by the grass, crops, or in some cases desert land, is less than that absorbed by the forests which were there before. More energy is reflected back into space, less moisture is evaporated and so less rain is produced and, of the rain which does come, more is run off the land. The inevitable increase of dust blown up from the surface can produce effects similar to those of volcanic eruptions or sandstorms.

From the point of view of food production, or general agricultural production, the most important effects have been a decline in rainfall in some areas. Tickell, again taking the Western historian's approach, cites the progressive aridity in historical times of the swathe of land from the Mediterranean to northern India. This area was once covered by dense forest and later became the site of successive civilizations; its increasing aridity seems to have been caused mostly by human destruction of the natural environment. Most of the grain-growing parts of the area which once supported large populations, including those of the Roman Empire, are now largely scrub or desert. Estimates are that about 7 per cent of the Earth's surface is man-made desert, and the process of progressive desertification is continuing. This inevitably contributes to some extent to changing weather and climate.

In more recent times other human activities have become important. For instance there are now situations in which human intervention occurs to modify the weather rather directly, for instance in the case of rainmaking by cloud seeding. The short-term local effects may be desired and beneficial. The effects elsewhere, or in the longer term, may be less obvious and may not necessarily be beneficial. Another human influence is the interference with the hydrology of the Earth. The creation of artificial lakes for irrigation or hydro-electric purposes will affect the local heat balance and the evaporation rate. Changing the flow of rivers for irrigation purposes, and the consequent change in land use and land cover, will also affect the local heat balance and evaporation rate as well. The development of large cities in recent years is also important. Large cities are well known to form heat islands with temperatures significantly higher than their surroundings. This arises from two factors: (i) the replacement of vegetated surfaces by more highly reflecting road and roof surfaces and (ii) the heat generated by human activities within the cities. Locally the cities' effects on the heat balance will affect the weather, although in global terms their contribution to the heat balance is relatively small.

#### 1.2.3.4 *Other greenhouse gases*

We have already mentioned the substantial increase in $CO_2$ concentration in the atmosphere, as a result of human activities, and the greenhouse effect associated with the atmospheric $CO_2$. In terms of concentration, $CO_2$ is by far the most important of the greenhouse gases, after water vapour. However, $CO_2$ is not the only important product of the combustion of fossil fuels. Other gases, principally $SO_2$ and oxides of nitrogen (denoted generally by $NO_x$) are produced by the combustion of fossil fuels. Some fuels are cleaner than others in terms of the pollution they produce. Thus gas and oil can be treated so as to remove sulphur before they are burned and therefore to prevent $SO_2$ being produced when they are burned; coal, being a solid, cannot be treated in this way before it is burned, and so expensive scrubbing facilities have to be used to remove the $SO_2$ from the gases produced by the combustion process. The production of oxides of nitrogen is more difficult to prevent because the nitrogen comes from the air used in the combustion process and not from the fuel.

The effect of $NO_x$ as far as $O_3$ is concerned actually appears to depend on the height in the atmosphere. At low altitudes the oxides of nitrogen favour the production of $O_3$ whereas at higher altitudes they lead to the destruction of $O_3$. Apart from their role as greenhouse gases, $SO_2$ and $NO_x$ are also undesirable products of combustion because through various reactions with water they produce mineral acids; these lead to acid rain which has catastrophic effects on the vegetation in some areas. As far as $SO_2$ is concerned, it should be pointed out that not all the gaseous sulphur compound emissions into the atmosphere arise from anthropogenic causes. $SO_2$ is liberated by volcanoes, $H_2S$ and DMS (dimethyl sulphide) are liberated by plants, and DMS is liberated by plankton in the ocean. It has been suggested that apart from producing acid rain the DMS emissions by plankton lead to sulphate particles that constitute hygroscopic nuclei for cloud drops and thereby significantly alter the cloud patterns and, thereby, the radiation balance (see Section 4.6).

#### 1.2.3.5 *Human activities and the greenhouse effect*

We have alluded to the greenhouse effect on several occasions in this section; the time has come to be a little more specific. First, from the point of view of life on Earth, the greenhouse effect is good, not bad. The average temperature of the surface of the Earth is 15 °C, and in the absence of any greenhouse gases in the atmosphere it would be reduced by 33 °C, i.e. to −18 °C. −18 °C may not sound too cold and many of us will have experienced that temperature briefly. But as an average temperature it would make the Earth a rather inhospitable place, and any semblance of modern agriculture would only be possible in a very few locations. In past ice ages the drop in average temperature has been less than 10 °C. However, what is worrying people at the moment is the possibility of a substantial global warming as a result of changes in the greenhouse effect induced by human activities. We have touched on many of these already in this section; they are the production of $CO_2$ and $N_2O$ (one of the oxides of nitrogen we denoted by $NO_x$) from combustion processes, and the release of CFCs – principally $CCl_3F$ (CFC-11) and $CCl_2F_2$ (CFC-12). In addition to this there are sulphur dioxide, $SO_2$, and methane, $CH_4$, which are produced by a whole variety of sources, some of which are natural and some of which are not. We shall see later that, for climate modelling, one simply represents all the greenhouse gases by an amount of $CO_2$ that gives rise to a warming equivalent effect to that of all the gases.

## 1.3 Natural factors affecting climate change

The interpretation of climatic changes centres on the effect of variations in atmospheric composition and relevant anthropogenic activity. Physical aspects of this problem include the influence of variations in the atmospheric composition on the greenhouse effect and on the radiative regime of the atmosphere, the contributions from cloudiness, aerosol and optically active minor gaseous components.

### 1.3.1 The atmosphere

The radiative regime of the atmosphere is known to be largely determined by such optically active components as water vapour, $CO_2$, $O_3$ and aerosols, while the greenhouse effect is thought to be one of the major mechanisms by which radiative factors influence climate. Most of the incoming solar radiation passes through the atmosphere and is absorbed by the Earth's surface (ocean, ice, rock, soil, vegetation, etc.). The evaporation of moisture and the direct heating of the surface lead to heat transfer between the surface and the atmosphere in the form of latent and sensible heat. The atmosphere transports this heat meridionally, mainly via transient weather systems with a timescale of the order of days.

The following factors are important in determining the behaviour of the atmospheric component of the climate system:

- turbulent transfer of heat, momentum and moisture at the surface of the Earth;
- the surface albedo (i.e. its reflectivity), which depends on the nature of the surface; this determines the proportion of reflected to incoming solar radiation;
- the release of latent heat when water vapour condenses; clouds, which play an important role in reflecting incoming solar short-wave radiation and in absorbing and emitting long-wave radiation;
- the radiative cooling and heating of the atmosphere by $CO_2$, water vapour, ozone and other trace gases;
- aerosols (such as volcanic dust), the orbital parameters, mountain ranges and the land–sea distribution.

These factors, however, are not all independent and, as we shall see later on, feedback between some of these processes and between these processes and the surface of the Earth is very important. The last two sets of items in this list of atmospheric processes include components that are strongly influenced by human activity. However, temperature changes induced by these two processes are likely to cause changes in the cloudiness and may therefore introduce further changes via the third process. This provides one example of the feedback which complicates climate studies. Let us suppose that an increase in concentration of $CO_2$, a greenhouse gas, occurs. This will lead to some global warming. As a result of this warming, the amount of water vapour in the atmosphere will increase. The water vapour is also a greenhouse gas and so we see that an increase in the concentration of one greenhouse gas ($CO_2$) leads to an increase in another greenhouse gas (water vapour) and thus provides a positive feedback mechanism. Calculations quoted in section 3 of chapter 3 of Houghton *et al.* (1990) indicate that a doubling of the concentration of $CO_2$ in the atmosphere would lead to an increase of 1.2 °C in the global-mean surface temperature, without feedback, and that if one includes water vapour feedback this would be increased to 1.9 °C.

The traditional approach to the greenhouse effect, and in particular to its variations, is based on an estimation of the contribution made by $CO_2$. One of the most popular hypotheses of climate change, both historically and at present, relates variations in temperature to those of $CO_2$ concentration, followed by variations in the greenhouse effect. However, the atmospheric absorption spectrum in transparency windows is known to be determined not only by $CO_2$ but also by such gaseous components as water vapour, $O_3$ and many other minor gases as well as by aerosols. From the viewpoint of the theory of climate changes, the leading role belongs to optically active components with long-term trends. In particular, chlorofluorocarbons (CFCs or freons) are atmospheric components of entirely anthropogenic origin which are steadily increasing in concentration.

To study the impact of the gaseous and aerosol components on the greenhouse effect, radiation fluxes for the models of the terrestrial atmosphere of different chemical compositions were calculated (Kondratyev and Moskalenko 1984, Wang and Mólnar 1985, Wang *et al.* 1986, 1991, Wuebbles and Edmonds 1988). The calculation schemes took into account many components affecting the absorption of radiation in the Earth's atmosphere, including water vapour, $CO_2$, $N_2$, $O_2$, $CH_4$, $NO_x$, CO, COS, $SO_2$, nitric acid vapour, ethylene, acetylene, ethane, formaldehyde, chlorofluorocarbons, ammonia and aerosol formations of different chemical composition and size distribution. Tables 1.1 and 1.2 illustrate the greenhouse composition of the atmosphere and its changes as well as relevant radiative effects.

Atmospheric chemistry plays a very important role in climate studies. Studies of changes in atmospheric chemical composition and associated chemical reactions are of particular concern (Wuebbles *et al.* 1989). The reliability of assessing the anthropogenic impact on the atmospheric chemical composition, and the resulting environmental changes, is strongly limited by inadequately studied physical, chemical and biological processes that determine the properties of the global environment. It is important to realize that on a time scale of decades, the atmosphere, the world's oceans and the continents are a single system. Physical and chemical processes in the atmosphere, being subject to an increasing anthropogenic impact, are closely involved in the formation of biogeochemical cycles that govern the living conditions on the planet.

**Table 1.1.** Greenhouse gases

| Gas | Common name | Surface concentrations[a] (ppmv) | Atmospheric trend[a] (%/yr) | Atmospheric lifetime (yr) | Primary man-made sources |
|---|---|---|---|---|---|
| $CO_2$ | Carbon dioxide | 345 | ~0.4 | ~500 (air–biosphere–oceans) | Fossil fuel burning; land use conversion |
| $CH_4$ | Methane | 1.7 | ~1 | ~7–10 | Domestic animals; rice paddies; biomass burning; gas and mining leaks |
| CO | Carbon monoxide | 0.12 | ~1–2 | ~0.4 | Energy use; agriculture; forest clearing |
| $N_2O$ | Nitrous oxide | 0.31 | ~0.3 | ~150 | Fossil fuel burning; cultivation and fertilization of soils |
| $NO_x$ ($=NO + NO_2$) | | $1–20 \times 10^{-5}$ | unknown | $\leqslant 0.02$ | Fossil fuel burning; biomass burning |
| $CFCl_3$ | CFC-11 | $2.0 \times 10^{-5}$ | ~5 | ~75 | Chemical industry product |
| $CF_2Cl_2$ | CFC-12 | $3.2 \times 10^{-4}$ | ~5 | ~110 | As above |
| $C_2Cl_3F_3$ | CFC-113 | $3.2 \times 10^{-5}$ | ~10 | ~90 | As above |
| $CH_3CCl_3$ | Methyl chloroform | $1.2 \times 10^{-4}$ | ~5 | ~6–9 | As above |
| $CF_2ClBr$ | Ha-1211 | $1 \times 10^{-6}$ | ~10–30 | ~12–15 | Fire extinguishers |
| $CF_3Br$ | Ha-1301 | $1 \times 10^{-6}$ | unknown | ~110 | Fire extinguishers |
| $SO_2$ | Sulphur dioxide | $1–20 \times 10^{-5}$ | unknown | ~0.02 | Coal and petroleum burning |
| COS | Carbonyl sulphide | $5 \times 10^{-4}$ | <3 | 2–2.5 | Biomass burning; fossil fuel burning |

[a] Corresponds to mid-1980s values.
ppmv = parts per million by volume.
After Wuebbles *et al.* (1989).

**Table 1.2.** Direct radiative effects and indirect trace gas interactions affecting climate

| Gas | Greenhouse gas? | Are its trophospheric concentrations affected by chemistry? | Indirect effects on: Tropospheric chemistry? | Indirect effects on: Stratospheric chemistry? |
|---|---|---|---|---|
| $CO_2$ | yes | no | no | yes, affects $O_3$ |
| $CH_4$ | yes | yes, reacts with OH | yes, affects OH and $O_3$ | yes, affects $O_3$, $H_2O$ |
| CO | no | yes, reacts with OH | yes, affects OH and $O_3$ | not significantly |
| $N_2O$ | yes | no | no | yes, affects $O_3$ |
| $NO_x$ | no | yes, reacts with OH, $O_3$ | yes, affects OH and $O_3$ | yes, affects $O_3$ |
| $CFCl_3$ | yes | no | no | yes, affects $O_3$ |
| $CF_2Cl_2$ | yes | no | no | yes, affects $O_3$ |
| $C_2Cl_3F_3$ | yes | no | no | yes, affects $O_3$ |
| $CH_3CCl_3$ | yes | yes, reacts with OH | no | yes, affects $O_3$ |
| $CF_2ClBr$ | ? | yes, photolysis | no | yes, affects $O_3$ |
| $CF_3Br$ | yes | no | no | yes, affects $O_3$ |
| $SO_2$ | yes, but weak | yes, reacts with OH | yes, but not significant to climate | yes, increases aerosols |
| COS | yes, but weak | yes, reacts with OH | no | yes, increases aerosols |

After Wuebbles *et al.* (1989).

Figure 1.4 illustrates the complicated set of phenomena and processes which establish the carbon cycle as one of the most important factors in the formation of the atmospheric composition and climate.

The heterogeneous nature of the processes in the atmosphere is its special feature (Schryer 1982). Water in all its phases is of key importance for the chemistry of the atmosphere: it is a source of OH hydroxyl (an important component of gas-to-particle reactions) and $H_2O_2$ (a critically important oxidizing agent in reactions between liquid components). Liquid water (clouds and precipitation) substantially affects the chemical transformations of various compounds and the tropospheric cycles of sulphur, nitrogen, carbon and halogens as well as other elements. An adequate understanding of the cycles is impossible without detailed studies of gas-to-particle conversions and of the role of clouds and precipitation in such conversions (Kondratyev *et al.* 1983a,b, 1985).

The emphasis has long since been placed on studies of the role of aerosols in atmospheric chemistry and as a weather and climate-forming factor (Kondratyev *et al.* 1973). In this connection, studies of, for example, the composition and optical properties of aerosols, the transformation of gaseous components of natural and anthropogenic origin to particles, and catalytic reactions of the particles are of great importance (Kondratyev 1991b, Jennings 1993).

Developments in the heterogeneous chemistry of the atmosphere are a key component of global atmospheric chemistry, especially in connection with the processes

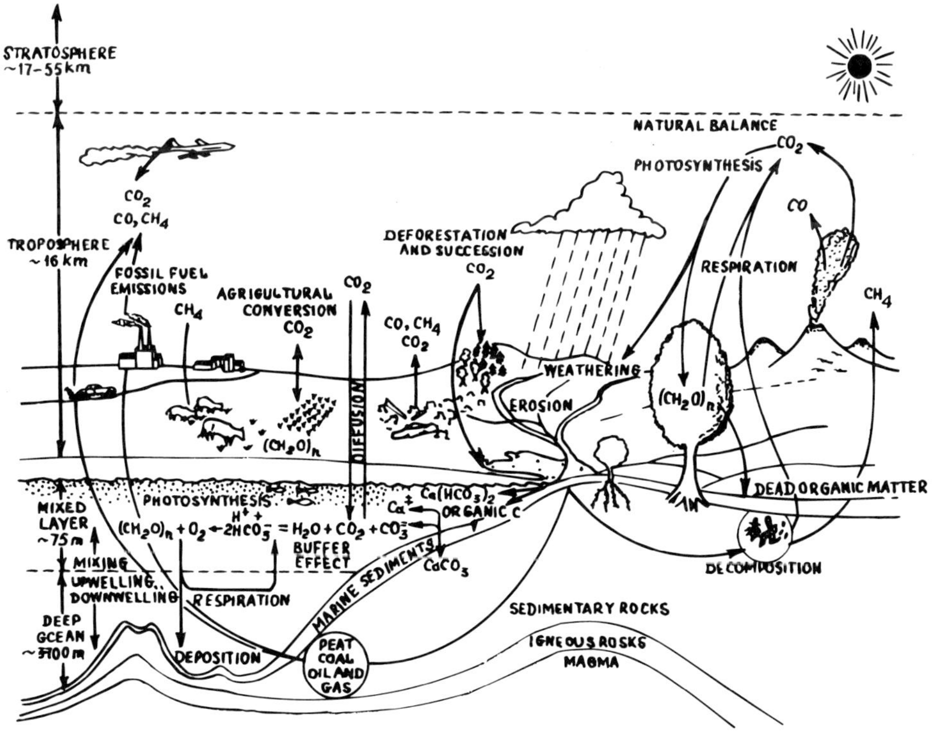


**Figure 1.4** Processes responsible for the formation of carbon cycle.

of formation of biogeochemical cycles, and they include the following major aspects:

- studies of basic heterogeneous processes responsible for the formation of cycles (feedback between homogeneous and heterogeneous processes, the importance of the rates of gas-to-particle reactions);
- determination of the role of heterogeneous processes in transport, physical and chemical transformation, and wash-out mechanisms, which govern the spatial and temporal variability of the key components of biogeochemical cycles (precipitating and non-precipitating clouds, aerosols, etc.);
- contribution from heterogeneous processes as factors responsible for changes in weather and climate.

When considering possible climatic changes due to anthropogenic variations in the chemical composition of the atmosphere, it is necessary to take into account the interactions among various climate-forming factors. For instance, an increase in chlorofluorocarbons can substantially change the content of atmospheric ozone. It is important to appreciate that variations in the ozone concentration are inhomogeneous vertically. Calculations have shown that the greenhouse effect due to ozone is very sensitive to variations in the vertical distribution of the ozone concentration and shows itself not only directly through the effect of the ozone but also indirectly through other atmospheric components. For example, Lacis *et al.* (1989) show that the surface air temperature (SAT) must increase both with decreased ozone content above 30 km and with its increase below 30 km. The trend of ozone content observed in the 1970s led to a surface air temperature decrease (mainly resulting from a decrease of ozone content in the lower stratosphere), reaching about half the calculated greenhouse warming due to increased $CO_2$ concentration.

The complicated nature of the greenhouse effect requires the regular monitoring of various optically active minor gaseous components of the atmosphere. Various techniques have been developed and applied, so that, for example, the content of minor gases and aerosols can be determined in the same way as air temperature at different heights, measuring from satellites the outgoing atmospheric thermal emission at different wavelengths.

There is also a more efficient way, which is by limb-sounding; this involves using satellite-borne instruments that measure the extinction of solar radiation on its way from the Sun to the satellite through the atmosphere, near to sunrise and sunset, with respect to the satellite (the so-called occultation measurements). Such an experiment on the retrieval of minor gases and aerosols was accomplished for the first time on the manned orbiting station Salyut-4, which carried a complex of Sun spectrometers that registered solar radiation at various wavelengths at sunrise and sunset with respect to the orbiting station. Measurements of the absorption of solar radiation by the atmosphere also provide, for instance, information about water vapour and ozone content in the atmosphere. Abundant information about stratospheric aerosol has been obtained from SAGE and Nimbus-7 occultation measurements. The limb-sounding technique (atmospheric thermal emission measurements) has become an important source of information.

Anthropogenic impacts on climate are numerous and different in their spatial and temporal scales. Putting aside the problems of micro-climate and meso-climate, we shall dwell on the global-scale impacts which result from changing the content of minor gaseous and aerosol optically active components in the atmosphere that affect the formation of what has become known as the atmospheric greenhouse effect. In this connection, the problem of increased $CO_2$ concentration and of an assessment of the consequential climate changes in the future are of particular concern.

These assessments include the necessity of forecasting industrial development and its impact on the biosphere in the forthcoming 50 to 100 yr, understanding the laws of the global carbon cycle and developing a reliable climate theory.

However, these aspects are characterized by many uncertainties. For instance, current research on the problem of the carbon cycle is complicated by contradictory estimates of the contributions made by marine and continental biota. As for the nature of the atmospheric greenhouse effect, studies performed during the last 10 to 15 yr have demonstrated the importance of a variety of minor gases and aerosols and their substantial contribution to the formation of the greenhouse effect (Kondratyev and Moskalenko 1984). All this testifies to the impossibility of reliable forecasts of anthropogenically induced climate changes for the next 50 to 100 yr. Various scenarios of climate change may be considered, however (Houghton *et al.* 1990, 1992).

### 1.3.2 The biosphere

Processes in the biosphere are a driving force in the composition of the atmosphere. Anthropogenic impacts on the biosphere therefore play a vital role in climate-changing factors. In this connection, studies of the global carbon cycle are relevant (Gorshkov 1984, 1986, 1987a,b, 1995, Gorshkov and Kondratyev 1988, Gorshkov *et al.* 1989, 1996).

Biospheric productivity is manifested through the synthesis and decomposition of organic matter. Stability of the chemical composition of the environment is maintained by the closed cycles of biospheric substances. As Gorshkov shows, with only small-scale anthropogenic disturbances of biospheric processes in the pre-industrial period, flows of substances due to the synthesis and decomposition in the biosphere had been mutually compensated to an accuracy of about 0.01 per cent. This had been achieved by natural selection of ecologically balanced communities of numerous types of living organisms.

In the case of anthropogenic impact, flows of substances due to the synthesis and decomposition of organic and industrial products are unbalanced. This determines the urgency of the following basic problems of theoretic ecology: (i) studying the laws that provide the closed nature of the cycles of substances in conditions of the undisturbed biosphere, (ii) determining the degree of disturbances that do not lead to irreversible changes in the biosphere and environment and (iii) obtaining observational data on the present state of the biosphere and environment and, in particular, on the degree of the closing of the cycles of various substances.

Based on an analysis of biospheric energetics, Gorshkov's study estimated the distribution of energy fluxes and nutrients from the size of organisms of less than 1 cm. Highly closed geochemical cycles are possible if less than 1 per cent of the energy flux is absorbed by organisms larger than 1 cm. These conclusions agree well with available empirical data for natural communities that are undisturbed anthropogenically.

Gorshkov also studied the closed nature of the present-day carbon cycle, using data on variations in the isotopic composition of carbon in tree rings. It is now established that the redistribution of carbon can mainly take place between the four reservoirs: atmosphere, ocean, fossil fuel and biosphere. Any leakage of carbon outside these reservoirs can be neglected. To determine changes in the carbon content of the ocean and biosphere, it is enough to assess variations for only one of these reservoirs, since a change in the content of carbon in the second reservoir is determined from a balance equation (the law of conservation of matter). It is essential here that the absorption of carbon by the ocean is characterized by two constants. The first is determined by the physico-chemical resistance to absorption with changing total content of carbon, including bicarbonate and carbonate ions. The second is determined by the absorption resistance due to variations in the concentration of dissolved $CO_2$ only and corresponds to variations in biological activity.

The presence of the 'buffer' effect in the absorption of the basic $^{12}C$ isotope and the absence of this effect in the absorption of rare isotopes of carbon (the Bolin–Erickson effect), mean that the two constants are introduced into the balance equation in different combinations. This makes it possible to determine two independent combinations of the constants using data on variations in the content of the two rare isotopes of carbon in tree rings, $^{14}C$ and $^{13}C$. In addition, empirical data are used for the buffer factor of resistance to the penetration of $CO_2$ across the atmosphere – ocean interface as well as the well-known corrections for the fractionation of the $^{13}C$ isotope. Data on the processes in, and the structure of, the ocean and biosphere are not required here.

The estimates by Gorshkov have shown that with the existing poor accuracy of empirical data, emissions of carbon from land biota are 2.3 times larger than those from fossil fuel burning. The biospheric emissions are absorbed by the ocean because of its varying biological activity. Thus, on the whole, the carbon content of the biosphere, including its near-surface and oceanic parts, remains almost stationary. Separate changes in the near-surface and oceanic parts of the biosphere and the respective flows of carbon through the atmosphere greatly exceed emissions due to fossil fuel burning.

Previous studies have not taken into account possible changes in the biological activity of the ocean, and therefore the estimates of emissions of carbon by land biota have been less than those by fossil fuel burning. If this factor is ignored, results corresponding to the previous models can be obtained. In this case an agreement with the equation for the carbon balance is provided with an error of less than 30 per cent. Naturally, these studies on the carbon cycle require a new approach to the problem of anthropogenic impacts on climate. The complexity of this problem necessitates further studies and discussion.

A study by Esser (1987) illustrates another approach to the problem of the global carbon cycle. It is well known that human impact on the biosphere during a period of more than a century (deforestation, agricultural activity, etc.) has led to global carbon cycle perturbations which are comparable to the contribution from fossil fuel burning. The anthropogenic transformation of vegetation cover reduces its age, and agricultural activity reduces net primary productivity (NPP). Since phytomass depends both on NPP and on vegetation age, it follows that there is a decrease of phytomass because of human activity.

Another important aspect of the problem is that, as a rule, an increase in $CO_2$ concentration has a stimulating effect on photosynthesis (development of natural vegetation) and, apparently, the NPP. This stimulation is determined by two mechanisms:

1. Increased quantum yield of photosynthetic reactions;
2. Increased efficiency in the use of water by vegetation (due to increasing stomatal resistance, with preserved input of $CO_2$ to the leaves) and decreased water loss on transpiration (Idso 1987, 1988, 1989, 1991a–d, Idso *et al.* 1989).

The NPP can also vary under the influence of climate (temperature and humidity change). In this connection, Esser (1987) assessed the sensitivity of the carbon cycle to the factors mentioned above, based on the use of a biospheric model (over a 2.5° by 2.5° grid on the continents except the Antarctic) developed in Osnabrück University (Germany) together with a multi-box oceanic model and a one-box model of the atmosphere. Data on regional soil use have been taken from the World Agricultural Atlas. Soil use dynamics have also been taken into account. The model shows that the initial value of $CO_2$ concentration (for the year 1860), which is equal to 290 ppm, is close to that found in the analysis of ice cores (287 ± 3 ppm). The post-1958 growth of $CO_2$ concentration agrees with the data from observations of Mauna Loa, but the observed interannual variability has not been reproduced.

By the end of the period considered (1860–1981), the biospheric reservoir of carbon turned out to be more or less balanced. Emissions due to deforestation were compensated by a stimulation of photosynthesis under the influence of increased $CO_2$ concentration. Before 1970, the biosphere functioned as a weak source of carbon, since $CO_2$ release due to forest cutting had not been compensated by a $CO_2$-caused enhancement of photosynthesis; later on, it became a weak sink for carbon.

If the effect of photosynthesis enhancement is not taken into account, the results of the numerical modelling of carbon reservoirs in the atmosphere and in the ocean strongly differ from those expected from the observational data. A biospheric cumulative release of carbon to the atmosphere (over the entire period mentioned above) amounted to 137 Gt (gigatonne), whereas an emission due to fossil fuel burning reached 166.6 Gt (a total release of 303.6 Gt). As a result, $CO_2$ concentration in the atmosphere grew by 50 ppm (this corresponds to the increment of carbon mass by 105.2 Gt), 64 Gt of carbon being assimilated by the ocean (11.8 Gt remained in the mixed layer).

Due to photosynthesis enhancement, the NPP of the terrestrial biosphere increased from 43.1 Gt of carbon in 1860 to 48 Gt of carbon in 1981. Since the NPP of agricultural fields is much lower than that of natural vegetation, a 17 per cent increase of $CO_2$ has led to only a 13 per cent increase of NPP. The phytomass reservoir is the most powerful in the biosphere, reaching about 600–700 Gt of carbon, but the global phytomass decreased by 17.5 Gt of carbon (the balance of phytomass is determined by its decrease by 137 Gt of carbon because of deforestation and by its increase by 199.5 Gt of carbon because of photosynthesis enhancement).

Different scenarios of global climate changes (annual mean temperatures and precipitation) for the period under consideration revealed their substantial effect on the balance of carbon in the biosphere and on $CO_2$ concentration in the atmosphere, with precipitation variations being the strongest. An account of regional climatic variability is even more important. A comparison of the obtained estimates of the global carbon cycle components with the data of other authors revealed considerable differences.

In this connection, an important question is whether it is possible, using observations of climate changes during the previous one hundred years, to detect (filter out) an anthropogenic signal (in particular the $CO_2$ signal). This is very difficult because the anthropogenic signal does not exceed the level of noise (natural climate variability). We shall turn to the consideration of this problem in Chapter 2.

For the moment we note that the questions that need to be considered in relation to the problem of the impact of $CO_2$ and other minor atmospheric gases on climate include:

- estimations of the possible fuel combustion in the forthcoming century, based on the consideration of population, economic development and prospects for the use of alternative sources of energy;
- possible dynamics of the global biosphere resulting from human industrial activity;
- further study of the carbon cycle and a redistribution of carbon among the basic reservoirs (especially the monitoring of the $CO_2$ dynamics in the atmosphere and ocean, estimating the dynamics of the continental phytomass and analysing an interaction among the cycles of the various components);
- more reliable and adequate estimates of climatic response to increasing concentration of $CO_2$ and other minor atmospheric gases;
- the effect of climate on natural ecosystems and human activity;
- complex long-term observations of the climate system to monitor the dynamics of the minor atmospheric gases and to detect an anthropogenic signal.

As has been pointed out, the simulation of the biosphere as a coupled component of the climate system has become a very important task. This has also stimulated interest in making assessments of the impact of climate changes on the biosphere dynamics. It was in this context that Claussen and Esch (1994) used the biome model developed by Prentice *et al.* (1993) to predict global patterns of potential plant formations, or biomes, from climatologies simulated by ECHAM, a model used for climate simulations at the Max Planck Institut für Meteorologie (Cubasch *et al.* 1991, 1992). The purpose of the modelling was twofold: a qualitative test of simulated climatologies and an assessment of the effects of climate change.

The first important result obtained by Claussen and Esch (1994) was that a good overall agreement between simulation and observation of global patterns of biomes has been found. There were also discrepancies for the Kalahari Desert, in Australia and for the middle west of North America, which can be traced back to failures in simulated rainfall as well as summer or winter temperatures.

As an example of potential vegetation changes in response to simulated changes in climate, global patterns of biomes have been computed from an ice age simulation 18 000 yr BP. These calculations indicate that North America, Europe and Siberia should have been covered largely by tundra and taiga, whereas only small differences are seen for the tropical rain forests. Since a new improved ECHAM-3-T42 model has become available it will be used for the continuation of similar modelling.

In accordance with the IPCC scenario A (Business as Usual) (Houghton *et al.* 1990) of the $CO_2$ concentration increase, a simulation has been accomplished to predict relevant biome dynamics. The results show that the largest changes occur for boreal biomes, whereas little change is seen for the Sahara and the tropical rain forests. Claussen and Esch (1994) noted, however, that since the biome model is not capable of predicting changes in vegetation patterns caused by a rapid climate change, this simulation has to be taken as a prediction of changes in conditions favourable for the existence of certain biomes, not as a prediction of a future distribution of biomes. Because the biome model does not take into account changes in vegetation due to changes in $CO_2$ and soil fertility, subsequent studies are necessary to couple a dynamic model of vegetation succession to a climate model. A further step in the study will be to use a nested approach to simulate smaller-scale dynamics. The sensitivity of a climate model to changes in global patterns of biomes will also be investigated.

### 1.3.3 The hydrosphere (the oceans)

A study of the large-scale atmospheric processes responsible for weather and climate changes requires a consideration of the interaction of the atmosphere not only with the land but also with the ocean, which is a gigantic heat reservoir. Over half of the solar radiation reaching the Earth's surface is first absorbed by the oceans, where it is stored and redistributed by ocean currents before escaping to the atmosphere, largely as latent heat through evaporation but also as longwave radiation. These currents have a complicated horizontal and vertical structure determined by the pattern of winds blowing over the sea and the distribution of continents and submerged mountain ranges.

The vertical structure of the oceans comprises three layers:

- the seasonal boundary layer, mixed annually from the surface, which is less than 100 m deep in the tropics and reaches hundreds of metres in the sub-polar seas (other than the North Pacific) and several kilometres in very small regions of the polar seas in most years;
- the warm water sphere (permanent thermocline), ventilated (i.e. exchanging heat and gases) from the seasonal boundary layer, which is pushed down to depths of many hundreds of metres in gyres by the convergence of surface (Ekman) currents driven by the wind;
- the cold water sphere (deep ocean), which fills the bottom 80 per cent of the oceans' volume, ventilated from the seasonal boundary layer in polar seas.

The atmosphere–ocean interaction, showing itself in the exchange of heat, water and momentum, is of key importance to the solution of the problem of long-range forecasts of weather and climate changes. A system of conjugated equations of thermohydrodynamics and the specially developed theory of perturbations was used by Marchuk and co-workers (Marchuk *et al.* 1987, Marchuk 1992) to study the processes of long-term weather anomalies and climate changes manifested as surface air temperature variations over large areas. These have shown that the short-range and long-range forecasts of temperature anomalies are determined mainly by processes taking place in different (depending on the short-range or long-range forecasts) regions of the world's oceans, called the energetically active zones of the ocean (EAZO) (Kondratyev and Khvorostyanov 1986, 1988, Marchuk *et al.* 1987, 1989).

This theoretical analysis has been verified by global maps of the world's oceans' heat budget, based on observational data as well as satellite observations of the Earth's radiation budget, and the empirical relations found previously between the state of the ocean and the resulting weather. This has confirmed the conclusion that the concept of the energetically active zones of the oceans is the most important factor determining long-term weather anomalies and climate changes. This conclusion is particularly important because the available observational means are still inadequate to obtain regular meteorological and oceanographic information on global scales. The First GARP Global Experiment (FGGE), with a 10-yr preparatory period, could not have been accomplished before the year 1979. This experiment has made it possible, for the first time, to obtain global information for a 1-yr period which can now be used to test theoretical models of the atmospheric general circulation and of climate.

Discovering the energetically active zones of the oceans has drastically changed the situation. The concentration of available observational means (primarily ships) on monitoring the processes taking place in the energetically active zones, including migration and variability therein, and also the use of satellite data, permit a deeper insight into the processes governing the long-term weather anomalies and climate changes, and provide long-range weather forecasters with much-needed information. That is why the 'Sections' programme has been initiated, with the major objective to study short-term (several years) climate changes. Based on the energetically active zones concept, the 'Sections' programme includes both field measurement in different key climatic regions and numerical experiments using observational data. We shall return to this programme in Section 5.3.

There are physical and biological mechanisms in the oceans which are important in controlling the concentration of $CO_2$ in the climate system. $CO_2$ is transferred from the atmosphere into the interior of the ocean by the physical mechanism caused by differences in the partial pressure of $CO_2$ in the ocean and the lowest layers of the atmosphere. The rate of gas exchange depends on the air–sea difference in partial pressure of $CO_2$ and a coefficient which increases with wind speed. The ocean branch of the carbon cycle involves a flux of $CO_2$ from the air into the sea at locations where the surface mixed layer has a partial pressure of $CO_2$ lower than that of the atmosphere and vice versa. The annual ventilation of the seasonal boundary layer from the surface mixed-layer controls the efficiency of the biological mechanism by which ocean plankton convert dissolved $CO_2$ into particulate carbon, which sinks into deep water. These two pumps are responsible for extracting $CO_2$ from the global carbon cycle for periods in excess of a hundred years. The mixed-layer partial pressure of $CO_2$ is depressed by enhanced solubility in cold water and enhanced plankton production during the spring bloom.

### 1.3.4 The lithosphere (geosphere)

The biological involvement of the land, insofar as its surface is vegetated, in the carbon cycle has been discussed in Section 1.3.2 already. The role of the oceans in the radiation balance at the surface of the Earth has also been considered in Sections 1.3.2 and 1.3.3. We now turn to the role of the land in its physical aspects, rather than its biological aspects, and the question of energy balance at the land surface. The land surface plays an important role in terms of its exchange of energy with the atmosphere through the reflection, absorption and emission of radiation. The land also plays an important part in the hydrological cycle. The processes involved concern the amount of fresh water stored in the ground as soil moisture (thereby interacting with the biosphere) and in underground reservoirs, or transported as run-off to different locations where it might influence the ocean circulation, particularly in high latitudes. The soil interacts with the atmosphere by exchanges of gases, aerosols and moisture, and these are influenced by the soil type and the vegetation, which again are strongly dependent on the soil wetness. Present knowledge about these strongly interactive processes is rather limited.

A Chapman Conference on the Hydrologic Aspects of Global Climate Change was held at Lake Chelan, Washington, in June 1990. The conference was motivated by recent interest in the description of land surfaces at scales much larger than the river basins traditionally studied by hydrologists and particularly by problems in the representation of the land surface hydrology in general circulation models. A collection of 13 papers presented at the conference appeared in an issue of the *Journal of Geophysical Research*, volume 97, number D3 (pages 2675–2833) which was published on 28 February 1992. A new source of information is a thematic issue of *Climate Change*, volume 31, numbers 3–4, published in 1995.

### 1.3.5 The crysophere

The cryosphere can be classified as follows.

- Seasonal snow cover, which responds rapidly to atmospheric dynamics on time scales of days and longer. In a global context the seasonal heat storage in snow is small. The primary influence of the cryosphere comes from the high albedo of a snow-covered surface.
- Sea ice, which affects climate on time scales of seasons and longer. This has a similar effect on the surface heat balance as snow on land. It also tends to decouple the ocean and the atmosphere, since it inhibits the exchange of moisture and momentum. In some regions it influences the formation of deep water masses by salt extrusion during the freezing period and by the generation of fresh water layers in the melting period.
- Ice sheets of Greenland and the Antarctic, which can be considered as quasi-permanent topographic features. They contain 80 per cent of the existing fresh water on the globe, thereby acting as long-term reservoirs in the hydrological cycle. Any change in size will therefore influence the global sea level.
- Mountain glaciers are a small part of the cryosphere. They also represent a freshwater reservoir and can therefore influence the sea level. They are used as an important diagnostic tool for climate change since they respond rapidly to changing environmental conditions.
- Permafrost affects surface ecosystems and river discharges. It influences the thermohaline circulation of the ocean.

Variations in the nature and extent of the ice cover clearly alter the energy fluxes at the Earth's surface and thereby influence the climate. The variation in the surface area covered by ice provides another example of a feedback mechanism in the climate.

The transport of ice, by floating icebergs, towards the equator also provides another possible positive feedback mechanism (see Ledley 1991a for example). Suppose that a general warming causes more ice to be released as icebergs in the polar regions. These icebergs move towards the equator leaving more unfrozen sea surface near the poles; this leads to a larger energy transfer from the sea to the atmosphere near the poles and this additional energy is spread around and leads to a general further warming of the atmosphere. We can look at it from the opposite viewpoint – if there is a warming of the Earth, leading to less snow and ice cover, then the surface of the Earth becomes (on average) less reflecting and more absorbing. This causes further warming and a further reduction in the snow and ice cover, i.e. this is an example of positive feedback. The real situation is, of course, more complicated and eventually the amplification of the warming by positive feedback (it is hoped) comes to a halt.

## 1.4 Nuclear war and climate

In recent years, assessments of the possible impact of a nuclear war on the atmosphere and climate have become of great concern (Chylek *et al.* 1983, 1987, Berger 1984, Chylek and Ramaswamy 1984, Robock 1984b, Sagan 1984, Cess 1985, Crutzen and Hahn 1986, Malone *et al.* 1986, Penner 1986, Thompson and Schneider 1986, Brühl and Crutzen 1987, Velikhov 1987, Kondratyev 1988b, Singer 1988, Vupputuri 1988b, Pittock *et al.* 1989). The complexity of estimating the possible climatic impacts of post-nuclear strong disturbances of the gaseous and aerosol composition of the atmosphere necessitates a thorough analysis of the contributions from numerous interacting factors and the role of numerous feedbacks, using not only simulation techniques but also possible natural analogues and data of observations after the nuclear tests of the late 1950s and early 1960s. In this connection, results should also be analysed from studies on possible climatic effects connected with increasing concentrations of $CO_2$ and other greenhouse gases as well as volcanic eruptions. Despite their extremely limited similarity, natural analogues contain signatures that help reduce the uncertainties of numerical modelling caused by simplifying assumptions.

The results from numerical modelling of the climatic impact of a possible nuclear war, which substantiate a well-known concept of nuclear winter, have lately been criticized in connection with the simplifications and assumptions which can seriously affect the final results. Basic assumptions involve: (i) the neglect of the atmospheric greenhouse effect and the initial patchy distribution of smoke aerosol, (ii) a very schematic simulation of the effects of coagulation and removal of smoke aerosol, (iii) the neglect of various feedback mechanisms between the effect of smoke aerosol on the wind field and the effect of atmospheric motions on the distribution of smoke and (iv) interaction between the atmosphere and the ocean.

A group of United States workers developed a programme of studies on possible climatic impacts of a nuclear war, aimed at the reduction of uncertainties in respective estimates, based on the nuclear winter concept (National Climate Program Office of Science 1985). Bearing in mind the insufficient reliability of both climate models and input data, they concluded that at present this concept of nuclear winter can be neither rejected nor accepted.

Without sharing the excessive caution of the US workers, we believe that the nuclear winter concept is realistic but one-sided. This conclusion is based on the following (Kondratyev 1988b): (i) results from an analysis of the effects on the atmosphere and climate of the nuclear tests in the late 1950s and early 1960s and (ii) theoretical estimates of the climatic effect of the post-nuclear atmospheric composition. Nuclear testing involved the formation of enormous amounts of post-nuclear $NO_2$, which leads to an increase of solar radiation absorption by the atmosphere and a global climate cooling that could reach 10 °C in the case of a nuclear war. Theoretical estimates of the transformation of the greenhouse effect in the zones of explosions show the possibility of warming and subsequent changes of the sign of climatic change.

Thus, it seems certain that a nuclear war will trigger a global ecological catastrophe which will manifest itself through a strong instability as well as spatial and temporal variabilities of climate. The catastrophic consequences will affect the ozone layer. All this will annihilate civilization. Further studies are needed for a more reliable and detailed substantiation of possible impacts of a nuclear war on the atmosphere and climate.

## 1.5 The World Climate Programme

The 1979 First World Climate Conference (WCC-1) resulted in the World Climate Programme (WCP). This programme has become the basis for subsequent multidisciplinary developments aimed at a deeper understanding of the laws of climate formation and changes, estimating the impact of climate on human activity and ecosystems, and analysing the socio-economic aspects of the problem. The scientific part of the World Climate Programme is the World Climate Research Programme (WCRP), which plays a key role in the prospects of further studies (Houghton 1984, 1986, Houghton *et al.* 1990, 1992, Kondratyev 1992a–d). Developments connected with the WCRP are divided into six sub-programmes having the following general aims: (i) substantiating the physical principles of long-range weather forecasts, (ii) revealing predictable aspects of global climate changes in time scales from months to several years and (iii) estimating the reaction of climate to natural and anthropogenic impacts during periods of several decades.

The six sub-programmes include the following.

1. Studies on atmospheric climate forecasts aimed at further improvements of general circulation models (GCMs) used to predict the weather, especially from the viewpoint of a more reliable consideration of heat flux divergences (diabatic factors) in the atmosphere and at the surface level. To achieve aim (iii) above, it is important first to develop interactive models of the ocean–atmosphere system and then subsequently to develop models considering the interaction between all the components of the climatic system. It is particularly important to increase the reliability of the forecasted cloudiness and precipitation and the effect of extended cloudiness on the radiative heat flux divergence.
2. Studies of the interacting boundary layers of the atmosphere and ocean which are still at the stage of conceptual developments.
3. The programme on studies of the tropical ocean and global atmosphere (TOGA) for a period of 10 yr (1985–1995), aimed at the analysis of predictability

of the interactive system 'tropical ocean–atmosphere' and its impact on the global atmospheric climate on time scales from several months to several years.

4. The World Ocean Circulation Experiment (WOCE) aimed at the diagnosis of oceanic circulation and heat transport in the ocean with the use of new (especially, satellite-based) observational means.
5. Studies of the processes in the cryosphere, oriented primarily toward analysis of the laws of seasonal and interannual variability of the ice cover as well as how the processes in the atmosphere and the ocean determine this variability. The final aim is to construct interactive 'atmosphere–cryosphere–ocean' models needed to forecast long-range climate changes.
6. Assessment of the climate sensitivity to various external factors (including astronomical factors) and physical mechanisms determining the respective impacts.

At a later stage a very important programme, the Global Energy and Water Cycle Experiment (GEWEX), has been initiated; it is discussed below.

The use of improved old and new observational means is decisively important for the accomplishment of the WCRP. It is of particular importance to raise the representativity of observations in the tropics by broadening the programme of conventional soundings and maritime meteorological observations (on ships and buoys) as well as by retrieving the wind field from satellite data. Still urgent is the problem of a combined assimilation of conventional and satellite data on sea surface temperature; ship data play the role of reference information, whereas satellite data are intended to fill the gaps in ship-based observations.

A substantial step is to obtain information about the global distribution of precipitation and sea ice cover from satellite microwave data. In the accomplishment of the WCRP a key role will be played by active microwave remote sensing from satellites (with radar altimeters and scatterometers) used to retrieve data on wind fields and wind shear near the ocean surface, as well as sea roughness and currents; the possibilities of retrieving the fluxes of latent and sensible heat will be investigated.

A principal aspect of the WCRP is the accumulation of complex and long-term observations up to a period of decades. Of great importance are observations that are limited in time and specialized (problem-oriented) at various key sites.

Two processes have first priority: (i) the formation, maintenance and dissipation of extended cloudiness as well as its interaction with the radiation field (the determination of the role of cloudiness in the formation of climate and its changes) and (ii) the effect on climate of land surface processes connected, first of all, with the water cycle (the key aspect is the problem of evapotranspiration). In the global system of observations, in the interests of the WCRP, monitoring of the dynamics of the carbon cycle, tropospheric and stratospheric aerosols and the Earth's radiation budget must play an important role.

Progress in palaeoclimatic studies points to the necessity to include them in the WCRP (Wigley *et al.* 1981, Adamenko *et al.* 1982, Manabe and Bryan 1985, Berger and Tricot 1986, Kutzbach and Guetter 1986, Zubakov 1988, Adem and Mendoza 1987, 1988, Bernhardt and Mäder 1987, Gerasimov and Velichko 1987, Mészáros 1987, Velichko 1987, Adem 1988, 1990, Borisenkov 1988, Crowley and North 1988, Hugget 1991, Jenkins 1993). Analysis of climate changes in the past is interesting from the viewpoint of assessing possible changes in the future. Although there are no close analogues to a possible climate warming due to increased $CO_2$ and other greenhouse gases, Webb and Wigley (1985) note that during the last 200 000 yr, a number of warmings of several degrees have taken place in vast regions of the Earth. In particular there were (i) the mediaeval warming (approximately during the period AD 900–1300, (ii) the Holocene warm period (4000–8000 yr ago) and (iii) the last interglaciation period (about 120 000 yr ago).

Data analysis of pollen, marine plankton and lake level for the Holocene warming has received the most attention. In the middle of the Holocene warming, during the hypsithermal period, a 1 °C global mean warming was observed (compared with the present climate). The climate of that period differed greatly from the present one; for example, in July in middle and high latitudes both positive and negative temperature anomalies had taken place (with respect to the present temperature field). Compared with temperature, precipitation has varied more strongly. In the mid-western USA, the annual sum of precipitation was found to be 10–20 per cent below the present one. This anomaly, together with more frequent and prolonged periods of dry air, shows a similarity with the climatic conditions in 1930 in the mid-western USA when severe dust storms occurred. Estimates of precipitation for Africa and India were 50–100 per cent higher than at present. Apparently, the present deserts of the Sahara and Rajasthan used to be dry savannahs and steppes. These data on the rainfall regime are very interesting; they point to the possibility of strong variations in precipitation and desert extent, with relatively small variations in the global mean temperature.

One more approach to the analysis of climate changes in the past is the comparison of climatic conditions in the periods of cooling and warming from observational data for the last hundred years. Comparisons of the surface air temperature data for the northern hemisphere unexpectedly showed that during the hemispherical mean warmings, the winter temperature over most of Europe had been lower and characterized by a stronger interannual variability (apparently due to blocking effects). Precipitation during the periods of warming decreased everywhere in spring and in summer but increased in autumn and in winter. On the whole, the field of increased temperature in the warm world had varied less during the whole year, but the structure of precipitation had been complicated and inhomogeneous, with vast regions of positive and negative anomalies. Observations in the northern hemisphere show that the level of warming averaged about 0.5 °C, but in the Arctic the warming reached 2 °C.

The Second International Conference on the assessment of the role of $CO_2$ and other greenhouse gases, together with aerosols and associated impacts on the biosphere and climate, was held in Villach, Austria, on 9–15 October 1985 and was sponsored by UNEP/WMO/ICSU. It was a milestone toward further consideration of climate changes (Houghton *et al.* 1990, 1992, Kondratyev 1992b–d). The participating countries totalled 29 (World Meteorological Organization 1986). The greenhouse gases were defined as the optically active gaseous components of the atmosphere that contribute significantly (together with $H_2O$ and $CO_2$) to the formation of the atmospheric greenhouse effect. These are chlorofluorocarbons (CFCs), methane ($CH_4$), nitrous oxide ($N_2O$) and tropospheric ozone ($O_3$).

The Statement to the conference participants included the following sentence: 'As a result of increasing concentrations of greenhouse gases, it is now believed that in the first half of the next century a rise of global mean temperature can occur which is greater than any in man's history.' The possibility of global climate warming requires an analysis of measures to provide successful socio-economic development in the future; this is all the more important because there is a close relation between climate changes and associated problems of sea level rise, the threat to the ozone layer, and acid rain. The speed and amplitude of climate warming could be substantially reduced by more economical energy and fossil fuel consumption, which would reduce the level of greenhouse gas emissions to the atmosphere.

The participants of the Villach Conference developed a consensus on the following: the atmospheric greenhouse gas concentration has increased (the $CO_2$ concentration has grown from 315 ppmv (parts per million by volume) in 1958 to 354 ppmv in 1984); estimates from data on $CO_2$ in the bubbles of ice cores gave a pre-industrial revolution concentration of $275 \pm 10$ ppmv. During the last 150 000 yr the $CO_2$ concentration has varied between 200 ppmv and 270–280 ppmv and has been correlated with climate changes.

The uncertainty of forecasting energy consumption (even on a scale of decades) leads to a wide range of estimates of possible $CO_2$ releases in the year 2050, from 2 Gt per year to 20 Gt per year (at present it is 5 Gt per year).

Although a better understanding of the global carbon cycle has been obtained, a balance cannot yet be reliably determined. Existing carbon cycle models must be revised, especially from the viewpoint of considering the marine biological pump that moves carbon from the continents to the ocean. In the participants' opinion, however, this does not seriously affect the estimates of future $CO_2$ concentrations.

With an assumed rate of growth in concentration not exceeding 0.5 per cent per year during the subsequent 40 yr, followed by slow growth by the end of the forthcoming century, the $CO_2$ concentration will be below 400 ppmv (i.e. it will exceed the pre-industrial level by less than 60 per cent). However, with the current rate of growth (1–2 per cent per year, starting from 1973) it will double within four decades. Different scenarios of the possible time period for doubling the concentration of $CO_2$ give periods varying from the mid twenty-first to the twenty-second century.

It is probable that, at the present time, about half of the possible greenhouse warming (as compared with that induced by $CO_2$) has been caused by the other greenhouse gases. If observed trends of increasing concentrations of greenhouse gases continue, their contribution, especially from CFCs, will be comparable in several decades to that of $CO_2$ (this will probably happen in the third decade of the twenty-first century). The assessment of the greenhouse gases' contribution to the greenhouse effect is hindered by an insufficient knowledge of the details of the biogeochemical cycles that govern the concentrations of $CH_4$, NO and $O_3$. In practical terms, there is no reliable basis on which to forecast a trend for aerosol content in the future.

The participants also concluded that estimates of the global mean climate warming due to a doubled $CO_2$ concentration, based on the use of the most sophisticated climate models, varied within the range 1.5–4.5 °C. It is mentioned in the Statement that '. . . because of the complexity of the climatic system and the imperfections of the models, particularly with respect to ocean–atmosphere interactions and clouds, values outside this range cannot be excluded'. This and further conclusions are in complete accordance with the IPCC Report (Houghton *et al.* 1990, 1992).

The participants emphasized that '. . . because of the model imperfections these numbers should for the time being not be interpreted as a prediction for the range of the eventual impact on climate. Rather they can be considered as a measure for the agreement among the different model approaches'. They added that, on the whole 'at present, and certainly for a number of years to come, climate change studies by means of models provide possible scenarios, not predictions of future climate'.

The ocean-induced delay in equilibrium warming would presumably take several decades. With such factors affecting the climate as variations in the aerosol content, extra-atmospheric solar radiation and vegetation cover, the role of the greenhouse gases in the next century will probably be most important. As for possible climate changes on a regional scale, results from numerical simulation cannot be considered reliable, though the following conclusions deserve attention: (i) the warming in high latitudes in late autumn and winter is stronger than in the tropics; (ii) an increase of the annual mean run-off in high latitudes is possible; (iii) dry summer seasons are most likely on the continents in high latitudes of the northern hemisphere; (iv) whereas the tropical warming is weaker than the global mean warming, its effect on humanity and ecosystems can have far-reaching consequences; (v) an increase in evapotranspiration in the tropics is possible as well as intensified rains in the regions of the humid tropics; (vi) a warming within 1.5–4.5 °C must raise the world sea level by 20–140 cm, mainly due to thermal expansion. However, substantial melting of the west Antarctic ice sheet is unlikely in the forthcoming century.

According to the conference Statement, based on analyses of observational data, the estimated increase in global mean temperature during the last 100 yr of 0.3–0.7 °C is consistent with the projected temperature increase attributable to the observed increase in $CO_2$ and other greenhouse gases, although it cannot be ascribed scientifically to these factors alone. Moreover, the present temperature-change signal predicted by climate sensitivity studies cannot yet be discriminated from the noise caused by natural fluctuations.

Forecasting global and regional precipitation and the regional structure of temperature anomalies are problems which are still unsolved. Models of ocean, cloud–radiation interaction and land surface processes should be revised. Of primary importance are the following developments: (i) the long-term monitoring of greenhouse gases and aerosols, solar radiation and sea level (with subsequent data interpretation based on the use of present model simulations), (ii) palaeoclimatic studies (especially from the viewpoint of analysis of the interaction between the oceans and ecosystems), (iii) studies of the effect of the interaction between atmospheric composition and climate changes (especially extreme variations) on sub-tropical and tropical ecosystems, boreal forest and the water regime, (iv) analysis of the response of agriculture to increased concentrations of $CO_2$ and other greenhouse gases (as a rule, an increased $CO_2$ concentration raises the vegetation productivity) as well as to climate warming and (v) assessment of the socio-economic consequences of rising sea level.

On the whole, the Report (World Meteorological Organization 1986) contains an objective analysis of the problem, but no doubt requires some comment. A solution to the problem that anthropogenic impacts on climate may be manifested through variations in atmospheric composition involves the following studies: (i) global cycles of $CO_2$ and other greenhouse gases, (ii) assessment of the contribution of various greenhouse gases to the greenhouse effect, (iii) analysis of climate changes from observational data and possibilities for revealing the $CO_2$ signal, (iv) numerical modelling of climate changes and (v) analysis of reliability of the climate forecasts. Although the authors of the assessment mention a consensus, the current state of the problem has produced quite different assessments. The status of the problem will be considered in Chapter 4, in the context of the Second UN Conference on Environment and Development held in Brazil in June 1992.

In January 1989 by the initiative of the University Corporation on Atmospheric Research (UCAR), a working meeting was held with the participation of more than 100 experts from the USA, which worked out the Climate System Modelling Programme (CSMP) aimed at speeding up progress in creating reliable methods to forecast changes in regional and global climates on time scales of the order of decades (Anthes *et al.* 1992). This programme, considered (in the future) international, will take into account six major interactive components, including: physics, dynamics and chemistry of the atmosphere; land surface processes; marine biogeochemistry; dynamics and thermodynamics of the ocean and land ecosystems.

The realization of the Climate System Modelling Programme will promote a better understanding of measures needed to remove (or overcome) harmful consequences of global changes. In this connection, of particular importance is an assessment of the role and relations of natural and anthropogenic climate changes on different time scales. The principal difference of the CSMP from other programmes is that it will: (i) be focused exclusively on one objective, the improvement of the climate systems dynamics forecasts, (ii) foresee an interdisciplinary approach to climate predictions on time scales from decades to centuries, (iii) be based on a broad cooperation of all experts interested in the cooperative use of technical facilities (first of all, computers) and (iv) release regular (every 3–4 yr) assessments of possible climate changes for governmental agencies.

The key directions of developments should include: (i) a deeper understanding and improvement of the reliability of simulating in climate models such phenomena and processes as clouds, precipitation, vegetation cover and turbulence, (ii) development of coupled subsystems of models (modules) describing various climate-forming processes, and substantiation of requirements for their resolving power, (iii) development of highly efficient software, (iv) application of new concepts of the system stability (for example, the theory of chaos) to study climatic predictability, (v) assessment of the role of biogeochemical cycles in the inter-decadal climate changes and the effect of the dynamics of ecosystems of regional climate, (vi) analysis of the ways of estimating the extreme changes and variations of climate variability as well as relations between natural and anthropogenic variabilities of climate, (vii) improvement of techniques to validate the climate numerical modelling results, (viii) substantiation of the strategy to improve the observational systems and (ix) estimation of uncertainties in climate forecasts.

The following organizational measures are envisaged under the Climate System Modelling Programme: (i) the organization of the Coordination Centre headed by Professor F. Bretherton at the University of Washington, (ii) financing (starting from 1992) the work of 4–6 qualified experts, in which the interdisciplinary problems are of the first priority, (iii) organization of a small core scientific group and coordinating group aimed at scientific and organizational coordination of all developments and (iv) support of projects focused on various concrete problems, which will form the basis for the CSMP.

In the period 1990–1991 efforts were focused on the strategy of realization of this Programme and on the substantiation of requirements for the supercomputer, ordered from industry through the University Corporation on Atmospheric Research (UCAR), to be used, in particular, in numerical modelling of regional climate changes. The problems already launched are as follows: ocean–atmosphere interaction, land–atmosphere interaction, the transport of pollutants and atmospheric chemistry, natural variability of the climate system (the diagnostics of climate from numerical modelling results and its validation), global forcings and regional climates, climatically significant processes (cloud–radiation interaction, etc.) and computer hardware and software.

In 1990 the first version of the Programme 'The Computer Hardware, Advanced Mathematics and Model Physics (CHAMMP)' was prepared, supported by the US Department of Energy and aimed at using new supercomputers and advanced climate models to provide more reliable climate forecasts for time periods from a decade to a century. During the next 10 yr new fast-operating computers are to appear, capable of (1–10) $\times$ $10^{12}$ floating-point operations per second (1–10 Tflops), which will be three to four orders of magnitude faster than those available now.

As Bader *et al.* (1992) note, an important problem is an adjustment of available climate models to a new generation of massively-parallel computers. This foresees: (i) a more adequate account of a complicated (interactive) totality of climate-forming physical, biogeochemical and ecological processes, (ii) comprehensive application of the possibilities of the architecture of current computers and software, (iii) analysis of the processes of climatic predictability and (iv) continuation of studies for a better understanding of the 'greenhouse' aspects of climate change.

The CHAMMP programme is planned to last 10–12 yr and includes three phases. The first phase (2–4 yr) has the major objective to adjust the available atmospheric general circulation models and oceanographic general circulation models to new computers and to start work on the analysis of climate predictability for a period of 10 to 100 yr. The second phase (3–5 yr) will be aimed at developing coupled models of the atmosphere–ocean system and their realization using computers with a fast-operation of 100 Gflops. The third phase will result in creating the coupled climate models realized with the 1–10 Tflops computers, which will make it possible to perform numerical modelling on a time scale of 10–100 yr with a regional spatial resolution.

In 1988 the Intergovernmental Panel on Climate Change (IPCC) was set up by the World Meteorological Organization (WMO) and the United Nations Environment Programme (UNEP) under the chairmanship of Professor B. Bolin. The Panel was charged with

1. Assessing the scientific information that is related to the various components of the climate change issue, such as emissions of major greenhouse gases and modification of the Earth's radiation balance resulting therefrom, and that needed to enable the environmental and socio-economic consequences of climate change to be evaluated.
2. Formulating realistic response strategies for the management of the climate change issue.

The Panel established three Working Groups:

I (Chair: Dr J. T. Houghton) to assess available scientific information on climate change;

II (Chair: Prof. Y. Izrael) to assess environmental and socio-economic impacts of climate change;

III (Chair: Dr F. Bernthal) to formulate response strategies.

In addition, it also established a Special Committee on the Participation of Developing Countries to promote, as quickly as possible, the full participation of such countries in the IPCC's activities.

The findings of Working Group I have been published (Houghton *et al.* 1990, 1992). In the preparation of the IPCC Scientific Assessment (Houghton *et al.* 1990, 1992) about 170 climatologists contributed and a further 200 were involved in the review of the draft report. The report was presented as an 'authoritative statement of the views of the international community' at the time of its preparation, while admitting that some minority opinions were not included.

A major event in the discussion of climate problems was the Second World Climate Conference (WCC-2) which was held in Geneva in the period 29 October to 7 November 1990 (Kondratyev 1992c). The participants in this Conference totalled 747 from 120 countries. The principal objective of the Conference was to summarize the results of the scientific studies for the previous decade and to discuss the perspectives of further studies as well as human-dimension problems; the latter involve necessary political decisions, which could prevent human economic activity and life from causing harmful (and even catastrophic) climate changes. In connection with this, the WCC-2 consisted of two stages: (i) scientific discussion based on consideration of the report prepared by the Intergovernmental Panel on Climate Change (IPCC) with the participation of about a thousand experts from 70 countries and summarizing the WCRP results and (ii) a conference at the level of ministers to work out political recommendations (this second stage distinguishes WCC-2 from WCC-1, which had been mainly a scientific conference). A consensus has been reached on the IPCC Report, which has been pointed out in both the final Statement of the participants at the scientific–technical stage of the Conference and in the Ministerial Declaration. These documents (the IPCC Report included) played an important role in the preparation of the Framework Convention on Climate Change, which was begun in Washington DC on 4 February 1991 and was completed during the Second UN Conference on Environment and Development in Brazil the following June (Kondratyev 1992d).

When opening WCC-2, Zou Jigmen (China), President of the WMO, emphasized that special attention to climate problems was explained by their social priority and the serious socio-economic consequences of possible future climate changes, and therefore increasing concentrations of greenhouse gases in the atmosphere and anthropogenic impacts on the ozone layer were of particular importance. The imperfection of climate theory, on account of the lack of an adequate account of interactive processes in the atmosphere–ocean system and the dynamics of cloud cover and its interaction with radiation, requires that further effort be made to develop more reliable methods of numerical modelling and climate forecasts.

It will be extremely important for a global climate observing system, with the use of both conventional and space-borne observational means, to be created as a result of close international cooperation.

One immediate comment is that despite all the rich information contained in the reports made at WCC-2 by the leaders of various international organizations, they have not suggested a clear enough concept of the long-term perspectives of studies and developments (e.g. on the scale of the next decades) within the World Climate Programme (WCP) and its scientific component, the World Climate Research Programme (WCRP), as well as other closely related programmes (first of all, the International Geosphere–Biosphere Programme (IGBP) (Kondratyev 1989b,c)). This situation seems to result from the stereotypes established during the first decade of the WCP accomplishment, connected with unjustified statements about the decisive role of anthropogenic factors in the present global climate change, i.e. increasing concentrations of greenhouse gases and subsequent climate warming during the last hundred years (some publications even contain conclusions about $CO_2$ signal detection from observational data (Budyko 1989a,b, Budyko and Groisman 1991, Budyko *et al.* 1991).

This circumstance has caused the overestimation of the significance of rather conditional calculations of climate changes resulting from spasmodic doubling (and sometimes quadrupling) of $CO_2$ concentrations to the detriment of some other numerical experiments aimed at the analysis of the sensitivity of the climate system to such factors as cloud cover dynamics, cloud–radiation interaction, atmosphere–ocean interaction, etc. Only recently have scientists undertaken, for example, a study of the climate sensitivity to a chosen scheme of cloud formation parametrization, which has led to very interesting results that have shown (as could be expected) that this sensitivity is very high (see below, Sections 4.1 and 4.8).

## 1.6 On the perspective concept of climate change studies

An important result from preparation of the IPCC report and WCC-2 has been a rejection of the stereotypes mentioned above, and a development of an understanding of the serious uncertainties of numerical climate modelling and the necessity of radical improvements of the existing climate models. The following key directions of further research deserve special attention.

The first aim is the achievement of a new stage of numerical modelling of the dynamics of the atmosphere–ocean–land–cryosphere–biosphere interactive climatic system. This is to take account of the fact that this system is open; that is, it is subject to external cosmic impacts (first of all, those of the Sun, because of variable solar constant and solar activity).

In this connection, the main point is to consider the biosphere as an interactive component of the climate system. The first steps in this direction have been taken; these involve approaches to the parametrization of the biosphere from the viewpoint of accounting for the heat and moisture exchange between the atmosphere and the surface covered with vegetation. Of great importance has been a realization of the programme of a complex field experiment, the First International Satellite Land Surface Climatology Programme (ISLSCP) Field Experiment (FIFE), together with a simple model of the biosphere developed by Sellers and Dorman (1987), Sellers and Hall (1987) and Sellers *et al.* (1988).

Next, the biospheric stage should consist of a generalization of climate theory by means of an interactive account of the global carbon cycle. Such a development of numerical climate modelling could serve as a sound basis for predictions of future variations in $CO_2$ concentrations. At present these predictions are reduced to a totality of rather conditional *a priori* scenarios of the $CO_2$ concentration growth; in the case of interactive simulation of the global carbon cycle, $CO_2$ concentration will become a predictant and not a predictor. Of course, the generalization of climate theory in question is a very complicated problem. Its solution has been favoured, however, by a number of key projects within the International Geosphere–

Biosphere Programme (IGBP). The first is the Joint Global Ocean Flux Study (JGOFS) which is aimed at studies of the global carbon cycle both on land and in the ocean (unfortunately, the single and most urgent problem of the global carbon cycle has been artificially divided into parts in the IGBP). The joint development of a climate model taking account of the carbon cycle can serve as a sound basis for the cooperation of the two large-scale programmes, WCRP and IGBP.

Further development of climate theory should establish an interactive account of the cycles of sulphur and nitrogen. Regarding the cycle of sulphur, of special interest is a simulation of gas-to-particle reactions of the transformation of gaseous sulphur compounds into sulphate aerosol. A variable emission of dimethyl sulphide (DMS) by the ocean (due to the dynamics of phytoplankton) puts forward quite a new problem of taking into account not only the gas-to-particle mechanism of the formation of sulphate condensation nuclei but also their subsequent effect on the dynamics and optical properties of cloud cover, which can become an important factor of the Earth radiation budget (ERB) variability (and hence that of climate). Here, again, there is a broad basis for concrete cooperation between the WCRP and IGBP within the problems of biogeochemical cycles, forming the thematical core of the IGBP. Thus the first component of the WCP perspective concept (and, mainly, WCRP) is a gradual realization of the biospheric stage of numerical climate modelling, which in its turn will ensure the development of realistic models of the biosphere.

The second key direction of research is the substantiation and realization of the Global Climate Observing System (GCOS). The WCC-2 participants reached a consensus on the necessity to develop the GCOS as a system for climate monitoring, detecting climate changes and monitoring the response (first of all, of ecosystems) to climate changes. Also, the objectives of the GCOS are: the provision of climate data to aid economic development, studies aimed at a deeper understanding, and a consequent improvement, of numerical modelling techniques and prediction of the climatic system dynamics. Major components of the GCOS should be (a) a proposed Global Ocean Observing System (GOOS), which would ensure observations of the characteristics of the physical, chemical and biological processes in the ocean, (b) a Global Terrestrial Observing System (GTOS), (c) an improved programme of the World Weather Watch (WWW) and (d) programmes of monitoring various key components of the climatic system, which would provide data on variations in the chemical composition of the atmosphere and surface ecosystems, cloud cover and water cycle, Earth radiation budget, ice cover and precipitation over the ocean.

A special place is occupied by the Global Atmosphere Watch (GAW) operating from 1989 under the aegis of the WMO and organized as a result of amalgamating and broadening the existing global ozone-observing system and network of the background monitoring of air pollution. The principal objective of the GAW is to provide regular observations of (a) the content of the most important greenhouse gases in the atmosphere (carbon dioxide, chlorofluorocarbons, methane, nitrous oxide, tropospheric ozone), (b) total ozone content, its concentration near the surface and its vertical profile, as well as the content of gases which are precursors of ozone, with the use of various techniques and instruments (spectrometers, sondes, lidars, microwave radiometers), including their installation on satellites, (c) surface radiation budget and its components, optical thickness (or transparency) of the atmosphere, meteorological visibility, UV-B solar radiation, total content of aerosol and water vapour in the atmosphere, (d) the chemical composition of liquid and solid precipitation, (e) the content of chemically active components, such as sulphur dioxide and reduced forms of nitrogen compounds, and carbon monoxide, (f) number, density and composition of aerosols (including mineral aerosols) and (g) the content of radionuclides, including krypton-85, radon and tritium, as well as the isotopic composition of some components of the atmosphere. The realization of the GAW should make an important contribution to studies of the atmospheric greenhouse effect dynamics, the global ozone layer and acid rain.

Since climate prediction is considered as the solution of a problem with initial data, it is important to obtain a time series of the statistical characteristics of climate for each region (with a spatial resolution of at least 5° latitude by 5° longitude) with an interval of a month, for a period of hundred years, with the use of regularly renewed intitial data. A simpler approach consists in prescribing a data base for one year (with a characteristic annual course). In this case it would be possible to release annual climate forecasts one hundred years in advance, similar to annual economic predictions published regularly. This similarity is not accidental, since data of the socio-economic predictions serve as the basis for assessing the contribution of anthropogenic impacts. The anthropogenic boundary conditions should contain data on the dynamics of land properties and water use, the rate of greenhouse gas emissions and atmospheric pollutants that react with them.

As for the limits of climatic predictability, they have been poorly studied and, apparently, depend on special features of the parameters to be predicted. Maybe, it is possible to make useful forecasts of the climatic system one hundred years in advance. Apparently, the accuracy of the forecasts will depend on the adequacy of the models and the reliability of the initial data. For the atmosphere, the data should include information on the content of greenhouse gases and stratospheric aerosols. Naturally, information about the ocean, the most slowly varying component of the climatic system, is of key importance (the whole oceanic thickness is meant). Thus, initial data should be based on (1) an improved system of land surface monitoring, (2) the renewed World Weather Watch (together with the GAW) and (3) quite a new system to monitor the world's oceans. If the errors of the initial state parameters are small enough and the errors in the prediction of the oceanic dynamics do not grow too fast, this will make it possible to make useful predictions. The estimates obtained have shown, for example, that for the signal to noise (S/N) ratio to be 5:1, it is necessary to estimate the atmosphere–ocean heat exchange, annually averaged for North America, to an accuracy not worse than $\pm 1\,\mathrm{W\,m^{-2}}$, which can be achieved only with the use of new measurement techniques (similar estimates made earlier had allowed an error of the order of $\pm 10\,\mathrm{W\,m^{-2}}$).

Although there is a need to resolve the mesoscale structure of the oceanic dynamics (synoptic vortices etc.), it is not practicable to create a network of stations uniformly distributed all over the world's oceans to make *in situ* observations; this necessitates the choice of key regions and parameters, among them being the following: the seasonal boundary layer, cycles and currents, areas of kinetic energy concentrations, as well as fluxes of heat, water and chemical components, and interbasin fluxes.

It is worthwhile to select the following regions where the observational means can be concentrated to study the seasonal boundary layer: (1) the Arctic Ocean (monitoring the processes that govern the ice cover dynamics), (2) the Norwegian and Greenland Seas (formation of deep waters), (3) the high-latitude North Atlantic Ocean (formation of deep waters), (4) the belt of the seas round the Antarctic (formation of bottom waters), (5) the equatorial zone of the Pacific Ocean (El Niño) and (6) the tropics of the Indian Ocean (the Somali current and monsoon processes).

Particularly interesting are studies of key mesoscale processes, such as (1) the interaction among the time-dependent vortices, permanent currents and bottom topography on horizontal scales of about several tens of kilometres and (2) interaction between stretched vortices in non-stationary jet streams and the biological pump on the same scales. So far, there has not been any technique to parametrize such processes. However, testing of the first high-resolution simulation models of the ocean (with a grid spacing of about 25 km and 70 levels in the vertical) has demonstrated that, on the whole, they succeed in reproducing the interaction among the vortices, permanent currents and bottom topography, though it would be expedient to obtain a higher resolution. An account of the non-linear dynamics of the vortices has provided a much more adequate simulation of the circulation in the ocean.

The most important new step towards a further development of the oceanic general circulation models consists in taking account of chemical and biological processes, with emphasis on the processes with participation of plankton and hydrosol which regulate the carbon flux between the atmosphere and the ocean. An important problem is to provide coordinated developments within the programmes such as GEWEX, TOGA and WOCE (World Ocean Circulation Experiment), as well as the IGBP core projects like JGOFS (Joint Global Ocean Flux Study) to investigate the oceanic component of the global carbon cycle and GOEZS (the study of processes in the euphotic zone). The joint programme of observations can be classified into five components: (i) temperature and salinity (hydrographic component), (ii) nutrients, the content of the dissolved oxygen, and geochemical tracers (chemical component), (iii) phase transformations of water–ice, precipitation, evaporation (water component), (iv) the external forcing of wind stress, the atmosphere–ocean heat exchange, insolation and heat flux across the sea bottom and (v) concentration of phytoplankton in the surface layer (biological component).

The urgency of the GCOS realization contrasts with a lack of serious (quantitative) substantiation of requirements for observational data from the viewpoint of the priority of parameters to be measured, allowed errors, required spatial and temporal resolution, optimal combination of conventional and space-borne observational means. The concept of energy-active zones as a methodology to reveal the global regions most significant for the formation of climate change and to optimize the combination of conventional and space-borne observational means has been considered by Marchuk *et al.* (1986, 1987, 1989). However, on the whole, the problem remains unsolved.

The difficulty of the situation can be illustrated by an almost total lack of numerical experiments on climatic sensitivity to various factors and processes, the results of which could be used to substantiate the priorities and to estimate permissible errors. The existing information is fragmentary and, as a rule, not very convincing. Of course, the numerical experiments under discussion are expensive and require much computer time. But much more substantial expenses are connected with the creation of space-borne observational means (the cost of the American programme of the Earth Observing System (EOS) will reach more than 10 billion dollars) and the lack of optimization will lead to more serious losses than the expenditure on developments of methodology to plan the observational system. Therefore the problem of extreme importance is to provide a scientifically justified approach to optimize the GCOS. The international cooperation in the realization of global systems to observe the environment is also not effective enough.

The third key direction of research is the analysis of the socio-economic aspects of the problem of climate changes. In this connection the following situation has occurred. The efforts of the second and third Working Groups of the IPCC have been dedicated to socio-economic implications of climate change, but these efforts have been concentrated only on a detailed analysis of the effects of possible climate changes on human economic activity (water resources, agriculture and food, ocean, fishery and coastal zones, energy production, land use and planning of urban development, forestry) and human health.

At the final stage of the first part of WCC-2 the emphasis was placed on the problem of greenhouse gas emissions and recommendations for the reduction of these emissions. But the key problem, which is the interactivity of the system of human society and the environment, i.e. the impact of socio-economic factors on the biosphere, environment and climate, has practically been ignored. The discussions (and the final 'Statement' and 'Declaration' made on their basis) were concentrated on the problem of reducing the emissions of $CO_2$ and other greenhouse gases.

A very important aspect of the problem is the assessment of climate change impacts and responses on the regional scale; we have already indicated that within a global trend there will be large regional variations. A new effort in this direction is a method called MINK which is discussed by Rosenberg (1993). We shall consider the question of regional variations in more detail below (Section 5.1).

CHAPTER TWO

# The observed regularities of climate

## 2.1 The $CO_2$ cycle and global-scale climate changes from observational data

### 2.1.1 The $CO_2$ cycle

Consideration of greenhouse-induced climate changes requires an analysis of observational data for a number of greenhouse gases such as $N_2O$, $CH_4$, chlorofluorocarbons (CFCs) and tropospheric ozone, in addition to the main contributors, namely water vapour and $CO_2$. Relevant summaries have recently been published by Houghton *et al.* (1990, 1992) and Viner and Hulme (1993a,b). The need to study anthropogenic sulphate aerosol formation and its impact on climate (greenhouse warming versus aerosol cooling problem), as well as tropospheric ozone dynamics, motivated the compilation of inventories for not only the above greenhouse gases, but also for $SO_2$ and tropospheric ozone precursors (Prinn *et al.* 1987, Prather 1988, Hao and Ward 1993, Law and Pyle 1993, Akimoto and Narita 1994, Klimont *et al.* 1994, Guntharatham and Young 1995 and many others). Here we shall concentrate on carbon dioxide.

Analysis of observational data gives evidence for an increase of $CO_2$ concentration in the atmosphere, caused by human industrial activity, and manifested mainly through fossil fuel combustion and through impact on the biosphere (burning and cutting of forests, soil cultivation, burning of timber, etc.) (Borisenkov and Kondratyev 1988, Rosenberg *et al.* 1989, Houghton *et al.* 1990, 1992, Handel and Risby 1992, Heimann 1995). This is illustrated in Figure 2.1. There was a reduction in the rate of $CO_2$ emissions during a period of economic depression in the 1920s and 1930s, an increase of $CO_2$ emissions before World War II, and then a decreased growth rate and even a drop in $CO_2$ emissions between 1941 and 1945, with a subsequent continuous increase of atmospheric $CO_2$ after World War II. Viner and Hulme (1993a) note that $CO_2$ concentration in the atmosphere has risen from a pre-industrial level of about 280 parts per million by volume (ppmv) to the 1991 level of 355 ppmv.

At present, the annual $CO_2$ emissions exceed 5 Gt (by carbon equivalent) or 18 Gt (by $CO_2$). As calculations show (Figure 2.2), not more than 50 per cent of the ejected carbon remains in the atmosphere. The rest is absorbed in the ocean. The $CO_2$ concentration has a pronounced annual cycle, with a minimum in winter. This

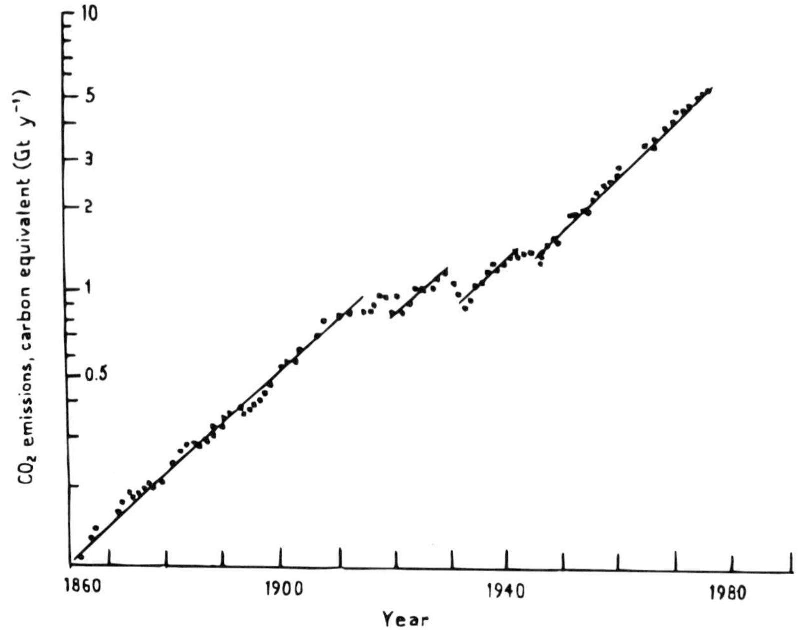


**Figure 2.1** The $CO_2$ input to the atmosphere due to fossil fuel combustion (GtC $y^{-1}$).

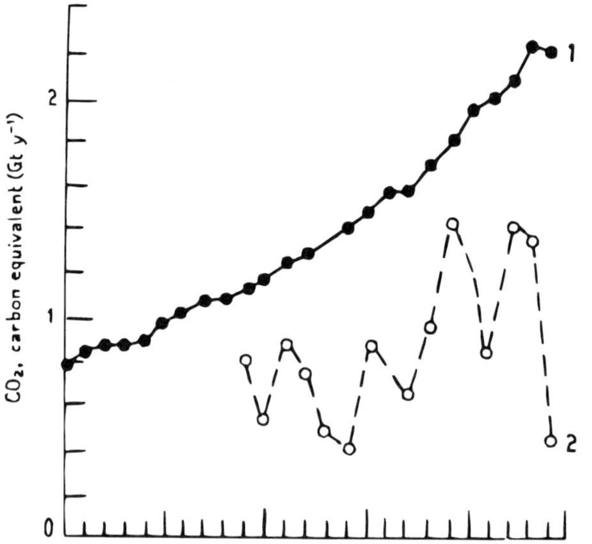


**Figure 2.2** The input of anthropogenic $CO_2$ and the share of $CO_2$ remaining in the atmosphere: (1) the input of anthropogenic $CO_2$ due to fossil fuel combustion; (2) the $CO_2$ amount remaining in the atmosphere.

is indicative of an active participation of the biosphere in the formation of the $CO_2$ annual cycle.

An analysis of 2419 air samples made by Keeling *et al.* (1985) between 1959 and 1981 at the oceanic weather station P (50° N, 145° W) revealed three important regularities in the temporal variability of $CO_2$ concentration; these are a well-reproduced annual change, an interannual variability which correlates with the dynamics of the general atmospheric circulation, and a long-term increasing trend which is approximately in proportion to global industrial $CO_2$ emissions caused by fossil fuel burning.

An increase was observed in the amplitude of the smoothed annual change (a difference between maximum and minimum values) from 13.3 ppm in 1969 to 14.5 ppm in 1981. This was apparently connected with an intensified vegetation productivity in a vast region of the northern hemisphere land surface, where the vegetation processes take place mainly in summer (this assumption is suggested by data on the observed phases of the $CO_2$ concentration annual change).

The interannual variability and its first derivative closely correlate with the dynamics of the Southern Oscillation. The first derivative lags 6 months behind the interannual variability, and this points to the possible existence of a remote oceanic or continental source-sink for $CO_2$ in the southern hemisphere tropics. Between May 1969 and June 1981, the $CO_2$ concentration increased from 324.9 to 340.8 ppm, which constitutes about 60 per cent of the increase which would take place if the whole amount of the $CO_2$ resulting from fossil fuel burning remained in the atmosphere and were uniformly distributed. The concentration of $CO_2$ at Station P, averaged over the period 1975–1981, turned out to be 0.8 ppm below that at Point Barrow (71° N, Alaska) and 0.9 ppm above that at Mauna Loa (19° N, Hawaii), which testifies to a north-to-south decrease of $CO_2$ in a broad latitudinal band (70–20° N).

The analysis of daily observations of atmospheric $CO_2$ concentration at the Mauna Loa observatory between 1952 and 1982 revealed an increase of the amplitude of the annual change of concentration (a maximum in May and a minimum in early October), reaching nearly 1 ppm, i.e. a considerable part of the average amplitude of the annual change (about 6 ppm). With a rate of concentration growth of about 0.7 per cent per year, the mean square error in linear approximation of the growth constitutes 0.09 per cent, i.e. the concentration increase in question is statistically significant. A thorough analysis of errors showed that the possibility of an instrumentally-induced trend could be excluded.

Since the annual change of $CO_2$ concentration in the northern hemisphere appears to be determined by the metabolic activity of land vegetation, one can assume that at least part of this increasing trend is connected with the dynamics of the continental vegetation cover. Other factors may be a stimulation of vegetation productivity by increased $CO_2$ and a change in climate that favours the growth of productivity.

The amount of anthropogenic carbon equal to 5 Gt per year released to the atmosphere corresponds to the natural accumulation of carbon in the Earth's crust during a millennium of the Earth's evolution (Borisenkov and Kondratyev 1988). For the natural carbon cycle, this is a serious disturbing factor. For this purpose, the carbon cycle must be analysed as well as the uncertainties caused by lack of knowledge of various mechanisms and parameters which govern the carbon cycle.

Table 2.1 gives some known parameters of the carbon cycle in the industrial

**Table 2.1** Some parameters of the carbon cycle and the degree of their uncertainty

| | Parameters | Estimates | Uncertainty degree |
|---|---|---|---|
| 1 | Amount of $CO_2$ due to fossil fuel burning (oil, carbon, gas) between 1861 and 1981 | 160 Gt | 140–180 Gt |
| 2 | Average $CO_2$ concentration in the atmosphere | 340 ppm | 339–343 ppm |
| 3 | Pre-industrial $CO_2$ content (early nineteenth century) | 260 ppm | 240–290 ppm |
| 4 | Amount of $CO_2$ absorbed by the ocean by the year 1981 (calculated for equivalent carbon) | 2 Gt $y^{-1}$ | unknown |
| 5 | Carbon in biomass at the present time (carbon content in global ecosystems and in harvests) | 560 Gt | 460–660 Gt |
| 6 | Annual loss of carbon by the biosphere due to soil cultivation, etc. | 2 Gt $y^{-1}$ | 2–5 Gt $y^{-1}$ |
| 7 | Present-day supplies of organic carbon in soil, peat, humus | 1450 Gt | 1200–1700 Gt |

epoch and characterizes their uncertainty. As can be seen from this table, there is considerable uncertainty in the quantitative characteristics, especially those which govern the exchange processes between various media. This calls into question the results of numerical modelling of the carbon cycle and explains the scatter of the estimates which have a bearing on final conclusions about possible climate changes.

Recent reliable observations of $CO_2$ in different global locations point to a considerable spatial and temporal variability of the concentration, which necessitates a thorough analysis of regional sources of $CO_2$. Interhemispherical and latitudinal differences do exist in the rate of $CO_2$ concentration increase. Variations in the amplitude of the annual change of concentration, which depends in particular on latitude, can be substantial.

In connection with these circumstances Marland *et al.* (1985) undertook a new inventory of the global distribution of $CO_2$ emissions due to fossil fuel burning over a 5° latitude by 5° longitude grid, based on UN data on industrial development for 1980 (without taking account of the contributions by the cement industry, fossil gas burning in plumes, and some others). Data in Table 2.2 characterize total emissions for 5° latitudinal belts in both hemispheres. The total global emission (including the indicated sources left out of account) amounted to 5.284 Gt of carbon. Ninety per cent of global emissions originate within the latitudinal belt 20°–60° N (95 per cent on the northern hemisphere), with maxima in the spatial grid cells where Frankfurt, London and Tokyo are located.

Tarko *et al.* (1992) estimated the contributions of various countries and regions to the formation of the $CO_2$ global budget, taking account of anthropogenic emis-

**Table 2.2** $CO_2$ emissions in 1980 for 5° latitudinal belts in both hemispheres

| Northern hemisphere latitude (°) | $CO_2$ emissions (kt C) | Southern hemisphere latitude (°) | $CO_2$ emissions (kt C) |
|---|---|---|---|
| 85–90 | 0 | 0–5 | 11 544 |
| 80–85 | 1 | 5–10 | 24 190 |
| 75–80 | 17 | 10–15 | 11 365 |
| 70–75 | 601 | 15–20 | 12 387 |
| 65–70 | 9 141 | 20–25 | 39 291 |
| 60–65 | 39 539 | 25–30 | 52 387 |
| 55–60 | 294 268 | 30–35 | 60 469 |
| 50–55 | 854 434 | 35–40 | 20 463 |
| 45–50 | 634 248 | 40–45 | 4434 |
| 40–45 | 846 680 | 45–50 | 742 |
| 35–40 | 726 033 | 50–55 | 336 |
| 30–35 | 578 493 | 55–60 | 0 |
| 20–25 | 140 728 | 65–70 | 0 |
| 15–20 | 76 712 | 70–75 | 0 |
| 10–15 | 45 891 | 75–80 | 0 |
| 5–10 | 36 094 | 80–85 | 0 |
| 0–5 | 25 744 | 85–90 | 0 |

sions. With 30 types of ecosystems considered, the globe was divided into cells of 4° latitude by 5° longitude. The model used for the carbon cycle took into account the dynamics of the biosphere from 1860 to 2000. The following parameters were calculated with a 1-yr time step: the carbon content in the biomass of living plants, soil humus (for each cell of the grid) and the carbon content in the atmosphere. Table 2.3 gives the resulting change in the content of carbon in the reservoirs of countries and regions (Gt C for a given period).

**Table 2.3** Changes in the content of carbon in the reservoirs of countries and regions (Gt C/period)

| Country | Phytomass | Humus | Phytomass + humus |
|---|---|---|---|
| | Period 1950–1980 | | |
| USSR | 5.51 | −1.37 | 4.14 |
| Europe (without USSR) | 3.82 | −0.43 | 3.39 |
| USA | 4.25 | −2.36 | 1.85 |
| Canada | 3.31 | −0.58 | 3.26 |
| Tropical forests | 0.97 | −8.19 | −7.22 |
| Globe | 31.64 | −26.8 | 4.84 |
| | Period 1981–2000 | | |
| USSR | 7.04 | −0.59 | 6.45 |
| Europe (without USSR) | 4.86 | −0.13 | 4.73 |
| USA | 5.31 | −1.58 | 3.72 |
| Canada | 4.27 | 0.25 | 4.52 |
| Tropical forests | 26.39 | −4.97 | 21.45 |
| Globe | 64.69 | −16.09 | 48.61 |

According to these results, the absorption of $CO_2$ took place in every region (despite a decrease in the content of humus), the former USSR contributing most to the absorption; these conclusions contradict the global estimates made by Gorshkov (1987a,b). As a result of deforestation, tropical forests between 1950 and 1980 became less effective as absorbers of $CO_2$. The northern hemisphere mid-latitude forest ecosystems play a very important role in $CO_2$ absorption. Their contribution over the entire period 1950–1980 over the territory of the former USSR reached 5.22 Gt. The land is likely to become a net source of $CO_2$ in the period 2000–2030 as a result of anthropogenic impacts.

Smith *et al.* (1992) performed numerical modelling to study changes in the carbon budget on the continents and, hence, in ecosystems, associated with a doubling of the concentration of $CO_2$ in the atmosphere. These estimates were based on climate scenarios obtained with four climate models: Oregon State University (OSU), Geophysical Fluid Dynamics Laboratory (GFDL), Goddard Institute for Space Studies (GISS), and the United Kingdom Meteorological Office (UKMO). The global distribution of vegetation cover was characterized using the Holdridge Life-Zone Classification (37 life-zones were taken) over a 0.5° longitude grid. The values in Table 2.4 characterize variations in carbon reservoirs, and those in Table 2.5 are for areas covered with major biome-types.

All four scenarios suggest an increase in area occupied by forests, although details of predicted patterns vary among the scenarios. There is a significant decrease in the global extent of desert. The decline in tundra observed under all scenarios is primarily due to a shift from tundra to mesic forest. Terrestrial carbon storage increased from 0.4 per cent (8.5 Gt) to 9.5 per cent (180.5 Gt) above estimates for present conditions. These changes represent a potential reduction of 4 to 8.5 ppm on elevated atmospheric $CO_2$ levels.

**Table 2.4** Changes in carbon storage (Gt) in above-ground biomass and soil for four climate change scenarios

| Scenario | Above-ground biomass | Soil | Total | Change[a] |
|---|---|---|---|---|
| Current | 737.2 | 1158.5 | 1895.7 | |
| OSU | 860.4 (16.7) | 1215.8 (4.9) | 2076.2 | 180.5 (9.5) |
| GFDL | 782.3 (6.1) | 1151.3 (−0.6) | 1933.6 | 37.9 (2.0) |
| GISS | 829.6 (12.5) | 1213.0 (4.7) | 2042.6 | 146.9 (7.7) |
| UKMO | 765.2 (3.8) | 1139.0 (−1.7) | 1904.2 | 8.5 (0.4) |

[a] Values in parentheses are percentage change from current.

**Table 2.5** Changes in the areal coverage ($10^3$ km$^2$) of major biome-types under current and changed climate conditions

| Major Biome-types | Current area | OSU | GFDL | GISS | UKMO |
|---|---|---|---|---|---|
| Tundra | 939 | −302 | −515 | −314 | −573 |
| Desert | 3699 | −619 | −630 | −962 | −980 |
| Grassland | 1923 | 380 | 969 | 694 | 810 |
| Dry forest | 1816 | 4 | 608 | 487 | 1296 |
| Mesic forest | 5172 | 561 | −402 | 120 | −519 |

Biosphere fertilization problems due to $CO_2$ increase have been discussed in detail by Idso (1991a–d).

Despite the uncertainty of the existing forecasts of $CO_2$ dynamics, a concern for $CO_2$ accumulated in the atmosphere has stimulated the development of different approaches to a solution to this problem, a key aspect of which is the dynamics of energy production. In this connection, Marchetti (1988) attempted an analytical approximation of the dynamics of the production of energy from various fuels during the last century or so based on available data, with an extrapolation of actual trends to the future. Figure 2.3 illustrates the results obtained (parameter $F$ characterizes the relative contribution of various kinds of energy production). Figure 2.4 shows estimates of the energy production dynamics, with a total for all kinds of fossil fuel (the estimates for nuclear fuel have been obtained on the assumption that the characteristic time of its adoption constitutes 75 or 100 yr). As can be seen, during the next 100 yr natural gas must become the prevailing fossil fuel. This conclusion is illustrated in Figure 2.5. In Figure 2.6 the data have been transformed into the curves of the dynamics of global $CO_2$ emissions. Proceeding from the assumption that nuclear energy production will prevail in the future, the scheme of hydrogen production as a chemical fuel, shown in Figure 2.7, can be considered the most promising. $CO_2$ can be withdrawn by, for instance, pumping it into used oil wells. It should be emphasized, of course, that the scenarios by Marchetti (1988) are conditional; they only illustrate one of the possible approaches.

### 2.1.2 Variations in surface air temperature and other parameters

According to the conclusions of some studies (Budyko 1985, 1989a,b), the results of meteorological observations contain indications of the beginning of a warming of

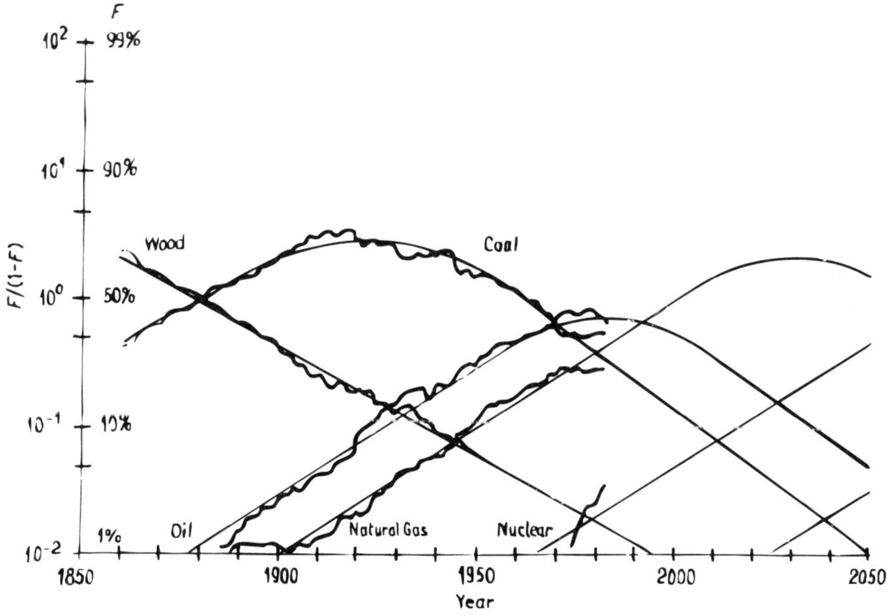


**Figure 2.3** Relative temporal variation of energy production from various fuels.

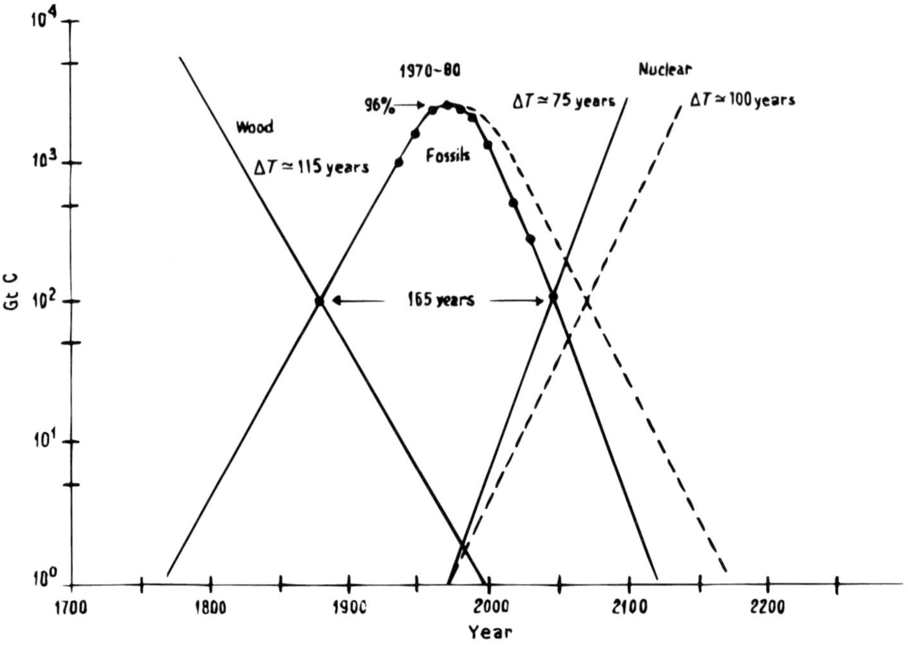


**Figure 2.4** Global energy production from various fuels.

the climate due to an increase in $CO_2$. However, this problem is still far from clear. Woodward and Gray (1993) published convincing results which demonstrated the need for great caution in the analysis of trends in time series data and predictions concerning future climate change. Important relevant considerations have been dis-

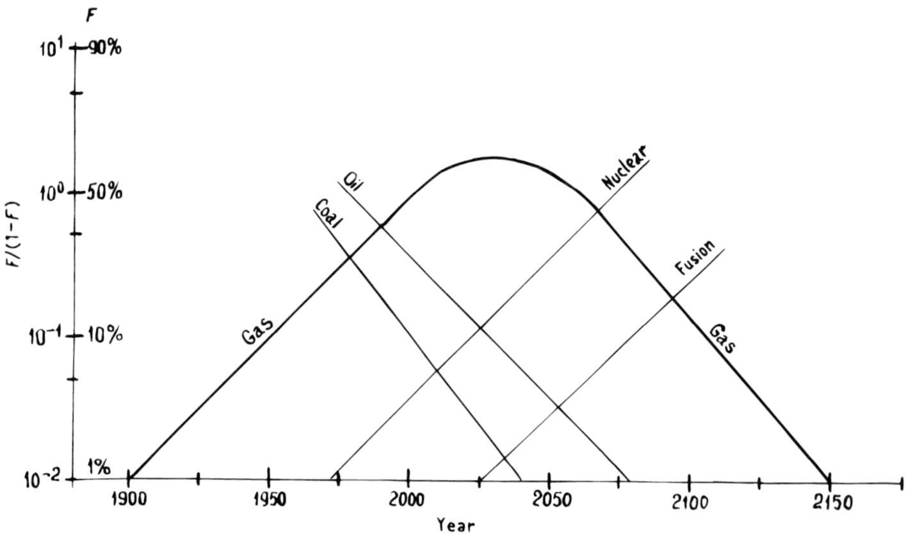


**Figure 2.5** Consumption dynamics for natural gas compared with other fuels.

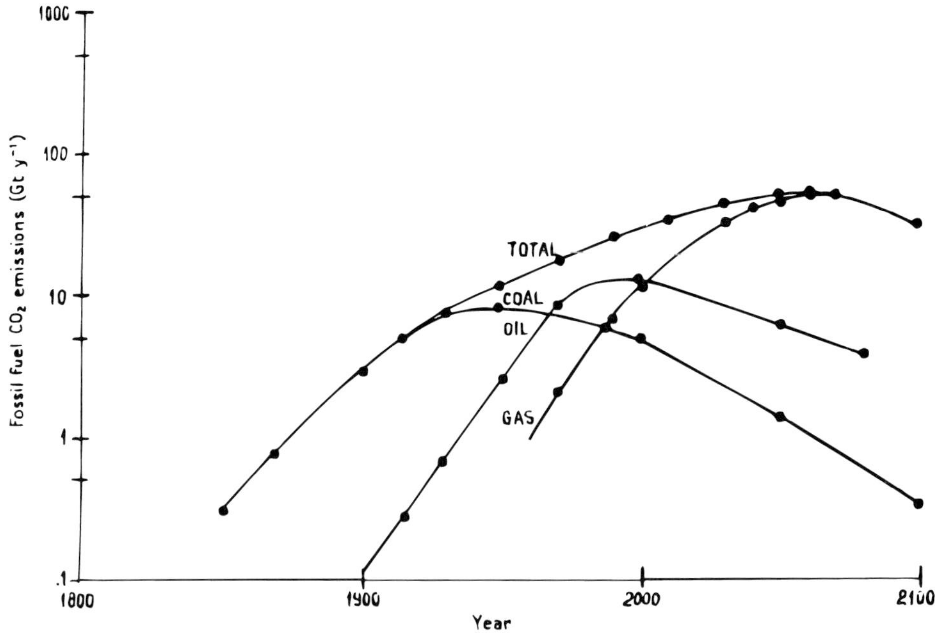


**Figure 2.6** The evolution of global $CO_2$ emissions. 1 tonne coal produces 3.5 tonnes $CO_2$ ; 1 tonne $CH_4$ produces 3 tonnes $CO_2$ ; $10^3$ $m^3$ $CH_4$ produces 2 tonnes $CO_2$ .

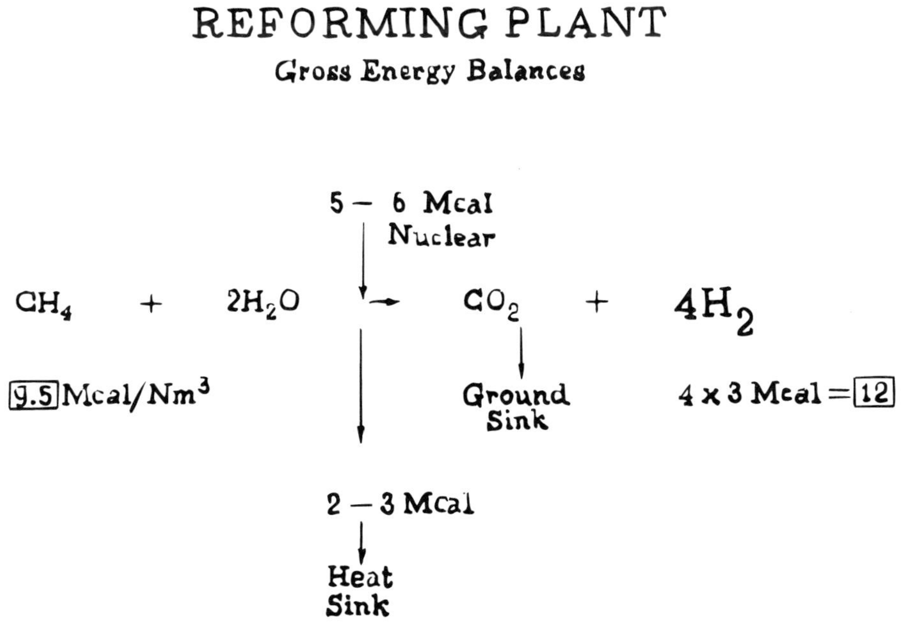


**Figure 2.7** The scheme of rational use of nuclear energy.

cussed by Kheshgi and White (1993), as well as by Nasrallah and Balling (1993). Hoskins and Pierce (1983) made a well-balanced conclusion in this context: 'There is no conclusive evidence that global warming, associated with the enhanced greenhouse effect, is occurring. We have far to go in our understanding of and ability to model the complex climate system before confident predictions [can be made] of the aspects of climate change that need to be known'. Similar assessments have been made by Dickinson (1989). Of especial interest is the analysis of tropospheric temperature observations in the Arctic, where the enhancement of greenhouse warming has been predicted by the results of numerical modelling. On this basis, Kahl *et al.* (1993) showed that absolute trends of 3 °C in thirty years, or larger, were found in the troposphere during the period 1958–1966 in the Arctic, with both cooling and warming tendencies observed in four layers: 800–700–500–400–300 hPa. The majority of trends, however, were not significant statistically, at the 90 per cent confidence level. Therefore Kahl *et al.* (1993) concluded that 'greenhouse-induced warming is not detectable in the Arctic troposphere for the 1958–1986 period'.

D'Arrigo and Jacoby (1993) analysed boreal tree rings recorded from high altitudes in North America, Scandinavia and Russia for 1682–1968, and their overall evaluation confirmed that high northern latitudes are now in an anomalously warm state relative to the past three centuries.

Gruza *et al.* (1989) showed that in the last 100 yr the increase of mean air surface temperature in the northern hemisphere reached 0.4 °C and global mean warming amounted to 0.55 °C. The positive trend in the northern hemisphere during this period included a maximum of warming between the 1920s and 1940s, when the $CO_2$ concentration increase was negligible. Also, in the years following the maximum warming, when an intensive increase of $CO_2$ concentration occurred, a cooling in the northern hemisphere was unexpectedly observed, together with a slight warming, especially after the 1960s, in the southern hemisphere. So far, this fact has not been convincingly explained. The estimates of climate cooling due to volcanic aerosol, given in some publications, cannot be considered convincing either. One can, for instance (by choosing empirical constants), explain observed surface air temperature variations by solar activity, without accepting any hypotheses of the effect of $CO_2$ or aerosol. Therefore, criticisms and doubts can increasingly be found in the literature with regard to the hypothesis that climate warming in the middle of this century can be explained by the effect of $CO_2$.

Angell (1986a,b) processed data from 63 aerological stations in the northern hemisphere to analyse the trends of surface air temperature and the temperature of the 100–50 hPa layer (16–20 km) and the 850–300 hPa layer (1.9–9.0 km) in the period 1958–1985 and to examine the impact of eruptions of the volcanoes Agung in 1963 and El Chichón in 1982 on the hemispheric mean surface air temperature and the temperature of the free atmosphere. The results obtained (Figure. 2.8) showed that in both cases a post-eruption stratospheric warming took place; in 1982 it was weaker because the eruptive aerosol was mainly above the 16–20 km layer and the stratospheric cooling took place due to other factors.

After the Agung eruption a decrease of surface air temperature and of the temperature of the 850–300 hPa layer was observed, though it was not unique. However, the El Chichón eruption was followed by a warming both near the surface and in the troposphere. The global mean temperature of the troposphere in 1983 turned out to be the highest since 1958. An examination of data on the temperature in both hemispheres after the Agung eruption showed that whereas surface air tem-

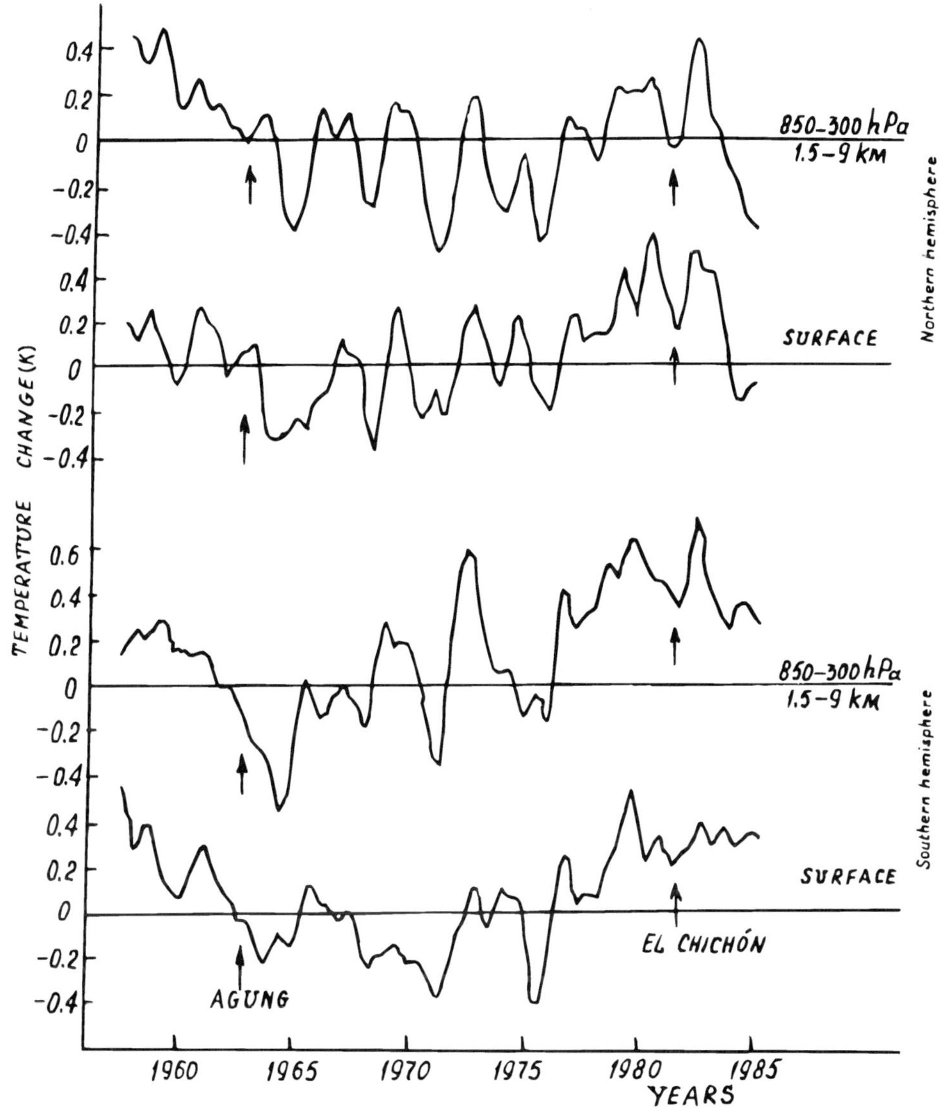


**Figure 2.8** Variations in air temperature near the surface and in the troposphere in the northern and southern hemispheres.

perature decrease in the northern hemisphere is clearly pronounced and suggests its volcanic origin, the tropospheric cooling is not as definite.

In the southern hemisphere the eruption-induced surface air temperature decrease was not so pronounced, since the cooling trend had taken place before the eruption and continued after it. In this case an analysis of observational results is seriously hindered by their fragmentary character. A combined analysis of the sea surface temperature near the equator (0–10° S, 80–180° W) and of the tropical troposphere temperature (30° S, 30° N) revealed (Figure 2.9) a high correlation (the correlation coefficient is 0.6, with a time shift of about 6 months). In this connection, it is very important that a few months after the El Chichón eruption an anomalously

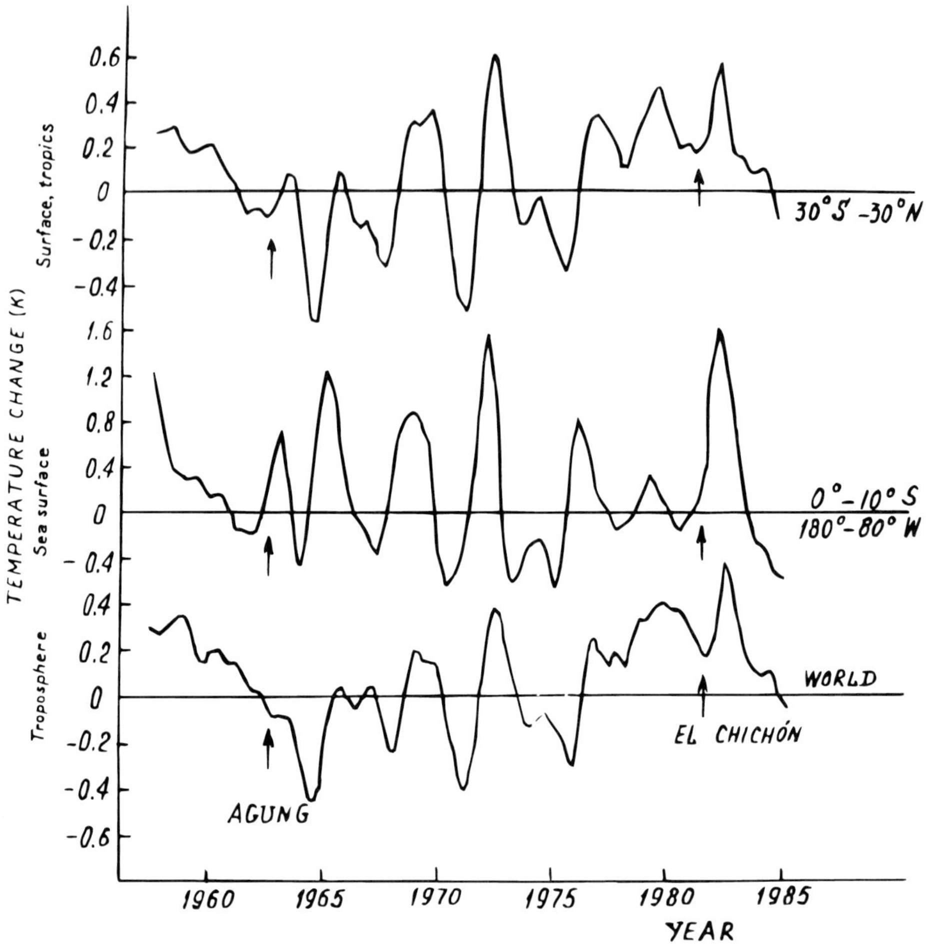


**Figure 2.9** A comparison of global, tropospheric air temperature variations with those of surface air temperature in the tropics and sea surface temperature in the eastern equatorial Pacific.

intensive El Niño event occurred (a strong positive sea surface temperature anomaly which occurs every few years in the Pacific Ocean) which must have caused a heating of the tropical troposphere, mainly due to the latent heat of condensation, with intensive cloud formation.

The prevailing effect of El Niño appeared to be the cause of the post-El-Chichón tropospheric warming, which showed itself not only in the tropics but also on a global scale. The El Niño event was also observed after the Agung eruption (such coincidences have brought forth a hypothesis of a volcanic origin of El Niño), but it was much weaker – its signal did not suppress the volcanic signal. Thus, to detect the volcanic signal the El Niño contribution should be excluded, which is very difficult, because the relations between various factors are non-linear and because of the delay of the sea surface temperature change following the surface air temperature anomalies in the tropics.

Although the sea surface temperature anomalies are twice as large as the surface air temperature anomalies and with a phase shift assumed to be two seasons, even an approximate estimate makes it possible to recognize reliably the volcanic signal in the variations of the surface air temperature and of the tropospheric temperature in the tropics after both eruptions (Figure 2.10), but an application of this approach to hemispheric and global estimates is not as convincing.

Later, Angell (1988a) made a further analysis of the data of global aerological observations at 63 stations for the 30-yr period 1958–1987 and the 15-yr period 1973–1987 in order to determine the laws of surface air temperature variability as well as mean temperatures of the 100–50, 300–100 and 850–300 hPa layers. The data considered (Table 2.6) reflect an increase of global mean surface air temperature and $T$ (850/300), i.e. temperature in the 850–300 hPa layer, at a 5 per cent significance level, of 0.08 °C and 0.09 °C, respectively, for 10 yr.

During a 5-yr period, both surface air temperature and $T$ (850/300) have decreased statistically to a negligible extent: −0.03 °C in 10 yr in the northern polar zone, 60–90° N and, respectively, −0.010 °C and −0.09 °C in 10 yr in the surface air temperature (mean-latitude northern hemisphere zone, 30–60° N), whereas global

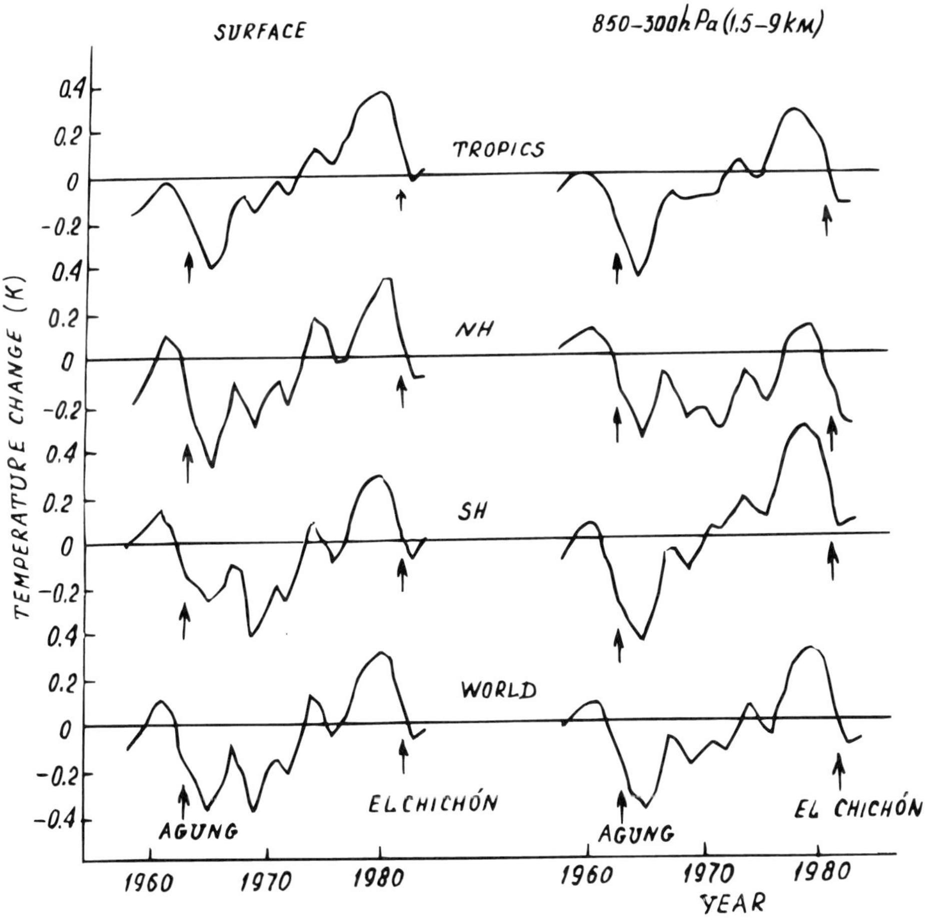


**Figure 2.10** Air temperature variations near the surface and in the troposphere, on average, for the hemispheres and over the globe, excluding the contributions of SST anomalies in the eastern tropical Pacific.

**Table 2.6** Changes in global mean (G) and hemispheric mean (SH and NH) temperature of different atmospheric layers (°C for 10 yr), from data of aerological observations during the periods 1958–1987 and 1973–1987

| | 1958–1987 | | | 1973–1987 | | |
|---|---|---|---|---|---|---|
| Layer (hPa) | G | NH | SH | G | NH | SH |
| 100–50 | | | | $-0.62 \pm 0.48$ | $-0.34 \pm 0.38$ | $-0.90 \pm 0.67$ |
| 300–100 | $-0.18 \pm 0.10$ | $-0.18 \pm 0.13$ | $-0.15 \pm 0.11$ | $-0.29 \pm 0.32$ | $-0.15 \pm 0.39$ | $-0.43 \pm 0.34$ |
| 850–300 | $0.09 \pm 0.09$ | $-0.02 \pm 0.10$ | $0.19 \pm 0.10$ | $0.24 \pm 0.26$ | $0.14 \pm 0.28$ | $0.26 \pm 0.28$ |
| Repeatability | $0.08 \pm 0.08$ | $0.08 \pm 0.09$ | $0.09 \pm 0.09$ | $0.28 \pm 0.17$ | $0.10 \pm 0.29$ | $0.34 \pm 0.20$ |

mean values of surface air temperature and *T* (850/300) have increased by 0.28 °C and 0.24 °C in 10 yr (the latter value is statistically insignificant). The global mean *T* (300/100) value has decreased by 0.18 °C in 10 yr (at the 1 per cent significance level). A decrease of *T* (300/100) is characteristic for all latitudinal belts considered (90–60° N, 60–30° N, 30–10° N, 10° N–10° S, 10–30° S, 30–60° S, 60–90° S), with a maximum decrease in the southern polar zone due to the ozone hole phenomenon. The *T* (100/50) value decreased substantially (−0.62 °C in 10 yr) during the period 1973–1987, a decrease in the southern polar zone reaching −2.04 °C in 10 yr. The 30-yr period 1958–1987 is characterized by an increasing meridional temperature gradient between equatorial and northern polar latitudinal belts, both near the surface and in the 850–130 hPa layer. However, in the southern hemisphere the gradient decreased near the surface and remained unchanged in the 850–300 hPa layer. The global mean climate warming, both near the surface and in the 850–300 hPa layer, was much stronger in March–May and in June–August than in December–February and in September–November.

From data for the period 1958–1987, a correlation (equal to 0.71 at the 1 per cent significance level) was observed between sea surface temperature in the eastern equatorial Pacific and *T* of the 850–300 hPa layer in the equatorial belt, with a time shift of one season. With increasing latitude, this correlation is reduced and the time shift increases.

The observational data under discussion could be considered as confirmation of climate warming due to the greenhouse effect, since (i) in 1958–1987 a statistically negligible increase was observed in global mean values of surface air temperature and *T* (850/300) and (ii) in 1983–1987 there was a considerable decrease of *T* (100/50). However, this conclusion is not correct, because there was (i) a small decrease of surface air temperature and *T* (850/300) in the northern hemisphere polar and mid-latitudes in the period 1958–1987 and (ii) a small decrease of surface air temperature in the northern hemisphere polar zones and in the southern hemisphere between 1973 and 1987. The fact that between 1958 and 1987 a global-scale cooling of the 300–100 hPa layer (the tropopause layer) was observed, can be considered neutral because numerical modelling, as a rule, does not predict a greenhouse cooling near the tropopause. An intensified increase of surface air temperature observed in winter in the northern hemisphere and southern hemisphere polar zones, compared with other seasons, agrees with results of calculations.

On the whole, a conclusion can be drawn about the unjustified categorical statements that the greenhouse impact on climate is now being observed. Major difficulties in the $CO_2$ signal identification are connected with global-scale forcings on the fields of temperatures of volcanic eruptions and sea surface temperature anomalies (such as the El Niño / Southern Oscillation – ENSO), as well as the formation of the ozone hole in the Antarctic. The latter, for instance, is responsible for a clearly manifested 1 °C decrease of *T* (100/50) in the southern hemisphere which was observed at the end of the phase of warming in the lower stratosphere and which was caused by the 1982 El Chichón eruption. The post-1985 increase of surface air temperature, *T* (850/300) and *T* (300/100) is, at least partly, explained by El Niño. The impact of El Niño reflects the similarity of temperature variations both in tropical and subtropical latitudinal belts. The 1963 Agung volcanic eruption did cause a substantial and prolonged cooling in the southern hemisphere. The effect of the anomaly on the field of tropospheric temperature determined the absence of the post-El-Chichón tropospheric temperature decrease.

As Ellsaesser (1985) notes, the empirical information which served as the basis for most of the investigators in assessing climatic changes during instrumental observations, includes almost no data from the equatorial zone of latitudes, about 20° wide, mainly over the oceans. At the same time, this zone covers 34 per cent of the hemispheric area. Some investigators have shown that the trend of temperature variations in this zone of the northern hemisphere is inverse to that in the Arctic. Therefore, if all the observational data were available and taken into account correctly, there is a possibility that temperature variations in the equatorial zone can either compensate for, or even change the sign of, the hemispheric mean temperature variations. The significance of data for the tropics has been demonstrated by Rind (1995).

A detailed analysis of regional features of the temperature trends in the northern hemisphere has shown that the variability of increases and decreases of temperature has sophisticated latitudinal–longitudinal structure and localization. On average, the amplitude of surface air temperature fluctuations increases with increasing latitude. In mid-latitudes, temperature variations for individual years and regions do not exhibit any clear latitudinal dependence. Therefore, large amplitudes of temperature changes in the Arctic, and the alternating sign of these changes in middle and low latitudes, determine an increasing trend of the amplitude of temperature changes in the Arctic, compared with mid-latitudes (in the case of mean values). Real features of large-scale temperature variations are complicated and not yet explained with theoretical models.

Note should be taken of the presence in real data of long-term periods of cooling and warming which have taken place in all periods, including the Little Ice Age (AD 1400–1850). So, for instance, temperature minima were observed in the 1810s, 1840s, 1910s and 1960s. Between these minima, temperature increases took place that were not connected with the $CO_2$ concentration increase. If we analyse these surface air temperature fluctuations and take into account that they have been observed either during small anthropogenic $CO_2$ emissions or without them, we may suppose that the warming of the 1920s to 1940s was an ordinary manifestation of such a temperature wave, i.e. it characterized the return of temperature to its norm, from very low temperatures of the Little Ice Age to the higher temperature regime similar to the small climatic optimum (AD 900–1300). The nature of these temperature waves is still unclear. However, there is considerable evidence that they resulted from a change of circulation epochs in the atmosphere and also, probably, in the ocean, which is illustrated by a sharp change of the types of circulation.

Empirical data show that at present the temperature in Greenland remains below the temperature of the period of colonization of the island by the Vikings during the small climatic optimum. The latter, however, had not been the warmest period in the Earth's history or even in the history of the Holocene, though at that time the temperature had been 0.5–1.0 °C higher than the present temperature. During the great climatic optimum (5000–8000 yr ago) the temperature was 2–3 °C higher than at present. Therefore, even if in the next decade the temperature rose by 2 °C, it would not necessarily point to a manifestation of the $CO_2$-induced greenhouse effect, because the climatic optimum must have been determined by the effect of other (natural) factors and, in particular, by changing parameters of the Earth's orbit, since not even a small increase of $CO_2$ at that period has been recorded (Berger and Tricot 1986).

But let us return to the problem of present-day climate changes. Following Ellsaesser (1985), if we exclude the 0.34–0.97 °C temperature increase during the period 1917–1938 from the temperature trend corresponding to the middle of the twentieth century (the choice of 0.34 °C or 0.97 °C values depends on the use of 15-yr running means or annual data) then no substantial trends are observed in the remaining temperature record. So if we want to assume the hypothesis of the effect of $CO_2$ increase on surface air temperature, we have to explain the amplitude of the warming trend that occurred without a $CO_2$ increase during the period 1917–1938.

Assuming that the atmosphere retained about 50 per cent of emitted carbon during the period mentioned above, the content of carbon ought to have increased by 43 Gt. This corresponds to a 4 per cent (11 ppm) increase of the atmospheric $CO_2$ concentration. With an assumption, according to the models, that a doubling in $CO_2$ causes a 3 per cent increase of temperature, and with logarithmic interpretation for this period, the $CO_2$-induced temperature increase should be only 0.17 °C, i.e. less than half as much as in reality.

Despite some uncertainty of the pre-industrial $CO_2$ content, this result can be partly reinforced. The pre-industrial level of atmospheric $CO_2$ is estimated at 260–290 ppm, and the $CO_2$ concentration in 1938 was 310 ppm. This corresponds to a 7–19 per cent growth of $CO_2$ (this seems to be an upper-limit estimate). With the same logarithmic interpolation, such a growth causes a $CO_2$-induced temperature increase of about 0.3–0.75 °C. The estimates obtained are close to those needed to justify the $CO_2$ hypothesis, but in this case such an explanation contradicts the temperature trend observed after 1938, when a cooling took place. This explanation does not prove to be correct either for the whole period of instrumental observations. So, for instance, in 1983 the content of $CO_2$ in the atmosphere reached 345 ppm, which corresponds to a 19–33 per cent increase of $CO_2$ over a hundred years (depending on the pre-industrial level of concentration).

With logarithmic interpolation, this corresponds to a possible 0.75–1.23 °C temperature rise. However, empirical data on surface air temperature variations between 1860 and 1984 show that the average warming (with an account of the trend of the 1930s) was about half as much. In this case we either have to assume that the doubled-$CO_2$-induced temperature increase should be not 3 °C, as deduced from modelling results, but 1.5 °C, or we have to acknowledge the non-synchronicity in atmospheric reaction to a $CO_2$ increase, as well as other factors affecting the climate. Since these factors had also acted in the pre-industrial period, there are no grounds to think that they do not act now nor that they do not govern and will not govern the present and future climate changes.

Another unexpected feature of available observational data is that in the northern and southern hemispheres, temperature variations are different, which is illustrated by the results in Table 2.7. This disagrees with early theoretical estimates which point to a $CO_2$-induced temperature rise towards the poles, independent of the hemisphere.

We are not trying to argue that there is no global warming resulting from anthropogenic causes, principally emissions of $CO_2$ into the atmosphere. But we are trying to argue that it is very difficult to separate the anthropogenically induced effects on the climate from the natural climatic fluctuations. If one is not very careful, predictions of climate change caused by $CO_2$ increase, based on the analysis of empirical data, may be hasty and unreliable. Even more unreliable are the forecasts of a catastrophic climate warming in the near future. This problem is far from being understood. A reasoned answer to the question of whether real data from

**Table 2.7** Average temperature characteristics for 5-yr periods between 1955 and 1975

| Hemisphere | Parameter | Region | Temperature change (°C) 1955–1960 | 1960–1965 | 1965–1970 | 1970–1975 | Average |
|---|---|---|---|---|---|---|---|
| Northern | TMT | 0°–90° | no data | | −0.068 | −0.324 | −0.196 |
| | SAT | 0°–90° | 0.088 | −0.204 | −0.068 | −0.068 | −0.068 |
| | TMT | 65°–90° | 0.556 | −0.944 | −0.300 | −0.208 | −0.224 |
| | SAT | 70°–90° | −0.760 | −0.428 | −0.184 | +0.412 | −0.240 |
| | TMT | 35°–90° | 0.328 | −0.316 | −0.620 | −0.160 | −0.192 |
| | TMT | 50°–90° | 0.476 | −0.476 | −0.584 | −0.264 | −0.212 |
| | SAT | 50°–90° | −0.156 | −0.072 | −0.308 | −0.324 | −0.018 |
| | SST | Northern Pacific | 0.304 | −0.492 | 0.168 | −0.512 | −0.128 |
| | SST | Northern Atlantic | −0.480 | −0.200 | −0.260 | no data | −0.313 |
| Southern | TMT | 10°–30° | no data | | 0.144 | −0.316 | −0.086 |
| | SAT | 0°–50° | 0.156 | −0.244 | 0.028 | −0.220 | −0.070 |
| | TMT | 0°–90° | no data | | 0.116 | 0.068 | 0.092 |
| | TMT | 60°–90° | no data | | 0.396 | 0.468 | 0.432 |
| | TMT | 30°–90° | no data | | −0.016 | 0.180 | 0.082 |
| | TMT | 10°–30° | no data | | 0.272 | 0.104 | 0.188 |
| | SAT | 0°–20° | 0.168 | −0.508 | 0.072 | −0.112 | −0.095 |

*Note*: TMT = tropospheric mean temperature;
SAT = surface air temperature;
SST = sea surface temperature.

observations contain evidence for climate change due to $CO_2$ increase will be considered below.

Setting themselves the task of discussing in detail the climatic changes observed during the last century, Ellsaesser *et al.* (1986) analysed all the available data from instrumental observations of climatic parameters and reviewed studies on global climate trends during the last century; a brief discussion was also made of some palaeoclimatic data. Information about the surface air temperature was considered particularly carefully (Figures 2.11 to 2.18). Summaries were also compiled on precipitation, the extent of snow and ice cover, world sea level and the advance or recession of continental glaciers.

The principal conclusion from this analysis was that existing information was not reliable enough for global climatic trends to be revealed, because of the complicated variability and causal nature of climate. Although data on the northern hemisphere air surface temperature constitute a major source of information for revealing global trends, they cover, in fact, less than half the globe. Moreover, according to Jones *et al.* (1986), these data are the least representative (from the viewpoint of global climate characteristics) compared with other observational series, such as data on surface air temperature for both hemispheres as well as data from ship observations of air temperature and sea surface temperature.

During recent years a number of interesting publications have appeared on climate diagnostics which change substantially the ideas and assessments of secular variations of surface air temperature during the last hundred years.

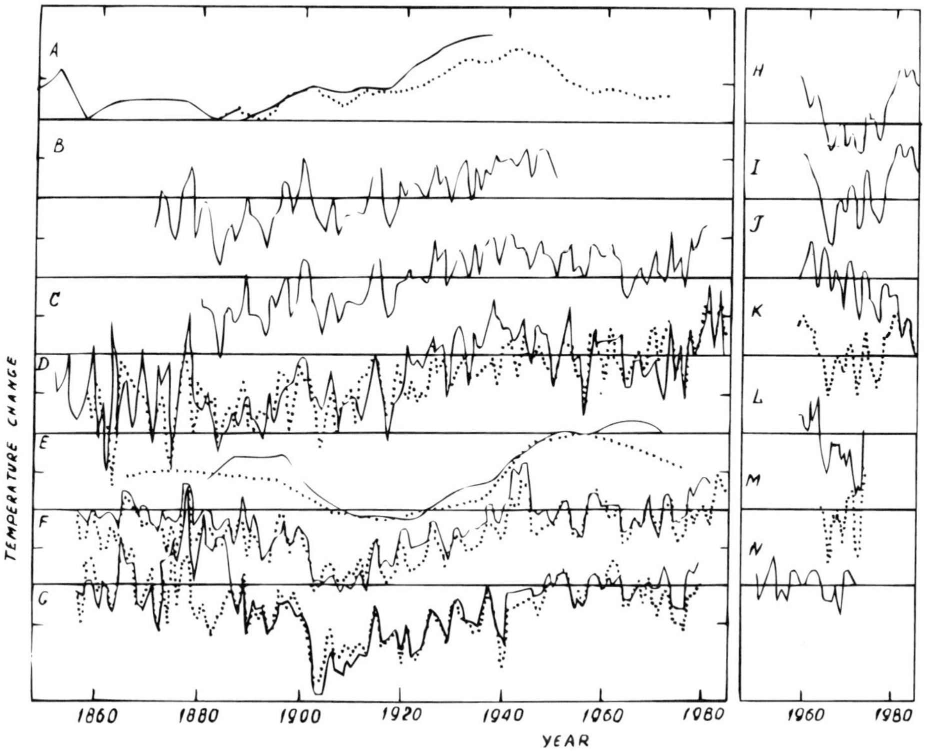


**Figure 2.11** Changes in global mean temperature (surface air unless otherwise stated). Scale: 0.5 K between lines. Curves A: solid curve, Willett (1950) as corrected by Mitchell (1961); dotted curve, Mitchell (1961) as updated by Lamb (1981); 5-yr mean departures from 1880–1984 normal. Curve B: Callendar (1961), annual mean 60° S–60° N departures from 1901–1930 normal. Curve C: Hansen *et al.* (1981), also private communication (1985), annual mean departures from 1881–1980 normal. Curves D: Jones *et al.* (1986), annual mean departures from 1951–1970 normals (solid curve, NH; dotted curve, SH). Curves E: solid curve, Paltridge and Woodruff (1981), 5-yr mean sea surface temperature (SST) and land surface air temperature departures using primarily ship data, renormalized to 1951–1960 period; dotted curve, Folland and Kates (1984), also private communication (1985), 5-yr mean ship SST departure from 1951–1960 normal with linear in time correction for switch from bucket to injection water temperatures. Curves F: Folland *et al.* (1984), also private communication (1986), annual mean ship (solid curve) near surface air temperature (NSAT) and (dotted curve) SST departures from 1951–1960 normals; NSATs corrected for ship size and SSTs for switch from bucket to intake water temperatures. Curves G: COADS (comprehensive ocean data set) annual mean ship (solid curve) SAT and (dotted curve) SST departures from 1951–1970 normals with corrections of Folland *et al.* (1984) and war years 1918–1919 and 1942–1945 omitted. Curve H: Angell and Korshover (1983) and CDIC (Carbon Dioxide Information Center), seasonal mean departures from 1958–1977 normal with 1-2-1 smoothing applied 12 times (approximates amplitude of annual mean). Curve I: Angell and Korshover (1983), for temperature of 850–300 hPa layer. Curve J: Angell and Korshover (1983), for temperature of 300–100 hPa layer. Curve K: Angell and Korshover (1983), for temperature of surface to 100 hPa layer. Curve L: Oort (1983), monthly mean departures from 1958–1973 monthly normals with 15-point Gaussian smoothing. Curve M: Oort (1983), for pressure-weighted 1000–25 hPa temperature. Curve N: Chen (1982), annual mean departures from 1949–1972 normal, combined land/sea data. After Ellsaesser *et al.* (1986).

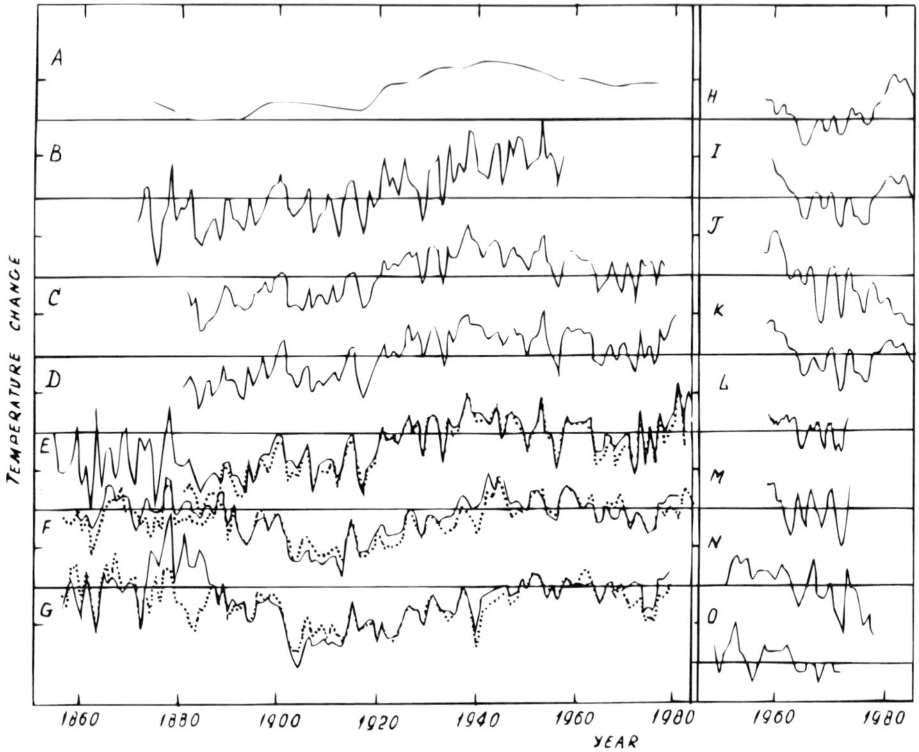


**Figure 2.12** Changes in northern hemisphere temperature (surface air unless otherwise stated). Scale: 0.75 K between lines. Curve A: Mitchell (1961) as updated by Reitan (1974), Brinkmann (1976), and NCAR, from Lamb (1981), 5-yr mean departures from 1880–1884 pentad. Curve B: Callendar (1961), annual mean departures from 1901–1930 normal. Curve C: Vinnikov *et al.* (1980), annual mean departures from 1881–1978 normal. Curve D: Hansen *et al.* (1981), also private communication (1985), annual mean departure from 1881–1980 normal. Curves E: solid curve, Jones *et al.* (1986), annual mean departure from 1951–1970 normal; dotted curve, Jones *et al.* (1982) and CDIC, same as above, renormalized to same period. Curves F: Folland *et al.* (1984), also private communication (1985), annual mean ship (solid curve) NSAT (near-surface air temperature) and (dotted curve) SST departure from 1951–1960 normals; NSATs corrected for ship size and SSTs for switch from bucket to intake water temperature. Curves G: COADS annual mean ship (solid curve) SAT and (dotted curve) SST departures from 1951–1970 normals with corrections of Folland *et al.* (1984) and war years 1918–1919 and 1942–1945 omitted. Curve H: Angell and Korshover (1983) and CDIC seasonal mean surface temperature departures from 1958–1977 normal with 1-2-1 smoothing applied 12 times. Curve I: Angell and Korshover (1983) and CDIC, same as for curve H but for temperature of 850–300 hPa layer. Curve J: Angell and Korshover (1983) and CDIC, same as for curve H but for temperature of 300–100 hPa layer. Curve K: Angell and Korshover (1983) and CDIC, same as for curve H but for temperature of surface to 100 hPa layer. Curve L: Oort (1983), monthly mean departures from 1958–1973 monthly normals, 15-point Gaussian smoother. Curve M: Oort (1983), same as for curve L but for pressure-weighted 1000–25 hPa temperature. Curve N: Namais (1980), annual mean 1000–700 hPa temperature departures from 1951–1978 normal. Curve O: Chen (1982), annual mean departure from 1949–1972 normal based on combined land/sea data. After Ellsaesser *et al.* (1986).

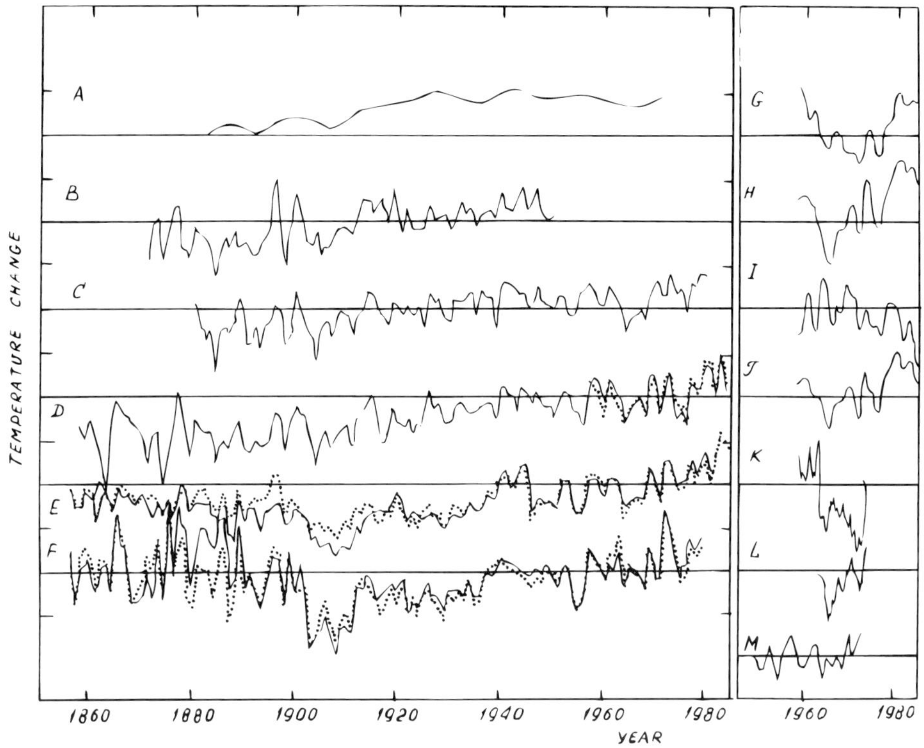


**Figure 2.13** Changes in southern hemisphere temperature (surface air unless otherwise stated). Scale: 0.75 K between lines. Curve A: Mitchell (1961) as updated by NCAR from Lamb (1981), 0–60° S 5-yr mean departures from 1880–1884 pentad. Curve B: Callendar (1961), 0–60° S annual mean departures from 1901–1930 normal. Curve C: Hansen *et al.* (1981), also private communication (1985), annual mean departures from 1881–1980 normal. Curves D: Jones *et al.* (1986), annual mean departures from 1951–1970 normals for (solid curve) SH60 (0–60° S) and (dotted curve) southern hemisphere total (SHT) (0–90° S) (after 1957 only). Curves E: Folland *et al.* (1984), also private communication (1985), annual mean ship (solid curve) NSAT and (dotted curve) SST departures from 1951–1960 normals; NSATs corrected for ship size and SSTs for switch from bucket to intake water temperatures. Curves F: COADS annual mean ship (solid curve) SAT and (dotted curve) SST departures from 1951–1970 normals with corrections of Folland *et al.* (1984) and the war years 1918–1919 and 1942–1945 omitted. Curve G: Angell and Korshover (1983) and CDIC, seasonal mean departures from 1958–1977 normal with 1-2-1 smoothing applied 12 times. Curve H: Angell and Korshover (1983), same as for curve G but for temperature of 850–300 hPa layer. Curve I: Angell and Korshover (1983), same as for curve G but for temperature of surface to 100 hPa layer. Curve K: Oort (1983), monthly mean departure from 1958–1973 normals with 15-point Gaussian smoother. Curve L: Oort (1983), same as for curve K but for pressure-weighted 1000–25 hPa temperatures. Curve M: Chen (1982), annual mean departures from 1959–1972 normal based on combined land/sea data. After Ellsaesser *et al.* (1986).

Having undertaking an overview of data on the secular trend of the global mean surface air temperature, Bloomfield (1992) came to the conclusion that the range of possible increase of surface air temperature should be estimated at 0.2–0.8 °C per 100 yr and noted that calculations made during recent years revealed the possibility

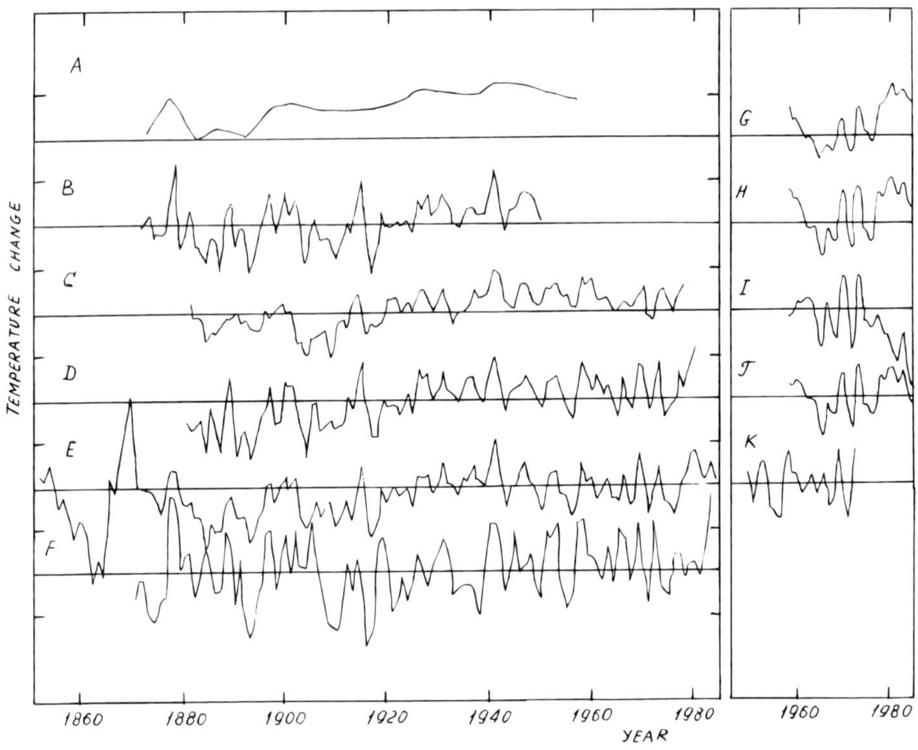


**Figure 2.14** Changes in tropical temperatures (surface air unless otherwise stated). Scale: 0.75 K between lines. Curve A: Mitchell (1961), 30° N to 30° S 5-yr mean departures from 1880–1884 pentad. Curve B: Callendar (1961), 25° N to 25° S annual mean departures from 1901–1930 normal. Curve C: Vinnikov *et al.* (1980), 17.5–37.5° N annual mean departures from 1881–1976 normal. Curve D: Hansen *et al.* (1981), also private communication (1985), 23.6° N to 23.6° S annual mean departures from 1881–1980 normal. Curve E: Jones *et al.* (1986), 0–25° N annual mean departure from 1951–1970 normal. Curve F: Angell (1981), also private communication (1985), seasonal SST departures from long-term mean in the equatorial Pacific area 0–10° S, 90–180° W (half scale, same as curve G). Curve G: Angell and Korshover (1983), 30° S to 30° N seasonal mean departures from 1958–1977 normal with 1-2-1 smoothing applied 12 times. Curve H: Angell and Korshover (1983), same as for curve G but for 850–300 hPa temperature. Curve I: Angell and Korshover (1983), same as for curve G but for 300–100 hPa layer. Curve J: Angell and Korshover (1983), same as for curve G but for surface to 100 hPa layer. Curve K: Chen (1982), 20° N to 20° S annual mean departures from 1958–1972 normal based on combined land/sea data. After Ellsaesser *et al.* (1986).

of practically total compensation of the greenhouse warming by the aerosol cooling due to the increased content of sulphate aerosol in the atmosphere.

Very interesting surface air temperature data for central England for the period 1659 to 1990 were analysed by Murray (1992), who found a strong variability of the annual mean surface air temperature values on a time scale of 10 yr before the beginning of the industrial revolution. So, for example, the surface air temperature amplitude between AD 1731 and AD 1740 constituted 3.7 °C (of course, this variability could have been only natural) and remained practically the same (3.8 °C) 332 yr later.

The estimates of the surface air temperature data reliability and other results presented at the Fifth Conference on Climate Change, organized by the American

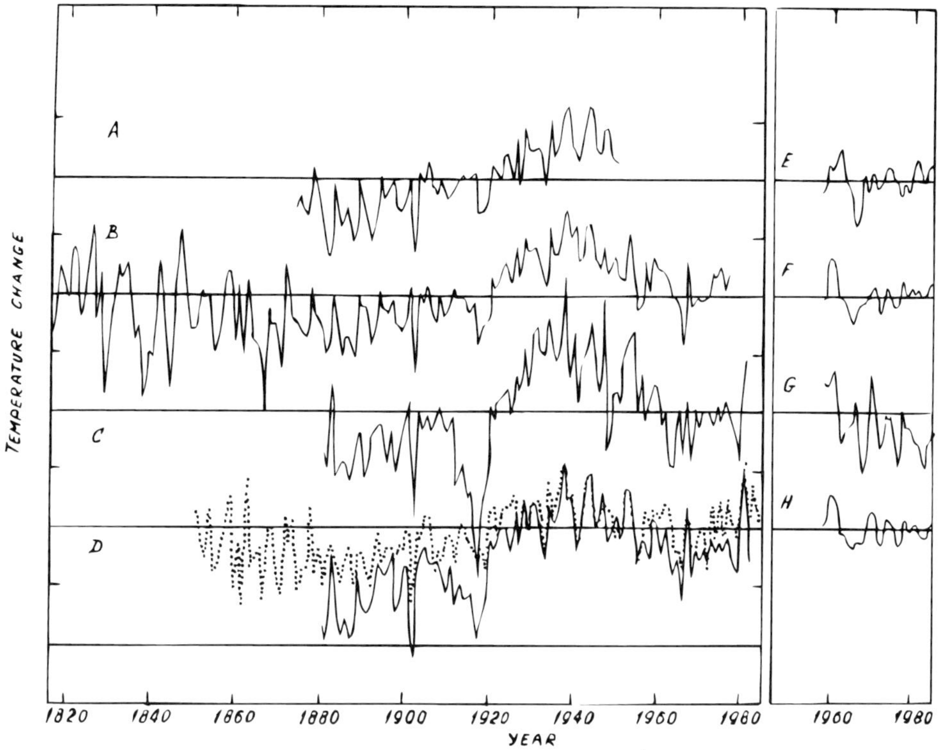


**Figure 2.15** Changes in temperature of the Arctic (surface air unless otherwise noted). Scale: 2 K between lines. Curve A: Callendar (1961), 60–90° N annual mean departures from 1901–1930 normal. Curve B: Self *et al.* (1981), 60–90° N annual mean departures from 1880–1977 normal. Curve C: Vinnikov and Lugina (1982), 72.5–87.5° N annual mean departures from 1881–1980 normal. Curves D: solid curve, Kelly *et al.* (1982) and CDIC, 65–85° N annual mean departures from 1946–1960 normal; dotted curve, 60–90° N annual mean departure from Jones *et al.* (1986) renormalized to 1946–1960. Curve E: Angell and Korshover (1983), 60–90° N seasonal mean departures from 1958–1977 normal, 1-2-1 smoother applied 12 times. Curve F: Angell and Korshover (1983), same as for curve E but for 850–300 hPa layer. Curve G: Angell and Korshover (1983), same as for curve E but for 300–100 hPa layer. Curve H: Angell and Korshover (1983), same as for curve E but for surface to 100 hPa layer. After Ellsaesser *et al.* (1986).

Meteorological Society (1991a) were also noteworthy. As Madden (1991) showed, one of the serious sources of errors in estimates of the global mean surface air temperature for the whole period of instrumental meteorological observations (from 1860 till now) was connected with evolution of the spatial representativity of the network. The proportion of land surface covered by observations increased from 20 per cent to 90 per cent between 1860 and 1965. Naturally, this gives rise to errors when calculating the global mean surface air temperature values from observational data, which can be estimated using results from numerical climate modelling.

Calculations with the use of the Community Climate Model (CCM) developed by scientists from the National Center for Atmospheric Research (NCAR) showed that the mean square deviations of true global mean surface air temperature values from those of the real network decreased from 0.224 °C in 1860 to 0.045 °C in 1950 and

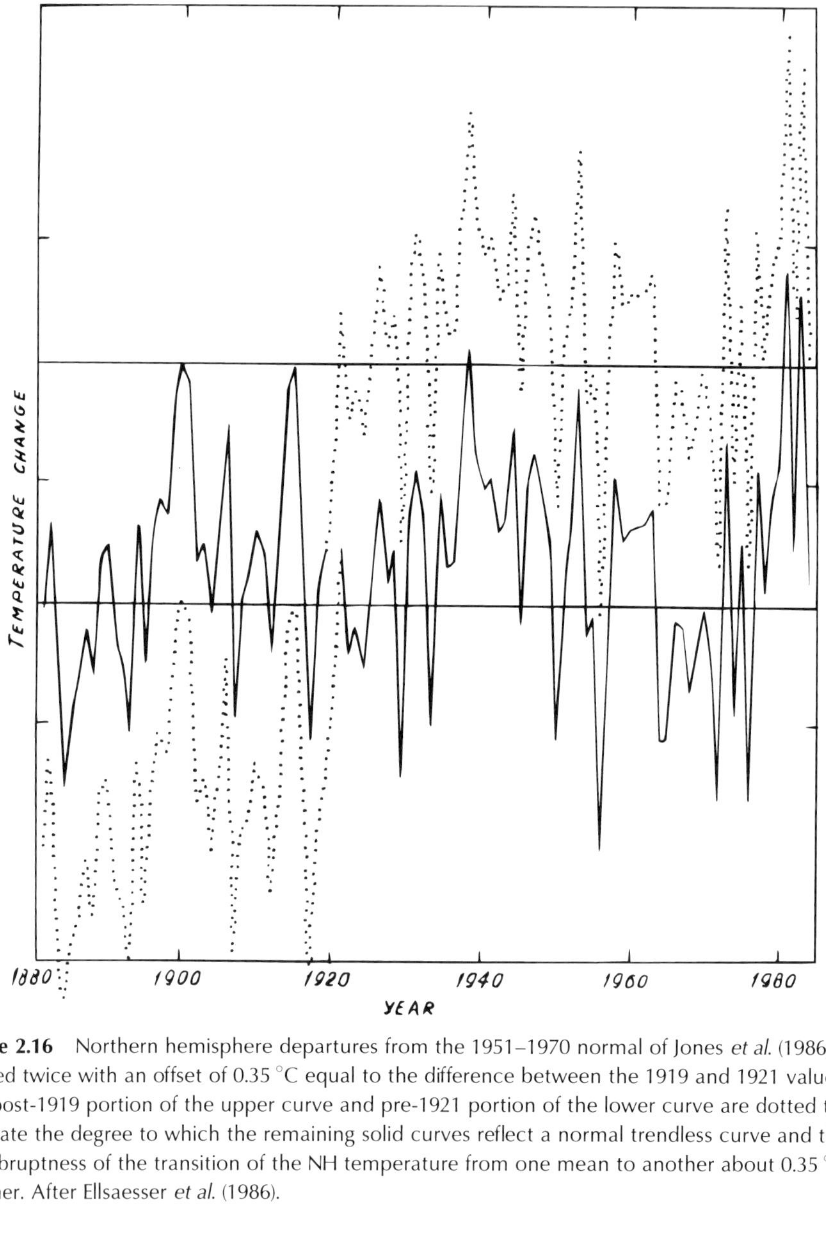


**Figure 2.16** Northern hemisphere departures from the 1951–1970 normal of Jones *et al.* (1986) plotted twice with an offset of 0.35 °C equal to the difference between the 1919 and 1921 values. The post-1919 portion of the upper curve and pre-1921 portion of the lower curve are dotted to illustrate the degree to which the remaining solid curves reflect a normal trendless curve and thus the abruptness of the transition of the NH temperature from one mean to another about 0.35 °C warmer. After Ellsaesser *et al.* (1986).

later. Therefore two techniques were suggested to correct the global mean surface air temperature values calculated from observational data, with the non-representativity of the network taken into account.

Philips *et al.* (1991) note that the analysis of the effect of adequacy of selected time series on the statistical characteristics of climate (first and second moments) which have been, so far, neglected, is complicated by the existing practice of data archiving, when only the results of calculations for time intervals not less than 6 hours are archived. In this connection, a simulation model was made of the

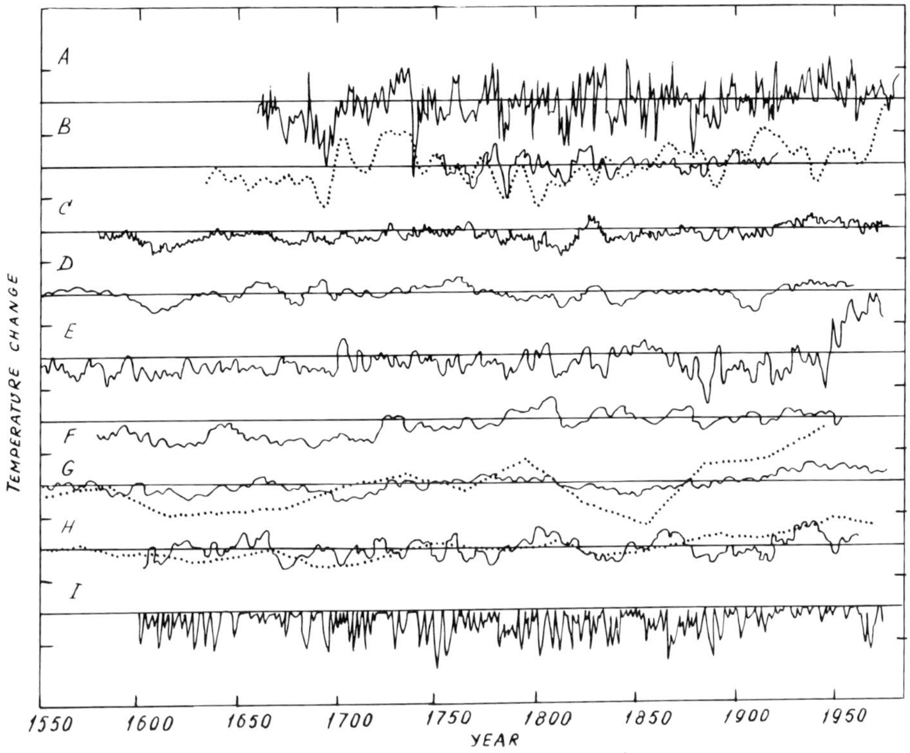


**Figure 2.17** Long-term and proxy records of northern hemisphere surface air temperature. Scale: 2 K between lines except where otherwise indicated below. Curve A: Manley (1974) and CDIC, central England departures from their mean. Curves B: solid curve, Humphreys (1940), extension of Köppen's (1873) weighted departures of NH stations from their mean; dotted curve, van den Dool *et al.* (1978), mean winter departures from mean at De Bilt with 1-2-1 smoothing applied 4 times. Curve C: Lauscher (1981), update of Groveman and Landsberg's (1979a, b) back extension of NH departures from 1881–1975 mean of Borzenkova *et al.* (1976). Curve D: Sirén (1961), Finland tree ring indices in arbitrary units (from Groveman and Landsberg, 1979a, b). Curve E: Gray (1974), reconstructed Tokyo winter temperature departures from 1881–1975 normal (from Groveman and Landsberg, 1979a, b). Curve F: Karlstrom (1961), Alaskan tree ring indices in arbitrary units (from Groveman and Landsberg, 1979a, b). Curves G: solid curve, Jacoby and Cook (1981), tree ring indices from 13 white spruce near Dawson, Yukon Territory, Canada, in arbitrary units; dotted curve, LaMarche and Stockton (1974), 30-yr mean tree ring indices from bristlecone pine averaged over nine sites in California, Nevada, Colorado, and New Mexico. Curves H: solid curve, Fritts and Lough (1985, also private communication, 1985), annual mean departures of US and SW Canadian surface temperature reconstructed from 65 arid-site tree ring chronologies; dotted curve, La Marche's 20-yr means of tree ring indices from White Mountains of California in arbitrary units (from Lamb, 1977). Curve I: Koch (1945), weeks per year with drift ice at the coast of Iceland in arbitrary units and reversed scale (from Lamb, 1977). After Ellsaesser *et al.* (1986).

minimum needed repeatability of data (frequency of observations) as well as relative contributions of intradiurnal and low-frequency variability when assessing the statistical characteristics of climate from numerical modelling data using the 19-level spectral (T-42) climate model developed at the ECMWF (European Centre for

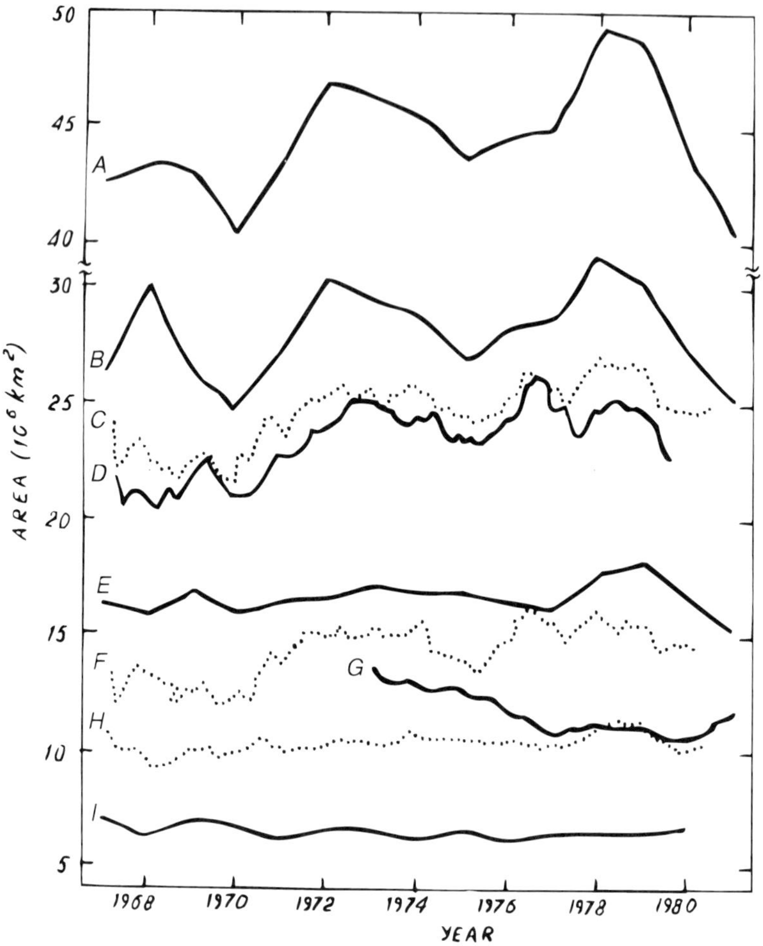


**Figure 2.18** Estimates of trends in sea ice and snow cover ($10^6$ km$^2$). Curve A: Foster and Rango (1982), NH average winter (December–February) snow cover. Curve B: Foster and Rango (1982), Eurasian average winter (December–February) snow cover. Curve C: Dewey and Heim (1982), 12-month running mean of NH weekly snow cover area. Curve D: Matson and Wiesnet (1981), 12-month running mean of NH monthly snow cover charts. Curve E: Foster and Rango (1982), North American average winter (December–February) snow cover. Curve F: Dewey and Heim (1982), 12-month running mean of Eurasian weekly snow cover area. Curve G: Zwally *et al.* (1983), 12-month running mean of Antarctic sea ice area. Curve H: Dewey and Heim (1982), 12-month running mean of North American weekly snow cover area. Curve I: US/USSR Workshop (unpublished manuscript, 1981), Arctic ocean summer (second half of August) sea ice cover. After Ellsaesser *et al.* (1986).

Medium-Range Weather Forecasts). The results of these calculations show that the choice of a 6-hour interval can be used for many climatic parameters, but it leads to a substantial loss of information in the case of such parameters as global solar radiation, sensible and latent heat fluxes near the surface and characteristics of convection processes, when a 3-hour interval is preferable.

Van Loon (1991) compared the trends of the annual mean surface air temperature values between 1880 and 1980 calculated as deviations from mean values for a long period (1946 to 1960) for the northern hemisphere and the Arctic (65–85° N) and found that (1) the variability of surface air temperature and its amplitude are higher in the Arctic than in the northern hemisphere, (2) between 1880 and 1940 the surface air temperature increased, then during the subsequent three decades it decreased and then, again, the surface air temperature increased, and this continues up to now, and (3) before 1950 the surface air temperature increase had not been monotonic, and the smoothed change of surface air temperature reflected rather a change in mean values from the first 40 yr to the next 20 yr of the period under consideration.

A detailed analysis of the surface air temperature variations for the period 1949 to 1972, when there was a decreasing trend of surface air temperature which could be analysed from data of aerological soundings, led to the conclusion that the hemispherical mean decrease of surface air temperature arose as a difference between large regional anomalies of different signs closely connected with variations in the atmospheric general circulation caused by the sensible heat transfer by quasi-stationary waves.

Space-derived meteorological information is becoming more and more useful. Satellite data series, which now cover more than thirty years, have made some climatological generalizations possible (there is, however, some disagreement concerning the interpretation of satellite data).

Analysis of the NOAA (National Oceanic and Atmospheric Administration) MSU (Microwave Sounding Unit) data, beginning from late 1978, revealed the long-term stability of channels 2 (53.74 GHz) and 4 (57.9 GHz), with weighting function maxima in the troposphere and lower stratosphere, respectively, which provided the possibility to use these data for climatological generalizations (Spencer and Christy 1991). Comparisons of data from six satellites for the period 1978–1991 revealed differences averaged over 10 yr in the brightness temperature ($T_B$) within 0.01–0.03 °C. The comparison of measured $T_B$ monthly means for channel 2 for 10 yr (1979–1988) with $T_B$ calculated from data of aerological measurements (averaged over squares 2.5° latitude by 2.5° longitude) revealed a very high correlation (with a correlation coefficient of 0.95–0.97). The observed $T_B$, with nadir viewing, correlates best with the 1000–200 hPa layer. The combination of data for several viewing angles provides the highest correlation with the 1000–400 hPa layer.

Using the NCAR climate model CCM-1, Christy *et al.* (1991) calculated the change of monthly-mean global-mean surface air temperature and mean tropospheric temperature between 1978 and 1987 with prescribed variations of the sea surface temperature over the grid 4.4° latitude by 7.5° longitude from observational data, which are quite scarce in the southern hemisphere (and completely absent south of 40° S).

Comparison of conventional (surface air temperature) and satellite (tropospheric temperature, MSU) observations revealed a low correlation; the correlation coefficient for seasonal surface air temperature values constituted only 0.32, which apparently resulted from inadequate parametrization of processes on the land surface. In the case of tropospheric temperature the correlation was higher (0.53). The model atmosphere can assimilate the heat from the ocean during strong positive anomalies of sea surface temperature (1979–1980, 1982–1983) but does not emit this heat to space so rapidly as the real atmosphere. On the whole, the results of numerical modelling show that the sea surface temperature anomalies can explain about 25 per cent of observed tropospheric temperature variability.

Data of surface observations of surface air temperature and marine observations of sea surface temperature are the most widely used source of information about the current variations of global climate. In connection with the urgency of analysis of the quality of the available data base, Christy and Trenberth (1991) assessed the reliability of the results of conventional marine observations, comparing them with the results of satellite observations made with the NOAA MSU.

The results of observations for six satellites from 1979 have given a homogeneous global data base over a grid of 2.5° latitude by 2.5° longitude. Data shown in Table 2.8 characterize the correlation coefficient of surface air temperature or sea surface temperature (conventional observations) and the results of satellite observations.

Analysis of the causes of low correlations, illustrated by the data of Table 2.8, requires first of all an estimation of the errors of observations. In the case of the MSU the noise level (0.005–0.023 °C$^2$) never exceeds 10 per cent variability (0.09–2.3 °C$^2$) and the standard error for the chosen grid varies from 0.07 °C (the tropics) to 1.5 °C (high-latitude continents). Estimation of the errors of conventional observations is more difficult.

Calculations have shown that in this case the variability of the errors reaches 1.39 °C$^2$ and the standard error of an individual observation is 1.2 °C (naturally it decreases by a factor of $\sqrt{N}$ for $N$ observations). The results obtained show that the previous conclusions about the temperature trends in the past should be treated cautiously, since they are very uncertain because of lack of acceptable representativity of the observational data.

Satellite data on total water content of the atmosphere obtained during the last 10–12 yr are of special interest for several reasons, as follows.

**Table 2.8** Coefficient of correlation between conventional and satellite observations of surface temperature

| Region | MSU2 | MSU2R | $\sigma_{MSU}$ | $\sigma_{SFC}$ | $S$ |
|---|---|---|---|---|---|
| Globe | 0.60 | 0.63 | 0.21 | 0.13 | 78 |
| Northern hemisphere | 0.52 | 0.65 | 0.21 | 0.19 | 90 |
| Southern hemisphere | 0.55 | 0.45 | 0.23 | 0.11 | 67 |
| North America | 0.90 | 0.95 | 0.69 | 1.32 | 100 |
| Eurasia | 0.77 | 0.89 | 0.40 | 0.95 | 96 |
| Australia | 0.50 | 0.78 | 0.47 | 0.49 | 92 |
| North Atlantic | 0.47 | 0.48 | 0.37 | 0.22 | 100 |
| South Atlantic | 0.30 | 0.23 | 0.31 | 0.24 | 77 |
| North Pacific | 0.37 | 0.49 | 0.40 | 0.32 | 100 |
| Indian Ocean | 0.56 | 0.56 | 0.30 | 0.22 | 95 |
| ESTPO | 0.82 | 0.83 | 0.54 | 0.58 | 80 |
| WSTPO | 0.10 | 0.17 | 0.27 | 0.18 | 93 |

*Note*: MSU2 and MSU2R denote, respectively, MSU channels 2 (the 1000–200 hPa layer) and a combination of MSU data for several viewing angles, which determines the temperature of the lower troposphere; $\sigma_{MSU}$ and $\sigma_{SFC}$ are mean square deviation values (°C$^2$); $S$ is the share (per cent) of the respective surface covered with observations; ESTPO and WSTPO are the east and west sectors of the Tropical Pacific Ocean.

1. The dynamics of water content reflects connections among evaporation, clouds and precipitation, as well as latent heat release, which is of key importance for the formation of climate variability in the tropics and for long-distance correlations between the processes in the tropical and extra-tropical latitudes.
2. Water vapour contributes most to the atmospheric greenhouse effect and determines the feedback functioning, whose account is of critical importance for assessing climate warming.
3. The Earth radiation budget formation is considerably determined by water vapour content both directly and through the effect on cloud formation. These circumstances have stimulated the appearance in 1990 of the programme GEWEX dedicated to studies of the role of water vapour in meteorological and hydrological processes. In this connection, Bates and Stephens (American Meteorological Society 1991a) performed an analysis of possibilities to retrieve the precipitable water from data of infrared and microwave observations using the following space-borne instruments: SSMR, 22 GHz channel (there are data for the period from 1979 to 1983); SSM/I, 22 GHz channel (from 1987 until now); HIRS/2, 6.7, 7.3 and 8.2 $\mu$m channels (from 1978 until now); AVHRR/2, 11.1 and 12.2 $\mu$m (from 1981 until now).

A comparison of observational results for the period 31 January to 5 February 1983 suggested the following conclusions. The infrared data are characterized by underestimated precipitable water (compared with microwave information) in the regions of overcast and broken cloudiness in the tropics and subtropics and by overestimated values in the subtropical regions of downward motions. Since the precipitable water field from infrared data is almost zonal (in contrast to that revealed by microwave observations), this illustrates serious differences between infrared and microwave information from the viewpoint of estimates of moisture transport from the tropics to mid-latitudes.

From the NOAA satellite data on the outgoing longwave radiation for 11 yr (1979–1989) in the 6.5 $\mu$m water vapour band, which serves as an indicator of water content in the upper troposphere, and for the channel 11 $\mu$m in the atmospheric window, Chesters and Neuendorfer (American Meteorological Society 1991a) considered the global dynamics of water content as a function of the low-latitude convection intensity characterized by the brightness temperature ($T_B$) at the wavelength 11 $\mu$m. Data from the infrared remote-sounding radiometer HIRS 6.7 and 7.2 $\mu$m channels were used to calculate $T_B$ corresponding to the centre of the 6.5 $\mu$m band using the relation

$$T_B(6.5) = 1.5\, T_B(6.7) - 0.5\, T_B(7.2)$$

This $T_B$ combination corresponds to a weighting function with a maximum near 350 hPa.

Analysis of global maps of the $T_B$ monthly means (for the whole 11-yr period) reveals the regions of $T_B(11)$ minima in low latitudes which reflect the presence of high-top powerful convective clouds (for example, in the regions of monsoons in the northern Indian Ocean or Central America in July–September and in Indonesia in January–February). As a rule, low values of $T_B(6.5)$ are observed, except for the regions of prevailing downward motions in the upper troposphere (the Middle East, the south Indian Ocean, the south-eastern sector of the Pacific Ocean in June–September; the central Pacific Ocean in the northern hemisphere in January–

February). The dynamics of the $T_B(6.5)$ global field contains hardly any signs of the effect of the El Niño phenomenon on the water content of the upper troposphere. The variability of $T_B(11)$ values averaged over the eastern and western hemispheres is characterized by a clear-cut semi-annual change with a phase shift of 180° in different hemispheres. The semi-annual change of $T_B(6.5)$ is weakly manifested.

From the data of Nimbus-6 and Nimbus-7 ERB (Earth radiation budget) wide-angle sensors, Ratan *et al.* (1991) drew global maps of monthly means of albedo and absorbed solar radiation for 140 months (from 1975 to 1987) over a grid 5° latitude by 5° longitude. For example, they drew a map of absorbed solar radiation in December averaged over 12 yr, which illustrated a comparatively smooth field with maxima in the tropical belts of the oceans. The map of standard deviation of the absorbed solar radiation for 12 yr shows maxima in the tropics of the Pacific Ocean, in the equatorial Atlantic Ocean and in the middle part of the Indian Ocean. A maximum of about 25 W m$^{-2}$ in the central Pacific is explained by the El Niño phenomenon taking place in 1982 and 1986.

Smith *et al.* (1991) analysed a 12-yr series of the absorbed solar radiation (ASR) monthly means, obtained from data from Nimbus-6 and Nimbus-7 wide-angle sensors. An expansion of the time series in empirical orthogonal functions showed that in 1983 and 1987 there were observed considerable anomalies of absorbed solar radiation due to the El Niño phenomenon. These two El Niño events were characterized by similarly distributed absorbed solar radiation, although, from the viewpoint of spatial and temporal variabilities, the 1987 absorbed solar radiation anomaly was stronger. The contribution of the first empirical orthogonal function to the retrieved absorbed solar radiation variability constitutes 0.9106 (those of the second and third constitute only 0.0283 and 0.0122, respectively) and, upon the filtering out of annual change, 0.102. The first and second orthogonal functions global maps calculated with the annual change being filtered out, reveal a similar dipole structure of the spatial distribution, with one pole over Indonesia, and the other over the equator near the International Date Line. These distributions are also characterized by some differences.

From the data of observations at 5328 stations (including the Antarctic), Diaz (1991) performed an analysis of precipitation variability on regional (North America, Australia) and global scales. The North-American and Australian continents were divided into regions taking account of the amplitude and phase of the annual change of mean monthly precipitation. Observational data point to different trends of precipitation in the tropical and extra-tropical regions of the northern hemisphere. (The same, but weaker, effect is observed in the southern hemisphere.) After World War II the precipitation intensified in high latitudes, while an opposite trend was observed in the tropics. The precipitation amount near the equator varied weakly.

To return to the analysis of surface air temperature, large-scale instrumental ground-based observations began in the late nineteenth century, near the time of minimum temperature in 1883. If it had been possible to have reliable data prior to 1850, then there probably would never have been such a concern about the climatic implication of a growth in $CO_2$ concentration. An account of additional data of observations and their thorough analysis performed by Jones *et al.* (1986) gave considerably higher surface air temperature values between 1881 and 1917 (in the Arctic in particular) and a lower amplitude of climate warming during the first half of the twentieth century. These data are illustrated in Figures 2.19 and 2.20 (World Meteorological Organization 1988g).

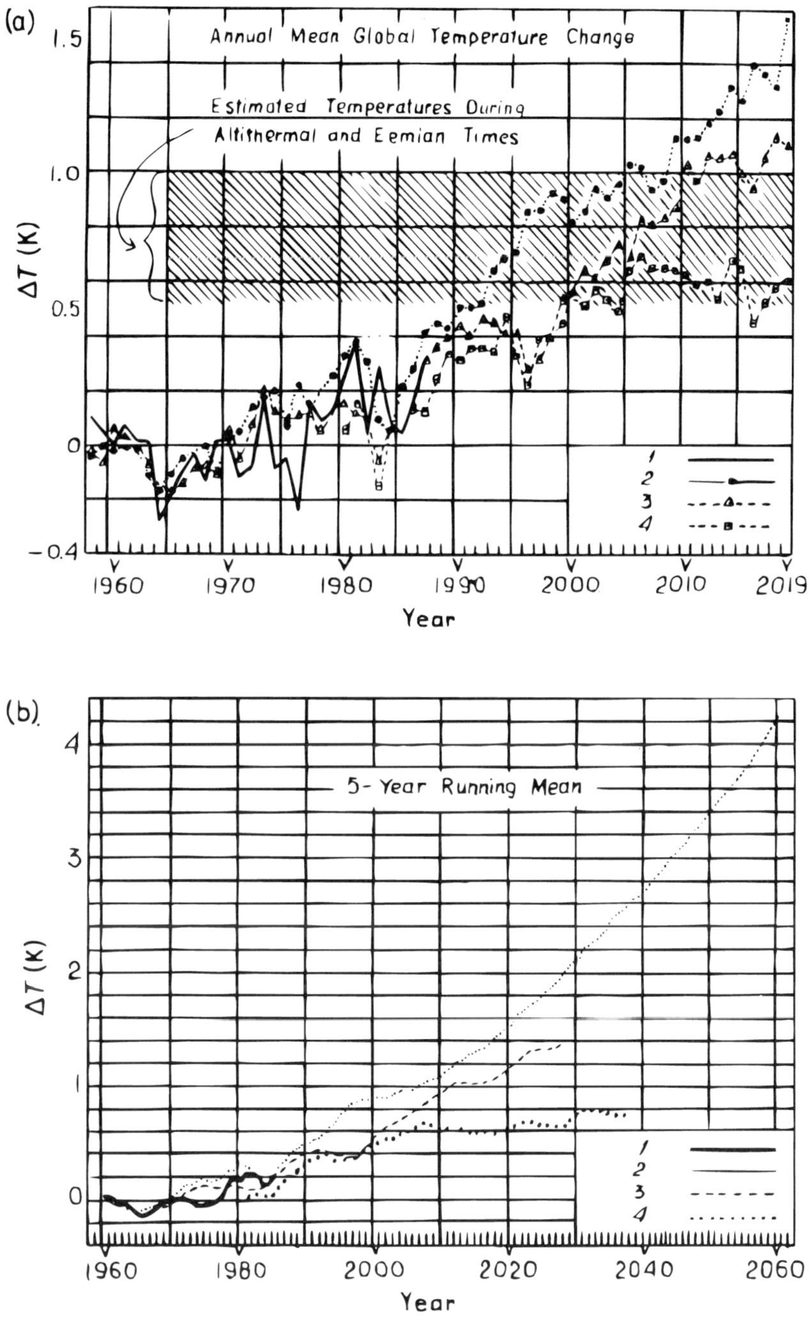


**Figure 2.19** Annual-mean global surface air temperature computed for scenarios A, B and C (see Hansen, 1988): (a) annual mean global temperature change, 1958–2019; (b) five-year running mean, 1960–2060. Curve 1 denotes observational data are from Hansen and Lebedeff (1987, 1988). Curves 2, 3 and 4 are for calculations based on scenarios A, B and C, respectively. The shaded range in (a) is an estimate of global temperature during the peak of the current and previous interglacial periods, about 6000 and 120 000 yr ago, respectively. The zero point for observations is the 1951–1980 mean (Hansen and Lebedeff, 1987); the zero point for the model is the control run mean. After Hansen (1988).

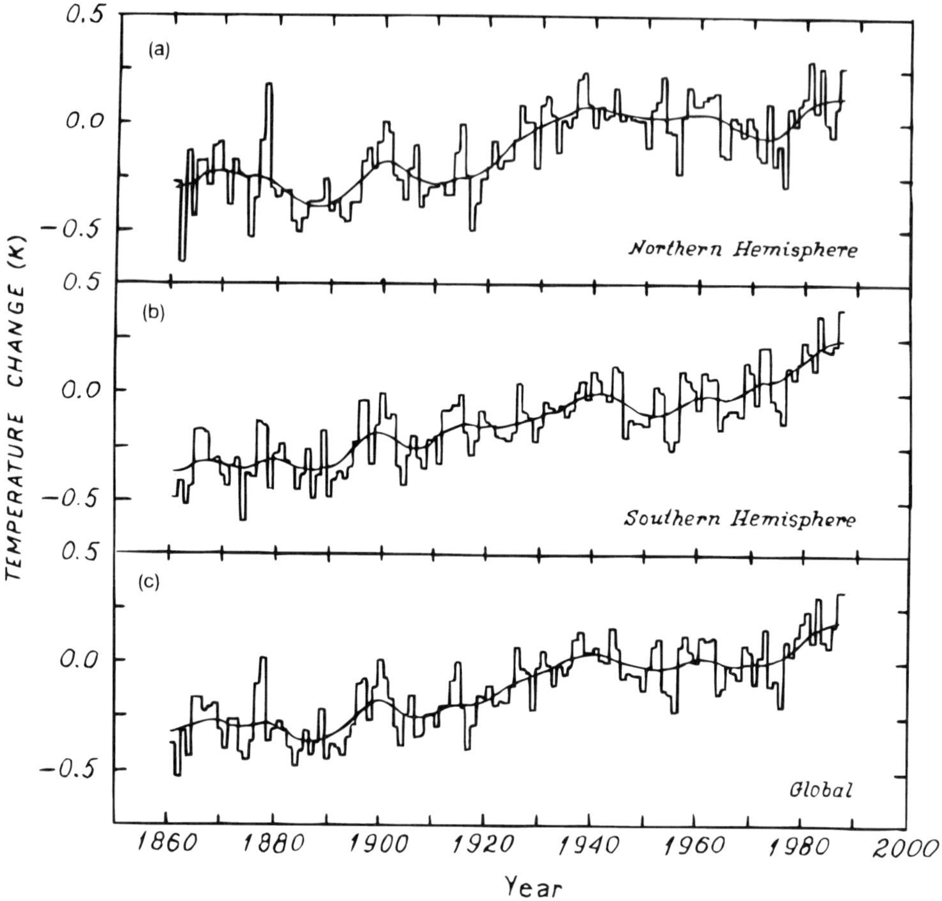


**Figure 2.20** Observed changes in temperature relative to the 1950–1979 reference period mean, for (a) the northern hemisphere, (b) the southern hemisphere and (c) the globe. The values shown are averages over both land and ocean areas. The smooth curves are obtained by applying a 20-yr Gaussian filter designed to show the longer time scale changes more clearly. Values prior to about 1900 are less reliable than those after 1900. Modified and updated from Jones *et al.* (1986).

There is no doubt that the global mean surface air temperature has been increasing (though not monotonically) from the time of the Little Ice Age and during the whole period of instrumental observations (starting from approximately 1750). This conclusion is verified both by data from direct observations and by indirect climatic indicators (retreat of glaciers, reduction of sea ice extent, shifting of the snow line and of the tree line in mountains, increase of the length of no-frost summer seasons, etc.).

Analysis of observational data has convincingly demonstrated that climate warming taking place during the last century has been followed by a very inhomogeneous distribution of regional climate changes over the globe. The 1980s, the warmest decade in the whole history of instrumental observations, has been no exception. So, for example, except for 1988 and 1990, 30 per cent of 2° latitude by 2°

longitude squares, covering the land surface, suffered a decrease of surface air temperature.

In this connection, Ropelewski *et al.* (1992) undertook a detailed analysis of spatial inhomogeneity of climate change during the 1980s, manifested in such parameters as surface air temperature monthly means, sea surface temperature, and snow cover extent (the latter was obtained from satellite observations). The surface air temperature anomalies assessed with respect to the data base for the period 1951–1980 were averaged over the decade 1981–1990.

Examination of the global distribution of surface air temperature anomalies has shown that with vast regions of positive anomalies in both hemispheres, negative anomalies in excess of $-0.5\,^{\circ}C$ are clearly observed in regions such as Greenland, Baffin Island and part of South America. Most substantial positive surface air temperature anomalies on land took place in central and eastern parts of Russia, Alaska and the western part of Canada, on the Antarctic peninsula and on the islands of the south-west Indian Ocean.

As for surface air temperature anomalies, they were usually weakly manifested and positive, except for the regions of the north Pacific and of the north Atlantic round Greenland. In all three oceans of the southern hemisphere the surface air temperature anomalies were positive, with an anomaly of more than $0.5\,^{\circ}C$ in the eastern Pacific, which was apparently connected with the El Niño events in 1982–1983 and, to some extent, in 1986–1987. The analysis of observational data suggested the conclusion that El Niño caused a modulation of global mean surface air temperature on a timescale of about 2–3 yr and, probably, also affected the 10-yr averages.

Observations show a trend of concentrations of surface air temperature and sea surface temperature anomalies in the winter and spring in the northern hemisphere, which determines the prevailing contribution of the northern hemisphere to the formation of global mean anomalies. There are considerable differences in temperature anomalies during the first (December–May) and second (June–November) parts of the year. So, for example, during the first 6 months, positive anomalies of more than $1\,^{\circ}C$ were observed in the regions of Alaska, west Canada and Russia and negative anomalies in Greenland. In contrast to this, in June–November weak negative surface air temperature anomalies occurred in Alaska and in west Canada, and those of sea surface temperature in the northern Pacific, whereas in north America the El-Niño-induced surface air temperature anomalies occurred.

The regions of Russia and Greenland are the only large territories of the northern hemisphere with considerable surface air temperature anomalies on a timescale of a decade not connected with either El Niño or other known modes of the interannual climate variability. During the second half of the year the spatial structure of the 10-yr surface air temperature anomalies was much weaker than in December–May when there were vast regions of positive anomalies in north Africa (here, at least partially, they resulted from a drought in the Sahel) and south Europe.

Analysis of the spatially inhomogeneous distribution of precipitation anomalies has been hindered by inadequate observational data and confined to the northern hemisphere. In December–February relatively arid conditions formed in the regions of south America (here an approximate conformity with the December–May positive surface air temperature anomaly took place) and south-east Asia. The spatial inhomogeneity was also characteristic of the snow cover extent variations. On the whole, the conclusion can still be drawn that it is not clear whether the climate changes considered are only manifestations of the internal variability of the climate system or whether they also reflect the existence of the trend of a transition to another climatic regime.

Ellsaesser *et al.* (1986) believed, however, that there were no grounds to ascribe the surface air temperature increase during the last century by about 0.5 °C to the effect of a $CO_2$ concentration increase alone, since that would entail neglecting the fact that the reference point in the analysis of the trend is a surface air temperature minimum, and, apparently, some part of the warming (probably 25–50 per cent) is determined by a return to the previously observed surface air temperature level. Explaining the warming solely as the effect of increasing concentration of $CO_2$ is unacceptable, given the existing differences between calculated and observed spatial and temporal variability of surface air temperature.

All observational data testify to a cellular spatial structure of the surface air temperature anomalies fields (both warmings and coolings). This situation took place not only in the periods of the northern hemisphere warmings in 1882–1940 and coolings in 1940–1960, but it was characteristic, for example, of surface air temperature variations happening in the twentieth century in the Indian subtropics sea surface temperature variations from the time of the last maximum glaciation to the present time and of variations in both of these temperatures during the last six thousand years.

Such a universal patchiness in surface air temperature variability points either to the inadequacy of the existing climate models or to radical differences between climate changes in the past and those expected from $CO_2$ increase. Although, according to numerical modelling data, an increase of warming with latitude is observed, no other features in common have been observed between calculated and observed surface air temperature variations. In particular, all attempts to reveal typical $CO_2$ signatures (correlations between the warming and other phenomena) have led to negative results. For instance, no cooling was observed in the stratosphere, which (according to calculations) ought to have followed an increase of surface air temperature.

For about forty years there has been not only an inversion of the sign of the northern hemisphere surface air temperature trend but (as recent studies have shown) also a climate warming in the southern hemisphere. This situation should be interpreted as being caused by the functioning of a so-far unidentified, natural mechanism for climate change, the role of which has turned out to be more substantial than that of $CO_2$. Naturally, until this mechanism is understood, all attempts to identify the climatic $CO_2$ signal and to forecast climate change due to $CO_2$ increase will fail.

This unidentified mechanism can, apparently, cause an increase of the equilibrium northern hemisphere surface air temperature by 0.35–0.4 °C over 2 to 4 yr, in the absence of the retarding effect of the ocean or any other retarding effect. An important question to be answered is whether the $CO_2$-induced greenhouse effect can determine the transition of the climate system from one equilibrium state to another through a step-like internal change, including such specific features, for instance, as details of the oceanic circulation, which govern the location of the polar front in the northern Atlantic.

Apparently, the local climate-forming mechanism had two results: (i) during the warming in the southern hemisphere no jump-like transition to a new equilibrium state took place and (ii) after a period of fast surface air temperature increase between 1917 and 1938 a pause occurred in the northern hemisphere temperature increase (and even the sign of the trend changed), but in the southern hemisphere a relatively continuous warming developed.

At present, the failure to perceive the climatic $CO_2$ signal in a hemispheric (northern hemisphere) or global climate warming, as predicted by equilibrium values in climate models, is usually explained by the retarding effect of the ocean. As Ellsaesser *et al.* (1986) note, this explanation faces at least two difficulties. One is connected with the fact that the warming of the ocean from the surface is similar to atmospheric cooling from the surface: as soon as this process begins, an increase of static stability brings forth an isolation of the surface layer from the bulk of the ocean. This complicates the conditions for warming the layers below the surface layer, which suffers forced mixing under the influence of local winds. If the $CO_2$-induced warming leads to wind attenuation (as predicted by numerous calculations), the thickness of the mixed layer could even decrease, compared with the present one.

The second difficulty is that the ocean can react to the disturbance via a redistribution of currents which determine the localization of the oceanic fronts or convergence zones, as could have happened, for instance, during the warming of the Arctic between 1917 and 1920. This reaction does not depend directly on the intensity of the vertical mixing and, though a delay in time may occur before the redistribution of the currents, the redistribution in itself can provide a rapid transition from one equilibrium state of the climate to another.

The tendency to ascribe the observed climate warming to $CO_2$ is disputable for other reasons, too. One of them is that the warmings in the past (altithermal and the mediaeval climatic optimum) happened without $CO_2$ interference. The second is that, of course, an increase of $CO_2$ concentration is not the only possible anthropogenic mechanism for the warming. There is no doubt that the imperfection of the schemes of parametrization of physical processes used in climate modelling leads to unreliability in assessing the effect of increasing $CO_2$ on the climate. There are grounds for believing that forecasts of climate warming due to a doubled $CO_2$ concentration have been overestimated. As Ellsaesser (1987a) notes, one of the causes of this overestimation is the incorrect parametrization of tropical convection.

The parametrization schemes used now do not take into account the fact that in the case of penetrating convection, the latent heat released from water vapour condensation is not input to the atmosphere at the level at which condensation takes place. Instead, it serves as an energy source for the buoyancy which drives the lifting of a convective bulb or plume to a level where it becomes gravitationally stable, acquiring the same density as the surrounding atmosphere. Though some further inertial lifting above this level can take place, this is only temporary.

This process shows itself in the formation of the so-called thunderstorm cloud anvils. Around the centres of the lifting air masses a more extensive zone of downward motion occurs, where (in the absence of water droplets) a dry-adiabatic lapse rate forms. It is these downward motions that determine the atmospheric heating in conditions of convection. Calculations of the $CO_2$-induced climate warming suggest that there is an intensification of the moisture cycle and an increase of atmospheric moisture content, which intensifies the greenhouse effect. However, as follows from the considerations mentioned above, an intensification of the hydrological cycle in the tropics, covering about half the globe, means an intensification of downward motions and, hence, a decrease of the tropospheric moisture content and an attenuation of the greenhouse effect. Thus, not positive but negative feedback occurs, caused by the moisture content variability.

The existence of such processes shows that in the tropical belt the intensification of the moisture cycle must lead to further drying of the arid regions and to moistening of the wet tropics. From this point of view, such climate changes as droughts in the Sahel region could be the first manifestation of the effect of $CO_2$ growth on climate (Ellsaesser 1987b). The second illustration of an overestimated climate warming is an incorrect account of the processes of interaction between the tropical atmosphere and oceans. It is difficult to see that the temperature of the ocean surfaces can rise by more than 2 °C, but calculations give 4 °C.

The problem of the transfer of the $CO_2$-induced atmospheric warming to the ocean has been discussed in detail by Schlesinger and Jiang (1988). They undertook a simulation modelling study of a period of 20 yr in order to analyse the physical processes determining the transfer of the $CO_2$-induced warming to the ocean on the basis of the Oregon State University model. Calculations of global mean values of the ocean heat-balance components have shown that, with a doubled $CO_2$ concentration, the rate of the heat input to the ocean constitutes about $3\,W\,m^{-2}$, which corresponds to the annual heating of the whole ocean by about 0.005 °C. It is important, however, that the warming begins near the surface and gradually penetrates the oceanic depth. By the end of year 20, the warming (in parentheses) of various layers of the ocean constituted: 0–50 m (1.2 °C); 50–150 m (0.9 °C); 250–750 m (0.5 °C); 750–1550 m (0.1 °C) and was practically equal to zero at depths exceeding 1550 m.

The heat input to the ocean is largely determined by the convection process. The above results can be compared with data of calculations based on the Geophysical Fluid Dynamics Laboratory (GFDL) interactive (but not global) atmosphere–ocean model for the case of a quadrupled atmospheric $CO_2$ concentration. This model primarily concerns the controlling role of convection as a process responsible for the heat input to the ocean. The comparison reveals many features in common, and even a similarity between quantitative estimates. However, the calculations in the GFDL model proceeded from a quadrupled $CO_2$ concentration, and the control integration using the Oregon State University model has not resulted in establishing an equilibrium climate.

Analysis of the annual mean zonal mean profile of the warming due to a doubled $CO_2$ concentration has demonstrated that the warming of the surface layer intensifies from the tropics towards the mid-latitudes in both hemispheres and gradually penetrates deep into the ocean, reaching greater depths in the subtropics and mid-latitudes than in the equatorial belt. The meridional profile of the warming turns out to be very similar to the latitudinal distribution of excess $^{14}C$ concentration observed (compared with its values before nuclear tests in the atmosphere).

Calculations of zonal mean profiles of variations in the ocean heat balance components have shown that the doubled-$CO_2$-induced warming of the ocean takes place largely through heat being transported downwards, whereas the meridional heat transport is of secondary importance. Although over the whole ocean surface the heat flux is directed deep into the ocean, latitude dependence has a substantial effect, giving maximum warming values near 60° N, 75° N, 0°, 40° S, and 60° S. The penetration of heat from the ocean surface deep into the ocean is at a minimum in the tropics and reaches a maximum in the polar latitudes of both hemispheres (the minimum in the tropics is determined by the cold water upwelling). Part of the heat released in the tropics is transported poleward by pole-directed branches of the cells of meridional circulation, which are formed (in connection with the upwelling) following the law of mass conservation. The downward transport of this heat in the subtropics is determined by downwellings. There are vortices (gyres) in the subtropical regions which represent anticyclonic horizontal circulations in the ocean; these are driven by westerlies in mid-latitudes of both hemispheres, as well as by equatorial easterlies. Due to friction in the oceanic surface boundary layer, the horizontal transport of water takes place in the form of gyres (the Eckman transport) and, following the law of mass conservation, downwellings appear. This gyre-caused downwelling is responsible for the transport of heat deep into the subtropical oceans, where it combines with the poleward heat fluxes from the equatorial belt.

In high latitudes, the suppressed convection serves as an important factor in heat transport deep into the ocean. If in usual conditions the ocean gets heat from the atmosphere in the tropics and delivers it to the atmosphere in high latitudes then, with increasing $CO_2$ concentration, the heating of the high-latitude ocean surface layer causes a persistent stratification and reduces the intensity of convective mixing. This, in its turn, reduces the heat release by the ocean to the atmosphere in high latitudes and contributes to the heat input from the atmosphere to the whole ocean. Further studies should be aimed at analysing the laws of geographical distribution of the heat input to the ocean and of the contributions of individual oceans and such zonally asymmetric phenomena and processes as gyres, boundary flows and regional centres of convection.

Thus, real causes of the global climate warming are still unclear. We now turn to a detailed discussion of the problem of detecting the $CO_2$ signal.

## 2.2 Detecting the $CO_2$ signal

The problem of detecting the greenhouse signal in global climate changes is quite urgent. Though the global mean increase of surface air temperature within the range 0.3–0.6 °C observed during the last century does not contradict estimates of the warming due to the enhancement of the greenhouse effect, other causes of the warming are possible, including the internal variability of the climatic system. The problem of recognition of the greenhouse effect from surface air temperature data alone is extremely difficult. The natural variability of climate can be filtered out more easily with the use of complex indicators in the form of the most sensitive set of parameters, which can be found by comparing the results of numerical modelling for real conditions and a doubled atmospheric $CO_2$ concentration ($2 \times CO_2$). Such a comparison has demonstrated that one of the most characteristic and persistent features of climate variability in the $2 \times CO_2$ case consists in tropospheric warming and cooling in the lower stratosphere, the warming being at a maximum in the upper troposphere in the tropics and in the high-latitude lower troposphere in winter. In this connection Karoly (1991) suggested the use of variations in the zonal mean air temperature at different altitudes and latitudes with doubled $CO_2$ as a complex indicator of climate change.

Based on the GFDL, NCAR and UKMO climate models, calculations of the annual mean values of this indicator were made, which gave very similar results, despite the substantial difference of the models. Table 2.9 gives the results of calculations of the correlation between calculated and observed temperature values (data

**Table 2.9** Correlation of calculated and observed air temperatures

| Period of averaging | Model | | |
|---|---|---|---|
| | GFDL | NCAR | UKMO |
| Year | $0.65^{10\%}$ | $0.74^{5\%}$ | $0.72^{5\%}$ |
| JJA | 0.51 | $0.63^{10\%}$ | $0.63^{10\%}$ |
| DJF | $0.45^{10\%}$ | $0.56^{5\%}$ | $0.60^{1\%}$ |

*Note*: JJA = June–July–August mean; DJF = December–January–February mean.

of radiosondes averaged over two 10-yr intervals: 1978–1988 and 1963–1973), showing its statistical significance (per cent) averaged over 1 yr and over 2 seasons (June–July–August, December–January–February).

As can be seen from this table, the correlation is positive; about half of the observed variability of the annual mean and about one third of the observed variability of the seasonal temperature values can be explained by the contribution of the greenhouse effect. An analysis of the observational data has not, however, revealed any enhancement of the warming in high latitudes and in the upper troposphere of the tropics, as follows from results of numerical modelling. Since an important property of the complex indicator should be its unequivocal property as an indicator of the greenhouse effect manifestations, an analysis has been made of the possible effect of such factors as the El Niño / Southern Oscillation (ENSO), total increase of sea surface temperature and decrease of the total ozone content in the stratosphere. This analysis has led to the conclusion that the effect of each of the factors mentioned above differs from the complex manifestation of the greenhouse effect, though, for example, simultaneous increase of surface air temperature and decrease of total ozone content can cause similar variations.

Although most of the available theoretical assessments of the impact of increased $CO_2$ concentration on global climate testify to an unavoidable substantial climate warming, an analysis of observational data has not, so far, permitted a reliable detection of the $CO_2$ signal. An International Workshop of experts was held in the USA on 8–10 June 1981 to discuss possibilities for early detection of the impact of $CO_2$ on climate, bearing in mind the three principal aspects of the problem (US Department of Energy 1982, MacCracken and Kukla 1985): (i) detecting (from observational data) variations in climatic parameters of biospheric characteristics, (ii) revealing the factors that can affect the biosphere and climate, causing these variations and (iii) filtering out the $CO_2$-induced variations.

As noted above, the most promising approach seems to consist of a search for a specific complex signal which characterizes variations in a set of parameters but not in a single parameter. Since the atmosphere is the most mobile component of the climate system, the emphasis must, naturally, be placed on analysis of its parameters. First of all, the climate models must be improved further, taking into account the interactive multi-layer ocean and the annual change of the depth of the mixed layer, the horizontal transport in the ocean, the annual change of insolation, the interactive cloud cover, the realistic topography and continuous temporal variations in $CO_2$.

In order to achieve reliable detection of the $CO_2$ signal, numerical experiments are needed on climate sensitivity to such factors as volcanic eruptions, extra-atmospheric insolation, the concentrations of aerosols and optically active gaseous components and variations in surface albedo, snow and ice cover extent, and the sea surface temperature. Atmospheric parameters to be analysed include the following:

- daily mean temperatures of the surface, lower troposphere and stratosphere,
- characteristics of the radiation budget (spectral outgoing thermal emission, total and spectral atmospheric emission, extra-atmospheric insolation, planetary albedo),
- total water vapour and liquid water content in the atmosphere and precipitation,
- soil temperature in arid and humid regions,
- sea ice cover extent,
- characteristics of the impact of volcanic eruptions on the atmosphere.

In the analysis of observational data, special attention must be paid to the spatial and temporal variability of the signal/noise ratio and available data for the summer polar stratosphere (to reveal the trend of its cooling). In studies of climatically sensitive polar regions, numerical modelling makes it possible to assess the impact of the $CO_2$-induced climate warming and aerosols on the snow cover in the transitional zone, variations in the mass balance of small polar caps during the warming and strong fluctuations of the ice cover.

The following observational data are of the first priority: surface albedo (especially along the sea-ice edge, along the southern boundary of the Arctic water basin in summer and in the northern hemisphere in spring), the extent of stratus and strato-cumulus clouds over pack ice in the Arctic, the 1000–700/300–100 hPa thickness ratio, the extent and thicknesses of sea ice and land snow covers (from satellite data), dates of freezing and breaking-up of the ice cover for a representative set of lakes, and characteristics of the radiative regime in the Arctic and Antarctic. These data must be used to analyse the trends of monthly mean surface temperatures, the ratio of temperature to soil moisture content and daily mean temperatures at various stations in the snow cover transition zone. Special emphasis must be placed on the oceans as a key component of the climate system. Numerical modelling must be aimed here at detecting an expected $CO_2$ signal from the oceanic properties, estimating the signal/noise ratio for pre-calculated characteristics of the ocean, assessing possible ranges of the $CO_2$-induced variability of the key oceanic parameters, analysing the interactive ocean–atmosphere–cryosphere system and developing models for assessing the conditional probabilities of detecting the $CO_2$ signal from specific sets of parameters.

An observational data base must include the following values: world sea level, sea surface temperature, $CO_2$ concentration in sea water and its alkaline property, and velocity of currents. Analysis of these and other parameters must include a study of the structure of their time series, estimation of the signal/noise ratios and determination of the most sensitive regions. An impact on the biosphere of both increased $CO_2$ concentration (stimulating photosynthesis) and climate warming is not in doubt. However, detecting the $CO_2$ signal is very difficult in this case.

As MacCracken (1983) notes, according to numerical climate modelling data, there is a 50 per cent probability that a doubled $CO_2$ concentration will raise the global mean surface air temperature by between 1.5 °C and 5 °C. Thus, since the

$CO_2$ concentration has increased from 290 to 340 ppm over the last hundred years, a climate warming should have taken place constituting 0.7($\pm$0.4) °C. To reveal the climate warmings that occurred in the past and may be likely in the future, it is necessary to fulfil the following conditions: climate data should be representative and adequate to permit a correct statistical analysis of climate variability, and information should also be available that allows us to identify and recognize the effects of various factors on climate change. Analysis of observational data made by MacCracken (1983) shows that at present these conditions cannot be satisfied. Some more optimistic conclusions have been made more recently (Monastersky 1995).

There are also uncertainties in the estimates of various external factors of climate variability: volcanic eruptions and solar activity. Thus, so far, a reliable determination of the $CO_2$ signal is impossible. This is verified by a number of other publications.

Many recent efforts have been made to reveal $CO_2$-induced climate changes from observations and numerical modelling data. Also included was an analysis of new palaeoclimatic (geological and archived) and instrumental data. Although these studies have not led to definite results, they are very important for a better understanding of the causes of present-day global climate changes. As Gates and MacCracken (1985) note, a major difficulty in filtering out the $CO_2$ signal is connected with the multi-factor internal variability of the climate system (climatic noise) in timescales up to hundreds of years, comparable to variations which can be determined both by external factors and by increased $CO_2$ concentration. According to palaeoclimatic data, the present interglaciation period began about ten thousand years ago, and a typical duration of such periods is about ten thousand years; therefore, assumptions have been made about the possible onset of the next glaciation during the forthcoming few centuries or millennia. Proceeding from the hypothesis that palaeoclimatic variability is determined by variations in orbital parameters, this must happen in four-to-five thousand years.

However, an enhancement of the atmospheric greenhouse effect due to increased $CO_2$ concentration and other optically active minor gas components must cause a climate warming which will exceed the natural cooling. After the period of the warmest climate (about six thousand years ago), a global mean cooling occurred, constituting, apparently, not more than 1 °C. During the last thousand years, regional-scale climate changes mainly took place (the Little Ice Age was the most substantial recorded anomaly in the seventeenth to nineteenth centuries).

In the analysis of numerical modelling results, it is important to bear in mind that the pre-calculated climate change is

$$\Delta c = f(\Delta c_r, \Delta c_m) \tag{2.1}$$

where the r and m indices correspond to real and modelled climate changes. The problem is that, as a rule, contributions from $\Delta c_r$ and $\Delta c_m$ are unknown and therefore cannot be identified. The problem can partially be solved by analysing the results obtained with different climate models. In this connection, the development of a sufficiently reliable climate model, which could be used as a reference, is decisive.

From the viewpoint of detecting the $CO_2$ signal, available observational data are inadequate and insufficiently representative and climate models are still imperfect. In these conditions the estimates of statistically significant variations in climate are unreliable. Therefore, at present the $CO_2$ signal cannot be detected reliably from observational data, though the data are consistent with the possibility of climate warming due to increased $CO_2$ concentration. Only further development of climate theory and obtaining more representative observational data will make it possible to understand the cause-and-effect characteristics of climate formation.

Wigley *et al.* (1985, 1989) summarized the results from analysis of the surface air temperature data and the causes of its variations, averaged over large regions (with emphasis on hemispheric means) on a time scale of decades and longer, bearing in mind, first of all, the problem of the impact of increased $CO_2$ concentration on climate. A combined analysis of data for the land and the ocean suggests that there has been an oscillating trend of warming in the northern hemisphere, of 0.4–0.6 °C during the last hundred years. (The cooling during the 1940s to 1960s, typical of land, looks weaker in the combined data base.) Fragmentary observations in the southern hemisphere reflect a trend of climate warming (0.3–0.7 °C) starting from the year 1900. Though this warming trend is consistent with a possible warming due to increased $CO_2$ concentration, analysis of empirical data does not permit a reliable determination of the impact of $CO_2$ on climate.

Analysis of data with characteristic time intervals of a decade and longer must serve as the basis for detecting climate changes due to increasing concentrations of $CO_2$ and other minor gas components (it is important here that the climate cooling between 1940 and 1970 still remains unexplained). Attempts to substantiate empirical models of these changes remain unsuccessful and this gives rise to diverging opinions on detecting the climate variations caused by increasing $CO_2$ concentrations.

Empirical models are based on a surface air temperature regression of the following type:

$$T_i = a_i V_i + b_i S_i + c_i C_i + \beta_i + e_i \tag{2.2}$$

where, $V_i$, $S_i$, $C_i$ are variations in the volcanic aerosol content, solar activity and atmospheric $CO_2$ concentration observed in the past (during one year), $\beta_i$ is the autoregression term, $e_i$ is the residual error, and $a_i$, $b_i$ and $c_i$ are the regression coefficients. The fact that $c_i \gg 0$ is sometimes considered to be testimony to an impact of $CO_2$ on climate, but this is incorrect (statistically unsubstantiated) since a reliable determination of the confidence interval for regression coefficients is impossible. The latter is determined by three circumstances: (1) uncertain $V_i$, $S_i$ and $C_i$ functions (in various studies they are prescribed rather controversially, but the results always agree well with observations); (2) specific statistics: with both predictors and predictants having similar medium-term and long-term trends, a good description of empirical data can be reached even when some of these trends happen to be well approximated, by coincidence; (3) apart from external influences, the internal variability of the climate system, especially that connected with the oceanic impact, is an important factor of climate variability.

No attempt has yet been made to take into account the role of the ocean in numerical modelling of the past climate. It is likely that observed inter-hemispheric differences in the hundred-year trends of the surface air temperature, averaged over decades and longer, will be the key to an assessment of the role of a regionally specific impact of the ocean on climate. Special attention must be paid to analysis of regional features of the $CO_2$ impact on climate, with continuous (but not stepwise) growth of $CO_2$ taken into account.

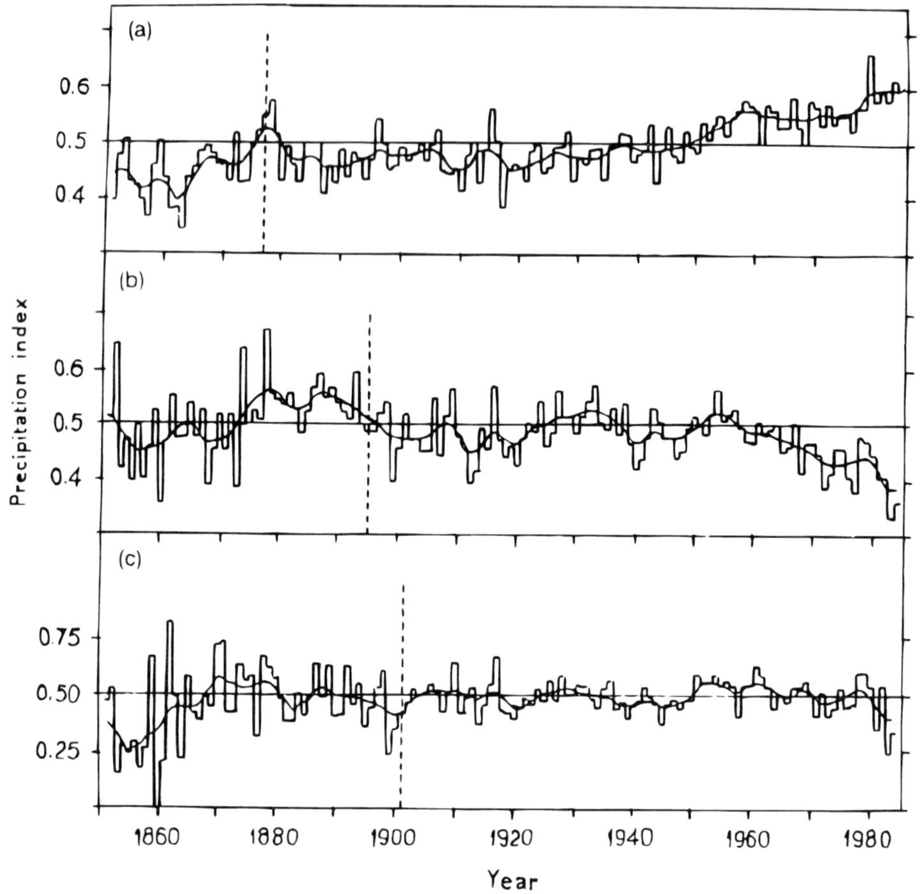


**Figure 2.21** Precipitation indices showing changes in area-averaged precipitation over the land areas in the zones (a) 35° to 70° N, (b) 5° to 35° N, and (c) 0° to 5° N (from Bradley *et al.* 1987).

Numerical modelling shows that the $CO_2$-induced climate warming must intensify the moisture cycle. A doubled $CO_2$ concentration must intensify precipitation and evaporation by 5–10 per cent. Apparently, precipitation will intensify most in low and high latitudes, whereas middle latitudes are characterized (depending on geographical coordinates and season) by both intensified and reduced precipitation. Evaporation must increase mainly over the mid-latitude continents in warm seasons.

As Barnett (1986) notes, all these results of numerical modelling depend substantially, however, on the specific nature of the climate model, and therefore they must be considered hypothetical, especially from the viewpoint of regional anomalies of precipitation. Comparatively brief and inhomogeneous observations of precipitation over land and their almost complete absence in the case of oceans exclude the possibility of reliably estimating global means of precipitation. Therefore, comparisons with calculations can only be made over some regions of land, but even in these cases serious difficulties arise because of patchy precipitation and large errors in observations.

Estimates of the river run-off dynamics for large water basins could serve as an alternative, but in this case an analysis of data is seriously hindered by anthropogenic impacts on the run-off as well as by the fact that the run-off is sometimes determined by the difference between precipitation and evaporation. Attempts to analyse the regional dynamics of precipitation gave no positive results; almost everywhere the interannual variability of precipitation and the trends of 10-yr means are such that it is impossible to detect a relatively weak $CO_2$ signal. Part of precipitation variability is determined by the effect of the El Niño / Southern Oscillation, as well as by other causes, whose combined effect cannot be filtered out reliably.

Thus, though the precipitation dynamics (see Figures 2.21 and 2.22) can be one of the most important indicators of the $CO_2$ signal, it is very difficult to detect this signal reliably. Broadening the data base on precipitation and improving its homogeneity over the oceans, in particular, to obtain the regionally-averaged estimates, must be an important goal of further studies. The emphasis must be placed on the problem of enhancing small variations in precipitation, manifested in various components of the biosphere. It was found that small variations in precipitation in the western USA cause substantial variations in the river run-off. The level and chemical composition of lakes, characteristics of vegetation canopy, etc. can also serve as respective indicators.

To assess a possible impact of increasing $CO_2$ on the processes in the oceans, the following are needed: (1) theoretical forecasts of expected variations (so far, there are

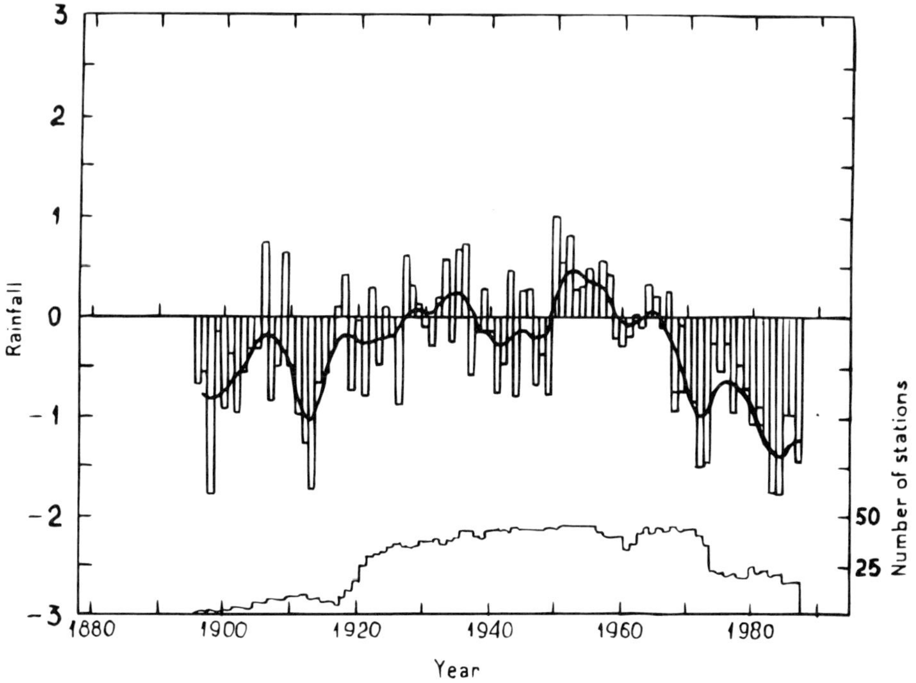


**Figure 2.22** Time series of May–October rainfall at stations in the western Sahel for the period 1896 to 1987. Rainfall is expressed as a normalized anomaly from the 1941–70 base period. The smooth line is obtained using a ten-term Gaussian filter, and the lower bar chart shows the number of stations included each year.

only obvious qualitative assessments of the warming of the upper layer of the ocean, which only permit an analysis of the $CO_2$ signal against the noise of the ocean) and (2) data on variations in the physical characteristics of the ocean in the past. Such information is rather fragmentary; there are long series of observations of the key parameters (sea surface temperature, salinity, sea level), but only for individual locations.

Barnett (1986) undertook a detailed analysis of such series from which it follows that, according to observations near the continental coastlines, in most cases sea level has risen by 10–25 cm over a hundred years.

The sea surface temperature data processing, hindered by almost uncontrollable changes in observational techniques (systematic errors reach 0.3–0.7 °C), reveals substantially different variations in the sea surface temperature. This confirms the importance of taking into account both the sea surface temperature and surface air temperature over the oceans when estimating the hemispheric and global mean values.

Globally averaged sea surface temperature values revealed a cooling by about 0.4 °C from the end of the nineteenth century until about 1910, with a subsequent warming by about 0.5 °C until the 1950s, and then a slight cooling (however, possible systematic instrumental errors must be analysed). As for variations in temperature, salinity and density at different depths, because of fragmentary evidence, their trends retrieved for the last 50 yr are, as a rule, an order of magnitude below the level of noise determined by internal variabilities of different origin.

To obtain information about the $CO_2$ signal manifested in the ocean, first of all, quantitative theoretical estimates of the expected signal are needed, which can serve as a reference point in planning the observational systems. In this connection, there is an urgent need for sea level observations in the southern hemisphere and on the islands in the centres of the oceans. Further analysis of the available oceanographic data base is needed to reveal the global-scale variabilities of such parameters as surface air temperature, sea surface temperature, wind speed, atmospheric pressure and cloud amount, based on the maximum possible filtering out of various measurement errors. An actual problem is to accomplish continuous and frequent (meaning the filtering of high-frequency variability) observations in individual key regions of the ocean (first of all, where long observational series are already available). The thoroughly selected information for such regions, together with the use of results of numerical modelling of the oceanic dynamics, will form the basis for $CO_2$ signal detection.

Global manifestations of the climatic effect on the variability of snow and ice cover can be subdivided into two categories, to which corresponds either a fast or a slow response of the cryosphere (US Arctic Research Commission 1989). To the first category belong changes in snow and sea ice cover and the second category refers to land ice, glaciers and ice sheets. Figures 2.23–2.25 illustrate the global ice cover dynamics (Gloersen and Campbell 1988). The effect of the cryosphere on climate is determined by high albedos of snow and ice cover, low heat conductivity and large thermal inertia. In this connection Barrie (1986) and Barrie and Chorley (1988) undertook a detailed overview of observational data on the variability of various components of the cryosphere. In particular, there is no clear-cut trend of snow cover extent on the northern hemisphere. The Arctic sea ice extent, having reached a minimum in the 1950s, has grown thereafter. Though the extent of the summertime ice in the Antarctic between 1930 and 1970 has been decreasing, nevertheless it has

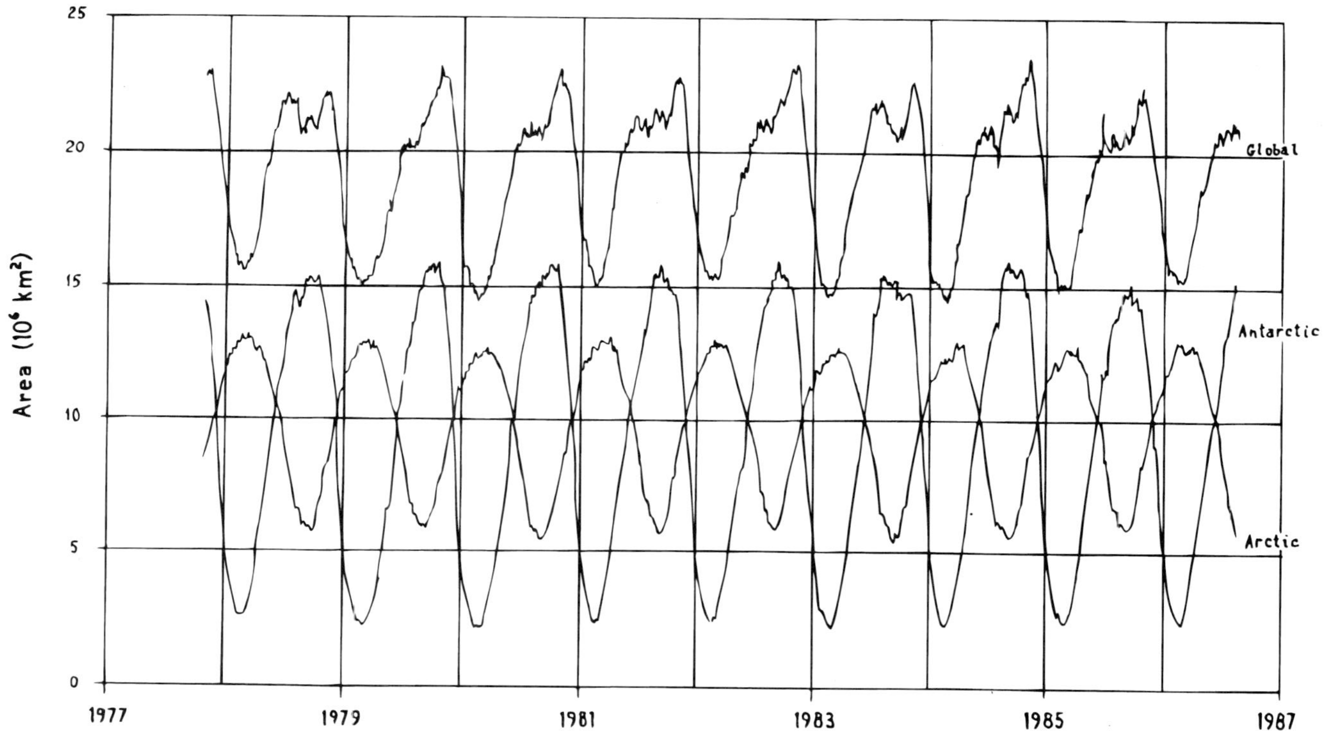


**Figure 2.23** Area coverage by sea ice versus time; the vertical grid lines correspond to 31 December of the year indicated.

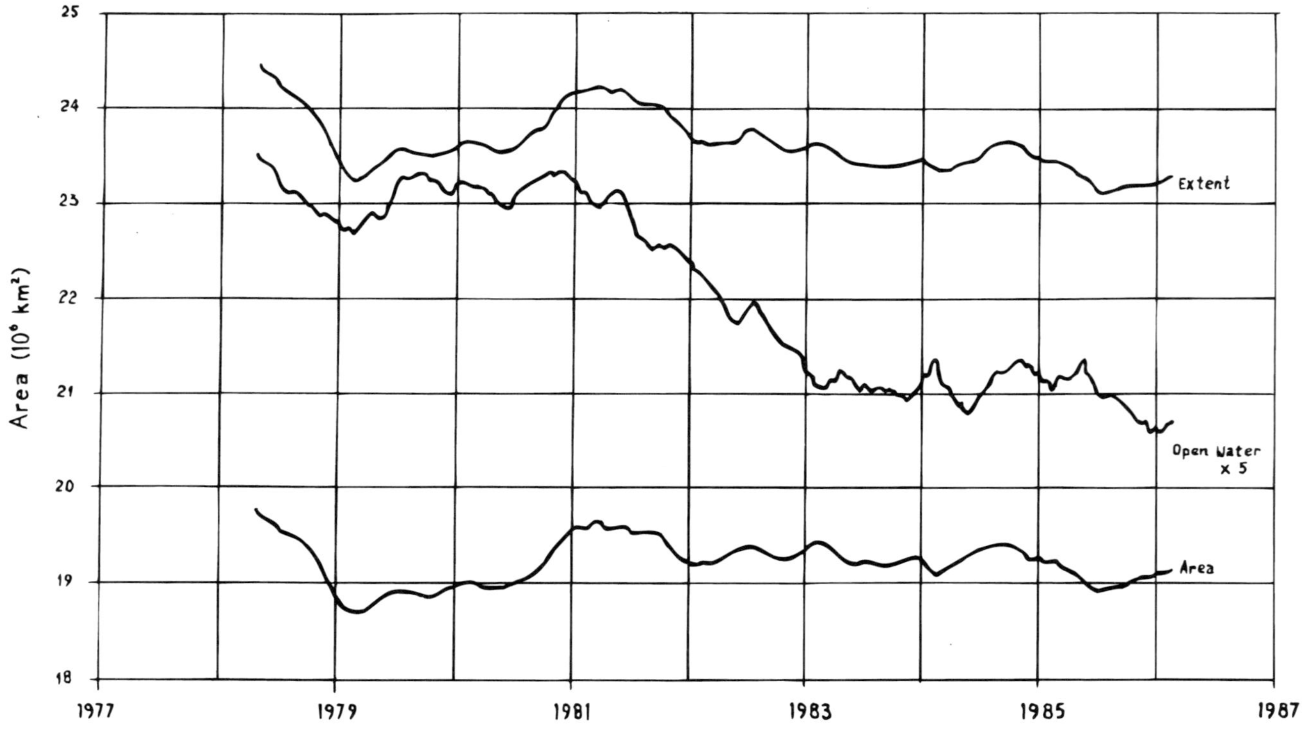


**Figure 2.24** One-year running mean of global sea ice versus time; the vertical grid lines correspond to 31 December of the year indicated.

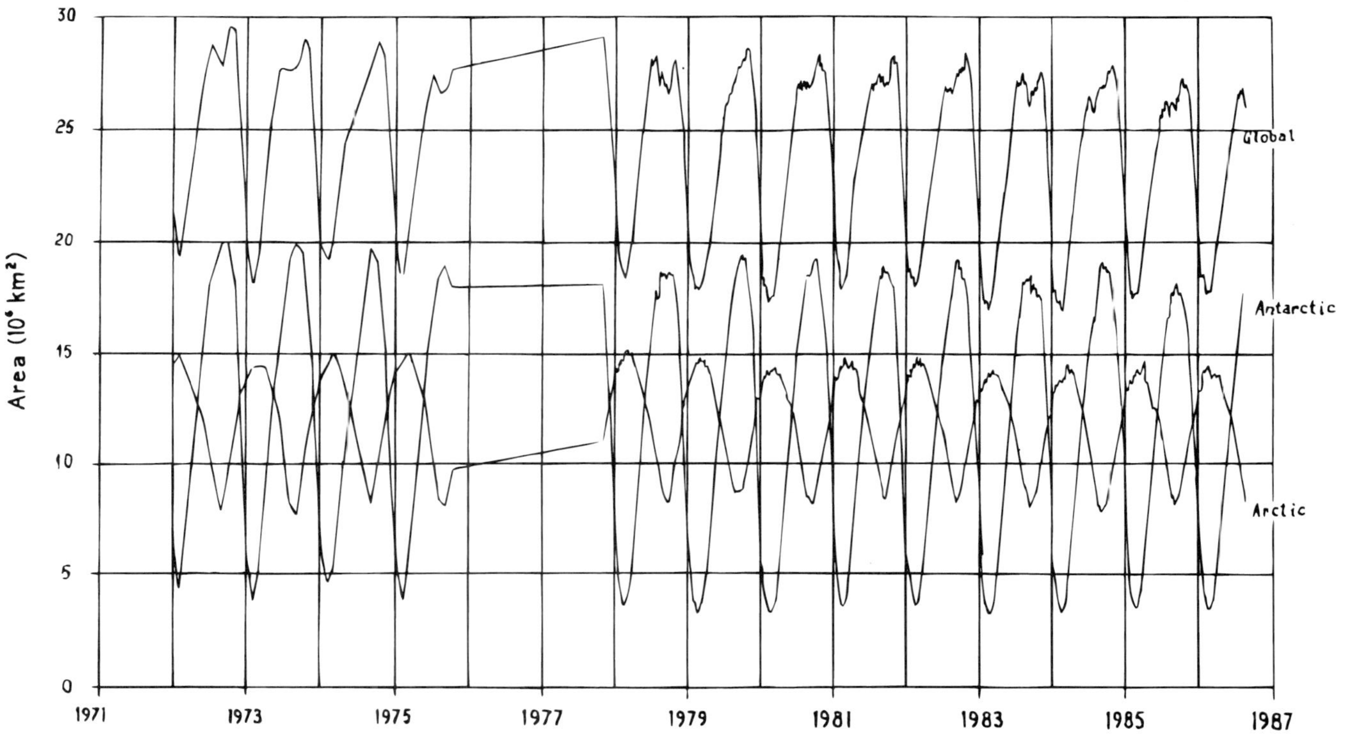


**Figure 2.25** Ice extent versus time (from combined ESMR and SMMR records); the vertical grid lines correspond to 31 December of the year indicated. There is a period of approximately two years, from late 1976 to late 1978, for which no data are available.

been determined by other factors. The requirements of the data of observations of the two categories of the cryosphere components mentioned above are substantially different (Table 2.10).

In connection with the very important role of albedo feedback, of primary importance are studies of the factors determining the long-term variability of the extent and thickness of sea ice cover as a function of the sensitivity of the Arctic pack ice to variations of the climatic, oceanic and hydrological regime. The latter determines the urgency of complex programmes of observations of the characteristics of the ice cover, ocean and cryosphere, as well as of the development of climate models; an account of the interactive system atmosphere–ocean–cryosphere would be useful in pre-calculating the changes in the extent and thickness of ice cover which can be caused by increasing $CO_2$ concentration.

Bearing in mind the multi-factor dynamics of the ice cover, it can hardly be expected that monitoring the variability of its characteristics during the decades to come will make it possible to detect the $CO_2$ signal. Of more importance may be satellite monitoring of the dynamics of freezing and ice-breaking of lakes, especially in North America and Eurasia. The corresponding data may become the most prominent crysophere $CO_2$ signal.

An analysis is needed of possible regional and seasonal responses of the extent, thickness and life-time of snow cover to increasing $CO_2$ concentration. A decrease of the extent and life-time of the northern hemisphere snow cover may become the second most important manifestation of the $CO_2$ signal in the cryosphere after the information on dynamics of lake ice in the decades to come. The problem of the

**Table 2.10** Data requirements for observations of cryospheric characteristics as applied to climate studies

| Characteristic | Objective | Accuracy desired/ needed | Horizontal resolution (km) | Temporal resolution |
|---|---|---|---|---|
| Snow cover extent | T, M | 3–5% | 250 | 3 days |
| Water equivalent of snow cover | T | 0.5–1 cm | 250 | 1 week |
| Surface albedo | T | 0.02–0.04 | 250 | 1 week |
| Ice cover boundaries | M | 20–50 km | 250 | 1 week |
| Ice cover concentration | T, M | 3–5% | 250 | 1 week |
| Sea ice drift | S | 2–5 km day$^{-1}$ | 500 | 3 weeks |
| Sea ice thickness | T, M | 10–20 cm | 250 | 1 month |
| Sea ice melting | S | ? | 250 | 1 week |
| Share of multi-year ice | M, S | 0.1–0.1 | 250 | 1 week |
| Glacier topography | M | 0.1–1.0 m | 2–250 | 1–10 yr |
| Glacier boundaries | M | 1–5 km | | 1–10 yr |
| Glacier thickness | M | 10–100 m | 250 | 1–10 yr |
| Glacier motion | S, M | (1–10)–100 m | m/yr at a point | 1 yr |
| Permafrost boundaries | M | 10–50 km | | 1–10 yr |
| Time-period of lake freezing | M | ? | individual lakes | 3 days |

*Note*: M = monitoring;
S = studying the processes;
T = testing the model reliability.

west-Antarctic ice sheet still remains unsolved, which necessitates further theoretical studies and extended observational programmes. A more adequate observational data base is needed to assess the mass balance and changes in the volume of the two largest ice sheets as well as a regular monitoring of glacier dynamics, bearing in mind calculations of the possible rise of world sea level due to climate warming. Of great interest is the monitoring of permafrost.

As noted above, estimates of the effect of volcanic eruptions on climate have led to controversial conclusions. It follows from some of these estimates that in order to explain the centennial trend of global mean surface air temperature, it is enough to take into account only the effect of eruptions. The estimates of the effect of solar activity are even more contradictory.

A combined account of increasing $CO_2$ concentration, volcanic eruptions and solar activity undertaken by Hansen *et al.* (1981) as well as by Gilliland and Schneider (1984) has led to substantially different conclusions. So, for example, according to Hansen *et al.*, the effect of $CO_2$ started showing itself in 1880, whereas according to Gilliland and Schneider it was after 1825. In the first study, the eruptions of the volcanoes Krakatoa in 1883, Suffrier-Santa Marie in 1902–1904 and Agung in 1963 had the strongest effect on climate, whereas in the second study it was those of Askania in 1875, unidentified eruptions in 1885–1886, Katmai in 1912 and Surtsei in 1963–1965.

The estimates of the effect of solar activity are also radically different. An analysis of the variability of such climate parameters as the stratospheric temperature, the extent of sea ice and land snow cover, and the world sea level, has not given convincing results either. No doubt, a climate warming has taken place during the last hundred years, which does not contradict a possible $CO_2$-induced warming, but there is no unique identification of the latter so far.

Ellsaesser (1985, 1987a) draws attention to the fact that available observational data do not reveal an increase of temperature near the surface and in the troposphere that is quasi-homogeneous and monotonically increasing with latitude, as is predicted by theoretical calculations of the impact of increasing $CO_2$ on climate. The most complete data for the northern hemisphere reveal a non-stationary and geographically varying temperature field, with peaks and troughs, of cellular structure (by latitude and longitude) and increasing with latitude. The latter determines the prevailing contribution of high latitudes to the formation of the hemispheric temperature trend. It is important, however, that the data considered refer to the continents and do not take into account the 0–20° equatorial belt, where temperature variations are in counter-phase to those in the Arctic and can compensate for the latter.

The interpretation that the climate warming trend observed during the last century has been caused by increasing atmospheric $CO_2$ concentration can be accepted only under the following four conditions: (1) the possibility that the warming is connected with the natural process of returning to a normal, milder climate such as that during the Little Optimum (AD 900–1300) after a temperature drop during the Little Ice Age (about AD 1550–1850) is excluded; (2) there are grounds for the supposition that $CO_2$ emissions reaching 100–200 Gt C due to changes in the biosphere (deforestation, agriculture) took place mainly before 1938, whereas emissions due to fossil fuel combustion (about 175 Gt C) were made after 1938; (3) there exist other climate-forming factors (stronger than $CO_2$) which may have led to warmings and coolings over periods of 30–50 yr, which suppressed a persistent trend of warming due to increasing $CO_2$ concentration; and (4) an assumption is made that the warming due to a doubled $CO_2$ concentration does not exceed 1.5 °C or that the time shift of the warming with respect to variations in $CO_2$ is less than 50 yr. Since the first two conditions cannot be guaranteed and the problem of the contribution by other factors remains unsolved, there are no grounds for ascribing the climate warming trend to the effect of increasing $CO_2$ concentrations.

Parker (1985) undertook an attempt to detect the climatic impact of increasing $CO_2$ by a combined analysis of air temperature variations at the 300 hPa level (where a warming must take place) and at the 50 hPa level (the level of cooling), based on aerological soundings in middle and high latitudes of both the northern and southern hemispheres during the last 25 yr, both in summer and in winter. The emphasis was placed on revealing an intensified cooling at the 50 hPa level compared with the 300 hPa level. The most substantial result of this analysis was the absence of the cooling trend both in summer and in winter. Some decrease in temperature difference at the 50 and 300 hPa levels was observed in the middle and late 1960s in summer, but later this difference was practically constant. This decrease can apparently be explained by several factors: errors in observations, variations in the regime of long waves in the troposphere, and volcanic aerosol since the 1963 Agung eruption could have caused some warming of the mid-latitude stratosphere.

With a 7 per cent increase of $CO_2$ concentration between 1958 and 1978 and a doubled $CO_2$ concentration assumed to enlarge the temperature difference between the 50 and 300 hPa levels by 10 °C, the temperature difference during this period should have reached 0.70 °C or much less (provided the effect of the ocean in reducing the response of the climate to $CO_2$ is taken into account). Since the threshold change in this difference, from the viewpoint of its statistical significance, is 0.55 °C (in winter) and 0.50 °C (in summer), it is clear that a trend of about 0.75 °C over 10 yr could be detected in not less than 35 yr, i.e. by about 2010, provided the rate of the $CO_2$ growth is preserved (3.5 per cent over 10 yr). A contribution comparable to the $CO_2$ contribution to the greenhouse effect from other minor greenhouse gas components (methane, CFCs, nitrous oxide, etc.) must also be taken into account, and this hinders the detection of the $CO_2$ signal.

Recommendations on the priority of various observational data as a source of information about climate changes are given in Table 2.11.

Some estimates suggest the conclusion that the $CO_2$ effect on climate predicted by numerical modelling ought to have been detected from data on the interannual variability of the surface air temperature at 60° N. However, since an analysis of the data for a 72-yr observation series did not allow the detection of such variations (mean temperature between 1958 and 1977 was not higher than during the 20-yr period 1906–1925), the following alternative possibilities can be assumed: (1) there are factors that compensate for the warming caused by an enhanced greenhouse effect, (2) a manifestation of the effect of the warming is delayed by the thermal inertia of the ocean or (3) numerical modelling overestimates the effect of increasing $CO_2$ on climate.

Using a one-dimensional box model of the ocean, Wigley and Schlesinger (1985) obtained new estimates of the delay in climate warming caused by a doubling of $CO_2$, explained by the thermal inertia of the mixed layer and by the heat transport to deep layers of the ocean. Calculations show that the transition process determining the response of the climate to stepwise variations in $CO_2$ concentration depends

**Table 2.11** Parameters needed for the early detection of anthropogenic climate changes

| Parameter | Grounds | Types of observations |
|---|---|---|
| 1. Air temperature from surface network and aerological soundings | Air temperature directly depends on atmospheric radiation budget and therefore on responses to increasing $CO_2$; model calculations point to a warming near the surface, intensifying with latitude, and to a stratospheric cooling | Network observations within the World Weather Watch programme |
| 2. Sea surface temperature and surface air temperature over the ocean | Response of the ocean upper layer is an important aspect of the warming; air temperature over the ocean is needed to obtain representative averages | Network observations; accumulation and distribution of data must be improved |
| 3. Global $CO_2$ concentration | A major potential factor of climate change | At several stations |
| 4. Global concentration of other long-lived minor greenhouse gases (CFCs, $CH_4$, $N_2O$, etc.) | A potentially important factor that intensifies the $CO_2$ effect | *In situ* sampling and remote sensing |
| 5. Global concentration of stratospheric aerosol, especially after powerful eruptions | The post-eruption stratospheric aerosol reduces solar radiation and causes a cooling near the surface, but a warming of the lower stratosphere | Specialized surface, lidar, aircraft and satellite observations |
| 6. Atmospheric transparency | It characterizes total reduction of solar radiation by tropospheric and stratospheric aerosol | Network actinometric observations; they must be broadened in low latitudes and in the southern hemisphere |
| 7. Solar constant and surface observations of solar activity (spots, flares, the Sun's diameter) | Variations in solar constant and solar activity can be an important factor of climate change | Specialized satellite and surface observations |

strongly on the level of the warming at a doubled $CO_2$ concentration and on the intensity of mixing in the ocean.

Thus, in view of inadequate understanding of the causes of present climate changes, we can state that at present it is impossible to detect reliably the $CO_2$ signal from observational data. With the temperature difference of 1 deg K as a threshold for detecting the $CO_2$ signal, we may expect that during the forthcoming

10–60 yr a $CO_2$-induced climate change can be detected. This may hold true if such factors as volcanic eruptions and solar activity do not mask a long-term variation of this order of magnitude.

As Hoffert *et al.* (1985) and Gaffin *et al.* (1986) note, attempts to select the contribution by external factors from observed interannual temperature variations have not given convincing results, because of arbitrarily prescribed correlations with disturbing effects of solar activity, volcanic eruptions and other factors. In this connection an analysis was made by Hoffert *et al.* (1985) of internal free (self-exciting) oscillations in the interannual variability of climate which appear in a system of interactive non-linear differential equations with a destabilizing feedback, which is significantly close to the basic state, and a stabilizing feedback which starts functioning when the amplitude of fluctuations becomes large. In this case a cyclic variability can be formed in the absence of external effects, as well as a more complicated regime of climate in the presence of these effects, often contradictory to intuition.

Hoffert *et al.* (1985) made calculations using a model of the SST ($T_a$), with the mixed layer $h_m = 100\,\mathrm{m}$ thick, the coefficient of turbulent mixing $k = 2000\,\mathrm{m^2\,y^{-1}}$ and a fixed temperature of the polar ocean $T_p = 1.1\,°\mathrm{C}$:

$$\frac{\mathrm{d}T_a}{\mathrm{d}t} = \frac{T_e - T_a}{\tau} + \frac{1}{h_m}\left[\mathrm{k}\,\frac{\partial T}{\partial z} + w(T - T_e)\right]_0 \tag{2.3}$$

where $T_e(t)$ is the equilibrium temperature and the relaxation time for the mixed layer, $\tau = 4\,\mathrm{yr}$ and $w$ is the varying velocity of upwelling. The expression in square brackets determines the total heat flux at the boundary of the mixed layer with the thermocline, which is zero at a state when diffusion compensates for upwelling. As a rule, $k\partial T/\partial z < 0$, and therefore an abrupt intensification of upwelling causes an increase of $T_a$. With the destabilizing feedback between the upwelling and $T_a$ expressed as

$$w = w_0 + \alpha\,\Delta T \tag{2.4}$$

where $\alpha > 0$ and $\Delta T = T_a - T_e$, the warming will intensify the upwelling, which in its turn will favour further warming. The stabilizing feedback is prescribed by adding a third non-linear term $\beta(\Delta T)^3$. In this case the resulting feedback becomes unstable at small $\Delta T$ ($\partial w/\partial t = \alpha > 0$) but stable at large $\Delta T$ ($\partial w/\partial t = -3\beta(\Delta T)^3 < 0$) at both positive and negative $\Delta T$. At $T_e$ = constant, this model illustrates the formation of unforced internal oscillations of $T_a$ on a timescale of 80 yr at acceptable values of the parameters $w_0$, $\alpha$ and $\beta$ and reproduces well the course of the globally averaged surface air temperature during the last century with the disturbing effect of increasing $CO_2$ concentration taken into account. This verifies the importance of an adequate account of the non-linear atmosphere–ocean interaction. Gaffin *et al.* (1986) demonstrated the possibility of internal self-excited oscillations with parameter-dependent periods of the approximate order of a hundred years exhibiting surface temperature changes of the order of $\pm 0.5\,°\mathrm{C}$.

As Luther (1983) noted, the most natural way to detect the $CO_2$ signal consists of monitoring the changes in the spectral distribution of shortwave and longwave radiation fluxes, which reflect respective radiative signatures of $CO_2$. The outgoing longwave radiation (OLR) spectrum, for instance, can serve as a signature. Calculations show, however, that even with doubled $CO_2$ a maximum decrease of the

OLR will constitute only 12 per cent in the 13–17 $\mu$m wavelength interval and, hence, early detection of the $CO_2$ signal is impossible.

In this connection, Luther (1983) calculated the spectral distribution of the downward thermal emission (DTE) flux in the atmosphere at different altitudes in the mid-latitude summer in order to find the spectral interval where the DTE is most sensitive to variations in $CO_2$ concentration. The results show that the DTE is most sensitive in the interval 700–800 $cm^{-1}$ at a height of several kilometres. The height of maximum sensitivity of DTE to increasing $CO_2$ concentration depends on latitude and decreases with increasing $CO_2$ concentration, which is caused by an accompanying decrease of the atmospheric water vapour content. The DTE sensitivity to $CO_2$ variations in high latitudes is three to four times greater. The maximum absolute DTE variations with $CO_2$ enhanced by 20–60 ppm is reached in the interval 755–760 $cm^{-1}$ and takes place at a lower height than maximum relative variation, constituting 9.63 per cent at a height of 8 km, with the $CO_2$ mixing ratio increasing from 340 to 400 ppm.

The results obtained by Luther (1983) suggest that high-mountain observations of variations in the spectral distribution of atmospheric emissions are more promising with respect to detecting the $CO_2$ signal than are satellite measurements of the OLR spectral composition. However, even in these most favourable conditions, the problem might only be solved after several decades. Early detection of the $CO_2$ signal can be improved with a spectral resolution better than 5 $cm^{-1}$. In view of the prolonged impact of $CO_2$ on climate, periodic observations are, apparently, sufficient.

Kiehl (1985) analyses the possibility of detecting $CO_2$-induced climate changes using data on spectral distributions in nadir of atmospheric emission and OLR radiances. Calculations of OLR are made for the wave number interval 0–2500 $cm^{-1}$, with emphasis on the interval 500–800 $cm^{-1}$ (the $CO_2$ 15 $\mu$m band), where the OLR is generated mainly by the stratosphere. The choice of OLR in the 15 $\mu$m band as an indicator of climate changes is determined by two circumstances: (1) the temperature variations due to a doubled $CO_2$ concentration are stronger in the stratosphere than in the troposphere and (2) the OLR formed in the stratosphere is independent of cloud conditions.

The OLR was calculated with $CO_2$ mixing ratios enlarged by 10 per cent, 25 per cent and 100 per cent with respect to an initial value of 340 ppmv, at a spectral resolution of 5 $cm^{-1}$. With doubled $CO_2$, the OLR decreases by 2.7 $W\,m^{-2}$ in the interval 500–800 $cm^{-1}$, provided all the vertical profiles of temperature and humidity remain constant, and by 2.2 $W\,m^{-2}$ if a $CO_2$ doubling is taken into account. This decrease constitutes 0.96 per cent with respect to total OLR (230 $W\,m^{-2}$).

But in some cases relative variations in OLR in the 5 $cm^{-1}$ spectral intervals exceed 10 per cent. In particular, the OLR variations at a wavenumber of 667 $cm^{-1}$, corresponding to one of the TIROS-N infra-red bands, constitute about 5 per cent, and can be detected in particular with observational data averaged in time, to reduce the instrumental noise. However, in view of the natural variability of OLR, only data for latitudes below 60° should be considered and only daily means for the six warm months should be analysed. Since stratospheric temperature variations can be caused not only by $CO_2$ but also by other optically active components (ozone, primarily), monitoring the OLR variations in various spectral intervals is needed. The necessity of continuous (long-term) monitoring of the spectral distribution of the OLR, the difficulties in the maintenance (or control) of space-borne instrument calibration, and the lack of synchronous data on radiative and meteorological characteristics near the surface seriously limit the possibility of using satellite-derived OLR data as a reliable source of information for early detection of the $CO_2$ signal. A new approach has been suggested in this context by Goody *et al.* (1995).

Summing up the discussions on detecting the $CO_2$ signal, MacCracken and Kukla (1985) noted that calculations based on the use of simple climate models suggest that the $CO_2$-induced global climate warming during the last 150 yr should have constituted 0.3–1.1 °C. Substantial uncertainty in estimates of warming can be caused by several factors: uncertain pre-industrial $CO_2$ concentration (260–280 ppm), different approaches to the role of the world's oceans, and different estimates of climate warming due to a doubled $CO_2$ concentration (1.5–4.5 °C).

It is also important that even without further $CO_2$ increase (by the year 1985 it had reached about 25 per cent), an additional warming by 0.1–1.0 °C must occur. A considerable contribution to the warming is made by increasing concentrations of some other greenhouse gases.

During the last few years, it has become increasingly clear how difficult it is to detect the real $CO_2$ signal, owing to numerous factors affecting the climate as well as to the spatial and temporal inadequacy of the available relevant observational data. A serious complication is the lack of reliable data about the pre-industrial climate, since the $CO_2$ concentration increase began almost immediately.

The detection strategy foresees three approaches: (1) accumulating and analysing a representative empirical data base containing various quantities (especially those whose variations are statistically most substantial), (2) determining and assessing the contributions from various natural and anthropogenic factors of climate change and (3) filtering out from observational data the $CO_2$ signal and subsequent comparisons with results of numerical climate modelling.

Analysis of various sources of information about the $CO_2$ signal shows that each is inadequate for a convincing verification of the conclusions drawn from numerical modelling of the effect of increasing $CO_2$ on climate, though these conclusions do not contradict the observational data.

Globally averaged observed climate changes are, on the whole, within the predicted estimates, especially when the oceanic contribution that delays a manifestation of the anthropogenic effect is taken into account. However, coolings on land and over the oceans in the northern hemisphere from the end of the nineteenth century until about 1910 and in the period 1940–1970 remain unexplained; here the internal variability of the climatic system could have played a substantial role. These and other circumstances do not permit a reliable detection of the $CO_2$ signal in the field of temperature.

The rise of world sea level observed during the last few decades qualitatively seems to be the result of the expected $CO_2$-induced climate warming, but it can be explained mainly by the effect of thermal expansion of the ocean. In this connection an analysis of oceanic salinity would be very useful in order to help distinguish between the contributions from thermal expansion and from the melting of glaciers. Opinions about the dynamics of precipitation as an indicator of the $CO_2$ signal are, so far, controversial.

On the whole, available observational data do not contradict the conclusion about $CO_2$-induced climate warming, but there is still no convincing evidence for an actual contribution by $CO_2$ and for its relative role in climate change compared with other factors. The inadequacy of observational data and imperfection of the climate theory raise doubts about any assumption that it will be possible to detect reliably the $CO_2$ signal by the end of the current century.

From the viewpoint of $CO_2$ signal detection, a combined application of different approaches, consisting of analysis of the signal/noise ratio from long series of observations, a reduction of the noise level (by filtering out the contribution from other factors), and a search for combined indicators (signatures) of the $CO_2$ signal are most promising. It is important to consider available information along with surface air temperature (sea surface temperature, precipitation, snow and ice cover extent, sea level, meteorological parameters of the free atmosphere, etc.).

In connection with an interdisciplinary programme planned by NASA to identify the long-term climate changes (and to reveal their causes) from satellite data, a workshop was held in Washington DC (Schiffer 1987). The objective of the workshop was primarily to analyse an optimal data set of satellite observations needed for early detection of climate change arising from the enhanced greenhouse effect. In this connection, of great importance are studies which would help to answer the following questions:

1. How soon would it be possible to identify the enhanced greenhouse effect in climate change?
2. Which climatic characteristics would be most informative?
3. What is the natural variability of such parameters?
4. What are the requirements of the accuracy and of the spatial and temporal resolution of the observational data?
5. To what extent do planned satellite observations meet these requirements?
6. How should climate models be developed to improve understanding of the causes of, and to identify (from observations), long-term climatic variability?

The most interesting results of numerical modelling of global climate change are the $CO_2$-induced air temperature increases near the surface and in the lower troposphere, as well as the stratospheric cooling and an increase of specific humidity and precipitation. Also, an intensification of cloudiness is predicted in the upper troposphere and a decrease of cloud amount in the lower troposphere. The reliability of quantitative estimates needs testing, however, due to imperfections in climate modelling; these imperfections include, in particular, the need to take proper account of the atmosphere–ocean interaction and the parametrization of the process of cloud formation.

In addition, care should be taken over the rather controversial assessments of regional climate variability. It is obviously necessary to monitor the dynamics of such regional climatic parameters as the extent and concentration of sea ice cover, the extent of snow cover, vegetation index, the amount and properties of clouds, and the radiation balance.

The key problem of climate theory, whose solution is vital for early detection of the enhanced greenhouse signal, is how best to assess the contributions of natural and anthropogenic climate changes. The present understanding of the internal (naturally caused) variability of the climatic system is not reliable enough to exclude the possibility of such natural climate changes during the next few decades, which could be caused by internal factors shifting the climate to a new quasi-persistent state, without any external interference.

Analysis of climate warmings in the past suggests that they cannot be explained simply by external non-natural effects. The contributors to Schiffer (1987) emphasize, therefore, that the greenhouse warming that has taken place or will take place in the near future cannot be considered reliable. This conclusion was reached using observational data which, although reflecting the presence of a statistically significant (at a 10 per cent level) increase of surface air temperature, do not reveal any statistically significant trends of air temperature in the free atmosphere, of sea surface temperature, or of pressure at sea level.

There is no doubt that greenhouse warming can be identified only on the basis of analysis of the variability of a combination of climatic characteristics which should be obtained with an account of maximum signal/noise ratio. The climate-forming factors to be specified include: (1) wet convection; (2) cloud processes; (3) transport and supply of the heat in the ocean; (4) soil moisture change; (5) vegetation cover dynamics; (6) sea ice and glacier dynamics. Of particular importance is the development of climate models, including an account of the interaction between the climate system components.

Based on the use of the Goddard Institute for Space Studies (GISS) climate model, with an account of 'co-evolution' (interactive time-dependent dynamics) of variations in $CO_2$ concentration and global mean surface air temperature, Hansen *et al.* (1988a) discussed prognostic estimates of SAT variations with three scenarios of the dynamics of greenhouse gases (GG): (1) exponential increase of GG concentrations; (2) the exponential increase is gradually reduced to a linear one; (3) by 2000, $CO_2$ emissions will halve, and those of other GG will cease. (The second scenario is the most likely.)

Data from the control experiment (for a period of 100 years) give the surface air temperature MSD of 0.1 °C, whereas, according to observational data for the last century, this is 0.20 °C (or 0.12 °C, with the trend excluded). To identify a global climate warming which would be statistically significant at a level $2\sigma$, the surface air temperature MSD would have to reach 0.25 °C (or 0.35–0.40 °C at a $3\sigma$ level).

For the period 1958–1985, neither observational data nor results of calculations reveal any statistically significant warming. Numerical modelling has shown, however, that by the end of the 1980s a global mean surface air temperature increase must occur at a level of $3\sigma$. This conclusion holds for a wide range of possible GG trends, provided no strong volcanic eruption takes place. In the 1990s, the 10-year mean warming should exceed the amplitude of the internal surface air temperature variability in many regions, while, by 2010, the global mean warming will be several times stronger than the interannual variability of the surface air temperature annual means.

The warming is expected to be strongest over land and in high latitudes (especially over sea ice). Individual regions, where the warming is particularly strong, are: (1) the low-latitude ocean (here the interannual variability is small and the mixed layer is thin), (2) the interior of the Eurasian continent (especially the region of Kazakhstan–Tibet–Mongolia) and (3) the ocean near the Antarctic and the North Pole (here the contribution of the albedo feedback shows itself).

It should be borne in mind, however, that identification of such sensitive regions depends on specific features of the climate model. In this connection, the difference between the models and observational data can be used to analyse the adequacy of an account of various climatic processes. None of the climate models could reliably simulate the observed combined cooling of the upper troposphere and the warming of the lower troposphere. If this observation is maintained in the future, it will testify to the still unknown defects of the models. Since the available numerical modelling data suggest that a greenhouse warming could be identified by the end of the current century, it is important to substantiate the requirements for observational data (the priority of climate indicators, acceptable uncertainties, spatial and temporal resolution, etc.).

## 2.3 A global system of observations

Before an optimal global system of observations can be substantiated, an analysis of priorities must be undertaken.

Analysis of the data from the meteorological services of 50 countries sent at WMO's request, suggested the following order of priority for data on the factors of interannual climate variability (Kellogg and Bojkov 1983): (1) the atmosphere–ocean interaction, (2) deforestation, (3) the variability of the snow and ice cover extent and (4) other factors, including urbanization, $CO_2$, tropospheric aerosols, desertification, stratospheric aerosols, and soil moisture. On a timescale of decades, the priority might be the following: (1) $CO_2$, (2) deforestation, (3) urbanization, atmosphere–ocean interaction and (4) other factors, including aerosols, solar activity, volcanic eruptions, stratospheric ozone, anthropogenic heat releases, and snow and ice cover.

The priority for atmospheric parameters, which must be presented in the form of global fields, is as follows: surface air temperature (absolute values and anomalies), atmospheric pressure at sea level (anomalies), precipitation (absolute values and anomalies), fields of winds or stream lines near the surface (in the tropics), cloud distribution, heights of 500 and 200 hPa levels (anomalies), relative topography 500–1000 hPa (anomalies), fields of winds or stream lines at the 200 hPa level (in the tropics).

The priority for oceanic parameters presented as time series is: sea surface temperature (anomalies), wind speed near the surface (averaged vector of wind stress), temperature of the ocean surface layer (in specific regions). Observations of the outgoing longwave and shortwave radiation (Earth's radiation budget) are of vital importance. The most important characteristics of the cryosphere are the extent of continental snow and ice cover and the fluctuations of glaciers.

Data are also needed on global mean concentrations of optically active atmospheric minor gaseous components (MGCs) – $O_3$, $CO_2$, nitrogen oxides, freons, $CH_4$, etc. – extra-atmospheric solar radiation, ocean surface level (in specific regions) as well as on such indices of the climate system as the Southern Oscillation index, the oscillation characteristics of the Atlantic and Pacific oceans in the northern hemisphere, the indices of zonal transport, blocking and trade winds, the amplitude and phase of the quasi-biennial oscillations in the stratosphere, and various indices describing the teleconnections (long-distance correlations) in the troposphere.

The emphasis must be placed on the following typical features of climate: jet streams, prevailing trajectories of cyclones, trade wind systems, intertropical convergence zone, and centres of action (the Aleutian and Icelandic minima, the subtropical high-pressure bands). Considerable progress in the development of satellite remote-sensing techniques will help the monitoring of vegetation cover, precipitation zones, stratospheric aerosols (of volcanic origin, in particular), polar ice thickness, oceanic currents, surface albedo, and insolation.

Data of the following physical properties of climate are required: the transport of latent and sensible heat at the surface–atmosphere interface, the latent heat releases in precipitation zones, the radiation–clouds interaction and interactions between the optically active MGCs and aerosols, the moisture cycle, atmospheric turbidity, etc.

Table 2.12 identifies the regions of the globe which are important from the viewpoint of monitoring the parameters and factors of climate.

Naturally, in the global system of climate monitoring, of decisive importance must be the use of space-based means of observation, though it should be emphasized that this by no means belittles the necessity of further development of conventional techniques (ground, aircraft and balloon observations). In this connection, of interest are recommendations worked out by the 1974 Stockholm Conference as well as plans for satellite monitoring of climate developed in the USA (Kondratyev 1982a,b, 1983a,b).

The monitoring system is an organic part of an extensive programme of climate studies within both national and international programmes, connected with the perspective of the World Climate Programme for the decade 1980–1990 (Kondratyev 1982a).

The WCRP Joint Scientific Committee recommends three different classes of observational programmes, to which three classes of data on climate correspond. These data must be used (1) to study the climate-forming processes and to develop techniques for their parametrization (these data of the so-called class P must be complex, detailed, but they can be confined to one or several small regions and short time intervals from several weeks to several years), (2) to test the adequacy of theoretical climate models (these data of the so-called class A require global-scale observations during long time periods and must contain information about numerous characteristics of the climate system) and (3) to perform long-term monitoring of slowly varying components of the climate system (the so-called class M), which is important for assessing the present global climate, understanding natural factors of the formation and variation of climate, as well as studying the possibilities of climate forecasting.

Table 2.13 gives a preliminary list of climatic parameters, requirements of their spatial and temporal resolution and measurement errors. These requirements are formulated according to the needs of numerical climate modelling, and they are therefore correlated, in particular, with the spatial grid step of climate models.

A considerable arbitrariness in the determination of acceptable errors in measurements is difficult to avoid, because it is impossible to formulate universal requirements of accuracy – since they depend on the specific character of the problems to be solved. The use of correlations among the various parameters of the climate system is important for an optimization of accuracy requirements. One of the most efficient ways to increase accuracy is to average data of frequent measurements (especially in the case of observations from satellites).

In testing the adequacy of climate models by comparing them with observational data, there is the problem of objective analysis. Experience has shown that application of various techniques for objective analysis gives (with identical input data) substantially different results. However, to check the adequacy, it has been found to be sufficient to have monthly averaged data bases.

Requirements for the accuracy of Earth radiation budget (ERB) measurements mean that instrumentation for the ERB experiment must provide an accuracy of not

**Table 2.12** Key regions with respect to monitoring the parameters and factors of climate

| Climate system components and climatic parameters | Key regions, first-priority processes and parameters |
|---|---|
| Atmosphere | 1 Regions of the most frequent formation of blocking cyclones: 160–120° W; 40–70° W; East Siberia – the Chuccha Sea along 60° N |
| | 2 Zones of jet streams |
| | 3 Regions of teleconnections: western Pacific Ocean, western Atlantic Ocean |
| | 4 Location of ridges and troughs of super-long waves: Greenland, Taimyr, Alaska, Aleutian Islands |
| | 5 Subtropical high-pressure bands: eastern and western Pacific and Atlantic Oceans |
| | 6 The Tibet maximum: from middle Asia to east Asia in summer |
| | 7 The wind field in the tropics, including flows across the equator over Africa, the Atlantic and Pacific Oceans |
| | 8 The South Pacific maximum, the Australian–Indonesian minimum |
| | 9 Studies on atmospheric composition at background stations ($CO_2$, etc.); lidar observations at individual locations |
| Ocean | 1 The Equatorial Eastern Pacific (El Niño): 180–90° W; 0–10° S |
| | 2 SST along 137° E in January and July, especially in the 0–7° N band |
| | 3 The Somali current in the Indian Ocean |
| | 4 The Kuroshio and Oyashio currents in the north-western Pacific Ocean |
| | 5 The central Pacific in the northern hemisphere |
| Cryosphere | 1 The Tibet plateau |
| | 2 Siberia |
| | 3 North America |
| | 4 Ice cover in the Chuccha Sea |
| | 5 Ice cover in the Arctic Ocean |
| Soil surface | Soil moisture in low and middle latitudes, especially in semi-arid regions |
| Precipitation | 1 The equatorial–tropical belt, especially from the Indian Ocean to the western Pacific Ocean |
| | 2 The semi-desert regions: Sahel, South Africa, Australia, Mexico, northern China |
| Radiation | 1 Global distribution |
| | 2 Deforestated regions |
| | 3 Regions affected by volcanic aerosols |

**Table 2.13** Requirements for observations of climate system parameters

| Type of observations | Climate system parameter | Class of data | Spatial resolution (km) | Temporal resolution (days) | Error |
|---|---|---|---|---|---|
| Basic meteorological parameters according to GARP requirements | Temperature | P, A, M | 500 (100–200 hPa) | t.b.s. | 1 °C |
| | Wind speed | P, A, M | 500 (100–200 hPa) | t.b.s. | 1–3 m $s^{-1}$ |
| | Relative humidity | P, A, M | 500 (200–300 hPa) | t.b.s. | 7–30% |
| | Surface pressure | P, A, M | 500 | t.b.s. | 1–3 hPa |
| | SST | P, A, M | 500 | t.b.s. | 1 °C |
| Radiation budget and its components, cloudiness | Solar constant | M | – | 1 month | 1–5 W $m^2$ |
| | Extra-atmospheric UV solar radiation | P, M | – | P: 1<br>M: t.b.s. | P: 10%<br>M: t.b.s. |
| | Earth's radiation budget | A, M | 500 | 15–30 | 2–5 W $m^{-2}$ |
| | Cloudiness | A | 500 | 5–15 (vertically) | 5% total<br>1 °C–(top boundary temperature) |
| | Global radiation | P, A | t.b.s. | 5 | 1–3% |
| | Net longwave radiation | P, A | t.b.s. | 5 | 1–3% |
| | Surface albedo | A | 500 | 5–15 | 0.01–0.03 |
| Oceanic parameters | SST | A, M | 500 | A: 5–10<br>M: 30 | 0.5–1.5 °C |

**Table 2.13** *Continued*

| Type of observations | Climate system parameter | Class of data | Spatial resolution (km) | Temporal resolution (days) | Error |
|---|---|---|---|---|---|
| | Upper-layer heat content (200 m) | A | 200 | 5–10 | 1–3 kcal $cm^{-2}$ |
| | Wind stress | A | 200 | 5–10 | 0.1–0.4 dyn $cm^{-2}$ |
| | Ocean surface level (to determine the speed of currents) | A | 200 | 5–10 | 2–10 cm (dynamic topography) |
| | Surface currents | A, M | Only critical regions | 30 | 2–10 cm $s^{-1}$ |
| | Deep-water circulation | A, M | 1000 | 5 yr | 0.1–0.5 cm $s^{-1}$ |
| Precipitation and hydrology | Precipitation over the ocean | A | 500 | 5–10 | 4 levels (1 mm/day) |
| | River run-off | A | Major river basins | 15–30 | t.b.s |
| | Soil moisture | A | 500 | 15 | 10% field water stress, 2 levels |

**Table 2.13** *Continued*

| Type of observations | Climate system parameter | Class of data | Spatial resolution (km) | Temporal resolution (days) | Error |
|---|---|---|---|---|---|
| Cryosphere | Snow cover extent | A, M | 50–100 | 5–15 | Presence/absence |
| | Sea ice extent | A, M | 50–100 | 5–15 | Presence/absence |
| | Ice thickness | A, M | 200 | 15–30 | 10–20% |
| | Sea ice melting | P | 50 | 5 | Yes/no |
| | Sea ice drifting | P | 400 | 1 | 5 km |
| | Thickness of glaciers | M | 200 | 1 yr | 0.1–1.0 m |
| | Deformation of glaciers | M | 200 | 1 yr | 1–10 m |
| | Variations in the glaciers' boundaries | M | 1–5 | 1 month | 1–5 km |
| Atmospheric composition | $CO_2$ | A, M | 2 background and 10 regional stations | 15 weeks | ±0.1 ppm |
| | Vertical profile of ozone concentration | P, A | 10 stations over the globe | 1 week | ±1 ppm |
| | Total ozone content | M | WMO network | 1 | 1–5% |
| | Global distribution of ozone | M | 500 km; 2 km (vertically) | 1 | ±0.5 ppm |
| | Tropospheric aerosols | M | WMO background stations | – | – |
| | Atmospheric turbidity | M | WMO background stations | 1 week | 1% |
| | Stratospheric aerosols | M | 2–4 background stations | – | – |

*Note*: t.b.s. = to be specified.

less than $\pm 5\,W\,m^{-2}$, which will make it possible to analyse the climatically substantial ERB variations on regional scales.

The problem of obtaining data on the three-dimensional cloud field remains unsolved. Information on cloud cover climatology must include the following data: global distribution of total cloud cover, types of clouds and the height of their upper boundary, interannual and seasonal variabilities of cloud cover, the diurnal change of low-level clouds and convective clouds, and statistical characteristics of cloud size. Achieving high sea surface temperature retrieval accuracy (bearing in mind the prevailing contribution of the ocean to the formation of climate and its changes) is a serious problem. The present retrieval accuracy of about 1–2 deg K permits neither reliable determination of the sea surface temperature nor diagnostic studies.

The uncertainty of such a critically substantial parameter as wind stress (0.01–0.04 $N\,m^{-2}$) is equivalent to that of wind speed, 3.6 $m\,s^{-1}$, which can be improved upon, since the error in scatterometric measurements from the Seasat satellite did not exceed $\pm 2\,m\,s^{-1}$. There is a need for an extensive programme of numerical simulation experiments to substantiate a set of parameters, and requirements for the accuracy of their measurement, as applied to various problems.

The existing and planned satellites permit remote sensing of the characteristics enumerated in Table 2.14, which also gives the requirements of spatial and temporal resolution and the accuracy acquired. The meteorological satellites of the TIROS-N NOAA series, first launched in 1978, make it possible to raise the accuracy of the estimation of sea surface temperature from 1.5–2.0 °C (for a spatial grid step of 160 km) to 1 °C, using the AVHRR, ensuring mapping over the spatial grid at intervals of 50 km. In this case the third of the requirements listed in Table 2.14 will be satisfied with an AVHRR field of view (FOV) ensuring a resolution of about 1 km.

A five-channel microwave radiometer developed for the Nimbus-7 and Seasat satellites enabled (for a grid step of 100 km) an accuracy in sea surface temperature (SST) retrieval of about 1.5 °C, with measurements made in all weathers. The infra-red instruments used earlier meet the requirements of measurements of the temperature of snow, ice and land surface. The presence in the latter case of a strong diurnal change necessitates the use of geostationary satellites. The utilization of the Along Track Scanning Radiometer (ATSR) on board ERS-1 and ERS-2 has resulted in a substantial improvement of SST accuracy to about 0.5 °C.

Use of microwave radiometers and radars on the Nimbus-7, Seasat, ERS-1 and ERS-2 satellites has opened up the possibilities of estimating the wind speed near the ocean surface (for Seasat) with an accuracy of $2\,m\,s^{-1}$ (with a grid step of 50 km and 369 h repeatability). Averaging over the area 200 km × 200 km for 5–10 days ensures meeting the requirements mentioned in Table 2.14. Estimating the wind speed up to $1\,m\,s^{-1}$ (with a grid step of 25 km) could soon be achieved.

Existing infra-red radiometers enable identification of the boundaries of sea currents and vortices, with temperature contrasts of about 2 °C and less. The GEOS-3 radiometer permitted an accuracy of sea-level measurements which meets the WCRP requirements. The required accuracy of 10 cm (the 20 km grid step, over 5–10 days) was achieved with the altimeter carried by the oceanographic satellite Seasat. Apparently, this accuracy can be further improved to 2–10 cm.

At present, the technique for analysing space-derived imagery has been thoroughly developed to study the spatial structure of ice cover (with clouds filtered out). Microwave remote sensing has made this an all-weather technique and opened up possibilities to distinguish reliably between open water, one-year and multi-year

**Table 2.14** WCRP requirements of surface characteristics measured using the techniques of remote sensing from space

| Parameter | Spatial resolution (km) | Temporal resolution (repeatability) (days) | Accuracy |
|---|---|---|---|
| SST | 200 | 5–10 | 0.5–1.5 °C |
| | 500 | 5–10 | 0.5–1.5 °C |
| | maximum possible | 30 | 0.5–1.5 °C |
| Land surface and ice cover temperature | 100 | 5 | 1.0 °C |
| | 500 | 5 | 0.1–3.0 °C |
| Wind stress near the ocean surface | 200 | 5–10 | 0.1–0.4 dyn cm$^{-2}$ |
| Sea currents and vortices | 200 | 5–10 | 2–10 cm (topography) |
| Ice cover extent | 100 | 5 | presence/absence |
| | 150 | 5–10 | presence/absence |
| | 50 | 5–15 | presence/absence |
| Ice cover thickness | 200 | 13–30 | 10–20% |
| Ice cover melting | 50 | 5 | yes/no |
| Ice cover drift | 400 | 1 | 5 km |
| Snow cover extent | 50 | 5–15 | presence/absence |
| Soil moisture | 100 | 5 | 10% field moisture content |
| | 500 | 15 | 2 gradations |
| Surface albedo | 100 | 5 | 0.1–0.3 |
| | 500 | 5–15 | ? |

ice, as well as to assess the ice cover thickness (Kondratyev *et al.* 1996a). Identification of the types of ice and the subsequent assessment of its thickness can be made from scatterometer data. Side-looking radars (SLR) are an important means of monitoring the ice cover. Synthetic-aperture radar (SAR) of a very high resolution (about 25 m) was carried by Seasat and later by satellites of the type 'Kosmos' and ERS-1 and ERS-2 (Paci and de Leffe 1983).

Brightness contrasts in the visible spectrum recorded with AVHRR and multichannel scanning radiometers on the Landsat and SPOT satellites make it possible to identify the zones of melting snow and ice cover. Microwave measurement data can also be used for this purpose due to a strong contrast between the emissivities of water and ice (or snow).

However, microwave thermal images have a very low spatial resolution for monitoring ice drift (icebergs, in particular). For this purpose, radar data are more useful. Another possibility is to monitor the drift of sea buoys.

The requirements for monitoring of the snow cover extent can be met with available data. The combined use of visible and near-infra-red imagery provides information on variations in the structure of snow cover and, in particular, snow melting. Previous results have revealed a good correlation between the dynamics of snow cover extent and river run-off. So far, no remote sensing technique has been developed to assess the snow cover thickness. Techniques for remote sensing of soil moisture are at an initial stage of development. To solve this problem, data from multichannel microwave measurements are the most useful. Techniques to estimate surface albedo from satellite data must be improved. Of principal importance is ensuring that all data obtained can be processed efficiently.

In connection with the climate problem, monitoring the sea ice cover is of particular importance. The WMO-organized working meeting of experts on the effect of sea ice on global climate discussed three problems: (1) available data on the ice cover, (2) natural climate change and the role of sea ice and (3) possible impact of human activity on climate and the consequences for ice cover dynamics. It was mentioned in the report (CAS 1977) that both ice and snow cover are the most sensitive indicators of climate change. This is manifested, in particular, through a strong annual change of surface albedo in the latitudinal bands 55–80° N and 60–70° S, which in the northern hemisphere is largely determined by variations of the continental snow cover and in the southern hemisphere by the seasonal variability of sea ice.

Regular observations of the global cryosphere were started only in 1966, following the launch of the meteorological satellites which were put into polar orbits. Available data are, so far, irregular, inadequate and, for periods longer than 100 yr, have been obtained, as a rule, from different indirect indicators (World Meteorological Organization 1989a).

The principal difficulty in the numerical modelling of the atmosphere–ocean–cryosphere system is in the parametrization of sub-grid processes, especially those connected with the existence of sea ice. Therefore, further studies are needed of sea ice dynamics and of the development of a respective parametrization technique, with an account of processes taking place in the ice cover and in the atmospheric and oceanic boundary layers, variability of the oceanic salinity regime, albedo, thickness of the ice cover and other factors.

A very important problem is the monitoring of ice and snow cover with the intention of meeting the requirements enumerated in Table 2.15.

Hooper and Sherman (1986) analysed the requirements for satellite-derived oceanographic information determined by the needs of governmental interest, as well as of interest for scientific research and commercial organizations. Data in Table 2.16 characterize general requirements which form the basis for the development of a scientific complex for the United States national system of oceanographic satellites in order to obtain operational geophysical information. This table contains summaries of the special requirements of spatial and temporal resolution of various parameters for different fields of applications.

For most users, in 70 per cent of the cases, a 6 h data repeatability is enough. The requirements of spatial resolution are more diverse. It is shown that to obtain a global data base on the wind field every 6 hours it is necessary to have an instrumental complex containing two types of scatterometers as well as three scanning multi-channel microwave radiometers. Monitoring of the quality of natural waters can be achieved with two spectrometers of the ocean colour. Global-scale information on the ice cover can be obtained with the use of one synthetic-aperture radar. It seems that about 70 per cent of the need for oceanographic information can be obtained from the ERS, DMSP and NOAA satellites.

A minimum programme of climate monitoring should include the gathering of information on the following (Kondratyev 1983a): (1) weekly microwave images of both polar regions, similar to those obtained with Nimbus-5 (19 GHz) and Nimbus-6

**Table 2.15** Requirements for monitoring the sea ice and snow cover parameters

| Parameter | Coverage | Type of observations | Resolution: Temporal (days) | Resolution: Spatial (km) | Accuracy | Duration | Priority | Technique | Remarks |
|---|---|---|---|---|---|---|---|---|---|
| *Ice* | | | | | | | | | |
| Location of boundaries | Global | Contours | 3 | 20 | 20 km | Continuous | 1 | AVHRR, IRSR, VISR | Applicable only in regions with clear ice edge |
| Concentration | Global | Part of area | 3 | 100 | 20% if $C > 0.8$, 8% if $C < 0.8$ | Continuous | 1 | AVHRR, SAR | Correlation with albedo is possible |
| Trajectory of motion | Arctic basin and selected Antarctic regions | Shifting of certain areas of ice cover | 1–3 | 500 | 2 km | Several years | 1 | AVHRR, MSR, dropped sea buoys | Monitoring the ice drift in the Greenland Sea using buoys with a grid step of 100–200 km |
| Albedo | Global | Areal characteristic | 3 | 100 | 0.03 | Several years from late spring to early autumn | 2 | Multispectral satellite-based radiometry and aircraft observations | Correlation with ice concentration is possible |
| Surface temperature | Global | Areal characteristic or point-to-point data | 3 | 100 | 1 °C | Several years | 2 | IR satellite radiometry | Higher accuracy near the ice edge is needed |

**Table 2.15** *Continued*

| Parameter | Coverage | Type of observations | Resolution: Temporal (days) | Resolution: Spatial (km) | Accuracy | Duration | Priority | Technique | Remarks |
|---|---|---|---|---|---|---|---|---|---|
| One-year/ multilayer ice | Global | Areal characteristic | 15 | 100 | Presence/ absence | Several years | 2 | ESMR, MSR | |
| Ice cover thickness | Global | Areal or profile characteristic | 15 | 100 | Thickness 0.5–1.0 cm part of area 2% | Several years | 2 | Satellite lidars (surface topography), sub-water lidars | |
| *Snow* | | | | | | | | | |
| Extent | Global | Part of area | 3 | 100 | 10% | Continuous | 1 | AVHRR | |
| Albedo | Global | Areal characteristic | 3 | 100 | ±0.03 | Several years | 2 | Satellite multispectral radiometry | 3-day temporal resolution only in period of rapid snow cover variability |
| Water content | Global | Areal characteristic | During winter maximum | 100 | 10% | Several years | 2 | | |

*Note*: AVHRR = Advanced Very High Resolution Radiometer; IRSR = Infra-red Scanning Radiometer; VISR = Visible Infra-red Scanning Radiometer; MSR = Multi-channel Scanning Radiometer; ESMR = Electronic Scanning Microwave Radiometer.

**Table 2.16** Requirements for operational oceanographic information

| Parameter | Error | Resolution: Spatial (km) | Resolution: Temporal | Delay in obtaining information |
|---|---|---|---|---|
| *Winds* | | | | |
| Speed | $2\,m\,s^{-1}$ | 25 | 12 h | 3 h |
| Direction | 10° | 25 | 12 h | 3 h |
| *SST* | | | | |
| Global | 1.0 °C | 25 | 3 days | 12 h |
| Local | 0.5 °C | 10 | 1 day | 12 h |
| *Roughness* | | | | |
| Substantial height of waves | 0.3 m | 25 | 12 h | 3 h |
| Direction | 10° | 25 | 12 h | 3 h |
| *Ice cover* | | | | |
| Extent | 15% | 20 | 3 days | 12 h |
| Thickness | 2 m | 50 | 3 days | 12 h |
| Age | Fresh, 1-yr or multi-year | 20 | 3 days | 12 h |
| Height of glaciers | 0.5 m variation | 10 | 1 yr | 1 month |
| *Water masses* | | | | |
| Chlorophyll content | 100% | 0.4 | 2 days | 8 h |
| Turbidity | Low, middle, high | 0.4 | 1 day | 10 h |
| *Subsurface currents* | | | | |
| Velocity | $5\,cm\,s^{-1}$ | 20 | 1 day | 1 day |
| Direction | 10° | 20 | 1 day | 12 days |

(37 GHz) (ESMR), (2) daily global images in the visible and infra-red spectral regions, (3) data of global scale objective analysis on the fields of wind, pressure, air humidity and temperature for every 12 hours, (4) daily information on the global distribution of cloud cover, (5) all available data for measuring the ERB components with the instruments carried by polar satellites and (6) selected data on surface and aircraft meteorological observations in polar regions.

NASA has undertaken the development of the 10-year programme of climate studies based on the preparation and realization of systems monitoring climatic parameters and further improvement of techniques for numerical modelling of climate and its changes, in accordance with the American general programme of climate studies aimed at foreseeing climate change and its national and international aspects, as well as revealing possible impacts of man on climate on regional and global scales (NASA 1977, 1987).

The climate is determined here as the state of weather near the Earth's surface averaged over time periods of a month and longer. Depending on time scales, four aspects of the problem have been considered. Climate A is defined as the present state of the climate, and the respective developments are aimed at obtaining actual data on the present climate and its analysis (for example, assessing water resources from the snow cover data). The principal objective of observations is to obtain information on the parameters which are important from the viewpoint of the production of food and fibre, assessment of water resources and solution of the problem of energy production.

Climate B is a regional climate on characteristic time scales of more than a month and less than a decade. The principal objective here is the forecast of natural variability (characteristic scales of hundreds of kilometres) of regional and global climates, with an account of previously obtained and continuously accumulated data, bearing in mind modelling of climate changes caused by such slowly varying characteristics as sea surface temperature, snow and ice cover extent, soil moisture, etc. An initial problem is understanding the nature of the annual change and interannual variability as well as studies of such regional-scale phenomena as monsoons.

Climate C is characterized by time scales of more than decades. The eventual objective is understanding the causes of natural variability of global climate, whereas initial stages are studies of the factors affecting the radiation budget (for example, volcanic eruptions, solar activity, etc.), of the interaction between radiation and cloud formation processes, of heat transport by sea currents, of ice cover physics, etc.

It should be emphasized that to determine a number of characteristics (precipitation, properties of clouds, evapotranspiration and water stress of vegetation, sea currents, ocean–atmosphere interaction, aerosol, cryosphere and solar-terrestrial relations), it is particularly important to synchronize satellite, aircraft and ground-based measurements, in view of the ambiguity of remote sensing data (Kondratyev 1983a).

The principal initial tasks for numerical modelling of atmospheric general circulation (AGC) and climate, which could be solved with the use of climate monitoring data, are the following: (1) major persistent features of AGC in middle and high latitudes (for example, prevailing trajectories of cyclones and anticyclones), (2) persistent features of AGC in the tropics (Inter-Tropical Convergence Zone (ITCZ), large-scale convection, sea surface temperature anomalies, oscillation of the Walker circulation in the southern hemisphere, tropical cyclones, Asiatic monsoon) and (3) large-scale interaction between circulations in the tropics, middle and high latitudes. Studies of climate on other planets (Venus, Mars, Jupiter) could help in studies of the Earth's climate (Kondratyev and Hunt 1982).

The urgency of assessing the anthropogenic impact on climate indicates numerous requirements for monitoring atmospheric pollutants (minor optically active gas and aerosol components). In this connection, an informative overview has been made (Committee on Atmospheric Sciences 1980, NSF-UCAR Long-Range Planning Committee 1987) of existing and future possibilities for detecting and monitoring environmental pollution based on the technique of remote sensing from space. Table 2.17 characterizes the atmospheric pollutants, objectives and requirements of measurement accuracy in those cases where the effect of pollutants on the environment has been relatively reliably established. Table 2.18 gives summaries of the components whose effect on the environment has been insufficiently studied.

One of the major sources of information in monitoring the atmosphere, ocean and land pollutants is multi-spectral images such as those obtained with the MKF-6 camera on Soyuz, Salyut and Mir manned orbital stations, as well as with multi-spectral instruments carried by Landsat, SPOT and Skylab.

Important data have been accumulated on the basis of the application of spectroscopic techniques used on the manned orbital stations Soyuz, Salyut and Mir, as

**Table 2.17** Requirements for measurements of atmospheric pollutants whose effect on the environment has been more or less reliably established

| Component and atmospheric region | Objective and importance of measurements | Accuracy |
|---|---|---|
| *Global processes* | | |
| $CO_2$<br>troposphere and stratosphere | Changes in increasing concentration affecting the climate | 0.5 ppm |
| $O_2$<br>upper troposphere and stratosphere | Formation of aerosol particles from $SO_2$ coming from the troposphere and ejected by volcanoes and supersonic aviation | 0.5 ppb |
| $O_3$<br>stratosphere | Study of the long-term trend to reveal causes, in particular the correlation with solar activity | Total content 1%; vertical profile of concentration 10% |
| $H_2O$<br>stratosphere | Study of the effect on:<br>(1) concentration of ozone<br>(2) stratospheric radiation budget<br>(3) aerosol particles size distribution in the sulphate layer | Total content 20% for (1) and lower accuracy for (2); vertical profile 0.5 ppm for (3) |
| $NO_x$<br>stratosphere | Study of the effect on ozone concentration | $NO_2$ and NO 10 ppb |
| *Regional processes* | | |
| $SO_2$ | (1) Damage to vegetation<br>(2) Formation of aerosol particles affecting precipitation acidity | 10 ppb |
| $H_2S$ | (1) Is oxidized to give $SO_2$<br>(2) Natural sources are uncertain | 0.1 ppb |
| $NO_x$ | (1) Damage to vegetation and toxicity at high concentrations;<br>(2) Photo-oxidation of hydrocarbons and the formation of particles;<br>(3) Precursors of PAN | 0.1 ppm for (1);<br>10 ppb for (2, 3) |
| ⟨HC⟩ | (1) Leads to the formation of particles due to photochemical processes;<br>(2) Determines the formation of toxic substances | 1 ppb (?) |
| $O_3$ | Exciting and destructive effects; product of photochemical processes with the participation of ⟨HC⟩ and $NO_x$ | 10 ppb |
| PAN<br>peroxyacetylnitrate | Toxic products of photochemical reactions | 2–10 ppb |
| Hg | Study of the transfer in the atmosphere and accumulation in the biosphere | $10^{-2}$ ppb |
| Thermal pollutants | A factor of regional climate changes | |

**Table 2.18** Requirements for measurements of atmospheric pollutants whose effect on the environment has been insufficiently studied

| Component and atmospheric region | Objective and importance of measurements | Accuracy |
|---|---|---|
| *Global processes* | | |
| CO<br>troposphere<br>stratosphere | (1) The effect of fuel burning on CO concentrations<br>(2) Study of the effect on the stratosphere | 10 ppb |
| $HNO_3$<br>stratosphere | The effect on ozone concentration and aerosol formation | 1 ppb |
| ⟨HC⟩<br>stratosphere | The effect on aerosol formation | ? |
| $CH_4$<br>troposphere and stratosphere | (1) Formation of particles due to photochemical reactions in the troposphere<br>(2) The effect on the distribution of water vapour and ozone in the stratosphere | Troposphere 0.5 ppm<br>Stratosphere 0.2 ppm |
| Chlorofluoromethanes<br>troposphere and stratosphere | An accumulation due to anthropogenic releases | 0.001 ppb (?) |
| $N_2O$<br>stratosphere | Photodissociation in the stratosphere | 50 ppb |
| CO | (1) Revealing the sources, sinks and life-time in the lower atmosphere<br>(2) Formation of particles due to photochemical reactions<br>(3) An indicator of processes in the oceans | 10 ppb |
| $H_2CO$ | (1) Formation of particles due to photochemical reactions<br>(2) The exciting factor for concentrations of about 1 ppm | 1 ppb |
| Halogens | (1) Toxicity and destructive effect<br>(2) Bromine as an indicator of oceanic bioproductivity | 1 ppb |
| $NH_3$ | Reacts with $H_2SO_4$ giving $(NH_4)_2SO_4$ particles | 10 ppb |

well as Nimbus-7 and the Stratospheric Aerosol and Gas Experiment (SAGE) satellites. In this connection, the development of techniques for correlation and interferometric spectroscopy is very important. Such techniques have been used during Shuttle flights to measure concentrations of such atmospheric components as CO,

$NH_3$, $H_2O$, $NO_2$, $CH_4$, $NO_x$, $SO_2$ and others. Techniques of remote sensing of oceanic pollution from space must be thoroughly developed. In this connection, the colour of the ocean is the most important characteristic.

To study the numerous and complicated chemical processes taking place in the troposphere, it is necessary to employ space-based observational means for their investigation, principally to assess the content of optically active minor gaseous and aerosol components (MGACs). Natural and anthropogenic surface sources of MGACs are as follows: volcanoes, biogenic processes in vegetation cover, soil and oceans, burning of fossil fuel and plant residuals, functioning of internal-combustion engines, etc. Tropospheric sources of MGCs include lightning, photochemical reactions induced by ultraviolet (UV) solar radiation, and numerous chemical reactions which transform atmospheric properties. The urgency of adequate information about MGCs is determined by their important role as factors of air quality, visibility, acid rain, ozone layer depletion and climate formation.

Experience in applying techniques for classical and correlation spectroscopy, spectral radiometry and lidar sounding testifies to the usefulness of remote sensing of MGACs on the basis of satellite observations. At a workshop in New York on 9–13 September 1987 (NASA 1987), the scientific and technical aspects of remote sensing of atmospheric composition were discussed in detail. Practically, the results obtained include: (1) retrieving the aerosol optical thickness of the atmosphere from measurements of the surface-atmosphere brightness in the visible and near infra-red spectral regions, made from operational (polar-orbiting NOAA and Geostationary GOES) meteorological satellites and Earth's resources Landsat satellites, and (2) assessing the content of aerosol, water vapour, $O_3$ and some other MGCs in the middle and upper troposphere from data of occultation soundings (satellites SAGE, Nimbus-7, ERBS (Earth Radiation Budget Satellite), orbital station Spacelab-3, the Shuttle).

Of particular interest is the successful testing of the interferometric (Fourier) spectrometer within the ATMOS programme (Spacelab-3) and the correlation gas-filter nadir-viewing radiometer (MAPS programme, Shuttle, 1981 and 1984). These instruments were planned to be used, initially, to estimate the content of CO and $CH_4$ in the troposphere.

Studies of the chemical composition of the troposphere from satellite observations are aimed at (in priority): (1) determination of continental and oceanic sources and sinks of such optically and chemically active MGCs as $CO_2$, CO, $CH_4$ and other hydrocarbon components, $N_2O$, $NH_3$, $(CH_3)_2S$ (DMS), $H_2S$, OCS and $SO_2$, (2) analysis of spatial and temporal variability (in particular, annual and latitudinal change, regionally and globally averaged trends) of various MGACs, in particular, $H_2O$, CO, $O_3$, $NO_2$, $SO_2$, acid and soot aerosols.

In this connection, it is important to emphasize that all first-priority components ($H_2O$, $O_3$, CO and $CH_4$) participate in chemical and photochemical reactions of the formation and destruction of hydroxyl OH. Thus, OH is the key chemically active component of the troposphere, taking part in reactions with almost all the other components and governing their residence time in the troposphere (NO, whose life-time is controlled by $O_3$, is the only important exception). The very short life-time of OH (about a second) and its very low concentration (about $10^5$ to $10^6$ molecules $cm^{-3}$) exclude any possibility of reliable direct measurements of its content in the troposphere, determined by the reaction

$$O(^1D) + H_2O \rightarrow 2OH \tag{2.5}$$

with excited atoms of oxygen resulting from ozone photolysis:

$$O_3 + \rightarrow O(^1D) + O_2 \tag{2.6}$$

The destruction of hydroxyl is controlled by two reactions:

$$OH + CO \rightarrow CO_2 + H$$

$$OH + CH_4 \rightarrow CH_3 + H_2O \tag{2.7}$$

Thus, $H_2O$ and $O_3$ are responsible for the photochemical formation of OH, whereas CO and $CH_4$ cause the chemical destruction of OH.

Four principal projects have been described in detail (NASA 1987) connected with the use of spectrometry, correlation radiometry, spectral radiometry and lidar sounding. Based on this discussion, plans have been formulated for studies and developments during the period 1986–1992 and later (1992–2000), including a programme of supportive investigations.

The first programme mentioned foresees the use of correlation radiometers to retrieve CO (three layers in the troposphere) and $CH_4$ (lower and upper troposphere), an interferometer spectrometer (Michelson interferometer) recording the outgoing thermal emission spectrum (the content of several MGCs in the lower, middle and upper troposphere) and lidar (aerosol sounding). In perspective, lidar soundings are to provide information not only about aerosols but also about several MGCs (water vapour, $O_3$, CO, $CH_4$ and, probably, $NH_3$).

Because of the numerous climatic parameters to be monitored and the difficulty of acquiring a complete system of satellites for climate monitoring, a minimum data base must be determined to solve the first-priority problem of studying the climate and its changes.

The report (COSPAR 1978) prepared by the COSPAR Working Group-6 contains results of discussions on the available possibilities and prospects of satellite observations for studying the physical processes that determine the climate which are still very important and informative. The discussion took place at the meeting of experts in NCAR (Boulder, Colorado, USA, July 1976) convened by the COSPAR WG-6 on the recommendation of the GARP Organizing Committee. The problems discussed were grouped as follows: (1) interaction of cloud cover and radiation, (2) the effect of land surface processes, (3) radiatively active MGACs of the atmosphere, (4) ocean–atmosphere interaction and (5) cryosphere and hydrosphere, and processes taking place there.

The main objective was to determine a set of climatic parameters and factors which can be remotely sensed from space, as well as to choose adequate observational means. Table 2.19 summarizes the results of discussions. Since, so far, it is impossible to substantiate the requirements for a satellite system of climatic observations, the related considerations and recommendations are strictly preliminary. This particularly concerns studies of ocean processes, since recent studies have put forward many new and, so far, unsolved problems (the role of mesoscale vortices, upwellings, large-scale sea surface temperature anomalies, etc.).

A set of stratospheric MGCs must be specified, which, apart from three basic components (water vapour, $CO_2$ and $O_3$) should include numerous compounds of

**Table 2.19** Possibilities for studying climate-forming factors and processes using observations from space

| Factors and processes | Effect on climate | Estimated parameter | Measured parameter | Techniques for processing and analysis | Accuracy and spatial and temporal resolution | Remarks | Observation technique |
|---|---|---|---|---|---|---|---|
| *Radiation and climate* | | | | | | | |
| Incoming solar radiation | Original source of energy for the climate system | Total insolation at the level of orbit and Earth's surface, in the range from UV to IR | Same | Non-specific | Desired absolute accuracy 0.1% but not below 0.7%; daily observations at present several times per year | Satellite and rocket-borne instruments are available. Instruments for shuttles and automated satellites must be further developed | Cavity radiometers calibrated against secondary electrical references |
| | | Insolation variability | Same | Same | Accuracy 0.1% for a period of Sun rotation daily | Instruments like those of Nimbus-6,7 can be used | Radiometers with a flat receiving surface calibrated against the Sun |
| Heating and photodissociation in the stratosphere and mesosphere | Controls the thermal regime dynamics and photochemical processes | UV solar radiation flux in the range 180–310 nm with a resolution of 5 nm | Same | Same | Absolute accuracy 10%, relative accuracy 1% for 5 nm intervals; daily | Accuracy is determined by reliability of references | UV radiometer, spectrometer |
| Atmospheric heating due to radiatively active gases | Heat divergence in the atmosphere | IR radiation flux in the range 0.8–4 $\mu m$ for 10 intervals | Same | Same | 1% or better | None, so far | IR filter radiometers |

**Table 2.19** *Continued*

| Factors and processes | Effect on climate | Estimated parameter | Measured parameter | Techniques for processing and analysis | Accuracy and spatial and temporal resolution | Remarks | Observation technique |
|---|---|---|---|---|---|---|---|
| Earth's radiation budget and its components | Principal energy source | Global distribution of outgoing shortwave and longwave radiation | Outgoing radiation flux | Schemes of the intensity-to-flux transition and account of diurnal change | Desired accuracy 1%, but 5% is acceptable for shortwave and 10% for longwave radiation; not less than 4 times a day | High-priority observations | Radiometers |
| Synoptic-scale cloudiness in connection with radiation balance | Effect of clouds on radiation balance | Type, amount, height of boundaries of cloud cover | Brightness in the visible and IR | Retrieval of the sought-for parameters from the brightness field | 5%, cloud amount; 1%, height of boundaries | Observations of clouds and radiation must be synchronous | Radiometers; stereo-images or lidar |
| *Surface* | | | | | | | |
| Evaporation, sensible heat exchange | Components of the atmospheric energy balance | Effective albedos of different types of surface | Reflected radiation | Empirical correlations | Several per cent, the scale is less than typical quasi-homogeneous regions | Satellite data serve, largely, to identify the type of surface and observation conditions | Multispectral images of the Landsat type |

**Table 2.19** *Continued*

| Factors and processes | Effect on climate | Estimated parameter | Measured parameter | Techniques for processing and analysis | Accuracy and spatial and temporal resolution | Remarks | Observation technique |
|---|---|---|---|---|---|---|---|
| *Optically active components* | | | | | | | |
| Radiation flux divergence to the atmosphere and latent heat flux | Effect on the thermal regime and dynamics | Water vapour vertical profile | Emission spectra in the IR and microwave regions | Retrieval algorithms | *Troposphere*: accuracy 0.25° per day of radiative cooling calculations in 3–5 km layers *Stratosphere*: 0.3–0.5 ppm for 3–5 km layers; daily; 500 km horizontally | Retrieving gives a low but, at present, acceptable vertical resolution | Multispectral IR and microwave radiometers; microwave limb scanning over the nocturnal pole |
| UV heating and IR cooling of the stratosphere | To calculate energy balance of the atmosphere and to test stratospheric models | Vertical ozone profile; $CO_2$ concentration in the stratosphere | Emission spectra with synchronous ground, balloon, and rocket observations | Same | 10% or 0.5 ppm for diurnal means; 2%, monthly means; 50 km horizontal, 3–5 km vertical | Ground-based network for ozone observations must be improved; balloon and rocket observations must be developed | UV radiometers and spectrometers; limb scanning in the IR and microwave intervals |
| Radiation transfer in the atmosphere | Controls the UV radiation in the biosphere | Total ozone content | Outgoing UV or IR radiation | Same | 1% global means; daily; 5% with 500 km averaging; daily; 1% with 500 km averaging over a month | Comparison of data of satellites and ground network is important | UV and IR radiometers (nadir); limb scanning |

**Table 2.19** *Continued*

| Factors and processes | Effect on climate | Estimated parameter | Measured parameter | Techniques for processing and analysis | Accuracy and spatial and temporal resolution | Remarks | Observation technique |
|---|---|---|---|---|---|---|---|
| Absorption and scattering of radiation | Effect on atmospheric energy balance | Aerosol properties (optical thickness, size distribution or average size, complex refraction index and lidar data) | Multi-spectral extinction and polarization, images and lidar data | Same | 0.3 μm (size), 0.1 (optical thickness); 0.02 (refractive index) | Revealing first-order effects and assessing atmospheric transfer function | Photometers, polarimeters, lidars |
| IR cooling of the atmosphere | Affects the atmospheric radiation | Total content or mean concentration of $CO_2$ | Concentration at selected ground stations | Absorption in a gas cuvette | Available accuracy is high enough to reveal the trends | Observations from space are not needed | IR radiometer |
| | | | Emission spectra as a function of height in the stratosphere | Spectroscopic calibration | Accuracy to be determined; several times per year | | |
| Chemical interaction with ozone | Controls the ozone concentration in the stratosphere | Nitrogen oxides, chlorofluorocarbons | Outgoing radiation or absorbed radiation (occultation experiments) | Retrieval algorithms | To be determined | Observations are of a search character | Various instruments to be tested |
| Aerosol formation processes | Determine the distribution and properties of aerosols | Sulphur compounds | Same | Same | Same | | |

**Table 2.19** *Continued*

| Factors and processes | Effect on climate | Estimated parameter | Measured parameter | Techniques for processing and analysis | Accuracy and spatial and temporal resolution | Remarks | Observation technique |
|---|---|---|---|---|---|---|---|
| *Ocean–atmosphere interaction* | | | | | | | |
| Ocean–atmosphere heat exchange | Energy component of the climate system | SST | IR and microwave outgoing emission | Same | 1% or better; 10 km resolution; daily | Correct account of atmospheric transfer function is a principal problem | IR and microwave observations |
| Meridional heat transport in the ocean | Energy component can determine the short-term climate variability | Heat content and velocity field in the upper layer of the ocean (before the seasonal thermocline) | Profiles of temperature and speed of currents | Direct measurements and tracing the buoys' trajectories | Tenths of of degree, 1 cm $s^{-1}$, 10 km resolution | Space-based monitoring technique is to be tested | Small buoys |
| Mass and momentum exchange between the ocean and the atmosphere | Connected with the ocean upper layer dynamics, affects the short-period climate variability | Wind Stress near the surface | Near-surface wind at a height of 20 m and at a level of trade-wind cumulus cloudiness | Empirical correlations | About 1 dyn $cm^{-1}$, resolution up to 10 km | Seasat scatterometer has been tested | Direct ship measurements of wind and space-based monitoring of clouds |
| *Cryosphere and hydrosphere* | | | | | | | |
| Processes of cooling, formation of polar bottom waters | Ice–ocean interaction in polar regions, affecting the energetics of the climate system | Salinity profile in the upper layer | Direct measurements of electrical conductivity and multispectral microwave reflectivity | Dependence of electrical conductivity on salinity; retrieval algorithms | About 0.1%; 200–500 km resolution | Actual local studies of the parametrization technique | Salinometers on buoys; remote sensing techniques are under development |

**Table 2.19** *Continued*

| Factors and processes | Effect on climate | Estimated parameter | Measured parameter | Techniques for processing and analysis | Accuracy and spatial and temporal resolution | Remarks | Observation technique |
|---|---|---|---|---|---|---|---|
| Formation of snow cover albedo, and of heat flux in the atmosphere | Heating of the Earth's surface and atmosphere | Snow cover extent | Images in the IR and microwave regions | Direct processing of images | 50–100 km for 5–10 days | Mapping of minimum brightness is useful | Scanning photometers and radiometers |
| Formation of ocean surface albedo, energy fluxes and divergences, effect of atmospheric dynamics on sea ice | Important role in the climatic system's energetics | Sea ice extent | Images in the visible, IR and microwave regions | Same | 50–100 km; 5–10 days; high resolution for special studies | Maps of relative reflectivity and analogy to maps of minimum brightness are useful | Scanning photometers and radiometers |
| | | Type | Multi-spectral microwave emission; radar data | Retrieval algorithms | 100–200 km; 10–30 days | Different age of ice | Passive and active radar technique |
| | | Concentration | Images in the visible IR and microwave regions | Direct processing | 100 km; 5–10 days | Important for studies of the heat budget and albedo of vast regions | Scanning photometers and radiometers |
| | | Drift and deformation | Images in the visible, IR and microwave regions | Direct processing | Drift to 5 km/day; 100–500 km | Average drift and deformation based on analysis of images | Scanning instruments; radars |
| Large-scale and persistent heat sinks | Continental glaciers as a factor and indicator of climate change | Glaciers extent | Images in the visible, IR and microwave regions | Direct processing | 1–5 km; month–season | Typical velocity | Scanning instruments |

**Table 2.19** *Continued*

| Factors and processes | Effect on climate | Estimated parameter | Measured parameter | Techniques for processing and analysis | Accuracy and spatial and temporal resolution | Remarks | Observation technique |
|---|---|---|---|---|---|---|---|
| | | Accumulation, thickness, topography | Multi-spectral microwave emission, radar-altimeter | Retrieval algorithms | Annual accumulation to 10 cm, 200 km | Indicator of climate and accumulation mechanism | Passive and active radar technique |
| | | Deformation | Images, ice buoys, topography | Direct processing | 1–10 m; 200 km; annually | Polar-orbit radioaltimeter is needed | Passive and active radar technique, buoys |
| Latent heat release, the hydrological cycle | Important factors of the climate system's energetics | Precipitation, water and moisture content | Images; attenuation of microwave emission, radar data | Retrieval algorithms | 4 levels of intensity; 100 km; 1–5 day averages | Data repeatability (several times per day) is very important | Scanning instruments in different ranges |
| Evapotranspiration processes | Surface–atmosphere interaction component | Moisture profile of a 1 m soil layer | Multi-spectral microwave emission, annual change of temperature, the state of vegetation | Same | To be determined | Search studies are needed | Passive and active radar techniques |
| Hydrological processes | Snow water content factor | Surface run-off connected with hydrological cycle | Images in the visible, microwave regions; radar data are possible | Same | 0.1–1 cm; 50–100 km; 5–10 day averages | Important in studies of water balance | Scanning instruments in different ranges |
| | Part of the hydrological cycle | River run-off | Data via satellite transmission of surface sensors' readings | Direct processing | To be determined | Use of experience in automated sensors' readings | Instruments for sensors' readings |

nitrogen ($N_2O$, $NH_3$, $NO_2$, NO, $HNO_2$, $HNO_3$), carbon ($CH_4$, CO), hydrogen (OH, $HO_2$), sulphur ($SO_2$, $H_2SO_4$, $H_2S$, $CS_2$, COS) and halogens ($CFCl_3$, $CCl_4$, $CF_2Cl_2$, HCl). Since there are data on an increasing amount of water vapour in the stratosphere, special emphasis must be laid on studies of the processes of water vapour formation due to methane oxidation.

Sherman (1985) reviewed the current state in the USA and western Europe of satellite systems which would provide information on the world's oceans and assessed future developments to the end of the century. The key component of these systems was to have been included in the ocean remote sensing system (N-ROSS), together with the Earth resources satellite ERS-1 developed by the European Space Agency, which, in total, would constitute an integral system; however, the N-ROSS project has not been realized. The so-called polar platforms (polar-orbiting satellites) combined with the permanent orbital manned station planned by the USA, had been planned to become a component of this system. All these incompletely realized projects have been transformed into an Earth Observing System (EOS) (NASA 1984, 1991, Kondratyev 1992e, 1994).

The global nature of the processes in the atmosphere, ocean and on land determines the need to consider the Earth as a single interactive system in order to understand and forecast the state of this system. On the other hand, the complexity of the internal processes taking place in each of the system's components necessitates their individual study. From the viewpoint of creating an observational system, the final objective is the development (with an account of the circumstances described above) of a system that would be able to provide information on the planet as a whole, and on its components, using both conventional and satellite observational means.

Of great interest are programmes to prepare the EOS polar platforms driven by the European Space Agency (ESA), Japan and the USA. The EOS-A satellite is planned for launch in 1998, followed by EOS-B in 2001. Both are designed to obtain data to study global climate change over a 15-yr period. A small part of the observational programme includes the problems of stratospheric chemistry, and geophysical studies of the Earth's solid shell (Dozier 1992).

The Consulting Council for the Complex of Scientific Instruments made decisions based on recommendations of the Intergovernmental Panel on Climate Change (IPCC) (Houghton *et al.* 1990), the US Environmental Protection Agency (EPA) and the Committee on Earth Sciences and the Environment. These decisions foresee the following priorities for development (NASA 1991, Dozier 1992):

1. Global water and energy cycles
    - 1.1. Formation, development and dissipation of clouds, and their radiative properties which affect the climate system's sensitivity to greenhouse forcing;
    - 1.2. Large-scale hydrological processes including evaporation and precipitation
2. The world's oceans
    - 2.1. Energy and mass exchange between the atmosphere and the ocean as well as between the upper and deep layers of the ocean;
    - 2.2. Sea ice cover dynamics; formation of bottom waters
3. Chemistry of the troposphere and lower stratosphere

3.1. Relations between the water cycle and ecosystems;
3.2. Transformation of greenhouse gases in the atmosphere and its effect on climate changes
4. Land surface processes (hydrology and dynamics of ecosystems)
4.1. Improvement of techniques to assess river surface run-off and the input of river water to the ocean;
4.2. Sources of, and sinks for, greenhouse gases;
4.3. Energy and moisture exchange between land surface and atmosphere;
4.4. Land surface changes (deforestation, desertification, etc.)
5. Glaciers and polar ice sheets: forecasts of the world's oceans' sea level and global water balance
6. Chemistry of the middle and upper atmosphere: chemical reactions; Sun–atmosphere connections; sources of, and sinks for, radiatively active minor gaseous components of the atmosphere
7. Earth solid shell: volcanism and its role in climate change.

In accordance with the enumerated priorities, the following problems are to be studied: (1) cloud cover, radiation, moisture and water content of the atmosphere (precipitation) with the diurnal change taken into account, (2) bioproductivity and general circulation of the ocean, ocean–atmosphere interaction, (3) greenhouse gas sources and sinks, greenhouse gas transformation in the atmosphere with account of the priority of the problem of the global carbon cycle, (4) changes in land use, land surface characteristics, primary production and the water cycle, (5) polar ice and ocean level, (6) interaction of the processes of formation and destruction of ozone with the dynamics of climate and biosphere and (7) volcanic eruptions and their impact.

The EOS scientific complex must ensure the continuation of the observation series (Dozier 1992): Earth's radiation budget (Nimbus-7 and ERBE), precipitation, snow, and ice cover extent, moisture and water content of the atmosphere (TRMM, SSM/I, DMSP), ocean colour (CZCS, SeaWiFS), topography of land and ocean (TOPEX/POSEIDON), winds near the ocean surface (NSCAT, ADEOS), land surface characteristics (AVHRR, SPOT), standard meteorological parameters (various meteorological satellites), chemical composition and dynamics of the upper atmosphere (UARS) and ozone content (TOMS, SAGE-II).

For the period 1997–2000, the following instruments are intended to be operational: Atmospheric Infra-red Sounder (AIRS); Advanced Microwave Sounder (AMSU-A); Microwave Humidity Sounder (MHS); Advanced radiometer to meaSure Thermal Emission and Reflection characteristics (ASTER) (ASTER permits one to obtain multispectral images of surface and clouds in a broad wavelength range, from the visible to the thermal infra-red spectral interval, and one of the channels can give stereo-images); the CERES complex to measure the ERB; MultI-spectral video Microwave Radiometer (MIMR); MultI-angle Spectroradiometer (MISR); mid-resolution video-spectrometer (MODIS-N); the MOPITT complex to measure tropospheric pollution; the well-known SAGE-II and the ocean colour radiometer SeaWiFS.

After 2000, the scientific complex will include a radiometer to measure the solar constant (ACRIM); a radar altimeter (ALT); a scanning spectropolarimeter (EOSP);

instruments for coordinate and geodetic observations (GGI); a laser altimeter (GLRS-A); a high-resolution limb sounder (HIRDLS); a high-resolution video-spectrometer (HIRIS); instruments to measure extra-atmospheric ultraviolet (UV) solar radiation (SOLSTICE); a scatterometer (STICKSAT) and an emission spectrometer to sound the troposphere (TES).

The possibility of using the microwave limb sounder, MLS, and the spectrometer SAFIRE for the far-infra-red region is now under discussion. However, the problems of financially supporting the development of the lidar wind sounder, LAWS, and the three-frequency synthetic-aperture radar (SAR) have not been resolved. Certain instruments have been excluded from the complex under development: the geodynamic laser range-sounder (GLRS-R); the geomagnetic complex (GOS); an instrument to measure the characteristics of ionospheric plasma (IPEI); the lightning video-sensor (LIS); a video-spectrometer (MODIS-T); the infra-red limb wind sounder (SWIRLS) and the X-ray video-spectrometer (XIE). Also under consideration is the possibility of installing the multi-channel photometer SAGE-II on the European polar platform.

Satellites of the EOS system will be launched in both the morning (am) and afternoon (pm) polar orbits. The first polar platform is due for a 1998 launch in the morning orbit, on the premise that this orbit will allow more reliable simultaneous measurements of the characteristics of cloud cover, outgoing radiation field and the surface, compared with the afternoon orbit. The major scientific complex of the satellite AM-1 (1998) will consist of ASTER, CERES, MISR, MODIS-N and MOPITT, initally to obtain data on radiative properties of clouds and aerosol, as well as energy and mass exchange between the stratosphere and the ocean. The intention is to launch the am and pm satellites at five-year intervals, so as to secure a continuous observation series to 2030.

The great interest in the effect of increasing concentration of greenhouse gases on climate has led to greater emphasis on the data for the lower stratosphere than for the upper stratosphere. In this respect, the first priority becomes the long-term monitoring of total ozone content (SAGE-II > HIRDLS, MLS, SAFIRE, where A > B,C indicates that instrument A is preferable to instruments B and C), global mapping of the total ozone content (HIRDLS > MLS, SAFIRE), and air temperature (HIRDLS > MLS, SAFIRE).

Of some urgency for attention are the problems of assessing the content of radicals: OH (SAFIRE, MLS), ClO (MLS), $NO_2$ (HIRDLS, SAGE-III), $NO_3$ (SAGE-III), OClO (SAGE-III), BrO (MLS), $HO_2$ (SAFIRE, MLS). The most significant reservoir components are: HCl (MLS, SAFIRE), HF (MLS), $ClONO_2$ (HIRDLS), $HNO_3$ (HIRDLS, MLS, SAFIRE); $H_2O_2$ (SAFIRE); HOCl (SAFIRE, MLS), $HBr_2$ (SAFIRE), $N_2O_5$ (HIRDLS, SAFIRE). The most substantial source gases are: $H_2O$ (HIRDLS, SAGE-III, MLS, SAFIRE), $N_2O$ (HIRDLS, MLS), $CH_4$ (HIRDLS), CFCs (HIRDLS). Measurements of the characteristics of aerosol and polar stratospheric clouds (SAGE-III) and extra-atmospheric ultraviolet solar radiation (SOLSTICE-II) are a necessary component of the observational programme.

The enumerated radicals and minor gaseous components play the most important role in the chemical and photochemical processes in the atmosphere, characterizing the following families: $O_x$, $ClO_x$, $NO_x$, $HO_x$. However, such quasi-conservative families like $NO_y$, $Cl_y$ and $Br_y$ also need consideration. MLS and SAFIRE are the instrumental priorities, from the perspective of problem solving.

As noted above, the EOS system of polar platforms should be supplemented by two small satellites to measure the most significant parameters of climate and climate-forming processes. The totality of these parameters can be classified into three groups (Hansen *et al.* 1990):

1. *Climatic forcings* – In this case, observations of greenhouse gases are needed first, e.g. $H_2O$, $CO_2$, chlorofluorocarbons (CFCs), $CH_4$, $N_2O$, as well as $O_3$, both in the troposphere and in the stratosphere. Other important parameters are: the contents of aerosols in the troposphere and stratosphere, extra-atmospheric solar radiation and surface albedo. The use of satellite remote sounding of minor optically active gaseous and aerosol components of the atmosphere should be based on the application of SAGE and EOSP together with *in situ* measurements.
2. *Climatic feedbacks* – In this context, of particular interest is the problem of the dynamics of cloud cover and its interaction with radiation. Most significant are the following characteristics of clouds: amount, height (temperature) of cloud top, optical thickness, effective size of particles (as an indicator of particle size distribution), and phase state of clouds. Of priority is information on the vertical profile of water vapour content in the atmosphere, snow and ice cover extent, heat transport in the ocean and ocean–atmosphere heat exchange, and types and state of vegetation cover. The operational meteorological satellites and instruments (SAGE and EOSP) are important sources of this information. Bearing in mind that much is unclear in respect of climatic feedback, information requirements must be carefully specified, and consequently consideration must be given to processes like the emission of dimethyl sulphide by the ocean, ejections of gas hydrates as a result of permafrost thawing, gas-to-particle conversion of anthropogenic sulphate aerosol, and so on.
3. *Climate diagnostics* – The solution of related problems, and first the water and energy cycle problems, requires various measurements with diverse parameters, including first priority values such as (i) the Earth radiation budget (ERB), (ii) the temperatures of the air, land surface, ocean (surface and deep water), and (iii) precipitation. Inevitably, in the interests of climatic diagnostics, conventional and satellite-derived information should be combined, including data from dedicated ERB satellites.

A minimal system of small satellites could include only two space platforms carrying the SAGE, EOSP and ERBE instruments. The total cost of such instruments would reach about \$100 million, whilst the launch of two satellites will cost below \$150 million, and the cost of one Delta rocket is \$25 million, with flight control and data analysis over 5 yr reaching about \$50 million. Thus the total programme cost might be estimated at \$350 million.

Of critical importance is the provision of calibration techniques which would guarantee the long-term stability of both the observational data and retrieved geophysical parameters. This particularly concerns the solar channels of various space-based instruments. It is also urgent to achieve complex programmes of sub-satellite control observations. In this connection it is worthwhile, in particular, to realize in the future a programme like the First GARP (Global Atmospheric Research Programme) Global Experiment (FGGE). Important components of such a programme should be theoretical studies aimed at a better understanding of the interaction of processes within the global-scale numerical models and sensitivity studies needed for an objective planning of the observational system. Retrieval algorithms must be further developed. From the viewpoint of developing instruments and techniques for retrievals, particularly important are efforts to raise the reliability of such retrieved data as soil moisture, rain rate, land surface temperature and emissivity, and wind vector.

Experience of data interpretation has shown (Jet Propulsion Laboratory 1985) that although it is possible to distinguish between three levels of soil moisture with a spatial resolution of 100 km and repeatability of once every 2 days, it is preferable to identify five levels at this resolution. The 21 cm channel is the best one to solve this problem, but in this case measurements are impossible when vegetation cover is thick.

A substantial increase in retrieval accuracy of the surface temperature data, and a resolution of about 1 km, can be achieved with the MODIS spectrometers with the following channels (the width of the channels, in nm, is given in parentheses): 3.750 $\mu$m (90); 3.989 $\mu$m (50); 4.05 $\mu$m (50); 8.55 $\mu$m (500); 10.45 $\mu$m (500); 11.03 $\mu$m (500) and 12.02 $\mu$m (500) with the combined use of MODIS and other remote sensing data. Problems to be solved include retrieving data on phytomass and mapping its global extent, as well as a search for techniques for retrieving surface heat balance components. Lidars and a six-channel radar for centimetre wavelengths are being developed to retrieve atmospheric pressure data near the ocean surface.

In accordance with the initial plans, the two categories of space-borne instruments were intended (1) to obtain information for the global-scale mapping of environmental and biospheric parameters, with a 2–3-day repeatability and a spatial resolution of several kilometres and (2) to obtain high-resolution data (tens of kilometres) with a limited spatial coverage for detailed monitoring of the processes in different regions. The measurement complex is to be carried by three satellites and include 37 instruments. Table 2.20 contains a brief outline of this complex. These plans have subsequently been changed (Kondratyev 1992e).

Curran (1987) briefly described the NASA lidar complex for EOS; this is intended for remote sounding of the atmosphere and for altimetric observations. Research should be directed at the following goals: (1) a deeper insight into the global water cycle based on analysis of data on water vapour and liquid water content in the atmosphere, (2) a better understanding of the role of aerosols and clouds as climate-forming factors, (3) obtaining more reliable information on the large-scale structure and dynamics of the atmosphere from remote sensing of its composition and structural parameters and (4) analysis of the laws of the distribution of snow and ice cover based on satellite altimetry. The Doppler lidar is also being developed to retrieve the wind field with an accuracy of several metres per second, with 100 km horizontal (1 km vertical) resolution. Unfortunately, there were many changes in the EOS planning, and the use of lidar soundings has been postponed.

An important component of the global-scale observational system is provided by developments within the International Satellite Cloud Climatology Project (ISCCP) which have led to the choice of an operational algorithm to retrieve cloud characteristics. The following ISCCP principal objectives have been approved, after the accumulation of a 5-yr global data base: (1) to assess the reliability of the cloud cover characteristics based on analysis of data of specialized regional complex sub-satellite

**Table 2.20** The measurement complex of satellites of the Earth Observation System

| Objective | Instruments | Basic characteristics | Parameters to be measured |
|---|---|---|---|
| Obtaining surface images and remote sounding | Slant-axis moderate-resolution video-spectrometer, MODIS-T | 0.4–1.1 $\mu$m; 64 channels; spatial resolution 1 km | Biogeochemical cycles; ocean colour |
| | Nadir-viewing video-spectrometer, MODIS-N | 0.4–1.1 $\mu$m; 36 channels; 0.5 km | Biogeochemical cycles; temperature |
| | High-resolution video-spectrometer, HIRIS | 0.4–2.2 $\mu$m; 200 channels; 30 m | Biogeochemical cycles; geology |
| | Thermal video-spectrometer, TIMS | 3–5, 8–15 $\mu$m; 9 channels; 30 m | Biogeochemical cycles; geology |
| | High-resolution multi-frequency microwave radiometer | 1.4–91 GHz; 6 channels; 1.1 km × 1.8 km at 91 GHz | Radiobrightness temperature of the ocean and land |
| | Lidar sounder and altimeter, LASA | A set of wavelengths; the receiving mirror is 1.25 m in diameter | Atmospheric correction; aerosols and clouds soundings; altimetry |
| | Advanced system for data collection and coordination referencing | Referencing accuracy; 2–5 km | Mainly, data from sea buoys |
| Active sounding in radiowave region | Synthetic-aperture radar, SAR | Bands C, L, X at different polarizations; 30 m | Characteristics of land, vegetation and ice cover |
| | Radio-altimeter, ALT | Height measuring accuracy 10 cm | Surface topography |
| | Scatterometer | Wind speed retrieval accuracy $\pm 2\,m\,s^{-1}$ | Wind share near the ocean surface |

**Table 2.20** *Continued*

| Objective | Instruments | Basic characteristics | Parameters to be measured |
|---|---|---|---|
| Monitoring of physical and chemical characteristics of the atmosphere | Correlation radiometer | | Retrieval of the vertical profiles of CO and $NH_3$ concentrations |
| | Nadir-viewing Fourier spectrometer | | Retrieval of $NO_3$, $CH_4$, $H_2S$ |
| | Lidar sounder–altimeter | | Profiles of temperature, pressure, water vapour and ozone in the troposphere |
| | Doppler lidar | | Tropospheric wind |
| | IR radiometer | | Limb scanning of $O_3$, $N_2O$, etc. |
| | Pressure-modulated radiometer | | Limb scanning for $CFCl_3$, etc. |
| | Sub-millimetre spectrometer | | Limb scanning for OH, HCl, etc. |
| | Microwave limb sounder | | Sounding the gas composition of the atmosphere |
| | Visible and IR spectrometers | | Limb sounding of CO, $O_3$, etc. |
| | Fabry–Perot interferometer | | Profiles of wind and temperature |
| | Cryogenic Fourier spectrometer | | Profiles of wind and temperature |
| Monitors | Solar constant; particles and fields; Earth's radiation budget | | |

observations, (2) to improve further the retrieval algorithms, (3) to study more thoroughly the cloud life-cycle and calculations of radiation fluxes to substantiate a more reliable parametrization of the cloud–radiation interaction and (4) to analyse satellite ERB information for a better understanding of cloud–radiation feedback. Mokhov (1993) and Mokhov and Schlesinger (1993) have demonstrated that even the results of satellite cloud observations are still rather controversial.

The USA and other countries are planning regional-scale observations to improve techniques for the parametrization of clouds in climate models. The ISCCP was planned to run from 1983 to 1988, but the necessity to obtain a more adequate data base determined the prolongation of the project.

Special attention has been paid to the problem of obtaining data on precipitation (in the interests of the WCRP) with an accuracy of $\pm 1$ cm and spatial resolution on land and over the oceans of 250 km × 250 km. Therefore a suggestion has been approved to accomplish a special Project on Global Precipitation Climatology, with the use of both conventional and satellite information.

The problem of accumulating information about the wind field near the ocean surface (in the tropics, in particular), retrieved from cloud cover dynamics, is still unsolved. Differences with conventional wind-sounding data reached 4.5–5.0 m $s^{-1}$ at the lower level and 9.0–10.0 m $s^{-1}$ at the upper level (in the latter, there is a systematic underestimation of the sub-satellite wind speed). Since January 1985, the Global Center of SST Data within the US Center for Climate Analysis has been undertaking a regular issue of global maps of sea surface temperature monthly means over a $2° \times 2°$ grid.

An important stage in the development of remote sensing is the operation of the European Earth resources satellite (ERS-1), which was launched in 1991 (Table 2.21). Later a French–American satellite TOPEX/POSEIDON was launched in a circular orbit 1320 km high (the inclination angle 63.1° provides a 10-day repeatability). The measuring complex includes an advanced radiometer (13.6 and 5.3 GHz) and a three-channel microwave radiometer (18, 21 and 37 GHz). The latter is designed to solve problems of atmospheric correction which ensures the retrieval of data on sea surface level with an accuracy of 1–3 cm.

The SAR (L-band) and multi-channel radiometer in the visible and infra-red spectral regions are the principal components of the Japanese Earth resources satellite JERS-1, which was launched in 1992. Of great importance are developments of microwave sounders for geostationary satellites (100–200 GHz).

Present developments suggest that by 2000 substantial improvements can be made in the accuracy and adequacy of satellite observational data used in the global objective analysis of the atmospheric state prescribed as an initial state in atmospheric general circulation (AGC) models. In the period 1990–1995, an important contribution to studies of the ocean has been made by oceanographic satellites equipped with microwave and optical instruments to retrieve, for example, sea surface temperature, wind speed and wind shear near the surface, as well as parameters of the ocean topography, etc. A discussion of the draft International Orbital Station Programme has concluded that the planned launching of a polar platform and a manned orbital station should be a priority.

As noted above, in connection with further development of studies within the WCRP and planning of the International Geosphere–Biosphere Programme (IGBP), a Global Energy and Water Cycle Experiment (GEWEX) has been proposed (World Meteorological Organization 1987b, 1988a). Its aims are: (1) the

**Table 2.21** Geophysical parameters that can be retrieved from ERS-1 data

| Parameter | Range | Error | Instruments |
|---|---|---|---|
| Wind speed ($m\,s^{-1}$) | 4–24 | $\pm 2\,m\,s^{-1}$ or 10% | Scatterometer and radar-altimeter |
| Wind direction | 0–360° | ±20° | Scatterometer |
| Significant wave-height (m) | 1–20 | ±0.5 m or 10% | Radar-altimeter |
| Wave direction | 0–360° | ±15° | SAR (synthetic-aperture radar) |
| Wavelength (m) | 500–1000 | ±20% | SAR |
| *Images*: ocean; coastal zones; ice cover; land | Minimum viewing bandwidth 80 km | Geometric radiometric resolution: (a) 30 m/3.5 dB; (b) 100 m/1.0 dB | SAR |
| *Satellite altitude*: over the ocean; over the polar glaciers | 745–845 km | 2 m (absolute) ±10 cm (relative) | Radar-altimeter |
| SST | Viewing bandwidth 500 km | ±0.5 K | Multi-channel radiometer scanning along the orbit |
| Water vapour content | Spatial resolution 25 km | 10% | Microwave radiometer |

description and analysis of the laws of the transport of water (as vapour, liquid and solid) as well as of energy fluxes in the global atmosphere and at the surface level and (2) the development of techniques to forecast natural and anthropogenic changes in the distribution of water in all three phases in the atmosphere and at the surface level. Meeting these objectives requires various advanced observational means, including space-borne instruments.

The accomplishment of GEWEX will make it possible to take the first step in studies of global changes taking place in the environment. The choice of the period 1995–2000 for GEWEX was determined by the following circumstances: (1) increasing urgency of the problem of global changes, whose key aspects necessitate studies of biogeochemical cycles, general circulation of the atmosphere and ocean and energy and mass exchange, (2) availability of new space-borne means of observation of the atmosphere and ocean by 1995–2000 (particularly referring to active lidar and radar sounding) and (3) expected progress in computers.

The existing observational system has a number of shortcomings, including (1) insufficient representativity of a global data base on the three-dimensional wind field (over the Indian Ocean, in particular), (2) serious uncertainties of data on precipitation over the land and practical lack of data over the ocean, (3) substantial inadequacy of global data bases on three-dimensional fields of water content and cloudiness, (4) a very limited volume of information on the radiation budget and optical properties of the surface on global scales and (5) scanty quantitative geographical information on land use, types and properties of vegetation cover, etc.

Requirements for data from satellite observations necessitate the use of geostationary and low-orbit satellites. In particular, five geostationary satellites are necessary which would carry: (1) multi-spectral instruments for the visible and infra-red spectral regions to obtain both high-resolution images and quantitative data on the brightness fields at various wavelengths and (2) medium-resolution sensors for the infra-red and microwave intervals to retrieve three-dimensional fields of temperature, water and moisture content, as well as other parameters.

As far as low-orbit satellites are concerned, at least two polar-orbiting (altitude nearly 800 km), Sun-synchronous satellites are needed, and a launch of an orbital station with individual platforms in an almost equatorial (30°) orbit has been planned. The equipment of the low-orbit satellites should include: (1) an operational meteorological unit to obtain images, to perform remote sounding and to collect information from various automated platforms, (2) a unit to obtain data on the cloud cover structure and on the wind field (lidar altimeter), a multi-spectral radiometer to obtain stereographic images of clouds, a scatterometer to retrieve winds and wind shear near the ocean surface and a Doppler lidar to obtain data on the wind in the cloudless atmosphere, (3) a microwave unit to obtain data on precipitation, comprising a multi-channel scanning microwave radiometer with broad spatial coverage, including a low-frequency channel in the S-band to retrieve soil moisture, a radar at frequencies 10–15 GHz and 35 GHz to measure rain rate with a high spatial resolution and two types of radar (a large-antenna radar on an orbital station providing a 200 km viewing band, with a high vertical resolution (0.5 km), and a compact radar on polar-orbiting satellites to identify the zones of precipitating clouds) and (4) a unit to measure the ERB components, including advanced ERB-sensors, narrow-band (in the bands of $H_2O$, $CO_2$, etc.) radiometers to assess the effect of cloudiness and greenhouse gases on the ERB, a lidar to retrieve data on the heights of cloud tops and the atmospheric boundary layer, a multi-spectral radiometer to obtain medium-resolution images and a multi-channel infra-red radiometer with several fields of view (FOVs) (along the track of the satellite) to retrieve sea surface temperature data.

Also, important information has already been obtained from the satellite GEOSAT (launched in 1985) and the Defense Meteorological Satellite Program (DMSP) Special Sensor Microwave Imager SSM/I (launched in mid-1986). The geostationary satellites Meteosat (Lennertz 1982) are an important component of the European Space Agency's programme aimed at using space-based observational means to study the environment and the Earth's resources. The three-channel (0.5–0.8 $\mu$m; 10.5–12.5 $\mu$m; 5.7–7.1 $\mu$m) scanning radiometers carried by these satellites provide a 5 km nadir spatial resolution, and a 2.5 km resolution with the use of duplicated sensors in the visible spectrum (one of these sensors functions alternatively with the sensor for the 5.7–7.1 $\mu$m channel).

Processing of data from Meteosat-1 (which was launched in November 1977, but from late 1979 suffered a malfunction of its scanning radiometer) and Meteostat-2 (which was launched in June 1981) has provided information on the following: cloud cover distribution, ocean surface and cloud top temperature, water content of the upper troposphere, the Earth's radiation budget and wind field (from the motion of clouds).

Two Skylab experiments have been planned based on the use of: (1) a metric camera to obtain images over an area of 188.5 km × 188.5 km (a 23 cm × 23 cm square, on a film 120–150 m long), with a resolution of 10–20 m, which will enable a thematic mapping on a scale of 1 : $10^5$ and (2) a SAR–microwave radiometer operating in various regimes (40° to the vertical). In the SAR regime, images of an area of 9 km × 2500 km, with a resolution of 25 m × 25 m, are to be obtained, and in the radiometer regime sea surface temperature is to be retrieved with an error of ±1 °C.

By the end of the 1990s, eight satellites should carry 19 new types of instrumentation (mainly for the microwave region): 3 or 4 scatterometers, 4 radar altimeters, 3 SARs, 5 microwave radiometers, 1 low-frequency sensor, 2 infra-red radiometers and 1 multi-channel visible radiometer.

The rate at which combined conventional and satellite information was added to the data base on the world's oceans in the period 1984–1989 (in thousands of reports per day) is given in Table 2.22. The increase in throughput has continued to the present day.

The ERS-1 scientific complex which was launched in 1991 (with the subsequent transition to a multi-satellite operational system) consists mainly of instruments for active and passive microwave sounding to provide the most adequate information on the sea surface wind field and sea roughness. Table 2.23 gives characteristics of ERS-1. As can be seen from this table, the satellite programme has been oriented, first of all, to monitoring the ocean and ice cover.

The on-board radar complex includes: (1) the 5.3 GHz S-band SAR at horizontal–horizontal polarization (the size of the antennae is 10 m × 10 m) to obtain images with a 30 m spatial resolution, within the 80 km swath, (2) a scatterometer to retrieve sea surface roughness (12 wavelength intervals, 100 to 1000 m) for 100 km bands along the satellite's track (resolution 5 $km^2$) at an angle of about 23° to the vertical and (3) a scatterometer to retrieve wind speed (error ±20°), at a spatial resolution of 50 km × 50 km within the 500 km band (25–55° from nadir). In the third case, three antennae have been used: 3.6 m × 0.3 m (two antennae) and 2.5 m × 0.3 m in size, directed forward, downward and backward.

The radar complex also includes the 13.5 GHz $K_u$-band radar altimeter (antenna 1 m in diameter), with a measurement accuracy not less than ±10 cm (±5 cm in the future), to retrieve the significant wave height within 1–20 m (error ±0.5 m), as well as a laser retro-reflector (to track the satellite from the ground). The ERS-1 spacecraft also carries the Along Track Scanning Radiometer (ATSR), which is a three-channel conical-scanning radiometer (3.7, 11 and 12 $\mu$m; resolution ±1 km; swath width 500 km), to retrieve sea surface temperature (with error 0.5 deg K for areas 50 km × 50 km, 8-point cloudiness) from data of two-angle measurements, as well as a two-channel (23.8 and 36.5 GHz) microwave radiometer to retrieve total water content in the atmosphere (two additional channels can be used at 90 and 150 GHz).

Japan has put an oceanographic satellite MOS-1 into a circular (909 km altitude) Sun-synchronous orbit, with an inclination of 99.1°, carrying the following:

**Table 2.22** Rate of accumulation of information on the world's oceans (thousands of reports per day)

| | 1984 | 1985 | 1986 | 1987 | 1988 | 1989 |
|---|---|---|---|---|---|---|
| Wind speed | 2–4 | 60 | 1000 | 2000 | 2000 | 4000 |
| Wind direction | 2–4 | 2–4 | 2–4 | 1–4 | 2–4 | 40–910 |
| SST | 30–40 | 30–70 | 30–70 | 30–70 | 30–70 | 950 |
| Sea state | 2–4 | 60 | 60 | 60 | 60 | 120 |

**Table 2.23** The ERS-1 radar complex

| Parameters | Range | Error | Horizontal resolution | Coverage and repeatability | Instruments |
|---|---|---|---|---|---|
| Wind | | | | | |
| Speed | 4–24 $m\,s^{-1}$ | $\pm 2\,m\,s^{-1}$; $\pm 10\%$ | 50 km × 50 km | Global; 3 days | Scatterometer |
| Direction | 0–360° | ±20° | 20 km × 20 km | Global; 3 days | Radar-altimeter |
| Sea state spectrum | | | | | |
| Direction | 0–180° | ±7°; 30% | | Discrete measurements in grid knots with | Scatterometer (2-D roughness |
| Spectral accuracy | | | 10 $km^2$ | a step of 900 km in longitude at the | spectrum) |
| Wavelength | 30–480 m | Spectral resolution 30 m | | equator; 3 days | |
| Ice cover thickness | | ±2 m | 10 km × 10 km | Global; 3 days | Radar-altimeter |
| Significant wave height | 1–20 m | ±0.5 m; ±10 per cent | 10 km × 10 km | Discrete measurements in grid knots | Radar-altimeter |
| High-resolution images | | Resolution 1 dB by brightness levels | 100 m × 100 m<br>30 m × 30 m<br>15 m × 15 m | A band 80 km wide; 3 days; global monitoring for 35 days | SAR |

1. a four-channel (0.51–0.59 μm; 0.61–0.69 μm; 0.72–0.80 μm; 0.80–1.10 μm) self-scanning radiometer with an instantaneous FOV of (54.7 ± 5.0) m rad$^{-1}$ and a swath width of 100 km;
2. a four-channel scanning visible and infra-red radiometer (0.5–0.7 μm; 6.0–7.0 μm; 10.5–11.5 μm; 11.5–12.5 μm) with instantaneous FOVs of 1 mrad (visible) and 3 mrad infra-red, and a swath width of 1500 km;
3. two-channel ((23.8 ± 0.2) GHz, (31.4 ± 0.25) GHz) microwave scanning radiometer with horizontal (23.8 GHz) and vertical (31.4 GHz) polarizations with the beam width, respectively, of 1.9° and 1.3° and a swath width of 317 km (the dynamic range is 30–300 K, with an error of 1 deg K at 300 K).

The use of MOS-1 data makes it possible to solve problems of retrieving sea surface temperature, sea surface wind speed, atmospheric water content and other parameters. In Japan, a launch had been conducted of the Earth resources satellite JERS-1 in the 570 km altitude, 99° circular Sun-synchronous orbit, whose measuring complex includes: (1) the SAR at 1275 MHz frequency, with horizontal–horizontal polarization, a swath width of 75 km (nadir angle 33°) and spatial resolution 25 m × 25 m and (2) a four-channel scanning radiometer for the visible and near infra-red spectral regions (0.45–0.52 μm; 0.52–0.60 μm; 0.63–0.69 μm; 0.76–0.95 μm), with an instantaneons FOV of 43.9 mrad and a swath width of 150 km. This satellite is designed to study the sea surface state in particular.

The launching of the MOS-2 satellite has been accomplished, for which the following instruments have been developed: (1) ocean colour and sea surface temperature scanner (0.43–0.45 μm, 0.50–0.53 μm, 0.54–0.58 μm, 0.65–0.67 μm, 0.75–0.79 μm, 3.7 μm, 10.5–11.5 μm, 11.5–12.5 μm), with a swath width of 1400 km, (2) the 14 GHz radar altimeter (error ±10 cm for the ocean level and ±0.5 cm for the significant wave height) with a spatial resolution of 2 km, together with a laser retro-reflector, (3) an advanced four-channel (31.4 GHz, 23.8 GHz, 10.7 GHz, 6.6 GHz) scanning microwave radiometer with 50 km spatial resolution (temperature resolution 1 deg K, swath width 1400 km) and (4) a 14 GHz scatterometer (vertical and horizontal polarizations) with spatial resolution of 50 km and a swath width of 500 km (errors in retrieving wind speed and direction, respectively, ±2 m s$^{-1}$ and better than ±20°). The ADEOS-1 satellite was launched in 1996.

At present, various data-processing projects are being realized in the interests of the World Climate Research Programme (WCRP). First of all, these are studies connected with the International Satellite Cloud Climatology Project (ISCCP). These studies are mainly focused on comparing and testing the reliability of algorithms to retrieve cloud cover characteristics from satellite data. In addition, complex field observation programmes to improve the parametrization schemes for clouds and radiation in climate models have been realized. The B1-level data (nominal spatial resolution 10 km) obtained within the ISCCP can be used to retrieve net radiation and rain rate over the oceans.

In connection with the realization of the TOGA and WOCE programmes, it was decided to continue the ISCCP until 1995, at least, to acquire further accumulation and processing of data for B1 and B3 levels in the formats available (a 5-yr data base on cloud climatology can be considered adequate). Achieving the aims of TOGA and WOCE, as well as an identification of the $CO_2$ climatic signal, requires satellite ERB information, which determines the significance of this aspect of the WCRP. The ERB observations should be prolonged until at least 2000.

The surface radiation budget data, obtained both by direct surface measurements and by retrieval from satellite data, are also very important for climate studies.

It is likely that net radiation averaged over 2° × 2° latitude–longitude squares, with an error of 5 W m$^{-2}$ in mid-latitudes and 5–10 W m$^{-2}$ in tropical latitudes, can be retrieved if B1-level data are used.

At present, the US Center for Climate Analysis regularly issues global monthly maps of sea surface temperature of two types: (1) those based on the use of data from buoys and ships and (2) those based only on satellite-retrieved sea surface temperature fields. The mean standard deviation (MSD) of the two sea surface temperature (SST) global data bases is 0.5 deg K, i.e. it meets the WCRP requirements but further efforts are needed to improve the retrieval accuracy, especially in the tropics. For this purpose, *in situ* sub-satellite SST measurements are also needed.

Of great importance is the development of techniques to retrieve heat losses due to evaporation from the oean surface and also rain rate over the ocean, which could be combined with the use of direct measurement data. Also urgent is the problem of retrieving the wind fields from cloud motion and from microwave and scatterometric observations. Radiometry can be applied to monitor the ice cover. Since the desired accuracy in retrieving numerous atmospheric parameters has not been achieved so far, various instruments have to be improved and special means for their testing must be developed.

As has been mentioned by Stowe (1988), high accuracy of satellite ERB observations is required, in particular, because potential global mean climate warming within 1.5–4.5 °C could happen if the ERB disturbance reaches only 2.3 W m$^{-2}$ (in this case, variations in each of the ERB components reach approximately 4.6 W m$^{-2}$) with regional anomalies varying from 20 W m$^{-2}$ in the tropics to 100 W m$^{-2}$ near the poles (the latter figure refers to the absorbed solar radiation variability). Hence, data from long-term (decades) ERB observations can be a very important source of information for the detection of anthropogenically-induced climate changes.

Table 2.24 contains a summary of basic requirements for ERB observational data. No agreement has been reached, so far, with respect to the choice of an optimal ERB measuring complex, but the following requirements must be mentioned: (1) active thermal control of the sensors, (2) their 'smooth' spectral sensitivity (with selectivity), (3) the possibility of computer modelling of instruments, (4) provision for in-flight and pre-flight calibrations and (5) lack of sensitivity to polarization. The combined use of scanning and non-scanning instruments, or (by priority) the use of either of them, raises no doubts.

Already at the first stage of observations with the use of meteorological satellites (the first meteorological satellite TIROS-1 was launched on 1 April 1960) it has become possible to obtain not only qualitative information in the form of visible and infra-red images of clouds and the surface, but also quantitative data on some parameters (the amount and top height of clouds, atmospheric water content, sea surface temperature, etc.) (Kondratyev 1983a). An extensive history of meteorological satellite programmes is given by Rao *et al.* (1990). Entering the fourth decade of satellite meteorological observations, it is of interest to analyse the possibilities of interpreting the data available from operational satellites, which is very important in connection with the growing interest in the problem of climate changes.

A NOAA report (NOAA 1991) contains such an analysis made with respect to the following instruments carried by American operational satellites: AVHRR (NOAA satellites), TOVS (NOAA satellites), SBUV (NOAA satellites), VISSR

**Table 2.24** Spheres of application and requirements for ERB data

| Application | ERB components | Spatial resolution (km) | Temporal resolution: Period of averaging | Temporal resolution: Repeatability | Error ($W m^{-2}$) |
|---|---|---|---|---|---|
| Input information for monitoring and numerical modelling of climate | Solar constant | – | 1 month | Daily | 0.2–1 |
| | OLR Global means | – | 1 month | 1–3 hours | 0.2–1 |
| | OSR Zonal means | – | 1 month | 1–3 hours | 1–3 |
| | SRB Regional means | 100–200 | 1 month to 1 day | 1–3 hours | 2–5 |
| | OLR Daily means | 100–250 | 1 month | 1–3 hours | 2–5 |
| | OSR Synoptic SRB studies | 100–250 | Without averaging | Synoptic periods | 10 |
| | SRB components | 100–250 | 1 month | 1–3 hours | 10 |
| Input data and validation of numerical weather forecasts | OLR, OSR, SRB: Regional means | 100–250 | Without averaging | 1–3 hours | 10 |

*Note:* OLR = outgoing longwave radiation; OSR = outgoing shortwave radiation; SRB = surface radiation budget.

(GOES satellites), SSM/I, SSM/T and SSMR (DMSP satellites). A detailed description of the instruments enumerated above can be found in a monograph by Kondratyev *et al.* (1992), which contains a discussion of the informational potential of the Soviet meteorological satellites. Following the report (NOAA 1991), we shall describe, below, the characteristics of the potential of the American operational satellites for various applications.

### 2.3.1 Earth radiation budget, clouds and aerosol

In all kinds of satellite observations, obtaining the observational series, the duration of which in some cases already exceeds at the present time two decades, plays the key role. This involves the necessity (when archiving the data) to ensure the reliability of calibration and to take into account the characteristics of instruments, orbital parameters, etc. For example, the needs of the problem of climate change are known to determine the need for data on the global mean ERB components with an error of not more than $1\,W\,m^{-2}$, whereas in the case of regional averages the error can increase to $5\,W\,m^{-2}$, and the errors in estimating the long-term trends must not exceed $1\,W\,m^{-2}$. Hence, very strict requirements have to be met for the stability of the instruments and the reliability of their calibration.

One of the efforts of NOAA has been the development of algorithms to retrieve the amount and radiative properties of clouds from the data of the five-channel AVHRR, as well as algorithms to retrieve air temperature and pressure at the cloud top level and cloud liquid water content from the data of the TOVS. Beginning from 1989, the aerosol optical thickness (AOT) of the atmosphere over the oceans has been retrieved regularly from the data of one shortwave AVHRR channel, and in the future the results of observations at 2–3 wavelengths are planned to be used to estimate not only the aerosol optical thickness but also the aerosol number density. An important problem in the future will be the further development and improvement of algorithms to retrieve: (1) the ERB components, including the spectral distribution of outgoing longwave radiation (OLR) and radiative heat flux divergence from the TIROS Operational Vertical Sounder (TOVS) data, (2) surface radiation budget component, (3) cloud cover characteristics (cloud amount, temperature and pressure at cloud top level, water and ice content) and (4) total number concentration and size distribution of aerosol (not only over the ocean but also over land). Of particular importance is meeting the requirements of the Global Energy and Water Cycle Experiment (GEWEX) and an experiment on studies of the tropical ocean and global atmosphere (TOGA) for observational data.

### 2.3.2 The stratosphere

In this area of observations the problem lies in estimating the vertical distribution and total ozone content (TOC), air temperature and UV solar radiation flux. The basic requirements for observational data reliability are that the errors in the total ozone concentration trends on a decade-long timescale do not exceed 1–1.5 per cent, that errors in the vertical profile of ozone concentration do not exceed 3–5 per cent, and that errors in the stratospheric temperature do not exceed 1–1.5 per cent. Since the dynamics of the ozone layer depend on the variability of extra-atmospheric UV solar radiation in the wavelength interval 170–240 nm (radiation at these wavelengths is responsible for the formation of atomic oxygen in the upper stratosphere), a particular problem is the monitoring of UV radiation variations (which can reach 20 per cent during an 11-yr solar cycle). A reliable retrieval of the natural variability of stratospheric ozone content forms the basis for reliable recognition of anthropogenically induced effects.

One of the most poorly estimated stratospheric parameters is its temperature; since the top height of radiosondes is confined, as a rule, to altitudes below 30 km, rocket sondes are rarely launched and remote sounding techniques (especially lidar) have not been adequately developed. Meanwhile, information about the trend of the atmospheric temperature is very important. As demonstrated by numerical modelling of climate changes caused by the enhanced atmospheric greenhouse effect, the stratospheric response has been manifested as a cooling. It is very important here to consider the interactive variations in the ozone content and temperature.

The NOAA-9 satellite launched in March 1985 was the first to carry the SBUV-2 (Solar Backscattered UltraViolet) instrument, which was designed to obtain total ozone content data. An improved version of this instrument was installed on NOAA-11, which was launched in January 1989. Though these observations have been planned as operational (following the observations, from 1978 onwards, with the Nimbus-7 SBUV complex), they cannot be considered climatic, because of problems of data interpretation. A new stage of ozonometric observations has been introduced with the Upper Atmosphere Research Satellite (UARS) and Meteor/TOMS instrumentation.

Initial data from remote temperature sounding of the atmosphere were obtained in 1969 via the Satellite Infra-red Spectrometer (SIRS) developed for the Nimbus series satellites. In 1978 the TIROS-N series of NOAA satellites began to be employed for operational remote temperature sounding of the atmosphere. The current TOVS complex comprises three units: the HIRS/2 (High Resolution Infrared Spectrometer), the MSU (Microwave Sounding Unit) and the SSU (Stratospheric Sounding Unit). Each instrument scans in a plane normal to the orbit, with a spatial resolution between about 17 km and 200 km. This ensures the retrieval of temperature vertical profiles in the 1000–0.4 hPa (0–54 km) atmospheric layer, with a vertical resolution of about 10 km. Of course, consideration must be given to the problems of retrieval error, and the reliable determination of trends is very important. Comparison with rocket sounding data shows that the differences between temperatures at levels 5, 2, 1, and 0.4 hPa reached, respectively and on average 6, $-3$ , $-7$ and 6 °C, with considerable temporal variability; for example, errors at the 2 hPa level varied between 0 and 6 °C.

*In situ* surface observations are an important way of excluding the drift of the space-borne instruments' sensitivity in ozone and temperature observations. To achieve this aim, atmospheric monitoring is under way (1988–1997) in the USA, in order to provide homogeneity of space-based observations via the use of surface observations at various locations; five stations are in place, including the tropics, mid-latitudes of both hemispheres, the Arctic and the Antarctic, with the use of reliably calibrated instruments. Lidar sounding is also a major source of control information about the stratospheric temperature. Simultaneous remote soundings in the UV, visible, infra-red and microwave intervals are planned.

Since SBUV/TOMS data make it possible to retrieve the content of $SO_2$ in the stratosphere, at least after volcanic eruptions when the $SO_2$ content is substantial, a new retrieval algorithm is being developed to facilitate the collection of both $O_3$ and $SO_2$ data.

### 2.3.3 The troposphere

From the viewpoint of monitoring climate dynamics, satellite observations of atmospheric general circulation (AGC) characteristics (including water vapour and cloud cover) are essential. In this respect, the collection of three levels of data is planned:

1. direct observation data on outgoing radiation (both shortwave and longwave), with special emphasis on the homogeneity of the observation series (removal of differences caused by special properties of different instruments is essential),
2. retrieved values of geophysical parameters, and
3. data of four-dimensional assimiliation.

Such data will allow reprocessing of observational results if this should prove to be necessary as a result of the development of more reliable retrieval algorithms.

Atmospheric remote sounding accumulated (archived) data from 1969 include observational results from the following instruments: SIRS-A (Nimbus-3 satellite, 11–15 $\mu$m spectral interval, spatial resolution in nadir 250 km, 1969–1972), VTPR (NOAA-2–NOAA-5, 12–19 $\mu$m, 1972–1978) and TOVS (TIROS-N, NOAA-6, October 1978 to date).

The data base obtained from these instruments includes the following information, with a nominal horizontal resolution of 60 km: vertical profiles of temperature and humidity, total ozone content, pressure at top level and effective cloud cover, sea and land surface temperatures, outgoing longwave radiation, cloud-induced longwave radiation forcing, rain rate, microwave emissivity of snow and ice cover; and wind field (from either cloud or inhomogeneous water vapour distribution motions).

In order to interpret data from tropospheric observations, one should consider the cloud–radiation interaction in the context of the global energy balance (Harshvardhan *et al.* 1989, Hartmann *et al.* 1991, 1992, Henderson-Sellers 1986, Marchuk *et al.* 1986, Kondrateyev *et al.* 1988a,b, Randall *et al.* 1989, Ramanathan *et al.* 1989b). This requires reliable information on water vapour content and cloud cover characteristics. Available space-derived data include information on total water content of the atmosphere above the ocean (SSM/I, SSMR), and relative humidity of the upper troposphere (6.7 $\mu$m channel, GOES satellite, and equivalent TOVS data). The use of data for channel 2 of the MSU instruments is very important for intercalibrations. Since instrument sensitivity could have changed over time, retrospective data must be reprocessed to secure observation series homogeneity.

### 2.3.4 Land surface characteristics

Vegetation snow and ice cover characteristics serve as integrators of the effects of weather and climate on the environment, and can be used as indirect indicators of thermal and hydrological regimes. The problem of the parametrization of land surface processes in numerical climate modelling is central (Henderson-Sellers and Robinson 1986). As with remote sounding discussed above, high-quality data from space-borne measurements of outgoing radiation at different wavelengths play a key role. The most important component of this data base is information on the vegetation index, and the extent, thickness and water equivalent of snow cover.

Information on the vegetation index is used to solve numerous problems connected with studies of biogeochemical cycles and ecosystems dynamics. These include pasture, semi-desert, and tropical forest monitoring, soil classification, estimates of photosynthetic activity and primary productivity, as well as gas emissions in biomass burning. To accumulate data to estimate land surface parameters, two categories of AVHRR data are being archived:

1. data with a spatial resolution of approximately 1 km, ensuring the solution of local problems (Local Area Coverage, LAC);
2. data with a resolution of approximately 4 km (Global Area Coverage, GAC).

GAC data have been regularly used to calculate vegetation indices (usually the Normalized Difference Vegetation Index, NDVI) and for vegetation index mapping, since February 1982. Also from the AVHRR data the snow cover distribution has been mapped every week. Although, in principle, there has been the possibility of global snow cover mapping from microwave data since 1978, only work on the development of methodology has been undertaken so far. The same is true for the use of microwave data to retrieve the thickness and water equivalent of the snow. Great interest is being shown in the development of retrieval techniques for the photosynthetically active radiation and surface radiation budget components.

### 2.3.5 The world's oceans

Information on the world's oceans, with respect to climate studies, should be classified thus:

1. monitoring, of which the parameters are: temperature and surface level of the ocean, ice cover extent, and albedo;
2. assessments of various forcings and validation of climatic numerical modelling, of which the parameters are: sea surface temperature, ice cover extent, speed and direction of near-ocean-surface winds, surface heat balance components, difference between evaporation and precipitation.

At present, NOAA compiles data archives for sea surface temperature. The GAC AVHRR data (resolution 4 km) form the global data bases for day-time and night-time, averaging over 8 km × 8 km squares. During day-time, sea surface temperature retrieval is sourced from data from channels 4 and 5, wavelengths 11 and 12 $\mu$m, whereas measurements for channel 2 (0.9 $\mu$m) are used to filter out clouds. When processing night-time observations, both these problems are resolved with data from channels 3 (3.72 $\mu$m), 4 and 5. The mean standard deviation (MSD) of sea surface temperature retrievals, estimated by comparison with data from buoy observations, varies within 0.3–0.7 °C. Daily results of global objective analysis are released as maps with a grid 1° latitude by 1° longitude, in the interval 70° N to 70° S. Twice a week regional maps are drawn (usually for coastal waters of the USA) with resolutions of 0.5° and 0.125°, as well as lower-resolution global maps. All sea surface temperature data available are generalized in the form of monthly mean

**Table 2.25** Climatic parameters which can be assessed with the instruments carried by the American operational satellites

| Climatic parameter | Instruments | | | | | | | |
|---|---|---|---|---|---|---|---|---|
| | AVHRR | TOVS | SBUV | Pre-TOVS | GOES | SSM/I | SSM/T | SSMR |
| *Improved available data* | | | | | | | | |
| Vertical temperature profile | | + | | + | + | | + | |
| Vertical profile of concentration and total ozone content | | + | + | | | | | |
| Solar radiation flux | | + | | + | | | | |
| Outgoing longwave radiation (OLR) | + | + | | | | | | |
| Earth albedo | + | | | | | | | |
| Cloud cover parameters | + | + | | | | | | |
| Total aerosol content | + | + | | | | + | | |
| Vegetation index | + | | | | | + | | |
| Snow cover extent | + | | | | | + | | + |
| SST | + | | | | | | | |
| *New data* | | | | | | | | |
| Data base on outgoing radiation | + | + | | + | | | | |
| Land surface temperature | + | | | | | | | |
| Snow cover depth | | | | | | + | | + |
| Snow cover water equivalent | | | | | | + | | |
| Forest fires | + | | | | | | | |
| Clear-sky outgoing radiation fluxes | + | + | | | | | | |
| Water content of clouds | + | | | | | + | | |
| Water content of individual atmospheric layers | | + | | + | | | | |
| Temperature of individual atmospheric layers | | + | | + | | | | |
| Water vapour climatology | | | | | + | + | | |
| Winds from water vapour drift | | | | | + | | | |
| Winds near the ocean surface | | | | | | + | | |
| Ice cover extent and concentration | + | | | | | + | | |
| Atmosphere–ocean heat fluxes | + | | | | | + | | |
| Aerosol parameters | + | | | | | | | |
| Outgoing radiation histograms | + | + | | | | | | |
| Multispectral OLR | | + | | | | | | |
| Radiative heat flux divergence in the atmosphere | | + | | | | | | |
| Atmospheric downward emission | | + | | | | | | |
| Surface albedo | + | | | | | | | |
| Radiation gradient index | + | | | | | | | |

global fields over a 2.5° × 2.5° grid. Other forms of data presentation are used: e.g. weekly sea surface temperature fields over a 1° × 1° grid, drawn using optimal interpolation. To reduce retrieval errors to 0.3–0.5°, it is necessary to increase the accuracy of measurements and their verification via buoy data, to improve calibration and geographic referencing, to increase atmospheric correction reliability (especially with respect to aerosol), to introduce corrections for the temperature difference between the surface and the upper layer of the ocean, to unify data processing techniques for day-time and night-time, and to realize separate drawings of monthly mean sea surface temperature fields for day-time and night-time.

Information about wind speed near the ocean surface comes from passive microwave observation data (SSM/I), whilst radar sounding data are needed to retrieve wind shear. The use of data from active microwave sounding using scatterometers and synthetic aperture radars remains in an experimental phase; considerable progress is expected from the processing of data from the Almaz-1, ERS-1 and ERS-2 satellites, whereas, through processing the SSM/I data, the USAF Oceanographic Center was the first to draw global wind fields. Attempts to retrieve the height of sea level from data from radar altimeters have remained experimental.

The difficult problem is to retrieve the components of the ocean surface heat balance (Kondratyev *et al.* 1988a). The use of AVHRR data to retrieve net radiation and albedo has been successful, but only the first steps have been made in the solution of the problems of retrieving the latent and sensible heat fluxes.

### 2.3.6 Conclusion

In Table 2.25 we list the climatic parameters which can be estimated using the results of observations with each of the instruments carried by the American operational satellites.

CHAPTER THREE

# The World Climate Research Programme (WCRP)

The World Climate Conference held in Geneva on 12–23 February 1979 made a statement on the problems of global climate (Houghton 1984), in which it was emphasized that, although humans are now able to take advantage of the present favourable climate, human activity is being strongly affected by climatic variability, especially by droughts and floods, and particularly in the developing countries located in either arid or semi-arid zones or excessively humid regions. The interdependence of the climates in different countries necessitates a coordinated global-scale strategy for a better understanding and rational use of climate.

At present, human activitity causes inadvertent local-scale and (to some extent) regional-scale climate changes. There is deep concern that further intensification of human activity may lead to considerable regional and even global climate changes, which determines the urgency of international cooperation in solving this problem. Climate changes caused by natural processes will continue, of course.

Although much progress has been achieved in understanding the causes of climate changes, the role of various factors is still unclear. However, an anthropogenic increase of $CO_2$ concentration in the atmosphere is likely to contribute to the gradual warming of the lower atmosphere, especially in high latitudes. Some regional-scale and global-scale anthropogenic effects will probably have been detected by the end of this century, and by the middle of the next century they will become substantial. Such anthropogenic effects as some aspects of agricultural activity (especially the increasing use of nitrogenous fertilizers) and forestry, as well as releases of halocarbon compounds (CFCs) etc., should be given serious attention.

From the viewpoint of land processes, variations in surface roughness and surface albedo, as well as thermal characteristics of the upper soil layer and the water-carrying capacity of soils must be taken into account. Estimates of the consequences of the growth of population and human activity suggest that there could be inadvertent anthropogenically-induced climate changes on the planet, which can be compared to natural variations over the life-time of several generations. Estimates of the present climatic trends show that they will be determined mainly by natural factors, at least until the end of this century.

The 8th World Meteorological Organization (WMO) Congress held in April–May 1979 in Geneva made the decision to develop and accomplish a World Climate

Programme (WCP) consisting of four programmes (Kondratyev 1982a): Climatic Data Programme (CDP), Applied Climatology Programme (ACP), World Climate Application Programme (WCAP) and World Climate Research Programme (WCRP).

The WCP stems from the following circumstances: the increasing role of climate as a natural resource and the increasing effects of climate on human activity as populations grow and achieve a higher standard of living, the possible substantial anthropogenic impacts on climate, the probability of climate changes in the future; and the improvement in various means of studying the climate and its interaction with human activity.

The WCP is aimed at: studies of climatic mechanisms (the determination of the relative roles of natural and anthropogenic factors of climate), an enhancement and improvement of climatic observational data bases, an application of climatic knowledge to planning human activities and studies on the impact of climate changes on various aspects of human activity.

Studies of applied climatology connected with food production (agriculture and fisheries), energy production, water resources, medicine and constructions, transport and communication, as well as the development of marine and coastal resources, are of high priority.

From the viewpoint of climate monitoring, the following data are of major interest: meteorological and aerological observations, information about minor gaseous and aerosol components of the atmosphere, oceanographic data (including information on sea ice) and hydrological data (including data on snow and ice cover). Measurements of the solar constant and the radiation budget at the lower and upper boundaries of the atmosphere, as well as data on stratospheric dust-loading and geomorphological data, constitute the geophysical data base.

Because studies of the atmosphere–ocean interaction are vital for the climate problem, regions of the world's oceans where this interaction is most active (the Gulf Stream, the Kuroshio Current, zones of unstable stratification near the ice cover edge, regions of upwelling and zones of monsoon formation) must be studied first. This has been substantiated by the Russian programme 'Sections' (Marchuk *et al.* 1989). The problem of the meridional (towards high latitudes) heat transport in the ocean and the transport of heat from one part of the ocean to another, varying under the influence of mesoscale vortices, must particularly be emphasized. Possible variations in ocean albedo resulting from variations in biological productivity must be analysed. Apart from meteorological, hydrological, geophysical, oceanographic and ecological information, respective data are needed on agricultural activity, land use and industrial production. An adequate 30-yr data base on the climate system (atmosphere–ocean–continents–cryosphere–biosphere) must be a major goal.

Particular attention must be paid to monitoring anthropogenic climatic factors and their manifestations. This information must be obtained by enhancing the systems of conventional and satellite observations.

## 3.1 Fundamental principles of the WCRP

Major objectives of the WCRP, aimed at climate studies on time scales from several weeks to several decades, are to determine the predictability of climate and the extent of human impact on climate. To achieve these, the following are needed: a deeper insight into the laws of regional (a scale of about 1000 km and more) and global climates, a more reliable assessment of the trends of regional and global climates, development and improvement of the physico-mathematical climate models able to simulate the actual climate and to assess its predictability and a study on climatic sensitivity to natural and anthropogenic effects (i.e. an assessment of climate changes caused by these effects).

Studies should have the following order of priority: the synthesis of observational data for regional and global climate systems, the development of various theoretical climate models, the role of the oceans in the climate system, the biogeochemical cycles of radiatively active atmospheric gaseous components, the formation of clouds and their spatial distribution and radiative properties, types of aerosol and their optical properties and effect on clouds, the hydrological cycle (evaporation, precipitation, etc.), the properties of, and the processes on, the land surface and in the cryosphere and their impact on climate, and palaeoclimatic reconstructions, which are important for understanding the regional and global climate changes. All the five components of the climate system (the atmosphere, ocean, land surface, cryosphere and biosphere) are objects for study.

A key problem is the study of relations between the contributions to observed climatic variability on various time scales: from rapid internal fluctuations in the atmosphere and a slower variability determined by the boundary conditions at surface level and at the top of the atmosphere (variations in the extent of snow and ice cover, sea surface temperature, extra-atmospheric insolation, etc.), as well as other external factors (volcanic eruptions, $CO_2$ increase). The solution to this problem requires, first of all, complex diagnostic studies of both the atmosphere and the ocean.

Based on observations, an analysis must be made of the balances of energy (in all its manifestations), momentum, water vapour and other minor gaseous constituents (MGC) on regional and global scales. The laws of the climate system can be understood only with the use of quantitative techniques to estimate the transport and transformation in the atmosphere of the most important components of atmospheric and oceanic energy balances.

It is still not clear which climate models will be most promising, and this complicates the planning of observations aimed at developing and testing the techniques for the parametrization of various processes. There are, however, processes whose importance raises no doubts: for instance, radiation–cloud interaction. Studies of the climate system demand rather high requirements for the accuracy of various quantities, which cannot be met. For example, extra-atmospheric insolation must be measured to a relative accuracy of about $3 \times 10^{-4}$, sea surface temperautre to a relative accuracy of 0.1 °C, and the transport of the heat in the ocean and precipitation to a few per cent.

Of particular importance are complex studies of land surface processes including albedo, soil moisture, surface roughness, and snow and ice cover, which also refer to the problem of anthropogenic impacts on climate, as well as the effect of aerosol on the atmospheric radiation regime and cloud formation. The development of some atmospheric aerosol models which could be used in numerical climate modelling is also very important (Kondratyev *et al.* 1983b, 1991, Jennings 1993). The effect of clouds on climate formation is decisive, and this requires that appropriate parametrization techniques be developed with all the most important feedbacks taken into account. Studies on the predictability and sensitivity of climate are also of vital importance.

As Gates (1979) notes, an analysis of observational data shows that in mid-latitudes the poleward (and vertical) transport of heat and momentum (and, to a lesser extent, moisture) is determined by large-scale moving vortices which in their turn maintain the west-to-east zonal transport due to a convergence of eddy fluxes. In low latitudes, this transport is largely determined by the contribution from mean meridional circulation, supported by the vertical transport of momentum and moisture due to tropical convection. From the viewpoint of energetics, the energy cycle of the atmosphere is based on the average available potential energy (APE) determined by the meridional distribution of temperature, which is 6–8 times larger than the eddy available potential energy, as well as average and eddy kinetic energy.

The gigantic heat capacity of the ocean and the ocean–atmosphere heat exchange (especially in winter when the atmosphere takes up from the ocean a large amount of heat accumulated in summer) determine the effect of the ocean on the atmospheric general circulation (AGC) which smoothes the summer–winter contrast. Since both in spring and in winter a considerable poleward heat transport takes place, this determines the global-scale reduction of the latitudinal temperature contrasts followed by intensified regional ocean–continent constrasts.

As Gates (1979) recommends, an analysis of the oceanic impact on the AGC (and, hence, on climate) should incorporate four extreme models of the atmosphere–ocean interaction: (1) the Earth without oceans, which gives rise to an intensive AGC with strong seasonal contrasts and, as a rule, a severe climate, (2) the Earth with the present-day distribution of the land and seas but with 'swamp' oceans unable to achieve horizontal heat transport (here the AGC is less intensive than in (1), but the climate is more severe than that observed), (3) a realistic AGC model, with the atmosphere–ocean interaction taken into account, which satisfactorily simulates the observed climate and (4) the Earth completely covered with oceans, which determines a less intensive AGC, a milder climate and reduced seasonal contrasts. So far, the first and the fourth (extreme) cases have not been considered in detail, though these entail greater sensitivity of the AGC to factors related to the ocean.

As for the conclusion that the oceans govern the meridional heat transport, a number of questions arise: What are the mechanisms that govern this transport? What are the characteristic features of the vertical and geophysical distribution of the transport in the world ocean? What is the variability of the oceanic heat transport and what is its sensitivity to the ocean–atmosphere interaction? What are the possibilities of parametrizing the heat transport in the ocean? Answers to these questions are vital for a better understanding of the effect of the oceans on the AGC and on the climate.

Assessing the role of the ocean in heat transport to high latitudes and in smoothing the seasonal variability due to an accumulation (or release) of heat by the upper layer of the ocean is the least studied aspect of the global heat budget. So far, the estimates obtained from observational data have been based on calculations of the heat budget components, where the ocean–atmosphere heat exchange is found as the remaining term representing a small difference between large quantities of an unknown accuracy. Available estimates show that the meridional heat transport in the ocean is at a maximum in low latitudes, whereas in the atmosphere the heat transport is at a maximum in mid-latitudes.

Variations in the heat supply of the upper oceanic layer must be taken into account. For instance, the heat spent on the warming of this layer in spring and early summer in mid-latitudes reaches 50–100 W m$^{-2}$, i.e. a substantial part of the heat due to solar radiation. In December to January the heat released by the upper layer of the ocean is also substantial.

Numerical experiments with the use of oceanic general circulation models (OGCMs) show that the annual mean meridional cell of circulation in the ocean (meridional overturning) is the most important mechanism for heat transport at low latitudes, whereas at high latitudes the meso-vortices and horizontal correlations of the speed and temperature of the currents in stable structures of the oceanic circulation (effect of gyres) play an important role. So, the poleward heat transport can take place owing to the southward motion of cold deep waters. The effect of gyres can be exemplified by the Gulf Stream warm current compensated for by the transport near the cold eastern boundary of the current.

A combined effect of thermohaline circulation and wind-induced overturning provides the meridional heat transport in the trade wind zone, while at higher latitudes there is a trend of mutual correlation of these factors, which results in decreased efficiency of the meridional heat transport in the ocean. The enumerated conclusions are based, so far, on the results of modelling with the use of a complete OGCM, which have been verified by direct observations of the oceanic currents. Only after the accomplishment of complex observational programmes will it be possible to obtain reliable estimates of the ocean's contribution to global heat balance.

As noted above, the WCRP is aimed at studying the predictability of climate and of the contribution of human impact on climate change. For this purpose it is necessary: (1) to get a deeper insight into the laws of regional and global climates, their temporal variability, and of the mechanisms responsible for the formation and change of climate, (2) to analyse data on substantial trends of regional and global climates, (3) to improve the physico-mathematical models for an adequate simulation and assessment of predictability of the climate system as applied to various spatial and temporal scales and (4) to study climatic sensitivity to various natural and anthropogenic forcings and to analyse climate changes due to some forcings.

For the purpose of concentrating the developments on key aspects, the WCRP has been subdivided into three directions: (1) problems of long-range (monthly-to-seasonal) weather forecasts, (2) the interannual variability and (3) studies of the processes on a time scale of decades and longer.

Although, from the viewpoint of the WCRP, time scales from several months to several decades are primarily important, palaeoclimatic reconstructions (especially to verify the reliability of climate models) are also very important. The spatial scales vary from regional (about 1000 km) to global. The WCRP foresees three approaches to the solution of related problems: (1) studies of climatically significant processes (2) development of physico-mathematical models of the climate system and (3) statistical analysis of the parameters of observed climate.

Although the WCRP has been formulated so that all its aspects are important, two of the numerous climatically key processes deserve special attention on account of their great importance in the formation of climate and because of the long time period needed to prepare respective complex observational programmes: (1) the controlling effect of clouds on the radiative regime of the climate system and (2) the prevailing effect of the properties and dynamics of the ocean on global cycles of heat, moisture and various chemical elements (carbon, in particular) in the climate system. It should be noted that the programme 'Sections' has been based on an interactive consideration of these processes (Marchuk *et al.* 1989).

Of key importance for an accomplishment of the WCRP are numerical experiments which foresee the following (Kondratyev 1982a): (1) development of numerical models for large-scale weather forecasts and for the interests of the WCRP, (2) parametrization of the processes in the atmosphere and on the surface, (3) development of interactive ocean–atmosphere models, (4) establishment of the strategy of the development of climate models, (5) assessment of climatic sensitivity to external forcings as well as of weather and climate predictability, (6) numerical experiments with the use of First GARP Global Experiment (FGGE) data and (7) solution of the problems connected with planning the observational systems and with the four-dimensional (4-D) assimilation problem.

An examination of the present status of climate models suggests that, in connection with developing more realistic models, the emphasis must be placed on interdependent problems of 'climatic drift' and 'systematic errors', since the reasons for these two phenomena remain unclear. They are apparently connected with an inadequate consideration of a number of processes. It is important, in particular, to study the role of such factors as the spatial and temporal resolution of models, orography, the upper boundary conditions (usually prescribed as the lack of mass and energy exchange), ocean–atmosphere interaction, land surface processes (especially evapotranspiration), radiation–clouds interaction and convection.

From the viewpoint of improving the models, an adequate parametrization of physical processes for various levels of resolution is of key importance. In connection with the development of interactive models, an analysis is needed of the ability of the models to reproduce momentum flux, latent and sensible heat, and the radiation budget at the interface. The emphasis should be placed on the sensitivity of climate to the sea surface temperature. The same refers to the problem of cloud–radiation interaction.

The test of 'tuned' simple schemes for parametrizing the clouds by a threshold level with respect to humidity gave acceptable results in the case of mid-latitude synoptic systems, whereas in other cases (clouds in the tropics, maritime cumulonimbus clouds) specific features of the processes of cloud cover formation should be taken into account. This emphasises the need to have reliable pre-calculated fields of relative humidity and an analysis of possibilities of forecasting the specific types of cloudiness.

An analysis of the climate-forming role of sea ice dynamics has shown that on time scales of 1–2 months this role is relatively unimportant.

Of increasing concern are studies of the role of the biosphere in the formation of atmospheric gas composition and of climate. The present chemical composition of the atmosphere, incuding numerous MGCs, is largely a product of the biosphere, which serves as both a sink and a source of MGCs. There is a very sophisticated and diverse interaction between chemical processes in the atmosphere and biosphere. As a rule, the biosphere-produced gases exhibit reducing or weakly oxidizing properties, whereas gases in the atmosphere, which eventually sediment onto the Earth's surface, have clear oxidizing properties due to chemical reactions in the atmosphere.

Despite a small MGC concentration, the biospheric-produced gases substantially affect the chemical composition and radiative regime of the atmosphere (Kaufman *et al.* 1991). Chemical processes and radiation transfer in the atmosphere are known to be governed by MGCs, but not by principal gases (oxygen and nitrogen, which are chemically inert and optically inactive). The MGCs contribute most to the formation of the atmospheric greenhouse effect and regulate the input of solar radiation to the troposphere. Table 3.1, compiled by Ehhalt (1986), gives characteristics of some MGCs in the troposphere. Since the development of industry and agriculture continues, the trend of increasing MGC concentration is likely to continue. Even if anthropogenic MGC releases are stabilized, their accumulation in the atmosphere will affect the atmosphere and biosphere for a long time to come (Houghton *et al.* 1990, 1992, 1996).

Discussions in the US Congress of the greenhouse gases problem in 1986 prompted a resolution on the necessity for further studies of the effect of greenhouse gases and a suggestion that the year 1990 be designated an International Year for studies of the atmospheric greenhouse effect. Reports prepared by the US Department of Energy conclude that, although at present there are numerical models useful for assessing global mean climate changes during the forthcoming 50–100 yr, they are still unreliable in forecasting regional-scale variabilities. Further progress in this direction has been achieved in connection with the activities of the IPCC (Houghton *et al.* 1990, 1992, 1996).

Further achievements in climate studies resulted from new satellite observations. The GEOSAT satellite, launched by the US Navy in March 1986 and carrying a radar-altimeter, offered a new stage in oceanographic studies. It has made possible a regular global mapping of the world sea level, topography and circulation in the sub-surface layer of the ocean. Substantial progress has been reached in numerical monthly climate forecasts based on the National Meteorological Center dynamic model of the long-range forecast. The existing models have been unable, so far, to forecast reliably the large-scale blocking events which develop during the forecast period. The National Meteorological Center has an operational model to predict processes in the ocean (with the monthly mean wind field near the surface as a disturbing factor), which is reliable for the fields of temperature and currents. Further improvements of the mixed layer parametrization will make it possible to forecast sea surface temperature with an error of about 1 °C.

The Earth Radiation Budget Experiment (ERBE) data processing from the systems flown on NOAA-9 and -10 gave the solar constant estimates as $1364.8\,W\,m^{-2}$ (error 0.2 per cent) and made it possible to analyse the forcing effect of cloud cover dynamics on the Earth radiation budget. Important results have been obtained from cloud cover data processing within the ISCCP: (1) the annual mean

**Table 3.1** Observed values of mixing ratio, life-time, total content in the global troposphere ($M$), $dM/dT$, and the contribution of anthropogenic sources for different MGCs

| MGC | Mixing ratio (ppm) | Life-time (years) | Total content ($10^6$ tonnes) | $dM/dT$ (% per year) | Anthropogenic sources (%) |
|---|---|---|---|---|---|
| $CH_4$ | 1700 | 10 | 3900 | 1 | 60 |
| $N_2O$ | 305 | 100 | 1900 | 0.2 | 30 |
| CO | 100 | 0.2 | 400 | 1(?) | 60(?) |
| $O_3$ | 40 | 0.1 | 270 | 1(?) | 50(?) |
| | variable (0.5) | 0.002 | (0.3) | 1 | 60 |
| $CF_2Cl_2$ | 0.35 | 90 | 6 | 5 | 100 |
| $CFCl_3$ | 0.20 | 70 | 3.9 | 5 | 100 |

cloud amount in the southern hemisphere is larger than in the northern hemisphere, (2) global mean cloud cover in the period April 1979 to October 1984 has gradually decreased (this is partly determined, however, by unstable calibration of satellite-borne instruments), (3) the hemispherical mean cloud amount is characterized by the annual change with a maximum in summer (of the respective hemisphere) and a minimum in winter, which reflects a more substantial efficiency of convection than advection (polar front) in the formation of cloud cover and (4) a more clearly expressed homogeneous annual change of cloud cover in the southern hemisphere, determining the semi-annual cyclicity of the global mean cloud cover. Recent data on satellite cloud cover climatology have been discussed by Minnis and Wielicki (1988), Wylie and Menzel (1989), Mokhov and Schlesinger (1993) and others.

A special place within the WCRP is occupied by the Global Energy and Water Cycle Experiment (GEWEX), noted briefly above. We shall now explore GEWEX in greater detail, as it has become a key component of the WCRP.

## 3.2 The Global Energy and Water Cycle Experiment (GEWEX)

Key parts of the on-going large-scale international programmes within ecological studies – the International Geosphere–Biosphere Programme (IGBP) and the World Climate Research Programme (WCRP) – are connected with studies of mass and energy exchange: for example the key global biogeochemical cycles of such elements as carbon, sulphur and nitrogen.

Our current level of understanding of the laws of the global water cycle remains paradoxically low. For example, the global mean difference between evaporation and precipitation (and hence the resultant water transfer between the ocean and the atmosphere) is only known to an accuracy of a factor of 2. More problematically, the value of the water volume contained in ice, snow and soil has been estimated with an accuracy of a factor of 2–4.

These and other considerations have determined the priority of GEWEX as a key component of the WCRP (Browning 1990, World Meteorological Organization 1991a,c, Chahine 1992). The WCRP's main objectives are:

1. a study of water cycle and energy fluxes from the data of the global system of conventional and satellite observations,
2. numerical modelling of the global water cycle and its effect on atmosphere, ocean, and land surface,
3. the development of techniques to predict changes in the regional and global hydrological processes, as well as the dynamics of water resources and their response to environmental changes,
4. the improvement of techniques of observation, processing and analysis of data, including the assimilation techniques most effective for the forecast of weather, climate and hydrological conditions.

The special scientific objectives of GEWEX are the following:

1. a study of the spatial and temporal variability of water and energy balances on continental scales (this is the objective, in particular, of the GEWEX continental-scale international project focusing on basins of large rivers, and beginning with the Mississippi),

2. the development and validation of macroscale hydrological models and relevant high-resolution atmospheric models and coupled atmospheric hydrological models,
3. the improvement and validation of remote sensing techniques that foresee the combined use of satellite and conventional information,
4. the assessment of the impact of potential future climate changes on regional water resources.

The planned launches of the American, European and Japanese polar platforms within the Earth Observing System (EOS) are essential for realizing the goals of GEWEX.

The principal strategy of GEWEX comprises the use of existing programmes and data bases and the application of advanced coupled numerical models of the atmosphere–land–ocean system to simulate the water and energy cycles. To achieve the observational component of the programme it is important to formulate requirements for future planned space-borne observational instruments. Of particular urgency is the need to promote greater international cooperation to realize the differing levels of observations and field experiments (at local scales from 1 km × 1 km to 100 km × 100 km) to study mesoscale processes, continental and global.

GEWEX is planned for 1995–2000, and the GEWEX Continental-scale International Project (GCIP) was planned for 1992–1998 for the Mississippi, with future studies of the Neva, Rhine and Zambesi.

The period 1992–1998 will see the first phase of GEWEX, and this will be followed by the operationalization of the international Earth Observing System (EOS) (Kondrateyev 1992a–c, 1993, 1994). To accomplish EOS, the following must be realized:

1. provision of systems for collection, processing and distribution of a hitherto unprecedented volume of observational data,
2. study (based on field experiments) of the processes of the system of atmosphere, vegetation cover and soil that governs moisture exchange and energy exchange,
3. improvement of simulation models of varying complexity and scale.

In this respect it should be emphasized that GEWEX is not an experiment in the traditional sense, it is rather a large-scale programme whose aim is to establish a deeper understanding of the processes of water exchange and energy exchange based on the integrating efforts that combine observations and numerical modelling.

### 3.2.1 Observational system

Accomplishing GEWEX, and operationalizing an improved global observation system, will facilitate the first step towards a comprehensive understanding of the global-scale changes in energy exchange and moisture exchange happening in our environment. Its time scheduling (1995–2000) is determined by:

1. the increasing urgency of the global change crisis, explicit in outcomes of studies of biogeochemical cycles, the general circulation of the atmosphere and cycles of ocean energy and mass,

2. the realization of new space-borne systems to observe the atmosphere and the ocean (especially active lidar and radar soundings),
3. the on-going and dramatic development of computing power.

Our current observation system suffers from numerous shortcomings, including:

1. inadequately representing the global database on the 3-D wind field (especially over the oceans),
2. making serious errors in data on precipitation over land and suffering from a practical lack of such data over the oceans,
3. suffering from substantially inadequate data bases on the 3-D fields of water content and cloudiness,
4. carrying a very limited volume of information on radiation budget and optical properties of the surface on a global scale,
5. having only scarce quantitative geographical information on land use and on the types and properties of vegetation cover.

When analysing the study requirements for climatology and hydrology, the core GEWEX problems, Barron (1991) formulated three categories of requirements for observational data, as follows.

1. We need to obtain enduring and homogenous data series on various parameters. The physical and biological parameters are: water vapour content, temperature, wind, amount and properties of clouds, distribution of vegetation and albedo, land use characteristics, and the extent and albedo of snow and ice cover. The fluxes of matter and energy to be estimated are: precipitation, evaporation and evapotranspiration, energy fluxes near the surface and radiation fluxes in the atmosphere, run-off, latent heat of phase transformations (formation of sea ice, snow thaw, etc.). Of particular interest are reservoirs of soil, snow, sea, and continental ice and inland waters.

2. We need to study processes which determine interactions between the major components of the climatic and hydrological systems.
   (i) *Land*:
      (a) impact of land surface properties on energy and moisture fluxes,
      (b) climate–vegetation interaction,
      (c) surface–atmosphere interaction from the viewpoint of its effect on precipitation and temperature,
      (d) the impact of land use on the formation of the hydrological cycle.
   (ii) *Atmosphere*:
      (a) factors which determine the properties of clouds, interaction of atmospheric circulation and clouds,
      (b) cloud–radiation interaction,
      (c) moist convection processes,
      (d) factors which determine the distribution of moisture in the atmosphere.

(iii) *Cryosphere*:
   (a) the role of sea ice in the formation of atmospheric circulation and regional climates,
   (b) snow cover metamorphism,
   (c) permafrost dynamics,
   (d) the role of the cryosphere in the formation of energy and moisture fluxes.

(iv) *Oceans*:
   (a) atmosphere–ocean interaction,
   (b) factors which determine the water balance and atmospheric variability over the ocean,
   (c) the impact of the hydrological balance on circulation in the ocean,
   (d) the impact of sea ice on the ocean density field.

3. We need an improvement of numerical modelling, especially on regional scales, and accumulation of global data bases for their verification.

The requirements for satellite data for GEWEX and in general are determined by the need to employ geostationary and low-orbit satellites. As noted above, five geostationary satellites are needed, carrying: (a) multi-spectral instruments for the visible and infra-red spectral intervals to obtain both high-resolution images and quantitative data on the fields of brightness at various wavelengths and (b) mid-resolution remote-sounding instruments in the infra-red and microwave intervals to retrieve 3-D fields of temperature, moisture and water content.

As for low-orbit satellites, no fewer than two polar-orbiting (at 800 km) Sun-synchronous satellites are needed. It is also planned to utilize near-equatorial-orbit low-inclination satellites (30°), e.g. the Tropical Rain Measuring Mission (TRMM) to study equatorial zone processes (Kondratyev 1992a,b). The instrumentation complex of the low-orbit satellite systems proposed includes the following.

1. A meterological operational unit, for obtaining images, remote sounding and collecting information from various automated platforms.
2. An instrumental unit to obtain data on the structure of cloud cover and wind field, including a lidar altimeter, a multi-spectral radiometer to obtain stereographic images of clouds, a scatterometer to retrieve either near-ocean-surface winds or wind shear and a Doppler lidar to obtain data on winds in a cloudless atmosphere.
3. A microwave unit to obtain data on precipitation including a multi-channel scanning microwave radiometer, with a broad swath width and a low-frequency channel in the C band to retrieve soil moisture, and one or more radars at frequencies of 10–15 GHz and 35 GHz to estimate the rainfall rate with a high spatial resolution. Two types of radar are expedient, one being a radar with a large antenna at the orbital station thus ensuring a swath width of 200 km with a high vertical resolution (0.5 km) and the other being a more compact radar on polar-orbiting satellites to recognize the zones of precipitating clouds.
4. An instrumental unit to measure ERB components, including advanced ERB sensors used earlier: narrow-band (in the $H_2O$, $CO_2$ bands, etc.) radiometers to assess the effect of clouds and greenhouse gases on the ERB, lidar to retrieve the

heights of cloud tops and the atmospheric boundary layer, a multi-spectral radiometer to obtain moderate resolution images and a multi-channel infra-red radiometer with several fields-of-view (along the satellite track) to retrieve sea surface temperature.

### 3.2.2 Studies of heat and water balances in the world's oceans

Proper consideration of the atmosphere–ocean interaction plays a key role in resolving the problems of diagnosis and prediction of climate. Thus observations of the characteristics of the oceanic upper layer (not considered under GEWEX) form an important source of information about the components of the heat and water balances of the ocean surface. We can use this kind of information to assess the role of buoyancy as a forcing impact on the oceanic circulation due to heat and fresh water inputs. The importance of fresh water inflow is particularly great at high latitudes, where the surface temperature is close to freezing. In such conditions the water salinity, not its temperature, controls the stratification of the upper layer of the ocean. Palaeoclimatic data show that low frequency salinity changes could have affected substantially the formation of the thermohaline circulation in the northeast North Atlantic. The 'cap' of fresh water can strongly limit the convection-induced cooling, isolating the deep layers from the processes taking place on the surface of the ocean. Sea ice transport from the Arctic Basin is an important source of fresh water, leading to the observed anomalies of salinity. In this respect, assessing the contributions of precipitation, river run-off, thawing, and sea ice transport to the fresh water balance in the North Atlantic is very important from the point of view of climate study.

Of the utmost interest is the retrieval from satellite data of the global field of precipitation. Browning (1990) notes that for climatic studies, monthly means of precipitation are required over a 500 km × 500 km grid, with an error of no more than 20 per cent (especially within the intertropical convergence zone (ITCZ)). This corresponds to 1 mm per day, and the energy equivalent is $30\,\mathrm{W\,m^{-2}}$, that is, it reaches approximately 10 per cent of the average surface radiation budget. On the other hand, this error is equivalent to the heating of the 10 km layer of the atmosphere by 0.2 °C per day. For accurate weather prediction, it is important to have data on instant precipitation, with a random error of ±50 per cent. 'Nowcasting' requires very high spatial (ca. 5 km) and temporal (ca. 10 min) resolution. In hydrology, data on precipitation are needed; these data should have temporal resolution from several hours (with thorough spatial details, for example, in flood predictions) up to and beyond 24 h. In oceanography it is necessary to estimate the rain rate with an error of about 1 mm per day. The retrieval of the global distribution of rain rate from satellite data is based on three techniques, as follows.

1. The first technique is the employment of data on minimum cloud top temperature determined from given values of brightness temperature in the infra-red, and they indicate rain rate. For example, the precipitation index calculated from geostationary satellites (GOES etc.) data allows us to determine the proportion of clouds within a grid cell 2.5° × 2.5° with cloud top temperature below −40 °C. Although the infra-red threshold technique permits only a recognition of precipitating and non-precipitating clouds for each grid cell, the experience of data processing for the GARP Atlantic Tropical Experiment (GATE) has indicated that

there is a high correlation between the precipitation index averaged over the area and the total precipitation for 6 h (though note that this only refers to precipitation from convective clouds). The combination of brightness temperature data and wind speed fields in the visible wavelength range has proven useful.

2. Techniques to retrieve rain rate from microwave data on absorption (in the frequency range 5–35 GHz) or scattering (35–183 GHz) have been used widely. In the former, precipitation is retrieved from intensified microwave emission of liquid water (under conditions of observations over the water surface, that is, with a low emissivity and a homogenous background). The minimum threshold of retrieved rain rate is about 2 mm $h^{-1}$. At a frequency of 19 GHz with a rain rate of 20 mm $h^{-1}$ a saturation occurs; the maximum measurable rain rate is much greater with measurements within 5–10 GHz. Radiometers, functioning via the scattering principle, are mainly sensitive to large ice particles near the tops of powerful convective clouds and therefore they permit only indirect estimates of rain rate near the surface. Such observations can also be conducted over land surfaces.

3. The employment of satellite radars functioning via the principles of backscattering or attenuation is intriguing. The advantage of the former is that it allows the use of the standard technique for radar data processing; its disadvantages are manifest in the need to prescribe the size distribution of precipitation, in order to determine relations between radar echo and rain rate, and in the need to filter out the effect on background signal attenuation by precipitation, which is substantially manifest at high frequencies. The advantage of the principles of attenuation is the presence of a linear relation between attenuation and rain rate, as well as independence of the size distribution. The possibility of very strong attenuation of the signal necessitates measurements at more than one frequency: for example, 14 GHz in the case of heavy rain and 35 GHz in the case of light rain.

Rain rate retrieval is hindered both by the complexity of the problem itself and by the strong spatial and temporal inhomogeneity of the physical nature of precipitation (size distribution, phase state); the latter complicates the estimation of respresentative averages. All this demands that we develop retrieval algorithms which take into account the observation conditions (convective or complex clouds, liquid or solid precipitation, etc.). To reach the vertical resolution required (ca. 0.5 km) with scanning in a plane normal to the sub-satellite trajectory ($\pm 15°$), a space-borne radar at a frequency of 14 GHz should have an antenna 8 m diameter for an orbit of 600 km altitude or 4 m diameter for an orbit of 300 km altitude.

Until such radars are available, we can use diverse subsidary information to retrieve the vertical profile of rain rate:

1. retrieval of water content (to estimate evaporation) from data for a 'split' transparency window,
2. cloud cover texture, which allows one to distinguish between convective and stratus clouds,
3. climatological assessment of evaporation,
4. large-scale 4-D assimilation models to assess precipitation and evaporation in the sub-cloud atmospheric layer,
5. modelling clouds with parametrized microphysics to analyse the 3-D distribution of precipitation and evaporation.

It is relatively simple to account for the horizontal inhomogeneity of precipitation in the calculation of averages with the rain rate distribution function given, while the required repeatability of observations (so as to provide the temporal representativity of averages) should be estimated as a function of the special characteristic time of autocorrelation for the rain rate field. For example, during the GATE period the autocorrelation time was such that, with a cloud cover area of several square kilometres, observations were repeated in less than 1 h; for a region 280 km × 280 km, they were repeated in about 10 h (with 12 h repeatability, the errors in the calculated monthly means constituted 10 per cent). The clear diurnal change of rain rate in many tropical regions demonstrates the uselessness of data from Sun-synchronous satellites, whereas data from geostationary satellites do not provide sufficient spatial resolution. Thus satellites with a relatively small angle of inclination must be employed.

### 3.2.3 The GEWEX Continental-scale International Project (GCIP)

The GEWEX Continental-scale International Project (GCIP) is an important part of GEWEX's scientific strategy. It is concerned with a study of the heat and water exchange between the atmosphere and land surface, and forms an intermediary stage in the realization of a system of collection, processing and distribution of global observational data. Accomplishing GCIP will make it possible to obtain adequate information for the Mississippi Basin for a 5-yr period; this is required for the testing of the reliability of atmospheric and hydrological models. The choice of the region is determined by the presence of an adequate network of meteorological and hydrological observations over US territory, including a system of Doppler radars (NEXRAD), wind profilometers and automatic meteostations. The subsequent broadening of the network of aerological and hydrological observations will serve as the basis for the study of the atmospheric–hydrological processes on the scales characteristic of climate models, as well as for the control of the reliability of space-based observations of precipitation, humidity, winds, temperature and other parameters. GCIP's major objective is to study the relation between processes on the scales characteristic of numerical modelling of (a) individual processes responsible for the formation of the water cycle on land, and (b) global-scale processes determining the formation of climate.

It appears that for numerical climate modelling we can parametrize (in terms of simple biospheric models) mesoscale energy and mass exchange processes and the effect of vegetation cover on evapotranspiration. An analysis of the HAPEX-MOBILHY field data has demonstrated that the physical inhomogeneities of geometry of the river catchment areas, surface properties, vegetation cover, precipitation and mesoscale atmospheric circulation contribute most to the variability on the scales 10–100 km. On these scales, a linear averaging of local inhomogeneities of about 1–10 km is possible, since in this case non-linear feedbacks caused by mesoscale circulation do not occur. This conclusion is very important for planning the observations within GCIP so as to focus on those processes on a scale of 10 km and larger. The following experiments are relevant here: HAPEX-SAHEL (1992), EFEDA (in Spain in 1994), BOREAS (in Canada in 1994) and a study of the forests of the humid tropics (in Brazil in 1996).

### 3.2.4 The study of cloud–radiation interaction

Our understanding of cloud–radiation interaction is a weak part of current climate theory (Kondratyev *et al.* 1996b). This makes observations of cloud cover characteristics and the Earth radiation budget vital for GEWEX. Processing space-derived data has demonstrated the potential for the reliable retrieval of surface albedo and net radiation monthly means. However, our prospects for retrieving the longwave components of the surface radiation budget remain unclear; here it appears that the combined use of satellite and surface data is required, as with the retrieval of the fluxes of sensible and latent heat.

Given the urgency of raising the reliability of retrieving the vertical profile of water vapour content and then analysing its role in climate-forming processes, it has been suggested that a GEWEX sub-project be dedicated to studies of water vapour. This sub-project's objectives would be:

1. a comparison of data from direct balloon observations of water vapour concentration with data from passive and active surface, aircraft and satellite remote sounding,
2. the construction of one or two ground-based stations with Raman lidar devoted to long-term monitoring of atmospheric water content,
3. a summary of all available satellite data on the water content.

Comparing space-derived data with the results of numerical modelling has proved very instructive, particularly with respect to Earth radiation budget data, which we can use not only in comparisons with the results of numerical climate model calculations, but also to test the reliability of parametrization of the cloud–radiation interaction and the water cycle in climate models.

### 3.2.5 The Global Precipitation Climatology Project (GPCP)

At present the Center for Climate Analysis in Washington, DC, releases maps of precipitation for the tropical band ($\pm 40^\circ$ latitude), retrieved from infra-red data from geostationary satellites. Such maps (over the grid $5^\circ$ latitude by $5^\circ$ longitude) are available from 1 January 1986 to 30 September 1990 for every 5 days. Now, with regularly available digital data, this work has become operational. For the band $\pm 50^\circ$ latitude there are maps of precipitation for a 2-yr period, drawn from microwave data (19 GHz, SSM/I). The Global Center for Precipitation Climatology, Offenbach, Germany, compiles data from ground-based observations.

We should also acknowledge the important role to be played by a comparison of the reliability of various algorithms for precipitation retrieval with the use of control data from a dense ground-based network of rain-gauge observations. Broad prospects are opened up by the TRMM satellites.

Thus we can look forward to achieving a greater and more complete understanding of the laws of climate change via the GEWEX programme.

CHAPTER FOUR

# Principal climatically-important processes

## 4.1 Cloudiness and radiation

Cloud cover is an important factor which determines the geographical distribution and the annual change of the climate regime. Analysis of the sensitivity of climate to cloud characteristics, based on numerical modelling and diagnostic studies, has now become of great concern. Such an analysis is aimed at: (1) studying the features of the respective feedbacks that govern the interaction among the fields of motion, clouds and radiation, (2) developing a technique that adequately simulates the evolution of clouds in numerical climate models, (3) studying the effect of clouds on the radiative regime and (4) testing the results on the basis of observational data. The development of a theory of the climate-forming role of cloud cover for use in climate studies is of paramount importance (Kondratyev 1983b, 1986b,c, 1987a, 1992b, Washington and Meehl 1983, 1984, 1989, 1993, Marchuk *et al.* 1986, Liou and Ou 1989, Liou and Zheng 1989, Liou 1992, Houze 1993, Doe 1995, Weare *et al.* 1995, Kondratyev *et al.* 1996b).

Both the observational and theoretical data testify to a mutual compensation of the effect of clouds on the shortwave and longwave components of the radiation budget of the surface–atmosphere system and the respective components of radiative heat flux divergence. This compensation is, however, partial (even for averaged quantities), and therefore radiative heat flux divergence depends on the amount, type and height of the upper boundary of clouds and on their spatial distribution.

Cloud anomalies can affect the natural variability of climate. Here it is important to take into account the specific influence of various types of clouds on the radiative regime in the interaction between radiative and dynamical processes which complicates the analysis of climatic sensitivity to different factors. In this connection the development of techniques to predict the formation of clouds and their effect on radiation fluxes is of the highest priority.

As Henderson-Sellers (1986, 1987a, 1992) notes, three independent sources of information on climate show that the increasing total cloud amount must cause a climatic cooling: (1) from satellite data, an increase of the cloud amount leads to a decrease of the Earth radiation budget (i.e. the albedo effect of clouds dominates the greenhouse effect), (2) it follows from surface meteorological observations for the last hundred years that in the periods of climate warming the cloud amount grew, but in the presence of clouds the surface air temperature was lower than it ought to have been in clear-sky conditions and (3) results of numerical modelling agree, on the whole, with observational data, although they are controversial.

Even an adequate understanding of the climate-forming role of cloud-radiation feedback does not, however, permit a reliable forecast of the future climate, since it is unknown for certain how climate changes affect the cloud amount and it is also impossible to predict a change in the types of clouds. The climate forecast requires that three questions be answered: (1) what is the sign of the effect of clouds on climate, (2) in what direction does the cloud amount vary (increase or decrease) and (3) what is the contribution of different types of clouds to their impact on climate.

In general, these questions can be answered with the use of 3-D climate models (e.g. Kobayashi 1993). But in practice this is impossible, in view of the inadequacy of the schemes of cloud parametrization used in climate theory (in addition, there is no complete observational data base, which is needed to check the parametrization schemes). Therefore, the existing 3-D models simulate only incomplete patterns of the effect of clouds on the formation of climate. From the viewpoint of testing the reliability of numerical modelling results, the accumulation of observational data from the realization of the International Satellite Cloud Climatology Project (ISCCP) and Earth Radiation Budget Experiment (ERBE) is very important.

The available atmospheric general circulation (AGC) numerical models use a rough parametrization of the dependence of water vapour condensation on relative humidity, which forms the basis for predictions of precipitation and assessments of latent heat. A description of clouds based on this parametrization does not permit a reliable consideration of cloud–radiation interaction in climate models. In this case the actual variety of cloud types to be parameterized must be taken into account. It is difficult to describe both realistically and on an interactive basis an extended stratiform cloudiness, powerful cumulus cloud clusters and various types of upper-level clouds, although this is absolutely necessary in adequate climate models. A simplified approach based on an approximate parametrization may be used only in the case of simpler models.

Numerical modelling of the atmospheric general circulation shows that an interactive consideration of clouds and radiation provides more realistic results and, in particular, reveals an increase of the eddy component of kinetic energy as well as substantial changes in the temperature and water regimes in some regions. It is important that adequate simulations of 3-D cloud dynamics, microphysics and chemical processes are made (see, e.g., Wang and Chang 1993a–c).

Schlesinger (1988a–d, 1989a,b) reviewed parametrization techniques for convective and stratiform clouds as well as considering specific features of microphysical processes in such clouds, bearing in mind possible ways of improving parametrization techniques on the basis of a more adequate simulation of respective physical processes.

The leading role of cloud formation is manifested through interactions between: (1) dynamical and hydrological processes at releases/expenditures of condensation/evaporation heat as well as the redistribution of heat and momentum, (2) radiative and dynamical–hydrological processes due to reflection, absorption and emission of radiation, (3) hydrological processes in the atmosphere and on the surface (in the case of precipitation) and (4) effects, on the atmosphere and the surface, of transformations of radiative and eddy transports at surface level.

An analysis of the available schemes of parametrization suggests that, from the viewpoint of cloud physics (a consideration of microphysical processes responsible for the formation of clouds and precipitation), they are only very rough approximations. In the case of cumulus clouds parametrization, one of two approaches is being used. The simplest approach proceeds from the assumption of a moist-adiabatic adjustment. If the temperature lapse rate (TLR) in a saturated air layer exceeds the moist-adiabatic lapse rate, the resulting convection is intensive enough to restore a neutral temperature lapse rate (i.e. $(\partial\theta_e/\partial z) = 0$, where $\theta_e$ is the equivalent potential temperature). The relative humidity (RH) never exceeds a critical value, $RH_c$; excess moisture is immediately compensated for and this provides a restoration of $RH_c$, as the condensed water immediately precipitates.

Since spatially averaged RH values are considered, and convective rains occur even with averaged RH < 100 per cent, a critical value $RH_c$ < 100 per cent is usually prescribed. The second approach is connected with the use of parametrization schemes developed by Arakawa and Kuo. A major drawback of these parametrization schemes is that they do not consider a mesoscale structure of cumulus clouds and arbitrarily simulate the cloud amount. So, a scheme developed in the European Centre for Medium-Range Weather Forecasts (ECMWF) gives zero cloud amount at RH < 50 per cent, and at a prescribed higher RH it increases with height to some level in the middle troposphere and then decreases.

Washington and Meehl (1993) have applied two versions of a global R15 AGC model coupled to a non-dynamic, 50 m deep, slab ocean to assess the impact of a penetrative convection that has the effect of pumping more moisture higher into the troposphere. Calculations show that the addition of penetrative convection increases tropospheric moisture, cloud amount and planetary albedo, but decreases net solar input at the surface. Cess *et al.* (1992, 1993) have studied in detail the shortwave surface radiation budget variability dependence on cloud conditions from both satellite and surface observations. However, the competing effect of increased downward infra-red flux (from increased tropospheric moisture) causes a warmer surface and increased latent heat flux.

Washington and Meehl (1993) pointed out that in the tropics the use of the prescribed cirrus albedo feedback decreases net solar input at the surface, since the cloud albedos increase in the regions of high sea surface temperature and deep convection. Downward infra-red radiation (from increased moisture) also increases, but this effect is dominated by the reduced solar input in the tropics. Therefore, the surface is somewhat cooler in the tropics, latent heat flux decreases, and global average sensitivity of temperature and precipitation/evaporation feedback to the doubling of $CO_2$ is reduced. A very important result is that the addition of penetrative convection produces global effects, as does the prescribed cirrus albedo feedback, although the strongest effects of the latter occur in the tropics.

The parametrization of non-convective clouds is based on the use of two criteria of cloud formation: the atmosphere must be stable $((\partial\theta_e/\partial z) \geqslant 0)$ and a modelled relative humidity must exceed a certain threshold quantity (usually <100 per cent).

Various schemes of parametrization are characterized by a wide range of prescribed cloud albedo: 0.38–0.70 in the case of low-level cumulus clouds and 0.13–0.80 for clouds of penetrative convection (a similar situation is observed in the case of non-convective clouds).

A serious drawback of the most widely used schemes of parametrization is that the description of the cloud–radiation interaction based on *a priori* radiative characteristics of clouds leaves out of account both the microphysical mechanisms for the formation, maintenance and decay of clouds and the dependence of the radiative characteristics of clouds on their microphysical properties.

Possible improvements to the parametrization of stratus clouds, providing a more accurate prediction of rain rate, may be based on a differentiated consideration of droplets (or ice crystals) both falling out of and remaining in clouds. Two approaches have been suggested by Schlesinger (1988a–d) to parametrize non-convective clouds. They provide a physically substantiated consideration of microphysical processes, one being based on a detailed consideration of the physical processes that govern the formation of clouds, the other on the use of a parametrization. The parametrization schemes developed by Sundqvist (used in the ECMWF 15-layer model) and by Roads and co-workers describe the water content and cloud amount as predictants, which allows forecasting of the radiative characteristics of clouds.

Schlesinger *et al.* (1988) developed a technique for parametrizing the rate $E(z)$ of precipitation/evaporation in an atmospheric layer between the bottom of the cloud layer and level $z$:

$$E(z) = C\,P^{\alpha}(0)\psi(z)$$

where $P(0)$ is the rain rate near the bottom of the cloud layer, $\alpha = 0.606$ and $C = 2.63 \times 10^{-2}$ are the values of the parameters if $P(0)$ and $E(z)$ are expressed in $\mathrm{mm\,hr^{-1}}$ and $\psi(z)$ is a function of altitude estimated on the assumption of the constancy of either relative or absolute humidity in the sub-cloud atmosphere.

Liou and Zheng (1989), using a one-dimensional climate model, assessed the effect of the interactive nature of cloud cover on variations in the vertical temperature profile caused by an external forcing prescribed as a doubled $CO_2$ concentration, a 2 per cent increase of the solar constant, or increased emissivity of cirrus clouds. An eddy mixing is formalized on the basis of the mixing coefficient, which makes it possible to reduce the problem to a system of differential equations in partial derivatives describing the balance of thermodynamical energy, water vapour transport, vertical motions and cloud formation.

It is assumed that evaporation in the atmosphere can be neglected (this means that the rain falls out immediately upon its formation) and cloud formation is determined by a threshold relative humidity value dependent on height:

$$\mathrm{RH} = \mathrm{RH_s}\left(\frac{P}{P_s} - 0.02\right)\Big/0.88 \qquad (4.1)$$

at prescribed $\mathrm{RH_s}$ values near the surface equal to 90 per cent, 85 per cent and 80 per cent ($P$ is atmospheric pressure). Clouds are assumed not to form in the lowest atmospheric layer of the model.

The temperature, humidity, vertical velocity and amount of clouds are calculated with the use of a closed system of differential equations determining these parameters. The solution of this problem with a one-dimensional model assists a detailed consideration of the vertical structure of the atmosphere (it provides a description of the dynamics of clouds at three levels: isotherms 20.2 °C and 10 °C determine the location of cloud bottoms) and an accurate parametrization of the radiative heat flux divergence. An average emissivity of 0.47 is taken for cirrus clouds.

A comparison of the results of numerical modelling with fixed and interactive clouds taken into account had shown that, in the latter case, atmospheric warming is intensified (due to latent heat determined by a 1.5 per cent increase in the amount of low-level clouds with a simultaneous decrease of middle- and upper-level clouds by 3.2 per cent and 2.0 per cent, respectively) and radiative cooling increases, thus providing a thermodynamical equilibrium. A consideration of the interactivity substantially reduces the sensitivity of climate to an external forcing. So, a doubling of $CO_2$ concentration at a fixed cloudiness causes an additional warming of the troposphere by about 2 °C (and a near-surface warming by 2.3 °C), whereas in the case of interactive cloudiness the warming decreases to 0.8–1.3 °C (about 1.5 °C on the surface). With a 2 per cent increase of the solar constant, the surface temperature rise is reduced from 2.9 °C to 1.7 °C. Possible variations in the emissivity of cirrus clouds lead to disturbances in the temperature profile very similar to those caused by the growth of either $CO_2$ concentration or the solar constant (this reflects the exceptional urgency of the problem of the optical properties of cirrus clouds).

The functioning of the cloud-radiation feedback is determined by the following sequence of processes, with a stepwise increase of $CO_2$ or solar constant: surface temperature increase → evaporation intensification → an increase of the water vapour flux from the surface to the atmosphere → latent heat release at cloud formation and atmospheric warming, and simultaneous intensification of the warming. Compared with these processes, an increase in the low-level cloud amount and the surface–atmosphere system albedo is, however, most important; the albedo effect exceeds the greenhouse effect. This tendency is also favoured by a decreased cloud amount, which reduces the greenhouse effect. Therefore, in total, the cloud–radiation interaction manifests itself as a negative feedback.

It has been noted (Curry 1986, Curry *et al.* 1988) that the inclusion in atmospheric general circulation models (AGCMs) or weather forecast models of the liquid water content of non-convective clouds as a predictor adds several advantages: (1) in this case, precipitation from non-convective clouds forms directly from the predicted cloudiness, which is also taken into account in the scheme of parametrization of radiative processes, (2) optical properties of clouds are not given *a priori* but are calculated from pre-calculated water content and (3) it is possible to parametrize the eddy mixing in terms of variables preserved during phase transformations of water in the atmosphere.

A parametrization has been considered of the phase transformations of water, taking into account both the sources and the sinks of either liquid water or ice in clouds, within the UK Meteorological Office AGCM (the ice forms at a temperature below −15 °C), as well as the sinks due to precipitation. The numerical modelling results have been obtained without an interactive account of the radiative properties of clouds. A comparison of the mean meridional profile of the cloud liquid water content averaged over 3 months (September to November) with data from satellite microwave observations of water content revealed a similarity, though satellite data refer to convective clouds and calculated estimates to non-convective clouds. The predicted water content fields turned out to be most sensitive to the parametrization of the rate of the transformation of cloud water into rainfall. In the future, the role of an interactive account of the radiative properties of clouds will be analysed.

Hu and Stamnes (1993) have developed a new parametrization technique to simulate radiative properties of water clouds in the visible and infra-red spectral regions. The analysis of calculated results (using the Mie formulae) for various optical characteristics (volume extinction coefficient $\beta$, single scattering albedo, $\tilde{\omega}$, asymmetry of the phase function $g$) within the broad range of equivalent particle radii, $r_e$, (2.5–60 $\mu$m) showed that cloud optical properties depend mainly on $r_e$ and are only weakly sensitive to droplet size distribution. This is why a new parametrization is based on the simulation of cloud optical properties dependent only on cloud liquid water content, LWC, and equivalent droplet radius $r_e$. Thus

$$\beta/\mathrm{LWC} = a_1 r_e^{b_1} + c_1 \tag{4.2}$$

$$1 - \tilde{\omega} = a_2 r_e^{b_2} + c_2 \tag{4.3}$$

$$g = a_3 r_e^{b_3} + c_3 \tag{4.4}$$

where coefficients $a_i$, $b_i$ and $c_i$ have been calculated separately for the three intervals of equivalent radii: 2.5–12 $\mu$m, 12–30 $\mu$m and 30–60 $\mu$m dependent on the wavelength (24 intervals within the visible and near infra-red region 0.3–4 $\mu$m, and 50 intervals for the infra-red region 4.5–150 $\mu$m). A comparison of calculated radiation fluxes and flux divergences with observations has shown good agreement.

Buriez *et al.* (1988) tested the reliability of the parametrization of the cloud formation in the ECMWF prognostic model by comparing the calculated cloud characteristics with satellite data. For comparison purposes, 2866 cases were taken from the NEPHOS cloud data archive. The ECMWF model uses a simple parametrization consisting of the supposition that clouds cover the whole square of the spatial grid, if relative air humidity reaches a threshold level.

As the comparison has shown, calculated estimates of the cloud amount are systematically underestimated (to 26 per cent), especially in the presence of stratocumulus clouds, which is explained by underestimated relative humidity in the atmospheric boundary layer. The underestimation of relative humidity determines also the inability of the model to simulate the formation of stratocumulus clouds in the zones of downward motions. The potential temperature lapse rate is a more reliable predictor of low-level cloudiness. Other studies of cloud parametrization have been described by Slingo and Slingo (1991), Briegleb (1992), Ramaswamy and Freidenreich (1992) and Senior and Mitchell (1993).

The analysis of the variability of the Earth's radiation budget (ERB) by Ardanuy *et al.* (1989, 1991, 1992) is of particular interest. Table 4.1 (from Ardanuy *et al.* 1992) illustrates the interannual variability of the ERB and its constituents: solar constant, absorbed solar radiation and outgoing longwave radiation, as well as the surface temperature, total cloud amount and aerosol optical thickness for the 1 $\mu$m wavelength. Maximum absolute outgoing longwave radiation (OLR) values (272 W m$^{-2}$,

**Table 4.1** Observed interannual variability[a] of mean global annual values of the Earth's radiation budget (ERB), with other parameters

| Parameter | Observed range of variability |
|---|---|
| ERB (W m$^{-2}$) | $\pm$0.5 |
| Solar constant $S_0$ (W m$^{-2}$) | $-$0.7 to $+$1.3 [$-$0.05 to $+$0.01] |
| Absorbed solar radiation (ASR, W m$^{-2}$) | $\pm$0.6 [$\pm$0.2] |
| Outgoing longwave radiation (OLR, W m$^{-2}$) | $\pm$0.4 [$\pm$0.2] |
| Total cloud amount (%) | $-$0.5 to $+$0.3 [$\pm$0.8] |
| Surface temperature $T_s$ (°C) | $\pm$0.1 |
| Aerosol optical thickness | $\pm$0.002 to 0.06 |

[a] Relative variability is shown, as a percentage, in brackets.

corresponding to an effective temperature of −10 °C) were observed in the tropical belts of the Atlantic, Pacific and Indian Oceans in the southern hemisphere, as well as in southern Arabia, while minimum values (corresponding to −23 °C) were observed over the Amazonian and Congo forests as well as over the monsoon regions of Asia and Indonesia. Surface-atmosphere system albedo varies from 19 per cent to 38 per cent in the tropics, but reaches 80 per cent at high latitudes. The ERB varies between maximum values in the tropics (ca. 80 $W m^{-2}$, save for the Sahara and North Africa, where the ERB drops to −20 $W m^{-2}$) and minimum values in the Arctic and Antarctic (around −110 $W m^{-2}$).

The interannual variability of the ERB, which is characterized by the mean square deviation, reaches maximum values in the equatorial belts of the eastern parts of the Atlantic and Pacific Oceans (4.5–4.8 $W m^{-2}$). There is a positive long-term trend for the ERB of the order of 0.2 $W m^{-2} yr^{-1}$ caused by a decrease of the outgoing longwave radiation. The planetary albedo is comparatively stable.

Although the time series considered (1979–1987) is quite short, the following approximation is possible ($[CO_2]$ = $CO_2$ concentration):

$$T_s = 0.26 + 0.76 S_0 + 0.11[CO_2] \quad (4.5)$$

which explains 32 per cent of the surface temperature variability, whilst the remaining 68 per cent is due to the climate system's internal variability. The latter in turn is significantly affected by volcanic eruptions. For example, during the Mount Pinatubo eruption of 1991 the global mean volcanic forcing of the ERB reached 2–4 $W m^{-2}$, which could result in a global climate cooling of the order of 0.5 °C for several years thereafter. Basset and Lin (1993) have assessed the breaking of temperature records after the Mount Pinatubo eruption; Hansen and Wilson (1993) have made a number of more general comments concerning future potential climate change.

The differences in calculated and observed values of the ERB components are usually connected with different amounts of clouds. Different radiative characteristics of clouds provide a partial compensation of differences in cloud amounts. Therefore, the radiation budget average difference is only 7 $W m^{-2}$, though the mean standard deviation reaches 46 $W m^{-2}$ (which is equivalent to 43 per cent of the observed mean standard deviation). The calculated albedo of the surface–atmosphere system turns out to be overestimated, and its dependence on the cloud amount smoothed, because the model does not take into account the increase in the optical thickness as clouds increase in quantity. Observed large-scale values of albedo and outgoing longwave radiation are characterized by strong non-linearity of their dependence on the cloud amount. This shows that in calculations of the radiation budget components for conditions of partial cloudiness it is incorrect to use mean weighted quantities based simply on the proportions of clear and cloudy parts of the sky.

Though the problem of cloud formation parametrization and of cloud–radiation interaction has become the key one in numerical modelling of climate, it is still unclear, for example, what is the extent and also the sign of climate change caused by cloud cover variations taking place when $CO_2$ concentration increases in the atmosphere.

In connection with the far-from-realistic parametrization of clouds in the existing climate models, Wilson and Mitchell (1986) undertook a numerical modelling of climate to assess the role of diurnal change of clouds and radiation fluxes based on the use of a one-dimensional radiative–convective model and of a complete five-layer AGCM developed by the UK Meteorological Office (time step 10 min, horizontal resolution 330 km).

The model takes an interactive account of clouds, soil moisture and snow cover and is based on prescribed annual variations of sea surface temperature, ice cover extent and total ozone content. The formation of three types of clouds (lower, middle and upper level) is foreseen, the amount of which is given by

$$C = (\mathrm{RH} - \mathrm{RH_c})^2/(100 - \mathrm{RH_c})^2 \tag{4.6}$$

if relative humidity $\mathrm{RH} > \mathrm{RH_c}$, and $C = 0$ at $\mathrm{RH} \leqslant \mathrm{RH_c}$. The threshold relative humidity $\mathrm{RH_c} = 70$ per cent (upper level) and 90 per cent (lower and middle levels). In calculations of the amount of lower cloud an account is also taken of the temperature lapse rate in the lowest model atmosphere, bearing in mind the parametrization of the formation of stratocumulus clouds over the sea.

Wilson and Mitchell (1986) analysed the results of two versions of numerical modelling for a period of 50 days with initial conditions set for 12 June of the fourth year of the previous multi-year integration (results have been considered averaged over the last 30 days of integration). The versions discussed correspond either to the control case with calculations of cloud amount and radiative heat flux divergence every 2 hours (respective estimates were considered constant during the subsequent 2 hours) or to the case when the time interval was 6 hours.

The results obtained show that an inadequate simulation of the diurnal change causes considerable distortions in the calculated climate. Detailed calculations for North Africa (6–36° N, 10° W–36° E) revealed a 26 W m$^{-2}$ decrease of global radiation due to the smoothed diurnal change (absorbed solar radiation decreases by 20 W m$^{-2}$), which leads to an attenuation of convergence in the lower troposphere and of precipitation intensified by the growth of low-level cloudiness. In order to avoid such distortions, it is necessary to calculate cloud amounts more often than four times a day.

Similar results have been obtained from numerical modelling for land in the tropics in July. An analysis of numerical modelling results has demonstrated that the strongest variation in precipitation does not coincide spatially with the positive anomaly region of absorbed solar radiation but is shifted southward, i.e. there is a remote forcing on convective precipitation. A further special feature of the conditions considered is that precipitation ceases simultaneously with increasing sensitivity of clouds to the relative humidity forcing.

The conclusions about the possibility of an adequate account of the diurnal change equally refer to numerical weather forecasts. The results favour the supposition made by Charney about the existence of biogeophysical feedback: an increase of albedo due to the destruction of vegetation cover and a respective decrease of the surface radiation budget are compensated by atmospheric heating caused by intensified downward motions; this, in turn, leads to the weakening of precipitation.

Despite the important climate-forming role of extended stratus cloudiness (ESC) in the Arctic, the interaction between turbulent mixing, radiative and microphysical processes responsible for the formation and evolution of extended stratus cloudiness remains poorly studied. The June 1980 experiment of studying extended stratus cloudiness in the Arctic (over the Beaufort Sea) made it possible to obtain an extensive data base on the structure of the atmospheric boundary layer, vertical profiles of radiation fluxes and the size distribution of extended stratus cloudiness particles. Curry (1986) undertook an analysis of related data (six flights of 'Electra', in total) to analyse the interaction between turbulence, radiation and cloud size distribution, with four cases of aircraft extended stratus cloudiness soundings as an example.

Studies of these data on the thermodynamical structure and fluxes of thermodynamical parameters in the cloud-topped atmospheric boundary layer (liquid water flux, in particular) have shown differences between observed and calculated quantities (with the use of a boundary layer model). These differences manifest themselves, for example, in the observed sharp inversion of humidity above the extended stratus cloudiness. Mixing in clouds and in the above-cloud layer does not reach the surface.

Mixing in clouds results not from convection at the surface level but from processes within the cloud itself, connected with radiative cooling and latent heat release. The vertical structure of clouds is multi-layered. In the Arctic, because of the low altitude of the Sun, even at noon, the extended stratus cloudiness is always characterized by total cooling. The infra-red cooling near the top of the clouds is at a maximum because of the large water content near the top, and the heating near the extended stratus cloudiness bottom is determined by a temperature contrast between this boundary and the surface. In each of the four cases, special conditions of balanced cooling near the cloud top form, but in all cases the cooling is an important factor of convection in the boundary layer.

The upward heat fluxes provide a balance for not more than 10 per cent of cooling, 16–71 per cent fall on eddy entrainment, and the remaining contribution is made by latent heat condensation. Turbulent entrainment does not penetrate deeper than 50 m with respect to an average cloud top and, therefore, does not directly affect the evolution of the size spectrum of droplets, except for a thin layer near the upper boundary. Observed vertical profiles of fluxes of buoyancy, entropy and water as well as data on the variability of vertical velocity show that radiative cooling substantially affects the formation of liquid water.

Water fluxes are quantitatively comparable to water vapour fluxes. The $0.2\,g\,m^{-3}\,h^{-1}$ vertical variation of water fluxes determines a substantial vertical inhomogeneity of the size spectra of droplets and liquid water content, as well as intensifying the coalescence of droplets in the lower part of the extended stratus cloudiness. Water fluxes near the cloud bottom are either negative or positive (but small), which reflects the practical absence of the contribution of evaporation from the surface to the extended stratus formation.

So far, the effect of cirrus clouds, which cover about 20 per cent of the globe, on the formation of the weather and climate remains poorly studied. Only recently have the data of high-altitude aircraft and lidar soundings made it possible to obtain information about the cirrus clouds' microphysical parameters, but the mechanisms for their formation and dissipation are still unclear. As is known, a major feature of cirrus clouds (in contrast to liquid water clouds) is that they contain a large amount of large non-spherical ice crystals and, as a rule, are optically thin. Therefore, their emissivity is well below 1.0.

Liou (1992) undertook a review characterizing the microphysical and radiative parameters of cirrus clouds, special features of their variability both in time and in space (especially climatological parameters), as well as their role in the formation of the temperature field and atmospheric circulation; a very important study of semi-transparent clouds on the basis of satellite remote sensing data has been achieved by Lin and Coakley (1993). Analysis of the data of geostationary meteorological satellites (infra-red images, in particular) has demonstrated the global-scale distribution of cirrus clouds (it is difficult, however, to identify thin cirrus clouds) observed at all latitudes (over the land and ocean) during the whole year and characterized by a considerable temporal variability of their properties (cloud amount, geographical location, thickness, parameters of the spatial structure).

There is no doubt that the formation, maintenance and dissipation of cirrus clouds are directly connected with large-scale synoptic processes and with disturbances (cyclones, fronts, etc.), as well as with processes in the upper parts of powerful convective clouds but, so far, respective quantitative estimates are absent. The problem of possible feedback between clouds and weather, as well as the climate-forming processes of different spatial and temporal scales, has not been solved either. The height of the bottom layer of cirrus clouds varies between 4 km and 15 km at a temperature below −20 °C. The cirrus clouds consist, as a rule, of particles in the shape of bullets, columns and plates, with the hexagonal structure prevailing. The relation between longitudinal $L$ and transverse $d$ sizes of particles can be parametrized as $d = aL^b$. The size distribution of particles is very variable, and the equivalent ice water content, IWC, is assessed from the following empirical dependence on temperature $T$:

$$\ln(\mathrm{IWC}) = -7.6 + 4\exp[-0.2443 \times 10^{-3}(|\bar{T}| - 20)^{2.455}]; \quad \bar{T} > 20 \tag{4.7}$$

Ebert and Curry (1992) presented a new parametrization of ice clouds with five spectral intervals in the shortwave and five in the infra-red, with ice cloud optical properties parametrized in terms of ice water path (IWP) and the effective radius $r_e$ of the ice crystal size distribution. This allows the flexibility of varying IWP and $r_e$ separately. The parametrized optical properties were used to calculate the bulk reflectivity, transmissivity and emissivity for cirrus clouds with realistic ranges for IWP and $r_e$.

A considerable contribution to cirrus clouds studies has been made by an application of active and passive remote sensing techniques. In particular, depolarization of the backscattering from a lidar signal can be used to identify the phase composition of clouds and the types of crystals and snow flakes. Satellite lidar sounding will make it possible to monitor cirrus clouds and to retrieve the top and bottom altitudes of them. However, it is still not clear whether lidar sounding can be used to retrieve the shape and size distribution of cirrus cloud particles.

In a search for passive sounding techniques, it has been found possible to retrieve the top height and thickness of overcast cirrus cover from data from four-channel measurements of outgoing radiation near 10 $\mu$m and from multi-channel measurements in the 15 $\mu$m $CO_2$ band, as well as in the 6.7 $\mu$m $H_2O$ band. Data from 0.83, 1.61 and 2.125 $\mu$m channels contain information about the size distribution and phase state of cirrus cloud particles.

Since cirrus cloud information is determined by sub-grid-scale processes, information on their formation, maintenance and dissipation in AGC and climate models can be realized only through a parametrization. So far, no attempts have been made at such a parametrization using available physical models of cirrus formation, although some schemes of cloud parametrization, in which the liquid water content is one of the predictants, can be used in cirrus parametrization. Since the emissivity, $\varepsilon$, of cirrus clouds is determined by their ice content, the $\varepsilon$ variability can be taken into account in climate models only on the basis of interactive prediction of ice content with the use of the model of cirrus cloud formation. Data from aircraft measurements and calculations have shown that $\varepsilon$ can vary within the range

0.1–0.9. There is no doubt that such variations can substantially affect the formation of cirrus clouds.

Numerical experiments within one-dimensional climate models have demonstrated that, depending on thickness (emissivity $\varepsilon$) and the height of the location, cirrus clouds can cause both the cooling and the warming of the atmospheric layers below. Calculations based on a two-dimensional thermodynamical model have shown that the ratio between cloud albedo, $r_s$, and emissivity, characterizing the contributions of the albedo and greenhouse impacts of cirrus clouds on climate, is as follows:

$$\frac{\partial r_s}{\partial \varepsilon} \ll 1 \text{ (thin clouds) or } \gg 1 \text{ (thick, almost completely black cirrus clouds)}$$

An increase of $\varepsilon$ from an average of 0.47 to 0.57 causes considerable tropospheric heating due to increasing atmospheric thermal emission and almost unvarying incoming shortwave radiation. The opposite effect of cooling causes a decrease of $\varepsilon$ from 0.47 to 0.32. Realistic changes in $\varepsilon$ are equivalent, in their effect, to a doubled $CO_2$ concentration or to a 2 per cent increase of the solar constant. If, for example, the cloud emissivity is fixed and equal to 0.47, and the height of the cirrus bottom layer varies by $\pm 100$ hPa with respect to the 300 hPa level, when cirrus clouds rise the troposphere and stratosphere become substantially heated (especially in the tropics), whereas when the clouds descend, the effect is the opposite.

Preliminary numerical experiments with the use of 3-D models of atmospheric general circulation and climate confirm the necessity for cirrus parametrization. We have discussed the results of Washington and Meehl (1993) above. All this emphasizes the importance of obtaining reliable information about the climatological characteristics of cirrus clouds. Specialized complex field programmes are needed to provide observations in different climatic conditions, to study real mechanisms of the interactive effect of cirrus clouds on the temperature regime and dynamics of the atmosphere. Analysis of the data of such observations will make it possible to substantiate the schemes of parametrization of the processes of formation and dissipation of cirrus clouds in the models of AGC and climate.

Woodbury and McCormick (1986) investigated the laws of global spatial and temporal variability of cirrus clouds, using the Stratospheric Aerosol and Gas Experiment (SAGE) data for retrieving vertical profiles of the concentration of minor gaseous and aerosol components of the atmosphere, for a period of 34 months (from February 1979 to November 1981). An identification of cirrus clouds was made, including the fact that they are observed more frequently near the tropopause. Therefore, the zone of cirrus cloud localization was determined as having the upper boundary at the tropopause level $H + 1$ km and lower boundary $0.7H$ or 8 km (depending on which of these values is greater). An anomalous increase of the extinction coefficient in this zone up to values more than $1.0 \times 10^{-3}$ km$^{-1}$ (the normal background value of the aerosol extinction coefficient is $2.0 \times 10^{-4}$ km$^{-1}$) is considered as an indicator of the possible presence of cirrus clouds (in general, the term 'cirrus clouds' in this case is conditional, since any clouds in this zone are identified as cirrus clouds).

Depending on the extinction coefficient value, clouds are classified into two categories: (1) optically thick (extinction coefficient $> 8.0 \times 10^{-3}$ km$^{-1}$, i.e. exceeding the background level 40 times) and (2) optically thin (extinction coefficient between $8.0 \times 10^{-4}$ and $8.0 \times 10^{-3}$ km$^{-1}$). Use of the characterized criteria makes possible a sufficiently reliable identification of cirrus clouds. It is important, however, that data for subsequent orbits cannot be interpreted as referring to the same clouds (even in high latitudes, where the orbits are the closest). The spatial resolution of the data considered is about $100\,km^2$.

An analysis of the results obtained showed that the optically thin cirrus clouds are most often formed in mid-latitudes and the tropics (the frequency of occurrence for clouds constitutes 50–60 per cent), with a clearly pronounced minimum in the 20–30° latitudinal band in both hemispheres (frequency of occurrence about 25 per cent). The observed features of the global distribution of cirrus clouds favour the above theory of their origin, based on information on the troposphere–stratosphere exchange. The optically thin cirrus clouds occur more seldom and, apparently, exhibit no marked annual change.

Comparison of data on the zonally averaged frequency of occurrence for cirrus clouds with results of observations with the Nimbus-5 selective-chopper radiometer revealed both an agreement between the qualitative features of cirrus cloud distribution and their annual change, and substantial quantitative differences. According to SAGE data, the frequency of occurrence for cirrus clouds is twice as great. These data are also of great interest from the viewpoint of the statistical assessment of possibilities for satellite occultation sounding of the troposphere. Results showed, for example, that in high latitudes obtaining information on the troposphere at altitudes above 7 km is possible in 60 per cent of the cases, and in the tropics only 30 per cent of the cases. Bearing in mind, however, that the tropopause in the tropics is located higher, the sounding of a 5 km layer under the tropical tropopause is possible in only 55 per cent of the cases.

Woodbury and McCormick (1986) reproduced maps showing the geographical distribution of frequency of occurrence for cirrus clouds for the whole period of observations and for the four seasons; these reflected the spatial and temporal variability of the cirrus cloud field and, on the whole, agreed with the previous observations. For example, well-known zones of cirrus cloud over deserts, in the tropics, etc. are clearly shown.

Yentsch (1985) analysed the role of albedo and cloud feedback in the formation of the annual mean global regime of climate and its sensitivity to the external forcings given as solar constant variations. The results obtained testify to the critically important role of the moisture cycle as a factor in the climatic system's stability and sensitivity to external forcings.

Whereas information on the albedo feedback only suggests the system's transitivity, when information on the cloud feedback is added it becomes non-transitive. The combined results have led to the conclusion that in this case the climatic system's sensitivity to the forcings weakens substantially. With only the albedo feedback, a regime of a 'white' Earth (complete glaciation) sets in with a 2 per cent decrease of the solar constant, whereas with the addition of the cloud feedback this occurs with only a 10 per cent decrease of the solar constant. Hence, the cloud feedback softens the sensitivity of the global climate to the albedo feedback.

In connection with the development of cloud parametrization schemes in climate models, it is important to test them by comparing calculated and observed characteristics of cloud cover and ERB components (for this purpose, the FGGE III-C level data are recommended for 12 hours on 21 January and for 12 hours on 11 June, 1979). Preliminary comparisons have shown (Stowe 1988) that models of similar complexity realistically simulate the basic features of the spatial distribution

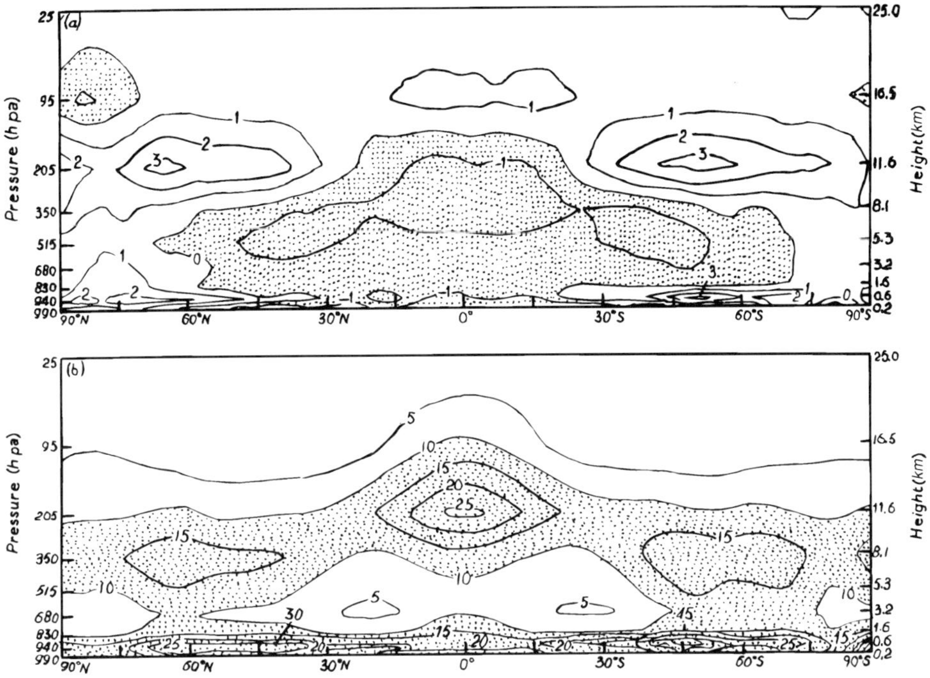


**Figure 4.1** Latitude–height distribution of (a) $CO_2$-induced change of zonal mean cloud amount (%), (b) relative humidity (%) obtained from the GVC model.

of clouds, with a high sensitivity of results to the modelled equilibrium distribution of moisture. Of great importance are present studies on comparisons of parametrization schemes for radiative transfer (and, hence, radiative heat flux divergence) in the atmosphere.

New developments in parametrization schemes for the atmospheric boundary layer over the ocean have led to the conclusion that an adequate parametrization of the atmospheric boundary layer is impossible without information on the interaction with radiation processes and with penetrative convection. The results of parametrization of clouds and radiation in the atmospheric boundary layer obtained with the use of various models turned out to be contradictory and insufficiently reliable from the viewpoint of forecasting the cloud cover.

On the basis of an intercomparison of the results of numerical climate modelling with a fixed cloudiness and with an interactive simulation of cloudiness, Wetherald and Manabe (1988) obtained new estimates of the role of cloud feedback in the formation of climate. The spectral nine-level global climate model with realistic orography simulating the processes in the atmosphere–ocean system (the ocean is approximated by a 50 m thick isothermic mixed layer) takes into account 15 harmonics (both in latitude and in meridian). Precipitation (snowfall, if the surface air temperature is below zero) takes place with a given threshold on its increase in the case of snowfalls. With the fixed threshold level of relative humidity at which clouds form (99 per cent), the global mean cloud amount turns out to be 50 per cent.

Since the model makes it impossible to predict the liquid water content of clouds, this excludes the possibility of taking into account the 'cloud-optical' feedback (changes of the albedo of clouds with their varying water content). The sea surface temperature is determined by heat balance conditions. The effect of horizontal advection of heat and of heat exchange with deep layers of the ocean is not considered. The measure of climatic sensitivity to cloud feedback is a change of equilibrium values (calculations have been made over 40 yr) of climatic parameters with a doubled $CO_2$ concentration. Analysis of numerical modelling results has shown (Figures 4.1–4.4) that in the case considered inclusion of data on cloud feedback causes an increase of climatic sensitivity to external forcing (doubled $CO_2$ concentration). The cloudiness intensifies near the tropopause, but the cloud amount in the upper troposphere decreases; hence the average temperature of its upper boundary and the outgoing longwave radiation decrease. The heat loss by the surface–atmosphere system is reduced, and the absorbed solar radiation varies. As a rule, the effect of the cloud cover decrease in the upper troposphere on the system's albedo dominates the effect of the intensification of clouds near the tropopause, which causes a decrease of the system's albedo and an increase in climate warming due to a $CO_2$ doubling.

The opposite situation (the growth of lower clouds, increase in the system's albedo and a reduction of warming) occurs only in high latitudes. However, the contribution of this negative feedback is small compared with the positive feedback mentioned above, determined by varying upper clouds. In the case of interactive simulation of cloudiness an increase of the global mean surface air temperature reaches 4.0 °C, whereas it is 3.2 °C for the fixed cloudiness model. Information on interactive cloudiness hardly affects the results of calculations of stratospheric cooling.

Wetherald and Manabe (1988) emphasized that, in view of a simplified parametrization of physical processes and neglecting several factors in the climate model considered, the results obtained should be interpreted not as realistic climate sensitivity assessments but as an analysis of the role of possible climate-forming mechanisms.

As Ramanathan (1987) notes, the variability of the ERB and its components (especially the outgoing longwave radiation) is of great importance as a still inadequately studied factor in the formation of climatic anomalies. Information on the global spatial and temporal variability of the ERB and its components can help in assessing the role of the cloud–radiation and albedo feedbacks as well as testing the

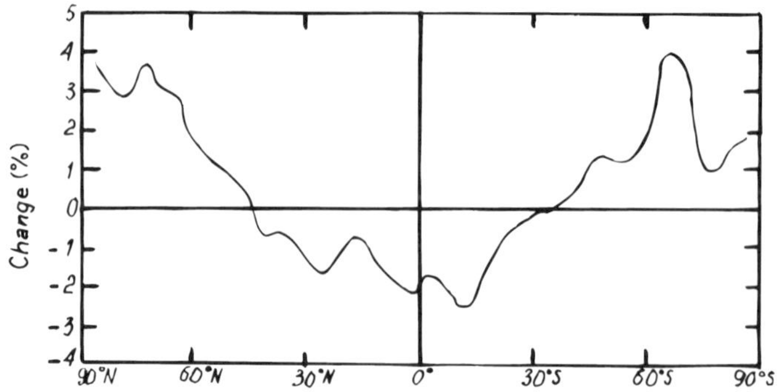


**Figure 4.2** Latitudinal distribution of the $CO_2$-induced change of zonal mean cloud amount (%) obtained from the GVC model.

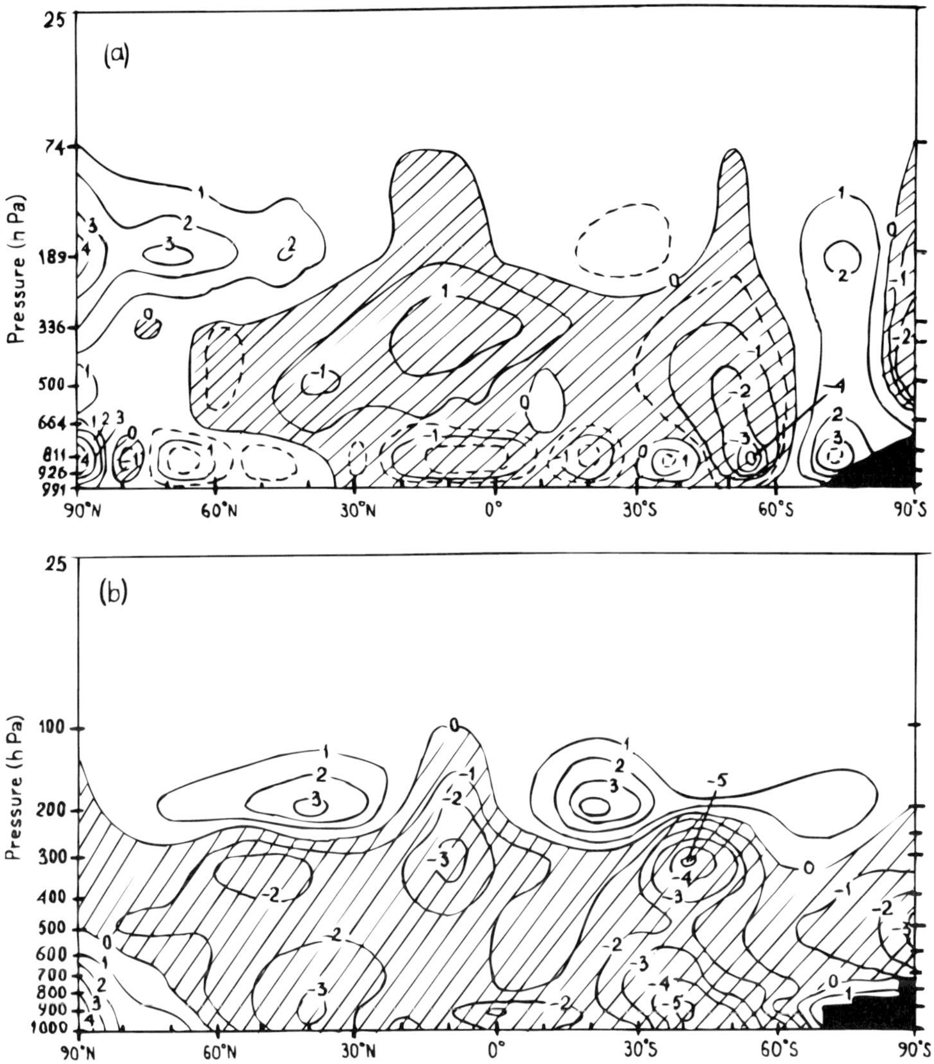


**Figure 4.3** Latitude–height distribution of the difference (annual mean change) in zonal mean cloud amount due to a doubling of $CO_2$ obtained from the (a) Washington and Meehl (1984) model, and (b) Hansen *et al.* (1984) model.

reliability of the results of climate numerical modelling. Figures 4.5–4.14 illustrate some results on ERB variability. Analysis of observational data has shown, for example, that the use of energy-balance climate models leads to a substantial overestimation of the role of albedo feedback.

As for assessments of the role of cloud-radiation feedback, they are rather contradictory. As the most reliable way (based on the ERB observational data) of obtaining such assessments, Ramanathan (1987) suggested calculating the cloud-radiative forcing due to longwave radiation using the relation $C_f(L) = F_c - F$, where $F_c$ is the outgoing longwave radiation value for clear skies and $F$ is that of real conditions of observations. Within the programme of the ERB experimental study, the $F_c$ values

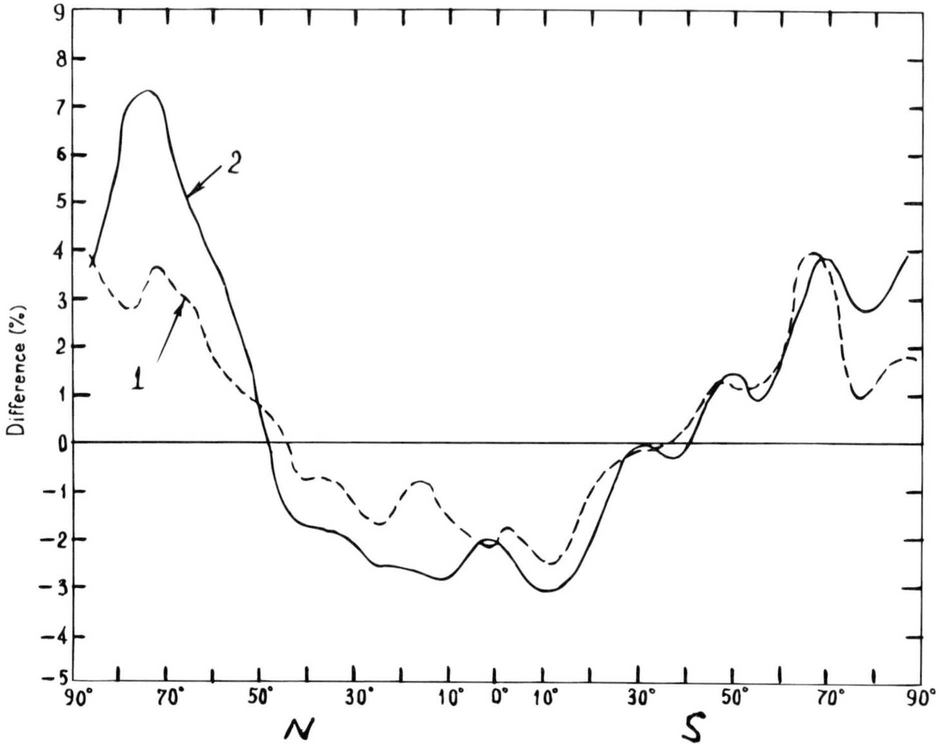


**Figure 4.4** Latitudinal distribution of zonal mean differences in total cloud amount obtained from $4 \times CO_2$ experiment (2, solid line) and $2 \times CO_2$ experiment (1, dashed line).

are estimated from data for $2.5^\circ \times 2.5^\circ$ squares by selecting maximum outgoing longwave radiation values and minimum albedo values for each resolving element (pixel). If a grid square is totally covered with clouds, the $F_c$ value can be found from data for the preceding and subsequent time moments, as well as from data for adjacent squares.

The forcing effect $C_f(S)$ due to shortwave radiation is determined in the same way. The total effect will then be

$$C_f(N) = C_f(L) + C_f(S) \tag{4.8}$$

Since, at that time, no estimates of $C_f$ had been obtained from satellite observations, Ramanathan (1987) considered the results of calculations of this parameter based on numerical modelling of climate. Later, Gupta *et al.* (1993) published important results on the seasonal variation of surface temperature and atmospheric cloud-radiative forcing over the globe, as derived from satellite data. These results show that (1) the longwave effect of cloudiness on the surface–atmosphere system is positive (warming), but the negative shortwave effect (cooling) is greater, especially in the southern hemisphere, and (2) there is a strong meridional gradient of the cloud-radiative impact on climate.

On the whole, the radiative effect of clouds on climate is similar to that of latent heat in the sense that in both cases a heat release takes place in the troposphere (due to solar radiation absorption). The amount of heat received by the troposphere (about 100 W m$^{-2}$) is comparable to the solar radiation absorbed by the surface and to the latent heat released in the troposphere. The longwave heat flux divergence in the troposphere is, in particular, one of the important factors determining the monsoon circulation.

The earlier neglected effects of decreasing solar radiation arriving at the surface in the tropics due to a thick vertically developed cloudiness, and the accompanying longwave heating of the troposphere, play an important role in the formation of the vertical thermal structure of the atmosphere. The introduction of $C_f(N)$ allows the following expression for the parameter $\beta$ of climatic sensitivity:

$$\beta = F_u \Big/ \left\{ \frac{\partial F_c}{\partial T_s} + \frac{\partial \alpha}{\partial T_s} - \frac{\partial}{\partial T_s} [C_f(N)] \right\} \tag{4.9}$$

Here $F_u$ is the outgoing longwave radiation for undisturbed conditions, $T_s$ is the surface temperature and $\alpha$ the clear-sky system's albedo. For realistic estimates of the $C_f$ quantities and analysis of the interannual climatic variability, a long series of observations is needed (at least 10–15 yr).

Only the availability of a long observational series will make it possible to study the long-term climatic trends and, in particular, to assess the anthropogenic effects on climate. For example, data of albedo for clear skies will allow analysis of the contributions of deforestation dynamics, ice and snow cover extent, and aerosol content variability to the radiative effect on climate.

Important circumstances connected with studies of the climate-forming role of the ERB are as follows: (1) direct measurements of the ERB components are needed using wide-angle sensors of shortwave and longwave radiation fluxes (the retrieval of total fluxes from data from narrow-angle scanners, sensitive in narrow spectral intervals, does not provide a reliable estimation of absolute values of the ERB components) and (2) data should be obtained on the spectral distribution of the ERB, which could help in recognition of the anthropogenic impact on the ERB.

Roeckner *et al.* (1987) assessed the climate-forming role of the cloud-radiation feedback caused by varying liquid water content (and therefore varying optical thickness and albedo) of clouds, with climate variations. The relevant calculations have been based on the use of a 3-D climate model with an interactive account of dynamics, radiation processes and moisture cycle, and with a fixed 2 per cent growth of the solar constant, which, from the viewpoint of climatic perturbation, is equivalent to a doubling of $CO_2$ concentration. The global finite-difference model (with a step of 11.25° on a spherical grid) includes: (1) a three-layer AGCM (radiation fluxes are calculated, however, for 8 levels), (2) an oceanic mixed-layer model and (3) a sea ice model. Land surface characteristics are either fixed (albedo) or predicted (temperature, humidity of soil, snow cover thickness).

An important feature of this climate model is the use of a new cloud-formation parametrization scheme, based on the numerical solution of the equation describing the liquid water content of broken cloud cover and microphysical parameters of clouds, including such processes as water vapour condensation, evaporation of cloud and rain droplets and coalescence of droplets. Radiative properties of cloud cover (albedo and emissivity) depend on the water content of clouds. Therefore, in contrast to other climate models, in this case the radiative properties of clouds turn out to be coupled with other climate characteristics through cloud dynamics.

Table 4.2 summarizes (in comparison with observational data) variations in the globally averaged climatic parameters calculated for the control case (integration

**Table 4.2** Observed and calculated globally averaged climate parameters

| | Surface temperature (°C) | Precipitation (mm $day^{-1}$) | Cloud amount (%) | Water content (g $m^{-2}$) | Earth albedo (%) | Cirrus emissivity (%) | Arctic ice extent ($10^6$ $km^2$) | Antarctic ice extent ($10^6$ $km^2$) |
|---|---|---|---|---|---|---|---|---|
| Observations | 14.9 | 2.7 | ~50 | – | 30.0 | – | 11.3 | 11.8 |
| Control | 14.7 | 2.5 | 53.4 | 43.2 | 27.8 | 42.4 | 11.9 | 12.0 |
| Perturbation control | 3.32 | 0.15 | 0.50 | 4.52 | 0.81 | 17.4 | −3.06 | −6.04 |

over 20 yr with the fixed solar constant equal to $1.370\,W\,m^{-2}$, with subsequent averaging over the last 10 yr) and for a 2 per cent growth of solar constant ($1.397\,W\,m^{-2}$).

As can be seen, the model satisfactorily reproduces the climate observed. A 2 per cent increase of solar constant causes a climate warming of 3.32 °C, which (as calculations show) is intensified with latitude. Most of the Arctic (about 25 per cent) and Antarctic (50 per cent) ice cover gets melted. The cloud amount increases slightly due to global increase of the upper clouds and to an increase of the lower clouds only in high latitudes (in other cases the cloud amount decreases).

A 10 per cent increase of the water content of clouds causes the growth of albedo and emissivity of cirrus clouds (emissivity of lower and mid-level clouds remains almost constant). Analysis of calculated ERB components has led to the conclusion that changes in the cloud cover characteristics intensify the atmospheric heating due to the greenhouse effect and lower the surface temperature (due to decreased solar radiation). The resulting effect of cloud feedback, determined largely by greater optical thickness of clouds, is to warm the surface–atmosphere system. Further numerical experiments of this kind have been performed by Ramaswamy and Chen (1993). They showed that, relative to the global annual mean value of the solar radiative forcing (a negative quantity for an increase of liquid water path of clouds), the forcing due to changes in the cloud liquid water path is more negative throughout the year in low latitudes and during the summer at high latitudes. By contrast, forcing is more positive than the global annual mean value during the winter season at high latitudes (poleward of 40°).

The complicated nature of the cloud optical feedback has resulted in contradictions in estimations even of its sign. In this connection Schlesinger (1988a–d) undertook new calculations of the contribution of cloud optical feedback (based on the use of a 1-D radiative–convective climate model) in order to estimate surface air temperature variations, $\Delta T_s$, taking place either with a 2 per cent increase of solar constant (2 per cent $S_0$) or with a $CO_2$ doubling ($2 \times CO_2$) with prescribed variations of the one-layer cloudiness optical thickness $\Delta\tau^*$ and emissivity $\Delta\varepsilon^*$.

The results of calculations are given in Table 4.3, where $\Delta N_t$ and $\Delta N_s$ are changes in radiation budgets at the tropopause level and surface level, respectively. Parameters $\Delta T_s^*$ have been calculated with a fixed adiabatic lapse rate ($6.5\,°C\,km^{-1}$) and moist-adiabatic lapse rate (in parentheses). As can be seen, in both cases (with and without cloud optical feedback) there is a climate warming, but in the first case it is stronger; hence, the cloud optical feedback is positive.

This situation can be explained by the following processes. The externally forced climate warming causes an increase of the water content of clouds at every level and of the emissivity of the upper clouds. The latter reduce the heat loss by Earth to space by an amount that exceeds the decrease of absorbed solar radiation. The resulting upper tropospheric heating reduces the convection and, consequently, the

**Table 4.3** Estimates of the contribution of the cloud optical feedback

| $\Delta\tau^*$ ($\tau^* = 1.1$) | $\Delta\varepsilon^*$ ($\varepsilon^* = 0.62$) | $\Delta N_t$ ($W\,m^{-2}$) | $\Delta N_s$ ($W\,m^{-2}$) | $\Delta T_s^*$ (2% $S_0$) (°C) | $\Delta T_s^*$ ($2 \times CO_2$) (°C) |
|---|---|---|---|---|---|
| 0.4 (0.65) | 0.16 (0.28) | 2.4 | −2.9 | 5.3 (2.6) | 5.3 (2.8) |
| 0 | 0 | 0 | 0 | 4.2 (2.2) | 4.4 (2.5) |

cooling of the Earth's surface due to evaporation. This results in an intensified warming of the surface air masses. Naturally, the estimates obtained are approximate and need to be specified with the use of a 3-D climate model.

Mólnar and Wang (1993) have undertaken a new analysis of cloud-optical feedback dependence on cloud optical thickness, $\tau$. This feedback is characterized by its intensity $S$, expressed as a percentage: $S = (\Delta T_\tau/\Delta T - 1)$, where $\Delta T_\tau$ and $\Delta T$ are changes of surface air temperature, with and without taking account of feedback. Earlier studies have indicated that increased $CO_2$ concentration results in variations of $S$ within a broad range of values, from small positive to large negative values. Such controversial results may be explained by differences in parametrization techniques for the simulation of cloud dynamics and cloud phase changes, as well as differences in control climates.

The value of $S$ may depend on a number of specific parameters which determine climate changes. For example, where assessing the influence of $CO_2$ concentration or solar constant increases, we should allocate an important role to variations of vertical profiles of radiative forcing. Since vertical radiative forcing distributions for $CO_2$ and other greenhouse gases ($CH_4$, $N_2O$, CFCs) are not similar, new assessments of the cloud optical feedback in climate change due to increases of one-layer and multi-layer cloud thicknesses have been obtained (Mólnar and Wang 1993).

For some cloud types, feedback parametrization is acceptable when based on the dependence of $\tau$ on cloud liquid water content through temperature (CT scheme). A CP scheme is also possible, where $\tau$ depends on liquid water content through latent heat due to precipitation formation. Mólnar and Wang (1993) calculated that on the basis of a one-dimensional radiative–convective climate model, stronger cloud optical feedback forms if $CH_4$, $N_2O$ and CFC concentration increase takes place than if the same were to occur for $CO_2$ (see Table 4.4). In Table 4.4 the difference between $CO_2$ and CFC is especially distinct; in the first case the cloud optical feedback is weakly negative, while in the second it is strongly negative. In using a CT scheme we can see that the transition from strongly negative to much weaker negative feedback occurs when one-layer cloudiness is substituted by multi-layer cloudiness.

Table 4.5 illustrates changes of cloud optical thickness, $\Delta\tau$, radiative flux divergence at the tropopause level, $\Delta F$, and cloud temperature, $T_c$, for various cloud conditions, where $CO_2$ doubling in the CT scheme has occurred. In the case considered, $S = -33$ per cent. As one might expect, the enhancement of upper level

**Table 4.4** The impact of cloud optical feedback (COF) on surface air temperature change (°C) and relative $S$ values

| Case | Without COF | With COF | |
|---|---|---|---|
| | | CP scheme | CT scheme |
| $CO_2$, 330–660 ppm | 2.88 | 2.49 [−13] | 1.68 [−42] |
| $CCl_3F$ (CFC-11), 0–2 ppbv | 0.57 | 0.39 [−31] | 0.34 [−41] |
| $CCl_2F_2$ (CFC-12), 0–2 ppbv | 0.57 | 0.39 [−31] | 0.34 [−41] |
| $N_2O$, 0.3–0.6 ppmv | 0.72 | 0.55 [−22] | 0.44 [−39] |
| $CH_4$, 1.7–3.4 ppmv | 0.48 | 0.58 [−20] | 0.33 [−30] |

*Note*: Data in brackets show relative $S$ values.

**Table 4.5** Changes of cloud optical thickness, $\Delta\tau$, radiative forcing, $\Delta F$, and cloud temperature change, $\Delta T_c$, in the case of $CO_2$ doubling (CT scheme)

| | Cloud levels | | |
|---|---|---|---|
| Parameter | Upper | Middle | Low |
| $\tau$ | 1.78 | 5.18 | 11.84 |
| $\Delta\tau$ | 0.22 | 0.58 | 0.87 |
| $\Delta F$ ($W\,m^{-2}$) | 0.25 | −0.70 | −0.84 |
| $\Delta T_c$ (°C) | 1.93 | 1.93 | 1.93 |

cirrus clouds results in weakening the feedback. In the case of upper level cirrus clouds, $\Delta F > 0$, that is, their contribution to feedback is positive. However, negative radiative forcing due to low and middle level clouds results in a significant negative $S$ value.

We need to take into account the latitude dependence of the contributions considered. In the tropics the role of cirrus clouds is naturally the strongest, which results in a weakening of negative feedback, whilst the opposite holds for high latitudes.

Satellite microwave sounding results confirm the existence of a correlation between $\tau$ and precipitation which has been used in the CP scheme. On the other hand, retrieval of cloud characteristics indicates that in low-latitude ocean regions a decrease of $\tau$ with the increase of temperature at low and middle level clouds occurs. This shows a deviation from adiabatic conditions. Strong cloud optical feedback sensitivity to vertical distributions of clouds and $\tau$ demonstrates the need to develop further the reliability of cloud cover climatology data, especially with respect to cloud liquid water content. We believe it is also important to obtain observational data on cloud microphysical parameters and on the 3-D structure of cloud cover.

The following circumstances necessitate a detailed development of the climatology of cloud cover and its interaction with the ERB (Kondratyev 1983b): (1) requirements to verify complete climate models by comparing calculated and observed cloud fields, (2) requirements to parametrize cloud-radiation feedback in simple climate models and (3) the importance of diagnostic analysis of climatic regularities from observational data. These problems are unlikely to be solved with the present status of cloud cover climatology. More reliable data are needed on the global distribution of clouds, the distribution of clouds by type and top height, as well as data on seasonal and interannual variability.

Starting in 1970, the US Air Force Meteorological Service has been processing satellite and conventional information on cloud cover in order to accumulate a multi-year data base on cloud climatology. The processing of satellite data, which serve as the basis for the 3DNEPH data base, resulted in data on cloud amount in 15 layers at a latitude of 60° (the hemispherical grid contains 162 × 144 points). Only in 5 per cent of cases is more than a three-layer cloudiness observed, and compared with the average scale (60 km) of ERB data, a reduced data base has therefore been accumulated about the amount of lower, middle and upper clouds with a horizontal spatial resolution of 160 km. The data of the reduced 3DNEPH data base agree well with available information on global cloud cover climatology –

with the exception of polar regions of the winter hemisphere, where these data are unreliable.

Liou and Zheng (1989) undertook a comparison of averaged 3DNEPH data with results of observations of the ERB components using the Nimbus-7 narrow-angle scanning radiometer (instantaneous field of view 0.25° × 5.12°; channels 0.2–4.8 and 4.5–50 μm). The transition to the fluxes of outgoing shortwave $F_s$ and longwave $F_{IR}$ radiation was made using an empirical transition scheme (by introducing the respective correcting coefficients). The comparison was aimed at assessing the sensitivity of the radiation budget $F_N$ and its components to the cloud amount, $n$. Table 4.6 gives the results obtained for the four regions for January ($r$ is the surface–atmosphere system's albedo).

Region 1 (eastern China, 22–30° N; 105–115° E) is characterized by prevailing persistent stratified low-level clouds, and therefore by maximum values of cloud amount and $F_s$. Small temperature contrasts between the surface and cloud top determine weak sensitivity of $F_N$ to $n$. There is a strong prevalence of the albedo effect of cloudiness over the greenhouse effect; the correlation between $F_N$ and $n$ is negative.

Region 2 (31–45° N; the longitudinal extent is about 10°) covers a mid-latitudinal band of storm-tracks in the Atlantic Ocean east of the US coastline and is characterized by a high sensitivity to cloud amount of both $F_s$ and $F_{IR}$, with maximum prevalence of the albedo effect. Rather special is region 3 (European territory of the former USSR, 49–63° N; 32–64° E), where the ratio $\Delta F_{IR}/\Delta n$ is the same as in region 2, but $\Delta F_s/\Delta n$ is much less, which determines the dominance of the greenhouse effect over the albedo effect. Such a situation can be explained by the effect of high snow-covered surface albedo. In this case, clouds absorb longwave radiation more than they reflect shortwave radiation. Region 4, located in the Sahara (18–31° N; 10° W–10° E), with its high albedo, has a small ratio $\Delta F_s/\Delta n$, but the albedo effect prevails nevertheless.

Table 4.7 illustrates the situation in July in the following four regions: (1) the western part of the Pacific Intertropical Convergence Zone (ITCZ) (0–13° N; 135–148° W), (2) the region of persistent stratus clouds in California (27–40° N; 120–137° W); (3) the eastern former USSR (38–70° N; 110–140° E) and (4) the Norwegian Sea (48–70° N; 10° W–10° E). Calculations of the zonal mean profiles of ERB sensitivity have led to the conclusion that the greenhouse effect dominates the albedo effect north of 60° N in January. Daily means of $\Delta F_N/\Delta n$ for the whole northern hemisphere in January and July constitute, respectively, $-8.1$ and $-50.9\ \mathrm{W\,m^{-2}}$ and agree well with the estimates obtained earlier.

From the viewpoint of information on cloud-radiation feedback, satellite observations of the ERB and its components are very important for analysis of the ade-

**Table 4.6** The sensitivity of the ERB and its components to cloud amount in January ($\mathrm{W\,m^{-2}}$)

| Region | $\Delta F_s/\Delta n$ | $\Delta F_{IR}/\Delta n$ | $\Delta F_N/\Delta n$ | $\Delta F_{IR}/\Delta r$ | $\Delta F_{IR}/\Delta r^a$ |
|---|---|---|---|---|---|
| 1 | 112.22 | −13.20 | −99.02 | −73.81 | −32.2 |
| 2 | 162.45 | −45.02 | −117.4 | −125.63 | −185.2 |
| 3 | 20.27 | −45.69 | 25.42 | 29.79 | 37.0 |
| 4 | 87.99 | −48.68 | −39.31 | −270.48 | −250.0 |

[a] After Hartmann *et al.* (1991).

**Table 4.7** The sensitivity of the ERB and its components to cloud amount in July ($W m^{-2}$)

| Region | $\Delta F_s/\Delta n$ | $\Delta F_{IR}/\Delta n$ | $\Delta F_N/\Delta n$ | $\Delta F_{IR}/\Delta r$ | $\Delta F_{IR}/\Delta r^a$ |
|---|---|---|---|---|---|
| 1 | 367.09 | −93.54 | −273.55 | −326.09 | −217.0 |
| 2 | 115.42 | −48.71 | −66.71 | −6.0 | −25.0 |
| 3 | 288.35 | −22.70 | −256.60 | −52.54 | −172.0 |
| 4 | 560.75 | −38.11 | −522.60 | −42.94 | −37.2 |

[a] After Hartmann *et al.* (1991).

quacy of climate numerical modelling. They also contributed much to the verification of previously calculated ERB data (Kondratyev, 1987a). Results showed that the Earth is much darker (it has a lower albedo, the albedos of the northern and southern hemispheres being almost equal, reflecting the prevailing contribution of clouds to the albedo formation) and warmer (higher outgoing longwave radiation) than followed from calculated data. The annual mean zero budget of absorbed solar radiation and outgoing longwave radiation is typical both of the Earth as a whole and of individual hemispheres (the latter result means that it is not necessary to transfer energy across the equator, from the viewpoint of ensuring the energy budget for the surface–atmosphere system).

To illustrate global mean regularities, we use the data from a review published by Ramanathan (1987). Figure 4.5 characterizes the components of the global mean annual heat balance of the Earth based on both calculations and observations. In connection with a study of the Earth's heat balance, of great importance is an estimation of the solar constant.

Mecherikunnel *et al.* (1988) undertook a comparison of solar constant values obtained from the ERB-satellite (ERBS), NOAA-9, NOAA-10, the Stratospheric Measurement Mission (SMM), Nimbus-7 and rocket soundings. From data for the first 2 yr of solar constant measurements on satellites ERBS and NOAA-9, the solar constant values averaged 1364.9 $W m^{-2}$. This value is 0.4 per cent lower than the Nimbus-7 estimate and 0.2 per cent below the solar constant value obtained with a cavity radiometer carried by SMM, and from the data of rocket soundings. The NOAA-9 and NOAA-10 measurement results averaged over the first 6 months gave a value of 1363.2 $W m^{-2}$. Crommelynck (1994) discussed further solar constant measurements made in 1992 and 1993 and compared them with previous observations. The solar constant values within the range 1366–1368 $W m^{-2}$ were obtained with a spread of less than 1 $W m^{-2}$ (see also Sklyarov 1994). Thus the data of the latest solar constant measurements agree reasonably well, differing within the error of the present-day pyrheliometric observations. However, the mutual calibration of satellite-borne pyrheliometers is needed to substantiate the pyrheliometric scale for space-based solar constant measurements.

The availability of measured outgoing longwave radiation data has made it possible to assess the planetary mean greenhouse effect. Since the surface emission (at a global mean temperature of 15 °C) constitutes 390 $W m^{-2}$, and the outgoing longwave radiation is 237 $W m^{-2}$, the atmospheric greenhouse effect reaches approximately 150 $W m^{-2}$. Figure 4.6 illustrates the spectral features. The data on the ERB meridional mean profile (Figure 4.7) were used to obtain estimates of the meridional heat transport in the atmosphere and ocean. Results showed that the maximum

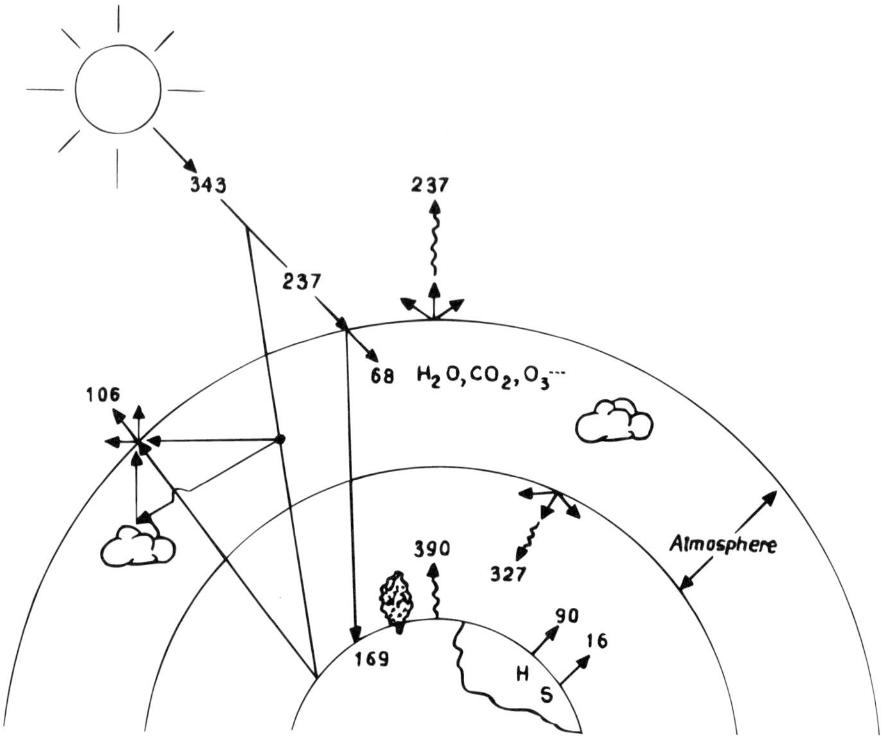


**Figure 4.5** The global energy balance for annual mean conditions. The top of the atmosphere estimates of solar insolation (343 W m$^{-2}$), reflected solar radiation (106 W m$^{-2}$) and outgoing longwave radiation (237 W m$^{-2}$) are obtained from satellite data. The rest of the components are computed: atmospheric absorption of solar radiation (68 W m$^{-2}$); surface absorption of solar radiation (169 W m$^{-2}$); and *H*, the latent (90 W m$^{-2}$), and *S*, the sensible (17 W m$^{-2}$), heat fluxes from the surface. After Ramanathan (1987).

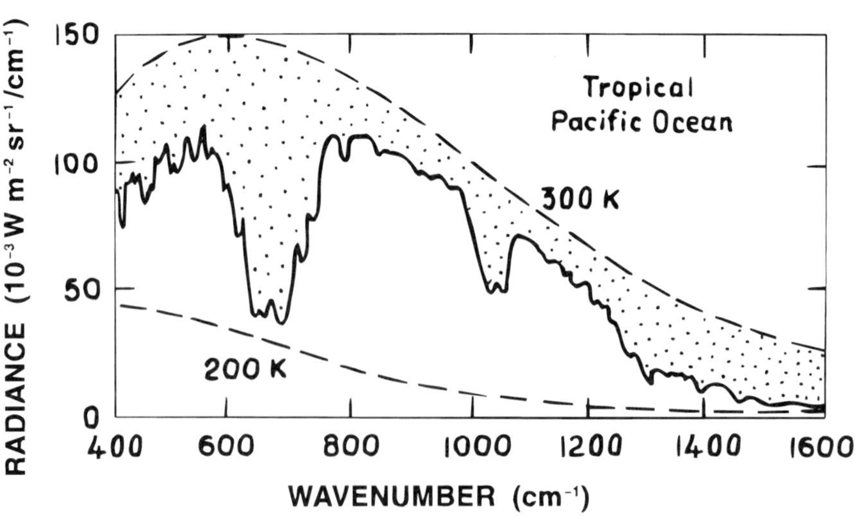


**Figure 4.6** Spectral distribution of outgoing longwave emission from Nimbus-4 IRIS measurements (the solid line is the measurement; the dashed lines are blackbody curves). After Ramanathan (1987).

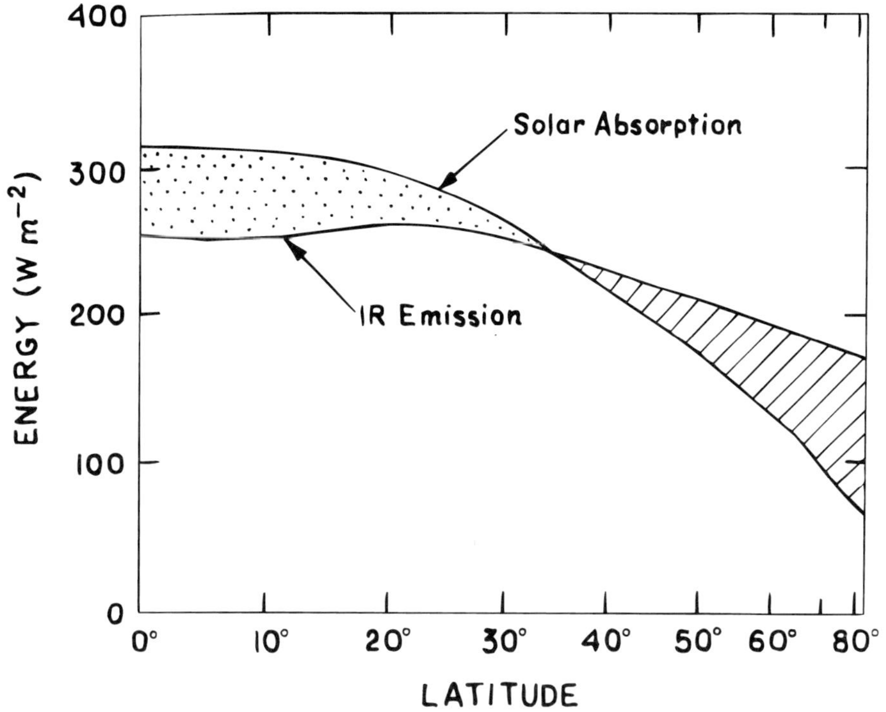


**Figure 4.7** Annual zonal mean estimates of absorbed solar radiation and outgoing longwave radiation flux. After Ramanathan (1987).

meridional transport was observed near latitudes $\pm 30°$, in the northern hemisphere the share of the heat transport by the ocean being about 40 per cent. Of great importance, constituting a still poorly studied factor of the formation of climatic anomalies, are the longitudinal variations of the ERB components (in particular the outgoing longwave radiation). Information about the global spatial and temporal variability and its components (Figure 4.8) is very interesting from the viewpoint of assessing the role of the cloud-radiation and albedo feedbacks and for testing the reliability of climate numerical modelling results (Figures 4.9 and 4.10). An analysis of observational data has shown that the energy-balance climate models substantially overestimate the role of the albedo feedback.

Borger and Vernekar (1988) undertook a comparison of calculated outgoing longwave radiation global fields, as a function of cloud conditions, with data from satellite observations. This comparison was made with the use of a modified (to eliminate the climatic drifting) AGCM developed at the Goddard Laboratory for Atmospheric Sciences (GLAS). The compared fields of outgoing longwave radiation are given in the form of an expansion in spherical harmonics (with 15 harmonics), which ensures a strictly quantitative comparison of the amplitudes and phases of the variability on various scales of averaging. The emphasis was placed on the analysis of information on the annual change, with the use of data from seven NOAA satellites for the period June 1974 to November 1983 (except for a 10-month gap from March to December 1978).

The nine-level GLAS model (uniformly subdivided along the sigma-coordinate) is realized on a $4° \times 5°$ grid. Two criteria of the cloud cover formation have been

(a)

(b)

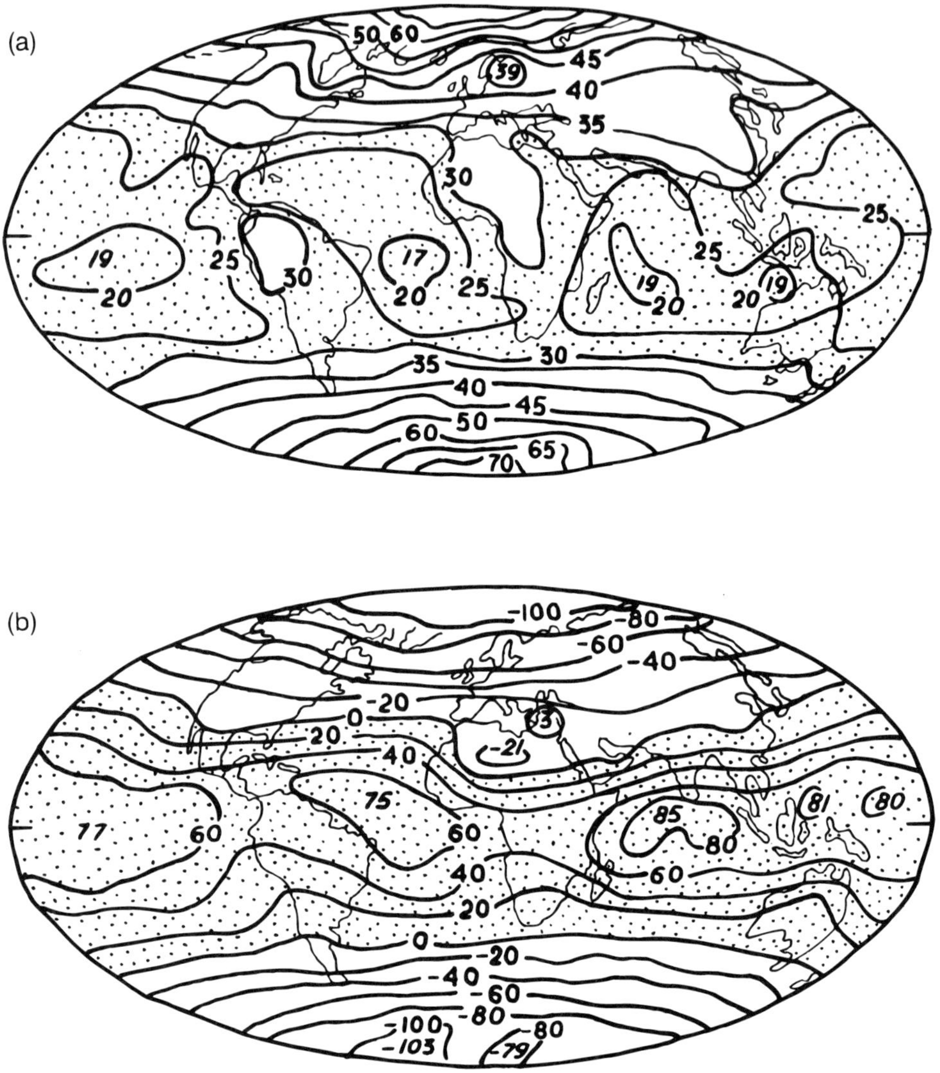


**Figure 4.8** Composite satellite estimates of annual mean regional radiation budget. (a) Albedo (%), regions with less than 30% albedo are shaded, (b) Net heating ($W\ m^{-2}$), i.e. absorbed solar minus outgoing longwave. Regions with net heating are shaded. After Ramanathan (1987).

prescribed: (1) more than 100 per cent relative humidity (within a grid cell) and (2) the buoyancy of the lower stratospheric layer with respect to that lying above, which determines the formation of cumulus clouds (however, since filling up the whole grid cell with cumulus clouds is unreal, these are left out in assessing the radiative effects). Optical thicknesses (varying from 2 to 32) and albedo (varying from 0.2 to 0.8) of different types of clouds were prescribed, as well as the annual change of sea surface temperature and the distribution of the surface and snow-ice cover albedos. The values of Student's $t$-criterion determined the statistical significance of differences between calculated and observed outgoing longwave radiation values.

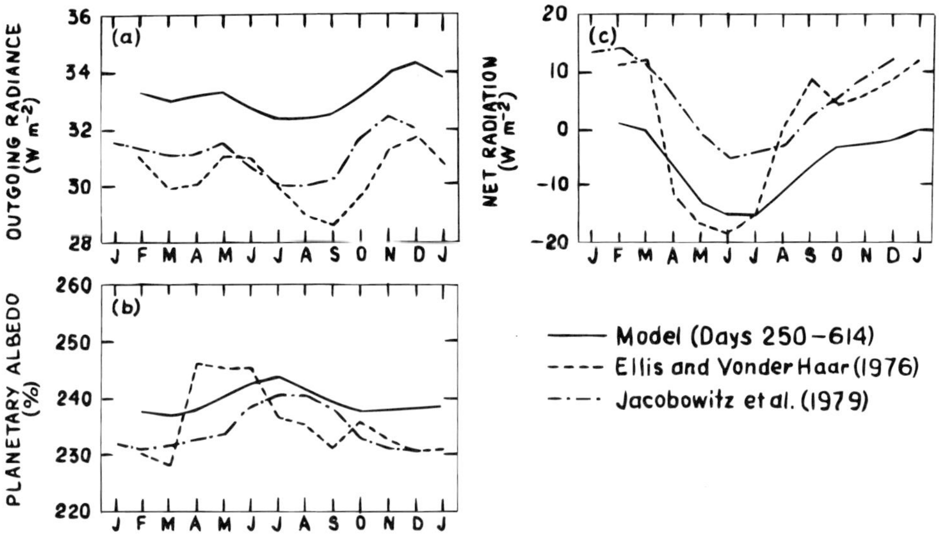


**Figure 4.9** Comparison of satellite radiation budget data with GCM results: global mean radiation budget. After Ramanathan (1987).

The calculated global mean outgoing longwave radiation value was found to be about $20\,\mathrm{W\,m^{-2}}$ lower than the observed one, which is explained by an overestimated cloud amount, and the calculated outgoing longwave radiation mean standard deviation exceeded that measured by $10\,\mathrm{W\,m^{-2}}$, which might be explained, largely, by an overestimated meridional outgoing longwave radiation gradient. The calculated annual change in global mean outgoing longwave radiation values and their variations agrees well with observational data. An analysis of the correlation

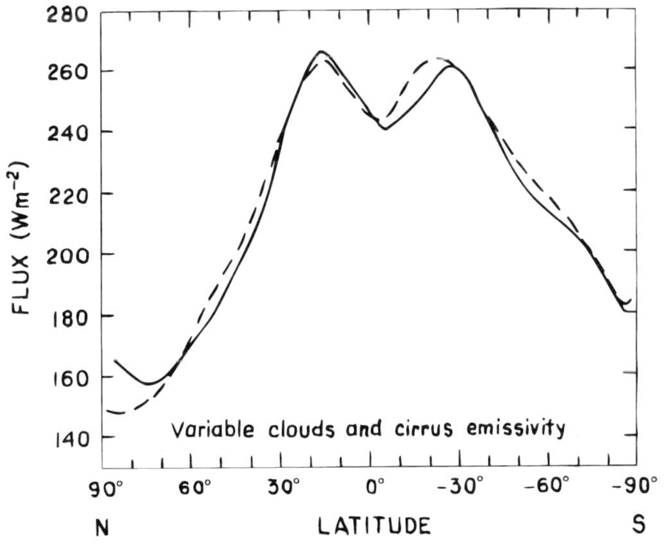


**Figure 4.10** Comparison of satellite radiation budget data with GCM results: zonal mean outgoing longwave radiation flux for January. —— observed – – – computed. After Ramanathan (1987)

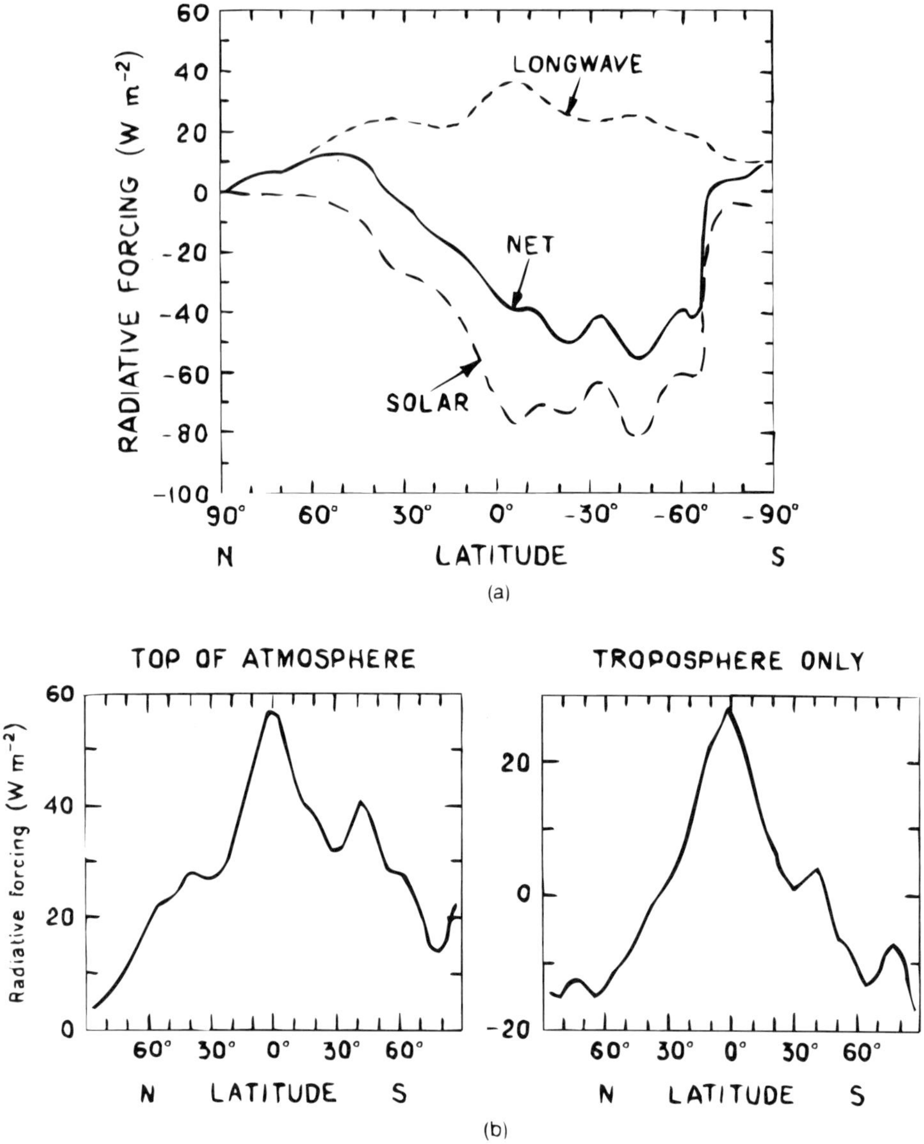


**Figure 4.11** Longwave, solar and net cloud-radiative forcing, inferred from NCAR GCM studies (January: 4 month average): (a) results obtained from Charlock and Ramanathan (1985), with severe underestimation of longwave forcing of clouds; (b) after Ramanathan (1987).

coefficients for observed and calculated outgoing longwave radiation, for different scales of averaging, has shown that the phase interrelations agree well in the case of large scales in January and July, but are merely satisfactory in April and October.

In the tropics, the differences between calculated and observed outgoing longwave radiation are close to zero, except for the regions with a strong convection (Asia, Amazonia, central Africa), when the cloud-radiation feedback is not taken into account. Very substantial differences appear in middle and high latitudes, espe-

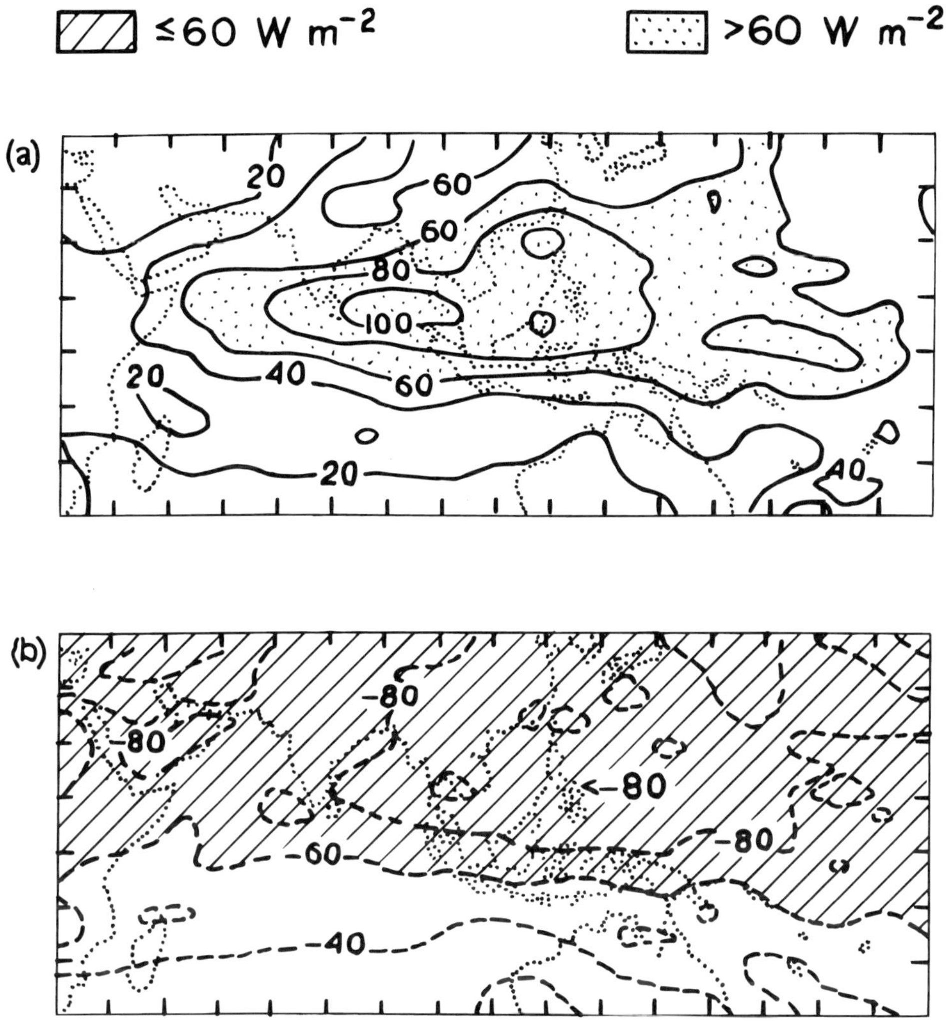


**Figure 4.12** Computed cloud-radiative forcing (W $m^{-2}$) for the surface–atmosphere column over the Indian subcontinent during the summer monsoon (June–July–August): (a) longwave; (b) solar. Solid lines indicate positive forcing (heating) and dashed lines indicate negative forcing (cooling). After Ramanathan (1987).

cially in the southern hemisphere. It seems that the greatest improvement in the AGCM would be connected with a more adequate parametrization of the transport of heat and water vapour due to vortices and meridional mean transport, and with information on the interaction between radiation and convective clouds.

The most important developments in the field of numerical climate modelling are as follows: (1) analysis of the causes of systematic errors in modelling (with emphasis on orographic effects), (2) comparison and testing of the reliability of results obtained with different models (this, in particular, concerns calculations of the ocean surface heat balance components) and (3) improvement in interactive atmosphere–ocean models.

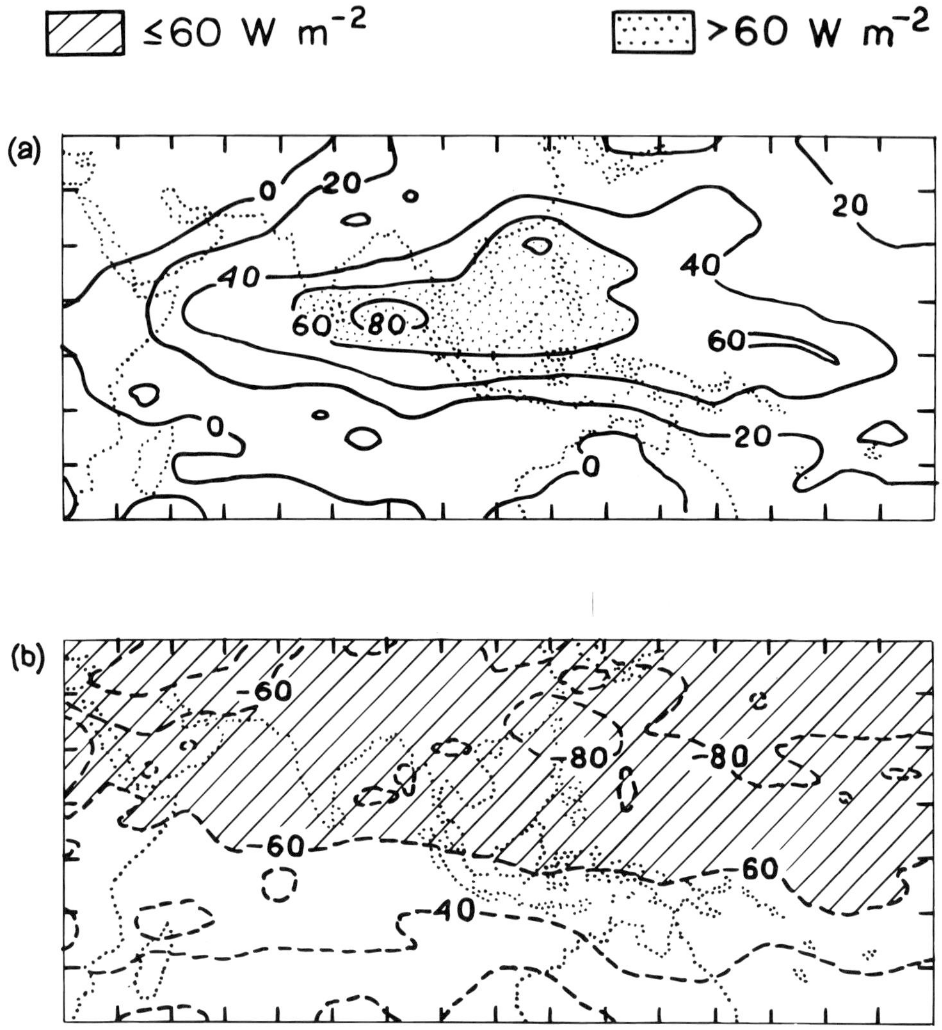


**Figure 4.13** Same as Figure 4.12 but separately for (a) tropospheric heating in the longwave and (b) surface heating in the solar. After Ramanathan (1987).

In connection with the realization of the TOGA programme, the following needs should be considered first: (1) advances in techniques for objective analysis and diagnostic studies of the tropical zone, (2) studies of 40-day to 60-day oscillations of atmospheric circulation in the tropics, (3) further analysis of the effect of minor positive sea surface temperature anomalies in the eastern Indian Ocean on the atmosphere and (4) improvements to interactive climate models.

Numerical modelling of the effect on the atmosphere of sea surface temperature anomalies caused by the 1982–83 El Niño event, undertaken by nine groups of scientists, revealed a satisfactory agreement of the results and showed that the principal effect of El Niño is the geographical shifting of a maximum of precipitation, but not its magnitude. Special features of the 1982–83 El Niño event were that a maximum of precipitation shifted eastward from the date line by about 20° longi-

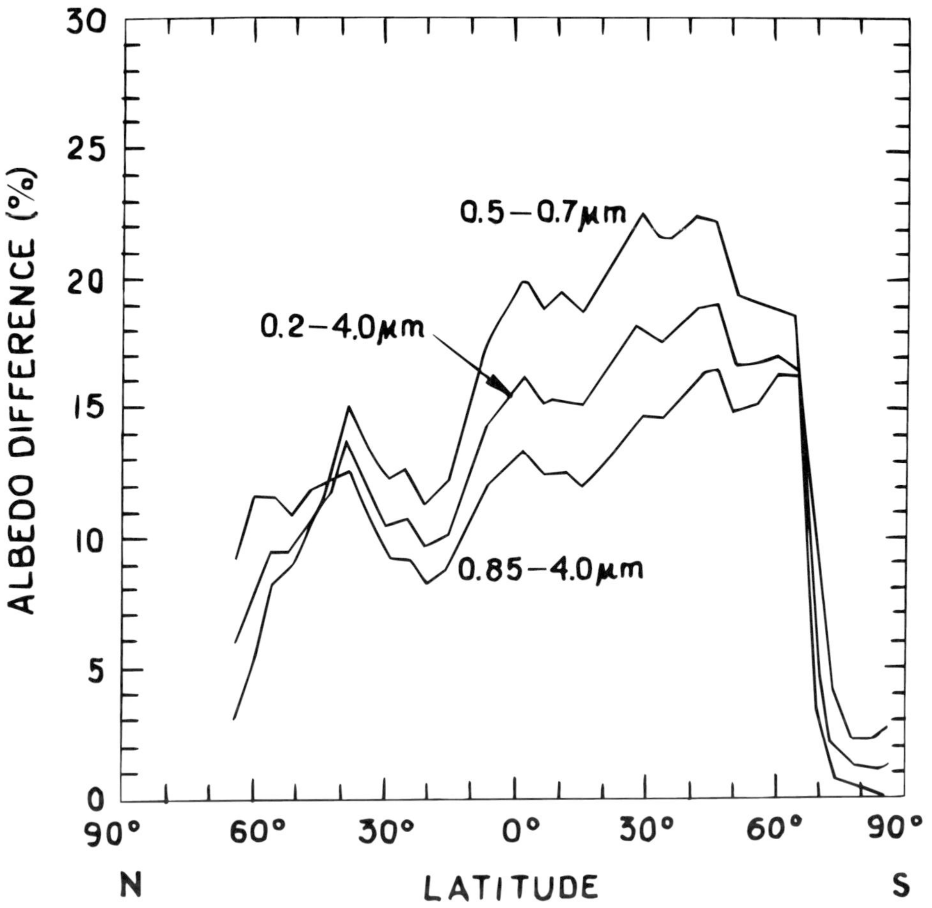


**Figure 4.14** Computed spectral variations in cloud-radiative forcing in the absorbed solar radiation for January conditions. After Ramanathan (1987).

tude. The mid-latitude atmosphere response to the sea surface temperature anomaly revealed by various AGCMs was rather variable and strongly dependent on the initial conditions and location of sea surface temperature anomalies.

The role of cloudiness is illustrated by results of calculations of global mean (in January) variability of absorbed solar radiation ($\Delta S$), outgoing longwave radiation ($\Delta L$) and ERB ($\Delta R$), for a doubled $CO_2$ concentration, made by Slingo (International Association of Meteorology and Atmospheric Physics 1989). The parameters were calculated with the use of three versions of the National Center for Atmospheric Research (NCAR) community climate model (Table 4.8). The parameter $\Delta n$ characterizes an increase of cloud amount (per cent) needed to compensate completely for the effect of a $CO_2$ doubling.

As can be seen, a compensation for the effect of $CO_2$ doubling should take place with increasing cloud amount only by 0.1–0.2 (such a change is almost within the error of ground-based visual estimation of cloud amount). The most convincing conclusions about the climate-forming role of clouds can be obtained from analysis

**Table 4.8** Calculated global mean (January) variability of ASR ($\Delta S$), OLR ($\Delta L$) and ERB ($\Delta R$)

| | Versions of climate model | | |
|---|---|---|---|
| | 1 | 2 | 3 |
| $\Delta S$ | 0.15 | 0.15 | 0.15 |
| $\Delta L$ | −2.20 | −2.27 | −2.34 |
| $\Delta R$ | 2.35 | 2.41 | 2.49 |
| $\Delta n$ | 19 | 14 | 14 |

*Source*: International Association of Meteorology and Atmospheric Physics (1989).

of the ERB satellite information and estimates of the cloud-radiation forcing, CRF:

$$\mathrm{CRF} = (Q - Q_{\mathrm{CLR}}) - (F - F_{\mathrm{CLR}}) \tag{4.10}$$

where $Q$ and $F$ are absorbed shortwave and outgoing longwave radiation (the index CLR indicates that the quantities refer to clear-sky conditions).

Although an estimate of $Q_{\mathrm{CLR}}$ and $F_{\mathrm{CLR}}$ is somewhat arbitrary (clearly, calculating them synchronously with $Q$ and $F$ from observational data is not feasible), the results obtained can apparently be considered sufficiently reliable, the more so that, as a rule, they agree reasonably well with the data of numerical modelling. The processing of satellite data gave the following global annual mean values:

$$Q - Q_{\mathrm{CLR}} = 48\ \mathrm{W\,m^{-2}} \tag{4.11}$$

$$F - F_{\mathrm{CLR}} = 31\ \mathrm{W\,m^{-2}} \tag{4.12}$$

i.e. cooling of the Earth due to the dominating effect of the cloud albedo constituted $17\ \mathrm{W\,m^{-2}}$. The effect of cloud dynamics on the variability of the ERB and climate is particularly manifested in the tropics, where latent heat release, maximally connected with cloud formation, takes place. A Hadley circulation cell results from upward motions in the Intertropical Convergence Zone (ITCZ), where latent heat, released in deep-convection clouds, and adiabatic cooling are major components of the energy balance (in the zone of the downward branch of the Hadley cell, adiabatic heating and longwave radiative cooling prevail). The radiative effect of the upper clouds in the tropics is determined by their water or ice content and particle size distribution. These clouds produce a heating of the surface–atmosphere system when their optical thickness is small (i.e. when the greenhouse effect dominates the albedo effect), but in the case of optically thick clouds a cooling effect is observed.

In recent years there has been a considerable amount of work done on the analysis of Earth radiation budget (ERB) data obtained from instruments flown on a number of polar-orbiting satellites. These include ERBE systems on Nimbus-6, Nimbus-7, ERBS (the Earth Radiation Budget Satellite), NOAA-9 and NOAA-10. Sets of data from these systems for various periods have been used in a variety of climate models by different research workers. This has involved the use of data from Nimbus-6 and Nimbus-7 to improve the input data on incoming and outgoing radiative flux for climate models (Ardanuy *et al.* 1991, Auer and Kao 1991, Smith and

Vonderhaar 1991) and for studying the incoming radiative flux particularly over the oceans with respect to the 1982–1983 El Niño phenomenon (Chertock *et al.* 1991) and then over the whole period from November 1978 to October 1985, using GOES data to validate the cloud-cover estimates (Chertock and Sud 1993). ERBE data from the ERBS, NOAA-9 and NOAA-10 satellites have recently been used in several ways. The effect of cloud type on the Earth's energy balance has been studied by Darnell *et al.* (1992) using ERBE data from 1983 and 1984 in conjunction with ISCCP (International Satellite Cloud Climatology Project) data to determine regional, zonal and hemispherical distributions of the downward and net flux components. Similar work has been done by Ockert-Bell and Hartmann (1992) and Hartmann *et al.* (1992) using data from 1985 and 1986. Darnell *et al.* (1992) claimed an accuracy of about $16\,W\,m^{-2}$ for the value of the average global flux. Wu and Chang (1991) studied the effect of clouds on the computation of longwave radiation budget parameters using two sets of cloud fields that were obtained from the HIRS2/MSU retrievals at the Goddard Laboratory for Atmospheres (GLA) and from the ISCCP at the Goddard Institute for Space Studies. Most discrepancies in the outgoing longwave radiation, using different cloud fields and geophysical parameters, were found to be less than $10\,W\,m^{-2}$. Discrepancies much larger than $10\,W\,m^{-2}$ were found in the downward flux values. The short-wavelength cloud radiative forcing at the top of the atmosphere, at the surface and of the atmospheric column has been studied by Laszlo and Pinker (1993) using ISCCP data.

ERBE data have been used in conjunction with image data from Meteosat and the AVHRR. Cheruy *et al.* (1991a,b) used Meteosat data from 1984 and 1985 to estimate the outgoing longwave radiance at times not sampled by the ERBE and to determine the longwave diurnal cycle. Li and Leighton (1991) and Ackerman *et al.* (1992) used AVHRR data, which has a much smaller instantaneous field of view than the ERBE data, to study the accuracy of the results obtained from ERBE data. Hartmann *et al.* (1991) used data from the ERBEs on ERBS and NOAA-9 (which pass over any given area at different times of day) to show that diurnal variations in cloud type and amount in many regions are sufficient to cause substantial errors in radiation budget quantities and cloud properties estimated from observations taken from a single Sun-synchronous orbit. The errors in estimated net radiation can be as large as $50\,W\,m^{-2}$ for oceanic stratus regions and for land regions during the summer.

Li and Leighton (1991) made comparisons between scenes identified from the ERBE algorithm and those deduced from AVHRR in the Arctic for 4 days in July 1985. It was found that they differ significantly from each other both in geotype and in cloud cover. For example, regions identified as clear fractional sea ice according to the AVHRR analysis were identified as being cloudy over open ocean by the ERBE analysis. Over the southern part of the area the ERBE-based values of net cloud forcing were as much as $50\,W\,m^{-2}$ smaller than the values determined when the scene identification was based on AVHRR data. At higher latitudes the ERBE-estimated cloud forcing was larger than that from the AVHRR although the magnitude of the difference between the estimates from the two sources was smaller. Wielicki and Parker (1992) attempted to test the accuracy of cloud parametrization obtained from low-spatial-resolution data by making comparisons with corresponding high-spatial-resolution satellite data. They used Landsat Thematic Mapper (TM) data at full resolution (i.e. 30 m) to provide a very high-spatial-resolution source of 'cloud truth'. The TM data were then spatially degraded to the spatial resolution of the data from AVHRR ($\sim$1 km resolution) and GOES (geostationary satellite data with 4 km and 8 km resolution).

Stephens and Greenwald (1991a,b) used Nimbus-7 and ERBE data to study the effects of atmospheric hydrology on the Earth's radiation budget. In clear-sky conditions this involved an observational study of the clear-sky components of the Earth's radiation budget and the relation of these components to the sea surface temperature and the microwave-derived water vapour amount using observations that are coincident in both space and time. In cloudy conditions this involved similar studies of the microwave-derived cloud liquid water vapour and cloud cover. Several studies have been made of the effects of using ERBE data to provide input data on cloud radiative forcing for various climate models (Hartmann and Doelling 1991, Minnis *et al.* 1991, Bony *et al.* 1992, Soden 1992, Cess *et al.* 1993, Graves *et al.* 1993, Li and Leighton 1993, Stephens *et al.* 1993) while the general question of evaluating clear-sky radiative fluxes in climate models in a manner that is consistent with the satellite data processing system has been considered by Cess *et al.* (1992). Their paper describes a clear-sky flux computation method that serves as an analogue to the data processing procedure and so provides a model diagnostic that is consistent with the processed satellite data.

During recent years, of great concern has been an analysis of the possible effect of aerosols on the optical properties of clouds, connected with increased concentration of the condensation nuclei caused by different factors, which leads to a decrease in the average radius of cloud droplets and, hence, an increase of cloud albedo, as well as an accompanying cooling effect which (from some estimates) may compensate for the warming effect of increasing $CO_2$ concentration. Dimethyl sulphide produced by marine phytoplankton and emitted to the atmosphere can be an additional source of condensation nuclei. The subsequent gas-to-particle reactions produce sulphate condensation nuclei (Kondratyev *et al.* 1973, 1983a, 1985, 1991, 1996a, Kondratyev 1983b, 1987b, 1989b, 1990a,b, 1991a,b, Jennings 1993, Wang and Chang 1993a–c).

Durkee (International Association of Meteorology and Atmospheric Physics 1989) performed a detailed analysis of the physico-chemical mechanisms for the formation of sulphate aerosol condensation nuclei, with the participation of dimethyl sulphide emitted to the atmosphere by the ocean, and attempted to reveal the effect of such condensation nuclei (CN) on clouds using aircraft and satellite AVHRR data for visible and near IR (3.7 $\mu$m) spectral regions. Calculations have shown that the albedo of clouds at $\alpha = 3.7\,\mu$m depends much on their particle size distribution. In this connection, aircraft observations were made of the albedo of clouds over the Pacific Ocean near San Francisco, in the presence of the plume of aerosol pollutants moving from the continent to the cloud zone. The observations revealed a marked albedo variability. Analysis of the global maps of aerosol optical thicknesses, retrieved from the AVHRR data, revealed a much higher albedo of clouds (by about 5 per cent) in the northern hemisphere, compared with the southern hemisphere, which can be ascribed to the effect of aerosols.

Shaw (1983, 1987, 1989b, 1995) drew attention to the non-linear effect of sulphate condensation nuclei on the optical properties of clouds and the climate. Therefore, the effects of both condensation nuclei and $CO_2$, equal in quantity, appear with a change in the aerosol mass, which is five orders of magnitude less than the respective change in the mass of $CO_2$. The global mean radiative cooling due to the changing cloud particle size distribution (albedo growth) can reach 13 $W\,m^{-2}$, whereas the greenhouse effect due to a doubled $CO_2$ constitutes only about

$2\,W\,m^{-2}$. The estimates have led to the conclusion that the global mass of dimethyl sulphide (DMS) in the atmosphere is $7 \times 10^{10}$ g, with the mass of a condensation nucleus (its diameter is $\sim 0.03\,\mu m$) being $4.5 \times 10^{-18}$ g. Due to gas-to-particle conversion, the condensation nuclei numbers can reach $2000\,cm^{-3}$, whereas an increase in concentration by only $95\,cm^{-3}$ ensures a complete compensation for climate warming due to $CO_2$ doubling. Thus, the biological mechanism of climate cooling due to cloud albedo growth can compensate for the warming caused by the growth of $CO_2$ and hence provide a global climatic stability (homeostatis). Such a hypothesis is verified by observed intensification of biogenic emissions of sulphur compounds by marshes and bogs when the temperature rises (the rate of emission increases by several per cent with a 1 °C increase of surface air temperature).

Rampino and Volk (1988) also discussed the hypothesis that there could have been an effect on the palaeoclimate through changes in the albedo of clouds over the ocean caused by variations in the amounts of DMS produced by marine plants and ejected to the atmosphere. DMS serves as a basis for the formation of condensation nuclei, since oxidation of DMS gives particles of submicrometre sulphate aerosol. This hypothesis is based on the events taking place about 66 million years ago, in the transition from the Cretaceous to Tertiary periods, when a mass extinction of plants and animals took place, including the destruction of nanoplankton and foraminifera (90 per cent).

Analysis of radiocarbon data has led to the conclusion that at that time (over at least several tens of thousand years, and, probably, up to a million years after the extinctions) the primary production of the ocean drastically dropped. The destruction of most of the marine phytoplankton ought to have resulted in a sharp decrease in the formation and emission to the atmosphere of DMS and, in its turn, a decrease of condensation nuclei concentration and an accompanying decrease in the number of cloud droplets and, hence, cloud albedo.

Approximate estimates show that a decrease of condensation nuclei concentration by more than 80 per cent and an associated decrease in the albedo of clouds over the ocean ought to have caused a global warming of about 6 °C or more. Isotopic analysis of oxygen in marine sediments from different regions of the global ocean revealed a substantial climate warming coinciding in time with the destruction of phytoplankton at the Cretaceous/Tertiary boundary. A decrease of the albedo of clouds over the ocean, and the resulting sea surface temperature increase, could have contributed much to maintaining the low-level bioproductivity of the ocean in the period following the destruction at the Cretaceous/Tertiary boundary.

In several studies, a hypothesis has been substantiated, according to which the atmospheric greenhouse effect had intensified at the Cretaceous/Tertiary boundary because of the growing concentration of $CO_2$ and/or $H_2O$, causing a climate warming. In fact, the nature of the warming has been more complicated. Since radiocarbon analysis revealed similar drops of bioproductivity in other transition periods (Precambrian/Cambrian, Devonian/Carboniferous, Permian/Triassic), it is not excluded that the episodes of cloud albedo decreases over the ocean and of global climate warming have been followed by other mass extinctions, revealed by geological data. The reality of such episodes needs testing by analysis of palaeotemperature data.

The observational evidence testifies to a long-term (on the scale of the age of the solar system, i.e. several billions of years) stability of global climate, in the sense that liquid water and life have always co-existed on Earth. It is clear that water has been a necessary condition for life, and the first direct evidence for the existence of liquid water refers to the time period about 3.8 billion years ago (detritus sediments found in Western Greenland testify to an active water cycle functioning at that time).

As Kasting (1989) notes, liquid water could have existed in a wide range of temperatures, from freezing point (273 K) to the critical point (647 K), since water vapour pressure, in the case of the totally evaporated ocean (with the Earth's mass assumed to be $1.4 \times 10^{24}$ g) could have amounted to 270 atm., which exceeds the pressure corresponding to the critical point (220.6 atm.). More severe restrictions for the upper limit of temperature are imposed by conditions for the existence of life. It is known, for example, that prokaryote microbes currently exist in hot sources at temperatures up to 90 °C.

The climate on Earth remained stable despite a considerable increase of extra-atmospheric insolation, which can be explained by a decrease of $CO_2$ concentration (and, hence, a reduction of the greenhouse effect) due to a negative feedback in the geochemical carbonate–silicate cycle. Such a negative feedback also determines the possibility of a habitable planet, like the Earth (i.e. a planet with liquid water), at a distance from the Sun as far as the orbit of Mars.

An initial $CO_2$ concentration in the terrestrial atmosphere could have been much higher than that needed to compensate for the cooling resulting from lower extra-atmospheric insolation, and hence the climate of the past could have been much warmer than it is now. This suggests, however, that as soon as the initial period of accretion had ended, a regime of stable climate ought to have become established, resistant to such factors as a runaway greenhouse effect and an accompanying complete evaporation of the world's oceans, or a rapid loss of water due to hydrogen dissipation as a result of water vapour photodissociation in conditions of high water content of the stratosphere.

Until recently, the long-term evolution of climate has been studied only by calculations using 1-D globally averaged models based on various assumptions, for example about the constancy of relative humidity or insignificance of the cloud-radiation feedback. A more reliable simulation modelling can be accomplished with the use of 3-D models, which will allow, in particular, studies of the problem of low-latitude glaciations in the late Proterozoic era and the performance of a deeper analysis of the problem of water loss on Earth and Venus.

Studies on cloud–radiation interaction, within the WCRP, must be based both on numerical modelling and on complex field observation programmes. In numerical modelling the following aspects are particularly important: developing techniques to forecast cloud formation in climate models for extended stratus clouds, powerful cumulus clusters, cirrus clouds and clouds of a complicated spatial structure; improving techniques to calculate radiation fluxes and their parametrization, with their preferred prediction as internal parameters, but not prescribed radiative characteristics in climate models, and analysing their reliability for the determination of radiation fluxes; comparing the calculated and observed radiation fluxes near the cloud cover boundaries as well as radiative heat flux divergences inside the cloud; and analysing the sensitivity of the surface–atmosphere system's radiation budget to cloud characteristics as well as the sensitivity of climate to the interaction between radiation and clouds.

Complex field observation programmes must be aimed at studies of dynamic, radiative and microphysical processes that determine the formation, maintenance and decay of all the types of clouds of interest from the viewpoint of analysing the role of cloud-radiation feedback in climate formation. Particularly important are (1) the improvement and testing (from observations) of the reliability of parametrization schemes for these feedbacks, (2) accumulating data on radiative properties of clouds, radiation budgets of the surface–atmosphere system and the surface and (3) accumulating data bases on climatic characteristics of clouds based on surface and satellite observations.

The successful accomplishment of complex field studies of microphysical and radiative properties of clouds can be exemplified by expeditions within the Global Atmospheric Aerosol Radiation Experiment (GAAREX) programme which took place in 1975–1979, particularly the Zaporozhye experiment (Kondratyev and Binenko 1984), which made it possible to study specific features of microphysical and radiative properties of polluted clouds (the observational programmes envisaged samplings and analyses of the chemical composition of cloud water), as well as the experiment 'Extended Cloudiness and Radiation' performed in the Arctic during the First Special Observational Period of the FGGE. Important data have been acquired from a complex of ship and aircraft observations (with the use of radiosondes, radars and other observational means) of microphysical and radiative characteristics of clouds within the Canadian–American programme Stratus Experiment (STREX) (studies of storms' movement and their effect on the atmospheric boundary layer). Experience of these complex observational programmes and numerical modelling confirmed the importance of aerological, radar and lidar soundings, surface meteorological and actinometric observations, direct aircraft measurements of water content and droplet size distribution in clouds, as well as vertical profiles of total and spectral radiative characteristics (irradiance and angular distribution of shortwave and longwave radiance). In the case of broken cloudiness, data on its three-dimensional spatial distribution are needed.

The final objective must be the acquisition of vertical profiles of radiation fluxes and radiative flux divergences together with information about the composition of the atmosphere and clouds, as well as their optical properties (optical thickness, single scattering albedo, phase function and transmission functions). Data on the optical characteristics of the surface (albedo and emissivity) are also important. A minimum data base must include aircraft measurements of radiation fluxes near the cloud boundaries, estimates of total liquid water content of clouds and results of aerological (and, if possible, radar) soundings. Observational programmes must be coordinated in the interests of the WCRP.

The problem of cloud–radiation interactions playing the key role in climate changes was in focus at the Second Symposium of the American Meteorological Society on Global Changes (American Meteorological Society 1991c), the results of which will be summarized below. Using the synchronous data of the Nimbus-7 wide-angle ERB sensors and data on cloud cover (of different levels) taken from the same satellite during the period April 1979 to March 1985 for the band 65° N to 65° S (93 per cent of the global surface), Sohn and Smith undertook calculations of total cloud-radiation forcing impact $C_f$ and its components determined by longwave (L) and shortwave (S) radiation for clouds of low (l), middle (m) and upper (h) levels:

$$C_{fl}(L) = \int_0^{A_l} \frac{\partial L^*}{\partial A_l} \, dA_l \tag{4.13}$$

$$C_{\mathrm{fm}}(\mathrm{L}) = \int_0^{A_\mathrm{m}} \frac{\partial L^*}{\partial A_\mathrm{m}} \, \mathrm{d}A_\mathrm{m} \tag{4.14}$$

$$C_{\mathrm{fh}}(\mathrm{L}) = \int_0^{A_\mathrm{h}} \frac{\partial L^*}{\partial A_\mathrm{h}} \, \mathrm{d}A_\mathrm{h} \tag{4.15}$$

where $A_\mathrm{l}$, $A_\mathrm{m}$ and $A_\mathrm{h}$ represent the amount of cloud for the three levels considered and $L^*$ is the outgoing longwave radiation. Similarly, the values of $C_\mathrm{f}$ are calculated for the case of absorbed solar radiation, $S^*$. The net cloud-radiation forcing is determined from the relation

$$C_\mathrm{f}(\mathrm{N}) = C_\mathrm{f}(\mathrm{S}) - C_\mathrm{f}(\mathrm{L}) \tag{4.16}$$

A comparison of averaged (over the 6 yr considered) meridional profiles of outgoing longwave radiation, absorbed solar radiation and ERB for real and clear-sky conditions has shown that $C_\mathrm{f}(\mathrm{L}) = -25.7\ \mathrm{W\,m^{-2}}$, $C_\mathrm{f}(\mathrm{S}) = -27.8\ \mathrm{W\,m^{-2}}$ and, thus, $C_\mathrm{f}(\mathrm{N}) = 2.1\ \mathrm{W\,m^{-2}}$; that is, the cloud cover results in a weak cooling of the Earth.

However, clouds affect much more the value of the meridional ERB gradients. In the absence of clouds the meridional ERB gradients would be much less in the tropics but greater in high latitudes, which of course is very important from the viewpoint of the formation of the meridional circulation and thermal regime. In this connection, three zones of the effect of clouds on the thermal regime can be specified: (1) warming over the tropics (0–20° latitude), (2) cooling in the zone of mid-latitudes (20–50° latitude) and (3) warming in high latitudes.

Assessments of the effect of cloudiness on the meridional circulation have shown that clouds can intensify the effect of the flux of the momentum vortex component and turbulent heat flux on the formation of Hadley circulation cells in low latitudes and those of Ferrel in mid-latitudes. Also, clouds bring about changes in the zonal available potential energy, which manifest themselves through the formation of the zone of zonal available potential energy generation in the band of latitudes less than 35° and two zones of zonal available potential energy destruction in the band of latitudes greater than 35°, where maximum cooling takes place. The formation of this zone in low latitudes is determined by the warming in tropical regions and by the cooling in mid-latitudes, which intensifies the meridional temperature gradient (the opposite situation is observed with respect to the bands of middle and high latitudes). In this connection, there is no doubt that the interannual cloud cover variability should affect general atmospheric circulation through the dynamics of the cycles of transformation of potential energy to kinetic energy. Of course, in the course of these interactive processes a reversed effect of clouds on circulation should take place.

Of importance is the fact that the cloud-radiation forcing depends not only on the amount but also on the height of clouds. The comparison of the meridional profiles of outgoing longwave radiation and absorbed solar radiation for real cloud conditions and clear skies performed by Sohn and Smith has led to the conclusion that, as a rule, in the case of upper clouds the longwave warming exceeds the shortwave cooling almost in all latitudes. By contrast, in the case of low and middle clouds the shortwave cooling prevails in middle and high latitudes. The net impact of longwave and shortwave effects is close to zero in the tropics. From the viewpoint of the generation of available potential energy, the upper clouds favour warming in the tropics and generation of zonal available potential energy in the zone between the tropics and mid-latitudes. As for low and middle clouds, they stimulate high-latitude cooling, intensifying thereby the generation of zonal available potential energy in a broad latitudinal band. Thus, the upper clouds in the tropics and middle and low clouds in middle and high latitudes intensify the previously existing meridional temperature gradients in the zones of generation and reduce the gradients in the zone of zonal available potential energy destruction.

Schmitt and Randall (1991) calculated the effect of sea surface temperature and cloud conditions on the ($2 \times CO_2$)-induced climate warming. For this purpose they used the Colorado State University (CSU) AGC model. This is the advanced model, initially developed at the University of California (Los Angeles) by Arakawa and others in the early 1960s. First of all, the parametrization of cloudiness (taking account of the three levels of clouds and penetrating convection) and radiation has been improved.

The finite-difference nine-layer (0–20 km) model in the $\sigma$-coordinate system was realized over a grid of 4° latitude by 5° longitude with time steps of 450 s (atmospheric dynamics and eddy fluxes at surface level) and 1 hour (all parametrized processes, including radiation). The effect of a $CO_2$ doubling (from 330 ppm to 660 ppm) was analysed by comparing the results of numerical modelling for the cases of $1 \times CO_2$ and $2 \times CO_2$. To reveal the effect of cloud cover, the climate was calculated with clouds neglected (method M2). The results of calculations for constant July conditions (boundary conditions for 15 July) were averaged over 60 days of integration (after the 30-day run of the model). In order to analyse the effect of sea surface temperature changes, the sea surface temperature anomalies were taken (compared with the norm for 15 July) to be equal to $-2$ deg K (case C112) and $+2$ deg K (C119). Case C112 was also used as the control to assess the effect of sea surface temperature increase by 4 deg K. In these conditions the surface warming constitutes about 4 deg K, reaching a maximum of 10.8 deg K near the 200 hPa level in the tropics. The water content of the atmosphere increases in all latitudes by about 50 per cent, the water vapour being concentrated mainly in the lower 2 km layer of the atmosphere.

In both cases (C112 and C119) the longwave cooling reached a maximum of 4.8 deg K per day, but, on the whole, was more intensive everywhere at a higher sea surface temperature, except for the atmosphere near the surface in the latitudinal bands 60–80° N and 25–30° S. The global mean outgoing longwave radiation constituted 220 W m$^{-2}$ (C112) and 234 W m$^{-2}$ (C119). In clear-sky conditions (M2) the cooling reached 6 deg K per day (C119), and the outgoing longwave radiation 265 W m$^{-2}$ (C119). The data in Table 4.9 illustrate the variability of the longwave balance at different altitudes when the $CO_2$ concentration is doubled. A comparison of results for real and clear-sky conditions makes it possible to assess the role of the effect of clouds on the longwave balance, reducing the effect of $CO_2$ doubling. The available data of numerical modelling of the longwave radiation fields for cloudy and clear-sky conditions have made it possible to analyse the contribution of clouds of different levels. The low and middle clouds prevailing in mid-latitudes intensify the ($2 \times CO_2$) effect throughout the whole troposphere, whereas the upper clouds in the tropics reduce this effect.

The sea surface temperature stimulates the $CO_2$-induced tropospheric warming; that is, a positive feedback takes place. When the sea surface temperature increases, the amount of cloud at all levels decreases, which intensifies the $CO_2$ effect (one

**Table 4.9** The (2 × $CO_2$)-induced variability of the LW balance (W m$^{-2}$)

| Cloud conditions | Type of experiment | Height | LW balance change: Global mean | LW balance change: Maximum | Location of maximum |
|---|---|---|---|---|---|
| Real | C112 | Near the surface | −1.24 | −2.2 | 90° N |
| | C119 | | −0.86 | −1.9 | 90° N |
| M2 | C112 | | −1.75 | −3.0 | 55° S |
| | C119 | | −1.20 | −1.8 | 60° S |
| Real | C112 | Tropopause | −4.33 | −5.0 | 15° S |
| | C119 | | −4.59 | −5.4 | 15° S |
| M2 | C112 | | −4.99 | −5.5 | 25° S |
| | C119 | | −5.18 | −5.6 | 25° N |
| Real | C112 | Atmospheric top | −4.04 | −4.7 | 15° S |
| | C119 | | −4.26 | −5.0 | 15° S |
| M2 | C112 | | −4.69 | −5.2 | 25° N |
| | C119 | | −4.84 | −5.3 | 25° N |

more positive feedback). Apparently, these two feedbacks should play an important role in the formation of cloud changes caused by increasing $CO_2$ concentrations. However, negative feedbacks should be taken into account, and this necessitates further investigations.

In connection with the study of the effect of cloud cover dynamics on the climate, Wylie *et al.* (American Meteorological Society 1991a) used the NOAA HIRS data for the period from July 1989 to assess the global statistical characteristics of cirrus clouds (results of observations for the periods June to August 1989 and December 1989 to February 1990 in the northern hemisphere were considered). Most clear cloud cover anomalies were observed in the ITCZ band and in the mid-latitude storm-band, where the cloud amount reached 4–8 points. The sub-tropical maxima of high pressure and deserts manifest themselves as cloud cover minima. Though the cloud bands are quasi-permanent, their boundaries move along latitudes all the year round. So, for example, a strong annual change is typical of the ITCZ boundaries in Africa, South America and the Indian–Indonesian region. A strong annual change of the cirrus cloud amount (from 8 points in winter to less than 4 points in summer) is observed over the Atlantic and Pacific Oceans in the northern hemisphere, which reflects the annual change of the cyclone intensity. An intensive annual variation of all types of cloud amount has been discovered in the Antarctic. Satellite-derived climatology of the ITCZ has been discussed by Waliser and Gautier (1993).

In continuation of the study of the role of cloud-radiation feedback in the formation of climate, manifested through changes in the optical thickness, $\tau$, of clouds, Tselonidis *et al.* (American Meteorological Society 1991a) performed an analysis of correlation between $\tau$ and low-cloud temperature (the top of which is in the 680–800 hPa layer) from the available global data base on cloud cover characteristics, and the results were used to assess the sensitivity of the surface temperature field $T$ to the optical thickness of low clouds, using the two-dimensional radiative–convective climate model. For this purpose, a numerical model was made with observed and halved values of $\tau$.

It follows from some previous results that the water content (and, hence, optical thickness) of clouds should increase with climate warming, which determines the formation of a negative cloud optical feedback. However, this conclusion has been drawn without taking account of two important factors: (1) the formation of cloud liquid water content as a result of a balance between water vapour condensation and water removal through precipitation and (2) the effect on $\tau$ of such parameters as the size of droplets and the vertical extent of clouds.

Analysis of observational data has shown that, on the whole, the optical thickness of low clouds decreases with increasing temperature (probably mainly due to more intensive precipitation than condensation). With this fact taken into account, the cloud optical feedback under discussion turns out to be positive and strongly varying with latitude. In the northern hemisphere, with $\tau$ halved, $T$ increases from 1.5 deg K in the tropics to 2 deg K in mid-latitudes. To characterize the cloud optical feedback, one can introduce the parameter $f = 1/\tau(\partial\tau/\partial T)$. The global means of $f$, calculated from observational data, were negative everywhere in the latitudinal band 15–55° N, varying from $-0.02$ in high latitudes to $-0.05$ in low latitudes. Numerical modelling of ($2 \times CO_2$)-induced climate changes with and without taking account of the cloud optical feedback has shown that this feedback produces an intensification of warming by 1–2.5 deg K. Further, more detailed studies of the role of this feedback are needed.

Wu and Chang (1992) performed calculations of global mean (over 10 days) longwave cloud radiation forcing from space-derived data obtained with HIRS-2/MSU for 1979–1980. Calculations have shown that the global mean cloud radiation forcing is practically constant all the year round, though the outgoing emission is characterized by a clear-cut annual course. The most substantial modes of cloud radiation forcing variability have been analysed with the use of techniques of principal components and empirical orthogonal functions: (1) the annual course over land (due to the annual change of solar warming), (2) the annual change over the ocean (due to sea surface temperature variations in the zone of trade winds) and (3) low-frequency oscillations in tropical and extra-tropical regions. The global mapping of the first and second empirical orthogonal functions, as well as calculations of temporal variations of the two first principal components, have demonstrated the known regularities of the east–west dipole in the precipitation field over the Indian Ocean and western Pacific Ocean, moving eastward; these are 40–60-day oscillations. An examination of the third and fourth empirical orthogonal functions reveals these oscillations, monsoon depression and the convergence zone in the southern Pacific. A high correlation of cloud radiation functions with such meteorological parameters as surface temperature, total water content in the atmosphere, etc., has been found.

One of the most important aspects of the problem of parametrization of cloudiness and cloud–radiation interaction is the dependence of the optical thickness of clouds on their size distribution, which can be connected, in particular, with the formation of a negative cloud optical feedback the temperature increase causes the increase of the optical thickness of clouds, and hence cloud albedo. Hahn *et al.* (American Meteorological Society 1991a) described a technique to retrieve the cloud particles' size distribution with the use of the data base of the International Satellite Cloud Climatology Project (ISCCP), based on the comparison of observed brightness of cloud cover at the wavelength 3.7 $\mu$m with the calculated one (data were used for warm clouds for November 1986). The single parameter of size distribution used was the effective radius of the droplets; values of 7–8 μm and 12–13 μm were used for continental and marine clouds, respectively. Attempts to determine the anthropogenic effect on the cloud particles' size distribution in the northern hemisphere and the effect of emission of dimethyl sulphide have proved abortive.

Since the transformation of liquid-water clouds into crystalline ones strongly changes their radiative properties, consideration of this phenomenon has an important role in the theory of climate. Chin *et al.* (American Meteorological Society 1991a) undertook studies of the effect of the phase of clouds on the vertical profile of diabatic heat flux divergence, determined by an ensemble of clouds, using a two-dimensional model of cirrus cloud formation as applied to the case of the squall-lines in mid-latitudes (an important feature of the model is a consideration of the microphysics of the phase transformation of clouds and radiative heat flux divergence).

Analysis of the budgets of heat, moisture and mass of the simulated system of the squall-lines has shown that, at least in the case of the system considered, an account of the ice phase does not lead to substantial changes of the net heat flux divergence and does not generate a feedback manifested through an effect on large-scale motions. However, the effect of the ice phase on the radiative properties of clouds influences not only the squall-line itself, but also the large-scale circulation at the expense of stratification that favours the growth of the life-time of anvils. Hence, further studies are needed of the effect of the ice phase on cloud-radiation feedback. However, in the parametrization of cirrus clouds the ice phase can be neglected, saving on computer time.

## 4.2 Processes in the ocean

As mentioned above, investigating the governing effect of the oceans on the climatic system is a critically important aspect of the WCRP. The effect is determined by the gigantic thermal and mechanical inertia of the oceans as well as by their effect on global cycles of water, heat and carbon. Basic features of the ocean as a component of the ocean–atmosphere–land–cryosphere–biosphere climatic system are as follows: (1) the larger extent of the oceans compared with the land (85 per cent of the southern hemisphere surface and 61 per cent of the northern hemisphere), (2) a comparatively low temperature (average temperature 3.5 °C) and larger heat capacity compared with the atmosphere (the heat capacity of the upper 3 m layer of the ocean is equivalent to the total heat capacity of the atmosphere), (3) oceanic stratification, at least in tropical and mid-latitudes, where a rather thin (several hundred metres) warm surface layer exists, which determines a stable stratification and the isolation of deeper cold layers from the atmosphere (leaving aside processes with a very long characteristic time period), (4) because of the large heat capacity and slow currents, variations in the ocean are much slower than in the atmosphere (the diurnal course of temperature in the open ocean cannot be determined, as a rule, and the annual course is observed only in the upper layer several hundred metres thick; in the regions of strong currents, far from the equator, even the annual course can be masked by temperature fluctuations due to passing vortices) and (5) oceans are major sources of moisture for the atmosphere.

The great thermal and mechanical inertia of the ocean results in high correlations between spatial and temporal variabilities of its various characteristics. Understanding the variability of large-scale dynamical processes in the ocean is important for assessing the relative role of the ocean and of the atmosphere in climate formation and is potentially essential for forecasting climate change.

The existence of the oceans on Earth facilitates climatic forecasts compared with hypothetical conditions on a planet without oceans. The transport to the atmosphere of heat accumulated by the ocean exerts a slowly varying influence on the atmosphere. Analysis of observational data has revealed correlations between the sea surface temperature anomalies and subsequent climate changes. It is likely that the interannual variability of the atmosphere is mainly governed by this slowly varying impact of the ocean (Kondratyev and Kozoderov 1984).

Major objectives of investigations are: (1) studies of processes in the ocean responsible for advection and accumulation of heat, as well as the ocean–atmosphere heat exchange and interannual variability of climate and (2) developing and testing the observation systems which combine conventional and satellite techniques to accomplish long-period observations of the ocean as the basis for understanding and forecasting climate changes.

These problems should be solved by observational and theoretical studies of weather-forming processes and climate-forming processes, and, first of all, processes of the large-scale ocean–atmosphere interaction in the tropics, where it is most strongly manifested and can show itself on a time scale of several days to several weeks, and where the seasonal and interannual variability is large. It seems that the redistribution of heat by sea currents is a major mechanism for large-scale heat anomalies in the equatorial ocean. Particularly important are studies of the equatorial zone of the Indian Ocean, where the monsoon is an example of the strongest large-scale and prolonged impact of the atmosphere on the ocean, and where the Somali current is the strongest in the ocean. Observational data point, in particular, to correlations between the sea surface temperature anomalies and precipitation over India.

Emphasis must be placed on the processes responsible for heat transport in the ocean, and on the effects on the global energy budget and associated fluctuations in global and regional climates. The meridional heat transport in the ocean and in the atmosphere must be studied. According to preliminary estimates, the oceanic contribution in the northern hemisphere prevails only in low latitudes, but in the southern hemisphere it is substantial even at 60° S. The monitoring of heat transport in the Atlantic Ocean across the 25° N latitude over at least several years can be a first step in further studies of this problem; for this purpose, the contribution of the Gulf Stream in the Florida Strait must first be measured, as well as the deep-water and surface heat transport in the remaining part of the ocean. Studies of low-frequency variations in the Antarctic circumpolar current and the formation of water masses under the influence of an intensive ocean–atmosphere interaction in the zones of the 40° and 50° stormy latitudes are important. The northern Atlantic and southern Pacific Oceans should be chosen as major targets for initial observations of heat transport.

Studies are required on mechanisms for heat transport in the ocean. These mechanisms include the transport by mean circulation in horizontal gyres (the gyre in the northern Atlantic exhibits the strongest effect on climate) and in vertical meridional cells of abyssal circulation, in the oceanic surface layer and by vortices. In mid-latitudes, as a rule, there is a heat transport from the ocean to the atmosphere, and near the eastern coastlines of Asia and Africa the temperature gradients in the zone of cold boundaries of the western warm currents create strong horizontal contrasts in atmospheric heating, which greatly affect its circulation, intensifying jet streams especially in the winter in the northern hemisphere. The important problem of the role of the ocean heat transport in climatic change has been discussed in detail by Covey and Barron (1988).

The problem of the transport of heat, moisture and momentum in the atmospheric boundary layer (especially in stormy conditions of the middle and high latitudes, where the storms cover about half the ocean surface) remains urgent, particularly the development of parametrization techniques for processes in the boundary layer. It is also very important to consider the processes governing the variability of sea ice distribution and its effect on seasonal and interannual variations in polar albedo, as well as the ocean–atmosphere interaction. The variability of the extent of ice cover and its concentration, which governs the variations in the albedo of polar regions, is the strongest indicator of the ocean–atmosphere interaction. In polar regions, where poorly understood processes governing the extent and concentration of ice cover are the leading climate-forming factors, a close correlation exists between heat flux and sea surface temperature anomalies. In this connection, the field experiments near the ice cover edge, with the major objective of studying the meridional heat transport as a factor that governs the distribution of sea ice, become particularly important. Special attention must be given to mechanisms of vertical and horizontal geochemical exchanges in the ocean, especially the $CO_2$ redistribution between the ocean and the atmosphere. All the above enumerated parts of the programme are interdisciplinary and require the joint efforts of oceanographers and meteorologists.

Compared with the atmosphere, the level of studies on numerical modelling of processes in the ocean is still rudimentary. Major objectives of the WCRP oceanographical component should be: improving models of oceanic processes, including prognostic hydrodynamical models (needed, in particular, to substantiate the forecasts of anthropogenic $CO_2$ increase in the atmosphere), accumulating an observational data base, developing new observational techniques useful for long-period (multi-year) observations at a sufficiently high spatial and temporal resolution (particularly satellite observations of the height of ocean surface level, wind stress near the surface, sea currents, sea surface temperature, evaporation, precipitation, etc.) and studying regional and global circulations in the ocean. There are a number of oceanographic programmes directly related to the WCRP; as has been mentioned above, the World Ocean Circulation Experiment (WOCE) and Tropical Ocean and Global Atmosphere experiment (TOGA) represent such observational programmes (Kondratyev and Kozoderov 1984).

The principal objective of WOCE consists of studying the annual variation of the upper oceanic layer circulation (at depths down to at least 1 km) using diagnostic models based on the observational data. This is to create the basis for the subsequent development of climate models with interactive information on the atmosphere–ocean interaction. The key element in obtaining information about the field of geostrophic near-surface currents is the use of satellite radar altimetry data. This information will then be used to retrieve the field of currents at great depths, based on relations for thermal winds. Scatterometric measurements are needed to provide the retrieval of wind speed near the ocean surface with an accuracy of up to $1\,m\,s^{-1}$ and direction 20°. The period 1987–1992 was planned for intensive satellite oceanographic observations in the interest of WOCE.

The WOCE experiment (World Ocean Circulation Experiment 1986), which started in 1987, has as one of its principal aims to obtain by 1995 the first global data base which would completely characterize the physical properties of the world's oceans. This will make it possible to analyse the long-term variability of the ocean and to test the reliability of numerical models of the oceanic general circulation (OGC), which are very important for the understanding and forecasting of global climate changes on a time scale of decades.

Important problems include: (1) analysis of the seasonal and interannual variability of the ocean on global scales, as well as its interaction with atmospheric variability (the results obtained will be important in planning the observational systems), (2) a better determination of boundary conditions on the ocean surface and a more adequate account of the ocean–atmosphere interaction, (3) studies of the interbasin exchange, (4) quantitative estimates of large-scale exchange of buoyancy and chemical components between the upper boundary layer and deep layers of the ocean, (5) estimates of oceanic heat transport and heat supply within the problem of the Earth's heat balance, (6) estimation of large-scale transport of passive tracers, (7) identification of the processes and balances which are important as factors of the oceanic general circulation variability and (8) improvements to numerical models for oceanic general circulation diagnostics and forecasting.

In connection with planning the optimal long-term observational system, the following problems are of the first priority: (1) identification of the most substantial parameters and indices and (2) development of an economical operational system of observations to describe the oceanic general circulation based on the combined use of conventional and satellite observation.

Theoretical and instrumental developments realized during the last few decades have created the real basis for planning the WOCE observational programme. A better understanding of the nature of oceanic general circulation makes it possible to make more reliable formulations of the requirements for the spatial and temporal resolution of data which can be obtained with existing instrumentation. Available numerical oceanic general circulation models (OGCMs) and high-speed computers will make possible adequate data processing and interpretation.

In the realization of WOCE, the use of satellite observational means, characterized in Table 4.10, plays the key role. As can be seen, the principal contribution to obtaining information about the ocean is made by both passive and active sounding in the microwave range. So, for example, the French–American satellite TOPEX/POSEIDON carries the US two-channel (5.3 and 13.6 GHz) altimeter and the French (13.5 GHz) radar altimeter, which operates with an accuracy of a few centimetres. An adequate interpretation of satellite data is possible, however, only with available data of control sub-satellite observations from ships and buoys.

The programme of regular observations foresees the priority of obtaining data on the following parameters: sea surface temperature, wind speed vector near the surface (or wind shear), atmospheric pressure at the surface level, air surface temperature and humidity, precipitation and the shortwave and longwave radiation budget. An assimilation of all this information will be achieved with the use of OGCMs.

From the viewpoint of studying the processes in the upper oceanic layer and the atmosphere–ocean interaction, the following four zones can be identified, each being characterized by special features: (1) equatorial regions, (2) subtropics, (3) sub-arctic and sub-antarctic regions and (4) high latitudes. In addition, sub-arctic and sub-

**Table 4.10** Satellite observational means needed for the World Ocean Circulation Experiment

| Satellite | Measuring complex |
|---|---|
| 1986 to 1992 | |
| TOPEX/POSEIDON | Radar-altimeter |
| ERS-1 | Side-looking radar (SLR), scatterometer, radar-altimeter, scanning IR radiometer |
| NROSS | Scatterometer, radar-altimeter, microwave radiometer to obtain images, IR radiometer |
| ERS-1 | Visible and IR scanning radiometer (VIRSR), microwave scanning radiometer |
| 1992 to 1995 | |
| MOS-2 | Scatterometer, radar-altimeter, VIRSR |
| ERS-2 | SLR, scatterometer, radar-altimeter, IR radiometer |
| RADARSCAT | SLR, scatterometer |

antarctic frontal zones require special consideration, as do regions of currents near the eastern boundaries of the continents, where the wind shear vortex is converted.

In each of these zones, specialized (problem-oriented) complex observational programmes should be set up. Universal problems for all regions are as follows: (1) balances of heat, mass and other characteristics, (2) transformation of water masses and (3) heat transport along the meridians and by the boundary currents, abyssal circulation, interaction between gyres and boundary currents, interaction between gyres, etc.

The provision of high-speed computers to solve problems of processing, archiving and analysis of observational data is of principal importance for the accomplishment of the WOCE programme. This would make it possible to work on eddy-resolving OGCMs, based on the use of the complete system of equations, on the scale of basins of individual oceans. It will then be possible to use global eddy-resolving (one-eighth degree) models, to include thermodynamical processes, but with low vertical resolution.

Of great interest are studies of the effect of the atmosphere on oceanic circulation. Wind shear near the ocean surface determines both the drift in the upper layer several tens of metres deep, directed to the right with respect to wind direction (Eckman circulation), and also deeper currents on the scale of the whole ocean (similar to the Antarctic circumpolar or to sub-tropical gyres) and is the principal forcing mechanism of circulation. In addition, the heat exchange between the atmosphere and the ocean, as well as changes in salinity under the influence of evaporation and precipitation, lead to variations in water density, which, in its turn, causes the formation of the thermohaline circulation. The latter manifests itself in the sinking of dense cold waters in high latitudes and in the horizontal transport under the warm-water layer to low latitudes. The presence of cold dense waters in deep layers of the ocean determines a very long period (about a thousand years) of the cycle of these abyssal waters, which apparently can be driven up to the surface only through mechanical vertical mixing. Much attention has recently been paid to the mechanism of exchange in the upper 1 km layer of the ocean due to mesoscale (synoptic) vortices.

Major aspects of the impact on the ocean of possible global climate changes, caused by increasing $CO_2$, are connected with revealing the oceanic response to variations in wind stress at the surface, air temperature and freshwater input (precipitation minus evaporation), occurring in different latitudinal belts. It is necessary here to take into account the effect of the ocean on the atmosphere, manifested through heat exchange and ice cover variability.

Characteristic time scales of the response of the global ocean system to the forcing effects vary widely. The slowest variations (characteristic scales of thousands of years) are connected with equalizing the properties of deep abyssal waters. Present ideas of the spectrum of climatic disturbances suggest that the distribution of the properties of deep oceanic waters never balances with climate. Hence, when studying the state of abyssal waters to calibrate climate models, it is necessary to include a time delay and to assume that the inert interior of the ocean has existed for centuries.

The present understanding of the dynamics of deep oceanic layers points to the fact that the oceanic general circulation component connected with the rate of formation of the deepest waters corresponds to the characteristic time of initial mechanical transfer processes on a scale of several decades, after which the mode of transfer processes prevails. This mode has been poorly studied, but probable scenarios of the development of these processes testify to the possibility of substantial climatic disturbances determined by this mode, especially in the northern Atlantic Ocean.

Possible effects of climate changes on global oceanic circulation include a 40 per cent attenuation of wind shear, combined with a several-degree warming and a possible decrease of the upper layer salinity in high latitudes. The consequences would be an attenuation of water circulation and a decrease of the slope of the constant-density surfaces, compared with that observed at present. During the transfer period, in the range of hundreds of years, surface waters, cooling in high latitudes, are expected to sink to intermediate depths (but not to form deep layers of the ocean), and the movement of surface waters from low latitudes, compensating for this process, will be changed, which will lead to a redistribution of heat transport and bioproductive components. All these processes can be very substantial in the Atlantic Ocean, where the fluxes across the equator are characterized by the features mentioned above.

The fundamentally important problem in the context of studying processes in the world's oceans is a development of climate models with adequate simulation of the interaction between the ocean and the atmosphere. Significant progress has recently been achieved in this field (Meehl and Washington 1986, 1988, Houghton *et al.* 1990, Meleshko 1991, Tucker 1991, Meehl 1991, 1992a,b, 1993, Cubasch *et al.* 1992, Trenberth 1992, Meehl *et al.* 1993a,b, Mitchell *et al.* 1995, Murthy 1995).

A significant motivation was a necessity to simulate the El Niño / Southern Oscillation (ENSO) phenomenon and its impact on the global climate. The atmospheric Southern Oscillation (SO) has been connected with the seasonally recurring ocean El Niño phenomenon by Bjerknes (1969) using dynamic and thermodynamic considerations. The implications of this El Niño / Southern Oscillation coupled ocean–atmosphere oscillation for tropical and global climate have led to many studies in recent years (e.g. Deser and Wallace 1987, Sud *et al.* 1991, Keppenne and Ghil 1992); its effects are also thought to be apparent in the winter in the northern hemisphere (Barnston *et al.* 1991). The simulation of the ENSO has become possible only on the basis of coupled models for the atmosphere–ocean system. A successful study of this kind has been accomplished by Latif *et al.* (1993) with the use of a coupled general circulation model (CGCM) developed in the Max Planck Institut für Meteorologie (Hamburg, Germany) to simulate long-term climate variability. Numerical modelling for a time period of 26 yr has been made on the basis of the utilization of a combined model consisting of a high-resolution ocean general circulation model (OGCM) and a low-resolution atmospheric general circulation model (AGCM). The OGCM is global over longitude (with 5° resolution) and covers a latitude range of $\pm 70°$. Latitude resolution is high in the tropical latitudes 10° S to 10 °N (0.5°) to simulate equatorial waves and the narrow band of equatorial upwelling, but then it decreases up to 1° at $\pm 20°$ latitude and 5° at $\pm 30°$ latitude.

A 17-level ocean model (to the bottom with prescribed topography) has been considered, with 10 levels embracing the upper 300 m layer. The ocean model is dynamically active only within the latitude belt $\pm 30°$. The atmospheric general circulation model is a simplified version of the 19-level spectral (T-21) model developed in the European Centre for Medium-Range Weather Forecasting (ECMWF).

The coupled climate model has been applied to simulate climate and its interannual variability in the tropical belt of the Pacific Ocean. The analysis of the results indicates the model's ability to simulate climate variability including annual and interannual variations, as well as irregular ENSO oscillations with prevailing periodicity of about 3 yr. The calculations revealed a key role of feedback under various time scales in climatic drift, annual and interannual variations. Although there is a certain similarity in the role of feedback in simple and sophisticated coupled models, the latter case is characterized by much more complex climate variability. An important result is that the presence or absence of interannual variability does not depend on climatic drift.

Significant features of the coupled general circulation model results are underestimation of trade wind intensity and, therefore, of equatorial upwelling, as well as exceedingly high sensitivity of the model's results to the intensity of equatorial upwelling. It is also very important that reliable simulation of the mean climate does not guarantee adequate simulation of an interannual variability which, in a number of cases, is the result of the mutual compensation of modelling errors. An important conclusion is that interannual variability is connected with two different interactive modes which may appear both separately and in combination: the 'delayed impact oscillator' and the 'slow varying sea surface temperature mode'.

A detailed survey of global coupled models (atmosphere–ocean–sea-ice) of varying complexity, with special emphasis on coupling strategies, has been published by Meehl (1992b). Among the basic phenomena considered as products of coupled interaction are: ENSO, low-frequency variability and the thermohaline circulation. In a later paper Meehl (1993) discussed the role of the ocean as a coupled air–sea biennial mechanism in the tropical Indian and Pacific regions and pointed out that variations in the upper-ocean heat content are part of a mechanism to explain biennial signals in the regions considered. The results suggest that changes in upper-ocean heat content contribute to the persistence of sea surface temperature anomalies important to the biennial mechanism, that both the Indian and Pacific oceans are actively involved in the ENSO and that the ENSO could be an amplification of the biennial cycle. The differential impact of 1984 and 1950 sea surface temperatures on rainfall in the Sahel has been modelled by Druyan and Hastenrath (1991).

A new important aspect of studying processes in the ocean is its biodynamics. The discovery of the ocean as a source of dimethyl sulphide (DMS) and carbonyl sulphide (OCS), which then may transform into aerosols through gas-to-particle conversion processes, results in the necessity to study oceanic input to climate change via biogenic production of atmospheric aerosols. This shows that the main biospheric component of the climatic system is also very important and so the problem of marine bioproductivity acquires some significance from the viewpoint of climate change apart from its role in ocean biology.

Deepening understanding of the coupled nature of climate change and biogeophysical cycles of minor gaseous components of the atmosphere, such as $CO_2$, CO, $N_2O$, DMS and OCS, makes it important to develop numerical climate modelling taking into account the possibility of using satellite remote-sensing data on relevant cycles. In this context Erickson and Eaton (1993) have considered the possibility of coupling computed geophysical fields from a three-dimensional general circulation model with the Coastal Zone Color Scanner (CZCS) satellite data on chlorophyll content of surface ocean waters to assess the photochemical lability of dissolved organic matter in the surface ocean. The areas with high chlorophyll estimates, as discerned from the CZCS data, are assumed to have high photochemical lability. Monthly estimates are made of the global ocean-to-atmosphere flux of carbonyl sulphide, with 2.8° latitude by 2.8° longitude spatial resolution on the basis of daily calculations every 24 hours for 5 yr.

The ocean-to-atmosphere flux of OCS can be computed via

$$F_{OCS} = K_{OCS}([OCS]_{so} - [OCS]_{eq}) \tag{4.17}$$

where $F_{OCS}$ is the flux of OCS from the ocean to the atmosphere, $K_{OCS}$ is the transfer velocity for OCS, $[OCS]_{so}$ is the surface OCS concentration and $[OCS]_{eq}$ is the expected oceanic OCS concentration in equilibrium with a 500 pptv atmosphere. It is assumed that the $[OCS]_{eq}$ is 10 $\mu$mol l$^{-1}$. The $[OCS]_{so}$ is computed from:

$$[OCS]_{so} = B[OCS]_{max} \tag{4.18}$$

where $[OCS]_{max}$ is the maximum potential concentration of OCS in the surface ocean based on a linear relation with the CZCS data:

$$[OCS]_{max} = 115[CHL] + 25 \tag{4.19}$$

where [CHL] is the retrieved chlorophyll concentration from the CZCS data. $[OCS]_{max}$ represents the maximum supply of organosulphur compounds in the surface ocean that are available for photo-oxidation. The value of $B = R_{sfc}/R_{max}$, where $R_{sfc}$ is the monthly mean-solar global radiation at every 2.8° latitude by 2.8° longitude region. $R_{max} = 375\ \mathrm{W\,m^{-2}}$ for the five seasonal cycles considered.

Erickson and Eaton (1993) note that the approach suggested allows the labile organic matter in the surface waters, as estimated from CZCS data, to be photo-oxidized in a way that is related to the amount of solar energy impacting the surface ocean.

The NCAR CCM1 general circulation model provides the surface radiation, surface wind speed and sea surface temperature every 24 hours for 5 yr of seasonal cycle runs. Year 4 of the integration is used for the OCS flux calculations. The results show that in January the concentrations vary from $10\,\mu mol\,l^{-1}$ in oligotrophic low solar energy flux areas to over $100\,\mu mol\,l^{-1}$ in highly productive regions that receive high surface insolation. In July, OCS concentrations greater than $75\,\mu mol\,l^{-1}$ are predicted for much of the North Atlantic and North Pacific. The surface ocean OCS concentration was never allowed to exceed $200\,\mu mol\,l^{-1}$ on account of the CZCS data being suspect at high chlorophyll concentrations because of saturation of the instrument.

The global annual area weighted mean surface ocean OCS concentration is $27\,\mu mol\,l^{-1}$, in close agreement with the average world ocean value of $32\,\mu mol\,l^{-1}$ (*in situ* observations). The analysis of the global ocean source terms, computed by Erickson and Eaton (1993) with general circulation model output and satellite data, shows that the values vary from just below zero (the ocean may be an OCS sink in some winter hemisphere regions) to 1500 nmol OCS $cm^{-2}\,h^{-1}$. Large areas of the North Atlantic and North Pacific have OCS fluxes in excess of 1000 nmol OCS $cm^{-2}\,h^{-1}$. The global area integrated flux is 0.25 Tg S (as OCS) $yr^{-1}$. This may be a lower limit of the global ocean OCS source.

Erickson and Eaton (1993) have pointed out that a similar approach may be used to study other global biogeochemical cycles, although in the cases of DMS and $CO_2$ there may be a much more complicated relation with CZCS data than in the case of OCS.

The primary bioproductivity of all the oceans except in the Antarctic is determined by nutrients rising from deep waters to the euphotic zone (the upper layer, where light from the Sun penetrates). This input is regulated by wind-driven mixing. Climate changes which affect oceanic circulation therefore influence the conditions for bioproductivity. Though the total effect on protein extraction (mainly by fishing) is difficult to assess, it may be supposed that an attenuation of mixing, and hence of primary bioproductivity, will lead to undesired consequences. However, in the tropical belt, intensified evaporation can strengthen the mixing, and the situation may reverse.

## 4.3 The Arctic, the Antarctic and climate

### 4.3.1 Introduction

There have been many studies of the climate-forming processes in the Arctic, whose principal feature is determined by the isolation of the ocean from the atmosphere by the ice cover (American Meteorological Society 1995a, Nagurny 1986, US Arctic Research Commission 1989, Kondratyev *et al.* 1996a). In this connection, of special interest is the programme of studies in the Arctic prepared by the US Interagency Committee on the Arctic Study Policy. This programme is based on the Act on the Arctic Study Policy approved by the US Congress in 1984. According to this Act: (1) the Arctic is of critical importance for national defence, (2) the Arctic is a natural laboratory to study the effect of extreme climatic conditions on man, which may play an important role in the solution of defence problems, (3) special atmospheric conditions in the Arctic determine its unique importance for testing telecommunication systems in high latitudes, which is extremely important from the viewpoint of defence problems and (4) developments of the arctic marine technologies are very interesting for economical extraction and transportation of energy resources, as well as for national defence. According to the Act, the Arctic is defined as the territories north of the Polar circle which belong to the USA and other countries, as well as all the US territories located north and west of the boundary formed by the rivers Porcupine, Yukon and Cuzcoquim, and by the adjacent seas, including the Arctic Ocean, the Bering, Beaufort and Chukcha Seas, as well as the Aleutian Chain.

The document on National Security Measures outlines the following aspects of US policy with respect to Arctic studies: (1) protection of vital interests in the Arctic region, connected with US security, including an easy access to the seas and adjacent air space, (2) favourable conditions for regional development, with a minimum negative effect on the environment, (3) support to field studies favouring either a better understanding of Arctic environmental conditions or the solution of other problems which can be most successfully realized in Arctic conditions and (4) favourable conditions for mutually beneficial international cooperation to solve the problems mentioned.

In this connection, the programme of the Arctic studies comprises three principal parts: (1) national security, (2) rational economics with minimum environmental impacts and (3) studies of phenomena and processes in the Arctic ('The Arctic as a natural laboratory'). Studies are planned in the field of meteorology and atmospheric physics (including the upper atmosphere), oceanography, permafrost, geology, medicine and engineering. Studies of the upper atmosphere largely concern the physics of the ionosphere and aurora.

A considerable part of the studies is dedicated to the conditions of operation of electro-optical systems in the Arctic. Serious effort will be made to study the Arctic haze and its impact on climate, as well as the gas and aerosol composition of the atmosphere in the regions of the broken polar tropopause, in order to develop an interactive climate model of the atmosphere–ocean–cryosphere system. Oceanographic studies will be focused on studies of underwater acoustics, processes in the zone of the polar ice edge, in the regions of polynyas, etc. Information obtained from the US defence meteorological satellites would be very useful.

The WMO discussions (World Meteorological Organization 1988a–f, 1989a–c, 1991a,b) on global climate modelling and the documents of the WCRP Joint Scientific Committee have emphasized that information has been accumulated on the effect of the Arctic Ocean on the global climate system, though, so far, different physical processes determining the ocean–atmosphere–ice-cover interaction remain poorly studied. In this connection, further realization of programmes of field experiments and monitoring is necessary as well as numerical modelling to study the processes of formation in the Arctic Ocean of stratification and circulation of water masses, advection of energy and fresh water, ice cover dynamics and overcast cloudiness–radiation interaction (Marchuk *et al.* 1986, Lappo *et al.* 1987, 1989, 1990, Matishov 1987, Kondratyev and Kotliakov 1991, Kondratyev 1992b, American Meteorological Society 1995a, IASC 1996). For these reasons, it was suggested that a specialized programme 'Study of the Arctic Climate System' be launched.

Of great interest is the dynamics of the ice sheets of Greenland and the Antarctic; the glaciers of Greenland can be the most sensitive indicator of global climate changes (US Department of Energy 1984). The problem of the gas-to-particle formation of condensation nuclei, and their effect on cloud cover, is full of uncertainties (Charlson *et al.* 1987, Kvenvolden 1988). The problem of the powerful enhancement of energy-exchange and mass-exchange in the polynyas and leads areas is still urgent (Nagurny 1986, Romanov *et al.* 1987, Romanov 1988, Meleshko 1991, Meleshko *et al.* 1991, Glowienska-Hense and Hense 1992). Special attention should be paid to the processes of formation of bottom waters in the high-latitude North Atlantic and, in this connection, the role of the Arctic Ocean in the global carbon cycle.

In the context of the problem 'Arctic and Climate', of great interest is the Programme to Study the Climate of Polar Regions developed in the Institute of the Arctic and Antarctic (1989) as well as programme developments such as Arctic Research for an Arctic Nation (US Arctic Research Commission 1989) and the Nansen Centennial Arctic Programme dedicated to the centenary of the expedition on *Fram* (Norwegian Research Council 1992).

In the conditions of the Arctic, the principal manifestations of the climatic effects of the processes in the polar oceans include: (1) the effect of heat transport in the ocean across the Greenland–Spitsbergen passage on the variability of pack ice, (2) the possible effect of the diversion of part of the Siberian rivers run-off on the fresh water influx to the Arctic basin and subsequent transformation of sea water stratification, (3) the impact of the ocean and pack ice on the formation and variability of summertime overcast stratus clouds, (4) intensification of climate warming in high latitudes due to increasing $CO_2$ concentrations and its possible consequences (more intensive $CO_2$ release by the ocean, the effect on ice melting, etc.), (5) cyclogenesis near the edge of the polar ice cover and (6) the influence of the atmosphere–ocean interaction in the Greenland–Norwegian Seas on the formation and variability of the North Atlantic deep waters (the state of balance between the output and input of these waters along the eastern coastline of the Atlantic; the impact of the Gulf Stream on Europe).

The unsolved problems of the Antarctic climate are as follows: (1) heat balance of the continent and adjacent seas (it remains unclear whether the heat flux can be directed northward from the Atlantic countercurrent in the southern Atlantic and southward in the Pacific Ocean), (2) the role of the meridional heat transport as a possible factor that determines the inadequacy of the classical concept of the oceanic circulation and affects the climate (here of importance is the interconnection of heat transport and sea surface temperature anomalies determined by the prevailing role of strong variations in the extent and concentration of the polar ice cover with its high albedo and heat-isolating effect in the formation of heat balance and climate: in the period of maximum development of the antarctic ice the size of the continent practically doubles, while in the Arctic the amplitude of the interannual ice cover variability reaches 50 per cent of the annual mean value), (3) the connection of sea surface temperature anomalies south of New Zealand with the ENSO, (4) the effect of the southern hemisphere oceans on the poleward heat transport in the atmosphere, (5) cyclogenesis near the ice cover edge and near the oceanic polar front, (6) fast 'barotropic' response of the antarctic countercurrent to changes in the zonal wind field and its connection with climate and (7) contribution of the 6-month oscillations in the atmosphere and ocean to climate variability. The factors of variability of the extent and concentration of the polar ice cover and the ice transport by sea currents have been studied inadequately, so far (in the Arctic, for example, the ice output from the Arctic basin east of Greenland is as important from the viewpoint of the formation of the regional heat balance as the heat transported by sea currents).

To solve the enumerated (and other) unsolved problems, first of all, it is necessary to expand the programme of observations in high latitudes, especially long-term observations in the key regions and special problem-orientated complex programmes (the atmosphere–ocean interaction at the polar ice edge, monitoring the properties of sea water under the ice cover and the ocean level, the meridional heat transport, etc.). For this purpose much effort is needed to improve the observational means available and to develop new ones (especially radar altimetry and acoustic tomography). Study of the climatic effect of the cryospheric properties (sea ice) and processes taking place there is one of the WCRP's first priorities (Houghton 1984, Houghton *et al.* 1990, Kondratyev 1992b, Rind *et al.* 1995). The most important problem is to develop the coupled climate models that would take into account the processes in the cryosphere, since: (1) the strong annual change and interannual variability of the ice cover are followed by respective variations in surface albedo as well as heat-exchange and water-exchange between the ocean and the atmosphere, (2) the ice cover dynamics intensifies the variability of the temperature regime, which is the result of a substantial positive albedo feedback in the global climate system and (3) the ice cover is an important indicator of climatic variability (in particular, that of anthropogenic origin).

The following four directions of research are of primary importance: (1) the establishment of more reliable techniques to parametrize the processes in the cryosphere (changes in albedo, heat-exchange and moisture-exchange, consideration of the contribution of polynyas and leads; cloud–radiation interaction, etc.), (2) study of interaction in the ocean–ice–atmosphere system at the ice cover edge, and the adaptation of climate theory to take account of this interaction, (3) analysis of the sensitivity of climate models to realistic changes in the extent and structure of the ice cover and (4) continuous monitoring of the ice cover characteristics from satellites, and especially at the ice cover edge. Apparently, of key importance is an account of sea ice dynamics in the Antarctic, characterized by the strongest seasonal variability.

Of particular interest are the following events in the Antarctic: (1) circulation in the Weddell Sea, (2) sub-ice sea stratification and (3) antarctic bottom waters, the main source of which is the Weddell Sea. The most important problem is the development of the coupled climate models which take into account (parametrize) the processes in the sea ice, based on diagnostic analysis of the observational data.

Table 4.11 illustrates the requirements on observational data that characterize: (1) the annual change of the extent and concentration of the ice cover averaged over a large time period, as well as the interannual variability, (2) drifting of the antarctic sea ice, (3) the distribution of the ice cover thickness over the Antarctic, (4) distribution and relative share of 'pancake ice', (5) the field of sea current, (6) fields of temperature and pressure in the atmosphere and (7) the annual change of the vertical heat flux. Of great importance is a realization of problem-orientated field experiments to study the most important aspects of the effect of sea ice on climate. Such experiments should be planned on the basis of respective simulation numerical experiments.

The first approximate climate models had taken into account only the albedo feedback determined by the formation (or melting) of sea ice. In fact, the ice cover is

**Table 4.11** Requirements on sea ice observational data for numerical climate modelling

| Parameter | Error | Spatial resolution (km) | Temporal resolution |
|---|---|---|---|
| Ice cover edge | 10 km | 50–200 | Averaged over 3 days, 1 week, and 1 month |
| Share of open water | 5–10% | 50–200 | same |
| Type (thickness) of ice | 5 levels[a] | 50–200 | same |
| Ice drift | 10% drift for 1 month | 50–200 | same |

[a] Open water; 0–10 cm; 10–70 cm; 70–200 cm; >200 cm.

a deforming barrier which reacts to the atmospheric forcing (and, in its turn, affects the atmosphere), the processes of interaction among the ocean, ice and atmosphere being very complicated. The ice cover variability determines an important contribution to the advective heat-exchange and salt-exchange between polar regions and the rest of the globe.

At present it has become clear that the ice cover should not be treated in climate theory as a fixed surface with varying albedo (depending on the presence of continuous ice cover, thawed patches, etc.). Besides, of critical importance are such characteristics as the thickness and concentration of ice cover, which control the motion of ice due to the interaction with the atmosphere and ocean, and the formation and disintegration of ice cover.

The contrasting geographical conditions of the two polar regions determine their marked thermodynamic and dynamic special features. The antarctic sea ice is characterized by seasonal variability; its thickness constitutes about 1.5 m, and its extent suffers great annual changes (from $2.5 \times 10^6\,km^2$ to $20 \times 10^6\,km^2$). The external ice boundary is not confined by the continent (as in the Arctic), which results in considerable meridional advection and free exchange with the ocean. The Antarctic Circumpolar Current, which ensures water transport of about $130 \times 10^6\,m^3\,s^{-1}$, is much more intensive compared with the currents in the Arctic Ocean. The meridional heat transport across the Antartic Circumpolar Current zone reaches $0.28 \times 10^{15}$ W and is, apparently, determined by gyres and not a regular convection (as in the East-Greenland current). The major Antarctic Circumpolar Current component is the gyre in the Weddell Sea and it is responsible for at least 40 per cent of the total meridional heat transport in the southern hemisphere ocean.

Multi-year ice about 4 m thick prevails in the Arctic, with a longitudinal asymmetry of distribution due to specific contours of the continental coastline compressing the arctic ice. The latter determines a much smaller amplitude of the annual change of the ice cover extent (from $7 \times 10^6$ to $14 \times 10^6\,km^2$) and the zone of the ice edge is not affected by winds and currents as much as it is near the Antarctic.

Characteristic features of the ice cover and relevant processes in the Arctic and Antarctic determine the necessity of special studies of the role of ice in the formation of climate models. Taking into account (parametrizing) the processes that determine the dynamics (growth, disintegration, splitting and melting) of sea ice is a long-standing problem; of particular importance is a description of the processes in the regions of polynyas and leads, as well as near the ice edge.

Estimates have shown that in the Arctic about half the heat release from the ocean to the atmosphere takes place in winter through polynyas, leads and ice less than 0.8 m thick, with total area less than 10 per cent of the whole ice cover extent. An adequate parametrization of the processes in the atmospheric boundary layer and the ocean is of critical importance. The turbulent heat exchange in the upper layer of the ocean is a major factor of the thermal dynamics and the dynamics of the ice-covered ocean. In particular, the ocean is seriously affected by salt release resulting from ice formation and causing variations in the upper ocean layer stability and advection of salt added to the ocean in the regions of ice formation and removed in other regions where the ice melts.

A major aspect of the problem of the impact of sea ice on climate is the heat exchange at the atmosphere–ice–ocean system interface. The decisive effect on the annual change of the ocean upper layer temperature is made by the latent heat resulting from ice formation or melting, the buoyant flow connected with salt exchange, and the atmosphere-to-ocean momentum transfer across the ice cover.

An upper limit on the climatic consequence of the melting of sea ice is achieved if all the sea ice on the planet were to melt; this limit has been estimated by Covey *et al.* (1991) using two versions of the National Center for Atmospheric Research Community Climate Model and by Rind and Chandler (1991) using the Goddard Institute for Space Studies (GISS) general circulation model. Rind and Chandler's work predicted a $+2\,^{\circ}$C rise in mean global surface temperature, with $+20\,^{\circ}$C in high latitudes and $+1\,^{\circ}$C rise near the equator.

Apparently, an important factor of the ice cover dynamics in the Antarctic is a still-poorly-studied heat input from the ocean, reaching $30\,\mathrm{W\,m^{-2}}$ (this can be a reason for a rapid disintegration of the ice cover in spring). In the arctic conditions, where the strong stability of stratification suppresses the vertical exchange, the role of the heat flux from the ocean is probably negligible.

The emphasis has been placed on such inadequately studied processes and phenomena near the ice cover edge as upwelling, convection, fronts, gyres and geostrophic flux along the edge. The construction of the coupled climate models requires a reliable substantiation of the parametrization of the ice cover dynamics taking account of ice rheology. A substantial contribution to the heat exchange in the ocean is made by ice advection, particularly strong near the ice cover edge.

The use of atmospheric general circulation models to describe the climate of polar regions revealed considerable differences from the observational data, characterized by: (1) the surface air temperature overestimated by $\lesssim 20\,^{\circ}$C in both hemispheres, especially in winter and (2) eddy heat flux towards the surface overestimated by $\lesssim 50\,\mathrm{W\,m^{-2}}$. In the models with prescribed ice cover, the arctic ice is often too thin ($\lesssim 1$ m) and the extent of the antartic ice is underestimated (there is no ice in summer and in winter its thickness is $\lesssim 50$ cm).

Apparently, major reasons for the differences are: (1) overestimated summertime total radiation due to the underestimated cloud cover, (2) overestimated downward sensible heat flux in winter and (3) overestimated diffusional horizontal heat flux in the ocean and underestimated Eckman transport to the equator across the 60° S parallel (the latter refers to the models which take into account the oceanic circulation). Before introducing the improved techniques of parametrization, it is expedient to adjust the AGCMs available, to reach an agreement with the observed extent and thickness of the ice cover. Apparently, the most serious defects of parametrization are the neglecting of the role of polynyas and leads in the Antarctic and the ice dynamics in the Arctic. To reveal the model components responsible for differences, a comparison with observational data on the surface heat balance is very important. From the viewpoint of the oceanic general circulation models, a description of heat flux towards the lower boundary of the ice cover in the Antarctic, controlled by the speed of upwelling and (or) the input of the warmer waters of halocline to the mixed layer, is very important. If these processes are determined by penetrating eddy mixing, a model will be needed of the mixed layer of varying thickness with buoyant fluxes depending on salinity, which is determined by the processes of melting and freezing. The problem of parametrization of the processes in the zones of leads and polynyas remains unsolved.

Simple thermodynamic models of a homogeneous ice layer should be specified, bearing in mind: (1) the ice dynamics (for example, the output of ice from the Arctic across the Greenland Sea, and the Eckman divergence round the Antarctic) and (2) the varying share of leads in each cell of the spatial grid.

The coupled AGCM/OGCM models can be tested by: (1) testing the reliability of calculated annual change, (2) comparing with data on the interannual variability and (3) simulating the annual change of palaeoclimate in the last glacial period (18 thousand years ago).

The most important impacts of sea ice on climate revealed through numerical modelling are as follows: (1) maximum climate warming with an increasing $CO_2$ concentration in the arctic winter due to the heat input from the ocean through the thinner (as a result of warming) sea ice, (2) the effect of albedo of the more extended sea ice 18 thousand years ago on the surface air temperature, comparable with the impact of continental glaciers and (3) possible reversal of the conventional relation between the amplitude of the annual change of temperature and depth of the oceanic mixed layer with the sea ice dynamics taken into account, since a thinner mixed layer favours the strengthening of the wintertime sea ice, which causes a delay of the springtime melting and results in a cooler summer. Recent data from numerical modelling generally confirmed these conclusions (Houghton *et al.* 1992).

A great contribution to understanding the problem under discussion has been made by the international programme MIZEX (MIZEX 1989) which is aimed at studies of the processes of mesoscale interaction among the atmosphere, ocean and ice cover in the marginal ice zones near the sea ice edge, since the marginal ice zones play a key role as factors determining the processes in the mid-latitude ocean and climate. The programme foresaw a study of such processes in two regions: in the Greenland Sea (Fram Strait), in deep waters, and on the continental shelf of the Bering Sea, with emphasis on observations in the Fram Strait since (1) this region contributes most to water-exchange and heat-exchange as well as the exchange of other materials between the North Atlantic and Arctic Oceans and (2) in this region various observational programmes have already been realized.

Analysis of the data available has shown that the processes of interaction of key importance prevail in the periods from mid-July to late August and from mid-February to late March. Therefore two observational experiments were undertaken, in summer (1983) and in winter (1985), each lasting 6 weeks during the respective period. The observational site (near the polar ice edge) covered a square of 200 km by 200 km with an equal area of ice cover and open water. The size of this site was large enough to monitor the most substantial processes in the marginal zone. So, for example, the longwave oceanic swell can penetrate the ice cover at a distance of 100 km causing ice crash near the edge within a band several tens of kilometres wide.

The thickness and size of the ice fields differ strongly from respective characteristics of the inner part of the ice cover, at least, at a distance of about 100 km from the edge. The diameters of the gyres marked by crashed ice are from 10 km to 60 km. A considerable transformation of the atmospheric boundary layer takes place within a distance of 30 km from the edge. The wind-driven motion of the ice takes place at a distance of 100 km from the edge.

The choice of a 6-week interval of observations ensures a representative coverage of the time scales of the processes in the atmosphere, ocean and ice cover. During this period several cyclones can pass across the test area, the gyres can form and disintegrate, oceanic fronts can deform and upwellings and near-edge currents can form.

The observational complex foresaw the use of a drifting station at the site, ships (ice-breaker, research ship of an ice-breaker type, conventional research ship), buoys, aircraft and satellites. Ships and aircraft played a decisive role in the accomplishment of the programme of observations of the atmosphere, ocean and ice cover (Table 4.12). In the sub-programme of meteorological observations (over pack ice and open water) the emphasis was placed on the determination of wind stress over the ice and measurements of the surface heat balance components, as well as the lidar profiling to retrieve the surface topography. The most important objects of the oceanographic sub-programme were gyres (their properties and dynamics), oceanic fronts, upwellings, tidal and inertial oscillations, small-scale processes (destruction of internal waves, convection), and roughness. In the ice cover observations the first priority is given to its dynamics and thermodynamics (inner stress, inertial effects, temperature field) and small-scale processes near the edge.

The programme of observations in the Bering Sea has many similar features but also specific features determined by the following: (1) small depths (20–150 m), (2) a weak effect of sea currents in the marginal ice zone (in contrast to the Greenland Sea) and (3) prevailing 1-yr ice (in contrast to the multi-year ice in the test area in the Fram Strait). The wintertime increase of ice cover in December–February with a mximum in March, and the springtime disintegration of ice in April–June characterize the dynamics of the ice cover.

One of the important and still unstudied features of the processes in the marginal ice zone is an appearance of divergence of the compact ice cover which leads to the deformation of the alternating bands of ice and water. In this connection, two major components of the summer (1985) and winter (1983) experiments were: (1) the study of ice dynamics in the original ice zone and (2) the observation of ice formation in the leeward polynyas. Analysis of satellite images of the ice cover has shown that the ice forms in the Bering Sea, as a rule, within several relatively small polynyas, where an extremely intensive heat exchange takes place between the water surface (releasing the heat) and the atmosphere with subsequent formation of the ice cover.

In late October–November the atmospheric circulation of rapidly-moving dry air masses, which forms in the Arctic, hinders the transport of water vapour and of industrial wastes to high latitudes; this causes the processes of atmospheric cleansing to be slowed down. In this period the amount of low cloud, which can cause the washing out of aerosol from the atmosphere, is small and the processes of dry deposition on the surface are impaired because of the formation of strong inver-

**Table 4.12** Basic parameters of ice cover and ocean, remotely sensed with airborne instruments (1, primary instruments; 2, auxiliary instruments)

| Parameter | Side-looking radar | Radar-altimeter | Microwave radiometer | Scatterometer | Lidar profilometer | Wind sensor | Camera | IR scanner | IR profilometer | Thermo-sondes |
|---|---|---|---|---|---|---|---|---|---|---|
| Ice cover concentration | 2 | 2 | 1 | – | – | – | 2 | – | – | – |
| Type of ice | 2 | – | 1 | 1 | – | – | 2 | – | – | – |
| Wind over ice | – | – | – | – | – | 1 | – | – | – | – |
| Wind over water / Roughness | – | 1 | – | 1 | – | 1 | – | – | – | – |
| SST | 1 | 1 | – | 1 | 2 | – | – | – | – | – |
| Temperature profile in the ocean | – | – | – | – | – | – | – | – | – | – |

sions. The inefficiency of the mechanisms of cleansing explains an accumulation of the polluting aerosol components.

As the data from surface observations discussed by Shaw (1988, 1989a, 1995) and Shaw *et al.* (1993) have shown, the aerosol concentration reaches its maximum in the period middle to end of winter. Only in April–May 1976 were the first aircraft aerosol observations made in the region of Point Barrow (Alaska) by aerosol sampling, whose analysis showed that the aerosol consisted mainly of soil particles from deserts of east China and Mongolia.

Therefore the conclusion has been reached that the arctic aerosol is apparently of natural origin. The subsequent analysis of ground-based aerosol samples taken beginning from the autumn of 1976 has given quite different results. These samples were characterized by higher contents of aluminium, vanadium and magnesium, compared with those from the spring of 1976, and are similar to the urban aerosol samples, which suggested the mainly anthropogenic origin of the arctic aerosol. Probably, by the spring of 1976 the anthropogenic aerosol had been removed from the atmosphere due to the processes of cleansing (as always in spring), and at the same time an unusual intrusion of the soil aerosol from the Asian deserts had taken place. Important indicators of the anthropogenic origin of aerosol are the soot component in it as well as the ratio of concentrations of vanadium and aluminium. If the vanadium to aluminium ratio corresponds to soil particles, the aerosol is deemed to be of natural origin. If the observed ratio of vanadium to aluminium exceeds that corresponding to soil aerosol by a factor of 1.5, this can be taken as the criterion for assuming a substantial contribution of the anthropogenic aerosol component.

The appearance after 1976 of a network in the Arctic to sample aerosols had led in 1981 to the following conclusions: (1) a considerable part of the aerosol haze observed in the wintertime Arctic consists of anthropogenic aerosol which comes, apparently, from Eurasia, (2) the formation of haze starts in late autumn and reaches its maximum in March–April, (3) the horizontal visibility in the presence of haze decreases to 3–8 km, (4) the anthropogenic component of haze consists mainly of sulphates (with an average concentration of $2\,\mu g\,m^{-3}$), organic carbon ($1\,\mu g\,m^{-3}$) and soot carbon ($0.3$–$0.5\,\mu g\,m^{-3}$) and (5) the aerosol that reaches the Arctic has travelled great distances and during this travel the aerosol survives strong ageing.

In March–April 1982 complex aircraft studies were made of aerosol, atmospheric chemistry and radiative characteristics of haze, using the flying laboratory WP-3D Orion on extended routes (up to transpolar). Visual observations revealed not only the presence of a multi-layer haze but also the brown-orange colour of the surface of the central Arctic ice cover. The haze covered the whole polar ice cap and propagated up to the flight ceiling (8.4 km). Actinometric observations revealed considerable (not less than 10 per cent) absorption of shortwave radiation by haze. The bottom layer of the overcast haze was at a height of 3 km, and above it there were observed broken layers at different altitudes, which looked like bands. The concentration of soot carbon in the haze particles was less by a factor of only 3–4 than in the region of Denver (USA) and only by an order of magnitude less than in New York, while it exceeded by several orders of magnitude that observed over the Pacific Ocean. Shaw *et al.* (1993) showed that the major perturbation to the radiation budget is a lowering of the albedo (heating) of the Earth–atmosphere system (due to soot particles).

The presence of soot particles can be substantial from the viewpoint of their effect on the albedo of the snow–ice surface and cloud cover properties. This requires a thorough analysis of the results obtained from the viewpoint of the impact of the arctic haze on the regional and global climate. The effect of depositing aerosol on the ecologically sensitive biomes of the tundra can be substantial, too.

An important aspect of climate predictions is the future of polar ice sheets. Numerous studies have produced rather controversial results on this problem. This is equally true for both palaeoclimatic and present-day climate variations. Weller (1993) has pointed out that, at present, considerable uncertainty exists as to whether the ice sheets are growing or shrinking. Ledley and Chu (1994) have emphasized that the question of what mechanisms are required to produce ice sheet growth has now become especially important in the context of global climate warming.

Miller and de Vernal (1992) presented geological evidence which suggested that it may be possible for ice sheets to grow under the conditions of greenhouse warming because: (1) warmer temperatures will increase the water-carrying capacity of the atmosphere and thus precipitation and snowfall rates and (2) the greatest warming is predicted to occur during the winter, having a relatively small effect on summer melting. Ledley and Chu (1994) have examined this possibility by using a coupled energy balance climate system: the atmosphere over the land, the atmosphere over the ocean, a mixed layer ocean, and a ground layer.

In the two present-day experiments with different ice thickness, the minimum land ice albedo is set to 0.4 and the initial land ice thicknesses are set to 0 m and 10 m in all zones. In the doubling-$CO_2$ experiments the initial land ice thickness is 10 m in all zones and the minimum land ice albedo is set to 0.4, 0.55, 0.6 and 0.65 to simulate different possible land ice surface conditions. The effect of doubling $CO_2$ in the atmosphere is simulated by increasing the radiative forcing at the tropopause by $4\ W\,m^{-2}$. Each experiment is run to equilibrium in the surface air temperature (the criterion for equilibrium is that the temperature change is less than 0.1 °C from one year to the next). The parameters used for the simulation of the present-day climate have been described by Ledley (1993).

The results obtained by Ledley and Chu (1994) indicate that if atmospheric $CO_2$ is doubled and the minimum ice surface albedo is set to 0.4, the increase in the snowfall will not be enough by itself, or, with an initial ice sheet 10 m thick, to overcome the melting away of the snow and ice during the summer, and start ice sheet growth. This is because the increase in summer temperatures produces an increase in the melt rate, and a decrease in the fraction of precipitation that falls as snow. These two factors dominate the increase in the snowfall during the winter due to warmer temperatures, to eliminate the possibility of the growth of an ice sheet.

If, however, the minimum albedo of the ice surface is relatively high, of the order of 0.65, so that absorbed solar radiation is reduced enough to maintain summer air temperatures below the freezing point, and if some mechanism can be found initially to maintain summer ice, as was simulated by specifying an initial 10 m thick ice sheet, then it may be possible to induce ice sheet growth under conditions of doubling atmospheric $CO_2$. Ledley and Chu (1994) have emphasized that such a situation may be realizable and, thus, the growth of an ice sheet may occur in the case of doubled concentration of $CO_2$.

### 4.3.2 Sea ice and climate

The most important large-scale problems connected with studies of the impact of sea ice on climate and necessitating an accomplishment of large-scale observational programmes are as follows: (1) study of the balance of ice masses in a given region determined by contributions of advection and ice formation and (2) revealing the physical processes that control the ice cover extent in conditions when it is not limited by the effect of land (the autumn and wintertime growth of the extent is largely determined by the ice formation and advection, whereas the effect of the summertime melting is a major factor of the sea ice retreat).

Since the annual change of the ice cover extent in the Arctic is characterized by a sharp maximum and flat minimum, it can be supposed that the prevailing winter conditions all year round will not cause further increase of the ice cover extent (the extent is already quasi-balanced), but the setting of a hypothetical round-the-year summer regime of climate can lead to total melting of the arctic ice cover. The relation between advection and ice formation as factors determining the location of the south boundary of the ice cover remains unclear. Still less studied and very complicated are small-scale and mesoscale processes in the zone of the ice edge, where the boundary layers in both the atmosphere and the ocean are transformed.

The dynamics of the polar ice cover is an important factor in the formation of climate (Crane 1986): the arctic pack ice extent varies annually from $8 \times 10^6\,km^2$ in the summer up to $15 \times 10^6\,km^2$ in the winter, and in the southern hemisphere the amplitude of the annual change of ice cover is much larger (within $(2.5–20) \times 10^6\,km^2$). The spatial and temporal scales of the effect of polar ice dynamics on climate are extremely large – from synoptic to decadal and longer, embracing the whole globe. Here both positive and negative climatic feedbacks occur. The role of polar cloud dynamics is also important, their amount varies from approximately 40 per cent in the summer to 80–90 per cent in the winter. From the viewpoint of the effect on the ERB in high latitudes, outgoing longwave radiation plays the leading role, whereas in low latitudes and mid-latitudes the albedo feedback prevails.

The interaction of ice cover and cloud cover in the Arctic is especially important. It means, for example, that the decrease of the polar ice cover induced by climate warming will intensify evaporation and increase cloudiness, which compensates for the albedo decrease taking place with the change to open water; thus, the nature of the known albedo feedback is much more sophisticated than was thought to be the case.

Calculations based on a one-dimensional thermodynamic climate model have shown that in the regions where there is the greatest interannual variability of the ice cover extent, the predicted ice cover thickness depends strongly on cloud conditions; this means that cloud variations may cause substantial variations in the ice cover thickness and extent. Reliable quantitative estimates of the role of these and other processes need more adequate meteorological and oceanographic information for polar regions than is now available.

On the basis of a one-dimensional thermodynamic climate model, Curry and Ebert (1989) have assessed the possible effect of the increasing pollution of the arctic atmosphere which may lead to an increase of the condensation nuclei concentration and a subsequent increase (and decrease of droplet sizes) of the cloud droplet concentration, as well as a change in the ice cover net radiation, which, in its turn, will cause an increase of the ice cover thickness. On the other hand, the ice thickness is sensitive to the repeatability and optical thickness of the low-troposphere ice clouds appearing in the cold period of the year.

In continuation of previous studies Curry *et al.* (1993) have investigated the impact of clouds on the surface radiation budget of the Arctic Ocean. They have emphasized that owing to changes in surface albedo and temperature associated with changing cloud properties, there is a strong non-linearity between cloud properties and surface radiative fluxes. The principal findings are as follows: (1) no simple relation exists between top of the atmosphere radiative fluxes and surface radiative fluxes, (2) clouds have a net warming effect on the surface, in contrast to clouds at lower latitudes and (3) for annual-averaged cloud fractions less than 0.4, the cloud radiative forcing is positive (warming) over the entire annual cycle.

It is clear that the best means to solve the problems mentioned lies in remote sensing from satellites. As experience has shown, in conditions of high latitudes, passive and active sounding in the microwave region is particularly important. Microwave passive sounding has made it possible, for example, to detect, within pack ice, vast regions of either open water or thin ice (the size of the polynya in the Weddell Sea reaches 300 km by 1000 km and remains so during the whole polar night) (Kondratyev *et al.* 1996a).

As has been pointed out, studies of sea ice must be of central concern, since during its maximum development it covers almost 10 per cent of the world ocean's surface. The climatic impact of substantial seasonal and interannual variations in the sea ice extent is determined by both its high albedo and heat and moisture exchange between the atmosphere and the ocean (the latter is particularly strong in the presence of polynyas). Recent studies have led to the conclusion that the heat transport across the ice cover in the Arctic is more intensive than had been considered previously. An adequate climate theory must be based on an interactive consideration of the processes of formation and dynamics of the ice cover. From the viewpoint of numerical modelling of climate, analysis of the response of ice to climate changes and the effect of ice on climate is very important (Sergin and Oreshko 1987).

The cryosphere is an important component of the climate system (Henderson-Sellers and McGuffie 1984). This is testified to by the data in Table 4.13, which reflect the spatial and temporal variability of the cryospheric components. As can be seen, the sea ice, land snow and permafrost play the leading role among the cryospheric components. A typical feature of the cryosphere is the rapid variability of its components during the course of the year. So the extent of sea ice varies from $20 \times 10^6\,km^2$ in September to $2.5 \times 10^5\,km^2$ in March in the southern hemisphere and within $(15\text{–}8.4) \times 10^6\,km^2$ in the northern hemisphere.

The variability of snow cover is more substantial; the difference between maximum and minimum values reaches 80 per cent. Rapid changes in the snow cover and ice cover due to melting constitute one of the key factors of climate formation, whose major manifestation is the albedo feedback and the thermal insu-

**Table 4.13** The spatial and temporal scales of the cryospheric components

| Component | Area ($10^6\,km^2$) | Time (years) |
|---|---|---|
| Ice sheets | 16 | $10^3$–$10^5$ |
| Permafrost | 25 | $10^4$ |
| Mid-latitude glaciers | 0.35 | 10–$10^3$ |
| Sea ice | 23 | $10^{-1}$–10 |
| Snow cover | 19 | $10^{-2}$–$10^{-1}$ |

lation of the ocean. The effect of snow cover has been modelled in the Goddard Institute for Space Studies (GISS) general circulation model by Cohen and Rind (1991). This showed that the snow cover caused only a short-term local decrease in the surface temperature. In the surface energy budget, the reduction in absorbed shortwave radiation and the increased latent heat sink of melting snow contributed to lower temperatures. However, all the remaining heating terms contribute to increasing the net heating over a snow-covered surface. The results emphasized the negative feedback which limits the impact of snow cover anomalies over longer time scales. Snow cover on sea ice leads to several competing effects which are described by Ledley (1991b). These include the following: (1) the snow acts as an insulator, thereby keeping the underlying sea-ice warm and thin, (2) the snow has a lower volumetric specific heat and volumetric latent heat of fusion than the ice, causing it to cool, warm and melt more easily than ice (this gives rise to longer periods of ice-free conditions during the summer and therefore a warmer climate) and (3) snow has a higher albedo than ice and so this leads to a reduction in the absorption of energy and a cooling of the climate. Calculations indicated that the albedo effect is dominant and this leads to an overall cooling. Loth *et al.* (1993) have described a snow cover model for global climate simulation work. Although the available information on permafrost is fragmentary, one may believe that this cryogenic component is the most sensitive indicator of climate.

Starting in November 1966, the National Oceanic and Atmospheric Administration (NOAA) (USA) has been issuing weekly maps of snow and ice cover distributions for the northern hemisphere, on a scale of 1:50 000 000 (in a polar stereographic projection). Dewey (1987) processed such data for the period November 1966 to October 1984 to draw monthly mean and annual mean climatological maps of the frequency of occurrence of snow cover over 89 × 89 grid points for the whole northern hemisphere, the tropics excluded. A grid cell was considered to be totally snow-covered if 50 per cent or more of the cell was covered with snow.

Table 4.14 characterizes the average snow cover extent per month and averaged over the whole year. Naturally, the climatological maps drawn from satellite data

**Table 4.14** Monthly mean snow cover extent ($10^6$ km$^2$) in the northern hemisphere during the period November 1966 to June 1985

| | Maximum | Minimum | Average | MSD |
|---|---|---|---|---|
| January | 51.0 (1979) | 41.2 (1981) | 46.2 | 2.6 |
| February | 52.8 (1978) | 40.1 (1970) | 46.4 | 3.0 |
| March | 43.5 (1971, 1978, 1979) | 37.0 (1970) | 41.0 | 1.9 |
| April | 36.3 (1979) | 27.6 (1968) | 31.7 | 2.3 |
| May | 24.6 (1978) | 17.2 (1968) | 21.2 | 2.0 |
| June | 15.5 (1976) | 7.5 (1984) | 11.8 | 1.9 |
| July | 8.5 (1967) | 3.6 (1982) | 5.5 | 1.3 |
| August | 5.3 (1980) | 2.7 (1970, 1984) | 3.9 | 0.8 |
| September | 9.9 (1972) | 2.6 (1970) | 5.2 | 1.6 |
| October | 31.6 (1976) | 7.4 (1968) | 17.7 | 5.7 |
| November | 37.6 (1973) | 29.5 (1979) | 33.5 | 2.0 |
| December | 45.6 (1973) | 37.1 (1969) | 41.9 | 2.5 |
| Year | 28.3 (1978) | 23.5 (1970) | 25.5 | |

provide a more detailed presentation of the variability observed (especially in high latitudes and in mountains) than the maps based on the data of surface observations. According to satellite data for the last two decades, the onset of snow cover in Asia and North America has taken place several weeks later than appears from the data of more recent conventional observations.

Analysis of space-derived maps revealed a persistent northward shifting of the southern snow cover boundary but a simultaneous southward shift of the isoline of the frequency of occurrence of 100 per cent snow cover. This reflects a narrowing by 25–50 per cent of the transition latitudinal zone, within which the frequency varies from zero to 100 per cent. Robinson *et al.* (1993), after updating previous global snow cover monitoring results, discovered a reduction of hemispheric snow cover over the preceding five years, particularly in spring.

### 4.3.3 Arctic and global climate

All hemispherical and global climate models take into account the processes in high latitudes, but relevant parametrization schemes are diverse. Numerical climate modelling for polar regions is seriously complicated by several factors, including the radical difference between the climates of the Arctic (an ocean covered with ice, whose dynamics is governed by temperature and convective mixing) and the Antarctic (a continental ice sheet up to 4 km thick) with sparse observational data. Another factor is the inadequacy of grid models near the poles (the existence of singularity) leading to serious distortions of results (spectral models are more useful here) (Kondratyev *et al.* 1996a).

Another difficulty is a reliable simulation of specific cloud conditions in high latitudes. With snow cover, the formation of extended cloudiness leads to surface heating (independently of the height of clouds), whereas in the case of the snow-free surface the low-level extended cloudiness causes a cooling and the upper-level clouds lead to a warming (the greenhouse effect of cirrus clouds).

The effect of the dynamics of permafrost on the hydrological regime is very important. Estimates show, for example, that with permafrost destroyed in the regions of the Lena and Enissey rivers the run-off for these rivers will be halved, which will affect the salinity regime of the arctic seas and the annual change of the extent of the arctic ice cover (the process of ice formation will slow down). Changes in the albedo of snow–ice cover caused by contamination can play a substantial role. This effect must be particularly considered in the analysis of the effect of a possible nuclear war on climate. More accurate consideration of the processes in polar regions is a fundamental aspect of further improvements on climate models.

A report was written (US Department of Energy 1984) to consider the problem of water exchange between the continental ice cover and the world ocean during the last century and the possibilities for forecasting climate changes and sea level changes during the next hundred years. This report analysed the grounds for the previous opinions that climate warming due to an increase of $CO_2$ and other greenhouse gases will cause continental ice melting and thermal expansion of the upper oceanic layer, which will lead to a considerable (possibly, catastrophic) rise of sea level. It concluded that sea level is rising but the speed of the rise is known only with an accuracy of a factor of 2 (from the beginning of this century it was 1–3 mm $yr^{-1}$). The poor accuracy arises both from the fragmentary observations (especially in the southern hemisphere) and from the difficulties in estimating the effect of tectonic and isostatic factors. Preliminary estimates of the continuous adjustment of the Earth to shrinking ice sheets (the Scandinavian and Laurentide ice sheets) gave values below 1 mm $yr^{-1}$ for the eastern coastline of the USA.

The sea surface temperature has increased from the beginning of the twentieth century, probably by $(0.6 \pm 0.3)\,^{\circ}C$, but this quantity can be distorted by systematic errors. A certain contribution to the rising of the ocean level is made by melting mountain glaciers and small ice caps, but it seems that the share of the Greenland ice shield in this contribution is negligible. As for the antarctic ice shield, most likely it is growing and taking water from the ocean. However, the rate of oceanic mass variations is practically zero.

It is still unknown whether the observed rise of the world sea ocean level may be explained only by thermal expansion of its upper layer. Calculations of thermal expansion are unreliable because of neglecting the deep-water processes. The limited observations of temperature, salinity and density of the oceanic water during the last few decades have not revealed any statistically substantial variations in individual regions (continuous observations are needed to draw reliable conclusions). North of 50° N, a slight freshening of the Atlantic water masses took place after 1972, but this could be temporary.

In assessing the balance between water masses and sea level for present conditions, a decrease in the thickness of the peripheral parts of the Greenland ice shield was discovered, as well as a thickening of its central part (Table 4.15). The uncertainty in the estimates for the Antarctic is explained by unreliable data on iceberg ejections, ice accumulation in some regions and ice melting beneath the shelf areas of the glaciers.

Physical processes governing the effect of climate changes on the antarctic ice shield are still inadequately studied. It follows from prognostic estimates that, apart from rainfall intensification (with supposed climate warming), an additional contribution to growing ocean water masses can be made by melting small glaciers and the Greenland ice sheet, but the total rise of sea level will not be more than a few decimetres by 2100 (Table 4.16).

Bromwich and Robasky (1993) noted that the precipitation rate over Antarctica appears to have increased by about 5 per cent over a time period spanning the accumulation means for the 1955–1965 to 1965–1975 periods, while over Greenland it has decreased by about 15 per cent since 1963 with a secondary increase over the southern part of the ice sheet starting in 1977. At the end of the 10-yr overlapping

**Table 4.15** Estimates of the balance between the water mass of the ocean and its level

| Ice mass | Equivalent change of mass (mm $yr^{-1}$) | Contribution to the change of the ocean level (mm $yr^{-1}$) |
|---|---|---|
| Glaciers and small ice caps | $-1.2 \pm 0.7$ | $+0.5 \pm 0.3$ |
| Greenland ice shield | $0.02 \pm 0.08$ | $-0.1 \pm 0.4$ |
| Antarctic ice shield | $0.02 \pm 0.02$ | $-0.6 \pm 0.6$ |

**Table 4.16** Estimates of the contribution by glaciers to changing the world ocean level as a result of climate warming caused by doubled $CO_2$ concentration by 2100

| Ice mass | Rate of change in level ($mm\,yr^{-1}$) | Total change in level (m) |
|---|---|---|
| Glaciers and small ice caps | 2–5 | 0.1–0.3 |
| Greenland ice shield | 1–4 | 0.1–0.3 |
| Antarctic ice shield | −3 to +10 | −0.1 to +1[a] |

[a] The range 0–0.3 is most likely.

period, the global sea-level impact of the precipitation changes over Antarctica dominates that for Greenland and yields a net ice-sheet precipitation contribution of roughly $-0.2\,mm\,yr^{-1}$. These conclusions are tentative, however. On the basis of climate sensitivity studies of the Greenland ice sheet using satellite AVHRR, SMMR, SSM/I data and *in situ* data Steffen *et al.* (1993) arrived at the conclusion that there was a consistent increase in accumulation rate for the northeastern part of the Greenland ice sheet from 1981 to 1986.

Further studies on sea level must be aimed at improving climate models, especially with regard to the role of sea ice and cloudiness, as well as processes in high latitudes, a more reliable determination of a global change of sea level from observations and a more accurate estimation of the contribution by volume thermal expansion of the ocean. Studies should also be concentrated on heat transport in the ocean across the continental shelf round the Antarctic (especially in the regions of the Ross, Amundsen and Bellingsgausen Seas as well as the Weddell Sea) and also on oceanic circulation beneath large masses of shelf ice, as well as the effect of a possible climate warming on these processes.

Since numerical modelling revealed an increased sensitivity of the climate in polar regions to anthropogenic impacts, the monitoring of polar climate changes is particularly interesting from the viewpoint of an early detection of global climate variations (Hall 1988). For this purpose, different indicators can be used. For example, the 2–4 °C permafrost temperature increase observed in the region of north Alaska is assumed to have taken place during the last century. A climate-warming indicator may be a retreat of small glaciers, which could have contributed markedly (up to 50 per cent) to the 10–15 cm rise of the world sea level which took place during the last hundred years. On the other hand, in some studies a possibility has been mentioned of intensified snowfalls in climate-warming conditions and, hence, of the growth of some glaciers on Alaska.

Though the climate warming should have resulted in globally decreased snow cover and sea ice cover extent, an analysis of satellite data accumulated during the last 20 yr has not revealed such a trend. The estimates obtained by glaciologists from conventional data show that the Greenland ice sheet has had a stable mass (and thus had been in an equilibrium), and the mass of the Antarctic ice sheet has been increasing. In this connection, satellite radar and lidar sounding is valuable as a source of information about the balance of the ice shield's mass. An analysis of the visible and near-infra-red images obtained from the Landsat and SPOT satellites permits monitoring of the glaciers' dynamics. Lake ice dynamics (the dates of the onset and melting of ice cover), resulting from the analysis of images in the optical and thermal regions, can serve as a sensitive indicator of regional climate variations.

As Carsey (International Association of Meteorology and Atmospheric Physics 1989) notes, the long-term satellite microwave observations, as well as processing of the data in the visible and infra-red spectral regions (including the information from Landsat and SPOT satellites) have provided the means to accomplish a regular global mapping of sea ice and land snow cover, to draw maps of sea ice concentration, and to study the detailed regional dynamics of the cryosphere. From microwave data for 1981–1987, no marked global trend of the ice cover extent has been observed (see also Kondratyev *et al.* 1996a).

Barry (International Association of Meteorology and Atmospheric Physics 1989) discussed a global database for the period 1979–1986 on sea ice concentration with a resolution of 50 km from observations with the Nimbus-7 Scanning Multichannel Microwave Radiometer (SMMR). The visible 0.6 km resolution imagery is being directly transmitted from the satellites of the US Defense Meteorological Satellite Program (DMSP). A system of arctic sea buoys supplements this information with data on ice drift and atmospheric pressure near the surface. The DMSP data have been used for a regular (every 3 days) mapping of the summertime ice cover in the Arctic. The following surface albedo values averaged over the Arctic have been obtained: 0.75 (May), 0.64 (June), 0.46 (July) and 0.41 (August). From mid-August, the albedo starts increasing in the central Arctic basin (due to snowfalls), but in some cases it may decrease (a minimum of the sea ice extent is usually observed in September). There is a considerable interannual variability of albedo (and of the absorbed solar radiation), which can explain the melting-induced variations in the ice cover thickness up to 0.66 m. Analysis of microwave data for 1979–1984 for Canada revealed in some regions a decrease of ice concentration amounting to 70–80 per cent, which is explained by the effect of quasi-stationary cyclones causing a divergence in the sea ice concentration of about 0.5 per cent per day. The appearance of an open-water surface must cause considerable intensification of the sensible heat exchange between the atmosphere and the ocean, which must be taken into account in the numerical modelling of the impact of the greenhouse effect and other factors on climate.

From the data of Cavallieri (International Association of Meteorology and Atmospheric Physics 1989), the annual change in ice extent in the northern hemisphere varies from a minimum of $8 \times 10^6\,km^2$ in September to a maximum of $15 \times 10^6\,km^2$ in March. In the southern hemisphere the ice cover reaches 54° S, and its extent varies from $4 \times 10^6\,km^2$ (February) to $20 \times 10^6\,km^2$ (September). An analysis of the SMMR data for nine seasons revealed a difference in the annual changes of the sea ice in both hemispheres, determined largely by the inhomogeneous distribution of land and oceans. The most important feature of these differences is an asymmetry of the annual changes, manifesting itself in a faster decrease in the southern hemisphere ice cover extent in the summer, caused by the more unstable thermocline in the Southern Ocean. A consideration of the dynamics of the cryosphere in five sectors in high latitudes of the southern hemisphere and in eight regions of the Arctic based on data for 15 yr revealed the existence of a substantial interannual variability (especially in the Antarctic) and asynchronous variations in different regions. For 9 yr of the period studied (1977–1987), there was no noticeable secular trend in the Arctic or the Antarctic of the variability of annual change of ice cover extent, but the global extent decreased by about 5 per cent, as a result of the change in the phase and shape of seasonal oscillations. The interpretation of the data on cryospheric dynamics, from the viewpoint of revealing the climatic variability, needs a very careful approach (see Kondratyev *et al.* 1996a,b).

Interesting results were presented at the Conference 'The Role of the Polar Oceans in Global Change' held in Fairbanks, Alaska on 11–15 June 1990 (Weller *et al.* 1991, Kondratyev and Kotliakov 1991). Some results of numerical climate modelling presented there are extremely important for analysis of observational data on climatic parameters aimed at revealing the anthropogenic climate signal. An integration for the period 200 yr for the atmosphere–ocean system with account of the deep-water circulation of the ocean demonstrated the practical absence in the Antarctic of not only the warming enhancement widely discussed but also any climate trend in general (so great is the effect of the gigantic thermal inertia of the ocean). Similar calculations for a period of 30 yr, which generally speaking is insufficient to reach a stabilized regime, revealed a vast zone of climate cooling in the region of the North Atlantic. Analysing the data of observations in Fairbanks, Bowling (1991) has shown that the signal-to-noise ratio (warming/interannual variability) is small, which will make any change hard to detect. In addition, the climatological data set at high latitudes is scanty and subject to most of the same problems as those in the temperate zone. In fact, the extreme ground inversions, low Sun angles, and seasonal polar night or continuous daylight conditions may lead to systematic errors with magnitudes much greater than would be predicted from mid-latitude experience.

In connection with the special role of coupling of various components of the climatic system in the Arctic, Walsh and Crane (1992) have undertaken an intercomparison of climate modelling results for high latitude with the use of five climate models which do not take account of the interaction between the oceans and the atmosphere. These models have been developed in the following institutions: the Geophysical Fluid Dynamics Laboratory (GFDL), Goddard Institute for Space Studies (GISS), the National Center for Atmospheric Research (NCAR), Oregon State University (OSU) and the UK Meteorological Office (UKMO).

Surface air temperature and surface pressure (geostrophic wind) have been chosen as the parameters which are most sensitive to interaction between the atmosphere and ocean (under Arctic conditions, the parametrization of ice cover dynamics is of special significance). In all cases, results of control integration ($1 \times CO_2$) have been averaged over time periods which were not less than three last years of integration (the intercomparison is complicated by differences in spatial and temporal resolution of the models). Table 4.17 illustrates the correlation between calculated and observed pressure fields using four of these models. As can be seen, there is a strong variability of the correlation coefficients. During the non-summer months, with the most active ice cover dynamics, the GFDL and UKMO models happen to be the most reliable, while during the summer months (especially in July and August) none of the models is adequate. Similar data for surface air temperature are given in Table 4.18. From the viewpoint of systematic errors the UKMO model looks the best (except for the summer months); this results from a reliable simulation of the surface heat balance. In the case of the GISS model there are differences of the order 2–4 °C all year-round, while the OSU model is characterized by the annual variations of the differences, with a maximum absolute error of 8 °C in November and December. The strongest deviations (up to −10 °C) from observations exist in the case of the GFDL model during the period from October to

**Table 4.17** Correlation coefficients between calculated and observed pressure fields, 70–90° N, for four models

| Period | UKMO | OSU | GISS | GFDL |
|---|---|---|---|---|
| DJF | 0.848 | 0.387 | 0.414 | 0.909 |
| MAM | 0.792 | 0.398 | 0.532 | 0.908 |
| JJA | 0.378 | 0.497 | 0.356 | 0.568 |
| SON | 0.743 | 0.169 | 0.550 | 0.760 |

*Note*: DJF = December, January, February; MAM = March, April, May; JJA = June, July, August; SON = September, October, November.

February, which should result in enhanced ice cover extent in winter. It is curious that the GFDL model, which is the best in simulation of pressure field, is the least adequate with regard to temperature simulation (a significant cause may be the lack of account of leads and polynyas).

A significant disagreement between calculated temperatures as well as between calculations and observations is a result of inadequate simulation of thermodynamic interaction between the atmosphere and the ocean in the presence of ice cover. Further development of climate models is necessary to take into account: (1) more realistic ice cover dynamics, better simulation of ice transport and the formation of leads and polynyas, (2) interactions in the atmosphere–ice-cover system which are responsible for dynamics and thermodynamics of the ice cover and (3) parametrization of extended stratus cloudiness and its interaction with radiation.

Gordon (1991) notes that the southern hemisphere ocean considerably contributes to the formation of waters which largely determine the present abnormally cold ocean. This situation is mainly determined by the existence of the antarctic circumpolar current which also serves as a significant factor of the antarctic glaciation. Two regions determine the formation of water masses affecting differently the global circulation of the ocean: (1) the antarctic zone of the formation of cold dense waters propagating along the sea bottom, intruding deep into the northern hemisphere, and (2) the polar frontal zone (within the circumpolar current) where the low-salinity water masses appear, moving northward near the low boundary and within the bottom part of the thermocline. The extent of the regions of water mass formation

**Table 4.18** Comparison between calculated and observed surface air temperatures (°C), 70–90° N, for four models

| | Average absolute error | | | | Average differences between model and observations | | | |
|---|---|---|---|---|---|---|---|---|
| | UKMO | OSU | GISS | GFDL | UKMO | OSU | GISS | GFDL |
| DJF | 4.26 | 8.42 | 4.28 | 10.10 | −0.74 | 4.67 | −3.03 | −10.07 |
| MAM | 2.54 | 6.13 | 3.16 | 5.46 | 0.31 | 3.66 | 0.99 | −5.46 |
| JJA | 3.15 | 2.09 | 2.91 | 2.58 | −2.15 | 0.17 | 2.38 | −2.39 |
| SON | 3.37 | 6.08 | 4.97 | 9.99 | −0.22 | 4.36 | −4.64 | −9.95 |

*Note*: DJF = December, January, February; MAM = March, April, May; JJA = June, July, August; SON = September, October, November.

depends strongly on the presence of polynyas and an intensive atmosphere–ocean interaction taking place in them.

### 4.3.4 Remote sounding

It is clear that remote sounding techniques are the key means of empirical studies of climate, and, as experience has shown, in conditions of high latitudes the use of both passive and active microwave sounding is of particular importance (Kondratyev and Melentyev 1986, Vaughan 1987, Cracknell and Hayes 1991, Kondratyev *et al.* 1996a). Passive microwave sounding has made it possible to reveal the existence within pack ice of vast regions of open water or thin ice (so, for example, the size of the polynya in the Weddell Sea reaches 300–1000 km$^2$ and remains during the whole polar night).

Turner (1991) processed data from satellite remote sounding (the NOAA TOVS instrumentation) and showed that the errors in retrieving the vertical temperature profiles and water content of the atmosphere in polar regions remained too large, preventing reliable retrievals of long-term trends. So, for example, the errors in temperature retrievals caused by insufficiently reliable filtering-out of the effect of clouds constitute about 2–3 °C, considerably increasing in the atmospheric boundary layer and near the tropopause. The processing of the brightness temperature ($T_B$) change data has not revealed any long-term trend either.

Fletcher (1991) undertook an analysis of the global data base of conventional meteorological observations (COADS), which led to a conclusion about the long-term variability of the atmospheric general circulation characteristics which did not agree with the known results of numerical climate modelling. So, for example, it was found that the observed global winds enhancement (this refers also to trade winds and the west–eastern transport) by about 20 per cent in the 1930s, compared with the period 1920–1970, does not agree with that calculated; during the last century three to four sudden changes of atmospheric general circulation took place. Long-term oscillations of two types have been observed: about 3–4 yr (this is connected with the El Niño / Southern Oscillation) and of the order of 130 yr (in this case the forcing appears, probably, in high latitudes in winter, more strongly apparent in the Arctic than in the Antarctic). Solar activity can be supposed to be a significant factor of long-term oscillations. Royer (1991) found an 18.6-yr periodicity of sea surface temperature and surface air temperature, which could be ascribed to the effect of lunar tides.

Having processed the data of the northern hemisphere aerological sounding to calculate monthly mean temperatures of various lower tropospheric layers for the last 35 yr, Herbert (International Association of Meteorology and Atmospheric Physics 1989) found the known effects of the Agung and El-Chichon eruptions on the temperature field, and the cooling-to-warming transition in 1978, with the warming taking place mainly in winter and in spring. In accordance with the numerical modelling data, the temperature variations in the 700–800 hPa layer were more substantial than in the 500–700 hPa layer.

Having processed the cloud data from the soviet meteorological satellites for the period 1971–1985, Mokhov (1993) found a general trend towards an enhancement of cloudiness all over the world, as well as in the northern and southern hemispheres as the surface air temperature increases, but against this trend, in some regions there was observed an anticorrelation of surface air temperature and cloud amount, especially in high latitudes of the northern hemisphere in winter and in spring (see also Weare *et al.* 1995). Stone and Kahl (1991) noted that the intensity of surface air inversions could be a good indicator of climate change in the zone of the Antarctic plateau. Bromwich emphasized the fact that precipitation in the regions of Greenland and the Antarctic could be an important factor in the world sea level variations. Radok and Wendler (1991) as well as Parish (1991) demonstrated the significant role of katabatic winds in the formation of tropospheric circulation in the Antarctic.

An analysis of requirements for sea ice information has led to the conclusion that the data are needed both with high resolution of the order of 100 m for spatial scales of about 100 km (routing of ships and other problems) and with 10 km resolution for sub-global and global scales (climate studies). In the case of snow cover the resolution and spatial coverage needed vary from 100 m and tens of kilometres (hydrological processes over the catchment area) to several kilometres for large scales, up to the global one (climate). The most important parameters (apart from the ice and snow cover extent) are as follows: the type (including the age) and concentration of ice cover and the state (e.g. melting) and water equivalent of snow cover. Major characteristics of glaciers are their extent, depth, topography and the rate of accumulation.

Continuous satellite microwave observations as well as processing of the visible and infra-red data (including information from Landsat and SPOT satellites) have made it possible to accomplish a regular global mapping of sea ice and continental snow cover, to draw maps of sea ice concentration and to study the detailed regional dynamics of the cryosphere. For the 1981–1987 period the microwave data do not reveal any marked global trend of the sea ice extent. The prospects for further development of remote sounding of ice and snow cover are connected with the use of space-borne radars, which will make it possible to obtain data with a resolution of tens of metres and better. In this connection, of great importance are the results of the Russian studies, especially those obtained with the synthetic-aperture radars carried by the satellites Kosmos-1500 and Kosmos-1870 (Almaz) as well as the ERS-1 and ERS-2 results (Kondratyev *et al.* 1992, 1996a).

The processing of the round-the-year data from acoustic sub-ice sounding from 13 nuclear submarines, covering practically the whole Arctic Ocean, has made it possible to draw maps of the ice cover thickness distribution in the Arctic and to characterize the fields of sea ice drift (Wadhams 1994). As McLaren *et al.* (1991) have shown, these data can be used also to retrieve the ice cover roughness.

The processing of a 10-yr (1978–1987) global SMMR data series by Gloersen and Campbell (1988) made it possible to characterize the statistical features of the global sea ice cover dynamics. The Fourier-analysis revealed not only the prevailing peak which corresponds to the annual change, but also considerable second and third harmonics, as well as marked fourth and fifth harmonics. The filtering-out of periodic variability had led to the conclusion about the existence of a negative trend of the extent of the arctic ice cover for 9 yr equal to $(1.9 \pm 1.3)$ per cent with a 90 per cent statistical significance at a level of $2\sigma$. (The ice cover extent was assessed in processing the satellite microwave data as a surface area outlined by the isoline of the ice cover concentration equal to 15 per cent.) In the case of the antarctic ice the trend is absent and its global mean value is $(-1.0 \pm 0.7)$ per cent. The area of the open water surface in the Arctic varied within $(1.6\text{–}3.2) \times 10^6\,\text{km}^2$, constituting about 10 per cent in the period of maximum ice cover extent. The respective values for the Antarctic are $(1.5\text{–}4.5) \times 10^6\,\mathrm{km}^2$ and 25 per cent.

The operation of the Geosat satellite has opened up possibilities to monitor the topography of ice shields and sea ice from the data of radar observations (Lingle *et al.* 1991), for example the annual change of the surface level in the zone of ablation in west Greenland. To reveal the climatic trends, the scientists of the US Geological Service started in 1978 a regular processing of Landsat data to monitor the dynamics of glaciers in different regions of the globe (Williams and Ferrigno 1991). Molnia (1991) emphasized, however, that the retreat of glaciers does not always serve as an indicator of climate warming; this indicates the necessity of a thorough analysis of the data of such observations.

The results of monitoring the ice cover dynamics from the DMSP SSM/I data at frequencies of 18, 37 and 85 GHz made it possible to obtain the ice cover concentration, the share of multi-year ice and the location of the ice edge in both polar regions. A joint processing of the satellite microwave information and the data from drifting buoys has opened up possibilities of studying the interaction between the atmosphere and the ice cover on synoptic scales in the Arctic. The ice cover microwave data interpretation becomes more informative with the use of the polarization ratio

$$\mathrm{PR} = (T_{37\,\mathrm{V}} - T_{18\,\mathrm{V}})/(T_{37\,\mathrm{V}} + T_{18\,\mathrm{V}}) \tag{4.20}$$

and the gradient ratio

$$\mathrm{GR} = (T_{18\,\mathrm{V}} - T_{18\,\mathrm{H}})/(T_{18\,\mathrm{V}} + T_{18\,\mathrm{H}}) \tag{4.21}$$

where the indices denote frequency (in GHz) and polarization state, taken as predictors. The combined use of the SMMR and AVHRR data (for 8 yr) has provided a reliable retrieval of the ice condition characteristics. Owing to the difference between the polarization ratio and the gradient ratio for one-year and multi-year ice, it is possible to determine the age of the ice. In the future, physical algorithms of retrieval are planned to be used.

Budd and Simmonds (1991) performed a numerical modelling of the greenhouse climate changes, prescribing (in July) a 2 °C sea surface temperature increase in the tropics and a 6 °C increase in high latitudes, as well as a decrease of the antarctic ice cover extent by 7° latitude and assuming different values of the relative area of leads: 1 per cent (control), 5, 20, 50, 80 and 100 per cent. Particularly marked were changes in the water cycle in the northern hemisphere high latitudes. Naturally, in the regions of the low ice cover concentration the fluxes of latent and sensible heat strongly increased. An intensified evaporation brought forth increasing precipitation in the coastal Antarctic and assimilation precipitation in the polar cap, which led to a decrease of sea surface level. With the rain rate per year assumed to increase by 30 per cent, it should cause a 2 mm decrease of the level.

Having compared the surface air temperature observation data from 30 antarctic stations for 30 yr with satellite data on the ice cover extent, Walsh *et al.* (1991) showed that, as a rule, large surface air temperature anomalies were followed by ice cover extent anomalies of the same sign, but this correlation was not the same in different regions, in particular, from the viewpoint of the phase shifts (the ice cover extent changes lag behind sea surface temperature anomalies but not in every case). In the Antarctic a complicated pattern of surface air temperature variations of different signs was observed, with an average trend equal to zero.

There is no doubt that the ice cover dynamics in the Antarctic (the area of seasonal ice is practically equivalent to the territory of the Antarctic) substantially affects the climate. However, the problem is that the southern hemisphere sea ice properties are still poorly known (ice cover thickness, texture characteristics, etc.)

As Lange (1991) notes, in contrast to the arctic ice, the one-year ice of the Antarctic is $(0.97 \pm 0.73)$ m thick and the thickness of the two-year ice is $(1.23 \pm 0.68)$ m. An analysis of observations in the Greenland Sea performed by Mysak *et al.* (1991) led to the conclusion that severe ice conditions took place in 1902–1920 and in the late 1940s, whereas the 1920s and 1930s were characterized by light ice conditions, as a rule.

Breitenberger and Wendler (1991) discussed the effect of the atmosphere–ice-cover interaction in the Antarctic on the formation of atmospheric general circulation, emphasizing the strong variability of the intensity, sign and phase of this interaction, which manifests itself most in the Ross and Weddell Seas.

Particularly substantial is the variability of the snow cover extent; the difference between maximum and minimum values reaches 80 per cent. Fast changes in the snow and ice cover, connected with thawing, are one of the key factors of climate formation, manifested mainly as the albedo feedback. Though the available information about the characteristics of permafrost is fragmentary, it can be assumed that this cryospheric component is the most sensitive indicator of climate.

In continuation of the analysis of the dynamics of snow cover thawing made by Foster (beginning in late 1960, the springtime thawing of snow cover in the tundra of North America started earlier than before, but the reasons for this trend are still not clear), Kodama *et al.* (1991) noted that the intensification of cloudiness in the process of global warming delays the snow thawing, while the surface albedo decrease and the increase of atmospheric thermal emission produce an opposite effect.

Naturally, the requirements for the observational data in the Arctic are not confined to information on the ice and snow cover or permafrost. The basic parameters of interest are as follows: the temperature and heat balance of the ocean surface, chlorophyll (phytoplankton) concentration, the area of the open water surface along the arctic coastline, the geographical distribution of the ice cover concentration and thickness, structure and extent of glaciers, the spatial and temporal variability of snow cover (including information about its thickness, density and water equivalent), the temperature and radiation budget of land surface, the extent of permafrost and zones of its instability, soil moisture, atmospheric transparency and trajectories of propagation of atmospheric pollutants, vertical profiles of radiative characteristics of the atmosphere, the amount and type of clouds, precipitation, stratospheric temperature, the content of ozone and other minor gaseous components.

### 4.3.5 Arctic and global changes

When discussing the problem of the role of polar regions in the formation of global changes, one should have in mind that two aspects of global ecology are of paramount importance (Gorshkov and Kondratyev 1988, Kondratyev 1987a, 1989a–c, 1990a,b, Kondratyev *et al.* 1990b, 1996a): (1) the anthropogenically included redistribution of the heat balance components of the Earth as a planet (with emphasis on the problem of the atmospheric greenhouse effect and its climatic impact) and (2) anthropogenically-induced breaking of global biogeochemical cycles (primarily carbon, nitrogen and sulphur). Polar regions are very specific as a component of the global ecosystem. From the viewpoint of Earth energetics, these are zones of a powerful sink of energy and their value in the formation of biogeochemical cycles is determined by the presence of high-latitude ecosystems highly sensitive to external forcings. The latter entails, however, that the dynamics of the processes in polar regions can be the most reliable indicator of global ecological changes.

Analysing the possibilities of studying data on sea ice dynamics as an indicator for early detection of anthropogenic climate changes, Parkinson (1991) notes that, in general, such indicators should meet the following four general conditions: (1) a considerable variability ensuring a reliable record of the signal, (2) the possibility of regular global-scale observations, (3) small and well-known natural variability (noise level) and (4) early changes with respect to changes in other climate parameters (for example, the surface air temperature).

From the viewpoint of the criteria mentioned, the sea ice dynamics cannot be considered as an adequate indicator, since, for example, the ice cover is characterized by strong interannual variability (both on local and hemispherical scales), and the observational series available are comparatively short. The phase relations in the variability of sea ice and other climate parameters have been inadequately studied.

From the data of Zwally *et al.* (1991), the amplitude of the regional interannual variability of the antarctic ice cover extent can reach 30 per cent while for the whole of the Antarctic it is about 10 per cent on average. The multi-year trend for the last 15 yr has turned out, however, to be negligible. A substantial interannual variability of the ice cover extent has also been recorded in the Arctic. From the data for the period 1966–1986 the area of ice in the Barents Sea decreased by 25 per cent in the end of the period, but no decrease of the maximum ice cover extent was recorded in late April for the same period (probably it was a result of the trend to decreasing ice cover thickness).

The most impressive manifestation of anthropogenic impacts on the antarctic atmosphere is the formation of the springtime minimum of total ozone content (the ozone hole) due to the specific combination of circumpolar circulation and low temperature of the stratosphere, polar stratospheric clouds and the input of chlorofluorocarbons (Kondratyev 1989a). Particularly interesting is the fact that the ozone hole event was not predicted on the basis of stratospheric simulation models which were presumed to be up to date.

Also important are the anthropogenically-induced changes in the arctic atmospheric chemistry, especially those connected with arctic haze formation. The destruction of ozone in the lower atmosphere caused by bromine and iodine compounds, apparently of marine origin, has been quite unexpected; the contribution of anthropogenic factors cannot be excluded (Barrie 1991).

As Schnell *et al.* (1991) have shown, a considerable photolytic destruction of ozone takes place around the summertime in the Antarctic in the atmosphere boundary layer. As for the arctic haze, its sources and ways of input to the Arctic, as well as the chemical composition and size distribution of aerosol, have become more clear, though the vital need for further studies remains.

Very interesting is the input of heavy metals to polar regions (Adamenko *et al.* 1991). Blanchet (1991) performed a numerical modelling of climate changes due to the arctic haze and showed that the effect of concentration of the soot component

$0.2\,\mu g\,m^{-3}$ (this exceeds the observed value by a factor of 5) was equivalent to the effect of a $CO_2$ doubling, the effect of the aerosol haze manifesting itself most in the regions of the Canadian arctic archipelago and the Greenland Sea.

The environment and the biosphere of the Arctic are characterized by a unique combination of such components as the polar ocean, ice and snow cover, glaciers, tundra, permafrost, boreal forests and wetlands, each being a sensitive indicator of global changes, subject to considerable variations under the slightest effects, manifesting themselves as variations in the solar radiation input, surface temperature, heat transport by the ocean, chemical composition of the atmosphere and the ocean, content and properties of atmospheric aerosol, etc. On the other hand, the processes in the Arctic can substantially affect climate changes. This refers, for example, to the role of permafrost and wetlands in the formation of biogeochemical cycles. Of particular interest are such problems as the role of $CH_4$ as a greenhouse gas and the biodynamics of the euphotic zones of the Arctic and Antarctic Oceans (this concerns, first of all, the dynamics of phytoplankton and, hence, the release of DMS to the atmosphere). Apparently, the high-latitude regions are parts of the globe where cosmic factors (mainly solar activity) significantly affect the environment and the biosphere.

An assessment of the role of the Arctic in the formation of global changes should be based on analysis of the interaction of the following factors: (1) nutrient fluxes, (2) carbon accumulation, (3) the dynamics of ecosystems and biological communities, (4) hydrology, (5) permafrost, (6) snow cover and glaciers, (7) coastal processes, (8) sea ice, (9) oceanic circulation and formation of bottom waters, (10) atmospheric dynamics and heat balance, (11) atmospheric composition and (12) solar and geomagnetic effects. The necessity to take into account the interaction of the factors mentioned determines the priority of seven inter-disciplinary problems, as follows.

1. The response of the arctic marine and continental ecosystems to external conditions. Interactive variations in the arctic ecosystems can manifest themselves, on the one hand, as the effects of changes in climate, world sea level and other environmental parameters on the ecosystems and, on the other hand, as the effects of the dynamics of ecosystems (for example, through changes in the carbon cycle) on the environment and, probably, global climate.
2. The role of the arctic wetlands in the formation of biogeochemical cycles can manifest itself, first of all, in varying carbon accumulation, which ought to have global consequences resulting from either assimilation or emission to the atmosphere of $CO_2$ and $CH_4$. In this connection, of great importance are studies of interaction between various cycles in the processes of biosphere–atmosphere exchange, in which the soil biology and permafrost play a special role (studies of connections between palaeoclimate and palaeoproductivity are urgent).
3. The biotic dynamics at the ice-cover–open-water interface. The high bioproductivity near the ice cover edge in the regions of polynyas necessitates a study of processes determining the accumulation of plankton and dynamics of the nutrient input, which leads to the blooming of plankton.
4. The effect of the arctic glaciers and ice shields on sea level in conditions of climate warming should be taken into account both from the viewpoint of its contribution to the rise of the world sea level and the effect on the coastal zone and plankton blooms.

5. Special emphasis should be placed on the water cycle in the Arctic, having in mind the unique nature of the hydrological ecosystem controlled by a complicated set of processes of thawing and freezing of water. The soil moisture, level of underground waters and river run-off significantly affect nutrient production, gas exchange, bioproductivity and other processes.
6. The chemistry and dynamics of the arctic atmosphere require serious attention in connection with global consequences which can cause changes in the concentration of greenhouse gases and stratospheric ozone, determined by a complex combination of chemical and photochemical reactions which depend on special features of the high-latitude atmospheric general circulation.
7. The urgency of the problem of interaction in the ocean–ice-cover–atmosphere system is determined by the key role of sea ice in the global-scale processes, especially from the viewpoint of the ERB variability and formation of bottom waters and hence climate.

In connection with the problems mentioned, the Arctic (like the Antarctic) is important as a source of palaeoclimatic information contained in ice cores and bottom sediments.

Of particular interest are studies of the role of polar regions in the formation of the global carbon cycle. The arctic ecosystems are known to contain a quantity of carbon as organic substance of soil, and, depending on climate changes in future, can be a powerful source of carbon or sink for carbon. The contribution of the tundra and boreal forests to the global reservoir of carbon on land constitutes about 30 per cent. The data of observations on the island Tulik (Alaska) considered by Oechel (1991) revealed the tundra-to-atmosphere carbon flux reaching $3\,g\,m^{-2}$ per day, which calculated for the whole Alaskan tundra is equivalent to 0.1–0.2 $Gt\,C\,yr^{-1}$, and for the whole of Alaska 55 $Gt\,C\,yr^{-1}$.

Bearing in mind the estimates of the contribution to the $CH_4$ global budget, Reeburgh and Whalen (1991) performed measurements of $CH_4$ flux at several locations of the Alaska tundra during 3.5 yr and in the taiga (for less than one year). In the period considered (the autumn of 1989) the taiga was always a sink for $CH_4$. The global estimates obtained from these data gave the emissions of $CH_4$ by the tundra as 19–33 $Gt\,yr^{-1}$, whereas the contribution of the taiga as a sink for $CH_4$ constituted 15 $Gt\,yr^{-1}$. The numerical modelling made by Bonan (1991) has shown that boreal forests of the inland part of Alaska should be a sink zone for carbon.

Kvenvolden (1988) analysed the role of a vast reservoir of hydrates of different gases, including the greenhouse gases (methane, nitrous oxide) frozen in permafrost (including the regions of shelf), which, in a condition of climate warming, can enter the atmosphere. The supplies of methane hydrate in permafrost have been estimated at 400 Gt C. On the whole, the atmosphere could gain up to 120 $Gt\,C\,yr^{-1}$ at the expense of the Arctic. Since sea transgression has taken place in eastern Siberia during the last 27 thousand years, it is possible that in this connection some amount of $CH_4$ has entered the atmosphere.

## 4.4 Land surface processes

The priority that the ocean and the atmosphere should have in studies of short-period climate variability makes land surface processes less significant, though still substantial. The urgency of the problem of parametrizing the atmosphere–land-surface interaction caused the WCRP Joint Scientific Commission to hold a working meeting in 1981 (Kondratyev 1985a). Some aspects of this interaction are connected with the atmospheric contribution: rainfall, global radiation and atmospheric thermal emission. In other aspects, the effects of both the atmosphere and surface are manifested (turbulent exchange of heat, moisture and momentum), whereas the absorption and reflection of radiation by the surface, river run-off and heat and moisture exchange in soils depend on specific topography, vegetation cover and soil properties. The principal circumstance is the strong spatial inhomogeneity of these processes, which determines the need for parametrization of subgrid processes and, in this connection, for a representation of input parameters on a subgrid-scale, depending on predicted quantities averaged over the scales of the applied spatial grid.

An analysis of available results of numerical modelling of atmospheric general circulation sensitivity to the parameters of land surface processes showed a strong atmospheric general circulation sensitivity to surface albedo and moisture in the upper soil layer. In this connection, the following values must be taken into account and parametrized: the spatial extent and vertical stratification of snow cover, the type and density of vegetation cover, the rate of growth of soil moisture due to water infiltration and the rate of soil drying caused by water evapotranspiration and infiltration. The interrelationship between moisture fluxes in the soil, soil moisture and density of vegetation cover is non-linear, and rainfall rate, soil and vegetation properties are characterized by considerable subgrid variability; thus calculations of averaged (including the grid step) moisture fluxes are the most difficult and most important problem in parametrization. The solution to this problem may be based on a one-dimensional parametrization of the respective physical processes depending on the observed physical parameters and an adequate technique for spatial averaging.

It would be important to undertake numerical experiments on the sensitivity of soil moisture and other essential parameters to a parametrization of hydrological processes, whose reliability must be verified from data of complex field observations on a scale of the catchment area of the size of the spatial grid cell. Although the present atmospheric general circulation models (AGCMs) are constructed so that the moisture balance in soils is included, they do not take into account the dependence of soil moisture capacity on the type of soil. Since this dependence is very strong, it must be considered in further improvements of the parametrization schemes.

In a very informative survey Garrat (1993) notes that an important condition of realistic climate modelling is the adequate parametrization of processes on the land surface and in the atmospheric boundary layer. Accomplishment of such a parametrization will allow: (1) the calculation of fluxes of momentum, heat and water vapour at the surface and in the atmospheric boundary layer, which also makes it possible to obtain information on relevant flux divergences, (2) the prediction of the boundary layer thickness and, thus, the vertical redistribution of heat and mass and (3) the prediction of an amount and thickness of cloud layer within the boundary layer, which interacts with the surface heat balance components.

The parametrization of land surface processes is supposed to take into account surface albedo (having in mind calculations of radiative transfer), surface roughness (momentum transfer) and hydrological parameters (transport of sensible and latent heat, precipitation). All these characteristics depend on such factors as specific features of vegetation cover, surface topography and horizontal inhomogeneity which have also to be taken into account. The parametrization of land surface areas for the purposes of climate models has been considered, for example, by Rowe (1993a,b), Matthews (1983, 1984), Oliver (1992) and Thompson and Pollard (1995). A survey has been made by Garrat (1993) of potential approaches to the parametrization of processes on the land surface and in the atmospheric boundary layer with the purposes: (1) to consider various parametrization schemes which are being used in contemporary climate models and (2) to analyse the results of sensitivity studies with regard to the choice of parametrization schemes.

The analysis of climate sensitivity to the choice of land surface processes parametrization schemes has been concentrated mainly on assessments of effects of changes of vegetation cover. The impacts of bare surface albedo and roughness variations have been studied; for instance there is a well-known tendency towards evaporation and precipitation decrease in cases of albedo growth and decrease of soil moisture. The results of calculations have shown, for example, that an albedo increase by 0.13 results in a precipitation decrease on land by about $1\,\mathrm{mm\,day^{-1}}$. Less attention has been paid to assessments of the role of roughness. It has been shown, for instance, that a decrease of precipitation in the Amazonia basin of about $1\text{–}2\,\mathrm{mm\,day^{-1}}$ can be expected to take place as a result of deforestation and a drop in the value of the roughness by three orders of magnitude. A combined parametrization scheme has been developed to take into account simultaneous changes of albedo, roughness and soil moisture which result from deforestation. In the case of parametrization of processes in the atmospheric boundary layer the most difficult is the problem of cloud dynamics parametrization which has been discussed in Section 4.1.

A very broad range of surface albedo variations from less than 5 per cent to around 90 per cent defines an important role of regional albedo variations in climate change. An especially significant contribution is made by high ice and snow albedo. In this context Colman *et al.* (1993) have applied the general circulation model (GCM) developed in the Australian Bureau of Meteorology Research Centre (BMRC) to investigate interacting climate feedbacks due to albedo and cloud parametrizations. Several numerical modelling experiments were carried out with the aim of quantifying the sensitivity of the model to albedo changes on the basis of the GCM run (for 210 days; the last 90 days were averaged to obtain the mean values) in perpetual April mode. It is expected that such results for April would be representative enough, since during other times the model would behave in a manner physically consistent with its behaviour in April, although April conditions represent an upper limit on the strength of simulated albedo–cloud interactions.

There were two pairs of numerical experiments carried out: (1) a perturbation in the form of a $\pm 2\,\mathrm{deg\,K}$ sea surface temperature change to simulate the associated top of the atmosphere (TOA) radiation flux changes as an analogue for climate change (in both cases sea ice extent and thickness were held fixed, but the surface albedo could vary: these are so-called variable albedo cases); (2) the surface albedo was held fixed at the final value determined in the initial $-2\,\mathrm{deg\,K}$ perturbation run, and the sea surface temperatures were once again perturbed by $\pm 2\,\mathrm{deg\,K}$ (fixed albedo cases). As usual, it is assumed that the change in the global mean surface temperature $\Delta T$ is proportional to the direct radiative forcing $G$ at the top of the atmosphere:

$$\Delta T = \lambda G \tag{4.22}$$

where $\lambda$ is a global climate sensitivity parameter. On the other hand

$$G = \Delta F - \Delta Q \tag{4.23}$$

where $\Delta F$ and $\Delta Q$ denote the changes in the emitted longwave radiation and net shortwave radiation, respectively. Thus,

$$\lambda = \frac{1}{\dfrac{\Delta F}{\Delta T} - \dfrac{\Delta Q}{\Delta T}} \tag{4.24}$$

In what follows, subscripts 's' and 'c' refer to conditions of fixed snow (or albedo) and clear sky, respectively. Since shortwave and longwave cloud radiative forcing may be defined by

$$\mathrm{CRF_{SW}} = Q - Q_c, \mathrm{CRF_{LW}} = F - F_c$$

and the total

$$\mathrm{CRF} = \mathrm{CRF_{SW}} - \mathrm{SRF_{LW}} \tag{4.25}$$

it is possible to introduce

$$\lambda/\lambda_c = 1 + \Delta\mathrm{CRF}/G \tag{4.26}$$

and

$$\lambda/\lambda_s = \frac{\Delta T}{\Delta T_s}(1 + \mathrm{SRR}/G) \tag{4.27}$$

where the snow radiation response is

$$\mathrm{SRR} = (\Delta Q - \Delta Q_s) - (\Delta F - \Delta F_s). \tag{4.28}$$

It is apparent from Equation (4.28) that SRR/$G$ is a direct measure of snow feedback, and a comparison of SRR for clear and cloudy conditions gives a measure of the amplification or reduction of the albedo feedback by cloud changes.

Table 4.19 illustrates some of the results. As can be seen, the chosen threshold of statistical significance shows all differentials to be significant at the 99 per cent level. The calculations indicate that, under clear sky conditions in the case of variable albedo, a dramatic increase of shortwave sensitivity (from 0.14 to 0.68 $\mathrm{W\,m^{-2}\,K^{-1}}$) takes place. Because of the large size of the change in $\Delta Q/\Delta T$ in comparison with $\Delta F/\Delta T$, almost the entire increase in $\lambda$ over $\lambda_s$ is due to the shortwave component: SRR $(\mathrm{SW})_c = 2.16\,\mathrm{W\,m^{-2}}$ whereas SRR $(\mathrm{LW})_c = 0.04\,\mathrm{W\,m^{-2}}$. The situation is quite different for overcast conditions.

Colman *et al.* (1993) noted that clouds somewhat mask the areas where surface albedo changes occur (in the variable albedo case, $\Delta Q/\Delta T$ decreases from 0.68 to 0.41 $\mathrm{W\,m^{-2}\,K^{-1}}$). For both clear and cloudy cases the snow radiative response for

**Table 4.19** Global climate sensitivity parameter $\lambda$ ($W^{-1} m^2 K$) as well as LW ($\Delta F/\Delta T$) and SW ($\Delta Q/\Delta T$) differentials ($W m^{-2} K^{-1}$)

| | $\lambda$ ($W^{-1} m^2 K$) | $\Delta F/\Delta T$ ($W m^{-2} K^{-1}$) | $\Delta Q/\Delta T$ ($W m^{-2} K^{-1}$) |
|---|---|---|---|
| *Mean global* | | | |
| Fixed albedo | 0.37 | 2.75 | 0.085 |
| Variable albedo | 0.46 | 2.8 | 0.62 |
| 99% threshold of statistical significance | | 0.07 | 0.09 |
| *Clear sky* | | | |
| Fixed albedo | 0.45 | 2.38 | 0.14 |
| Variable albedo | 0.59 | 2.37 | 0.68 |
| 99% threshold | | 0.03 | 0.04 |
| *Overcast sky* | | | |
| Fixed albedo | 0.35 | 3.06 | 0.21 |
| Variable albedo | 0.41 | 2.85 | 0.41 |
| 99% threshold | | 0.1 | 0.14 |

*Source*: Colman *et al.* (1993).

short wavelength radiation is 2.16 $W m^{-2}$. This means that although shortwave feedback dominates the sensitivity parameter in the change from fixed albedo to variable albedo, it has little impact on the change in sensitivity caused by clouds.

As far as the longwave sensitivity is concerned, its variability is fairly small. But, on the other hand, in going from clear to cloudy conditions (either in fixed albedo or variable albedo cases), it is the long wavelength sensitivity change which dominates the change in global sensitivity. In both fixed albedo and variable albedo experiments, the presence of clouds reduces the climate sensitivity by about 20 per cent.

Table 4.20 contains data on the mean global surface fluxes for fixed and variable snow experiments (in all cases the positive direction is defined to be into the surface), and the surface energy balance, $N$, is given by

$$N = \mathrm{SW} + \mathrm{LW} + \mathrm{LH} + \mathrm{SH} + \mathrm{S} \tag{4.29}$$

where SW is the short wavelength flux, LW is the long wavelength flux, SH is the sensible heat flux, LH is the latent heat flux and S represents the flux into the surface over land (heat flow from the subsoil) or over sea ice (heat flow from the underlying ocean). Over the ocean, S is zero.

Colman *et al.* (1993) pointed out that, at the surface, the heat balance change for perturbed SSTs is dominated by changes in latent heat flux and downward longwave radiation. If albedo is considered, however, the major contrast lies in the change in surface reflected shortwave radiation, amplified by changes in downward shortwave radiation caused by cloud amount changes. A detailed analysis of this kind has also been published by Randall *et al.* (1992).

The results considered demonstrate the complex nature of the interactions induced by the interplay between albedo and cloud feedbacks. Further studies are necessary on both short-wavelength and long-wavelength feedbacks to clarify all the mechanisms operating between the surface changes and the atmosphere.

**Table 4.20** Surface heat balance components and albedo, precipitation and precipitable worker

| Components | Variable albedo, −2 K | Fixed albedo, −2 K | Change | |
|---|---|---|---|---|
| | | | Variable albedo | Fixed albedo |
| N | −5.91 | −4.83 | −11.25 | −12.57 |
| SW | 159.81 | 160.92 | −0.63 | −3.30 |
| LW | −53.65 | −55.07 | 6.20 | 7.15 |
| R | 106.16 | 105.85 | 5.57 | 3.85 |
| LH | −85.98 | −84.66 | −15.58 | −14.91 |
| SH | −19.77 | −19.89 | 0.59 | 0.08 |
| LW (down) | 326.22 | 324.97 | 30.27 | 30.54 |
| SW (down) | 184.98 | 186.53 | −2.62 | −3.94 |
| LW (up) | −379.87 | −380.04 | −24.07 | −23.38 |
| SW (up) | −25.16 | −25.61 | 1.98 | 0.64 |
| R (TOA) | 4.84 | 5.42 | −9.70 | −11.60 |
| S | −6.32 | −6.13 | −1.83 | −1.59 |
| P | 2.95 | 2.89 | 0.56 | 0.55 |
| PW | 16.85 | 16.80 | 6.20 | 6.01 |
| A | 13.60 | 13.60 | −0.90 | 6.00 |

*Note*: down = downward fluxes; up = upward fluxes; up = TOA = top-of-atmosphere. All energy fluxes are expressed in $W m^{-2}$; precipitation P in $mm d^{-1}$; precipitable water, PW in mm and albedo A as %. R is the radiation budget.
*Source*: Colman *et al.* (1993).

An insufficient consideration of snow cover in most atmospheric general circulation models shows itself in an inadequate description of the heat balance of the soil–snow-cover system in the regions with a seasonal snow cover, run-off due to snow melting, solar radiation absorption by snow, and snow albedo. Available atmospheric general circulation models do not sufficiently consider the effect of vegetation on land surface processes, though generally there is a need for interactive consideration of the soil–vegetation–atmosphere system.

As for the physical processes and their parametrization, the unsolved problems of primary importance are the following: (1) developing statistical techniques to describe spatial variability of the forcing parameters (rainfall, in particular) and parameters characterizing physical land surface processes, (2) improving the parametrization of hydrothermal processes in snow cover (especially snow albedo), (3) developing techniques to establish correlations between substantial hydraulic and thermal properties of soil and classes of soil texture shown on global maps of soil (properties of the first priority are efficient porosity and hydraulic conductivity of water-saturated soil, the dependence of the conductivity logarithm on soil moisture, heat conductivity and specific heat capacity of soil), (4) establishing a similar correlation between the depth of the root zone of plants, soil moisture and vegetation classes (including the vegetation phase) shown on the global maps of vegetation cover and (5) developing techniques for parametrizing the state of natural and cultivated vegetation canopy (with and without irrigation) from the viewpoint of the type and density of the canopy, albedo, rate of transpiration from moisture and roughness of soil surface, depth of the root zone and vegetation phases.

In establishing the parametrization of processes in soil and vegetation, the specific biology of plants as well as interactions among the atmosphere, soil and vegetation must be taken into account. Of great importance are further numerical experiments on the assessment of atmospheric general circulation models' sensitivity to the parametrization of land surface processes, bearing in mind the following order of priority of the effects of (1) the choice of precipitation models on horizontal scales typical of soil moisture anomalies (and, generally, atmospheric general circulation models and parametrization of hydrological processes in the soil), (2) regional anomalies of snow-free surface albedo (including soil moisture and vegetation feedbacks), (3) winter snow cover, its interannual variability and extent, (4) anomalies manifested through albedo and the respective effect on soil moisture, (5) the sensitivity to variations in surface irregularities, especially those connected with deforestation and (6) the heat content in soil on diurnal and annual changes.

Under the auspices of several international organizations (COSPAR, IAMAP, UNEP), an International Satellite Land Surface Climatology Project (ISLSCP) has been worked out, with the major objective of using satellite data to improve and apply techniques for retrieving land surface parameters that would be interesting for studies of climate and its changes, as well as the impact of climate variations on the surface (soil, vegetation, ice cover) (Henderson-Sellers 1992, 1993, Sellers *et al.* 1992). The International Satellite Land Surface Climatology Project (ISLSCP) began in 1983 when a group of interested scientists from the remote sensing, climate and ecological communities came together to discuss the future of global environmental studies based on state-of-the-art satellite systems and general circulation models linked to interactive biosphere models. Inspired by the ideas generated by the nascent International Satellite Cloud Climatology Project, S. Ichtiaque Rasool, then with IBM Paris, and Hans-Jurgen Bolle, then with the University of Innsbruck, organized two small workshops. Plans were developed to assure the calibration of satellite sensors, to develop and validate algorithms to obtain biological and climatically important parameters from satellite observations, to study the problems in scaling from point measurements to the scale of satellite pixels and then to the scale of global circulation models and to use satellite data and models 'to better understand' the biosphere's role in the climatic system.

It was an exciting time in the development of the study of global change. Several factors combined to reorient the scientific community's approach to the study of the Earth's environment. These factors included advances in remote sensing, remote sensing technology and climate modelling. General circulation models had reached a point of sophistication where it became important to treat land surface properties in a realistic manner, allowing for the dynamic, biologically controlled response of the soil and vegetative cover in the partitioning of incoming solar radiation into latent and sensible heat.

To achieve the ISLSCP objectives, plans were made to organize retrospective studies of variations in the surface in key areas, using available data from surface and satellite observations (albedo, vegetation canopy, surface temperature), as well as to search for the key areas where surface variations are most sensitive to climate variations by making complex (satellite, aircraft and surface) observations in these areas. Bolle and Rasool (1985) performed a detailed analysis of the possibility of retrieving some of the parameters enumerated in Table 4.21, and emphasized the important role of combined satellite and ground observations at key sites. Substantial difficulties in data interpretation are connected with a solution to the problems of filtering out partial cloudiness entering the radiometer's field of view and of atmospheric correction.

The criteria for the choice of key sites to carry out complex observational programmes within the ISLSCP are: (1) high sensitivity of the respective regional climates to the ongoing or possible changes of surface properties, (2) the presence of a considerable anthropogenically-induced large-scale transformation of the surface, (3) a strong natural variability of regional climates, (4) a variety of surface types and their significance for agriculture and (5) good knowledge of the site.

Complex observations on two scales were planned: (1) local (largely to provide a satellite data calibration) and (2) covering large territories (to substantiate various parametrization techniques). Local key sites have been chosen in north-eastern Brazil, Senegal, south Sudan, north Africa, Argentina, the Canadian tundra, different regions of the USA, France, Austria, India and others. Large-scale key sites have been suggested in the northern (Canada) and southern (USA) parts of the Great Plains, central Niger, Sonoran desert (near the USA–Mexico frontier), a region near the Amazon river, etc.

The WCRP Joint Scientific Committee Steering Group has discussed the problems and ways to accomplish the programme of studying the role of land surface processes in the formation of climate and its changes (such processes are defined as phenomena determining the exchange of heat, moisture and momentum between land surface and the atmosphere) (World Meteorological Organisation 1995). For a complete comparison of the results of numerical modelling of climate (from the viewpoint of land surface processes parametrization), it would be worthwhile to develop a one-dimensional model which would provide an interactive description of processes on the surface and in the atmosphere boundary layer, and of cloud formation, as well as a simulation of the established diurnal and annual change, independent of initial conditions.

Considerable progress has been made in all the areas outlined in the original ISLSCP plans from 1983. Critical satellite data were preserved through effective inter-agency and international cooperation. Existing data and algorithms were used in the ISLSCP Retrospective Analysis Program (IRAP) leading to two special issues of journals: *Forest Agriculture and Meteorology*, volume 52, Nos 1 and 2, edited by G. Asrar, August 1990, and *Climatic Change*, volume 17, Nos 2 and 3, edited by G. Asrar, December 1990.

The first major experiment of the ISLSCP was the First ISLSCP Field Experiment (FIFE). The major experimental phase was conducted in 1987, just four years after ISLSCP began. However, it was quickly realized that a second field season was needed to compensate for an unexpectedly wet summer and this was accomplished in 1989. The experience gained in running that programme has now been used to help in planning other more recent experimental campaigns. A special issue of the *Journal of Geophysical Research*, volume 97, part D17, November 1990, has been produced with papers on the results from FIFE (see also American Meteorological Society 1995b).

The reliability of various parametrization techniques could be tested based on the realization of the Hydrological Atmospheric Pilot Experiment (HAPEX) aimed at (1) a better understanding of the land surface hydrometeorological processes and meteorological processes, on horizontal scales from several kilometres to characteristic grid sizes of climate models, to improve parametrization techniques, (2) testing the reliability of parametrization techniques of heat and moisture exchange, soil

**Table 4.21** Land surface parameters to be monitored and modelled

| Parameter | Spatial resolution (km × km) or (km) | Repeatability | Period of observations | Error: Absolute | Error: Relative | Remarks |
|---|---|---|---|---|---|---|
| *A Global modelling and monitoring of climate* | | | | | | |
| Vegetation cover | 250 × 250 | Three times per month | 1 year | ±0.057 | ? | Part (%) of major types of vegetation |
| Dry-soil albedo | 250 × 250 | Annual mean | 3 yr | ±0.05 | ±0.02 | Dry-soil albedo as a standard for comparison |
| Averaged surface albedo | 250 × 250 | 1 month | 1–3 yr | ±0.05 | ±0.02 | Parametrization in climate models |
| IR emissivity | 250 × 250 | Annual mean | ? | ±0.02 | ±0.01 | An account of the dependence of the type of surface |
| *B Validation of global data bases* | | | | | | |
| Surface temperature (diurnal and nocturnal) | 250 × 250 | 1 month | 1 yr | ±1 °C | ? | Data on diurnal change are needed |
| Noon global radiation | 250 × 250 | 1 month | 1 yr | $\pm 25\ \mathrm{W\,m^{-2}}$ | $\pm 10\ \mathrm{W\,m^{-2}}$ | |
| Soil moisture | 250 × 250 | 1 month | 1 yr | | | Five gradations of water content in the 0–10 cm layer |
| Precipitation index | 250 × 250 | ? | ? | ? | ? | Present retrieval error is 10–25% |
| Extent and depth of snow cover | 250 × 250 | 1 week/1 month | 1 yr | ? | ? | |
| Noon radiation budget | 250 × 250 | 1 day/1 week | 1 yr | $\pm 10\ \mathrm{W\,m^{-2}}$<br>±10–20% | | |
| *C Meso-scale modelling and monitoring of climate* | | | | | | |
| Surface topography (height, slope) | 10–20 | Fixed | From diurnal change on some days to a week round-the-year | ±5–10% | ? | |

**Table 4.21** *Continued*

| Parameter | Spatial resolution (km × km) or (km) | Repeatability | Period of observations | Error | | Remarks |
|---|---|---|---|---|---|---|
| | | | | Absolute | Relative | |
| Basic types of surface | 10–20 | 1 week | | ±10% | ? | |
| Share of vegetation, % | 10–20 | 1 week | | ±10% | ? | |
| Roughness | 10–20 | 1 week | | ±10% | ? | |
| Albedo | 10–20 | 1 week | | ±0.01–0.03 | ±0.02 | |
| Height of plants | 10–20 | 1 week | | ±5–10% | ? | |
| Leaf area index | 10–20 | 1 week | | ±0.05% | ? | |
| Emissivity | 10–20 | 1 week, 1 month | | ±0.01–0.02 | ? | |
| Soil structure | 10–20 | 1 season | | ±10% | ? | |
| *D Validation of meso-scale data bases* | | | | | | |
| Surface temperature | 5 × 5 | 1 month | 1 month/1 yr | ±0.5° | – | Also, soil moisture and surface wind |
| Evaporation | 5 × 5 | 1 month | 1 month/1 yr | ±10–20% | – | |
| Global radiation and radiation budget | 5 × 5 | 1 month | 1 month/1 yr | $\pm 10\ W\ m^2$ | – | |
| Precipitation | 5 × 5 | 1 day | 1 month/1 yr | $\pm 0.1\ mm\ day^{-1}$ | – | |
| *E Additional parameters to monitor the effect on climate* | | | | | | |
| Length of vegetation period | 250 × 250 | 1 week | Continuously | ±0.05 | ±0.02 | |
| Variations in the share (%) or type of vegetation | 50 for a search monitoring | ? | | | | |
| Soil moisture variations | 250 × 250 | ? | Continuously | ±0.05 | ±0.02 | |
| Soil erosion: | Several km | Twice a day | Continuously | ? | ? | |
| (1) Wind: dust storms | – | 1 day/1 week | 5 yr | ? | 0.1 | 1 km |
| (2) Water: hydrological network | 50 | 1 month | 5 yr | ? | ? | |
| Permafrost zone | 50 | 1 month | 5 yr | ? | ? | |

moisture and run off, applied in climate models and (3) a complex interpretation of remote sensing results with the use of direct measurement data. For an adequate account of the spatial and temporal variability of the processes, it is necessary to accomplish programmes of the HAPEX type in different climatic conditions at test sites with areas from 50 km × 50 km to 500 km × 500 km, covering a certain watershed territory, for at least one year (with periods of observations every hour to obtain data on diurnal change).

Scientifically substantiated by the WCRP plan, the requirements of observation accuracy necessitate an estimation of monthly mean energy fluxes over the 2° by 2° latitude–longitude grid (about 200 km × 200 km), with an error of 10 W m$^{-2}$, which is equivalent to an error of 10 mm per month for precipitation. For the territory 100 km × 100 km, this means that the errors of estimation of total precipitation and run-off must not exceed $10^8$ m$^3$ per month and 40 m$^3$ s$^{-1}$, respectively. To achieve the HAPEX objectives, errors of 10 W m$^{-2}$ for hourly mean surface heat balance components and about 5 per cent for monthly mean precipitation should be considered acceptable.

The complex of parameters to be measured includes: (1) sensible fluxes of heat and moisture, (2) heat flux in soils, (3) radiation budget components (including surface albedo and temperature), (4) precipitation, (5) standard meteorological quantities (temperature, relative humidity, wind speed, cloud amount) near the surface and in the atmospheric boundary layer, (6) hydrological properties of the soil (moisture content, groundwater level), (7) the run-off within the watershed basin and (8) surface characteristics (topography, vegetation cover, type of soils, land use). An adequate account of land surface processes in climate models is determined by simulation accuracy for: internal atmospheric dynamics, radiative regime, atmospheric boundary layer structure (from the viewpoint of the profiles of temperature and wind, as well as heat fluxes, moisture and momentum) and hydrological processes. An accomplishment of HAPEX will make it possible to obtain information about all these aspects, with the exception of the first.

One of the most important input parameters in numerical modelling of atmospheric general circulation and climate is surface albedo. Correct prescriptions (and variability parametrization) of albedo are, however, very difficult to achieve, because of a wide range of albedo variations and a limited data base of albedo observations, which determines the urgency of surface albedo retrieval from satellite observations.

Table 4.22 contains a summary of albedo prescriptions in different atmospheric global circulation and climate models (Henderson-Sellers and Wilson, 1983). This table illustrates a diversity of approaches, of great concern being differences in techniques for the parametrization of snow and ice cover albedos, since an account of the albedo feedback is one of the principal features of the present models of climate.

Though the retrieval of surface albedo from satellite data is very interesting (also because of natural spatial averaging close to the scales of the grid used in numerical modelling), there is no doubt that satellite information on its own is not enough (for example, it is impossible to estimate surface albedo in overcast conditions using only satellite information). The significant differences between the global maps of albedo drawn from satellite and conventional data have various causes – mainly differences in retrieving surface albedo. All this suggests that the available data on surface albedo should not be considered reliable, although recently successful attempts have been made to accumulate data on albedo and to substantiate global data bases on surface albedo.

**Table 4.22** Surface albedo prescribed in AGC climate models

| Example centre | Model | Snow-free and ice-free surfaces | Snow-covered and ice-covered surfaces |
|---|---|---|---|
| Atmospheric Environment Service (Canada) | AES | no specific details but implied geographical distribution based on Posey & Clapp (1964) | follows Holloway & Manabe (1971), equation (3): see GFDL |
| Australian Numerical Meteorology Research Centre | ANMRC | latitudinal variation based on Posey & Clapp (1964) | snow albedo prescribed as latitudinal variation of $\alpha$. $\alpha$ of sea ice = 0.07 |
| Computing Centre, Siberian Academy of Sciences | CCSAS | $\alpha = 0.2$, bare ground<br>$\alpha = 0.1$, ocean | snow: $\alpha - 0.2 + 0.4d_{sw}$<br>$\alpha \leqslant 0.6$<br>(same as NCAR)<br>ice: $\alpha = 0.6$ |
| Geophysical Fluid Dynamics Laboratory (Holloway & Manabe 1971) | GFDL | geographical distribution based on Posey & Clapp (1964) | snow: equation (3)<br>$\left.\begin{array}{ll}\alpha = \alpha_1 + (0.6 - \alpha_1)d_{sw}^{1/2} & d_{sw} < 1\,\text{cm} \\ \alpha = 0.6 & d_{sw} \geqslant 1\,\text{cm}\end{array}\right\} \quad (3)$<br>poleward of 75° lat., albedo for land and pack-ice 0.75 |
| (Manabe & Stouffer 1980) | GFDL | geographical distribution based on Posey & Clapp (1964) | sea ice: 0.5 for lat. <55°,<br>0.7 for lat. >66.5°,<br>0.45 if top melting |
| Goddard Laboratory for Atmospheric Sciences | GLAS | Feb.: geographical distribution based on Posey & Clapp (1964)<br>Aug.: Charney *et al.* (1977) vegetated land 0.14, desert 0.35, ocean 0.07 | snow and ice 0.70<br>Holloway & Manabe (1971) equation (3): see GFDL |
| Goddard Institute for Space Studies | GISS | land: 8 vegetation types have seasonally varying albedos for <0.7 μm and ≥0.7 μm | snow-free ice: 0.45 ocean, 0.5 land<br>snow albedo: $\alpha_s = 0.5 + e^{-a/5}$<br>ground partly snow-covered albedo: $\alpha_1 + (\alpha_s - \alpha_1)(1 - e^{-dsw/dmask})$<br>two spectral regions |
| Meteorological Office, U.K. | UKMO | 5-level model: snow-free land values vary with latitude range 0.150–0.223 | snow:<br>$\alpha = \alpha_1 + 0.38d_{sw}^{1/2}$<br>$\alpha \leqslant 0.6$<br>similar to Holloway & Manabe (1971), equation (3): see GFDL<br>sea ice and permanent snow cover:<br>$\alpha = 0.8$; $T_0 < 271.2\,\text{K}$<br>$\alpha = 0.5$; $T_0 \geqslant 271.2\,\text{K}$ |
| | UKMO | 11-level model: land 0.2, sea (where effective) 0.06 | transient snow cover 0.5, permanent snow cover, land ice and sea ice 0.8 |
| National Center for Atmospheric Research | NCAR | originally:<br>geographical distribution based on Posey & Clapp (1964)<br>third-generation model:<br>two spectral regions <0.7 μm and ≥0.7 μm; dependence on vegetation type and extent | originally:<br>snow or ice:<br>$\alpha = 0.2 + 0.4d_{sw}$ and<br>$\alpha \leqslant 0.6$<br>third-generation model:<br>snow albedo function of age and depth of snow: two spectral regions |
| Oregon State University | OSU | geographical distribution based on Posey and Clapp (1964) and model's 9 surface types | fixed value |
| Rand Corporation | RAND | geographical distribution based on Posey & Clapp (1964) | |
| University of California at Los Angeles | UCLA | bare soil 0.14, ocean 0.07 | snow-covered 0.7, ice-covered soil or sea water 0.4 |

Symbols: (i) $d_{sw}$ is the water equivalent depth of snow (centimetres); (ii) $a$ is the snow age (days); (iii) $d_{mask}$ is the vegetation snow masking depth equivalent thickness of water (centimetres); (iv) $\alpha_1$ is the snow-free land albedo, and (v) $T_0$ = surface temperature.
Source: Henderson-Sellers and Wilson (1983).

One of the most important climate-forming factors is evapotranspiration. Mintz *et al.* (1983) noted that, although the effect of evapotranspiration on the weather and climate is difficult to assess from observational data, it is reliably proved by numerical modelling of the atmospheric global circulation to play a substantial role. A reduction of heat expenditure through evaporation off the land surface, in comparison with the ocean, is approximately balanced by increased sensible heat loss. However, the latter causes a local heating only of a relatively thin atmospheric boundary layer, whereas latent heat released at water vapour condensation leads to a heating of the whole atmosphere and, due to advection, can manifest itself after a time lag in remote regions. This difference between the features of the spatial and temporal manifestation of two types of heat flux divergence determines the sensitivity of thermally-induced large-scale atmospheric global circulation to evapotranspiration.

The Goddard Laboratory for Atmospheric Sciences AGCM model has been generalised by Mintz *et al.* (1983) through an account of the atmosphere–biosphere interaction manifested through evapotranspiration. The water and energy transport through the surface depends on the morphology (spatial structure) and physiology of the vegetation cover. In the biospheric model considered, the vegetation morphology is determined by dependences of specific leaf area on height, $L_D(z)$, calculated per unit volume, and of specific length of the root system on depth $RT_D(z)$, which is determined as total length of the roots per unit volume, represented in discrete form for each of the cells of the AGCM model grid. The simplest version of the biospheric model foresees precribed, for each cell, values of $L_D(z)$ and $RT_D(z)$ depending on season, from data of phenological observations. In a more complicated model, an interaction is considered between these quantities and precalculated parameters of the atmosphere and soil moisture. Therefore, on time scales of phenophases, $L_D(z)$ and $RT_D(z)$ depend, for example, on winds and temperature extremes.

A more advanced model foresees a differentiation of vegetation cover (tropical forests, mid-latitude forests, deserts, grass cover, etc.) and its interaction with conditions in the atmosphere and soil. Specific leaf area affects: (1) the aerodynamical resistance to latent and sensible heat transport, (2) radiation transfer through vegetation cover and (3) evaporation from the surface and precipitation. The effect of the internal structure of roots, stems and leaves on water and energy transport is a major physiological factor. The climate model, incorporating variables of the biosphere, is planned to be the first, using FGGE data as initial information, with a fixed sea surface temperature. At a later stage, an interactive atmosphere–ocean–biosphere model will be developed, including the parameter of heat transport in the deep layers of the ocean. The realization of such a model is of particular importance for studying anthropogenic impacts on climate.

Dickinson (1982, 1984, 1985, 1986a,b, 1987, 1989) proposed a technique for parametrization of evapotranspiration in 3-D climate models, which permits a separate account of contributions to evaporation from the soil surface and vegetation cover. The soil moisture is assumed to be concentrated in the upper 1 m layer, but the 10 cm surface layer is also considered separately. In order to test the reliability of such a parametrization, comparisons have been made with a multi-layer model of evaporation from the soil surface. Parametrized evaporation from the soil is determined as a possible rate of evaporation or maximum rate at which water can diffuse to the surface, depending on which of these rates is less. The maximum rate is estimated empirically from soil-hydrological parameters.

Evapotranspiration of vegetation cover is incorporated as evaporation of water accumulated on the surface of vegetation cover elements, or as transpiration of water extracted by the roots from the soil. The water vapour flux from the external surface of vegetation cover elements to the atmosphere over the vegetation cover is determined by the difference between water vapour concentrations near the surface and in the atmosphere, as well as by the water input through the vegetation cover molecular boundary layers and through the vegetation cover as a whole (aerodynamical resistance). Transpired water suffers an additional resistance in the process of its transition from the interior to external parts of vegetation cover elements, whose temperatures and saturating water vapour pressure are calculated with the use of the vegetation cover heat balance model.

A rough account of precipitation infiltration to the soil and of the surface run-off is made, since it is apparently insignificant. Soil moisture determines the maximum rate of water extraction by the roots from the soil and, if evapotranspiration forces roots to exceed this maximum rate, the roots become narrowed until the respective equilibrium is reached. A detailed simulation of the technique for land surface evapotranspiration parametrization is needed for the realistic reproduction of the diurnal change of surface temperature and evapotranspiration in meso-scale and global climate models. The resulting increase in reliability of numerical modelling should manifest itself first in calculations of the surface, and in the estimates of climatic parameters of the surface and soil. This increase of reliability is possible, however, only under conditions of a reliable parametrization of the vertical transport in the atmospheric boundary layer and of the processes of precipitation formation, cloud cover formation and the radiation budget.

Practical application of the technique discussed is complicated by its cumbersome nature, determined by having to take into account small-scale processes, by lack of reliable input parameters (hydrological characteristics of the soil, vegetation cover properties), which should be prescribed at grid points, and by the necessity of simultaneous adequate parametrization of some other processes. In this connection, detailed numerical sensitivity studies are needed which would allow analysis of the priority of various factors.

Improved climate models necessitate a parametrization of not only heat and moisture fluxes from the vegetation-covered surface but also of the processes of dry deposition onto the surface of various gas and aerosol components. Of particular interest is, for example, an account of the fluxes of $CO_2$, $SO_2$ and $NO_2$, depending in particular on stomatal resistance in the leaves of plants as determining the surface sinks of various components of the atmosphere. The parametrization of related processes could be achieved, based on the use of schemes similar to that described.

As Dickinson (1984) notes, various changes are possible in the vegetation cover characteristics under the influence of climate changes which can cause albedo–climatic feedbacks. Changes which lead to an albedo increase are as follows, for example: the southward retreat of the northern boundary of forests (where areas of coniferous trees change into tundra), transformation of pastures into deserts and tropical forests into savannahs.

Although during the last few thousands of years, an anthropogenically induced decrease of forest areas has occurred, it seems that an associated change in albedo has not substantially affected the global climate. Probably, the desertification processes have only led to comparatively small albedo variations. Therefore, numerical modelling within the Climate of the Past Programme (CLIMAP) naturally revealed only a 0.22 per cent decrease of the global mean annual albedo of the surface–atmosphere system during the last 18 thousand years (from the time of the Wisconsin glaciation), which is equivalent to a 0.75 W $m^{-2}$ decrease of reflected radiation.

Shukla and Mintz (1982) performed numerical experiments to analyse climatic sensitivity to changes in evapotranspiration, based on the use of the GLAS global nine-level atmospheric general circulation model. The model with resolution 4° latitude × 5° longitude has the following prescribed parameters: sea surface temperature, topography, surface roughness and albedo, and soil moisture content as resources of evapotranspiration.

Two cases of limitations with regard to evapotranspiration have been considered: (1) when its quantity is equal to potential evapotranspiration modelled for conditions when the soil is wet and totally covered with vegetation and (2) when evapotranspiration is absent. The first case (wet soil) corresponds to an Earth totally covered with vegetation (with adequate moisture content) and the second case (dry soil) corresponds to the Earth without vegetation.

Shukla and Mintz (1982) discussed the results of integration of the atmospheric general circulation model equation system for the cases of wet and dry soil for 60 days, with the initial state corresponding to the 15 June atmosphere. These results characterize the July atmospheric general circulation when, in the northern-hemisphere extra-tropical latitudes, evapotranspiration is at a maximum. An analysis of data for the subsequent 15 days (1–15 August) has shown that under fixed boundary conditions, the July atmospheric general circulation in both these cases almost reaches an equilibrium. The results obtained show that in the case of wet soil, precipitation over Eurasia and North America is close to potential evapotranspiration, whereas in dry-soil conditions it is either small or completely absent (except for south-eastern Asia and India, where a strong moisture transport from the ocean takes place). The ocean moisture input provides a substantial prevalence of precipitation over evapotranspiration (wet soil) in the tropical latitudes of South America and Africa.

In the case of dry soil, a strong surface temperature increase takes place north of 20° S due to (1) a decrease of heat loss on evaporation (in the case of wet soil the heat loss in the belt 20° S to 60° N averages 126 W $m^{-2}$) and (2) an increase of global radiation from 172 to 257 W $m^{-2}$ (on average, for the belt 20° S to 60° N) caused by decreased cloud amount. In the case of dry soil, an increased radiative heating of the surface is almost entirely spent on sensible heat loss to the atmosphere (169 W $m^{-2}$, compared with 21 W $m^{-2}$ for wet soil), which leads to the formation of strong thermal low-pressure centres over the continents. The pressure drop over the continents by 5–15 hPa is compensated by a pressure increase, non-uniformly distributed over the oceans; if the pressure over the North Atlantic is almost constant, in mid-latitudes of the northern Pacific Ocean the pressure rises strongly.

Evapotranspiration takes place under three conditions: (1) when the soil is wet, (2) when vegetation releases the moisture to the atmosphere and (3) when the radiative heat is enough for the moisture to evaporate. Since the surface radiation budget depends on albedo, and the latter is determined by plant features, the respective feedback must be taken into account. If the plants favour an evaporation, the clouds play the role of the centres of water vapour condensation. In the extra-tropical latitudes, with a characteristic strong annual change, the soil functions as an ocean, accumulating the moisture in winter and evaporating it in summer, whereas the ocean accumulates the radiative heat in summer to return it to atmosphere over the ocean in winter. Hence, the problem of soil moisture monitoring should be urgently considered.

Of great importance is revealing the consequences of the impact of anthropogenic forcings on vegetation cover and soils (deforestation or reforestation, melioration, soil erosion, etc.), from the viewpoint of their effect on climate.

Based on the use of the National Center for Atmospheric Research (NCAR) Community Climate Model, Wilson *et al.* (1987) performed numerical climate modelling to answer the following two questions: (1) what is the effect of land surface processes on climate and (2) what is the priority of model parameters and parametrization schemes for these processes? Calculations were aimed at analysis of the interaction between the variables of vegetation cover, precipitation and leaf area index, soil texture, snow cover and latent and sensible heat fluxes.

The model unit responsible for land-surface processes parametrization takes into account the interaction between the atmosphere and soil–vegetation cover system (Biosphere–Atmosphere Transport Scheme, BATS) which consists of a two-layer soil (its upper layer containing the root system interacts directly with the atmosphere) and a one-layer vegetation cover. This model of interaction is not very logical, since the atmospheric characteristics are prescribed; these include a sinusoidal change of surface air temperature and global radiation, precipitation, $3\,m\,s^{-1}$ wind speed, as well as relative humidity which varies as a function of the type of ecosystem considered.

Numerical modelling has been largely aimed at assessing the climate sensitivity to parameters of soil characteristics and their variability. Calculations were made for five types of ecosystem: (1) low-latitude evergreen forests, (2) low-latitude sand deserts, (3) high-latitude coniferous forests, (4) high-latitude tundra and (5) prairies. Each of the ecosystems is considered for a certain season. In the case of tundra, snow melting is taken into account and the grass-cover model considers an accumulation of snow. Sensitivity studies focus on analysis of the role of soil-texture variations from a rough texture typical of sand to a moderately rough structure of loam and finely dispersed clay. Climatic sensitivity to the structure of stratified soil and a parametrization of the dependence of soil albedo on moisture and snow cover thickness have been studied.

Calculations have shown that the BATS scheme is most sensitive to soil texture variability caused by variations of hydraulic conductivity and characteristics of diffusion. The effect of the albedo dependence on soil moisture turned out to be weak. Snow cover is very significant due to an associated change of surface roughness. As for the effect of soil stratification and soil-layer depth, the variability of the upper-layer thickness is most substantial. This is determined mainly by the effect of the upper-layer thickness on the moisture saturation of the upper layer, which affects hydraulic processes. Soil moisture variations can also be caused by variations of such characteristics of vegetation cover as, for example, stomatal resistance, which affects the moisture transport through vegetation cover, and hence the radiation and water balances of the soil. On the whole, climatic sensitivity to texture variations can at least be compared to the effect of variations of vegetation cover characteristics, but it may be even more important (so far, averaged characteristics of soils on the scale of the climate model grid have been poorly studied).

Study of the global distribution of vegetation is of great importance. In this connection, Matthews (1986) has established a global data base on the distribution of vegetation cover, land use and types of surface for a 1° latitude by 1° longitude grid, based on data of about a hundred publications, supplemented by results of processing satellite imagery. This data base is a set of input parameters on global distribution of albedo for the numerical climate modelling.

The whole data base, archived at NCAR (USA), consists of six parts. The first part (VENTYPE) is characterized by the distribution of natural vegetation using the classification scheme adopted by UNESCO. The second part (COLTINT) contains data on land use represented as a share (per cent) of cultivated soils for each grid cell. The remaining four parts represent global data bases (also in the form of maps) of total land surface albedo for the four seasons (January, April, July, October) in the absence of snow cover (except for the regions of continental ice with permanent snow cover). Information on albedo was obtained incorporating both natural and cultivated vegetation. The scheme of data formats on magnetic tapes is given by Matthews (1986), together with brief descriptions of albedo maps for the USA territory. A total of 32 types of formation was incorporated, with the annual change of albedo, either weak or completely lacking. For example, a constant value of 75 per cent was assumed for snow–ice-surface albedo. A recent compilation of global data sets has been made by Meeson *et al.* (1996).

Starting from April 1982, the NOAA-7 AVHRR data have been processed regularly, and weekly maps have been drawn of the global normalized difference vegetation index (NDVI) determined by the relation

$$\mathrm{NDVI} = (B_2 - B_1)/(B_2 + B_1) \tag{4.30}$$

where $B_1$ and $B_2$ are the satellite-derived brightnesses of the surface–atmosphere system for AVHRR bands 1 (0.58–0.68 $\mu$m) and 2 (0.73–1.10 $\mu$m), respectively, for the spatial resolution 5 km × 3 km. In order to minimize the effect of cloudiness, maximum NDVI values for a week have been chosen for each resolving element (they also correspond to minimum nadir angles) in order to calculate global vegetation index averages. Only NDVI values were taken into account; these vary from $-1$ to $+1$, though the range of positive values from $+0.1$ to $+0.45$ corresponds to vegetation. Since clouds, water and snow are characterized by a spectral change of brightness opposite to vegetation, NDVI $< 0$ corresponds to them. Thomas and Henderson-Sellers (1987) undertook an analysis of data on the global vegetation index for three regions (Table 4.23), to assess their usefulness in the parametrization of land surface processes in numerical climate models.

Although it has been shown that a 3–4 week sampling lowers the level of contamination of data by clouds and snow, and reduces the effect of viewing geometry, as well as illumination conditions, in this case the results are not unique. The table reveals a substantial ambiguity in the identification of even basic types of vegetation by their GVI, which cannot be used to distinguish between the classes. The GVI values for a given ecotype vary widely from region to region, which excludes the possibility of global scale classification. The conclusion is that the available GVI data base is useful only for a qualitative analysis and cannot be used as the input information in numerical climate modelling.

The global water cycle plays a key role in climate formation and biospheric functioning. In this cycle, water takes part in its three phases due to the unique position of the Earth in the solar system, which determines the observed thermal regime. The most important problems in hydrology are: (1) the development of new techniques

**Table 4.23** Characteristic global vegetation index values for various types of vegetation

| Region | Type of vegetation (season) | Estimates | | | | |
|---|---|---|---|---|---|---|
| | | | Quartiles | | Extrema | |
| | | Average | Upper limit | Lower limit | Upper limit | Lower limit |
| Africa | Wet tropical forests (May, June) | 0.309 | 0.331 | 0.283 | 0.355 | 0.264 |
| | Wet tropical forests (October, November) | 0.276 | 0.298 | 0.265 | 0.333 | 0.158 |
| | Savannah (May, June) | 0.150 | 0.215 | 0.085 | 0.278 | 0.076 |
| South America | Wet tropical forests (May, June) | 0.293 | 0.323 | 0.264 | 0.330 | 0.242 |
| | Wet tropical forests (October, November) | 0.323 | 0.341 | 0.306 | 0.345 | 0.295 |
| | Savannah (May, June) | 0.227 | 0.246 | 0.207 | 0.250 | 0.187 |
| North America | Tundra (May, June) | 0.009 | 0.042 | −0.025 | 0.054 | −0.036 |
| | Tundra (October, November) | 0.218 | 0.248 | 0.187 | 0.258 | 0.170 |
| | Boreal forests (May, June) | 0.246 | 0.294 | 0.197 | 0.325 | 0.182 |
| | Boreal forests (October, November) | 0.329 | 0.385 | 0.273 | 0.412 | 0.239 |

for assessing soil moisture and evapotranspiration on a scale from individual fields to continents, (2) a quantitative estimation of precipitation and its distribution over the catchment area, (3) global monitoring of snow cover and its water equivalent, (4) the development of testing of reliability of hydrological models, coordinated with possibilities of obtaining information by satellite remote sensing (spatial and temporal resolution, repeatability), for events of different scales (from floods on small rivers to general circulation models) and (5) further studies of the laws of the global water cycle.

Certain requirements have been analysed by Rango (1984) for observations of soil moisture, evapotranspiration, precipitation and snow cover (Table 4.24). These requirements can be met only with the use of remote sensing instrumentation, which provides measurements in a wide wavelength range, from visible to microwave spectral regions.

**Table 4.24** Requirements for observations of hydrological parameters

| Parameter | Spatial coverage | Scale of averaging: Spatial (km) | Scale of averaging: Temporal (days) | Spatial and temporal resolution of observational data: Horizontal scale (km) | Spatial and temporal resolution of observational data: Repeatability (days) |
|---|---|---|---|---|---|
| Soil moisture | Sub-continental | 10 | 3 | 10 | 1–3 |
| | Irrigated regions | 1 | 3 | 0.1 | 1–3 |
| Surface temperature | Sub-continental | 10 | 3 | 1 | daily |
| Type of vegetation and biomass | Sub-continental | 10 | weekly | 1 | 3 |
| Albedo | Sub-continental | 10 | 3 | 1 | 3 |
| Rain rate | Sub-continental | 1 | 0.5 | 1 | 0.5 |
| | Oceans | 50 | daily | 5 | daily |
| Precipitation amount | Sub-continental | 1 | daily | 1 | daily |
| | Oceans | 1–4 | daily | 1–4 | 1 hour |
| Extent | Sub-continental | 0.2 | weekly | 0.1 | 3–5 |
| Water equivalent (depth) | Sub-continental | 1–5 | weekly | 1–5 | 3–5 |
| | Samples | 1–5 | weekly | 0.5 | 3–5 |
| Liquid water content | Sub-continental | 1–5 | weekly | 1–5 | 3–5 |
| | Samples | 1–5 | weekly | 0.5 | 3–5 |

## 4.5 Optically-active minor gaseous components

The formation of climate and its changes is determined by a complicated interaction among the components of the atmosphere–hydrosphere–lithosphere–cryosphere–biosphere system. Solar radiation is practically the only source of energy, and has an important role in the atmosphere in the formation of the greenhouse effect, due to numerous optically active minor gaseous and aerosol components (Kondratyev *et al.* 1982, 1991, 1996a, Kondratyev and Moskalenko 1983, 1984, Kondratyev 1985c, 1987a,b, 1988a, 1992b, 1996, Houghton *et al.* 1990, 1992, Schönwiese and Runge 1991). The cloudiness interaction is a key dynamical climate-forming mechanism (Marchuk *et al.* 1986). This determines the decisive importance of chemical and photochemical processes responsible for atmospheric composition as climate-forming factors.

The components of atmospheric chemical composition, directly affecting the most substantial atmospheric phenomena (Table 4.25) have been considered by Clark (1982, 1983, 1985). The significance of these components can vary, depending on the specific problems under consideration. It is also important to take into account numerous indirect impacts (for example, the effect of halocarbon compounds on ozone) and synergism, determined by interactions among various chemical processes in the atmosphere and biosphere.

The chemical composition of the atmosphere is affected by various factors (Table 4.26). Table 4.27 illustrates a synoptic assessment of the impacts on the atmosphere.

As has been mentioned above (see Chapter 2), an increasing amount of $CO_2$ in the atmosphere is of great concern because of its possible effect on climate, which is likely to show itself in a tropospheric warming and stratospheric cooling. So far, however, possible feedback mechanisms, which can complicate the nature of the effect of $CO_2$ on climate, have not been considered adequately. For example, the

**Table 4.25** Major impacts of atmospheric chemical components on the most important atmospheric phenomena

| Components | UV radiation absorption | Greenhouse effect | Photochemical formation of oxidants | Acidification of precipitation | Drop of visibility | Corrosion of materials |
|---|---|---|---|---|---|---|
| $O_3$ | + | + | + | | | |
| C (soot) | | + | | | + | |
| $CO_2$ | | + | | | | |
| CO | | | | | | |
| $CH_4$ | | + | | | | |
| $C_xH_y$ | | | + | | + | |
| $NO_x$ | | | + | + | + | |
| $N_2O$ | | + | | | | |
| $NH_3/NH_4^+$ | | | | + | | |
| $SO_x$ | | + | | + | + | + |
| $H_2S$ | | | | | | + |
| COS | | | | | | |
| Organic sulphur | | | | | | + |
| Halocarbons | | − | | | | |
| Other halogens | | | | | | + |
| Minor components | | | | | | + |

*Note*: + signifies a positive effect of a given component on the respective phenomenon.

**Table 4.26** Principal factors affecting atmospheric chemical composition

| Components | Oceans and estuaries | Vegetation and soil | Wild animals | Wetlands | Biomass burning | Crop production | Domestic animals | Petrol burning | Coal burning | Industrial processes |
|---|---|---|---|---|---|---|---|---|---|---|
| $O_3$ | | | | | | | | + | | |
| C (soot) | | | | | + | + | | + | + | |
| $CO_2$ | + | + | | | + | + | | + | + | |
| CO | + | + | | | + | | | + | + | + |
| $CH_4$ | | + | + | + | + | + | + | | | |
| $C_xH_y$ | + | + | | | + | + | | | | |
| $NO_x$ | + | | | | + | + | | + | + | + |
| $N_2O$ | + | + | | | + | + | | + | + | |
| $NH_3/NH_4^+$ | | + | + | + | + | + | + | | + | |
| $SO_x$ | | | | | | | | + | + | + |
| $H_2S$ | + | + | | + | | + | | | | |
| COS | + | + | | + | | | | | | |
| Organic sulphur | + | + | | + | | | | | | |
| Halocarbons | | | | | | | | | | + |
| Other halogens | | | | | | | | + | + | + |
| Minor components | | | | | + | | | + | + | + |

*Note*: + signifies a positive effect of a given component on the respective factor.

**Table 4.27** A synoptic assessment of impacts on the atmosphere

| | Atmospheric property | | | | | |
|---|---|---|---|---|---|---|
| Source | UV energy absorption | Thermal radiation budget alteration | Photochemical oxidant formation | Precipitation acidification | Visibility degradation | Materials corrosion |
| Oceans and estuaries | II-1 | | | | | II-3 |
| Vegetation and soils | | II-2 | III-2 | III-1 | | |
| Wild animals | | III-1 | | | | |
| Wetlands | | III-1 | | | | III-2 |
| Biomass burning | II-2 | II-2 | | | II-2 | |
| Crop production | I-1 | II-2 | | | | |
| Domestic animals | I-2 | | | | | |
| Petroleum combustion | II-2 | III-3 | IV-3 | II-3 | II-3 | II-3 |
| Coal combustion | III-2 | III-3 | I-3 | III-3 | III-3 | III-3 |
| Industry | III-3 | II-2 | | II-3 | II-3 | II-3 |

*Note*: Potential importance: I, some; II, moderate; III, major; IV, controlling.
Assessment reliability: 1, low; 2, moderate; 3, high.

tropospheric warming will affect the distribution of water vapour, cloudiness, winds, precipitation and sea ice. The total effect of the respective feedbacks cannot be known beforehand: it might lead to mutual compensation or intensification of the $CO_2$-induced effect. A similar situation is typical of the stratosphere, where a change in the temperature regime can affect photochemical processes and lead to a redistribution of ozone.

So far, the effect of $CO_2$ on climate is not at all well understood. Moreover, the radiative regime of the atmosphere is affected by many other optically minor active gaseous constituents (oxides of nitrogen, CFCs, etc.). Therefore, there is a need for the following: (1) a reliable parametrization of the radiative processes of various gases in climate models and a determination of the priority of their inclusion (recent studies revealed, for example, the importance of assessing the climatic consequences of increased ozone content in the troposphere), (2) the development of atmospheric models which characterize the typical features of the fields of concentration of the more important minor gaseous constituents, (3) continuous observations to reveal the trends of such components, (4) the development of photochemical blocks of climate models whose application will make it possible to make the concentrations of optically active minor gaseous constituents the internal variables of climate models, (5) the continuation of laboratory studies to specify data on the rates of various photochemical reactions and to enhance data bases on the spectral line parameters for various minor gaseous constituents and (6) an improvement of climate models used to assess climate sensitivity to varying atmospheric composition, by interactive consideration of processes in the ocean and with the ice cover contribution taken into account.

Progress achieved during recent years in the development of observational techniques and in a deeper understanding of the laws of chemical and physical processes in the atmosphere has made it possible to reveal substantial changes in the concentration of numerous chemically and optically active minor gaseous constituents of the atmosphere. Theoretical calculations have led to the conclusion that the contribution of increased concentrations of such components as methane, nitrous oxide, CFCs, etc. to the formation of the atmospheric greenhouse effect (and respective climate changes) turns out to be comparable to the effect of increased $CO_2$ concentration, an interactive effect (non-additive effect) of various components on the greenhouse effect with climate playing the substantial role. So, for example, a temperature increase would be followed by an intensification of natural processes of the formation of methane affecting the concentration of CO and $O_3$. In forecasting possible trends of minor gaseous constituent concentrations in the future, and associated changes in the atmospheric composition and climate, a thorough account of the complicated interaction between various processes is needed.

The simplicity and computational economy of 1-D models of the interaction between the processes establishing a radiative–convective equilibrium and the kinetics of chemical reactions have determined their wide use in assessing possible climate changes. There is an obvious need for more complete numerical modelling based on the use of 2-D and 3-D models incorporating real variability of minor gaseous constituent concentrations (instead of their supposed stepwise growth). In this connection, Wuebbles (1988, 1991) and Wuebbles *et al.* (1984, 1989, 1991) undertook a determination of scenarios of a possible variability of minor gaseous constituent concentration both in the past and in the future, with the following aims: (1) to establish a minimum of needed initial data including their acceptable approximate prescription, the use of which will make it possible to provide compatability of numerical modelling results (the choice of components is determined by the global-scale manifestation of their effects), (2) to limit as much as possible the number of scenarios, (3) to provide an interactive account of the effect of various components on climate, considering the data of minor gaseous constituents concentration in total but not for individual components, (4) to reveal characteristic trends both in the past and in the future, based on extrapolation of observational data but including the necessity to consider some extreme situations, (5) to analyse the effects of differences in data on changes in the chemical composition of the atmosphere and determination of alternative scenarios and (6) if possible, to update data bases with new information.

Tables 4.28 and 4.29 summarize data characterizing the scenarios discussed. In the future, however, it will be necessary to take into account such minor gaseous

**Table 4.28** Summary of proposed scenarios of changes in concentrations ($C$) of various components with time $t$ (years)

| $t_0$ | $t$ | Concentration |
|---|---|---|
| *Carbon dioxide* (ppm) | | |
| January 1850 | January 1958 | $C_{CO_2} = 270\exp[0.00141(t - t_0)]$ |
| January 1958 | January 1983 | $C_{CO_2} = 270 + 44.4\exp[0.019(t - t_0)]$ |
| January 1983 | January 2100 | (a) $C_{CO_2} = 341.4 + 1.539(t - t_0)\exp[0.09173(t - t_0)]$ |
| | | (b) $C_{CO_2}^{max} = 341.4 + 1.081(t - t_0)\exp[0.0258(t - t_0)]$ |
| | | (c) $C_{CO_2}^{min} = 341.4 + 1.82(t - t_0)\exp[8.34 \times 10^{-5}(t - t_0)]$ |
| *Methane* (ppm) | | |
| January 1850 | January 1983 | $C_{CH_4} = 1.0 + 0.65\exp[0.035(t - t^*)]$ |
| January 1983 | January 2100 | (a) $C_{CH_4} = C_{CH_4}$ (January 1983)$\cdot(1.01)^{t-t_0}$ |
| | | (b) $C_{CH_4}^{min} = 1.72$ |
| *Chlorocarbon compounds* | | |
| January 1950 | January 1983 | See Table 4.29 |
| January 1983 | January 2100 | (a) Constant emissions at the 1983 level |
| | | (b) Maximum emissions: a 3% annual increase up to 2100 with subsequent constancy at the 2100 level |
| | | (c) Minimum emissions: a 3% annual decrease with subsequent constancy at the 2100 level |
| *Nitrous oxide* (ppb) | | |
| January 1850 | January 1983 | $C_{N_2O} = 285 + 14.0\exp[0.04(t - t^{**})]$ |
| January 1983 | January 2100 | (a) $C_{N_2O} = C_{N_2O}$ (January 1983)$\cdot(1.0025)^{t-t_0}$ |
| | | (b) $C_{N_2O}^{min} = 302$ |
| *Nitrogen oxides* $NO_x$ | | |
| January 1850 | January 2100 | See Table 4.30 |
| *Carbon monoxide* (ppm) | | |
| January 1980 | January 2100 | (a) No changes |
| | | (b) $C_{CO}^{max} = C_{CO}$ (January 1980)$\cdot(1.01)^{t-t_0}$ |

*Note*: $t_0$, $t$ is the time interval to which the respective approximation refers.
$t^*$ = January 1980; $t^{**}$ = January 1978.

**Table 4.29** Typical trends of emissions of chlorocarbon compounds for the period $t_0 \leqslant t \leqslant t_1$

| Compounds | 1 January $t_0$ | 1 January $t_1$ | Emission rate ($10^6$ kg yr$^{-1}$) |
|---|---|---|---|
| $CFCl_3$ (freon-11) | 1850 | 1938 | 0.0 |
| | 1938 | 1943 | 0.1 |
| | 1943 | 1945 | 0.2 |
| | 1945 | 1950 | $0.40 \exp[0.50(t - t_0)]$ |
| | 1950 | 1954 | $5.64 \exp[0.285(t - t_0)]$ |
| | 1954 | 1957 | $17.7 \exp[0.195(t - t_0)]$ |
| | 1957 | 1960 | 31.3 |
| | 1960 | 1964 | $38.2 \exp[0.213(t - t_0)]$ |
| | 1964 | 1974 | $88.7 \exp[0.128(t - t_0)]$ |
| | 1974 | 1977 | 323.5 |
| | 1977 | 1980 | 325.5 |
| | 1980 | 1983 | 264.5 |
| $CF_2Cl_2$ (freon-12) | 1850 | 1931 | 0.0 |
| | 1931 | 1934 | 0.1 |
| | 1934 | 1940 | $0.214 \exp[0.377(t - t_0)]$ |
| | 1940 | 1946 | $2.05 \exp[0.248(t - t_0)]$ |
| | 1946 | 1948 | $11.2 \exp[0.427(t - t_0)]$ |
| | 1948 | 1955 | $32.9 \exp[0.0928(t - t_0)]$ |
| | 1955 | 1964 | $46.9 \exp[0.127(t - t_0)]$ |
| | 1964 | 1970 | $148.2 \exp[0.123(t - t_0)]$ |
| | 1970 | 1974 | $300.5 \exp[0.874(t - t_0)]$ |
| | 1974 | 1975 | 443.5 |
| | 1975 | 1976 | 434.6 |
| | 1976 | 1978 | $433.9 \exp[-0.0373(t - t_0)]$ |
| | 1978 | 1981 | 388.6 |
| | 1981 | 1983 | 412.2 |
| $CCl_4$ | 1850 | 1911 | 0.0 |
| | 1911 | 1916 | $0.96 \exp[0.388(t - t_0)]$ |
| | 1916 | 1922 | $0.67 \exp[0.132(t - t_0)]$ |
| | 1922 | 1925 | 16.0 |
| | 1925 | 1943 | $17.1 \exp[0.1067(t - t_0)]$ |
| | 1943 | 1948 | 116.8 |
| | 1948 | 1957 | $116.9 \exp[-0.0306(t - t_0)]$ |
| | 1957 | 1962 | 84.8 |
| | 1962 | 1975 | $84.8 \exp[0.0449(t - t_0)]$ |
| | 1975 | 1983 | 131.2 |
| $CH_3CCl_3$ | 1850 | 1951 | 0 |
| | 1951 | 1958 | $0.14 \exp[0.794(t - t_0)]$ |
| | 1958 | 1977 | $19.64 \exp[0.173(t - t_0)]$ |
| | 1977 | 1983 | 454.5 |
| $CFCl_2CFCl_2$ (freon-113) | 1850 | 1960 | 0.0 |
| | 1960 | 1977 | $2.5 \exp[0.1916(t - t_0)]$ |
| | 1977 | 1980 | $65.0 \exp[0.1346(t - t_0)]$ |
| | 1980 | 1983 | 97.3 |
| $CF_2ClCF_2Cl$ (freon-114) | 1850 | 1960 | 0.0 |
| | 1960 | 1977 | $5.0 \exp[0.072(t - t_0)]$ |
| | 1977 | 1978 | 18.0 |
| | 1978 | 1979 | 21.0 |
| | 1979 | 1983 | 18.0 |
| $CF_2ClCF_3$ (freon-115) | 1850 | 1970 | 0.0 |
| | 1970 | 1977 | $2.9 \exp[0.0628(t - t_0)]$ |
| | 1977 | 1983 | 4.5 |
| $CHF_2Cl$ (freon-22) | 1850 | 1950 | 0.0 |
| | 1950 | 1961 | $0.13 \exp[0.299(t - t_0)]$ |
| | 1961 | 1975 | $3.3 \exp[0.184(t - t_0)]$ |
| | 1975 | 1980 | $46.0 \exp[0.0893(t - t_0)]$ |
| | 1980 | 1983 | 71.9 |

constituents as water vapour, tropospheric and stratospheric ozone, nitrogen oxides, hydrocarbon compounds, as well as tropospheric aerosols (sulphate- and carbon-induced) and stratospheric aerosols. Table 4.30 gives levels of nitrogen oxides (as $NO_2$) against height (from aircraft-data).

To assess an equilibrium climate with a step-like change of atmospheric composition, the following model has been recommended (the change of the concentration of respective components is given in brackets): $CO_2$ (340–680 ppm), $CH_4$ (1.7–3.4 ppm), $N_2O$ (302–375.5 ppb), $CCl_3F$ (F-11) (0–744.8 ppt), $CCl_2F_2$ (F-12) (0–1789.5 ppt). The scenarios discussed are preliminary because of the inadequate understanding of global cycles of the respective components and uncertain predictions of economic development (anthropogenic emissions). Thus, further information is needed before the scenarios can be specified more precisely. Relevant efforts have been undertaken in the course of preparation of the IPCC Report (Houghton *et al.* 1990, 1992, Watson *et al.* 1992, Wigley 1993).

The important role of halogenized hydrocarbon compounds in the atmosphere as the sources of chlorine atoms stimulating the catalytic destruction of $O_3$, and greenhouse gases (in view of the resulting strong absorption in the 8–13 $\mu$m transparency window), has raised interest in observation of the content of hydrocarbon compounds that are totally ($CF_2Cl_2$, $CFCl_3$, $CF_4$) and partially ($CH_3Cl$, $CHF_2Cl$) halogenized. In this connection, Zander *et al.* (1987) considered the results of retrieving the vertical profiles of the mixing ratios for six compounds ($CH_3Cl$, $CF_2Cl_2$, $CFCl_3$, $CHF_2Cl$, $CCl_4$, $CF_4$) in the 10–30 km layer from data of occultation observations in April–May 1985 using the Spacelab Fourier-spectrometer ATMOS (these data refer to the latitudinal band 30° N to 47° S). The high spectral resolution of the instruments, about 0.01 $cm^{-1}$, has made it possible to record the fine structure of infra-red absorption spectra of the atmosphere in the ranges 600–1200 $cm^{-1}$ and 1100–1900 $cm^{-1}$ with a vertical resolution of about 4.1 km. Table 4.31 gives the

**Table 4.30** Levels of nitrogen oxides (as $NO_2$) against height (from aircraft data) (kt $yr^{-1}$)

| Height (km) | Year 1955 | 1975 | 1990 |
|---|---|---|---|
| 5 | 0.0 | 0.0 | 0.0 |
| 6 | 1.3 | 7.5 | 35.1 |
| 7 | 2.6 | 15.1 | 70.2 |
| 8 | 5.1 | 30.0 | 111.1 |
| 9 | 14.8 | 86.8 | 254.0 |
| 10 | 23.9 | 140.6 | 452.6 |
| 11 | 20.9 | 122.9 | 445.5 |
| 12 | 8.5 | 48.8 | 195.4 |
| 13 | 0.6 | 3.4 | 29.1 |
| 14 | 0.05 | 0.3 | 6.9 |
| 15 | | | 7.1 |
| 16 | | | 12.6 |
| 17 | | | 14.1 |
| 18 | | | 9.1 |
| 19 | | | 2.8 |

**Table 4.31** Retrieval of the mixing ratio for six halogenized hydrocarbons from ATMOS data (errors are given in brackets)

| Atmospheric pressure (hPa) | Height (km) | $CH_3Cl$ ($10^{-12}$) | $CF_2Cl_2$ ($10^{-10}$) | $CFCl_3$ ($10^{-10}$) | $CHF_2Cl$ ($10^{-11}$) | $CCl_4$ ($10^{-4}$) | $CF_4$ ($10^{-11}$) |
|---|---|---|---|---|---|---|---|
| 273 | 10 | 6.3 (0.63) | 3.2 (0.32) | 1.60 (0.24) | 5.5 (0.50) | 1.10 (0.16) | 5.7 (0.57) |
| 185 | 12.5 | 6.2 (0.74) | 3.1 (0.31) | 1.35 (0.20) | 5.4 (0.52) | 1.00 (0.15) | 6.1 (0.60) |
| 122 | 15 | 5.6 (0.78) | 2.9 (0.29) | 1.30 (0.20) | 5.5 (0.55) | 0.85 (0.13) | 5.6 (0.55) |
| 81 | 17.5 | 4.9 (0.78) | 2.7 (0.33) | 1.18 (0.21) | 5.4 (0.65) | 0.65 (0.13) | 5.7 (0.68) |
| 54 | 20 | 3.8 (0.68) | 2.4 (0.28) | 0.90 (0.18) | 5.0 (0.70) | 0.35 (0.09) | 5.9 (0.88) |
| 37 | 22.5 | 2.7 (0.54) | 1.8 (0.25) | 0.30 (0.13) | 3.8 (0.61) | 0.10 (0.04) | 6.0 (0.90) |
| 25 | 25 | 1.6 (0.40) | 1.3 (0.30) | 0.23 (0.07) | 3.3 (0.58) | – | 6.1 (0.95) |
| 17 | 27.5 | 0.5 (0.25) | 0.8 (0.20) | 0.06 (0.03) | 2.8 (0.70) | – | 5.9 (0.94) |
| 12 | 30 | – | 0.4 (0.20) | – | 2.6 (0.80) | – | 5.7 (1.2) |

observational results characterizing zonal mean values of the volume mixing ratio at 30° N.

The principal source of errors in retrieved values is the insufficiently reliable information about the parameters of the fine structure of absorption spectra (especially from the viewpoint of their temperature dependence) and about continuum absorption. The recorded transmission spectra characterize the contribution of these gases in the formation of the atmospheric greenhouse effect.

Of special interest are data on $CH_4$ which is one of the principal greenhouse gases responsible for the radiative forcing change, due to its concentration growth since pre-industrial times (1765) of about $0.4\,W\,m^{-2}$; it is second only to $CO_2$ (Osborn and Wigley 1994). Unfortunately, there is still very high uncertainty concerning sources and sinks of $CH_4$ which makes it very difficult to predict future $CH_4$ concentration. Assessments of total $CH_4$ emissions vary within the range 330 to $880\,Tg\,CH_4\,yr^{-1}$ (see Table 4.32) with a best guess of $515\,Tg\,CH_4\,yr^{-1}$ (Watson *et al.* 1992). Apart from the principal $CH_4$ sink (ca. 90 per cent) due to eventual oxidation to $CO_2$ and $H_2O$ by reaction with the hydroxyl radical in the troposphere, there is the $CH_4$ removal resulting from uptake by soils (ca. $30\,Tg\,CH_4\,yr^{-1}$), and reaction with OH in the stratosphere (ca. $10\,Tg\,CH_4\,yr^{-1}$). The chemical lifetime of $CH_4$ determined by the atmospheric chemical sinks is of the order of $(10 \pm 2)$ yr or perhaps slightly higher.

Osborn and Wigley (1994) have developed a simple $CH_4$ model to assess the global $CH_4$ dynamics. The key feature of the model is that $CH_4$ has a variable lifetime, of which changes are expressed as a function of $CH_4$ concentration and emissions of $NO_x$, CO and non-methane hydrocarbon compounds. The parametrized model which was fitted to results from five more complex atmospheric chemistry models and to the 1990 IPCC concentration projections (Houghton *et al.* 1996) (the data for these models are fitted quite satisfactorily) is used to estimate the history of past $CH_4$ emissions, lifetime changes and OH variations.

As Osborn and Wigley (1994) noted, the pre-industrial lifetime is estimated to be 15–34 per cent lower than today (this means that 23–55 per cent of past concentration changes are due to lifetime increases). Pre-industrial emissions are found to be much higher (220–330 $Tg\,CH_4\,yr^{-1}$) than the best estimate of present natural emissions (155 $Tg\,CH_4\,yr^{-1}$). The change in emissions since pre-industrial times is estimated to lie in the range 160–260 $Tg\,CH_4\,yr^{-1}$, compared with the current best guess for anthropogenic emissions of 360 $Tg\,CH_4\,yr^{-1}$. Thus, these results show either that current estimates of anthropogenic emissions are too high or that there have been large changes in natural emissions.

According to the 1992 IPCC emission scenarios (Houghton *et al.* 1992), uncertainties of future concentration and lifetime changes are large, but the relevant climate change consequences are small; in terms of future radiative forcing and global-mean temperature changes over 1990–2100 they correspond to uncertainties of at least $\pm 0.2\,W\,m^{-2}$ and $\pm 0.1\,°C$, respectively. Table 4.32 demonstrates some modelling results.

Hao and Ward (1993) have assessed emissions of $CH_4$ from various sources of biomass burning for tropical, temperate, and boreal regions. They show that 85 per cent of the total $CH_4$ is emitted in the tropical area, which is mainly the result of shifting cultivation, fuel-wood use and deforestation. Changes in land use practices and population growth in the tropics are possible causes of the increase of atmospheric $CH_4$ concentration. $CH_4$ emitted from biomass burning may have increased

**Table 4.32** $CH_4$ concentrations $C$, in 2100, $\tau$, lifetime changes from 1990 to 2100 and radiative forcing $\Delta Q$ from 1900 to 2100 (direct effect of effect of $CH_4$ only) for the IPCC 1992 emissions scenarios

| Scenario | C(2100) | $\tau$ (2100)/$\tau$ (1900) | $\Delta Q$ (2100–1990) ($W\,m^{-2}$) |
|---|---|---|---|
| IS92a,b | 3790 (3280–4380) | 1.15 (1.06–1.25) | 0.62 (0.49–0.76) |
| IS92c | 2250 (2000–2560) | 1.05 (1.02–1.11) | 0.19 (0.10–0.28) |
| IS02d | 2160 (1900–2380) | 0.97 (0.94–0.99) | 0.15 (0.07–0.23) |
| IS92e | 4180 (3710–4730) | 1.09 (1.03–1.16) | 0.72 (0.60–0.84) |
| IS92f | 4890 (4180–5690) | 1.20 (1.09–1.34) | 0.88 (0.72–1.05) |

*Note*: Figures in parentheses give lower and upper values.
The $CH_4$ emissions in the six IPCC scenarios IS92a–IS92f are identified in Table 2 of Houghton *et al.* (1992).
*Source*: Osborn and Wigley (1994).

by at least 3 Tg yr$^{-1}$ to a total of 34 Tg yr$^{-1}$ by the late 1980s because of increases in tropical deforestation and demand for fuel-wood.

The greenhouse effect of the atmosphere ($G$) may be defined by the difference between the thermal emission from the surface ($E$) and the outgoing longwave radiation ($F$):

$$G = E - F \tag{4.31}$$

Assessments of global/annual average values on the basis of satellite observations (Harrison *et al.* 1990) led to the following figures: $E = 390\,W\,m^{-2}$ and $F = 235\,W\,m^{-2}$, so that $G = 155\,W\,m^{-2}$. The clear-sky greenhouse effect ($G_A$) may then be defined as

$$G_A = E - F_{clear} \tag{4.32}$$

with a global annual average value of $F_{clear} = 265\,W\,m^{-2}$ (Harrison *et al.* 1990), so that $G_A = 125\,W\,m^{-2}$. Although it may seem very small, the cloud longwave radiative forcing is thus equal to only 30 $W\,m^{-2}$.

Following Raval and Ramanathan (1985), let us introduce normalized quantities to remove the strong temperature dependence

$$G = E/F_{clear} \tag{4.33}$$

There is a very simple possibility to assess qualitatively the influence of atmospheric temperature and humidity on the clear-sky greenhouse effect. Obviously,

$$F_{clear} = \varepsilon_A \sigma T_A^4 + (1 - \varepsilon_A)\sigma T_S^4 \tag{4.34}$$

where $\varepsilon_A$ is the atmospheric emissivity, $\sigma$ is the Stefan–Boltzmann constant, $T_A$ is the mean atmospheric temperature and $T_S$ is the mean surface temperature. Now, from Equation (4.33), we have

$$G' = T_S^4/[\varepsilon_A \sigma T_A^4 + (1 - \varepsilon_A)\sigma T_S^4] \tag{4.35}$$

Or, it is possible to introduce a parameter $g$ (Raval and Ramanathan 1985) where:

$$g = \varepsilon_A[1 - (T_A/T_S)^4]. \tag{4.36}$$

Hence

$$G = \frac{1}{1 - g} \tag{4.37}$$

$$= \frac{1}{1 - \varepsilon_A[1 - (T_A/T_S)^4]} \tag{4.38}$$

Here it is assumed that the emissivity $\varepsilon_A$ is proportional to the total column moisture content, $w$. Although it is a clear simplification, Equations (4.37) and (4.38) illustrate the essential result that there is a separation between the effects of the atmospheric humidity, as measured by $w$ and represented by $\varepsilon_A$ in these equations, and the temperature, as represented by $T_A$ (Webb *et al.* 1993). As is seen, the increase of $w$ and, consequently, of $\varepsilon_A$ leads to the enhancement of $G'$ or $g$, while the increase of $T_A$ results in the decrease of temperature contrast between the atmosphere and the surface and, thus, to the $G'$ decrease. It is quite obvious that the roles of $w$ and $T_A$ have to be geographically specific.

On the basis of the qualitative analysis, Webb *et al.* (1993) have pointed out that, for instance, since the temperature field in the tropics is comparatively homogeneous, the dominating role in the greenhouse effect variability belongs to $w$, which is controlled by sea surface temperature. The opposite situation occurs in middle and high latitudes where air temperatures are highly variable and therefore exert much more powerful control over $G'$ than at low latitudes.

The analysis of satellite data on the Earth radiation budget (Earth Radiation Budget Experiment, ERBE) and the total column moisture (Special Sensor Microwave/Imager, SSM/I) as well as the results of radiative transfer simulations made by Webb *et al.* (1993) has confirmed the conclusions of the simplified qualitative assessment. At low latitudes the clear-sky greenhouse effect varies mainly due to $w$ changes (which, on the other hand, are controlled by the sea surface temperature) and seasonal variations are small. In contrast, at middle and high latitudes both $G'$ and $w$ exhibit strong seasonal variations. The clear-sky greenhouse effect variations are controlled by the seasonal changes in atmospheric temperatures, which are strong enough to overcome the opposing effect of the moisture variations. There are strong seasonal variations of the greenhouse effect which has its maximum in winter (when surface–atmosphere temperature difference increases, which leads to the enhancement of the greenhouse effect, in spite of small moisture content) and its minimum in summer. The combined impact of both temperature and column moisture results in the formation of the meridional profile of the greenhouse effect which is characterized by the decrease of $G'$ towards the winter pole at a much slower rate than it does towards the summer pole.

As has been mentioned above, the estimates obtained during recent years have shown that the contribution to the formation of the greenhouse effect by increased concentrations of such optically active gaseous components as $CCl_3F$ (F-11), $CCl_2F_2$ (F-12) and $N_2O$, and by resulting changes in $O_3$ concentrations, is comparable to the effect of increased $CO_2$ content; in particular, stratospheric cooling due to a decrease of $O_3$ in the stratosphere caused by increased concentrations of CFCs can be compared to the respective cooling due to $CO_2$.

The data in Table 4.33 characterize variations in radiative heat flux divergence with increasing concentrations of minor gaseous constituents, obtained with the use of a one-dimensional model (calculations were made for a standard atmosphere).

**Table 4.33** Variations in radiative heat flux divergence due to the greenhouse effect caused by different MGCs

| | | Variations in radiative heat flux divergence ($W m^{-2}$) | | | |
|---|---|---|---|---|---|
| Component | Change of concentration | Stratosphere | Troposphere | Surface | Surface–troposphere system |
| $CO_2$ | 330 to 660 ppm | −2.64 | 3.72 | 1.16 | 4.88 |
| $N_2O$ | 0.30 to 0.60 ppm | −0.20 | 0.55 | 0.52 | 0.07 |
| $CH_4$ | 1.7 to 3.4 ppm | −0.06 | 0.42 | 0.29 | 0.71 |
| $CF_2Cl_2$ | 0 to 2 ppb | 0.17 | 0.00 | 0.87 | 0.87 |
| $CFCl_3$ | 0 to 2 ppb | | | | |
| $O_3$ | −25% | −0.15 | −0.27 | −0.08 | −0.35 |
| | +25% | −0.21 | 0.12 | 0.17 | 0.29 |
| $H_2O$ | 100% | −3.47 | 3.58 | 0.00 | 3.58 |

Calculations have also been made of the increase in surface temperature due to increased concentrations of $N_2O$, $CH_4$, $CF_2Cl_2$ and $CFCl_3$ with the use of a 2-D (zonal) radiative-dynamic climate model. Since this model does not take into account chemical processes, only the direct radiative effect of minor gaseous constituents has been considered; indirect effects due to the redistribution of $O_3$ have been ignored. Calculations revealed substantial differences in the effect of minor gaseous constituents and $CO_2$ on the greenhouse effect, determined by the difference in transmission functions.

A comparison of results of calculations of temperature with the use of 1-D and 2-D models has demonstrated that the second model leads to more realistic results, agreeing better with the data of 3-D models; with a doubled $CO_2$ concentration, the surface temperature increased by 3.7 deg K.

The surface temperature increase due to different minor gaseous constituents that was calculated using a 1-D model (with a critical temperature lapse rate of 6.5 deg K $km^{-1}$) turned out to be substantially overestimated. However, with an assumed wet-adiabatic lapse rate, the results are reliable enough. Table 4.34 shows the estimates of surface temperature increase by 2010 obtained with a 2-D model, and prescribed probable rates of the growth of concentrations of various minor gaseous constituents are shown in Table 4.35.

As can be seen, the hemispherical mean surface temperature increase with a zero growth of concentrations of freons (CFCs) after 1980 should constitute 1.2 deg K, with 60 per cent of this increase arising from $CO_2$. It is important to note, however,

**Table 4.34** Increase of surface temperature between 1980 and 2010

| | | $\Delta T$ (deg K) | | |
|---|---|---|---|---|
| Case | Components to be considered | 0–30° N | 30–90° N | 0–90° N |
| A | $CO_2$ alone | 0.67 | 0.83 | 1.75 |
| B | $CO_2 + CH_4$ + CFC (0% increase) + $N_2O$ | 1.08 | 1.31 | 1.20 |
| C | $CO_2 + CH_4$ + CFC (3% increase) + $N_2O$ | 1.17 | 1.44 | 1.30 |

**Table 4.35** Probable increase of mixing ratios for different MGCs during the period 1980–2010

| Component | Rate of growth (% $yr^{-1}$) | Mixing ratio (ppm) | |
|---|---|---|---|
| | | 1980 | 2010 |
| $CO_2$ | 0.5 | 330 | 383 |
| $N_2O$ | 0.2 | 0.30 | 0.32 |
| $CH_4$ | 1.2 | 1.70 | 2.40 |
| $CF_2Cl_2$ | 0 | $0.30 \times 10^{-4}$ | $0.78 \times 10^{-4}$ |
| | 3 | | $1.07 \times 10^{-4}$ |
| $CFCl_3$ | 0 | $0.17 \times 10^{-4}$ | $0.41 \times 10^{-4}$ |
| | 3 | | $0.61 \times 10^{-4}$ |

that the estimates discussed have been obtained without taking into account the thermal inertia of the ocean, whose effect can slow down the manifestation of warming by decades. The uncertainty of the estimates is also largely determined by only a schematic account of the effect of cloudiness (water cycle) on the formation of climate. An indirect impact on the greenhouse effect through a redistribution of $O_3$ with changing minor gaseous constituent concentration should be also considered.

Ramanathan (1987, 1988a,b) and Ramanathan *et al.* (1976, 1987, 1989a,b, 1991) suggested that the climatic response to an enhancement of the atmospheric greenhouse effect could be characterized by the climatic feedback parameter

$$\lambda = \frac{dR}{dT} \tag{4.39}$$

where $R$ is the Earth radiation budget and $T$ is surface temperature. According to the current estimates using 3-D atmospheric general circulation models, the $\lambda$ value varies within 0.9–1.2 $W\,m^{-2}\,K^{-1}$ which corresponds to a global mean warming by 1.1–4.5 deg K with a doubling of $CO_2$ concentration.

The reliability of estimates of the sensitivity of the global climate to a perturbation due to an enhancement of the greenhouse effect is determined by the adequacy of the account of various feedbacks and of the atmosphere–ocean interaction.

With an account of contributions of increased concentrations of all greenhouse gases during the period 1850–1985, estimates show that a climate warming by 0.6–2.4 deg K should take place during a period of 10 yr, and in the following 50 yr (with greenhouse gas emissions at the present level) by 1.2–5.0 deg K. An account of the effect of the world's oceans' thermal inertia reduces the level of climate warming that ought to have taken place to 0.5–1.1 deg K.

Although the warming observed between 1850 and 1985 constituted about 0.5 deg K, the closeness of these estimates does not prove that the greenhouse effect is the cause of the climate warming, since the real warming occurred not smoothly but step-by-step. As has been mentioned above, the processing of radio-sonde data revealed a stratospheric cooling in the period 1965–1980 (−0.14 deg K for 10 yr for the 16–20 km layer and −0.33 deg K for the 16–24 km layer), but this also cannot be considered as a confirmation of the greenhouse hypothesis, because the data series is too short.

An identification of the causes for global climate variability is complicated by an inadequate account of the various feedback mechanisms. As has been mentioned above, a supposition has been made during recent years about the substantial role of cloud optical feedback, which manifests itself in the growth of the optical thickness (and, hence, the albedo) of clouds when absolute humidity increases. Biogenic emissions of dimethyl sulphide by the ocean, leading in particular to a greater concentration of condensation nuclei and to respective changes in the optical properties of clouds, can add to the complicated nature of climate-forming processes. The consequences of the increase in methane concentration, from the viewpoint of its effect on water vapour content in the stratosphere and on tropospheric ozone, are still uncertain.

Brasseur and DeRudder (1987) discussed the results of numerical modelling based on the use of a 1-D time-dependent model with an interactive account of chemical, radiative and dynamic processes, which show that an increase of the concentration of such optically active minor gaseous constituents as freons, $CO_2$, $CH_4$ and $N_2O$ will have a substantial effect on the vertical profiles of ozone concentration and stratospheric and tropospheric temperature.

A substantial increase in concentration of active chlorine due to photodissociation of F-11 and F-12 will lead to a strong decrease of $O_3$ content and temperature in the upper stratosphere. If emissions of F-11 and F-12 are held constant at the 1980 level (309 kt of F-11 and 443 kt of F-12), the calculated decrease of $O_3$ content (assuming that other changes are absent) will be about 60–70 per cent (with respect to the pre-industrial level) at a height of 40 km, and total $O_3$ content will decrease by 8.7 per cent (taking account of temperature feedback) or 5.5 per cent (neglecting this feedback).

Calculations have also led to the conclusion that a $CO_2$ concentration doubling will raise the total $O_3$ content by 1.8 per cent and the surface temperature by 2 deg K. In this case the $O_3$ content at 40 km will increase by 15 per cent, and the temperature will decrease by 8 deg K. A doubled $CH_4$ concentration will raise the total $O_3$ content by 2.4 per cent, and a 20 per cent increase of $N_2O$ concentration will lower the total $O_3$ content by 1.4 per cent.

Calculations of the temporal variability of $O_3$ content and temperature (on the assumption that a 3 per cent annual increase of the production of F-11 and F-12 will continue until it reaches a limit which exceeds 1.5 times the level of production in 1985) suggested that a maximum of total $O_3$ content decrease should constitute 4–5 per cent if the rates of increase in the concentrations of $CO_2$, $CH_4$ and $N_2O$ reach, respectively, 0.5 per cent, 1.0 per cent and 1.25 per cent annually.

In the case of the scenario of minor gaseous constitutent dynamics considered above, the maximum cumulative decrease of total $O_3$ content is to take place in 2070. A small increase after 2070 depends on the rate of increase of $CH_4$ concentration. If the F-11 and F-12 production grows continuously by 3 per cent per year (without being limited to the threshold level mentioned) and if the concentration of other minor gaseous constituents is neglected, then the total $O_3$ content must decrease by more than 10 per cent not later than 2050.

Data of the model calculations discussed here revealed an almost linear interrelation between the total $O_3$ content decrease and the chlorine content, until the mixing ratio for active chlorine remained less than for active nitrogen. The sensitivity of $O_3$ content to active chlorine depends strongly, however, on the content of nitrogen oxides. If the content of active chlorine is comparable to, or more than, the content of active nitrogen, then the total $O_3$ content decrease rapidly intensifies as the chlorine content in the atmosphere grows. Although during the last decade the understanding of chemical processes in the atmosphere has deepened, many uncertainties still remain. The most substantial source of errors is unreliable forecasts of minor gaseous constituent emissions. The estimates of total $O_3$ content and temperature variabilities obtained by Brasseur and DeRudder (1987) should be considered as a possible upper limit. It is clear, however, that qualitative variations should consist in a decreasing $O_3$ content in the upper stratosphere (as a result of emissions of CFCs) and in an increasing $O_3$ content in the troposphere (due to increasing concentrations of $CH_4$ and nitrogen oxides). The total effect on total $O_3$ content cannot be calculated reliably because of uncertain scenarios of minor gaseous constituent emissions. A detailed consideration of global ozone dynamics problems has been published by Kondratyev (1989a).

The observed increase in the concentrations of several optically active minor gaseous constituents in the atmosphere should enhance the greenhouse effect and the radiative heating of the atmosphere. In this connection, the prediction of anthropogenically-induced warming of regional and global climates is correct, and is likely to be detected as statistically significant signals during the next one or two decades. The most significant source of uncertainties in the climatic forecasts is an inadequate account of the cloud–radiation interaction in the process of formation of the greenhouse effect in the atmosphere, as well as the role of the ocean in a delayed greenhouse effect on climate.

Ramanathan (1988a,b) emphasized that climate–chemical feedbacks are particularly important. So, for example, chemical reactions in the troposphere between CO, $CH_4$, $O_3$ and $H_2O$ control the concentration of hydroxyl, which in its turn governs the rate of CO and $CH_4$ oxidation. This determines a strong interaction between CO, $O_3$ and $H_2O$; thus the climate-induced variations in the $H_2O$ content can affect the concentration of such minor gaseous constituents as $O_3$ and CO. Similarly, an increase of CO can lead to an increase of $O_3$ in the troposphere.

There is one more potentially significant climate–chemical feedback determined by the interaction between $CH_4$ and stratospheric $H_2O$. $CH_4$ oxidation leads to $H_2O$ production, and this process is a major source of water vapour for the middle and upper stratosphere. Apparently, the pre-industrial concentration of $CH_4$ had constituted approximately 0.8 ppm, whereas the present concentration is 1.7 ppm. Such an increase of $CH_4$ concentration could have led to an increase of the mixing ratio for water vapour by about 1.8 ppm (since two molecules of $H_2O$ can be produced from one molecule of $CH_4$). The growth of stratospheric water content (above 25 km) could have intensified the atmospheric greenhouse effect up to 20–40 per cent (with respect to an increase due to $CH_4$ concentration growth).

Since the equator-to-pole diabatic circulation of the atmosphere provides the $H_2O$ transport from the middle and upper stratosphere to the lower polar stratosphere, the moisture content of the polar stratosphere could have increased. In the conditions of the cold Antarctic lower stratosphere, the 1 ppm increase of the mixing ratio for $H_2O$ is strong enough for polar stratospheric clouds to form (especially in winter), which strongly affect both the thermal regime and $O_3$ layer dynamics.

Table 4.36 contains an assessment of the role of various biogeochemical feedbacks made by Lashof (1989) and may be considered as a summary characterizing the importance of minor gaseous constituents in climate formation. Here the gain is $g = (w - i)/w$, where $w(i)$ is the output (input) signal defined by analogy to an elec-

**Table 4.36** Estimated gain from climate and biogeochemical feedbacks

| Feedback | Average gain | Possible range | Source |
|---|---|---|---|
| *Geophysical* | | | |
| Water vapour[a] | 0.39 | (0.28–0.52) | |
| Ice and snow | 0.12 | (0.03–0.21) | Dickinson (1986c) |
| Clouds | 0.09 | (−0.12–0.29) | |
| Sub-total[b] | 0.64 | (0.17–0.77) | |
| *Biogeochemical* | | | |
| Methane hydrates | 0.1 | (0.01–0.2) | After Revelle (1983) |
| Trospospheric chemistry | −0.04 | −(0.01–0.06) | Hameed and Cess (1983) |
| Ocean chemistry | 0.008 | | $\partial \ln(p_{CO_2})/\partial T = 4\%$ |
| Ocean eddy-diffusion | 0.02 | | $K \sim (\partial T/\partial Z)^{-2}$ |
| Ocean biology and circulation | 0.06 | (0.0–0.1) | After Sarmiento and Toggweiler (1984) |
| Vegetation albedo | 0.05 | (0.0–0.09) | After Hansen *et al.* (1984), Dickinson and Hanson (1984) |
| Vegetation respiration | 0.01 | (0.0–0.3) | Flux = 0.5 $Pg\,yr^{-1}\,K^{-1}$ |
| $CO_2$ fertilization | −0.02 | −(0.01–0.04) | 15% biomass increase for 2 × $CO_2$ |
| Methane from wetlands | 0.01 | (0.003–0.015) | Lashof & Fung, unpublished |
| Methane from rice | 0.006 | (0.0–0.01) | After Holzapfel-Pschorn and Seiler (1986) |
| Electricity demand | 0.001 | (0.0–0.004) | After Linder and Inglis (1989) |
| Sub-total[c] | 0.16 | (0.05–0.29)[d] | |
| Total | 0.80 | (0.32–0.98)[d] | |

[a] Includes the lapse rate feedback and other geophysical climate feedbacks not included elsewhere.
[b] Based on 1.5–5.5 deg K for doubling $CO_2$. The individual values do not sum to these values: see Dickinson (1986c) for details.
[c] Based on selected biogeochemical feedbacks which might occur together during the next century.
[d] Ranges are combined using a least-squares approach, i.e. by letting $C - L = [\Sigma(c_i - l_i)^2]^{0.5}$, where $C$ is the central estimate for the total, $c_i$ is the central estimate for feedback $i$, $L$ is the lower uncertainty bound for the total, and $l_i$ is the lower uncertainty bound for feedback $i$. A similar calculation is performed for the upper uncertainty bounds.
*Source*: After Lashof (1989).

tronic amplifier. The amplification (often referred to as the feedback factor) is

$$f = w/i = \frac{1}{1-g}$$

There are negative ($g < 0$) and positive ($g > 0$) feedbacks.

Some recent numerical modelling results show that variations of total $O_3$ content and its vertical profile may constitute a significant climate forming process (Kiehl *et al.* 1988). As Schwarzkopf and Ramaswamy (1993) have pointed out, observations from satellite and ground-based instruments indicate that significant changes in atmospheric ozone concentrations have taken place during the 1980s. Decreases in total ozone in middle and high latitudes have been measured by the TOMS instrument, while SAGE data indicate ozone depletion in the lower stratosphere at all latitudes, including the tropics. On the other hand, observations at several locations in the middle-latitudes of the northern hemisphere have shown significant decadal increases in tropospheric ozone. The earlier calculations led to the conclusion that the decadal changes in lower stratospheric ozone produce substantial negative radiative forcing of the surface–troposphere system at middle and high latitudes. The magnitude of the ozone forcing is sensitive to both total ozone loss and the altitude profile of the ozone depletion.

Schwarzkopf and Ramaswamy (1993) have undertaken new calculations of the sensitivity of radiative forcing to ozone changes with the use of SAGE I and II data on ozone vertical profiles above 17 km (below this level the presence of clouds complicates the remote sounding) as well as data on the increase of tropospheric ozone. The following parameters have been fixed: vertical temperature and humidity profilcs, cloudiness, surface temperature and albedo. Three scenarios for annual-averaged profiles within the 17–50 km layer for decadal ozone losses in tropical (4.5° N) and middle (40.5° N) latitudes (similar in each latitude belt) have been chosen:

1. no $O_3$ depletion below 17 km (S1);
2. a linear decrease with altitude of the percentage of $O_3$ depletion between 17 km and the tropopause, where the depletion becomes zero (S2);
3. the depletion percentage between the tropopause and 17 km is equal to the SAGE depletion percentage at 17 km (S3).

An additional 'standard' profile (S4) has been also used which employs a percentage ozone depletion adjusted so that the total ozone content is equal to that obtained using profile S1.

Tables 4.37 and 4.38 illustrate some of the results obtained by Schwarzkopf and Ramaswamy (1993). These results show that the amount of stratospheric $O_3$ radiative forcing during the 1980s depends principally on two factors: the total $O_3$ content depletion and the altitude profile of the $O_3$ change. In the tropics, where stratospheric $O_3$ change profiles are available almost to the troposphere, the prin-

**Table 4.37** Decadal stratospheric column ozone change, stratospheric ozone forcing, and stratospheric radiative forcing gradient (SRFG) for 'standard' tropical (January, 4.5° N) profiles

| Profile | S1 | S2 | S3 | S4 |
|---|---|---|---|---|
| Column change (DU) | −7.34 | −8.07 | −8.79 | −7.34 |
| Forcing ($W\,m^{-2}$) | −0.052 | −0.071 | −0.090 | −0.052 |
| SRFG ($W\,m^{-2}\,(DU)^{-1}$) | 0.0071 | 0.0087 | 0.0102 | 0.0071 |

*Note*: The SRFG derived here is valid over a stratospheric column ozone decrease of 0–10 DU.
*Source*: Schwarzkopf and Ramaswamy (1993).

**Table 4.38** Same as Table 4.37, except for mid-latitude (January, 40.5° N)

| Profile | S1 | S2 | S3 | S4 |
|---|---|---|---|---|
| Column change (DU) | −10.38 | −15.55 | −20.72 | −10.38 |
| Forcing ($W\,m^{-2}$) | −0.035 | −0.084 | −0.134 | −0.078 |
| SRFG ($W\,m^{-2}\,(DU)^{-1}$) | 0.0034 | 0.0054 | 0.0065 | 0.0075 |

*Note*: The SRFG derived here is valid over a stratospheric total ozone content column ozone decrease of 0–30 DU.
*Source*: Schwarzkopf and Ramaswamy (1993).

cipal question is the quantity of lower stratospheric $O_3$ loss. An important conclusion is that the magnitude of the decadal northern hemisphere stratospheric $O_3$ radiative forcing would increase significantly if $O_3$ depletion has occurred in the tropics, as implied by SAGE observations.

In mid-latitudes, the stratospheric $O_3$ forcing depends strongly on the altitude profile of decadal $O_3$ change in the mid-latitude lower stratosphere, especially in the layers nearest the tropopause (accurate measurements of $O_3$ change in the vicinity of the tropopause are necessary to arrive at a precise estimate of the stratospheric $O_3$ radiative forcing). If the observed increases of tropospheric $O_3$ in various regions of the northern hemisphere are hemispherically representative then one can conclude that a substantial positive tropospheric $O_3$ radiative forcing has occurred during the 1980s, of magnitude possibly comparable to that produced by decadal stratospheric $O_3$ decreases. Although the results considered require further confirmation (the uncertainty in both the magnitude and the sign of the decadal total $O_3$ forcing is rather high), it may be concluded that changes of total $O_3$ content and vertical $O_3$ profiles (both in the stratosphere and troposphere) are significant climate-forming processes.

Since the recent analysis of $O_3$ data for middle and high latitudes of the northern hemisphere has led to conclusions about the trends of $O_3$ depletion (during the last few decades) in the lower stratosphere and the growth of tropospheric $O_3$ (especially in the upper troposphere), Wang *et al.* (1993) have conducted calculations of the impact of such an $O_3$ change on the radiative forcing for the surface–troposphere system, including its shortwave and longwave components. Monthly mean $O_3$ profiles have been prescribed on the basis of ozone-sonde observation data for seven stations in middle and high latitudes of the northern hemisphere. Earlier calculations showed that the observed lower stratospheric $O_3$ depletion during the 1980s may have induced a significant radiative forcing (cooling) of the surface–troposphere climate system in contrast to the positive radiative forcing due to the increase of greenhouse gases during the same time period. Wang *et al.* (1993) have rightly pointed out that, however, the calculations at the middle and high latitudes of the northern hemisphere did not include the effect of observed tropospheric $O_3$ increases. It is in this context that new calculations have been made to intercompare radiative forcing values for clear sky conditions for the following time periods:

1. 1971–1980 and 1981–1990 (Hohenpeissenberg, Payerne, Goose Bay and Resolute data);
2. 1981–1985 and 1986–1990 (Churchill, Edmonton and Tateno data).

Two approaches have been used to calculate radiative forcing resulting from lower stratospheric $O_3$ depletion: for fixed temperature (FT) and for temperature adjusted according to a fixed dynamical heating (FD) assumption.

Table 4.39 illustrates the comparative impacts on radiative forcing of $O_3$ changes (including separate assessment for tropospheric $O_3$ increase) and increases of greenhouse gas concentrations ($CO_2$, $CH_4$, $CCl_3F$ (CFC-11), $CCl_2F_2$ (CFC-12) and $NO_2$ have been taken into account). For fixed temperature treatment, the total radiative forcing at Hohenpeissenberg is calculated to be 0.79 and 1.05 $W\,m^{-2}$, respectively, for January and July, while the effect due to $O_3$ changes contributes 0.41 and 0.57 $W\,m^{-2}$ , respectively. For the fixed dynamical heating treatment, the total radiative forcing for January and July is calculated to be 0.71 and 1.03 $W\,m^{-2}$, respectively. For both treatments it is quite clear that the tropospheric $O_3$ increase

**Table 4.39** Comparison of changes in radiative forcing ($W\,m^{-2}$) between $O_3$ changes and the combined $O_3$ changes and increases of greenhouse gases $CO_2$, $CH_4$, $CCl_3F$ (CFC-11), $CCl_2F_2$ (CFC-12) and $N_2O$

| | $O_3$ changes | | | | Combined $O_3$ and other greenhouse gases | |
|---|---|---|---|---|---|---|
| | Total | | Troposphere | | | |
| | FT | FD | FT | FD | FT | FD |
| Hohenpeissenberg (47° N, 11° E) | | | | | | |
| January | 0.410 | 0.327 | 0.222 | 0.136 | 0.794 | 0.710 |
| July | 0.570 | 0.548 | 0.490 | 0.467 | 1.050 | 1.027 |
| Payerne (47° N, 7° E) | | | | | | |
| January | 0.356 | 0.190 | 0.242 | 0.074 | 0.797 | 0.630 |
| July | 0.397 | 0.136 | 0.097 | −0.167 | 0.879 | 0.617 |
| Resolute (75° N, 95° W) | | | | | | |
| Winter | −0.038 | −0.085 | 0.012 | −0.036 | 0.230 | 0.183 |
| Summer | 0.194 | −0.009 | 0.052 | −0.153 | 0.620 | 0.415 |
| Tateno (36° N, 140° E) | | | | | | |
| Winter | 0.239 | 0.096 | 0.084 | −0.060 | 0.491 | 0.348 |
| Summer | 0.304 | 0.287 | 0.295 | 0.279 | 0.585 | 0.569 |

*Note*: FT = fixed temperature; FD = fixed dynamical heating.
*Source*: Wang *et al.* (1993).

contributes substantially to the total radiative forcing. An important aspect of the results is that they are sensitive to stratospheric temperature responses associated with $O_3$ changes. Wang *et al.* (1993) have emphasized that the large sensitivity of the $O_3$ radiative forcing to lower stratospheric temperatures demonstrates the need for studying the climatic implications of $O_3$ changes using general circulation models in which the interactions between radiation and dynamics are properly incorporated.

Significant changes of the total $O_3$ content in the atmosphere after explosive volcanic eruptions has attracted serious attention to careful analyses of combined observational data from satellite, ozone-sondes and ground-based instrumentation as well to the search for potential causes of $O_3$ changes. Although similar changes took place earlier during non-volcanic periods, it is quite natural that various attempts to explain post-eruption observations have been concentrated on revealing volcanically-induced mechanisms of total $O_3$ variations, especially in connection with possible enhancement of hetereogeneous chemical reactions on surfaces of volcanic aerosol particles which lead to $O_3$ destruction (as in the case of the Antarctic spring $O_3$ minimum (ozone hole) formation). It is in this context that Schoeberl *etal.* (1993) have analysed Nimbus-TOMS data of equatorial total $O_3$ content change after the Mount Pinatubo eruption.

The data show a decrease of up to 6 per cent compared with climatological average within the belt $\pm 12°$ latitude. $O_3$ losses began approximately a month following the eruption, consistent with the time required for the $SO_2$ to be converted to sulphuric acid aerosol. $O_3$ values remained below the climatological average until

December 1991. Following the eruption, retrieved total $O_3$ content values increased from 15 June through the first week of July by about 10 DU. Subsequently, total $O_3$ declined by about 20–30 DU, reaching a minimum shortly before November, then levelling off. The net decrease was about 10 per cent from the peak value during this period. This anomaly is, however, confined to the tropics. No anomalous minimum values were seen in the 12–18° N latitude band, but the 12–18° S band did show a weak signal.

As Schoeberl *et al.* (1993) have pointed out, under usual conditions equatorial total $O_3$ follows the phase of the quasi-biennial oscillation (QBO) in lower stratospheric equatorial winds. During the easterly phase, $O_3$ amounts are lower, while during the westerly phase the amounts are higher as a result of the secondary circulation associated with the quasi-biennial oscillation. Since Singapore 50 hPa winds switched to easterly in early 1991, it is possible that the tropical $O_3$ decrease is simply due to the quasi-biennial oscillation. However, the $O_3$ decrease following the Mount Pinatubo eruption amounts to a 5–6 per cent decrease below the climatological average of easterly phase quasi-biennial oscillation years.

New aspects of global $O_3$ dynamics and their impact on climate are connected with such phenomena as effects on stratospheric $O_3$ from high-speed civil transport (Pitari *et al.* 1993, Weisenstein *et al.* 1993) and the response of stratospheric $O_3$ to increasing levels of $CO_2$, $N_2O$, $CH_4$ and CFCs (Schneider *et al.* 1993, Law and Pyle 1993).

## 4.6 Aerosols and climate

### 4.6.1 Introduction

One of the most important factors in climate change is atmospheric aerosols, the product of a complicated totality of physical and chemical processes taking place in the environment as well as the results of direct injection of dust, ash or smoke particles into the atmosphere. Part of these processes is determined by the growing anthropogenic loading on the environment. Human's impact on atmospheric aerosols shows itself through changes in the intensity of formation (or disintegration) and in the physical and chemical properties of aerosol particles, which, in its turn, affect climate (Kondratyev 1978a, 1983a,b, 1991a,b, 1992b,c, 1996, Hobbs and McCormick 1988, d'Almeida *et al.* 1991, Götz *et al.* 1991, Husar *et al.* 1991, Jennings 1993). The three principal mechanisms through which aerosols affect climate are as follows.

1. The direct effect of aerosols on the radiation budget of the surface–atmosphere system through a re-distribution of shortwave solar radiation and thermal infrared emissions in this system, caused by scattering and absorption by aerosol particles non-uniformly distributed in the atmosphere (Galindo 1965, 1975, 1978, 1984, 1992a,b, Kondratyev 1972, 1973, 1976, 1978a, 1985b, 1988b, 1992b, 1996, Grassl 1973, 1974a,b, 1975, 1988, 1992, Kondratyev *et al.* 1973, 1983a,b, 1985, 1996b, Gerber and Deepak 1984, Coakley and Cess 1985, Hobbs and McCormick 1988, d'Almeida *et al.* 1991, Ackerman and Chang 1992, Deepak 1992, Hobbs 1993).
2. The crucial role of aerosol particles in the phase transformations of water in the atmosphere and in particular in cloud formation, which is more important for

the energetics of the atmosphere than the direct effect of aerosols (Kondratyev 1973, 1978a, 1991b, Ivlev 1982, Ivlev and Andreev 1986, d'Almeida *et al.* 1991, Jennings 1993). In this connection, of great concern is the role of gas-to-particle reactions of aerosol formation (condensation nuclei, first of all) connected with natural and anthropogenic emissions to the atmosphere of compounds such as dimethyl sulphide and sulphur dioxide.

3. Hetereogeneous chemical processes, in particular the processes of disintegration of $O_3$ molecules on the surface of aerosol particles, which affect the change in the atmospheric gas composition (Finlayson-Pitts and Pitts 1986, Kondratyev 1989a,b).

Complex field programmes of the aerosol–radiation experiments accomplished in various global regions within the CAENEX and GAAREX programmes (Kondratyev 1972, 1976, 1991b, Jennings 1993) have shown that the aerosol can substantially absorb the solar radiation and greatly affect the structure of the atmospheric radiation budget. The urgent problems are those of the impact of dust outbreaks from the arid regions on the radiative regime of the atmosphere over the continents and the oceans, the transport of aerosols to high latitudes leading to the formation of the arctic haze and also the extended stratified cloudiness in the Arctic (Kondratyev 1992a,b, Kondratyev *et al.* 1996a,b, Shaw 1995). A special role belongs to aerosol perturbations due to volcanic eruptions.

An extremely complicated problem is the consideration of the role of the radiation impact in the formation and evolution of the atmospheric aerosol field. It is necessary to study the relations between the radiative, convective and turbulent heat exchange and mass exchange in the processes of generation, formation, chemical composition, size distribution and optical properties of aerosols (Bolle 1982, Husar *et al.* 1991, Kondratyev 1991b).

### 4.6.2 Sulphate aerosols

In this context, the numerical modelling of the formation of the global distribution of aerosol is extremely urgent. In this connection, the sulphate aerosols have recently been of great concern (Charlson *et al.* 1987, 1991, 1992, Bates *et al.* 1992a, Fried *et al.* 1992). So, for example, based on the use of a three-dimensional global model MOGUNTIA of the transport and chemical transformation of minor atmospheric components, Langner and Rodhe (1991) and Langner *et al.* (1992) assessed the changes in the distribution of tropospheric sulphate aerosol and the deposition of non-marine sulphate aerosol on the underlying surface, which took place during the industrial revolution. The 10-layer model (up to the 100 hPa level) has been utilized over a 10° longitude grid with the prescribed monthly means of the meteorological parameters.

The concentrations of dimethyl sulphide (DMS), sulphur dioxide ($SO_2$) and non-marine sulphate aerosol ($SO_4^{2-}$) precursors have been calculated. The final product of the transformation of all these gaseous components is assumed to be $SO_4^{2-}$. For comparison purposes, calculations have been made with the prescribed emissions taking place before the industrial revolution and at the present time.

The data of Table 4.40 characterize the emissions of sulphur compounds ($Tg\,S\,yr^{-1}$) in both hemispheres. The limits of possible variabilities are given in parentheses. An important feature is a serious uncertainty in the assessments of natural

**Table 4.40** Natural and anthropogenic emissions of sulphur compounds in the northern and southern hemispheres

| Source of emissions | Global emissions ($Tg\,S\,yr^{-1}$) | Northern hemisphere | Southern hemisphere |
|---|---|---|---|
| Industry ($SO_2$) | 70 (62–80) | 64 | 6 |
| Biomass burning ($SO_2$) | 2.5 (2.3–2.8) | 1.1 | 1.4 |
| Oceans (DMS) | 16 (8–51) | 6.9 | 9.1 |
| Volcanoes ($SO_2$) | 8.5 (7.4–9.4) | 5.8 | 2.7 |
| Soils and vegetation (DMS, $H_2S$) | 1 (0.2–4.3) | 0.6 | 0.4 |

emissions of DMS by the ocean, which leads to the uncertainty of the total emissions. The emission of DMS by the ocean is characterized by strong spatial variability and annual change, especially in high and middle latitudes. Naturally, such a variability is also typical of the DMS concentration near the surface, reaching a maximum in summer over the ocean in high latitudes and (throughout the year) in the equatorial belt, and a minimum over the continents where the biogenic emissions of DMS are small. The calculated data, on the whole, are in satisfactory agreement with the available observational data.

Maps of the global distribution of the concentration of sulphates in the lower 500 m layer of the atmosphere, drawn from numerical modelling data, reveal distinctly the effect of anthropogenic emissions in the northern hemisphere with maxima of more than 1 ppb in the regions of Europe, North America and China. The sufficiently long life-time of $SO_4^{2-}$ (about 5 days) allows the possibility of long-range transport of sulphate aerosol, which shows itself east of North America and China, as well as east and south-east of Europe. In the southern hemisphere the anthropogenic impact can be observed in Africa, South America and Australia. In case of the absence of anthropogenic emissions, a maximum of the concentration of sulphates takes place in the tropical belt, where the natural emissions of DMS are most intensive. The calculations of the ratios of the pre-industrial and present concentrations of sulphates in the layer 0.5–1.5 km, where the lower-level clouds form (mainly stratus clouds), have led to the conclusion that in the most polluted European regions the concentration of sulphates has increased by two orders of magnitude, and in the northern hemisphere the overall increase of the concentrations in July (January) reached 51 per cent (0.8 per cent).

The dry and wet sulphate aerosol deposition has been intensified, respectively: twice and more over 58 per cent of the northern hemisphere of the surface and more than 16 per cent over the southern hemisphere surface. About half the $SO_2$ deposits onto the surface and a large part of the remaining $SO_2$ gets oxidized in cloud droplets. Only 6 per cent of the total anthropogenic emissions ($72.2\,Tg\,S\,yr^{-1}$) can be transformed into aerosol particles (condensation nuclei) which limits the possibilities of indirect impact of $SO_2$ emissions on the microphysical and optical parameters of clouds. Reliable quantitative estimates are, so far, impossible; however, Rodhe and Langner (1993) have undertaken a similar study concerning DMS. The existence of gas-to-particle reactions in the formation of sulphate aerosol from DMS emitted to the atmosphere by the ocean causes great concern for assessments of the varying (depending on the dynamics of the oceanic bioproductivity) emissions of DMS and the role of a possible impact of sulphate aerosol on climate.

Based on the use of a global 3-D model MOGUNTIA of the transport and transformations of minor gaseous components of the atmosphere and aerosols, calculations have been made by Rodhe and Langner (1993) of the contribution of DMS emitted by the ocean to the global sulphur cycle. Since the DMS oxidation giving $SO_2$ and the subsequent gas-to-particle reactions of the formation of sulphate aerosol are in proportion to the concentration of hydroxyl radicals, the initial task consisted in simulation of the global field of OH concentration with the consideration of the totality of chemical reactions for the system $CH_4$–CO–$NO_x$–OH–$O_3$. The scale of $SO_2$ oxidation to sulphate in cloud droplets is assumed to be proportional to their number concentration, wet and dry deposition of $SO_2$ and sulphate.

In connection with the problem of aerosol cooling versus greenhouse warming, great attention has recently been attracted by the above-mentioned hypothesis of dimethyl sulphide (DMS) climatic impact which is based on the assumption that phytoplankton could affect cloud albedo by producing DMS, that the latter is a precursor to aerosols and cloud condensation nuclei (CCN), and that cloud albedo could in turn affect the productivity of the phytoplankton, thus creating a climatically significant feedback cycle (e.g. Shaw 1983, Charlson *et al.* 1987, 1991, 1992, Foley *et al.* 1991, Idso 1992, Putaud *et al.* 1993). Lawrence (1993) has undertaken an empirical analysis of the strength of this phytoplankton–DMS–cloud–climate cycle by considering available data on the relations between individual components of the feedback and developing an empirical model of the cycle as a whole which allows an assessment of the strength of the cycle for thermostatic regulation of the climate.

The feedback considered by Lawrence (1993) includes three components: (1) the coupling between phytoplankton, DMS and cloud condensation nuclei, (2) the effect of changes in cloud condensation nuclei levels on albedo, incident irradiance at the surface and surface temperature and (3) the response of phytoplankton to changes in incident irradiance and seawater temperature. The first and third components of the cycle have been derived primarily from observations of their respective chemical, physical and biological constituents. The second component was based on the estimate by Charlson *et al.* (1987) that a 30 per cent increase of cloud condensation nuclei in the area covered by marine stratiform clouds would cause a decrease in the global average surface temperature of 1.3 °C, as well as on the conclusion of Charlson *et al.* (1992) that enhanced sulphur emissions should cause roughly the same magnitude of cooling via the direct aerosol backscatter effect as via the indirect cloud albedo effect.

The first step of the consideration accomplished by Lawrence (1993) has been the substantiation of the relation between phytoplankton population and seawater DMS concentration $[DMS]_{SW}$. The following relation for the DMS flux has been chosen:

$$F_{DMS} = K_L\left([DMS]_{SW} - \frac{[DMS]_{BL}}{H}\right) \tag{4.40}$$

where $K_L$ is the piston velocity, an empirical constant, $[DMS]_{BL}$ is the boundary layer DMS concentration and $H$ is the Henry's law constant for DMS.

Since for the range of atmospheric and oceanic DMS concentrations which were observed, the second term of Equation (4.40) is negligible, one can simply use

$$F_{DMS} = K_L[DMS]_{SW} \tag{4.41}$$

where the area-weighted global annual mean $K_L$ value (which, in general, depends on meteorological conditions) may be prescribed as equal to $(2.5 \pm 0.2)\,\mathrm{m\,d^{-1}}$.

The first step of the empirical parametrization is the choice of the following relation between $F_{DMS}$ and aerosol number density (condensation nuclei concentration) ([CN]):

$$[\mathrm{CN}] = A_1 \ln(F_{\mathrm{DMS}} + 1) + B_1 \tag{4.42}$$

If $F_{DMS}$ is expressed in $\mu\mathrm{mol}^{-2}\,\mathrm{d}^{-1}$ and [CN] in $\mathrm{cm}^{-3}$, then the values of the constants are $A_1 = (184 \pm 34)\,\mathrm{cm}^{-3}\,(\mu\mathrm{mol\,m}^{-2}\,\mathrm{d}^{-1})^{-1}$ and $B_1 = (43 \pm 61)\,\mathrm{cm}^{-3}$. The final step is the parametrization of the growth of CN, condensation nuclei, to CCN, cloud condensation nuclei:

$$[\mathrm{CCN}] = A_3[\mathrm{CN}] \tag{4.43}$$

where $A_3 = (0.46 \pm 0.24)$ (number densities are expressed in $\mathrm{cm}^{-3}$).

For sea water DMS concentration the following formula has been determined

$$[\mathrm{DMS}]_{\mathrm{SW}} = \frac{\mu\beta P_{\mathrm{p}}}{\left(K_{\mathrm{L}}\dfrac{A_{\mathrm{B}}}{V_{\mathrm{B}}} + L_{\mathrm{WC}}\right)} \tag{4.44}$$

where $P_p$ is the phytoplankton population density, $\mu$ is the phytoplankton specific growth rate, $\beta$ is the mean DMS content per phytoplankton cell, $A_B/V_B$ is the ratio of the area of the bloom at the ocean surface to the volume of the bloom, and $L_{WC}\,[\mathrm{DMS}]_{SW}$ represents various losses of DMS in the water column (microbial degradation etc.), apart from zooplankton grazing and, to a lesser extent, senescence which are considered as basic factors responsible for the release of DMS into surrounding waters.

Since $L_{WC}$ is significantly less light dependent and temperature dependent than $\beta$ in Equation (4.44), the simplified expression may be used for making assessments of changes due to incident irradiance and temperature variations:

$$[\mathrm{DMS}]_{\mathrm{SW}} = \eta\mu \tag{4.45}$$

where $\eta = 2.3\,\mathrm{nmol\,DMS\,l^{-1}\,d}$ (the potential range of $\eta$ values is from 1.7 to $21\,\mathrm{nmol\,DMS\,l^{-1}\,d}$).

The dependence of the phytoplankton specific growth rate ($\mu$) on the incident irradiance and temperature has been parametrized on the basis of existing empirical models. The first-order feedback strength, FS, is computed as:

$$\mathrm{FS} = 100\left(\alpha_d \frac{[\mathrm{CCN}]_{\mathrm{p}} - [\mathrm{CCN}]_0}{[\mathrm{CCN}]_{\mathrm{c}} - [\mathrm{CCN}]_0}\right) \tag{4.46}$$

where $[\mathrm{CCN}]_p$ and $[\mathrm{CCN}]_0$ are the predicted and background values, respectively, $[\mathrm{CCN}]_c$ is the value necessary to counteract the initial perturbation and $\alpha_d$ is a parameter which has been introduced to take into account not only the feedback through indirect (cloud albedo) cooling, but also direct cooling due to enhanced sulphur emissions (enhanced backscattering), which has approximately the same value (therefore $\alpha_d = 2$).

Calculations of FS values made by Lawrence (1993) for several different cases gave the expected FS (for an initial temperature of 15 °C) equal to 23 per cent, which indicates that the marine biogeochemical cycle could be playing a significant role in the global climate change (for example, in aerosol cooling versus greenhouse warming). It is important to note that an FS of 20 per cent represents a change in [CCN] of only 3 per cent for a temperature perturbation of 1.3 °C. Thus, the climate is extremely sensitive to the changes in aerosol and cloud condensation nuclei levels. The potential range of FS values is from 10 to 50 per cent. It is obvious that a 50 per cent value implies significant revisions of global climate change assessments, but, of course, the degree of uncertainty is very high and this shows the necessity for further studies. Lawrence (1993) has pointed out that two areas in particular require to be studied in order to reduce uncertainties: (1) phytoplankton ecology and the concomitant DMS production; (2) the relation between DMS emissions and the concentration of condensation nuclei.

Various aspects of the anthropogenic-aerosol–climate problem have recently been studied. An important finding by Kim and Cess (1993) has been that the satellite data (low-level cloud albedo) demonstrate enhanced cloud albedo near the coastal boundaries where sulphate concentrations are large. Similar trends are absent over ocean regions of the southern hemisphere that are far away from anthropogenic sulphate sources. Kaufman *et al.* (1991) have assessed the impact of fossil fuel and biomass burning on climate. They have pointed out that although coal and oil emit 120 times as many $CO_2$ molecules as $SO_2$ molecules, each $SO_2$ molecule is 50–1100 times more effective in cooling the atmosphere (through the effect of aerosol particles on cloud albedo) than a $CO_2$ molecule is in heating it. Kaufman *et al.* (1991) concluded that the cooling effect from coal and oil burning presently ranges from 0.4 to 8 times the heating effect. Within this large uncertainty, it is presently more likely that fossil fuel burning causes cooling of the atmosphere rather than heating. Biomass burning associated with deforestation, on the other hand, is more likely to cause heating of the atmosphere than cooling since its aerosol cooling effect is only half that of fossil fuel burning and its heating effect is twice as large.

Important studies of biogenic sulphur aerosol in the Arctic troposphere have been made by Li and Barrie (1993) on the basis of long-term observations from 1980 to 1990 of aerosol methane-sulphonate (MSA), sulphate, sodium and other related chemical species as well as a shorter time series of aerosol sulphur isotopic composition ($\delta^{34}S$) at Alert, Northwest Territories, Canada. The principal aim of the studies was to determine the contributions of sea salt, biogenic sources and pollution to aerosol $SO_4^{2-}$. The utilization of two different processing techniques gave similar values of MSA/biogenic $SO_4^{2-}$ mass ratio within 0.2 to 0.9 for the summer months and much lower in October to March ($<0.08$). The analysis of the isotopic composition data show that the aerosol $SO_4^{2-}$ in summer is 25 to 30 per cent biogenic and 1 to 8 per cent sea salt and the rest is anthropogenic in origin. At other times of the year it is $<14$ per cent biogenic and 1 to 8 per cent sea salt, while the rest is anthropogenic in origin.

Hunter *et al.* (1993) have studied seasonal, latitudinal and secular variations in temperature trend with the purpose of identifying the influence of anthropogenic sulphate. These data indicate that pronounced minima in the rate of temperature increase in summer months in northern hemisphere mid-latitudes are consistent with the latitudinal distribution of anthropogenic sulphate and changes in the rate of $SO_2$ emissions since the industrial revolution.

An estimation of the global distribution of the climatic (cooling) impact of sulphate aerosol due to DMS emissions (made by Rodhe and Langner 1993) revealed, as was expected, a maximum over the equatorial belt of the ocean (up to $1\,W\,m^{-2}$) and a vast region of large values (more than $0.5\,W\,m^{-2}$), stretching from the equatorial Atlantic across north Africa and towards the northern Indian Ocean. A strong effect over the land under consideration is determined by the clear weather prevailing here, which favours the prolongation of the sulphate aerosol life-time.

Bearing in mind that the assessment of the mean-climatic impact of the northern hemisphere anthropogenic sulphate aerosol gave $-1.1\,W\,m^{-2}$ and the positive greenhouse effect constituted about $2.5\,W\,m^{-2}$, it becomes clear that the effect of DMS is rather substantial. To obtain more reliable estimates of the role of DMS, further studies are needed of the DMS flux across the ocean–atmosphere interface, the mechanisms for DMS oxidation in the atmosphere, the processes of $SO_2$ oxidation in cloud droplets and in aerosol particles and washing out of sulphate aerosol from the atmosphere by rains.

From the viewpoint of global aerosol impact on climate the chemical composition and size distribution of particles, which determine their radiative characteristics, is a very important problem. In recent years a number of studies of atmospheric aerosol and its radiative properties have been published. However, most of the publications cannot be used adequately because the data on aerosol properties, which form the basis for calculations of the radiative characteristics, are fragmentary and therefore hardly representative. No doubt the estimates for anthropogenic aerosols, mostly sulphates, are more reliable than for natural aerosols (especially in the difficult-to-access regions of the world's oceans and continents). This very significantly affects the reliability of the assessment of relations between the effects of various aerosols on the climate, although the fact of human impact on the relevant processes, and on the cycles of sulphur, carbon and nitrogen in particular, is not in doubt.

### 4.6.3 Modelling aerosol properties

The accomplishment of various climatological calculations and assessments requires a consideration of the aerosol characteristics, which it is not always possible to measure directly. It is important to note that such characteristics vary strongly, depending on various environmental conditions. For example, the variability of the optical properties and size distribution of aerosol particles even within a certain climatic zone is well known (Mészáros 1981, Deepak 1992, Kondratyev *et al.* 1983b, Ivlev and Andreev 1986, d'Almeida *et al.* 1991). On the other hand, there are relatively stable aerosol characteristics subject to certain laws, such as particle size distribution and chemical composition of the aerosol substance (Mészáros 1985, 1987, 1988, Stowe 1988, d'Almeida *et al.* 1991, Götz *et al.* 1991, Kondratyev 1991b, 1992b). The difficult problem of substantiating the representative models is of fundamental importance.

The difficulties in the modelling of the characteristics and properties of aerosol particles are connected with the large number of parameters which influence their variability and with the absence of reliable criteria for the choice of the values of the various parameters. So, for example, the spatial and temporal variability of aerosol properties is determined by a complicated sum of the processes of generation, transformation and evolution of an ensemble of aerosol particles, their propagation from the sources, and the removal of the particles from the atmosphere. An isolated consideration of these processes, which is often practised, is rather arbitrary, since all of them are interconnected.

The problem of aerosol modelling may be divided into two parts, including:

1. global spatial distribution of aerosols (the field of the aerosol mass concentration), with consideration of the average characteristics of generation, propagation and removal of particles from the atmosphere (Ivlev 1982, Kondratyev *et al.* 1983b, Kondratyev 1991b, 1992b, Jennings 1993);
2. local properties of aerosols resulting also from the effect of the processes of transformation of the spectra of atmospheric aerosols of different origin (Grassl 1974a,b, 1988, 1992, Mészáros 1981, Ivlev and Andreev 1986, Götz *et al.* 1991, Galindo 1992a,b).

The processes of the generation of particles determine, first of all, their physical and chemical properties and their concentration in the atmosphere and, hence, the subsequent processes of transformation of the spectra of size distribution and the rate of removal of the particles from the atmosphere. The following types of natural aerosol particles can be identified:

1. products of the disintegration and evaporation of sea spray,
2. mineral dust, wind-driven to the atmosphere,
3. products of volcanic eruptions (both directly ejected to the atmosphere and resulting from gas-to-particle conversion),
4. particles of biogenic origin (both directly ejected to the atmosphere and resulting from the condensation of volatile organic compounds, for example terpenes, and chemical reactions taking place between them),
5. smoke from biomass burning on land,
6. products of natural gas-to-particle conversion (for example, sulphates formed from sulphur compounds supplied by the ocean surface).

The principal types of anthropogenic aerosols are particles of industrial emissions (soot, smoke, cement and road dust, etc.), as well as products of gas-to-particle conversion. In addition, some aerosol substance results from heterogeneous chemical reactions, in particular, photocatalytic reactions.

As a result of the sufficiently long life-time of aerosol particles, their transport, mixing and interaction with each other and with gaseous compounds of various substances, differences in the physical and chemical properties of particles from various sources are smoothed out and the particles form what are called the background aerosols.

In different atmospheric layers the characteristics of the background aerosols suffer substantial changes. In particular, due to the weak mixing of tropospheric and stratospheric aerosols, the particles in the stratosphere differ strongly in their composition and size distribution from the tropospheric ones. The troposphere can contain regions with particles of different types, in view of a relatively short life-time of the particles in the troposphere, especially in the surface layer, as well as some of the sources' markedly dominating the intensity of the remainder.

The processes of transport and removal of aerosols from the atmosphere are determined by both the meteorological factors and the properties of the aerosol particles themselves. The processes which depend on meteorological conditions include organized convective and advective transports of particles as well as mixing due to the turbulent diffusion. The individual properties of particles determine, in particular, such processes of their removal from the atmosphere as sedimentation on obstacles.

In the processes of transport and removal of particles from the atmosphere, their size (mass) and composition are transformed due to coagulation and condensation growth as well as to heterogeneous reactions. These processes are responsible for the spectrum of particle sizes and determine the value of the complex refractive index and the particles' shapes, that is their optical properties (d'Almeida *et al.* 1991, Ivlev and Andreev 1986).

The most substantial gas-to-particle reactions in the formation of aerosol particles are as follows:

1. reactions of sulphur dioxide with hydroxyl radicals which eventually lead to the formation of sulphuric acid molecules, which are easily condensed in the presence of water molecules;
2. reactions of hydrocarbon compounds with ozone and hydroxyl radical with the subsequent formation of organic nitrates through the reactions of primary products with nitrogen oxides (e.g., the well-known peroxyacetylnitrate).

All the processes of formation and evolution of aerosol systems depend on periodical variations of solar radiation in the atmosphere and at the surface. The processes of atmospheric heating near the surface, water evaporation and chemical reactions of aerosol formation have a clear-cut diurnal change resulting in variations of the aerosol characteristics, which in some cases are stronger than the annual change (Galindo 1965). Strong daily variations of anthropogenic aerosols have been documented.

The cycles of aerosol matter are closely connected with the hydrological processes in the atmosphere. On the one hand, clouds and precipitation play an important role in the formation, transformation and removal of aerosol particles from the atmosphere, while, on the other hand, the aerosols are condensation nuclei whose physico-chemical properties determine the microphysical processes in clouds. It is not accidental that both water vapour molecules and aerosol particles have approximately equal life-times. Therefore a deeper knowledge of the interactions of aerosols and cloud elements is needed to obtain a better understanding of the processes of the formation of aerosols and of the variability of aerosol properties in time and space.

### 4.6.4. Climatic impacts of aerosols

The principal problem connected with studies of the climatic impact of aerosols consists in the consideration of their influence on the longwave and shortwave radiation transfer in the atmosphere from the viewpoint of climate change that can take place due to regional and global-scale variations in the spatial distribution of the concentration, size distribution and chemical composition of aerosols (Grassl 1973, 1974a,b, 1975, 1988, 1992, Kondratyev 1980, 1987a,b, 1988b, 1992b, 1996, Galindo 1992a,b). In this connection, of great concern was the problem of a possible climatic impact of soot particles ejected to the troposphere and the stratosphere by fires caused by nuclear explosions in the atmosphere (Kondratyev 1988b). During recent years, the assessments of the climatic implications of fires on the oil fields of Kuwait have been of serious concern (Parungo *et al.* 1992).

#### 4.6.4.1 *Tropospheric aerosols*

The development of numerical climate modelling necessitates a reliable consideration of the effect of aerosol on climate, based on the development of adequate and realistic aerosol models and assessment of climate sensitivity to various characteristics of atmospheric aerosols (Kondratyev and Ter Markaryants 1976, Coakley and Cess 1985; Ramaswamy and Kiehl 1985, Hobbs and McCormick 1988, Hansen *et al.* 1992; Jennings 1993, Pollack *et al.* 1993; Rodhe and Langner 1993, Schlesinger *et al.* 1993).

The following aspects of the problem are of the first priority:

1. impacts of aerosols on regional and global processes,
2. identification of the types and properties of aerosols, most important from the viewpoint of their climatic implications, and
3. assessing the contribution of impacts on the regional and global climate of the variability of aerosols, minor gaseous components ($H_2O$, $O_3$, $CO_2$, CO, etc.), clouds and surface albedo.

The development of models of the dynamics of formation, transformation and transport of aerosols in interaction with the processes of climate formation is a substantial priority for the consideration of aerosol effects in climate theory. Numerical experiments on climate sensitivity to aerosol effects should be accomplished to reveal impacts of aerosols on the radiation budget of the surface–atmosphere system, heat flux divergence in the atmosphere, heat balance and the climatic system as a whole. Radiative–convective equilibrium models are useful for approximate assessments of the effects of aerosols on climate. In particular, they have revealed a strong effect of aerosols on the vertical temperature profile, when the optical thickness of aerosols is large, and a weak effect if their optical thickness is below 0.5. Depending on the surface albedo, optical properties of aerosol and aerosol spatial distribution, either cooling or warming of the climate by several degrees can take place (Kondratyev 1980, 1991b, 1992b, Charlson *et al.* 1987, 1991, 1992, Hobbs and McCormick 1988, Jennings 1993, Lawrence, 1993, Penner *et al.* 1993, Pollack *et al.* 1993).

A consideration of the aerosol effect on surface temperature in the zonal model of atmospheric general circulation (AGCM) suggests the conclusion that a temperature decrease of several degrees is possible. In the initial calculations with the use of three-dimensional atmospheric general circulation models and the aerosol characteristics for the regions of deserts, the following has been discovered:

1. a surface temperature decrease of up to 2.5 °C, provided the aerosol absorbs only the solar radiation,
2. a surface temperature increase of up to 3.5 °C, provided the aerosol impact on the thermal radiation transfer (greenhouse effect) has been taken into account,

3. considerably intensified atmospheric stability near the Earth's surface and growing instability of the free atmosphere (in the upper part of a dust-loaded atmosphere),
4. a marked transformation of the temperature field beyond the dust-loaded part of the atmosphere (west Europe, Asia and tropical Africa for the cases of dust outbreaks from the Sahara).

An assessment of the effect of the background aerosol in the troposphere on the radiative regime shows that the aerosol can determine a decrease of the shortwave radiation budget at the surface level by an order of $5.0\,W\,m^{-2}$, and at the atmospheric top level by $3.5\,W\,m^{-2}$: that is, it causes approximately the same values (but opposite in sign) of the radiative disturbances as a $CO_2$ doubling (of course, the effect of aerosols on the thermal emission field should be taken into account). In contrast to the global comparative homogeneity of radiative disturbance due to $CO_2$, a specific feature of the aerosol effect on the radiative regime is its spatial inhomogeneity connected with the spatial variability of the aerosol content and surface albedo as well as with variations in the Sun's elevation.

A comparison of the results of calculations of surface air temperature variations with the use of atmospheric general circulation models and an energy-balance climate model revealed serious differences; the data of calculations with the energy-balance model can give variations stronger by an order of magnitude. This can be explained by the following: the energy balance model considers only the albedo feedback, characterizes the mean-annual conditions and provides an equilibrium state, which in the case of atmospheric general circulation models depends strongly on the prescribed sea surface temperature.

An analysis of calculated zonally averaged temperature fields shows that in the troposphere (primarily, near the surface), the heating due to shortwave radiation absorption by aerosols is compensated, mainly, by non-radiative factors. Therefore the temperature field does not change substantially, whereas in the stratosphere the heating is compensated by an intensified longwave cooling.

The geographical distribution of the aerosol effect on climate for every season has two characteristic features. One of them consists of a considerably decreased solar global radiation due to desert aerosol. Both in the Sahara and on the Arabian Peninsula, the decrease of solar global radiation can reach $60\,W\,m^{-2}$. The second feature refers to the outgoing longwave radiation field and consists in the fact that in the belt from the north-eastern coastline of Brazil to Texas the cloud amount decreases, which causes an abnormal growth of the outgoing radiation in the centre of this zone.

It is very important to appreciate that numerical modelling to assess the contributions of various types of aerosols to climate change revealed the non-linear nature of the interaction among the contributions; this means that the sum of the effects of various types of aerosols is not equal to the effect of the sum of the contributions. The strongest spatial homogeneity of the effect of aerosol on climate is observed in the case of stratospheric aerosols. The stratospheric aerosol determines some suppression of the Hadley circulation as well as slowing down the zonal transport in the upper troposphere of the tropics and the subtropics.

An intensified emission of $SO_2$ to the atmosphere, with the subsequent formation of sulphate aerosol, taking place during the second half of the twentieth century has driven serious attention to the assessment of the aerosol effect on climate. This can manifest itself mainly through an increase of the albedo of the surface–atmosphere system due to the following two processes:

1. intensification of backscattering with an increasing aerosol content,
2. increase of cloud albedo as a result of increasing concentration of condensation nuclei, which leads to the growth of the small droplets concentration and, hence, of the albedo. Naturally the effect of anthropogenic aerosol on climate should be the strongest in the summer (with maximum insolation) and in the regions of high concentration of industry.

With the direct and indirect effects of aerosol assumed to be proportional to the content of aerosol in the atmosphere, Engardt and Rodhe (1993) undertook an analysis of surface air temperature trends in the summer and in the winter in different northern hemisphere regions. They considered monthly mean surface air temperature anomalies with respect to surface air temperature monthly means for the period 1950–1979. Table 4.41 illustrates the variation ($\Delta T$) between 1971 and 1990, as compared with the period 1926–1945, in winter (December, January, February) and in summer (June, July, August). The right-hand columns give standard deviations. $\Delta F\, SO_4^{2-}$ is the increase of the sulphate aerosol content in an air column, according to the previously obtained estimates. As can be seen from Table 4.41, in the regions with polluted atmosphere containing more than 6, 8 or 10 mg m$^{-2}$ $\Delta F\, SO_4^{2-}$, a summertime decrease of surface air temperature took place within 0.2–0.4 °C, compared with the remaining part of the northern hemisphere. In winter a strong warming took place (0.5–0.9 °C) in the polluted regions (e.g. Galindo 1992a).

An analysis of the maps of the global distribution of $\Delta T$ showed that in summer, in most of the southern hemisphere and in many regions of the northern hemisphere a warming took place, but in the vast continental regions of the northern hemisphere there was a cooling. The most substantial cooling trend in the polluted regions was recorded between 1960 and 1970, which agreed with the period of intensified emissions of sulphur compounds. Subsequently there was a gradual decrease of these emissions.

Despite the complexity of the interpretation of the results obtained, the following conclusions can be drawn:

1. the summertime cooling in the heavily polluted regions agrees qualitatively with the effect of sulphate aerosol, though other factors may also be important (especially the growth of cloudiness in these regions),

**Table 4.41** Temperature trends in different regions of the northern hemisphere

| Increase of the aerosol content $\Delta F\, SO_4^{2-}$ (mg m$^{-2}$) | $\Delta T$ (°C) summer (JJA) | $\Delta T$ (°C) winter (DJF) | Share of the northern hemisphere area (%) |
|---|---|---|---|
| $>10$ | $-0.33 \pm 0.39$ | $0.62 \pm 0.70$ | 4 |
| $<10$ | $0.05 \pm 0.13$ | $0.11 \pm 0.24$ | 96 |
| $>8$ | $-0.13 \pm 0.26$ | $0.88 \pm 0.69$ | 12 |
| $<8$ | $0.06 \pm 0.13$ | $0.01 \pm 0.21$ | 88 |
| $>6$ | $-0.09 \pm 0.23$ | $0.62 \pm 0.60$ | 21 |
| $<6$ | $0.07 \pm 0.13$ | $-0.02 \pm 0.22$ | 79 |

2. the wintertime warming in the polluted regions can be interpreted as a manifestation of the greenhouse effect of pollution,
3. the summertime cooling trend in the North Atlantic could have been influenced by both atmospheric pollution and other causes. Since the statistical significance of the results discussed is low, they should be considered preliminary.

Schlesinger *et al.* (1993) performed an analysis of the sensitivity of global mean surface air temperature variations due to a $CO_2$ doubling ($\Delta T_{2x}$) to the effect of two external forcings prescribed as variations in the radiation budget of the surface–troposphere system at the level of the tropopause at the expense of variations in the content of greenhouse gases and anthropogenic sulphate aerosol. The values of $\Delta T_{2x}$ were calculated with the use of the energy-balance model for the atmosphere–ocean system (in a 40-layer ocean divided into the polar regions, where the bottom waters form, and non-polar regions, the processes of upwelling and diffusion also being taken into account) as deviations from surface air temperature averages for the period 1861–1990, starting from 1765. The scenarios of the greenhouse gas dynamics have been prescribed from the data of the Intergovernmental Panel of Experts on Climate Change (IPCC) and the aerosol-induced radiative disturbance in 1978: $\Delta F\,SO_4^{2-}$ (1978) $= (-1.1 \pm 0.9)\,W\,m^{-2}$; that is, it can vary from $-0.2$ to $-2.0\,W\,m^{-2}$.

Since the estimates of this effect are very approximate, the numerical modelling has been made with prescribed variable $\Delta F\,SO_4^{-2}$ consisting of two components, corresponding to clear and cloudy skies:

$$\Delta F\,SO_4^{2-}\,(1978) = \Delta F\,SO_{4\,\text{clear}}^{2-}\,(1978) + \Delta F\,SO_{4\,\text{cloud}}^{2-}\,(1978) \tag{4.47}$$

The results of calculations of $\Delta T_{2x}$ were compared to the observational data for the period 1861 to 1990. The temporal variation of the aerosol effect is given as

$$\Delta F\,SO_4^{2-}(t) = \frac{\Delta F\,SO_4^{2-}\,(1978)\,[QSO_2 - S(t)]}{[QSO_2 - S\,(1978)]} \tag{4.48}$$

where $QSO_2 - S(t)$ is the emission of sulphur in the form of $SO_2$ to the atmosphere from the data for the period 1860 to 1977 and $S$ (1978) is the emission of sulphur as $SO_2$ in 1978. Approximate estimates, taking account of the uncertainty of the input parameters and the level of emissions, have led to a value of

$$\Delta F\,SO_{4\,\text{clear}}^{2-}\,(1978) = (-0.58 \pm 0.42)\,W\,m^{-2} \tag{4.49}$$

Estimates of the cloud component $\Delta F\,SO_{4\,\text{cloud}}^{2-}$ are still less reliable since they are connected with the necessity to consider the indirect effect of sulphate aerosol as condensation nuclei on the microphysical and radiative characteristics of clouds. An advection of anthropogenic condensation nuclei from the continents to the oceans is assumed to take place, equal to $10^{19}\,s^{-1}$, which leads to a 20 per cent intensification of nucleation when marine stratus clouds form (with respect to that induced by nature). This leads to a decrease of the radiation budget at the level of the tropopause by about $1\,W\,m^{-2}$ in the northern hemisphere (in the southern hemisphere this effect is considered to be absent).

Then $\Delta F\,SO_{4\,\text{cloud}}^{2-}$ (1978) $= (-0.5 \pm 0.5)\,W\,m^{-2}$, and the total effect is equal to that given above ($-1.1 \pm 0.9\ W\,m^{-2}$). The calculations of surface air temperature anomalies for the period from 1765 (when $\Delta T$ (1765) $= 0$) to 1990 were made with various values of $\Delta F\,SO_4^{2-}$ (1978) and $\Delta T_{2x}$.

The results of calculations can be approximated with the following regression equation:

$$\frac{1}{\Delta T_{2x}} = 0.7042 + 0.4584x - 0.1005x^2 - 0.06847x^3 \tag{4.50}$$

where $x = F\,SO_4^{2-}$ (1978). Thus $\Delta T_{2x}$ varies from 1.39 °C for $x = 0$ (without the aerosol effect) to 5.92 °C for the nominal value of $\Delta F\,SO_4^{2-}$ (1978) $= -1.1\,W\,m^{-2}$. Apparently, the most likely interval will be $1.39\,°C \leqslant \Delta T_{2x} \leqslant 3.00\,°C$, that is $\Delta T_{2x} = (2.2 \pm 0.8)\,°C$, which corresponds to the interval $0 \geqslant \Delta F\,SO_4^{2-}$ (1978) $\geqslant -0.74\,W\,m^{-2}$.

To reduce the uncertainty of the estimates of $\Delta F\,SO^{2-}_{4\,\mathrm{clear}}$ it is important, first of all, to have more reliable data on the efficiency of aerosol scattering calculated per unit mass, the life-time of $SO_4^{2-}$ in the atmosphere and the aerosol backscattering. The problem of improving the reliability of the estimates of $\Delta F\,SO^{2-}_{4\,\mathrm{cloud}}$ is much more complicated. Probably this value is very small indeed.

#### 4.6.4.2 *Stratospheric aerosols*

The effect of stratospheric aerosols on climate is a quite different problem which is relevant, first of all, to the impact of volcanic eruptions on climate (Kondratyev 1985b, Pollack *et al.* 1993). To substantiate the most essential properties of stratospheric aerosol affecting the climate, Lacis *et al.* (1992) calculated the global-mean vertical temperature profile based on the use of a one-dimensional radiative–convective model with the cloud amount assumed to be 50 per cent, and vertical temperature lapse rate constituting $6.5\,°C\,km^{-1}$. In this connection, calculations were made of the extinction coefficient $Q_{ext}$, the single-scattering albedo, $\varpi_c$ and the coefficient of asymmetry of the phase function $\cos\theta$ in the wavelength region 0.3–30 $\mu$m for the droplets of a 75 per cent water solution of $H_2SO_4$, as well as spectral reflection, transmission and absorption with an optical depth $\tau$ at the wavelength 0.55 $\mu$m equal to 0.1 and 1.0.

The size distribution has been prescribed for the May and October stratospheric aerosol observed, respectively, 1.5 and 6.5 months after the El Chichón eruption (the special feature of the May aerosol is the presence of a large number of large particles with radius $\geqslant 1\,\mu$m, which results in a considerable enhancement of the absorption of the thermal emission by the eruptive cloud).

As a measure of climate forcing by stratospheric aerosol one can consider a change of the radiation budget $\Delta F_{net}$ and its components $\Delta F_{solar}$ and $\Delta F_{IR}$ at the tropopause level (i.e., for the surface–troposphere system). Calculations of $\Delta F_{solar}$ and $\Delta F_{IR}$ as a function of the effective radius of particles, $r_{eff}$, have shown that the infra-red forcing depends strongly on the radii of the particles, increasing with increasing radius, within the interval 0.5–3.5 $\mu$m, whereas the albedo for the shortwave radiation is almost independent of the radii of particles, provided they are larger than the effective wavelength (0.5 $\mu$m). Therefore the aerosol with a critical radius of particles $r_c \geqslant 2\,\mu$m causes a strong heating of the troposphere (due to the attenuation of the longwave radiative cooling).

It turns out that $r_c$ depends weakly on the composition and shape of particles. Though the small aerosol particles ($r_{off} < 0.05\,\mu$m) also determine the warming, the effect is weak because of the small content of such aerosols. The width of the particle size distribution characterized by the parameter $v_{eff}$ markedly affects $\Delta F_{net}$. Thus it can be reliably concluded that the impact of net radiative forcing on the troposphere depends mainly only on two parameters, $r_{eff}$ and $v_{eff}$, and the aerosol size distribution can be approximated by a $\gamma$ distribution. With low values, $\tau < 0.2$, the simplest approximations are justified:

$$\Delta F_{net} \sim 30\tau \text{ W m}^{-2} \quad \text{and} \quad \Delta T \sim 9\tau\,^{\circ}\text{C} \tag{4.51}$$

Of great importance is the presence of admixtures in the droplets of the $H_2SO_4$ aerosol, causing changes in the single-scattering albedo. Even for the critical value $\varpi_c = 0.99$ the stratospheric cooling due to backscattering is halved, as compared with $\varpi_c = 1$, and the temperature of the stratosphere increases substantially. However, this does not impact on the troposphere.

The principal conclusion is that the radiative forcing of the stratospheric aerosol on the troposphere can be simulated precisely enough with only one parameter taken into account – the optical depth of stratospheric aerosol, with the exception that it is necessary to consider the effect of particles sizes, when they exceed 1 $\mu$m. An enhancement of the large-particle-induced greenhouse warming can be comparable to or even exceed the albedo effect of cooling, provided $r_{eff} > 2\,\mu$m.

To assess climate changes, it is necessary to substantiate the minimum combination of optical parameters of atmospheric aerosols needed. The calculation schemes are usually based on the following optical characteristics: the volume coefficient of attenuation (extinction), $\sigma_e$, (the availability of the data on the vertical $\sigma_e(z)$ profile makes it possible to calculate the optical depth of the atmosphere $\tau$), the volume coefficient of scattering, $\sigma_{sc}$ (or the single-scattering albedo $\varpi_c = \sigma_{sc}/\sigma_e$), and the volume phase function, which is approximately simulated by introducing the coefficients $g$ of asymmetry of the scattering function. The representative characteristic of the effect of aerosols on climate is the ratio of the coefficients of the aerosol attenuation of radiation in the longwave ($\lambda = 10\,\mu$m) and shortwave ($\lambda = 0.55\,\mu$m) spectral regions. Some of these parameters for different types of aerosols are shown in Table 4.42.

To obtain reliable data on the aerosol optical characteristics, it is necessary to have either complete information on the distribution and chemical composition of aerosol particles (number density, size distribution, complex refractive index of the aerosol matter and shape of particles), or a large statistically substantiated data base of the optical measurements.

Though during recent years a great number of studies have been conducted on the optics of atmospheric aerosols, the results of which have been broadly discussed in various publications (Kondratyev 1980, 1988b, 1992b, Mészáros 1981, Bolle 1982, Ivlev and Andreev 1986, Zuev and Kabanov 1987, Hobbs and McCormick 1988, d'Almeida *et al.* 1991, Götz *et al.* 1991, Deepak 1992, Jennings 1993), the available information should be considered inadequate.

The results of *in situ* measurements of the size distribution and chemical composition of aerosols, which contain rather reliable information, cannot be considered representative for large spatial and temporal scales. Therefore new sets of results of ground, aircraft, balloon and satellite observations should form the basis for the climatological aerosol models, with emphasis on the accomplishments of integrated programmes with the use of all observational means available. Also, it is necessary to develop both theoretical and empirical models of the processes of formation,

**Table 4.42** Typical values of the optical parameters of different types of aerosol

| | Wavelength | | | | | |
|---|---|---|---|---|---|---|
| | 0.55 $\mu$m | | 10 $\mu$m | | Horizontal | Vertical |
| Aerosol origin | $\tau$ | $\varpi_c$ | $\tau$ | $\varpi_c$ | variation | structure |
| Desert | 0.054 | 0.890 | 0.0052 | 0.44 | Yes | ED in ABL at $H = 3$ km |
| Stratospheric | 0.045 | 1.000 | 0.0049 | 0.06 | No | Homogeneous |
| Oceanic | 0.033 | 0.988 | 0.0083 | 0.76 | Yes | ED in ABL at $H = 1$ km |
| Continental | 0.031 | 0.890 | 0.0030 | 0.44 | Yes | ED in ABL at $H = 1$ km |
| Tropospheric | 0.030 | 0.890 | 0.0029 | 0.44 | No | Homogeneous |
| Urban | 0.008 | 0.647 | 0.0003 | 0.15 | Yes | ED in ABL at $H = 1$ km |
| Volcanic | 0.007 | 0.943 | 0.0002 | 0.15 | No | Homogeneous |

*Note*: $\tau$ = optical thickness $\varpi_c$ = single scattering albedo; ABL = atmospheric boundary layer; $H$ = scale height; ED = exponential decrease.

evolution and removal from the atmosphere of the aerosol particles, as well as various models of their size distribution and optical characteristics. In this respect a new Atmospheric Radiation Measurement (ARM) programme, which has been recently started in the USA on the basis of a broad international cooperation, opens new horizons.

Of particular importance are models with few parameters, based on both physical ideas about the processes of formation of aerosols and reliable observational data. Values of the optical parameters used in the various models are given in Table 4.42. In particular, very important are further studies to determine models based on a consideration of two easily measured input parameters, the relative humidity $H$ (per cent) and the meteorological visibility $S_m$ (km).

In the context of identification of climate signals due to volcanic aerosols there is a strong need to analyse coupling with the El Niño / Southern Oscillation (ENSO) phenomena. Such an analysis has been recently accomplished by Portman and Gutzler (1995) who considered regional surface air temperature changes for US territory under the impact of both volcanic eruptions and the ENSO on the basis of observational data for 1219 meteorological stations. Time series have been analysed with the separation of certain time periods for various eruptions (Table 4.43). This problem has been discussed in detail in the monograph by Kondratyev and Galindo (1997).

The intensity of eruptions has been charactcrized by various parameters such as the dust veil index, sulphate content, etc. As ENSO criteria the Southern Oscillation Index and sea surface temperature anomalies in the eastern sector of the tropical Pacific have been chosen.

Table 4.44 contains the list of observed ENSO events (those ENSO events which have not been considered are mentioned in brackets). A technique of superimposed epochs (moving averages) has been used to analyse observational data with the consideration of time series for various eruptions and for single stations. Mean

**Table 4.43** Volcanic eruptions

| Year | Volcano | Latitude | Month |
|---|---|---|---|
| 1902 | Mount Pele | 15° N | May |
| | Soufriere | 13° N | May |
| | Santa Maria | 15° N | October |
| 1907 | Ksudach | 52° N | March |
| 1912 | Katmai | 58° N | June |
| 1963 | Agung | 8° S | March |
| 1982 | El Chichón | 17° N | April |

monthly values have been calculated as deviations from 6-yr averages calculated for every calendar month which made it possible to filter out annual variations and long-term trends.

The analysis of the results for five eruptions has demonstrated an enhancement of sea surface temperature anomalies with latitude. In northern and central belts of the USA, stable negative anomalies of about 1–2 °C persisted during the first 6 months after an eruption. During the next 6 months in the eastern and northern parts of the country, positive anomalies were observed, whereas in the western sector of the central and southern parts negative anomalies prevail. Subsequently (during the second year after an eruption) negative anomalies were dominant again in the whole US territory (up to −2 °C in the northern belt).

The consideration of seven ENSO cases has led to the discovery of a number of significant anomalies including positive (negative) with an interval of 6–17 months after the onset of the ENSO in northern (southern) belts which have also been observed earlier. All volcanic eruptions considered (except Agung in 1963) took place during a warm ENSO of at least moderate intensity.

Christy (1993) has analysed a combined impact of the Mount Pinatubo eruption and ENSO 1991–1993 on temperature fields in the troposphere and lower strato-

**Table 4.44** El Niño / Southern oscillation phenomena (years)

| | | | | |
|---|---|---|---|---|
| A. Warm phase (very strong–strong) | | | | |
| 1911 | 1917 | 1925 | 1932 | 1940 |
| 1957 | 1972 | [1982] | | |
| B. Warm phase (moderate) | | | | |
| [1902] | 1904 | [1907] | 1910 | 1914 |
| 1923 | 1930 | [1939] | 1943 | 1951 |
| 1953 | 1965 | 1969 | 1976 | |
| C. Cold phase | | | | |
| [1903] | 1906 | [1908] | 1916 | 1920 |
| 1924 | 1928 | 1938 | 1942 | 1949 |
| 1954 | [1964] | 1970 | | |

sphere on the basis of the NOAA satellite microwave sounding unit (MSU) for channels N2R (mean tropospheric temperature 0–10 km) and N4 (100–50 hPa layer of the lower stratosphere, 15–20 km). The analysis of the time series for global and zonal averages of lower stratosphere temperature during 1979–1992 distinctly revealed impacts of volcanic eruptions: Nyamarugira (December 1981) and El Chichón (April 1982) as well as Mount Pinatubo (June 1991).

It is important to consider two effects of eruptive sulphuric acid aerosol:

1. absorption of the upward flux of longwave (thermal) radiation which leads to greenhouse warming of the stratosphere;
2. transformation of shortwave radiation fluxes, of which the characteristic feature is an almost complete mutual compensation between the attenuation of direct solar radiation (about 25 per cent) and the increase of the downward flux of scattered radiation.

During the first year after a powerful eruption a decrease of global radiation did not exceed 3 per cent, which led to certain cooling of the troposphere and the surface. The lower stratosphere in sub-tropical belts of both hemispheres reacted most rapidly to Nyamarugira and El Chichón and Mount Pinatubo but relevant warming was not zonally symmetrical. In the case of Nyamarugira and El Chichón it was stronger in the northern than in the southern hemisphere, whereas after Mount Pinatubo the opposite was true.

In a year after Nyamarugira and El Chichón, near global air surface temperature increased by 1.0 °C, but during the subsequent 18 months it decreased at the rate of 0.0025 °C per day. Within the equatorial belt the rate of air surface temperature increase was 0.01 °C per day (Nyamarugira and El Chichón) and 0.015 °C per day (Mount Pinatubo). Relatively stable mean global air surface temperature during 1984–1990 was about 0.5 °C lower than before the eruptions (in this connection a suggestion has been made that volcanic aerosol stimulates the washing-out of $O_3$ which leads to stratospheric cooling).

The surface air temperature time series in the cases considered were quite different. Rather rapid warming (0.013 °C per day) after Mount Pinatubo (June–September 1991) reached 1.3 °C and was five times stronger than the warming after Nyamarugira and El Chichón, but the rate of subsequent cooling was approximately equal.

Analysis of tropospheric temperature data has also convincingly revealed effects of both ENSO and volcanic eruptions. For instance, the 1982–1983 ENSO produced global tropospheric warming of the order of 0.5 °C (in comparison with the background level) and by the end of 1986 the tropospheric temperature increased by 0.6 °C, this warming being stable until September 1988. However, the onset of the cold ENSO phase in 1989 led to a drop of tropospheric temperature by 0.6 °C. The impact of the 1991–1992 ENSO was insignificant.

It is possible that the trend of tropospheric temperature decrease in 1982–1986 was enhanced by the Nyamarugira and El Chichón impact. If one can neglect the 1982–1983 ENSO contribution, then the tropospheric temperature global average drop during 4–5 yr reached 0.3 °C. The strongest anomaly during the whole period considered was a rapid cooling in the northern hemisphere from June 1991 till August 1992 (global mean tropospheric temperature decrease reached 0.73 °C). This cooling was especially strong over the land surface. For the whole Microwave Sounding Unit time series for a 4-yr period, tropospheric temperature minimum was observed in June–August 1992. It has happened to be very difficult to accomplish a cause and effect analysis of observational data to assess contributions from ENSO and volcanic eruptions because the amplitudes of high frequency and interannual tropospheric temperature variability were approximately equal.

Minnis *et al.* (1993) demonstrated that an unambiguous identification of volcanic climatic signal is possible on the basis of the analysis of ERBE (Earth Radiation Budget Experiment) satellite data, which has been accomplished for the case of the Mount Pinatubo eruption in June 1991. Such an analysis has discovered for the first time large-scale volcanic forcing on the basis of observations. A strong cooling was observed immediately after the eruption, which increased through September 1991 during the shortwave forcing increase relative to the longwave forcing. The primary aerosol effects were a direct increase in albedo over mostly clear areas and both direct and indirect increases in the albedo of cloudy areas; during the first 5 months after the eruption the stratospheric aerosol optical depth increased by up to two orders of magnitude between 40° N and 40° S.

Minnis *et al.* (1993) have characterized the strength of the volcanic forcing through the examination of zonal variations of monthly means of reflected shortwave fluxes ($M_{\mathrm{SW}}$) and outgoing longwave radiation ($M_{\mathrm{LW}}$) relative to the 5-yr means. The increase of $M_{\mathrm{SW}}$ after the eruption, between 5° N and 5° S in August 1991, reached $10\,\mathrm{W\,m^{-2}}$, while the longwave volcanic signal was barely detectable above the noise of the interannual variability. Therefore a net cooling of almost $8\,\mathrm{W\,m^{-2}}$ occurred in August 1991, which is twice the value of any other monthly anomaly.

This trend continued for the zone between 40° N and 40° S with a net volcanic forcing of $-4.3\,\mathrm{W\,m^{-2}}$ for August 1991. Since the standard deviation for the 5-yr period (as an indicator of the volcanic forcing accuracy) is $1.5\,\mathrm{W\,m^{-2}}$ the anomaly considered is statistically significant at the 95 per cent confidence level. Globally averaged volcanic forcing (if one can neglect volcanic impact outside the 40° latitude belt) was $2.7 \pm 1.0\,\mathrm{W\,m^{-2}}$ during August and September 1991.

The distribution and amount of observed volcanic forcing closely followed the spread of the volcanic aerosol cloud. For approximate assessments very simple linear regression may be utilized:

$$\Delta M_{\mathrm{SW}} = (17.1 \pm 5.6)\tau - 0.7 \qquad (4.52)$$

$$\Delta M_{\mathrm{LW}} = (-7.9 \pm 5.1)\tau + 1.9 \qquad (4.53)$$

$$\Delta M_{\mathrm{NET}} = (-11.5 \pm 4.4)\tau - 0.5 \qquad (4.54)$$

where $\tau$ is the zonal mean optical depth. However, such correlations are correct only for an initial period after the eruption. At the later stage, transformation of aerosol properties (primarily, size distribution) becomes an important factor.

Observational data demonstrate very pronounced spatial inhomogeneity of volcanic forcing distribution even within a given latitude zone. In general, Mount Pinatubo's effect was most evident over areas that tend to be cloud-free, especially over dark surfaces (that is why the apparent lack of significant volcanic forcing signal over the Sahara is not surprising).

In a few months after the eruption a substantial penetration of volcanic aerosol into the troposphere (primarily in the vicinity of tall convective storms and in the tropopause folds) took place. This is why lidar observations during August 1991 discovered significantly enhanced tropospheric aerosol loading as far north as 40° N. Such a penetration created favourable conditions for an interaction between volcanic aerosol and clouds. Since volcanic aerosol particles can serve as cloud and ice nuclei this could lead to an increase of the number and a decrease of the effective radius of cloud particles which resulted in cloud albedo increase. Although such an indirect effect is most noticeable for deep clouds ($M_{LW} < 220\,W\,m^{-2}$) it may also occur for other cloud types.

Pollack *et al.* (1993) have applied the Climate Model II developed at the Goddard Institute for Space Studies (the nine-level model with an 8° latitude by 10° longitude grid was run for 55 simulated years to reach a statistically steady state for the volcanic problem) to simulate the response of the climate system to a spatially and temporally constant forcing by volcanic aerosols having an optical thickness of 0.15. These climate changes are compared with those obtained with the same climate model for cases of doubled $CO_2$ concentration and when the boundary conditions associated with the peak of the last ice age were used. In all three cases, the absolute magnitude of the change in the globally averaged surface air temperature is approximately the same, ~5 deg K. Surface temperature change, $\Delta T_s$, in the case of volcanic perturbation does not sharply increase with increasing latitude, while $\Delta T_s$ in the lower stratosphere is positive at low latitudes and negative at high latitudes.

To some extent, the climate response in all three different forcings is similar: direct radiative forcing accounts for 30 per cent and 25 per cent of the total $\Delta T_s$ in the volcano and doubled $CO_2$ runs, respectively. The remaining part of the change is mainly due to water vapour feedback which is positive (as well as albedo feedback at high latitudes) in all three cases.

But in some respects, the climate response to the three forcings is quite different; the latitudinal $\Delta T_s$ profiles for the three runs differ considerably (due to differences in the latitudinal profiles of the primary radiative forcings). Partly for this reason, changes in eddy kinetic energy, heat transport by atmospheric eddies, and total atmospheric heat transport are quite different in the three cases. For example, atmospheric heat transport acts as a positive feedback at high latitudes in the volcano run and as a negative feedback in the other two runs (these results illustrate limitations of energy-balance climate models with regard to simple parametrization of atmospheric heat transport).

In connection with the problem of interdecadal climate changes, Crowley *et al.* (1993) have undertaken an analysis of the acidity of the ice core from Greenland for the last 1400 yr (see also Crowley and Baum 1994). The results show that changes in volcanism may have significantly influenced climate on this time scale. However, the ice core acidity record consists of volcanically-induced spikes superimposed on background acidity of non-volcanic origin that varied significantly in amplitude (the minimum acidity threshold to identify significant eruptions has been prescribed at the level of $0.6\,\mu EH^+\,kg^{-1}$). Thc principal conclusions made by Crowley *et al.* (1993) are that: (1) on decadal time scales, volcanism has varied by a factor of three to four over the last 1400 yr, but the climate volcanism correlation is considerably less impressive ($r = 0.23$) than previously concluded ($r = 0.52$) and (2) such a difference may be explained by the contribution of background acidity level variations that show a significant Little Ice Age increase which may reflect changes in ocean productivity (sometimes maximum acidity reached $1.2\,\mu EH^+\,kg^{-1}$, which was 50 per cent higher than after the Krakatoa eruption).

### 4.6.5 Conclusions

From the viewpoint of further studies of the properties of atmospheric aerosols and their impacts on climate, the following are most urgent:

1. improving the standard radiative models of the atmosphere to simulate more adequately the characteristics of various types of aerosols, including an account of the effect of humidity on their size distribution, vertical concentration profiles, etc.;
2. obtaining more adequate observational information about the spatial and temporal variability of aerosols on a global scale;
3. developing techniques to identify the anthropogenically induced trends of the content and composition of aerosols on regional and global scales (in particular, monitoring the dynamics of sulphate and soot components is especially urgent);
4. studying the dependence of the optical properties of real aerosols and their chemical composition and size distribution with an account of the effect of variability of the air humidity;
5. special comparisons of techniques to measure the aerosol characteristics (remote sensing versus *in situ* measurements, etc.);
6. observational and theoretical studies of the internal structure of aerosol particles as multi-component formations; the role of aerosol particles as condensation nuclei.

The chemical processes in the troposphere and the stratosphere are an important climate-forming factor (Finlayson-Pitts and Pitts 1986, Kondratyev 1992b, Kondratyev *et al.* 1996a,c). In this connection, the problem of stratospheric $O_3$, closely connected with the problems of stratospheric heterogeneous chemistry, has been intensively discussed (Kondratyev 1980, 1989a, 1996).

Apparently, most important are the following mechanisms affecting the heterogeneous reactions of the formation and disintegration of $O_3$ molecules:

1. changing the radiative regime in the stratosphere in the presence of scattering aerosol particles;
2. heterogeneous reactions taking place with an absorption by aerosol particles (polar stratospheric clouds) of gases, taking an active part in reactions with $O_3$ molecules (hydroxyl, nitrogen oxides, CFCs, etc.).

In the troposphere the following phenomena are connected with heterogeneous processes: formation of acid rain, removal from the atmosphere of various polluting gases, reduced visibility, fogs, clouds and, a special phenomenon, arctic haze. In these phenomena the effect of anthropogenic factors manifests itself, in particular, as a change of the physical and chemical properties of cloud systems: their optical characteristics and stability (Jennings 1993, Kondratyev *et al.* 1996b).

All the atmospheric climate-forming processes are closely coupled, forming a chain in which there are direct and inverse, positive and negative feedbacks. In some situations, certain links of this chain of atmospheric processes, conditions appear for unstable equilibrium; the connection between the links is such that coupling with one link, often energetically insignificant, brings forth more substantial changes in other links, i.e. a trigger mechanism works. The aerosols are most likely candidates for such trigger mechanisms in the atmosphere. In this context it is specially important to study thoroughly the physico-chemical processes that govern the dynamics of the atmosphere as a colloidal system. The urgency and the complexity of problems considered determine the particular importance of international cooperation in this field, closely connected with a study of global inadvertent and advertent impacts on the weather and climate.

## 4.7 Solar–terrestrial interrelations

For the characteristic spatial and temporal scales determined by the WCRP objectives, there is statistical evidence for solar–climatic correlations (Gille *et al.* 1984, Labitzke *et al.* 1986, Labitzke 1987, Labitzke and van Loon 1988, 1989, 1992, Wuebbles *et al.* 1991, Hood *et al.* 1993, Hoyt and Schatten 1993, van Loon and Labitzke 1993) but convincing physico-chemical mechanisms explaining these correlations have not been discovered yet (Kondratyev 1992b). Numerous unanswered questions in the problem of solar–climatic relations necessitate serious studies in this direction (i.e. to discover the potentially substantial role of extra-terrestrial factors on climate, including solar activity).

The available data related to the possibility of long-term solar constant variability show that the problem of solar constant variations directly concerns the WCRP. Without dwelling on the results discussed earlier (Hunt 1981, Kondratyev 1986b,c) we shall consider the subsequent studies. The most recent results have been discussed by Kondratyev and Nikolsky (1995a,b).

It is obvious that only a systems approach (interactive consideration of the impact of solar activity) will lead to a solution. This has recently been demonstrated by the results obtained by Jenkins (1993) who considered the sensitivity of a general circulation model (the NCAR CCMOA model was chosen) to a faster rotation of the Earth, enhanced $CO_2$ concentration and reduced solar constant. Four sensitivity studies have been performed under the assumptions of zero land fraction and 10 per cent reduced solar constant conditions: (1) increased rotation rate of the Earth (shorter day length), (2) eight times the present concentration of $CO_2$, (3) a combination of 1 and 2, (4) 3 with a 15 per cent reduction in the solar constant.

These numerical modelling results directed towards palaeoclimatic assessments indicate the following changes in the global mean sea surface temperature compared with the control simulation: (1) +6.6 deg K, (2) +12 deg K, (3) +18.5 deg K and (4) +6.6 deg K relative to the control's global mean temperature of 264.7 K. The warm sea surface temperatures are associated with a 20 per cent reduction in clouds for the rotation rate experiments and higher $CO_2$ concentrations in the other experiments. A very important conclusion is that, in contrast with previous studies that have used energy balance and radiative–convective models, the new results require a much smaller atmospheric $CO_2$ increase to prevent an ice-covered Earth. This demonstrates a potentially important role of cloud feedback for the climate dynamics of the early Earth as well as leading to the conclusion that very large atmospheric $CO_2$ concentrations may not be necessary to counterbalance the lower solar constant that existed early in the Earth's history.

Nesme-Ribes *et al.* (1993) have used an approach to reconstruct the solar constant over the past four centuries (with the use of magnetic solar cycle variations as proxy data) and then accomplished climate change simulation experiments on the basis of an atmospheric general circulation model developed at the Laboratoire de Météorologie Dynamique (LMD AGCM) with a grid of 7.5° longitude by 3.2° latitude and 11 layers. The response of the model to magnetoconvective solar forcing during the Maunder Minimum, when the mean value of the drop in solar constant reached 0.5 per cent, has been studied to simulate the Little Ice Age phenomenon. The simulated cooling for a solar constant decrease by 1 $W m^{-2}$ is compatible with observational data. An important conclusion is also that variations of globally homogeneous external forcing parameters, like solar constant or greenhouse gas loading (of equivalent forcing), yield climate responses with very similar geographical patterns. These responses are distinguishable only in the stratosphere. This means that such responses are determined mainly by the internal modes of climate variability, its amplitude only being modulated by the amplitude of the external perturbation.

Nesme-Ribes *et al.* (1993) pointed out that there are a few conspicuous features of the results: the stratospheric warming from mid-latitudes to poles, contrasting with tropospheric cooling, which may be attributed to a secondary greenhouse effect by water vapour; relatively strong warming of the upper tropical troposphere, due to a strongly positive cloud feedback in this version of the model, and the unexpectedly weak seasonal contrast in high-latitude cooling. The global average cooling is nearly 1.5 °C with little seasonal variation.

The most important conclusion made by Nesme-Ribes *et al.* (1993) is that the strong sensitivity of the results to parametrization approaches (especially in the case of cloud microphysics which makes the model highly sensitive), the possibility of other mechanisms acting simultaneously (like volcanic perturbations) and the presence of internal climate variability due to the transient component of ocean motion (omitted in the numerical modelling considered) should be a reminder of the complexity of climate system dynamics and the difficulty of reaching definite conclusions.

There are three basic solar impacts on climate: (1) the effect of solar constant variations on the Earth's radiation budget, and therefore on climate, (2) the orbital forcing effect (so called 'Milankovich mechanism' due to variations of the orbital parameters) and (3) solar activity influence. We shall start from the consideration of the first impact.

### 4.7.1 Variations in the value of the solar constant

The Solar-Maximum Mission satellite launched in February 1980 within the Programme on Studies of Maximum Solar Activity in an 11-yr cycle carried three similar precise cavity pyrheliometers developed at the Jet Propulsion Laboratory of the University of California to measure the solar constant. Two major objectives of measurements were (Willson 1984, Willson and Hudson 1988): (1) regular measurements of the solar constant with an accuracy of $\pm 0.1$ per cent which should be extended for a 22-yr solar cycle to study the solar constant long-term variability and (2) obtaining information about the solar constant short-term variability on a time scale from several minutes to several months, with maximum possible accuracy (see the survey by Sklyarov 1994).

Each of the three pyrheliometers has a 5° field of view, being independent and having a system of electric self-calibration. The cavity's absorptivity is 0.9995. Each pyrheliometer has a shutter which makes it possible to take measurements in different regimes. The phase operation of the instruments ensures an accuracy within 0.1 per cent for at least a year. A planned comparison with rocket- and Shuttle-carried sensors will make it possible to maintain this accuracy for a longer time.

Willson (1984) and Willson and Hudson (1988) discussed the results of the solar constant measurements using pyrheliometer A during the first 153 days, reduced to an average Earth-to-Sun distance, with solar constant variations averaged over one orbit, with respect to an average overall 153 days equal to 1368.31 W m$^{-2}$ (with an accuracy of $\pm 0.5$ per cent). The number of counts for an orbit, when the shutter was open, averaged 24, which corresponds to a period of about 52 minutes in each of 16 daily orbits.

From data for the first 112 days of observations (when the solar constant substantially decreased), a high correlation between the solar constant and the photometric sun-spot index (PSI) was found:

$$\mathrm{PSI} = a \sum \mu_i s_i \frac{(3\mu_i + 2)}{2} \tag{4.55}$$

where $a = 0.33$ is the normalization coefficient, $\mu_i = \cos\,\theta_i$ ($\theta_i$ is the angle between the axis of the instrument and the direction to the centre of the Sun) and $s_i$ is the area of the spot. The summation is carried out for the whole disc of the Sun. If the area of sun spots is expressed in terms of millionths of the disc's area, then PSI characterizes the solar constant decrease in millionths. Outside this period, without substantial solar constant variations, the correlation between solar constant and PSI substantially worsens (during the first 35 days it was completely absent).

The 5-day decrease near 13 May did not reveal any correlation with solar activity. Apparently, there are other mechanisms responsible for solar constant variations. An effect of faculae on solar constant variations is also possible. Analysis of the frequency spectrum of solar constant variations revealed the absence of periods shorter than approximately seven days. Such an interval is close to the characteristic time of formation and development of the groups of sun spots and associated zones of faculae.

Hoyt and Eddy (1983) proposed a model of interdaily variability of the solar constant which is part of the model of solar constant variations for the period 1974 to 1981 described above. Since neither faculae nor bright rings around sun spots can balance the blocking effect, an 11-yr modulation of solar constant is observed, with its relative decrease down to 0.07 per cent between 1969 and 1981. It is shown that although the solar constant variability determined from data from the Solar Maximum Mission satellite can be simulated with the use of different models, only the blocking model is adequate and can explain, in particular, an 11-yr cycle of solar constant.

Hoyt and Schatten (1993) examined several different solar indices measured over the past century that are potentially proxy measures for the solar constant. Three solar indices allow the solar constant to be modelled for the 1770–1992 period: (1) the length of the solar cycle, (2) the normalized decay rate of the solar cycle and (3) the mean level of solar activity (for all the proxy models considered, the solar constant varies by less than about $\pm 0.2$ per cent in the last century; this is why it is difficult to detect solar constant variations). A comparison with the northern hemisphere surface air temperature departures for the same period show that approximately 71 per cent of the decadal variance in the last century can be modelled with those solar indices, although this analysis does not include anthropogenic or other variations which would affect the results. Over the entire three centuries, ~50 per cent of the variance is modelled. In accordance with what has been mentioned before, Hoyt and Schatten (1993) note that, to understand our present climate variations, we must place the anthropogenic variations in the context of natural variability from solar, volcanic, oceanic and other sources.

The self-calibrating cavity radiometer (ACRIM) carried by Nimbus-7 provided regular (48-second series in each orbit) measurements of the solar constant, starting from 16 November 1978. An analysis of preliminary (engineering) data of solar constant observations has led to the following conclusions: (1) drops of solar constant within 0.2–0.4 per cent took place during periods of several days to weeks, (2) these drops corresponded to maxima of the simplest indices of solar activity, (3) a long-term trend of solar constant decrease of about 0.02 per cent per year was recorded and (4) the solar constant for the period November 1978 to January 1983 averaged 1372.6 $W m^{-2}$ with the variability within 6.2 $W m^{-2}$ (0.45 per cent) and mean standard deviation 0.95 $W m^{-2}$ (0.07 per cent).

The subsequent more detailed processing of observational data has led to the conclusion that the ratio of preliminary solar constant estimates to final ones averages 1.0012, with variations within 0.9997–1.003. An analysis of the series of final data for three years (16 November 1978 to 31 October 1981) for 764 days has shown that the solar constant averages 1371.0 $W m^{-2}$, with mean standard deviation 0.78 $W m^{-2}$ (0.057 per cent).

The daily data variability ranges within 1367.8–1373.0 $W m^{-2}$ (0.38 per cent) with respect to an averaged solar constant, and the long-term trend is −0.02 per cent per year. The preliminary processing of data for the period November 1980 to April 1983 did not reveal any similar trend and gave 1370.0 $W m^{-2}$ for the solar constant. A comparison with synchronous data of the SMM satellite revealed a systematic 0.2 per cent underestimation of these data, but an agreement of the short-term solar constant variability was good enough.

Fröhlich (1994) analysed the data of solar constant observations using the instrumentation carried by the satellites SMM and Nimbus-7 for the period November 1987 to August 1991 (in September 1991 a new solar constant observation series began with the satellite UARS). These data show that the solar constant decreasing trend was followed in the late 1980s by the trend of solar constant increase. Analysis of the cavity radiometer ACRIM data (the satellite SMM) for 8.5 yr had led to the conclusion that the sun spots dominate the solar constant variability at shorter periods by decreasing the irradiance, whereas increases of the radiation by the faculae and magnetic network take over for longer periods up to the 11-yr solar cycle when, at the maximum, the highest intensity is observed.

From the data of solar constant measurements by the satellite ERBE, the solar constant averages (1366.5 ± 3.6) $W m^{-2}$. The solar constant decreasing trend (till September 1986) constituted −0.037 per cent per year and the subsequent increase +0.03 per cent per year (including the year 1989). A maximum of the amplitude of the solar constant short-term variations reached 0.3 per cent, and the long-term variability did not exceed 0.1 per cent during an 11-yr cycle. The most recent solar constant measurement results have been summarized by Crommelynck (1994) and Sklyarov (1994).

An analysis of the results of recent continuous satellite observations of total and spectral solar constant has shed light on the solar activity-determined physical nature of short-term solar constant variability. The common supposition is that large sun spots or groups of spots cause a decrease of solar constant measured from satellites, when they (as the Sun rotates) appear on the visible side of the disc. The 25–27-day periodicity of solar constant follows from this effect, which corresponds to the equatorial period of the Sun's rotation.

Smith *et al.* (1985) showed that the variability of the solar constant spectral distribution from Nimbus-7 data can be explained as corresponding to a shift of the absolute blackbody radiation spectrum, with large spots appearing on the visible side of the disc. In the absence of this modulation under the influence of a simple thermodynamic process due to changing temperatures in the regions of spots, any deviation from blackness manifests itself most in the long-wavelength region, in the visible and near infra-red spectrum (see Kondratyev and Nikolsky 1995a).

It is very important that the revealed disturbances, especially seen in radiometer channels 6–9 (wavelength intervals 395–508 nm, 344–460 nm, 300–410 nm and 275–360 nm) are connected, as a rule, with disturbances of total solar constant. The total solar constant variability coincides in phase with variations in the index of projection area of sun spots. Since the spectral solar constant variations are in phase, it follows that solar constant variations for individual channels are limited by the level of total solar constant variations, determined as a weighted mean superposition of the spectral solar constant variations. More substantial solar constant fluctuations appearing sometimes for individual channels should be ascribed to the influence of various sources of errors.

Heath and Schlesinger (1984) discussed the results of observations of the spectral distribution of extra-atmospheric solar radiation in the wavelength interval 160–400 nm (spectral resolution 0.1 nm, but observations have been made with a wavelength step of 0.2 nm) made during five years (starting from November 1978) with the Nimbus-7 spectro-radiometer to measure backscattered ultraviolet radiation. A double monochromator recorded the spectrum of solar radiation reflected from the diffusor's plate for at least one orbit a day, when the satellite was near the northern hemisphere terminator.

From data of two series of solar constant measurements from satellites SMM (270 days) and Nimbus-7 (1445 days), Fröhlich (1982, 1994) performed an analysis of the frequency spectrum of solar constant variability for frequency intervals 0.06–4.0 $\mu$Hz. A 27-day periodicity of solar constant was observed, not connected with sun spots, with a modulation amplitude of about $\pm 0.012$ per cent. Almost half of this amplitude can be explained by the effect of the same factor, responsible for the ultraviolet radiation modulation in the interval 200–300 nm, with an amplitude from 4 to 1 per cent. To reveal the reasons for this variability, further studies are needed. The solar constant variability observed testifies to a possibility of a 3-fold faster rotation of the Sun's core than its surface.

According to Fröhlich (1982, 1994), analysis of available balloon and satellite data on the total solar constant revealed variations with periods from minutes to several hours with an amplitude within the range of millionths to hundredths of 1 per cent, as well as variations with characteristic time from several days to months with an amplitude reaching several tenths of 1 per cent. The latter is a manifestation of solar activity connected with variations of solar spots and flares.

The entire spectral range of variability covering frequencies (periods) from 70 nHz (150 days) to 5 $\mu$Hz (3.3 min) can be divided into three major intervals. The solar constant variations in the interval 70 nHz to 2 $\mu$Hz (5.8–150 days) are mainly determined by solar activity whose intensity varies by an order of magnitude during a cycle of solar activity. The spectrum of variations is characterized by prominent peaks at periods 51.4, 23.4 and 7.09 days, which probably exist independently of solar activity. In the periods of high solar activity the integrated power of variations reaches $7.5 \times 10^{-7}$ (46 ppm mean amplitude), decreasing to $1.5 \times 10^{-7}$ (21 ppm) during a weak solar activity.

The second spectral interval of solar constant variations covering $\mu$Hz (time periods from 14 hours to 5.8 days) follows the law $\nu^{-2}$ ($\nu$ is the frequency), which may be due to so-called g-modes. The integrated power in this range is $3.1 \times 10^{-9}$ (3.7 ppm mean amplitude). Within the third interval, 20 $\mu$Hz to 5 $\mu$Hz (14 h to 3.3 min), the spectrum follows a $\nu^{-1}$ law, with a broad peak around a 5 min period which is due to p-modes yielding amplitudes of several ppm. The integrated power is $1.5 \times 10^{-8}$ (0.5 ppm).

The linear structure of the variability spectrum within the third interval, revealed by balloon photometric observations, shows a spatial coherence of granulations and supergranulations, at least during several hours. As for the long-term trends, the observational data reliably reveal the decreasing trend for the solar constant of 0.18 per cent per year after 1980, whereas during the late 1960s–1970s an increasing trend of about 0.029 per cent per year was observed for the solar constant. The long-term trends and their causes can reliably be determined only on the basis of continuous satellite observations of the solar constant.

Solar ultraviolet radiation forms in very thin external parts of the solar atmosphere. Though the ultraviolet radiation at wavelengths shorter than 500 nm constitutes only about 1 per cent of the total energy of the solar constant, its variability affecting the processes in the upper atmosphere of the Earth is of great interest. From the data of Lean (1987), the maximum increase of ultraviolet radiation recorded during cycle 21 for characteristic scales of the Sun's rotation (days, months) constitutes: 1 per cent (300 nm), 2.5 per cent (250 nm), 6 per cent (200 nm), 16 per cent (150 nm), 40 per cent (the Lyman-alpha line, 121.6 nm), 60 per cent (30.4 and 120.6 nm) and 200 per cent (28.4 nm). The amplitude of a longer-term variability (for instance, 11-yr) of ultraviolet radiation is still unknown, despite long observation series. Available data show that the extra-atmospheric ultraviolet radiation has been increasing during cycle 21 with growing solar activity by 1 per cent (300 nm), 1.5–4 per cent (250 nm), 2.5–9.3 per cent (200 nm), 16–100 per cent (150 nm), 80–155 per cent (the Lyman-alpha line, 121.6 nm), about 100 per cent in chromospheric emission lines in the extreme ultraviolet range (102.6, 58.4, 30.4 nm, etc.) and from 100 per cent to an order of magnitude in coronal emission lines of the extreme ultraviolet range. Though 11-yr variations of the frequency of occurrence of floccules with intensified ultraviolet radiation reflect a possible 11-yr modulation of the extra-atmospheric ultraviolet radiation, the amplitude of such variations has not been estimated. Probably, therefore, in fact, this modulation is of a longer period.

The data from satellite daily observations of the ultraviolet solar constant allowed a detailed analysis of its short-period variability. The time series of the data are characterized by a high autocorrelation and instability with non-deterministic cycles of 27 and 13 days. Observations confirmed a consistency between the short-period variations of ultraviolet radiation and the evolution and rotation of active regions (floccules) on the Sun's disc. The emissions of the upper photosphere and chromosphere have, apparently, similar temporal structures which differ, however, from the variability of coronal emissions or 10.7 cm thermal emission. As for the indicators of short-period variations of ultraviolet radiation formed beneath the corona and determined from ground-based observations, the best are the emission of Ca II K and an equivalent width of the 1093 nm line.

The determination of the long-period variability of extra-atmospheric solar ultraviolet radiation is hindered by a limited volume of reliable observational data. The accuracy of measurements varies usually within 10–30 per cent – that is, within the same order of magnitude as the variability of ultraviolet radiation within its long-wavelength region during an 11-yr cycle. There are, however, grounds to suppose much larger systematic errors. These circumstances determine the hypothetical character of mechanisms for the long-term variability of the extra-atmospheric solar radiation.

Variations in solar magnetic activity are usually supposed to be a reason for 11-yr cycles, but these correlations have not been estimated quantitatively. The factors of other cycles (apart from the 11-yr one) are still less determined. An accuracy of several per cent for extra-atmospheric ultraviolet radiation measurements is very difficult to achieve. In developing the satellite instrumentation which will operate during cycle 22, the emphasis has been placed on the reliability of the pre-flight absolute and in-flight calibrations of these instruments. Absolute errors within 5–10 per cent with relative errors not more than 1 per cent are supposed to be reached in the wavelength interval 100–300 nm. The simultaneous use of several instruments, after a thorough absolute calibration, will make it possible to assess the reliability of measurements.

An analysis of data on the temporal variability of the ratio of emissions in the centres of lines MgII h and k to the continuum emission interpolated from the zone of line wings has shown that variations of radiation flux during the Sun's rotation are largely determined by single active longitudes, though time periods were observed when, in several orbits, disturbances with a period of 13.5 days took place. The presence of such 13.5-day and 27-day effects and sharp transitions from one regime of variability to another makes it possible to recognize reliably the Sun-induced and internal oscillations of the atmosphere. The amplitude of variations, on the whole, decreases with wavelength.

An analysis of several indices of solar activity suggests that most representative are F-10.7 (radio emission at 10.7 cm) and He-1083 (emission in the helium line). The observed dependence of Sc(MgII) – the ratio of fluxes in the centres of lines MgII h and k to continuum emission – on these solar activity indices has been used as the basis for extrapolation of variability of extra-atmospheric ultraviolet solar radiation in an 11-yr cycle.

Index F-10.7 is the best indicator of the slowly varying component of the spectral distribution of ultraviolet radiation in the interval 160–400 nm. Therefore, for Sc(MgII) estimates averaged over 27-day periods and F-10.7 monthly means, the following regression equation is valid:

$$\mathrm{Sc(MgII)} = 0.2585 + 1.142 \times 10^{-4}\ \text{F-10.7} \tag{4.56}$$

With an account of observational data for the maximum year of cycle 21 (1979), an extrapolation to F-10.7 = 70, from this equation, corresponding to a minimum of solar activity, leads to a change of solar constant value, which exceeds by 1.6 times the variations due to the Sun's rotation (emission CaK serves as the best index of short-term variability). This coefficient preserves the same value for other wavelengths. The results can be used to assess the effect of ultraviolet radiation variability on climate.

### 4.7.2 Solar activity impacts

Numerous publications of recent decades contain diverse arguments favouring the variability of the weather and climate caused, for instance, by 11-yr and 22-yr cycles of solar activity, which suggests correlations between the weather and radiation variations, depending on various phenomena on the Sun. For instance, an 11-yr cyclicity of precipitation and a 22-yr periodicity of various indicators of weather (the wintertime and summertime temperature, droughts) are clearly seen (Scuderi (1993) has established 125-yr periodicity on the basis of analysis of tree ring data for the past 2000 yr). There are numerous manifestations of correlations between the short-period weather variability and the magnetic field, most substantial in the auroral zone. Though the evidence for the effect of the auroral zone on the lower atmosphere raises doubts, some of the Sun–weather relations are supposed to be determined by high-energy particles intruding into the polar-latitude atmosphere.

The conclusions based on analysis of the Sun–atmosphere correlations and the estimates of the statistical significance of these correlations are rather controversial. Therefore, a search for physical mechanisms for the Sun–weather–climate interrelations gains importance. Studies of correlations between solar activity, weather and climate are, first of all, directed to revealing correlations for three spatial and temporal scales: (1) climate changes from centuries to millennia due to long-period variations of solar activity, (2) climate changes connected with a 22-yr cycle and (3) weather changes on a scale of several days due to short-period phenomena on the Sun. Though in these cases there is not enough convincing evidence for the respective correlations, there are abundant data to necessitate further studies on the Sun–atmosphere interrelation (Kondratyev and Nikolsky 1995b).

As for the long-term climatic variability, it may partially be caused by variations of solar luminance followed by marked changes in the solar constant. The nature of the Sun–weather correlations ought to be different. In this case, sensitive correlations may be looked for in the system Sun–solar-wind–magnetosphere–stratosphere–troposphere; that is, we seek evidence that small variations of the incoming energy can stimulate relatively large-scale consequences based on the trigger mechanisms (Kondratyev 1992b).

Estimates of energy connected with manifestations of solar activity (i.e. corpuscular radiation) show small quantities of such energy compared with the extra-atmospheric solar insolation and the energy needed to affect typical tropospheric weather-forming systems.

The difference between the quantities of corpuscular radiation and total insolation decreases, however, in high latitudes where the annual mean insolation is at a minimum (in winter it is absent altogether) and corpuscular radiation is at a maximum due to the effect of the Earth's magnetic field.

Available estimates show that in high latitudes a strong local increase takes place in the ratio of corpuscular emission to insolation. It follows that the effect of solar activity on the weather and climate must be a maximum in the high-latitude winter. The types of radiation which can manifest themselves at various latitudes in polar regions are as follows: (1) galactic cosmic rays modulated by solar activity – usually a basic factor of electrical properties of the atmosphere at altitudes 5–30 km, (2) solar proton events which are relatively rare but which can intensify the ionization near 300 km altitude by several orders of magnitude, (3) precipitating relativistic electrons, which are energetically less substantial but more frequent and can increasingly affect the ozone layer and (4) very frequent local precipitation of electrons in the zone of northern lights, which leads to the production of X-ray emission.

The transformation of the energy of electrons to the energy of X-ray photons ensures the penetration of radiation deeper into the atmosphere, resulting thereby in the effects of solar activity in the layers below the levels reached by electrons responsible for X-ray emission. Simultaneous observations of stratospheric ozone and particle intrusion reveal a correlation between them. So, for instance, during the solar protons events a decrease of ozone takes place at altitudes below 60 km. After the solar protons event in August 1972 the daily-mean zonal-mean $O_3$ content in high latitudes decreased by about 20 per cent. There was an increase of $O_3$ concentration near 50 km caused by X-ray emission. The resulting changes in the absorption of solar radiation by stratospheric $O_3$ and the stratospheric heating can affect the weather in the troposphere. All this suggests that the stratosphere acts as a buffer zone in the transfer of Sun-induced perturbations from the upper atmosphere to the troposphere. The magnetically-caused variations in the weather necessitate an analysis of the geophysical phenomena which vary during a solar cycle and are affected by the Earth's magnetic field (cosmic rays, for instance). It was found that the X-ray induced ionization at the 500 hPa level decreases by a factor of four in the period from minimum to maximum solar activity.

With the supposed correlation between ionization and thunderstorm activity, there is a possibility of the Sun-induced modulation of climate. Changes of lower atmosphere conductivity connected with the solar cycle can affect the electric field, cloud formation, temperature, thunderstorm activity and precipitation. Ozone is an important atmospheric component, varying during the solar cycle as well as at very short-term variations (solar protons events) which has produced special interest in the ozone mechanism of solar–atmospheric interrelations.

Besides, various other mechanisms have been suggested: (1) the effect of the energetically weak phenomena in the upper atmosphere on the lower atmosphere in close-to-resonance conditions, (2) the manifestation of an 11-yr cycle in the troposphere as a total effect of the short-period variability connected with either solar flares or geomagnetic storms, most frequent during a maximum of sun spots, (3) the effect of water vapour formed under the influence of solar wind, (4) the possible effect of the solar constant variations during a solar cycle, (5) a cyclic variability of the solar-wind–magnetosphere interaction when the direction of the interplanetary magnetic field changes and (6) the variability of the reflection of upward gravity waves by the upper atmosphere.

During recent years the emphasis has been placed on the mechanisms of Sun–atmosphere interrelations manifested through the effect of solar and galactic cosmic rays on the chemical composition of the atmosphere (Kondratyev and Nikolsky 1978).

One of the most likely mechanisms of the effect of solar activity on the weather and climate is connected with the consideration of an additional formation in the stratosphere of nitrogen oxides (NO and $NO_2$) due to galactic cosmic rays and solar protons events. The ozone-destructive effect of the galactic cosmic rays and solar protons events reduces the ozone content, which decreases the absorption of solar radiation by the stratosphere, causes its cooling and determines several processes which can change the weather and climate. The precipitation of high-energy particles of radiation belts during magnetic perturbations may also be a factor in the formation of nitrogen oxides.

The estimates of the effect of modulation of the stratospheric chemical composition due to galactic cosmic rays and solar protons events on the vertical temperature profile have shown that the content of nitrogen oxides $NO_x$ formed due to galactic cosmic rays is not large (compared with background concentrations) at all the altitudes. Above 20 km the time for the doubling of $NO_x$ concentration due to galactic cosmic rays exceeds 11 yr, and even at 15 km it constitutes 2–4 yr. Therefore, it is difficult to suppose that the $NO_x$ production due to galactic cosmic rays can cause variations of the concentrations of $O_3$ and temperature.

The effect of solar protons events can be more substantial. Available estimates show that one strong solar protons event can lead to the formation of a larger amount of nitrogen oxides than do galactic cosmic rays during the whole year.

Brasseur *et al.* (1987) have suggested a theory of the effect of solar activity on the ozone layer and stratospheric temperature field manifested as periodic variations of the extra-atmospheric solar ultraviolet radiation, bearing in mind the subsequent comparison of theoretical estimates with the data of new satellite observations. Calculations of the effect of short-term ultraviolet radiation variations on the stratospheric chemical composition (including ozone) and temperature were based on the use of a one-dimensional chemical-radiative time-dependent model (0–100 km), in which 40 components were taken into account (oxygen, carbon, hydrogen, nitrogen and chlorine compounds). The external forcing was taken as a sinusoidal variation of solar ultraviolet radiation with periods of 27 and 13.5 days (the second case simulates the existence of two active areas on the opposite sides of the Sun). The amplitude of solar radiation changes varies depending on wavelength, according to the data of the Nimbus-7 spectrometer measuring the solar backscattered ultraviolet radiation (the SBUV instrumentation). More complete simulation modelling has been developed by Balachandran and Rind (1995).

According to numerical modelling data, the sensitivity of temperature to ultraviolet radiation grows with altitude. The calculated amplitudes turned out to be much below those determined by the effect of the atmospheric dynamics. So, for instance, in the case of the periodicity of perturbations equal to 27 days the temperature at the 1 hPa level increases by 0.12 deg K with a 1 per cent increase of solar radiation flux at 205 nm wavelength. The calculated phase shift (at the 1 hPa level the temperature increase lags behind the ultraviolet radiation increase) constitutes 1.5 days. Apparently, the effect of solar activity on the thermal regime shows itself not directly (through radiative heating) but indirectly (through the dynamical processes, including planetary waves).

Calculations have shown that maximum sensitivity of $O_3$ to solar activity is observed near the 3 hPa level, with the amplitude and phase of $O_3$ variability strongly depending on the ozone–temperature feedback taken into account. The interaction between the temperature field and $O_3$ content is realized so that it causes a strong phase shift in the variability of $O_3$ in the upper stratosphere and mesosphere due to the effect of negative temperature feedback. Maximum ozone in the 0.2–0.5 hPa layer is reached several days earlier than maximum solar ultraviolet radiation.

As a rule, theoretical conclusions agree with observational data. However, the calculated phase shift in the $O_3$ response is 2–4 times less than that observed, which points to the existence of processes not considered by the model. According to the theory, a strong negative sensitivity of $O_3$ and positive sensitivity of temperature must be observed in the mesosphere to ultraviolet radiation variations caused by the absorption of solar radiation by molecular oxygen in the Lyman-alpha line.

It follows from calculations that variations of the nitric acid mixing ratio in the stratosphere must be twice as large as the respective (opposite in sign) variations of $O_3$ concentration and, hence, more easily detectable from observational data. With solar radiation flux at 205 nm wavelength supposed to vary within 2–15 per cent during an 11-yr cycle, the calculations give the limits of variability of $O_3$ and temperature shown in Table 4.45. As can be seen, the maximum variation in $O_3$ content is 5 per cent, temperature 2 deg K, but most probable are the limits 2.8–3.5 per cent and 1.3–1.6 deg K, respectively (with assumed variability of 205 nm radiation 8–10 per cent). Note that the estimates of the limits of the long-term variability of total $O_3$ (0.3–2.6 per cent) agree with the data from the Nimbus-4 SBUV spectrometer measuring backscattered ultraviolet radiation.

In later studies by Brasseur (1993), Huang and Brasseur (1993) and Rasch *et al.* (1995) 2-D and 3-D chemical–dynamic–radiative models of the middle atmosphere have been developed to investigate the potential changes of temperature, $O_3$ and other chemical constituents in response to variations in the solar ultraviolet flux, associated with the solar rotation (27 days) and the solar cycle (11 yr) (see also Ko *et al.* 1993). The observations showed that the variation between the early 1980s (solar maximum condition) and the mid 1980s (solar minimum condition) is of the order of 50–60 per cent at Lyman-$\alpha$, 5 per cent at the wavelength 180–200 nm and 6–7 per cent at the wavelength 205 nm; this is often used as an index to represent the intensity of the solar forcing on the stratosphere (these are Solar Maximum Mission satellite data).

Brasseur (1993) has pointed out that this model reproduces satisfactorily the response (amplitude and phase) to the 27-day forcing of ozone and temperature in the stratosphere, but does not properly explain the ozone and temperature responses of opposite sign observed near 70 km altitude. The change in the ozone column abundance associated with the 27-day solar forcing is estimated to be less than 0.5 per cent. Variations in middle and upper atmospheric ozone concentrations and temperatures induced by solar variations on the 11-yr time scale are not negligible compared with changes produced as a result of human activities over the same time period. The calculated change in the total ozone from solar minimum to solar maximum conditions is of the order of 1.1–1.3 per cent in the tropics and increases with latitude, especially in winter, to reach up to 1.5–1.7 per cent in the polar regions. Brasseur (1993) has emphasized the presence of large unexplained differences between the calculated and observed stratospheric responses. These discrep-

**Table 4.45** Possible variations of ozone and temperature during an 11-yr cycle

| Pressure (hPa) | Approximate amplitude (km) | Ozone variation (%) | Temperature variation (deg K) |
|---|---|---|---|
| 0.5 | 54 | 0.16–0.9 | 0.32–2.4 |
| 1.0 | 48 | 0.20–1.7 | 0.32–2.1 |
| 2.0 | 43 | 0.40–3.2 | 0.32–2.1 |
| 5.0 | 36 | 0.7–5.1 | 0.20–1.3 |
| 10.0 | 31 | 0.6–4.2 | 0.16–1.0 |

ancies could involve chemical and radiative, as well as dynamical, processes which are not yet fully understood. In this context three-dimensional numerical modelling is specifically required. It should be mentioned in this connection that according to Hood *et al.* (1993), a positive solar cycle variation of independently measured ozone and temperature occurs with maximum amplitude near the low-latitude stratosphere. This conclusion has been made on the basis of analysis of the Nimbus-7 SBUV data.

Of principal importance for understanding the Sun–climatic relation are studies by Hunt (1981) and Hunt and Gordon (1988, 1989) in which a numerical modelling of the effect of solar activity on atmospheric general circulation has been made, based on the hypothesis that the mechanism of this effect is connected with the formation of nitrogen oxide due to solar protons events, when the polar stratosphere is bombarded with charged particles. This leads (through the respective catalytic cycle) to a local destruction of $O_3$ in high latitudes which causes a disturbance of radiative heat flux divergence followed by changes in the temperature and wind fields. The latter changes the conditions for the transport of the energy of long waves from the troposphere to the stratosphere. Since the wave energy appears in synoptic systems of mid-latitudes, any variations of energy flux have an influence on synoptic systems. Thus, a direct connection must exist between solar protons events and weather changes.

Calculations have been made with the use of an eighteen-level hemispherical atmospheric general circulation model for annual mean conditions. The model does not take into account the effect of diurnal change and topography, but considers such factors as the hydrological cycle and radiative heat flux divergence (with a fixed climatological cloudiness). The surface temperature is determined, on the basis of a fixed swamp boundary condition (the heat capacity is considered to be zero and the source of water vapour is determined only by saturated water vapour pressure at the surface temperature).

Previous numerical modelling of atmospheric general circulation revealed a strong sensitivity of the model to the troposphere–stratosphere interaction, so this model can be considered useful in assessing the effect of solar activity on atmospheric general circulation. Such assessments have been obtained for two Sun–weather experiments, SWE-1 and SWE-2, which differ only in that their initial meteorological conditions are 15 days apart (this has been done to study the effect of meteorological conditions on the functioning of the Sun–weather mechanism).

In the consideration of each of the Sun–weather experiments started after an integration for the control case for several hundred days, the radiative heat-flux divergence was calculated, for which a changed content of $O_3$ in the high-latitude stratosphere was considered, with an account of satellite data obtained after a strong solar protons event on 4 August 1972 (a decrease of total $O_3$ in high latitudes reached 6 per cent in this case, with a 50 per cent decrease in the upper stratosphere). It is assumed that the $O_3$ content had decreased during 2 days and returned to the normal magnitude during the subsequent 3 days.

The SWE-3 (similar to SWE-2, but with a fixed $O_3$ deficit near 40° N and 45° N) and numerical modelling have been accomplished with an introduction in the lower layer of the model of a temperature noise within $\pm 1$ deg K (noise experiment). As applied to SWE-1, calculations have been made with an introduction of noise at the 550 hPa level (tropospheric noise experiment). The noise level $\pm 1$ deg K corresponds to supposed changes in temperature due to decreased $O_3$ content.

Hunt (1981) mainly considered the results of calculations of time series of zonal mean values averaged over 4-day intervals from SWE-1 data. A comparison of temporal variations of four principal components of the atmospheric energy balance (zonal and eddy kinetic energy, $K_z$ and $K_e$; zonal and eddy available potential energy $P_z$ and $P_e$) for SWE-1, the noise experiment and the control case revealed a typical oscillating regime of variability of the parameters considered. However, variations in $K_z$ and $P_z$ after approximately day 405 differ substantially for SWE-1, compared with the control case and the noise experiment, testifying to the presence of response of the atmospheric global circulation model to solar protons events.

It is important to note that it had taken 20 days before the response could be noticed. Though this time constant is natural from the viewpoint of the Sun–weather mechanism in question, it exceeds a shift of about 1–2 days, which is introduced in studies of empirical correlations between the index of vorticity area and the moment of crossing the sector boundary of the solar magnetic field.

A consideration of the vertical profiles of the difference of zonal mean wind speeds (SWE-1 and the noise experiment) for different 4-day periods at 45° N (it is assumed that mid-latitudes are a major source of wave energy propagating to the stratosphere) revealed a persistent intensification of the response of SWE-1 to solar protons events (especially in the upper troposphere) during the sixth period after solar protons events. In particular, the wind speed at a height of 10 km is observed to grow by $4.5\,m\,s^{-1}$, which constitutes approximately 25 per cent with respect to the control case ($17.8\,m\,s^{-1}$). It is essential that the effect of a solar protons event shows itself not in an appearance of a new mode of variability but in variations corresponding to the noise experiment.

An analysis of similar latitudinal profiles did not give any clear results (from the viewpoint of latitudinal selectivity of the response to the forcing), although, during the sixth period, the winds grew in a wide range of mid-latitudes. In comparison of data for SWE-1 and the noise experiment, a substantial increase of the meridional temperature gradient is observed as well as a small increase of temperature at 500 hPa in mid-latitudes in the case of SWE-1 (this result testifies to the uselessness of an empirical search for Sun–weather effects based on the analysis of zonal mean temperature profiles).

During the sixth period, $K_e$ at 45° N for SWE-1 exceeded the noise value ($1.45 \times 10^6\,J\,m^{-2}$) by $0.27 \times 10^6\,J\,m^{-2}$, pointing to a substantial intensification of atmospheric global circulation, with a strong increase of $K_e$ in the middle stratosphere after the sixth period, which again points to a substantial effect of solar protons events on synoptic activity whose nature is connected, apparently, with the effect on large-scale eddy heat exchange and the flux of relative angular momentum.

In contrast to all expectations, analysis of instant synoptic maps did not reveal a definite effect of solar protons events, and it follows that there is no simple way to recognize Sun–weather interrelations using a synoptic analysis. However, a local effect of solar protons events on the vertical temperature profiles is observed. An analysis of data for SWE-2 gave almost the same results as for SWE-1. This similarity should be interpreted as a strong argument in favour of the existence of a special mechanism for the effect connected with solar protons events. As has been expected (bearing in mind the high-latitudinal nature of the discussed interaction between the troposphere and stratosphere), data for SWE-3 did not reveal the effect of solar protons events.

The results obtained suggest that the solar protons event-induced variations of stratospheric temperature serve as a trigger-mechanism stimulating the development of atmospheric dynamics, leading to the tropospheric disturbances considered above. The prevailing mechanism for dynamical interaction between the troposphere and stratosphere is connected with a large-scale eddy flux of geopotential ($\overline{\omega'\phi'}$ are eddy fluctuations of the vertical velocity and geopotential, respectively), which is at a maximum in the zone of the high-latitude tropopause.

This high-latitude belt is similar to a waveguide for the interaction mechanism. Only in the case when the $O_3$ content in the region of the waveguide decreases is a clear response observed of the zonal wind field in the troposphere to the forcing. Despite the convincing results obtained, the presence of solar protons event-induced changes of the flux $\overline{\omega'\phi'}$ cannot be considered proved. The statistical significance of the sample used to calculate the geopotential flux remains unproved.

Although the comparison of meridional sections of $\overline{\omega'\phi'}$ for the control case, the noise experiment, SWE-3 and SWE-2 (period 6) shows that in the latter case a response to solar protons events is clearly seen, a comparison with data for SWE-1 does not reveal such a response; this should be attributed to statistical non-significance. Nevertheless, it should be borne in mind that in order to stimulate the effect, small variations in the geopotential flux are enough.

With the atmospheric general circulation model considered adequate, the conclusion can be drawn that for the observed effect to be revealed, it should be of a very long duration, but even in this case the effect will be temporary. Apparently, in most cases, the response to the forcing turns out to be hidden within the level of natural atmospheric noise, and thus part of the noise can be ascribed to the effect of solar activity.

In assessing the results obtained by Hunt (1981), on the whole, it should be borne in mind that (1) only one of the possible Sun–weather mechanisms has been taken into account, which manifests itself in decreased $O_3$ content, with a rare case of a superstrong solar protons event prescribed, and (2) the simplified nature of the atmospheric global circulation model requires estimates with the use of more adequate models. In order to find the final answer to the question of whether solar activity affects the weather and climate, it is necessary to realize coordinated, complex numerical experiments and observational programmes, which could be made during the decade to come.

Based on the use of a 3-D five-layer global climate model with a $10^\circ \times 10^\circ$ grid, Sellers (1982) estimated the effect of solar constant variations on climate. The two-layer stratosphere is proposed to be in a state of radiative equilibrium and the time step to be 5 days. The cloud amount in each of the three tropospheric layers is parametrized as depending on relative humidity near the surface and on the ratio of condensation rate to atmospheric water content. Calculations are made for a period of 20 yr with prescribed solar constant values 1360, 1347 and 1333 $W\,m^{-2}$. Since the model takes into account the ocean, a 20-yr period is not enough to establish thermal equilibrium. Therefore, the results of numerical modelling for 10 and 20 yr are considered.

An analysis of results shows that a 2 per cent solar constant decrease causes only a 0.8 deg K decrease in global mean surface temperature, but in all three cases a weak temperature increase takes place from year 10 to year 20 of integration, which points to an incomplete thermal equilibrium. A considerable decrease of global mean precipitation (and evaporation), taking place at the solar constant decrease, plays the most important role. It causes an increase of energy available to warm the air near the surface from 35.6 to 36.5 $W m^{-2}$, provided the solar constant decreases from a maximum to a minimum. The negative feedback is a major reason for decreasing climatic sensitivity to solar constant variations. Most sensitive to solar constant variations are precipitation (or evaporation) and the extent of continental snow cover and sea ice; a 2 per cent decrease of solar constant causes a 4.8 per cent decrease in global-scale rainfalls and 5.0 per cent, 7.0 per cent and 21.2 per cent increases in the extents of snow cover and sea ice in the northern hemisphere and sea ice in the southern hemisphere, respectively. The unstable ice cover in the southern hemisphere is one of the most interesting features of the climate model under discussion; it shows that the extent of ice cover in the southern hemisphere can be considered the most sensitive indicator of climate changes.

Ramanathan (1982) emphasized that the available results make it possible to perform a useful analysis of the climatic reaction to possible radiative disturbances with the time constant of the climate system taken into account. This aspect of the problem is particularly important in studies of the effect of solar activity, since it is either periodic or episodic. Here a consideration of the atmosphere–ocean interaction, which smoothes the cyclonic effects, is a key aspect. Therefore, the amplitude of surface air temperature variations in the case of impacts with periods from a few years to a decade is 50–75 per cent smaller than in the case of the equilibrium regime.

With the time scale of extra-atmospheric insolation variations ranging from weeks to a month, a change in the tropospheric mean temperature will be an order of magnitude smaller than in equilibrium conditions. This result must be taken into account in assessing the effect of short-term and episodic variations in extra-atmospheric insolation. The results of numerical modelling of variations in the atmospheric chemical composition and cloudiness are very important for assessing the effect of solar activity on climate. They show, for example, that not only total $O_3$ content but also its vertical profile of concentration must be taken into account. In studies of clouds, their amount, optical properties, altitude and geographical location must be considered.

Mólnar and Mika (1992) assessed the sensitivity of global mean surface air temperature, $T$, to solar constant variations, $S_0$ :

$$\beta = \Delta T/(\Delta S_0 = 1 \text{ per cent}) \tag{4.57}$$

with the use of available theoretical estimates of climate changes. The estimates are typified (being taken with different weights) depending on the reliability of the climate model determined with account of the following factors: the parametrization of horizontal heat exchange, the horizontal spatial resolution, the number of the climate system components considered in the energy balance model, vertical resolution, the parametrization of hydrological processes and latent heat, time averaging and taking account of time-dependence. An analysis of different climate models for their reliability suggested the priority of the GFDL model. Calculations of sensitivity to a $CO_2$ doubling gave a warming by 2.3 deg K, which agrees well with the results obtained with the most complicated models.

Hoffert *et al.* (1988), using a 1-D climate model for the atmosphere–ocean interaction system, assessed the changes in global mean surface temperature, caused by solar constant variations observed between 1980 and 1984, from data of a cavity pyrheliometer. (According to Willson (1984) and Willson and Hudson (1988), there was a solar constant decreasing trend of about 0.01–0.02 per cent per year, which

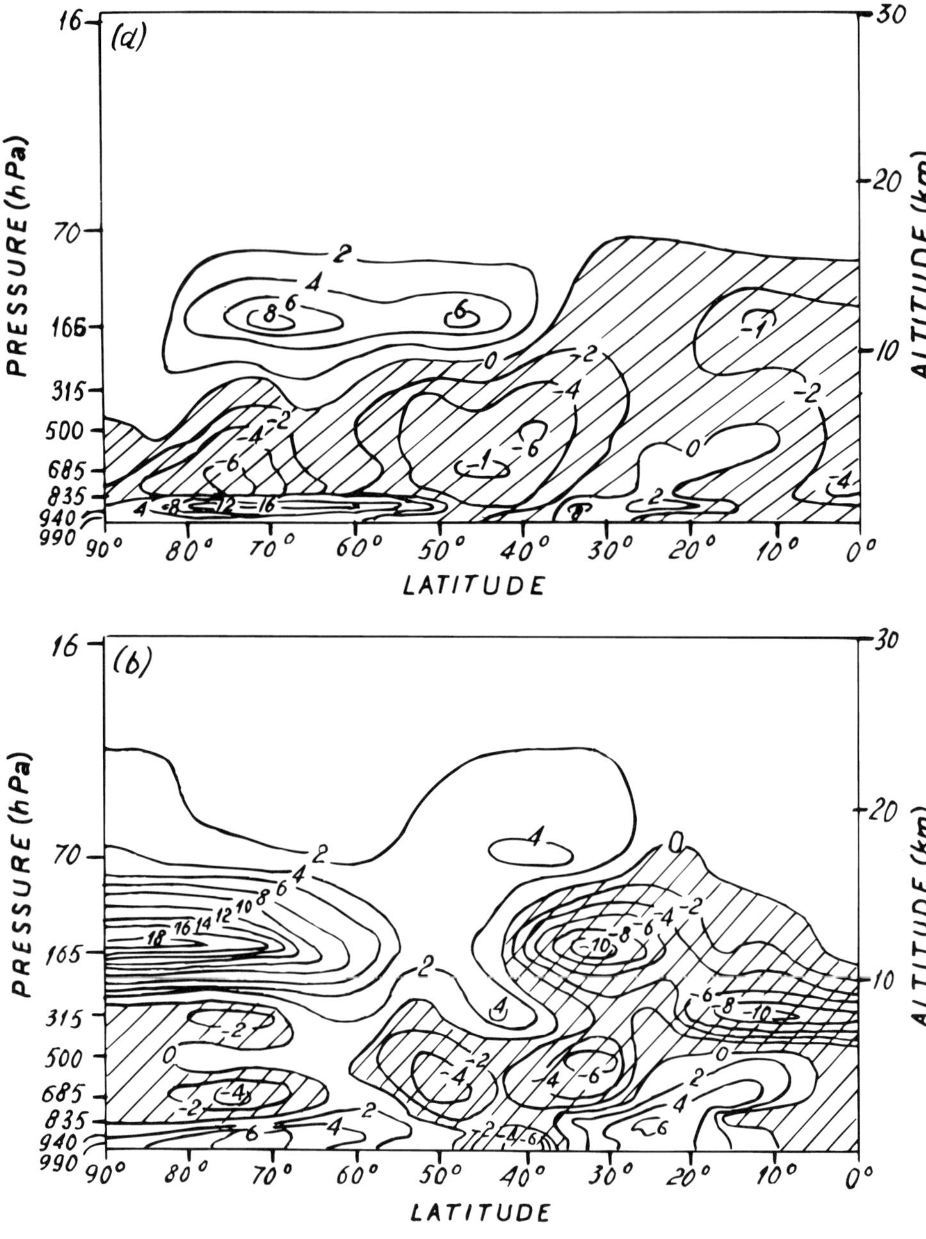


**Figure 4.15** Latitude–height distribution of the difference in zonal mean (a) cloud amount and (b) relative humidity obtained from a 6 per cent increase of the solar constant. Shaded areas indicate negative values. Values are in per cent.

reflected the presence of a positive correlation between solar constant monthly means and solar activity.) The calculated variability of equilibrium temperature turned out to be small ($\sim 0.1\,^{\circ}C$) but not negligibly small. This is clear if we take into account that a 0.1 per cent decrease of solar constant is equivalent to a global mean decrease of absorbed solar radiation by $0.24\,W\,m^{-2}$, whereas a decrease of outgoing longwave radiation due to the increase of $CO_2$ concentration by 6 ppm observed between 1979 and 1984 amounted to 90.26 W m$^{-2}$.

Use of a model of the solar constant secular change proposed by Schatten (1988) has led to the conclusion that the maximum amplitude of temperature variability due to solar constant variations can be only 0.05 °C (it is assumed that for zero solar activity $S_0 = 1366.5$ W m$^{-2}$). Thus the estimates obtained with the use of available satellite data on the solar constant reveal the impossibility of substantial short-term variations in surface air temperature due to solar constant change (these estimates have been obtained, however, without account of non-linearities in the climatic system). To reveal possible long-term variations, satellite-based solar constant observations are needed over several decades.

Wetherald and Manabe (1988) performed conditional calculations of the variability of global cloud cover (cloud amount) and relative humidity with a prescribed 6 per cent increase of solar constant (Figures 4.15, 4.16). Figure 4.17 shows, for comparison, a case of a quadrupled $CO_2$ concentration.

As has been mentioned above, an analysis of observational data shows that solar constant variations during the last few decades have not exceeded 0.2 per cent. Satellite measurements revealed, however, considerable variations in spectral solar constant in the wavelength interval below 300 nm, with the amplitude growing with decreasing wavelength. In this connection, Borucki *et al.* (1980) assessed the effect on total $O_3$ content and climate of solar constant sinusoidal variations in the ultraviolet spectral region (prescribed from observation results), with total solar constant remaining constant (the latter was reached by a small compensation for homogeneous variations in solar constant in the wavelength interval above 290 nm).

The results of calculations with a one-dimensional radiative–convective–

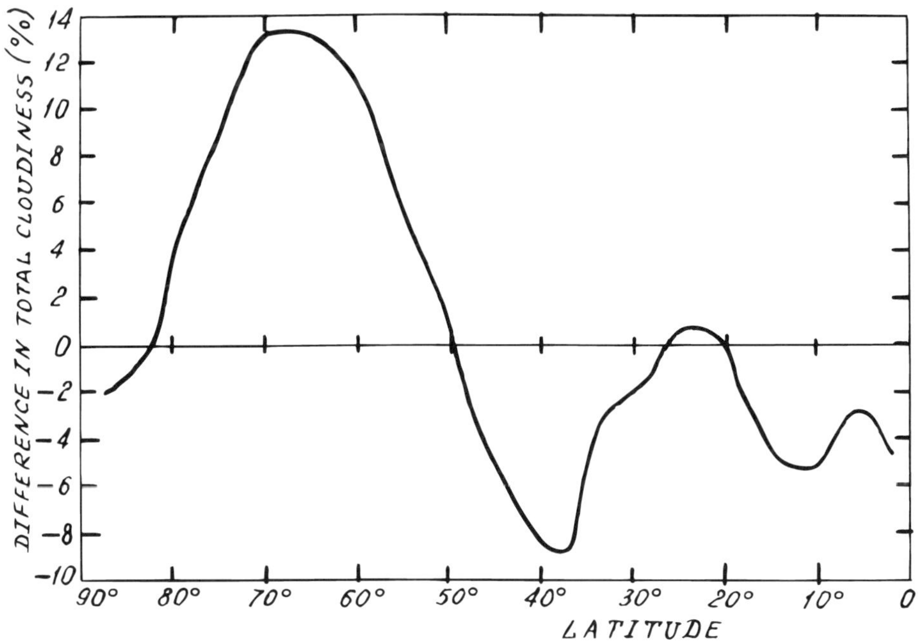


**Figure 4.16** Latitudinal distribution of zonal mean difference in total cloudiness obtained from a 6 per cent increase of the solar constant.

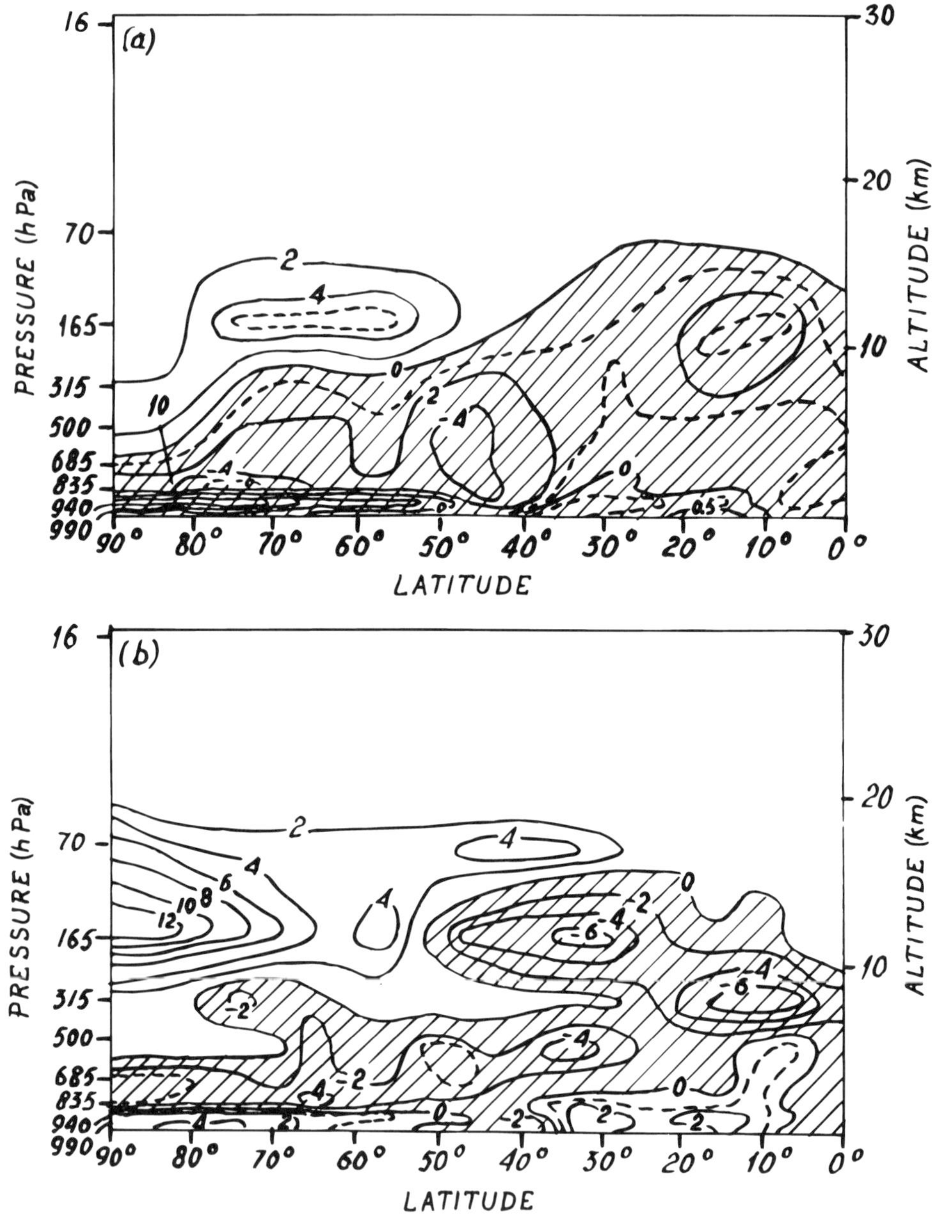


**Figure 4.17** Latitude–height distribution of the difference in zonal mean (a) cloud amount and (b) relative humidity obtained from a quadrupling of $CO_2$ content. Shaded areas indicate negative values. Values are in per cent

photochemical model show that ultraviolet radiation variations during a 27-day period of the Sun's rotation (a 5 per cent change at 175 nm wavelength) are not strong enough to cause noticeable changes in total $O_3$ content and temperature. Assuming an 11-yr variability of spectral solar constant, based on satellite observations (the ratio of maximum ultraviolet radiation to minimum is 2.5 at 175 nm), a 13 per cent change in total $O_3$ content is found, which does not agree with results of surface total $O_3$ content observations. According to these, total $O_3$ content variations did not exceed 3.5 per cent, the temperature at a height of 48 km increasing by 12 °C and near the surface decreasing by 0.06 °C.

Variations in the spectral composition of the extra-atmospheric solar radiation (with total solar constant being constant) had probably contributed much (about 0.2 °C) to the decreasing trend of global mean temperature in the 17th and 18th centuries (the Maunder Minimum), but such being the case, the dependence of ultraviolet solar radiation flux on the number of sun spots should have been opposite to that observed during an 11-yr cycle (temperature must decrease with increasing ultraviolet radiation, which leads to an increase in total $O_3$ concentration). Substantial climate changes may happen due to spectral solar constant variations, connected with a 22-yr magnetic solar cycle.

Having analysed possible mechanisms for disturbances taking place in the upper atmosphere due to solar activity and affecting the stratosphere and troposphere, Geller (1992) reaches the conclusion that one of the most probable mechanisms consists in changing conditions for the upward propagation of planetary-scale gravitational waves, which is connected with the effect of ultraviolet radiation variations on temperature, causing variations in the field of mean zonal winds at heights below 30–35 km. Since the characteristic time for probable manifestations of the mechanism should be about 2 weeks, this mechanism cannot explain the observed correlation between the vortex index and sector boundaries of the solar magnetic field. For reliable substantiation of this mechanism, further studies are needed of the meteorological regime and stratospheric composition (especially the content of ozone and nitrogen oxides) as well as continuous measurements of extra-atmospheric ultraviolet solar radiation. Data of such complex observations will serve as the basis for developing and testing a more adequate model of the effect of solar activity on the stratosphere and troposphere.

As far as further studies are concerned, continuous precise satellite measurements are needed of total and spectral solar constant. Of primary importance are further searches for mechanisms of solar–climatic feedbacks (Labitzke 1987, Labitzke and van Loon 1988, 1989, 1992). The emphasis is placed here on studies of modulation of the stratospheric $O_3$ content under the influence of varying extra-atmospheric ultraviolet solar radiation as well as modulation of the stratospheric $NO_2$ content, caused by variations in galactic cosmic rays and solar protons events. The estimates show that the $NO_2$ content increasing with intensified galactic cosmic rays and solar protons events can cause an additional absorption of total solar radiation by the stratosphere reaching 1 per cent.

Summing up studies of solar–terrestrial interrelations, the following key questions can be formulated (ICSU 1988).

1. What is the long-term variability of the incoming solar electromagnetic and corpuscular energy and what are the mechanisms for the Sun-induced variability?

2. In what way do variable effects impact on the upper atmosphere from above (the Sun, cosmos) and from below (the lower atmospheric layers) and what are the respective processes and feedbacks that determine these effects?
3. What is the role of the middle atmosphere in the transformation of minor gaseous constituents of tropospheric–biospheric origin and to what extent do natural effects (solar activity, volcanism) mask the anthropogenic ones?
4. What are the processes determining the interaction between the upper and lower atmospheric layers and their effects on the formation of weather and climate?

Palaeoclimatic data can be an important source of information about the effect of extra-atmospheric insolation variability on climate. An analysis of the ice core, deep water, bottom and soil cores revealed a high correlation between variations in the parameters of palaeoclimate and the Earth's orbit. In particular, this refers to the change of ice epochs in the Pleistocene period. Therefore, climate variations can be supposed to have been mainly caused by insolation variations due to unstable inclination angle and precession of the orbit. To explain the prevailing 100 000 year cycle, one has to suppose the existence of an unknown non-linear mechanism for the intensification of the cycle determined by variations in the orbital parameters.

At the present stage of development of the astronomical model of palaeoclimate changes, when it has been successfully tested statistically and serious ideas have appeared with respect to physical mechanisms, the time has come to investigate its reliability using numerical climate models. Berger *et al.* (1990) undertook such a study with the use of a 2–3-dimensional time-dependent climate model which took into account atmospheric feedbacks, processes in the upper layer of the ocean, sea ice cover and in the lithosphere. On the time scale of the order of one thousand years, the latitudinal variability of the seasonal insolation fields is supposed to play the most important role (from the viewpoint of the climate-forming effect of variations in the orbital parameters).

The calculations made with this model have made it possible to obtain information about changes in global ice volume for the last 125 000 years which, on the whole, agrees with the geological data available. This shows that an account of variations in the parameters of the orbit makes it possible to explain the principal features of the variability of the glacial–interglacial periods. In the model considered, the glacial cycle is very sensitive to changes in the heat balance of the surface of ice sheets (due to the low thermal conductivity of ice and snow, small variations in the heat input can bring about substantial changes in the heat balance).

The underestimated volume of the North-American glaciers given by the model should be partially ascribed to the ocean level variability which was not taken into account and, as a result, the propagation of glaciers over the continental plateaux and towards the Arctic Ocean turns out to be impossible. The difference between the model calculations and observations for the last six thousand years is explained, apparently, by neglecting some physical processes that affect the formation of the surface heat balance (it refers, for example, to variations in $CO_2$ concentrations). The prospects for further development of the climate model should be connected with taking into account the processes in the southern hemisphere and deep circulation in the ocean, as well as improving the model of the continental ice–lithosphere system and a more detailed test of the model's reliability.

## 4.8 Development, verification and application of climate models

Studies of the climatic system including the atmosphere, oceans, cryosphere, land surface and biosphere, are extremely complicated since they require a consideration of numerous physical, chemical, photochemical and biological processes, many of which are interrelated, in a wide range of spatial and temporal scales. Climate models play a leading role in understanding the laws of climate and its changes. In a most complete climate model all the variables and parameters should be considered interactively and be predictable explicitly and more or less continuously by integrating a system of equations. Such a model is still an abstraction and therefore the climate models can only be realized with certain simplifications, depending on the goals of the problem to be solved. Even the most complicated three-dimensional models are based on a parametrization of the numerous processes that cannot be considered explicitly.

### 4.8.1 Climate models and weather forecast models

In recent years a great deal of effort has been put into numerical modelling, and huge computing resources have been devoted to the running of such models. We have already, at various points in this book, quoted some results from numerical modelling work to illustrate particular points and we shall continue to do this in the remainder of the book. However, this is not a textbook or treatise on climate modelling; the reader who is interested in the details of the subject of modelling is referred to other, more appropriate, texts, of which perhaps the most up to date at present is the one by Trenberth (1992), but there are various others. We shall do two things in this section. One is to make a few general remarks about climate modelling and the other is to try to indicate, in very general terms, how climate models are used. Later in this book we shall draw heavily on the results summarized in the IPCC Report (Houghton *et al.* 1990, 1992) and one or two other more recent sources. The IPCC Report was very much concerned with trying to summarize the state of knowledge, from whatever source it arose, at the time that the report was written. That report is widely available and we do not intend to reproduce large quantities of material from it; we shall simply note some of its main features and refer the reader who wishes to have more details to the relevant sections of the IPCC Scientific Assessment Report and its supplement (Houghton *et al.* 1990, 1992). We also make no attempt to give comprehensive citations of references on this topic.

The idea of numerical modelling of climate may be regarded as a development from numerical weather forecast models. One can ask the question as to whether or not it is possible to predict the climatic consequences of some given forcing mechanism. Now it is widely accepted that this is possible, but in the early days of climate modelling it was not regarded as self-evident. We have already discussed the question of climate predictability, in general terms, in Section 1.1. To assess the possible climatic predictability, in a modelling context, Zwiers and Boer (1987) analysed the results of numerical modelling of climate 20.5 yr ahead performed at the Canada Climate Centre. Their results showed that a considerable contribution to the observed predictability can be connected with a time-dependent variability caused by the internal dynamics of the atmosphere. Climatic regimes can be identified (such as that prevailing during a large anomaly of the pre-calculated climate) when seasonal mean values can usually be predicted reliably. However, this was only demon-

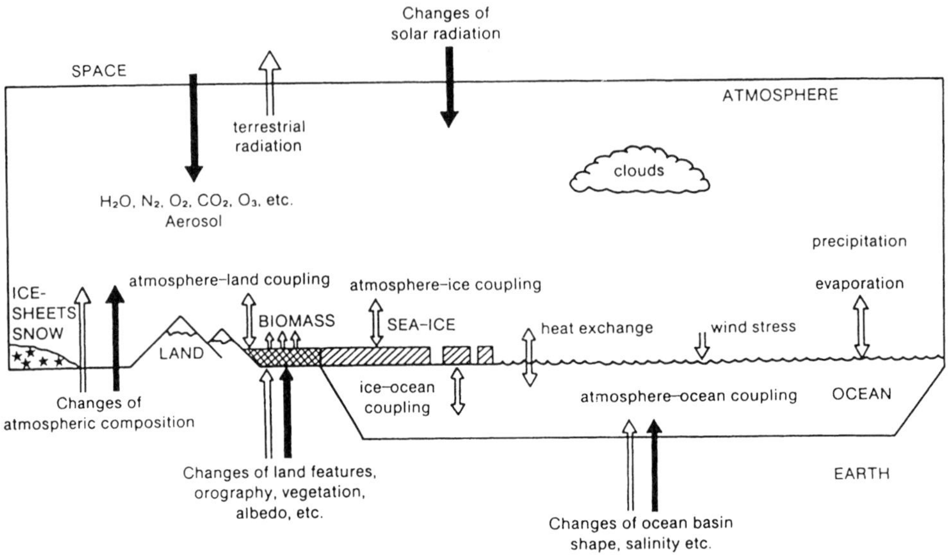


**Figure 4.18** Schematic illustration of the components of the coupled atmosphere-ocean-ice-land climatic system. The full arrows are examples of external processes, and the open arrows are examples of internal processes in climatic change (from Houghton, 1984).

strated for the period from December to February. The predictability was found to be higher in the southern hemisphere than in the northern hemisphere. If such regimes do exist, however, they happen seldom, judging at least from the data of simulation modelling. Therefore, the question of the possible predictability of inter-annual variability caused by internal dynamics still remains open. Further studies are needed of (1) the existence of such climatic regimes that can be predicted on a seasonal scale and (2) the repeatability of such regimes and the possibility that they can be revealed by numerical climate modelling.

We should, perhaps, say a few words about numerical weather forecast models since they were the precursors of numerical climate models. The objective of a weather forecast model is to be able to predict the weather at a particular place and time. The idea is that one writes down the mathematical equations that describe the various physical processes that occur, see Figure 4.18, and tries to solve those equations for some future time. In an atmospheric model the behaviour of the atmosphere is represented by the values of appropriate parameters specified on a three-dimensional grid of points, see Table 4.46. The UK Meteorological Office uses a global model operating with a grid with spacing of 90 km (i.e. of the order of 1° of latitude or longitude) in the horizontal and also a limited area model with a grid spacing of about 40 km; in each model there are about 20 levels in the vertical. The physical processes involved include the following:

- horizontal momentum equations
- the hydrostatic equation
- the equation of continuity (i.e. matter is neither created nor destroyed)
- the equation of state
- the thermodynamic equations

along with parametric descriptions of the following processes:

- evaporation
- condensation
- cloud formation and dispersal
- radiation and convection in the atmosphere
- the exchange of momentum, heat and water vapour at the Earth's surface.

It is also necessary to take into account effects or processes that occur on a scale which is smaller than the grid spacing that is used in one's model. We suppose that the results of meteorological observations allow the value of the atmospheric parameters to be specified at some starting time, $t = 0$. The forecasting problem then is to solve the various equations describing the physical principles and processes that we have just mentioned for a subsequent time $t$. That is, one seeks to determine new values of these atmospheric parameters at a subsequent time, in practice up to a few days ahead. Because there are several of these equations, and especially because they are non-linear equations, the problem has to be tackled by numerical methods using a computer. Hence the terms 'computer model' or 'numerical model'.

The operational use of numerical models at the UK Meteorological Office began in 1965. Although their usefulness at that stage was limited, there have been rapid developments since then and within ten years the models were able to provide better forecasts of the basic motion field than could be achieved by an unaided human forecaster. These improvements have come about for two reasons. First, the power of the computing facilities available has increased enormously. Secondly, the amount of data available to describe the present atmospheric conditions has been greatly increased in recent years. In addition to conventional surface measurements and radio-sonde measurements, there are now also satellite observations including satellite soundings and satellite-derived wind speeds obtained from cloud tracking from geostationary satellite images.

In using an atmospheric general circulation model for weather forecasting purposes the temperature over the surface of the Earth (land, sea, ice and snow) provides a very important set of boundary conditions on the model. It is therefore very important to have accurate values of surface temperatures, especially sea surface temperatures (given the large area of the oceans) as input data for the model. This is also true for climate modelling but there is one very important difference: in a weather forecast model the surface boundary conditions can be regarded as constant for the period for which the model is being run, i.e. for the period for which a forecast is being made. While the atmosphere reacts very rapidly to changes in its forcing (on a time-scale of hours or days), the ocean reacts more slowly on time-scales ranging from days (at the surface layer) to millennia in the greatest depths. The ice cover reacts on time-scales of days for sea ice regions to millennia for ice sheets. The land processes react on time-scales of days up to months, while the biosphere reacts on time-scales from hours (plankton growth) to centuries (tree growth). However, in making calculations of predicted climate change one is dealing with much longer periods of time, so that circulation of water and heat in the oceans is important. Temporal changes (but not circulation) in the land, ice and snow also need to be taken into account. For climate modelling, therefore, one cannot assume a constant set of boundary conditions at the Earth's surface. One

**Table 4.46** Characteristic input and output parameters of atmospheric general circulation models

| Input parameters |
| --- |
| *Initial conditions* |
| (Basic meteorological parameters, corresponding to input parameters) |
| *Boundary conditions* |
| (mainly at the surface level) |
| Height and type of surface (water, ice, land use character) |
| Albedo |
| SST and ice cover |
| Insolation (at the upper boundary level) |
| *Physical parameters* |
| cloud albedo, mixing coefficients, soil moisture capacity, atmospheric composition. |
| **Output parameters** |
| *Basic variables* |
| Pressure, temperature, humidity, wind, cloud amount, soil moisture, snow cover extent. |
| *Auxiliary variable processes* |
| Precipitation, evaporation, surface radiation budget and its components, sensible heat exchange, surface run-off. |

needs to consider the variations in the ocean surface conditions and one also needs to include time variation (seasonal changes) in the boundary conditions at the land surface. The consideration of the atmosphere–ocean interaction cannot be restricted to a study of only the surface of the oceans; since the surface is affected by processes below the surface within the ocean, one needs to use a general circulation model for the ocean as well as a general circulation model for the atmosphere. Moreover, one has to allow these models to be coupled together by exchanging information about the sea surface temperature, the ice cover, the total (latent, sensible and net long-wave radiation) heat flux, the solar radiation and the wind stress. Because including ocean circulation as well as atmospheric circulation is so much more complicated than using only an atmospheric general circulation model for weather forecasting, the spatial resolution and temporal resolution used in practice in climate modelling tend to be coarser than those used in weather forecasting. Grid spacing of 90 km or even 40 km in the horizontal surface are quoted for numerical models for weather forecasting. For climate modelling typical models have a horizontal resolution of 300 km to 1000 km and between 2 and 19 vertical levels. Such resolution is sufficient to represent large-scale features of the climate, but it allows only a limited interpretation of results on the regional scale. Thus, one very important problem that arises in climate modelling is that of the parametrization of subgrid-scale features, such as spatial variations in the surface albedo, and processes, such as cloud–radiation interactions and (relatively) small-scale vortices on the ocean surface. High-resolution models, which can resolve eddies, are enormously demanding in terms of computer power. As far as time-scales are concerned, 24 hours is a long time in weather forecasting terms whereas the time-scales of climate models range up to the equivalent of thousands of years.

### 4.8.2 Parameters and boundary conditions used

In the development of models of different complexity, two problems claim primary attention: (1) a parametrization of subgrid processes (first of all, cloud–radiation interactions as well as an interactive account of usually fixed parameters such as albedo) and (2) a consideration of atmosphere–ocean interaction and processes in the ocean (the latter is complicated by the fact that most of the ocean's energy is concentrated in small-scale moving vortices as well as by the large inertia of the ocean). To simplify a comparison of models, three stages of numerical modelling should be borne in mind: (1) an assessment of the geographical distribution of the time-averaged major climatic parameters (pressure, temperature, wind, precipitation, clouds, radiation budget), (2) calculations of the variability of these parameters and (3) calculations of covariations, transport, energy budget components, and related diagnostic studies.

Numerous studies suggested the conclusion that a considerable part of climatic variability in time-scales up to several months is explained by a combined effect of the weather-forming systems, which are individually predictable only for about a week. However, some specific climatic features can be predicted for longer periods (for example, the geographically excited modes, blocking the mid-latitude cyclonic trajectories, and the phase of the dominating mode of wind oscillations in the equatorial zone). Climate models must become a major tool for further studies of the predictability and sensitivity of climate to various factors.

A fundamental aspect of the WCRP is a study of anthropogenic impacts on climate, proceeding from the fact that even if two asymptotic states (the control state and the disturbed one) are predicted inaccurately, their difference can be reliable enough. However, since each asymptotic state can vary, the problem of statistical significance arises in the analysis of their difference. Reliable results can be obtained only after filtering out the climatic noise and differentiating the variability caused by various anthropogenic factors. Therefore, an accomplishment of numerical experiments on sensitivity and analysis of their results must be carefully made, especially bearing in mind the difficulty (or even impossibility) of their independent testing.

The current, most adequate, models of atmospheric general circulation make it possible to simulate realistically the observed large-scale features of the fields of temperature, pressure and wind. However, so far, they are not so reliable with respect to the variability of cloud cover and precipitation in view of the highly simplified nature of the parametrization of the respective physical processes. The main factors that affect the climate were discussed in Section 1.2. The problems involved in modelling them are discussed in detail by Houghton *et al.* (1990) and Trenberth (1992), for example. We simply note a few salient points here and refer the reader to these, or other, texts for more details and for references.

#### 4.8.2.1 *Radiation and clouds*

The parametrization of radiation is possibly the most important issue for climate change experiments, since it is through radiation that the effects of the greenhouse gases are transferred into the general circulation. A radiation parametrization scheme calculates the radiative balance of the incoming solar radiation and the outgoing terrestrial long-wave radiation and, as appropriate, the reflection, emission and absorption of these fluxes in the atmosphere. Absorption and emission are calculated in several broad spectral bands (for reasons of economy) taking into account the concentration of different absorbers and emitters like $CO_2$, water vapour, ozone and aerosols.

One sensitive part in any radiation scheme is the calculation of the radiative effect of clouds. We have already given an extensive discussion of the various effects of clouds in Section 4.1; the situation is very complicated. In early work with numerical general circulation models, clouds were prescribed using observed cloud climatologies (fixed cloud experiments) and were not allowed to alter during the experiments with (for example) changed $CO_2$ concentration. Later schemes contained interactive cloud parametrizations of various degrees of sophistication, but mostly based on an estimate of the cloud amount from the relative humidity experiments. Only the most advanced schemes involve calculating the variation of cloud optical properties.

The seasonal variation of the solar insolation is included in almost all experiments, but a diurnal cycle is omitted in many simulations. Climate experiments run without a seasonal cycle are limited in scope, and their reliability for climate change experiments is therefore doubtful. The inclusion of the diurnal cycle improves the realism of some feedback mechanisms and therefore the quality of the climate simulations.

#### 4.8.2.2 *Subgrid-scale transports*

Most of the solar radiation absorbed by the climate system is absorbed at the surface. The energy becomes available for driving the atmospheric general circulation only after it has been transferred to the atmosphere through the planetary boundary layer, primarily by small-scale turbulent and convective fluxes of sensible and latent heat, but also by net longwave radiative exchange. On the other hand, the general circulation is slowed down by frictional dissipation which basically takes place in the planetary boundary layer through vertical transport of momentum by turbulent eddies.

In most general circulation models the turbulent fluxes of heat, water vapour and momentum at the surface are calculated from empirical 'bulk formulae' with stability-dependent transfer coefficients. The fluxes at the planetary boundary layer top (at a fixed height, generally) are either neglected or parametrized from simple mixed-layer theory. In general circulation models that resolve the planetary boundary layer, the eddy diffusion approach is generally employed. Considerable efforts are made to incorporate into the planetary boundary layer parametrizations the effects of cloud, vegetation and subgrid-scale terrain height.

Cumulus convection in a vertically unstable atmosphere is one of the main heat-producing mechanisms at scales which are unresolvable in general circulation models. A common procedure is to adjust the temperature and water vapour profile to a conditionally stable state (Moist Convective Adjustment). The second class of cumulus parametrizations often employed in general circulation models is based on a moisture convergence closure. Other general circulation models use Penetrative Convection schemes to mix moist conditionally unstable air from lower model layers with dry air aloft. The question of how sophisticated convective parametrizations in general circulation models need to be, and how much the sensitivity of climate change experiments depends on their formulation, is still an open question.

#### 4.8.2.3 *Land surface processes*

Another important parametrization is the transfer of heat and water within the soil, for instance the balance between evaporation and precipitation, snow melt, storage of water in the ground and river run-off. This parametrization is of extreme relevance for climate change predictions, since it shows how local climates may change from humid to arid and vice versa depending on global circulation changes. It furthermore reflects, in some of the more sophisticated schemes, the changes that could occur through alterations in surface vegetation and land-use.

Most soil moisture schemes used to date are based either on the so-called bucket method or the force–restore method. In the former case, soil moisture is available from a single reservoir, or thick soil layer. When all the moisture is used up, evaporation ceases. In the latter method, two layers of soil provide moisture for evaporation, a thin near-surface layer which responds rapidly to precipitation and evaporation and a thick deep soil layer acting as a reservoir. If the surface layer dries out, deep soil moisture is mostly unavailable for evaporation and so evaporation rates fall to small values. However, in the presence of vegetation, realistic models use the deep soil layer as a source of moisture for evapotranspiration.

At any given grid-point over land, a balance between precipitation, evaporation, run-off and local accumulation of soil moisture is evaluated. If precipitation exceeds evaporation, then local accumulation will occur until saturation is achieved. After this, run-off is assumed and the excess water is removed. The availability of this run-off as fresh-water input to the ocean has been allowed for in ocean models only recently (Cubasch *et al.* 1991, 1992, Jentsch 1991a,b). Most models differ in the amount of freshwater required for saturation, while few treat more than one soil type.

#### 4.8.2.4 *Boundary conditions and ocean models*

To determine a unique solution of the model equations, it is necessary to specify a set of upper and lower boundary conditions. These are:

- input of solar radiation (including temporal variation) at the top of the atmosphere
- orography (mountain topography) and land sea distribution
- albedo of bare land
- surface roughness
- vegetation characteristics.

The lower boundary over the sea is either prescribed from climatological data or, as this is not very appropriate for climate change experiments, it has to be calculated from an oceanic global circulation model as we have already mentioned above. The typical oceanic general circulation model used for climate simulations follows basically the same set of equations as the atmosphere if the equation defining the water vapour balance is replaced by one describing salinity. As with atmospheric general circulation models, one specifies appropriate surface boundary conditions (i.e., fluxes of heat, momentum and fresh-water) either from observations (uncoupled mode) or from an atmospheric general circulation model. The vertical and horizontal exchange of temperature, momentum and salinity by diffusion or turbulent transfers is parametrized. Solutions are obtained by using numerical methods. A critically significant aspect is so-called 'flux adjustment'.

### 4.8.3 The use of climate models

There are many uncertainties and problems associated with any given model. Thus it would be rather risky to rely on absolute predictions made with one of these models. We should therefore consider the manner in which these models are used, and this is basically done by attempting to eliminate the effects of the uncertainties or errors. Thus a model is normally used to investigate the sensitivity of climate to internal and external factors and to predict the climate change by first carrying out a control integration with the parameters set for the present climate in order to establish a reference mean model climatology and the necessary statistics on modelled climatic variability. The climate change (perturbation) run is then carried out by repeating the model run with appropriately changed parameters (a doubling of atmospheric $CO_2$ concentration for example) and the difference between this and the parallel control run examined. The difference between the control and the perturbed experiments is called the response to the change in the values of the parameters used. The significance of the response must be assessed against the model's natural variability (which can be determined from the control run) using appropriate statistical tests. These enable an assessment to be made of the confidence that the changes obtained represent an implied climatic change, rather than simply a result of the natural variability of the model. We shall consider the use of climate models to study the effect of doubling the concentration of atmospheric $CO_2$ in Section 6.1 in some detail.

The period for which the integration in a model is carried out ranges typically from 5 to 100 yr, depending on the nature of the investigation. Until now, the greatest effort has been devoted to studying the response to increased levels of greenhouse gas concentrations by determining the equilibrium response of climate to doubling of $CO_2$, using atmospheric models coupled to slab ocean models. In an equilibrium response experiment, both simulations – the control experiment and the perturbation experiment – are run for a sufficiently long time to achieve the respective equilibrium climates.

## 4.9 The comparative climatology of planets

There is no doubt that a comparative study of the atmospheres and climates of various planets is urgent. Experience shows that studies of the atmospheres of other planets (Venus, Mars, Jupiter, Saturn) are useful for a deeper insight into the laws of the processes taking place in the Earth's atmosphere (Goody and Walker 1972, Kondratyev 1978a,b, 1990c, Kondratyev and Hunt 1982, Kondratyev and Moskalenko 1983, 1984, Kondratyev *et al.* 1987). In this connection, studies of the evolution of the atmospheres and palaeoclimates are very important.

Analysis of the factors of climatic variability of various planets requires an account of: solar constant variations due to varying orbital parameters, the gas and aerosol composition of the atmospheres, the volatiles' reservoirs, the surface properties and the internal compositions of planets. Probably, except for the last, all other factors have played a leading role in the evolution of the climates of terrestrial planets. Data compiled by Pollack (1979) illustrate the effect of insolation (solar constant) and other factors on the climates of terrestrial planets (Table 4.47).

Since the solar constants for the Earth and Mars are comparatively small, almost all the water vapour and $CO_2$ released from the solid shell are concentrated in the surface reservoirs and not in the atmospheres; the opposite situation is observed on Venus. The inclination of the orbit to the equatorial plane and the eccentricity (the orbital parameters determining the solar constant variability) could have varied within 22.1–24.5° and 14.9–35.5° and 0–0.06 and 0–0.14, respectively, on the Earth and Mars.

Despite the supposed lower emission of the Sun in the past, the climates of the Earth and Mars had warmer periods at early stages of their evolution, which phenomenon was connected with the intensifying of the greenhouse effect due to varying compositions of the atmosphere.

Calculations of the effect of variations in atmospheric pressure and composition on the thermal regime of the Martian atmosphere (Pollack 1979) show that periods of temperature increases can more easily be explained by an account of a primitive, not now present, oxidized atmosphere. A weaker gravitational field, the possible effect of varying orbital parameters and relatively more intensive solid-shell outgassing on its formation are, apparently, factors of more episodic periods of a milder climate on Mars than on the Earth.

With the 35° inclination of the Martian orbit's plane to that of the equator ($\delta = 35°$) the summer temperature of the polar-zone bare soil reaches 273 K. Such a high temperature seriously affects the cycles of volatile components ($H_2O$ in particular) and processes taking place in polar regions. The effect of such geological processes as volcanism has led to a decrease of $\delta$ resulting from changes in rotation

**Table 4.47** Climatic factors for the terrestrial planets

| Planet | Solar constant ($KWm^{-2}$) | Global albedo | IR emissivity[a] | Surface temperature (K) | Atmospheric pressure (hPa) | Gravity acceleration ($m\,s^{-2}$) | Distance to the Sun (AU) | Gas composition | Atmospheric aerosol |
|---|---|---|---|---|---|---|---|---|---|
| Venus | 5.17 | 0.77 | $8 \times 10^{-3}$ | 750 | 90.000 | 8.72 | 0.72 | 0.95 $CO_2$<br>$10^{-3}$ $H_2O$<br>$10^{-4}$ CO | $H_2SO_4$ clouds |
| Earth | 2.70 | 0.28 | 0.6 | 290 | 1.000 | 9.80 | 1.00 | 0.77 $N_2$<br>0.21 $O_2$<br>0.009 Ar<br>0.01 $H_2O$<br>$3 \times 10^{-4}$ $CO_2$ | Water clouds; aerosols; volcanic, soil photochemical sea salts |
| Mars | 1.16 | 0.24 | 0.9 | 218 | 0.006 | 3.72 | 1.52 | 0.96 $CO_2$<br>0.025 $N_2$<br>0.015 Ar<br>$10^{-4}$ $H_2O$ | Dust and ice ($H_2O$ and $CO_2$) |

[a] ε (IR emissivity) is determined from the formula $(1 - A)S_0 \pi R^2 = \sigma \varepsilon T_s^4 \times 4\pi R^2$ where $A$ is albedo; $S_0$ is the solar constant; $\sigma$ is the Stephan–Boltzman constant; $T_s$ is the surface temperature; $R$ is the planet's radius.

rate and moment of inertia. Therefore, in the past, values of $\delta$ could have been $(33.5 \pm 11.4)^\circ$. An increase of minimum $\delta$ value affects minimum atmospheric pressure and, hence, nitrogen dissipation. At $\delta = 45^\circ$ the summer surface temperature of the polar zone reaches to 300 K and the mid-summer mean daily temperature of the whole planet exceeds 273 K.

As a result of high temperatures in the polar regions, the atmosphere can acquire large amounts of water vapour, which promotes further temperature increase due to the greenhouse effect. This can easily explain the fluvial structures of the Martian relief in the equatorial band. However, low insolation, low atmospheric pressure and small temperature lapse rates strongly restrict the greenhouse warming of Mars.

The high surface temperatures and the dense atmosphere of Venus had resulted, apparently, from a higher (almost twice as high) level of insolation than on the Earth. This could have provided a runaway greenhouse effect on Venus, i.e. the concentration of all volatiles in the atmosphere and not on the surface (in the form of reservoirs of liquid or solid components). However, if it had formed with small amounts of water or without any, Venus ought to have had an oxidized atmosphere. In this case a decreased luminance of the Sun could have caused moderate surface temperatures at early stages of the evolution of Venus.

Quasi-periodic oscillations of eccentricity and inclination of the orbit to the plane of the ecliptic could have contributed to climate changes on the Earth during the glacial and interglacial periods of the Pleistocene as well as to the formation of stratified deposits in polar regions of Mars. Two combined mechanisms are supposed to have determined the formation of stratified soil on Mars: (1) dust particles served as nuclei for water vapour and $CO_2$ condensation, resulting in large-scale particles of dust, $H_2O$ and $CO_2$, with a higher rate of sedimentation compared with dust particles or liquefied dust particles; therefore, dust and ice deposited mainly in polar regions, forming stratified sediments and (2) changes in the angle of inclination of the orbit to the plane of the ecliptic caused variations in atmospheric pressure and surface temperature in polar regions, followed by changes in the concentration of dust particles and ice in the atmosphere. Drifting of the continents on Earth could have caused substantial climate changes on individual continents and favoured glaciations during the Pleistocene because of the movement of the continents to polar latitudes. Although no drifting of the continents has taken place on Mars, tectonic deformations of its lithosphere under certain conditions could have changed the inclination of the orbit to the plane of the ecliptic and, hence, could have caused substantial climate changes.

Atmospheric aerosol, substantially changing the radiative regime of planets, is an important climate-forming factor on all the planets considered. On the Earth, volcanic aerosol has been an important factor of climate changes on time scales from 1 to $10^6$ yr, and, for the future, of great importance is an account of possible impacts of anthropogenic aerosol on climate. The thermal regime of the Martian atmosphere is radically changing during global-scale dust storms. The $H_2SO_4$ clouds on Venus play a decisive role in the thermal balance of that planet. Intrusions of comets onto Venus could have provided a substantial additional source of water vapour and sulphur-containing gases in the Venusian atmosphere, which must seriously affect the properties of clouds.

Kasting *et al.* (1988) supposed that, initially, the three terrestrial planets (Venus, Earth, Mars) had been similar in many features, having a similar mineral composition of surface layers of the solid shell and a similar composition of the atmospheres (including the contents of $CO_2$ and water vapour). The climate on all three planets had permitted the existence of liquid water on the surface. Later on, the climates on the planets had drastically differed because the processes determining the carbon cycle between the solid shell and the atmosphere had become substantially different.

On the Earth, an almost constantly moderate climate had been determined by the special formation of the carbon cycle (by the presence of a negative feedback); the amount of $CO_2$ had increased with decreasing temperature (and vice versa), which had ensured the stability of the atmospheric greenhouse effect. The cooling on Mars had been determined by the absence of a mechanism for $CO_2$ to return to the atmosphere, whereas on Venus, an opposite situation had taken place, namely the absence of a mechanism for $CO_2$ removal from the atmosphere.

A negative feedback between the $CO_2$ content in the atmosphere and surface temperature observed on the Earth is largely determined by the geochemical cycle of carbonates and silicates, whose contribution to the $CO_2$ exchange between the Earth's crust and the atmosphere constitutes about 80 per cent (on a time scale of more than 0.5 million years). The initial phase of this cycle is connected with the solution of atmospheric $CO_2$ in rain water with the subsequent formation of carbonic acid ($H_2CO_3$). Precipitation causes the processes of water erosion of rocks containing calcium silicate minerals (compounds of calcium, silicon and oxygen), during which the carbonic acid reacts with rocks. As a result, calcium and bicarbonate ions ($Ca^{2+}$ and $HCO_3^-$) are released to surface waters. These waters eventually (mainly due to river run-off) reach the ocean, where plankton and other organisms provide an inclusion of ions to the composition of shells of calcium carbonate ($CaCO_3$). Dead organisms fall to the sea bed, forming carbonate sediments. Thousands of years later, carbonate compounds reach the edges of continents, where the sea bed slides down under the continental rocks. The resulting increase of temperature and pressure favours the reaction between calcium carbonate and silicon (quartz), ensuring the formation of reduced silicate rocks (a process called carbonate metamorphism). There is also a release of gaseous $CO_2$ to the atmosphere due to volcanic eruptions in the regions of joints between tectonic plates or ridges of the middle parts of the oceans. If, for example, a climatic cooling takes place under the influence of decreasing solar constant and, hence, a decrease of the ocean temperature, it means a decrease of evaporation and precipitation, which weakens the processes of water erosion. In such conditions, the rate of income of $CO_2$ from the atmosphere to the Earth's crust is reduced, but the generation of $CO_2$ from carbonate metamorphism remains at the same level, due to which the input of $CO_2$ to the atmosphere intensifies, hence intensifying the greenhouse effect, which compensates for the climatic cooling.

Since about 20 per cent of $CO_2$ is removed from the atmosphere by photosynthesizing plants, the role of the biosphere in the formation of the global carbon cycle is clear. It manifests itself through the effect of living organisms on the cycles of carbonates and silicates. It should be considered, however, that physical rather than biological processes play a major role in the formation of the global carbon cycle.

As far as Mars is concerned, its palaeoclimate is considered to have been warm enough to provide the possibility of the existence of liquid water on the planetary surface (for this purpose, an atmospheric pressure from 1 to 5 atm is needed). The mechanism for the input of $CO_2$ from the atmosphere to the solid shell has probably been the same as on the Earth, but the mechanism for the return of $CO_2$ to the atmosphere has been very different (on such a small planet as Mars, plate tectonics is impossible).

It is assumed that there have been oceans on Venus in the past, which vanished in the course of evolution due to the wet greenhouse effect, determined by an initially high concentration of water vapour (more than 20 per cent) near the planetary surface. The existence of an ocean is possible, provided that the solar constant is about 1.1–1.4 times its value on the Earth, but the ocean would vanish within several hundred million years due to hydrogen dissipation. Apparently, the hypothesis of the wet greenhouse effect is more correct than the previous concept of the runaway greenhouse effect. As in the case of Mars, a high temperature on Venus is determined not by its proximity to the Sun, but by a strong greenhouse effect due to the $CO_2$ composition of the atmosphere, which is explained by the disappearance of the ocean and cessation of carbonate formation. Apparently, the critical interval of distance to the Sun, which permits a climate fit for life on a planet, constitutes 0.95–1.5 AU. The conclusion suggests itself that the climate on the Earth is not unique. Similar conditions can exist in other planetary systems.

It is probable that the original atmosphere of the Earth had disappeared as a result of the collision of the Earth with a very large celestial body. In this connection, Pepin (1988) analysed possible scenarios of the evolution of the primary atmosphere resulting from accretion. The loss of initial volatile components at the time of a gigantic collision may explain, in particular, why 15 per cent of carbonate chondrites – one of the components in the process of accretion – had not left any signature in the present distribution of noble gases.

According to present models of accretion of gases from the protoplanetary nebula, the Earth ought to have captured more gases than it has done, and this means that the initial atmosphere ought to have been lost. It appears that the subsequent primordial atmosphere has formed as a result of another accretion of gases from the protoplanetary nebula. Hence, the episode of gigantic collision must have taken place before a dissipation of the nebula, i.e. approximately 1–10 billion years ago, and the content of captured gases must have been 50–500 times less than in the initial atmosphere. A decrease of the efficiency of gas capture can be explained by a higher temperature and a faster rotation of the Earth. The conclusion can be drawn that the gravitational condensation of gases from the protoplanetary nebula is quite possible as the process continued of the input of noble gases of the solar composition to the primary atmospheres of the Earth and Venus. This process has been, however, of low efficiency on Mars.

Fegley and Prinn (1988) emphasized that the bombardment of the Earth with large extra-terrestrial bodies has led in the course of its evolution to collision-induced heating and to a change of the chemical composition of the atmosphere and, in particular (in conditions of the present oxidizing atmosphere), to the formation of a large amount of $NO_x$ (respectively, $NO_2$ and acid rains of $HNO_3$), which, with a sufficiently high concentration of nitrogen oxides, could have explained various dangerous ecological consequences, including a decrease of total ozone content, strong attenuation of solar radiation by the atmosphere (and hence a decrease of its income to the Earth's surface), intensification of chemical weathering of continental rocks, and neutralization of acidity of the uppermost layer of the world's oceans, with subsequent increase of $CO_2$ concentration in the atmosphere.

On the other hand, the consequences of collision-induced heating for the hypothetical earlier atmosphere with reducing properties (and a low content of oxygen) ought to have consisted in the synthesis of a large amount of HCN, which is a critically important component for the synthesis of complex organic compounds. The abiotic formation of oxygen ought to have been an important result of the collision-induced heating and chemical conversion of the composition of the Earth's initial atmosphere. Photodissociation of $H_2O$ and $CO_2$ has been one of the sources of oxygen in the atmosphere. It was shown, however, that the principal contribution to the formation of oxygen in the early atmosphere had been made by collision-induced heating and chemical conversion of the atmosphere.

As Gautier (1987) noted, four giant planets within the solar system have a similar structure: analysis of data of their gravity moments had led to the conclusion that all the Jovian planets have cores (consisting, apparently, of rocks and ice with a total mass of about 10–15 times that of the Earth) surrounded with thin atmospheres largely of hydrogen and helium. Since the masses of Jupiter, Saturn, Uranus and Neptune are, respectively, 318, 95, 14.5 and 17 times the mass of the Earth, there is no doubt that the atmospheres of Jupiter and Saturn are the most thick, and those of Uranus and Neptune are relatively thin.

The atmospheres of all the Jovian planets have recently been supposed to be of the same composition as the protoplanetary nebula, from which the Sun and the planets were formed, but during the last few years it has become clear that in the period of formation of the planets and their subsequent evolution, the relative content of various gases and the protoplanetary nebula could have changed. The existing scenarios of the formation of planets can be classified into two groups: models of unstable gas and nucleation models.

In the first case it is assumed that under the influence of gravitational instability of the protoplanetary nebula, a local condensation of substance had taken place within extended volumes, whose sizes were thousands of times greater than the present radii of the Jovian planets. The subsequent hydrodynamic collapse of protoplanets and their slow compression to a state of hydrostatic equilibrium have led to the formation of the present planets (their interior stratification in the layers of different composition has resulted from chemical processes in the presence of thermochemical equilibrium). Since molecular hydrogen, helium and neon cannot condense in the external parts of the atmospheres of the Jovian planets, their composition should be the same as that of the protoplanetary nebula (and the Sun). Respectively, the concentrations of such elements that prevail in space as carbon, nitrogen and oxygen (mainly in $CH_4$, $NH_3$ and $H_2O$) in the atmospheres of the Jovian planets must be solar (with respect to hydrogen) except for cold upper tropospheres, where these gases can condense.

The nucleation model is based on the fact that particles of refractory materials (silicates, iron, etc.) and solidified gases ($H_2O$, $NH_3$, $CH_4$) or mixed clathrates and hydrates of volatiles ($CH_4$, $NH_3$, $N_2$, CO, Ar, etc.) have emerged into the gas protoplanetary nebula.

Gradually, an agglomeration of such particles into planetosimals took place, and the subsequent processes of accretion led to the formation of the planetary cores. Upon the formation of a large enough core, it began capturing gases from the protoplanetary nebula surrounding it, forming, thereby, the atmosphere. In this case, the concentrations of uncondensed components ought to be enriched, compared with solar ones, due to an evaporation of volatiles in the upper atmosphere.

Table 4.48 gives the observed concentrations of carbon and nitrogen in the atmospheres of the Jovian planets. The observed surplus of carbon contradicts the model

**Table 4.48** The contents of carbon and nitrogen in the atmospheres of Jovian planets

| Content | Sun | Jupiter/ Sun | Saturn/ Sun | Uranus/ Sun | Neptune/ Sun |
|---|---|---|---|---|---|
| C/H | $4.7 \times 10^4$ | 2.3 | 3–6 | 20–25 | 25? |
| N/H | $9.8 \times 10^5$ | 2 | 2–4 | ? | ? |

of an unstable gas, but agrees with the model of nucleation (for instance, the increase of this surplus as the planet moves away from the Sun). The content of helium is an important indicator of the evolution of the planetary atmospheres. According to both scenarios of evolution, the ratio $He/H_2$ in the atmospheres should be the same as in the protoplanetary nebula, since these elements cannot condense, but observations do not confirm this conclusion. So, for example, the relative content of helium in the atmosphere of Saturn ($0.06 \pm 0.05$) is much less than on Jupiter ($0.18 \pm 0.04$), whereas on Uranus the share of helium ($0.267 \pm 0.048$) is close to the proto-solar one (0.27–0.28). It seems that the evolution of Jupiter and Saturn has been followed by a differentiation of helium with respect to the core of metallic hydrogen. Measurements of the composition of the atmospheres of the Jovian planets, using automated interplanetary stations, will be very important for the solution of the problem of the evolution of planets.

CHAPTER FIVE

# The internal variability of the climatic system

## 5.1 Climate diagnostics

Processes of climate formation are complicated, diverse and almost always interdependent. These processes may be classified into two categories. The first determines the internal variability of the climatic system due to such processes as interactions between the atmosphere and the ocean, clouds and radiation, etc. The second involves external natural and anthropogenic effects, including numerous quasi-external forcings. The latter are determined proceeding from two principles: the possibility of considering certain climatic factors to be unchanged within limited time periods (when sea surface temperature, snow and ice cover extent, soil moisture, etc. may be prescribed as boundary conditions) and the difficulty of attributing some climatic factors to internal ones (for example, the problem of biogeochemical cycles, especially global cycles of carbon, sulphur and nitrogen).

Despite recent progress in studies of the factors determining climate and its changes, the understanding of these factors remains limited. So, for example, differences in atmospheric circulation, snow cover distribution and other characteristics of normal and severe winters have been well documented, but the reasons for these differences are still poorly understood. Though an account of external forcings on the climatic system (volcanic eruptions, anthropogenic factors, solar activity, etc.) is undoubtedly important, it should be borne in mind that the climatic system can be considered to be relatively closed for long time periods, exhibiting the characteristic interacting climate-forming processes.

In this connection, dynamic processes simulated by non-linear equations of dynamics and thermodynamics of the atmosphere (World Meteorological Organization 1988b) are particularly important. Even in conditions of a quasi-closed climatic system, substantial fluctuations can occur (at least, on regional scales) due to the effect of feedbacks between the components of the climatic system. The effect of non-linearities is a substantial factor of the unpredictability of the climatic system and itself can explain the interannual variability of climate (though there are other mechanisms, too).

In order to analyse the principal manifestations of the climatic system's non-linearity (as applied, mainly, to the problem of the interannual climatic variability),

Hunt (1988) used the simplest scheme, based on the fact that climate changes are determined by a superposition of its annual change and non-linear disturbances (see also Hameed *et al.* 1989, Palmer 1993). The latter are considered to be caused by an advection and, therefore, to be of a purely dynamic nature. The possibility of an interaction between the externally forced annual change and internal disturbances connected with non-linear dynamics is excluded. Any climatic parameter $Z$ is described by the expression:

$$Z_n = C\cos\left(\frac{2\pi n}{365.25}\right) + N_n \tag{5.1}$$

Here $C$ is the relative amplitude of the annual change; $n$ the number of days; $N_n$ the non-linear term determined from the following recurrent expression:

$$N_n = x_{n+1} = a(3x_n - 4x_n^3) \tag{5.2}$$

suggested earlier by Lorenz (1990). An analysis of Equation (5.2) has shown that at $1/3 < a < 2/3$, the recurrence relation is stable (there is a conversion to the so-called fixed point). If $a < 0.767$, then there are periodic doublings of the cycle period, whereas at $a > 0.767$ a chaotic attractor appears. In the latter case, the $a$-values, remaining limited, are characterized by changes in the sign and amplitude (with successions of different length) if $a > 0.866$. Since at $a = 0.866$ the non-linearity manifests itself strongly, this value of $a$ has been taken in the calculations considered.

Although Equation (5.1) cannot be considered physically substantiated as a climate model, it enables simulation of the interannual climate change and makes it possible to analyse the laws of real climate changes. With this aim in view, Hunt (1988) performed calculations with a time step of 1 day. To take into account the effect of an 11-yr cycle of solar activity on climate, a term $B\cos[2\pi n/(365.25 \times 11)]$ was added to the right-hand side of Equation (5.1), where $B$ is an arbitrary coefficient for which different values were fixed.

Calculations have shown that the effect of solar activity is strong, if the $B$-value exceeds a certain threshold level, when the solar signal can be noticed against the non-linearly induced climatic noise. The effect of non-linear interactions results in the occasional breaking of Sun-induced modulations. This result can be considered an argument in favour of solar–climatic relations, which have been rejected because from time to time an 11-yr climatic cyclicity gets broken. The latter may be connected with the effect of not only non-linear dynamics but also such factors as the El Niño / Southern Oscillation (ENSO), etc. Calculations based on the use of Equation (5.1) revealed considerable interannual variations in the amplitude of the annual change, as well as changes in the values for any fixed month determined by non-linear dynamics.

Calculations for a very long period (100 000 years) have led to the conclusion that long-term climatic variations can be explained only by the effect of non-linearity which causes, for example, typical anomalies every 100 and 500 yr. Simulations of the effect of a long-term external forcing (such as the growth of $CO_2$ concentration), intensifying by a factor of $10^{-5}$ for each time step, have shown that the disturbance clearly manifests itself, against the internally induced variability of the climatic system, only 40 yr later. There is no doubt that studies of non-linear effects on climatic dynamics remain one of the most urgent problems of climate theory.

Droughts are one of the most important manifestations of interannual climatic variability. In this connection, Gordon and Hunt (1987, 1991) used a two-layer (250 and 750 hPa levels) spectral (21 harmonics) climate model to analyse the formation of the global distribution of soil moisture for fixed values of surface albedo (in the case of snow the dependence of albedo on the snow cover thickness was taken into account) and prescribed (from climatological data) sea surface temperature (see also Hunt and Gordon 1988, 1989, Trenberth 1992). The model makes it possible to forecast the formation of two types of clouds. An initial integration (in order to provide the stabilization of climate) was made for two years. Results of calculations for the next 10 yr were used to analyse the interannual variability of climate.

A concern for soil moisture is determined by its important role in the redistribution of energy between the fluxes of latent and sensible heat. The results obtained have made it possible to reveal not only the features of the global distribution of soil moisture but also a considerable interannual variability. So, for example, the appearance of severe local droughts has been demonstrated, which illustrates the usefulness of numerical modelling from the viewpoint of drought studies. Despite neglecting numerous feedbacks and some external forcings (for example, sea surface temperature anomalies, volcanic eruptions, anthropogenic disturbances), a real-time simulation can be made of the interannual variability of the moisture cycle, except for the region of Brazil.

The results considered can be interpreted as evidence for the important role of internal variability (non-linear dynamics) in the climatic system as a factor of climatic evolution. It means, in particular, that an integration over 12 months, with a specific initial climate, will not lead to a simulation of an identical climate – rather the contrary, since interannual variability is not an exception but a rule. Numerical experiments are a reliable way to analyse the role of such internal variables as cloud amount, snow and ice cover extent and soil moisture anomalies in the formation of the interannual variability of climate and specific features of its geographical distribution.

The importance of non-linear mechanisms of climatic variability can be illustrated by a consideration of data on the rate of convective heating of the atmosphere for two Januaries (years 6 and 8 in a 10-yr period of integration), which is connected with the intensity of evaporation and, hence, soil moisture (note that the effect of sea surface temperature anomaly is also manifested through convection). The year-to-year variability of convective heating must be a leading factor of the interannual variability of climate.

An analysis of the two January global distributions of convective heating revealed (with their similarity in basic features) numerous local and regional differences, reaching 1 deg K per day. These differences determine the specific features of atmospheric dynamics, which manifest themselves, for example, in the field of the meridional component of wind speed at the 700 hPa level. In their turn, the disturbances appearing in the wind field affect the distribution of convective heating, which illustrates the interactive nature of the processes responsible for the formation of the interannual variability of climate.

From the practical point of view, the most important result of the numerical modelling under discussion is the possibility of a realistic simulation of droughts as a consequence of only internally-induced variability of the climatic system. The possibility is not excluded, however, that the numerical model under discussion overestimates the importance of the non-linearly induced droughts. More complicated models are needed which would take into account both the non-linear internal dynamics of the climatic system and the effect of sea surface temperature anomalies.

Using the UK Meteorological Office 11-level global atmospheric circulation model, Rowntree and Sangster (1986) performed a numerical modelling in order to analyse the role of feedbacks determined by land surface processes in the formation of droughts in the Sahel region. Within a 4-yr control experiment, calculations were made for a period of 9 months (starting from 1 March) for two different years with a prescribed soil moisture content in Africa, south of 10° N, reduced from 15 cm to 1 cm. This is equivalent to the substitution of an almost bare soil, thinly covered with vegetation, for the deep-root multi-year vegetation. Such a substitution led to the calculation of a substantial decrease of rainfall over the western Sahel and an intensified precipitation in Sudan and Ethiopia.

In the next pair of experiments, apart from reduced evaporation, an albedo increase was prescribed to take place in the band 10–17.5° N to the values corresponding to semi-desert conditions. In this case, the calculated decrease of precipitation was much more spatially homogeneous (in the Sahel region) and exceeded somewhat that observed during the last drought. The calculated effect on precipitation and atmospheric circulation was not confined to the Sahel region. So, for example, there was a calculated rainfall reduction as far south as the equator (especially in September and October), reflecting the effect of advection from the north on the atmospheric water content. There was a similar effect in the Atlantic zone, where it was followed by a southward shift of the Intertropical Convergence Zone (ITCZ) rainfall belt and by substantial changes in tropospheric circulation (all these changes were statistically significant). The latter closely correspond to circulation variations observed in the transition from arid to moist periods. This suggests that variations in circulation have been the consequence of, but not the reason for, the droughts.

Thus, the results of the numerical experiments under discussion point to a positive feedback manifested in that a reduction of rainfall in the Sahel leads to the loss of multi-year vegetation cover and this favours further intensification of the drought. This feedback can intensify owing to such factors as a rainfall reduction due to sea surface temperature anomalies and vegetation destruction by domestic animals and humans. Until recently, one of the mysterious natural phenomena in the region of the Great Plains (USA) has been a well-known trend of repeated periods of prolonged droughts every 20 yr (1910s, 1930s, 1950s, 1970s), when a 15–20 per cent reduction of rainfall was observed (with respect to the norm), and in July–August (the period which is critically important for the crop yield) the reduction has been up to 25–50 per cent, followed by a strong decrease of soil moisture and of the underground run-off.

In this connection Hameed and Currie (1986) undertook an analysis of tree rings (with the use of maximum entropy technique and Fourier analyses) for western Canada (British Columbia and Alberta) and the USA (Texas), which revealed a cyclicity with a period of 18.6 yr. Such a cyclicity may be supposed to be caused by lunar tides with a period of 18.613 yr affecting the atmosphere, to which the following control dates of the twentieth century correspond: 1917.5, 1936.1, 1957.7, 1973.3. Observational data for the eastern USA show that, starting from 1700, maxima of the droughts have been in phase with these epochs, whereas in western Canada they have been phase shifted by 180°.

The heat balance of polar regions is one of the key aspects of the formation of climate (World Meteorological Organization 1988f). In this connection, Nakamura and Oort (1988) performed calculations of four basic components of the heat balance of the atmosphere in the southern hemisphere and northern hemisphere polar regions from data of satellite and radio-sonde observations: (1) the rate of change of energy, $\Delta E/\Delta t = F_{RAD} + F_{WALL} + F_{SFC}$, (2) the radiation budget of the surface–atmosphere system, $F_{RAD}$, (3) poleward energy fluxes in the atmosphere at the surface level across the 70° latitudinal belt, $F_{WALL}$ and (4) upward energy flux in the atmosphere at the surface level $F_{SFC}$. The last component is calculated as a residual term, and the energy flux across the 70° S latitude circle was calculated using an atmospheric general circulation model, in view of fragmentary observational data.

An analysis of the results has shown that the role of heat exchange between the atmosphere and the ocean–cryosphere system is more substantial in the Arctic than in the Antarctic. However, in the case when polar caps reach the 70–60° latitudinal belt, various components of the heat balance in the Arctic and the Antarctic become closer in value. The north-polar cap is characterized by a strong annual change in $F_{SFC}$, with an amplitude of about 50–80 W m$^{-2}$, whereas in the region of the south-polar cap this amplitude constitutes only 20–30 W m$^{-2}$. Apparently, this difference is determined by the lesser intensity of the formation processes of snow and ice cover on the south polar cap.

The structure of the annual mean heat balance of the atmosphere in both polar regions is similar and determined by the $F_{RAD}/F_{WALL}$ ratio, and the value of $F_{SFC}$ is small. However, the annual change in the heat balance components is quite different. In the summertime north-polar cap zone, $F_{RAD}$ is small and the meridional heat flux coming to this zone is largely balanced by heat losses on snow and ice melting, as well as on the ocean heating. The meridional input of heat to the south polar cap is half as much and is, apparently, equally spent on heat loss due to $F_{RAD}$ and on snow and ice melting. In the wintertime at the north polar cap, the heat loss on emission ($F_{RAD}$) is very large and two thirds of it are compensated by the meridional heat flux, and one third by the heat flux through the surface. On the south polar cap the heat loss in the winter is almost equalled by the meridional heat flux, whereas the contribution of heat exchange through the surface is negligible. A major reason for the contrast between the polar regions is the presence in the southern hemisphere of a huge ice cap, whose height reaches 3 km, as well as such a peculiar feature of the south-polar atmosphere dynamics as slope (katabatic) winds.

Some diagnostic studies revealed the effect of ENSO events on the fluctuations in the ocean–atmosphere system, a sea surface temperature anomaly in the eastern equatorial Pacific being a critically important parameter (World Meteorological Organization 1988b). Of particular interest are correlations between sea surface temperature variability at the key point of 2.5° S, 130° W and variations in the global thermal regime, including mean temperatures of the atmosphere in the southern and northern hemispheres. The whole Pacific Ocean east of 170° E in the 20° S to 20° N band suffers an increase of temperature if the sea surface temperature at the key point increases (Figure 5.1). The sea surface temperature in the equatorial Atlantic and Indian Oceans varies in phase. However, the correlation with the mid-latitude sea surface temperature is negative.

In comparison with the key-point monthly mean sea surface temperature and southern hemisphere and northern hemisphere air temperature variations shown in

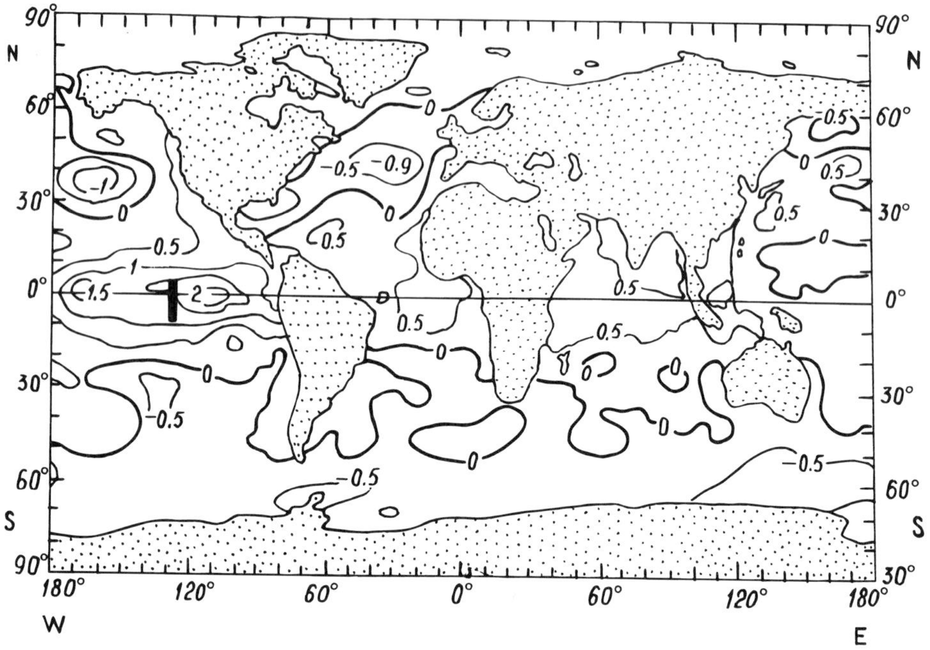


**Figure 5.1** The geographical distribution of the December–February SST difference (deg K) for the cases when, at the key point 2.5° S, 130° W, an increased (1963/1964, 1965/1966, 1968/1969, 1969/1970, 1972/1973) and decreased (1964/1965, 1966/1967, 1967/1968, 1970/1971, 1971/1972) SST was observed.

Figure 5.2, one can see that for the northern hemisphere the correlation coefficient, $r$, reaches 0.6 (Figure 5.3) when the phase shift (with respect to the key point) reaches 6 months. The correlation with atmospheric temperature of the southern hemisphere is weaker, which can be explained by the fragmentary observational data. The period of maximum variations between 1958 and 1973 is about 3 yr. Near the equator the temperature changes lag behind the sea surface temperature variations by several months, while the farther from the equator the greater is this shift.

These results show that, at least during the 15 yr under consideration, the basic short-period climatic variability in the atmosphere was connected with the sea surface variations in the key eastern sector of the tropical Pacific. Apparently, on such time scales the role of other disturbing factors (for instance, the albedo feedback in high latitudes) is secondary (see also Palmer and Branković 1989).

Lau (1985b, 1986) discussed observational data from the viewpoint of their use to test the results from numerical climate modelling, especially information on the spatial distribution of prevailing modes of atmospheric variability on time scales longer than a month, and a possible role of disturbances in boundary conditions in the formation of such anomalies.

An analysis was made of the geographical distribution of the mean standard deviation of the 50 hPa height in the northern hemisphere, calculated from data for the unfiltered time series which contains fluctuations from 1 day to 1 season and for the filtered time series (by eliminating the high-frequency fluctuations) which describes fluctuations from 10 days to 1 season. This analysis shows that a considerable contribution to the geopotential variability is made by slowly-varying processes

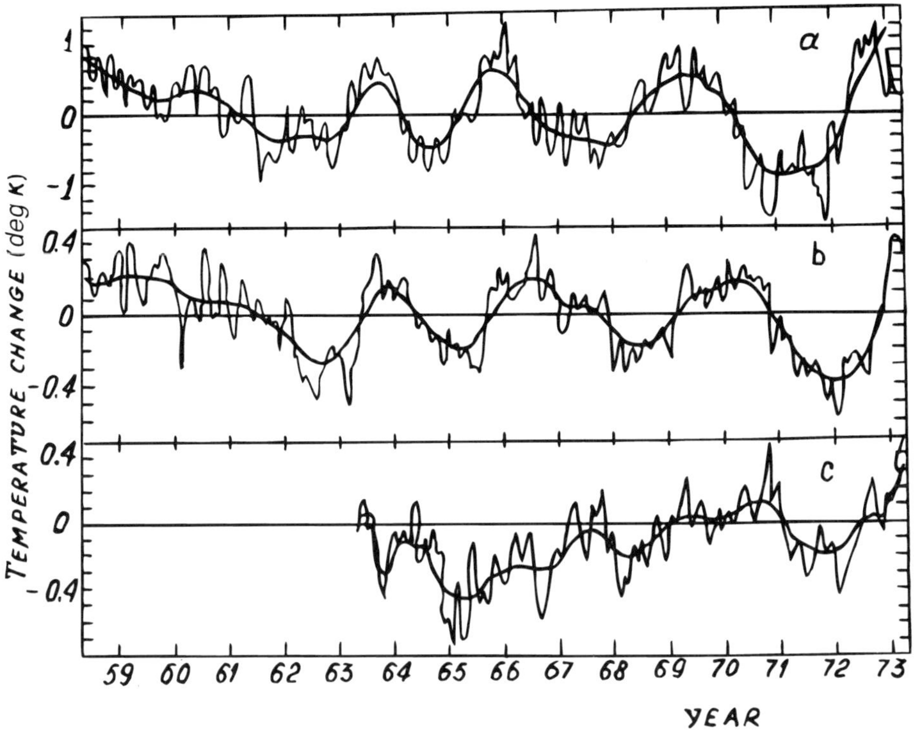


**Figure 5.2** Time series of the variability of the SST monthly means at the key point 2.5° S, 130° W (a) and of air temperature averaged over the vertical column of the atmosphere in the northern (b) and southern (c) hemispheres, for the period May 1958 to April 1973 (the annual change of temperature averaged over the period 1963–1973 is excluded).

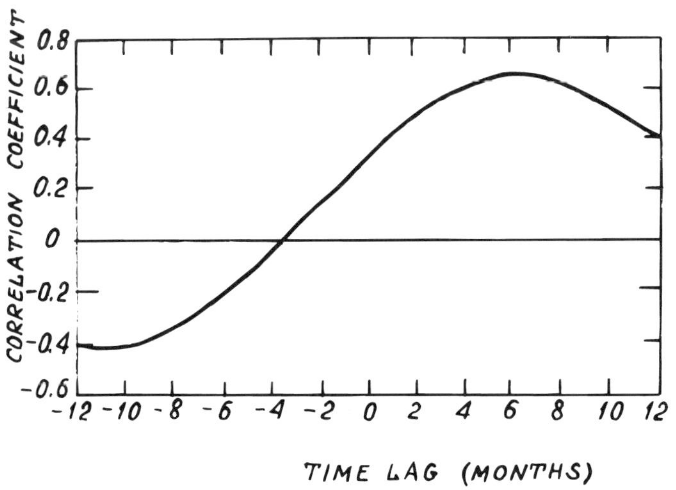


**Figure 5.3** The correlation coefficient between SST at the key point 2.5° S, 130° W and air temperature averaged over the whole northern hemisphere atmospheric mass. Positive values show that the SST anomalies at the key point were followed by the air temperature anomalies in the northern hemisphere.

(with a characteristic time scale of 10 days and longer). Similar results were obtained for the southern hemisphere as well as for other levels. Naturally, a thorough analysis of such variability, concentrated in the low-frequency range, plays an important role in forming a strategy for seasonal climate forecasts.

This analysis is also important because the variability considered has a specific spatial distribution, each mode often showing through variations in several meteorological parameters.

In this connection, the North Atlantic Oscillations, manifested through the meridional swing anomaly and a pressure field at sea level (Figure 5.4), are of interest. This anomaly is connected with a zonally oriented contrast (swing) in winter temperature between Greenland and northern Europe as well as large variations in the spatial structure of trade winds and precipitation. Analysis of the North Atlantic Oscillations-related observational time series revealed fluctuations in time scales of several years, with statistically substantial maxima corresponding to quasi-biennial and 6-yr periodicities. Similar North Pacific Oscillations are manifested as counterphase changes in pressure in the northern hemisphere sub-tropical Pacific and in Alaska and northern Canada (Figure 5.5), connected with characteristic spatial structures of the fields of temperature, precipitation and trade wind circulation. The drifting of the Aleutian Minimum is, apparently, the dominating factor of the North Pacific Oscillations.

The Southern Oscillation, covering the sub-tropical bands of the Pacific and Indian Oceans, is another illustration of the oscillation events. The Southern Oscillation, with a characteristic time of several years, is followed by variations in numerous meteorological parameters, both at sea level and in the whole atmosphere. Some of the Southern Oscillation related variations can be described as intensifications and attenuations of the Walker circulation which maintains the east-to-west mass exchange in the near-equatorial band (Figure 5.6).

The experience of long-range weather forecasts has long since revealed persistent anomalies of the geopotential field in the middle and upper troposphere, with a

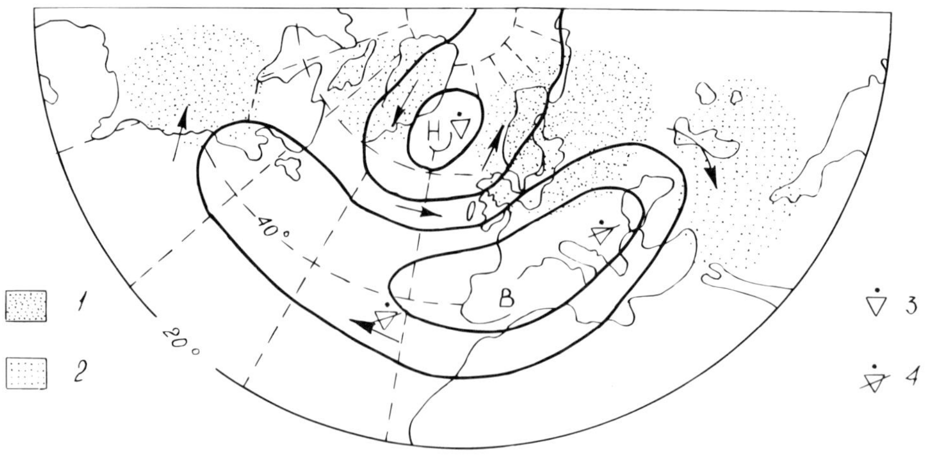


**Figure 5.4** A scheme of the North Atlantic Oscillation. The contours and arrows characterize the anomalies of pressure at sea level and of atmospheric fluxes, respectively. Anomalies of heat (1) and cold (2) are shaded. The zones of intensification (3) and attenuation (4) of precipitation are marked by respective symbols.

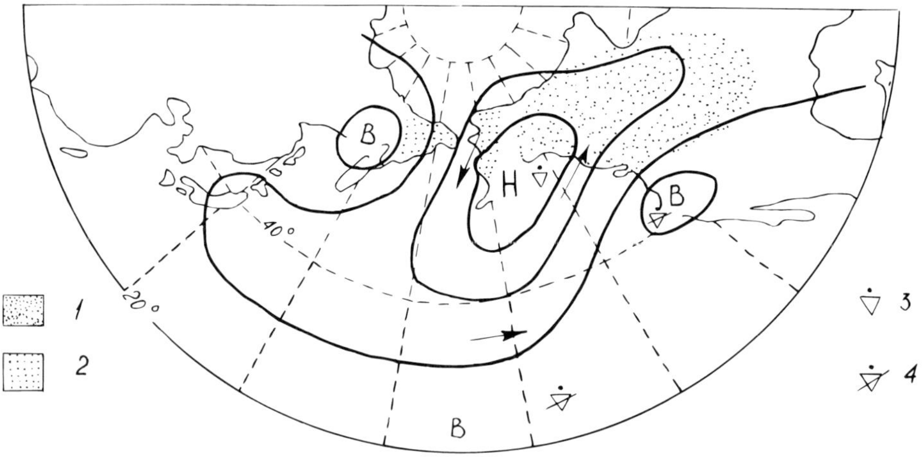


**Figure 5.5** A scheme of the North Pacific anomaly. Legend is as for Figure 5.4.

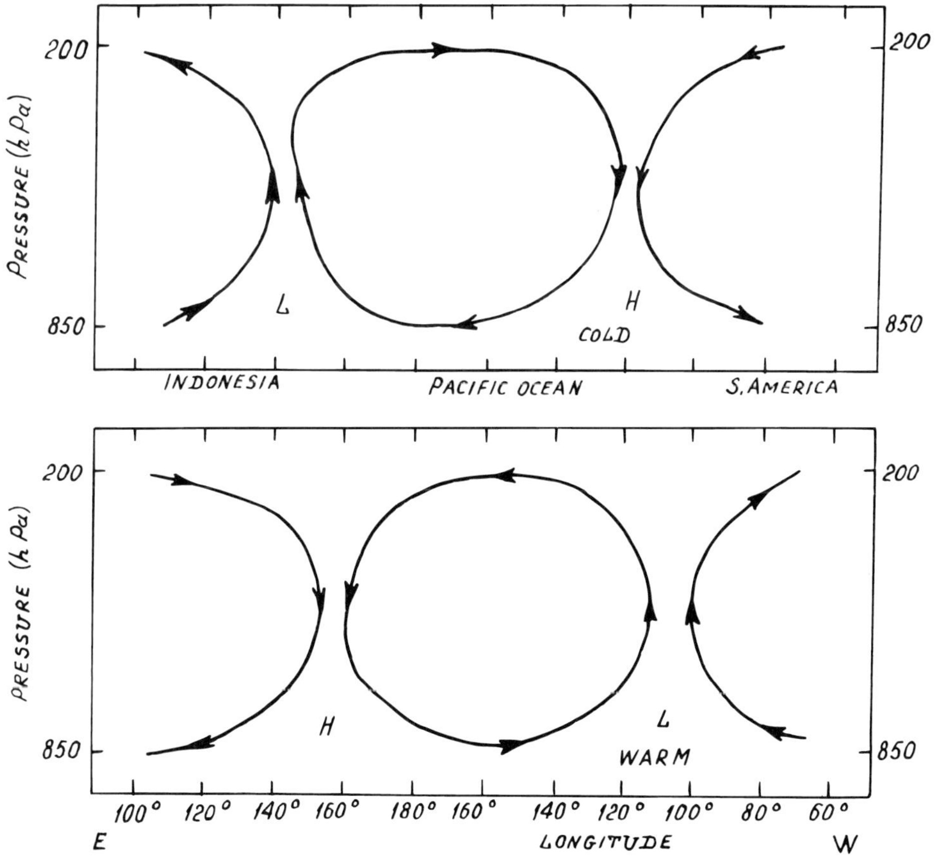


**Figure 5.6** The structure of the anomalous components of the Walker circulation in the zonal equatorial plane during the opposite phases of the Southern Oscillation.

typical wave structure and several centres. There are five major centres in the winter northern hemisphere (Figure 5.7), many of them coinciding with the regions of intensification of the low-frequency variability mentioned above. The combined effect of the anomalies shown in this figure explains 75–90 per cent of local normalized variability in the active zones (the northern regions of the Pacific and Atlantic Oceans and the north of the former USSR). Analysis of asynchronous cross-correlative statistical characteristics of the fields of meteorological parameters showed that the centres of action connected with the processes in the region of the Pacific Ocean and North America correspond to the variability of the Rossby–Blinova wave trains, propagating from the tropical Pacific, i.e. a remote forcing takes place here. The western Pacific anomalies centred over the oceans (Figures 5.7 and 5.8) are apparently connected with the North Atlantic Oscillations and the

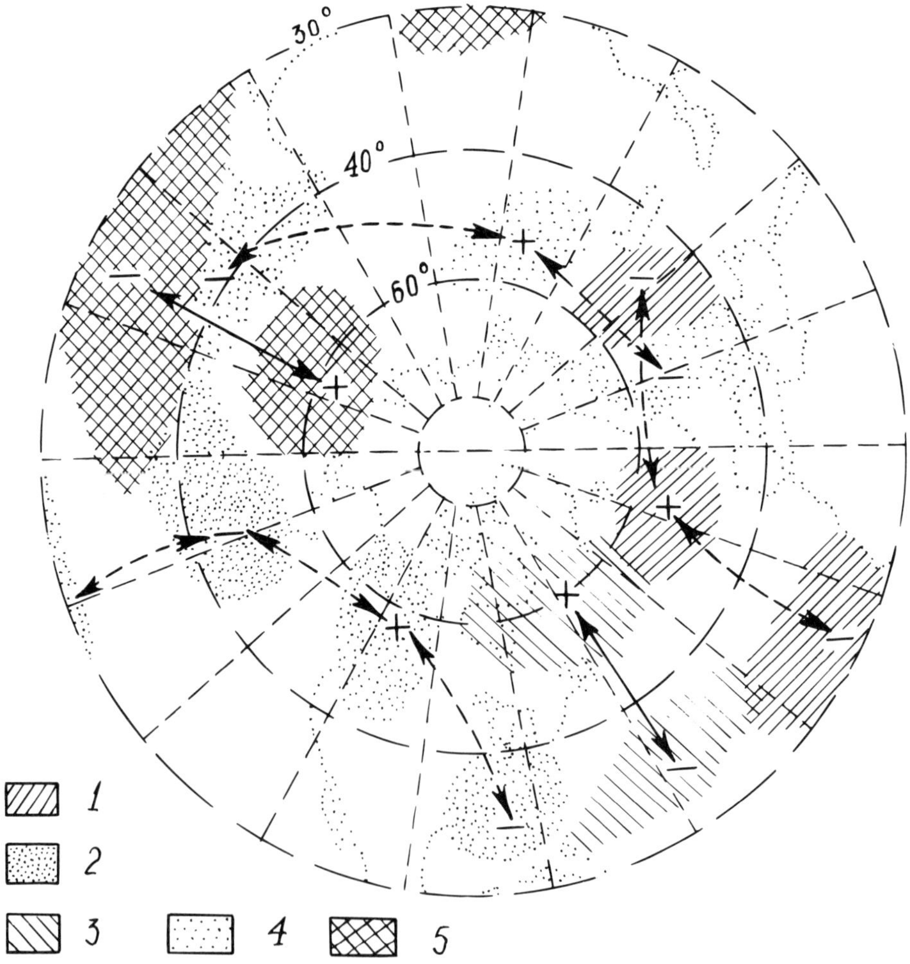


**Figure 5.7** A scheme of the five major centres of action: east Atlantic (1), North American Pacific (2), west Atlantic (3), Euroasiatic (4) and west Pacific (5). The relative polarity of the centres is marked by pluses and minuses.

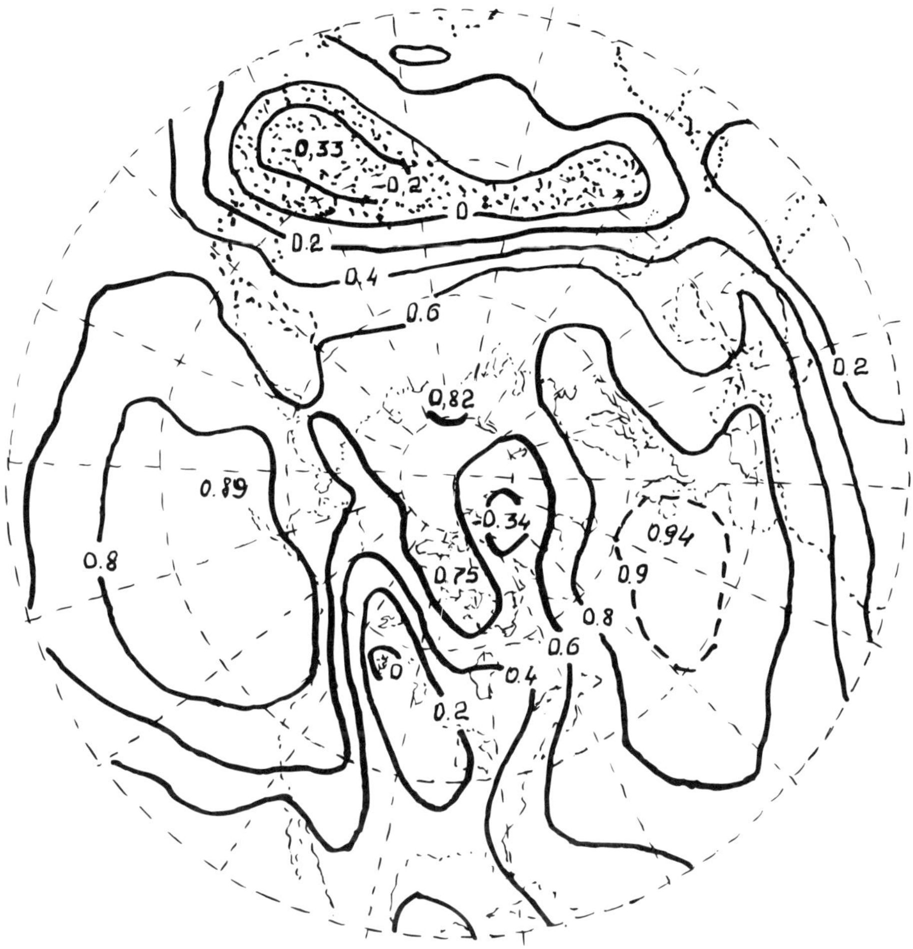


**Figure 5.8** The geographical distribution of the coefficients of time correlation between deviations of the level of geopotential surfaces 500 and 1000 hPa, winter.

North Pacific Oscillations. Thus, the conclusion can be drawn that major modes of anomalies have a global scale and a coherent three-dimensional pattern.

Rasmusson *et al.* (1983) analysed conditions for the atmosphere–ocean interaction in the equatorial Pacific and the subsequent large-scale sea surface temperature anomalies in this region. With respect to long-period weather forecasts, it is particularly important that abnormally warm waters near the Peru coastline usually move westward and, in 6 months, reach the central part of the equatorial Pacific. At this stage, sea surface temperature anomalies remotely affect the mid-latitude atmospheric circulation. To improve understanding of these processes, an extensive programme is needed of observations in the key regions of the world's oceans, aimed at continuous monitoring of sea level, the pattern of the temperature and current fields, the wind stress near the ocean surface and the heat and moisture budgets.

An analysis of data of observations of the low-frequency variability of atmospheric pressure at sea level in the region of the southern Pacific ITCZ had led to the conclusion that the sea surface temperature anomalies in the ITCZ are connected with the Southern Oscillation. An assumption has been made that sea surface temperature anomalies, which, in their turn, are connected with the preceding wind anomalies in the same region, can induce a regional response of the atmosphere, which favours the formation of an abnormal west-to-east transport over the largest part of the tropical Pacific in the southern hemisphere as well as heat anomalies in the Southern Oscillation. To substantiate such an assumption, von Storch and van Loon (1986) undertook calculations for a period of 10 yr with the use of the ECMWF spectral AGCM T-21, with prescribed sea surface temperature anomalies in the ITCZ of $+2$ deg K and $-2$ deg K. The results of calculations show that a positive sea surface temperature anomaly generates a persistent negative anomaly of pressure over the whole south-western Pacific in the presence of a ridge stretching from the northern coastline of Australia to the point 50° S, 150° W. A negative sea surface temperature anomaly determines a regular anomalous spatial structure of the pressure field. A specific impact of positive and negative sea surface temperature anomalies shows itself in the fields of other meteorological parameters.

As Lau (1985a) noted, during the last few years the emphasis in medium-range and long-range forecasts has been placed on the impact of the processes in the tropics on the development of weather-forming systems in mid-latitudes (of special interest was an analysis of the effect of the Pacific Ocean temperature anomalies on the climate of North America). However, the interaction between the tropical and middle latitudes is not isolated. For example, this interaction is subject to monsoon cold air outbreaks over the continent of Asia, which may intensify a convection in the region of Indonesia.

In this connection Lau (1985a) undertook diagnostic studies of the factors which determine the interaction between atmospheric circulation and tropical convection as applied to medium-range variability (10–90 days). The correlation between the fields at the 500 hPa and 200 hPa levels and the dipole anomaly of the convection-induced heating in the regions of Indonesia and the central equatorial Pacific Ocean was taken as an example. The maps of geopotential over the 4° latitude by 5° longitude grid obtained by the US National Meteorological Center for two 150-day periods (November to March) in 1974–1977 were used as initial data, and the convection intensity was estimated from satellite-observed outgoing longwave radiation (twice a day, 2.5° by 2.5° grid). In both cases the annual change was filtered out.

An analysis of observational data has shown that in the northern winter 40–50-day oscillations of dipole heating, contributing most to medium-range variability, are closely connected with an excitation in extra-tropical latitudes of the Rossby wave packets in certain geographical regions. In this connection, two groups of observational data are of particular concern. One of them shows that a convection intensification in the region of Indonesia results from the group of waves over Eurasia (propagating against the prevailing flow), whose orientation and evolution point to the prevailing energy transport from middle to tropical latitudes. Then groups of waves appear which propagate in the direction of flow in the Pacific–North-American region and reflect the energy transport both northward and to the equator.

The second group of observational data shows that a convection in the central equatorial Pacific Ocean follows a strong geopotential anomaly over the north-eastern Pacific, which excludes the blocking situation and synchronously correlates with the group of waves propagating along the flow over the Pacific–North-

American region, where the northward energy transport takes place. Thus, the extra-tropical anomaly is followed by an intensification of convection at both poles of the dipole structure of heating, testifying to the fact that in the case of medium-range variability the convection is governed, at least partially, by weather-forming mid-latitude systems. It does not, however, mean that the processes in the tropics play a passive role in this interaction.

The presence of the energy flux from the central equatorial Pacific Ocean to the north is an indicator of the participation of the tropics in the feedback mechanism which regulates the interaction of processes in the middle and tropical latitudes. Analysis of the interaction morphology during the whole cycle of oscillations reveals the possibility of correlation between the convection in the central equatorial Pacific Ocean and the mode of atmospheric general circulation covering the whole zone of mid-latitudes in the northern hemisphere. The results of diagnostic analysis are confirmed, on the whole, by theoretical studies of teleconnections between the processes in the middle and tropical latitudes. A major conclusion is that in order to understand the mechanisms of feedbacks, an account should be taken of the effect of both the moving groups of waves and unstable normal modes. Since the tropical convective systems seem to play an important role in other regions too (for example, in the Atlantic Ocean), the diagnostic studies are to be continued.

As satellite observations show, the extent of continental snow and sea ice cover is characterized by a considerable interannual variability. Having studied both synchronous and phase-shifted correlations, Walsh and Incek (1981) came to the conclusion about remote forcing on the topography of the 700 hPa surface, similar to that for the Pacific–North-American region. Therefore, an account of the extent of snow cover as a predictor improves the reliability of the monthly mean temperature forecasts for the USA territory, whereas the Eurasian snow cover affects the atmospheric circulation in China, where the thickness of snow cover is an important parameter.

The stability of the anomalies of the extent of ice cover is much higher than that of the meteorological anomalies. Therefore, major modes of the ice cover anomalies are manifested through the variability of meteorological parameters with a time shift of several months. The structure of the North Atlantic ocean-related circulation reflects the leading modes of ice variability in the North Atlantic Ocean. Thus, it is important to monitor the concentration, age and thickness of ice cover.

The problem of the effect of soil moisture anomalies on atmospheric circulation also claims attention. Its solution is hindered by the fragmentary character of observational data. On the whole, the conclusion can be drawn that correlations and large-scale circulations have several centres, sometimes located far from the regions of disturbances. Nevertheless, an analysis of long-range numerical modelling results shows that even without such disturbances the level of variability of the monthly means determined by external dynamics can be compared to that observed, i.e. there are long-lived circulation anomalies in middle latitudes.

A different situation is observed in the tropics where the simulated variability constitutes only a small part of the observed one, and some major modes of anomalies (the Southern Oscillation, for example) cannot be found at all. Apparently, this defect of numerical modelling of the tropical climate should be ascribed to the neglect of non-seasonal sea surface temperature disturbances, with the subsequent conclusion that the ocean surface conditions more strongly affect the tropical than the mid-latitudinal atmosphere.

Successful application of numerical interactive models to quantitative estimates of the role of the atmosphere–ocean interaction in the formation of short-period climate variations depends much on empirical analysis of the variability of heat content in the upper ocean layer, the heat transport by the ocean (contributions from horizontal and vertical advection) and the ratio between latent and sensible heat fluxes. In this connection, Namias (1983) analysed several characteristic cases of the effect of the atmosphere–ocean interaction in the North Pacific on large-scale atmospheric processes in North America observed during the last few decades.

Of primary importance is the fact that the upper oceanic layer as a heat source is characterized by intermonthly, interseasonal and interannual variabilities. For several years the temperature of this layer can be either above or below the normal (this situation creates the basis for climate forecasts). Observations for the North Pacific reveal a high spatial coherence of sea surface temperature anomalies and the temperature of the middle troposphere over the ocean.

With correlations calculated for the point 40° N, 150° W the structures of sea surface temperature anomalies and 700 hPa surface height are similar: large (small) values of heights correspond to high (low) sea surface temperature. Such anomalies can cover one third to one half of the North Pacific basin and remain for a month, a season and even several years. Sea surface temperature anomalies often reach depths of 100, 200 and 300 m and hence accumulate an abundance of heat which (under certain conditions) can be taken up by the atmosphere. These anomalies are most likely generated and maintained, first of all, by an anomalous atmospheric circulation.

Apart from spatial coherence, there exists also a temporal coherence of the anomalies of sea surface temperature and atmosphere, with sea surface temperature anomalies being an order of magnitude more persistent in time (and thus able to be considered a cause of climate changes). Anomalies of heat and cold near the surface, penetrating deep into the ocean, can remain deeply hidden in the summer (at a shallow thermocline) but are transported onto the surface by mixing processes in the autumn and winter. Despite a long life-time, anomalies can sometimes rapidly appear (for a month or a season). The atmosphere–ocean interaction is so strong that with prescribed sea surface temperature anomalies, one can calculate the anomalies of pressure topography (or pressure at sea level), using techniques of either multiple regression or eigen vectors. Thus, the possibility of forecasting sea surface temperature anomalies increases the predictability of atmospheric circulation.

Although, at present, reliable long-range forecasts can be made only for a month or a season, it is of interest to analyse the possibility of longer-range forecasts, taking into account the thermal memory of the ocean, which can affect the atmosphere for several seasons or even a year. An examination of monthly and seasonal sea surface temperature anomalies in the Pacific and Atlantic Oceans of the northern hemisphere reveals a slow evolution of the atmospheric circulation structure and a succession in the development of processes in time scales of months and seasons. Therefore, the possibility of forecasting the sea surface temperature evolution improves the likelihood of reliable forecasts of atmospheric circulation.

Having formulated the requirements on observational data needed to forecast short-range climate changes, Namias (1983) compiled a summary of some parameters (Table 5.1). During recent years, satellite observations of the Earth radiation budget components have become a new important source of information for diagnostic studies of climate (Kondratyev *et al.* 1988a).

**Table 5.1** Some important parameters needed to forecast short-range climate changes (from a month to a decade)

| Parameter | Region | Spatial resolution | Temporal averaging |
|---|---|---|---|
| *Parameters of primary importance* | | | |
| SST | Ocean basins | $5^\circ \times 5^\circ$ lat.–long. | 1 week to 1 yr |
| 700, 500 and 200 hPA surface heights | Hemisphere | $5^\circ \times 5^\circ$ lat.–long. | 15 days to 1 season |
| Pressure at sea level | Hemisphere | $5^\circ \times 5^\circ$ lat.–long. | 15 days to 1 season |
| Surface air temperature | Continent | $10^2$ km | 15 days to 1 season |
| Precipitation | Continent | $10^2$ km | 15 days to 1 season |
| Snow and ice cover extent (thickness, desired) | Hemisphere | $10^2$ km | 15 days to 1 season |
| *Parameters desired* | | | |
| Surface wind speed | Hemisphere | $5^\circ \times 5^\circ$ lat.–long. | 15 days to 1 season |
| Cloudiness | Hemisphere | $5^\circ \times 5^\circ$ lat.–long. | 15 days to 1 season |
| Soil moisture | Continent | $10^2$ km | 15 days to 2 seasons |
| Field of currents in the upper layer of the ocean | Ocean basins | $5^\circ \times 5^\circ$ lat.–long. | 15 days to 1 season |
| Heat supply in the upper layer of the ocean | Ocean basins | $5^\circ \times 5^\circ$ lat.–long. | 15 days to 1 season |
| Radiation and solar activity indices | World | $5^\circ \times 5^\circ$ lat.–long. | 15 days to 1 season |
| Atmospheric transparency indices (effect of volcanic eruptions) | World | $5^\circ \times 5^\circ$ lat.–long. | 15 days to 1 season |

The results obtained by Lau and Chan (1983) illustrate the possibilities of such an approach. In order to reveal the effect of sea surface temperature anomalies in the tropical Pacific on atmospheric circulation, not only in the tropics but also in mid-latitudes, these authors undertook an analysis of the variability of the spatial structure of convective clouds in the tropics, connected with sea surface temperature anomalies, based on satellite measurements of outgoing longwave radiation in the period from early 1969 to early 1978.

In the middle of this period (1976–1977), a positive sea surface temperature anomaly was observed in the eastern equatorial Pacific. After World War II, 3 large and 3–5 small heat episodes occurred, when sea surface temperature anomalies, averaged over the water basins of the central and eastern sectors of the equatorial Pacific, reached approximately 1 deg K. These anomalies had formed irregularly, had existed for a year and longer and had been connected with the ENSO event. From NOAA-4 data for the whole period under consideration, global maps have been drawn ($\pm 45^\circ$ latitude) of zonal mean total outgoing longwave radiation values retrieved with the use of direct outgoing longwave radiation measurements in the interval 10.5–12.5 $\mu$m, as well as similar maps of outgoing longwave radiation anomalies for individual seasons with a spatial averaging over a $5^\circ$ by $5^\circ$ grid (the outgoing longwave radiation anomaly maps reflect reasonably well the changes in the spatial structure of convective cloudiness as outgoing longwave radiation minima). These maps have been compared with time series of sea surface temperature anomalies and monthly mean outgoing longwave radiation anomalies at different locations of the eastern tropical Pacific.

The comparison has shown that a positive sea surface temperature anomaly was followed by simultaneous strong changes in outgoing longwave radiation. In the period of the anomaly, the area of convergence (and intensive convection) over Indonesia shifts eastward and combines with the well-developed ITCZ. The South Pacific convergence zone also combines with the Indonesian convergence zone, but develops more slowly and reaches a maximum intensity only a year after a sea surface temperature increase. By this time, the ITCZ returns to the state characteristic of the period prior to the sea surface temperature anomaly, and the South Pacific convergence zone shifts eastward, whereas the westward-shifting Indonesian convergence zone remains.

A numerical modelling of the effect of the equatorial sea surface temperature anomaly on atmospheric global circulation made by Webster (1984) (calculations were made based on the linearized complete system of equations in spherical coordinates) revealed that in this case the band of positive vertical velocity, similar in shape to the South Pacific convergence zone, moves northward and eastward from the region of the anomaly. Bearing in mind that during the years without any anomaly, the highest sea surface temperature is observed in the eastern Pacific, it can be supposed that the South Pacific convergence zone originates from this sector. During sea surface temperature anomalies the maximum sea surface temperatures are located far to the east; and this can explain the eastward motion of the South Pacific convergence zone in these periods. It should be emphasized, however, that convective cloudiness changes most, not in the region of maximum sea surface temperature anomaly, but on the opposite side of the Pacific Ocean.

These results testify to the great importance of a long series of radiation observations for understanding the nature of the interannual climate variability in the tropics. An analysis of the dynamics of the fields of the Earth's radiation budget and its components, which is very sensitive to variations in atmospheric general circulation, has opened up possibilities for a comparatively direct revealing of the forcing on the tropical atmosphere.

## 5.2 The theory of short-term climate changes

As Schlesinger (1988b, 1989a,b) noted, two basically different approaches are possible to climate modelling and climate forecasts. The analogue technique consists of developing a scenario for climatic dynamics in the future, based on reconstructions of the climates in the past and extrapolation of the results obtained into the future with the use of some mathematical approaches. One of the advantages of the analogue technique is that it is free of surprises, i.e. it can reproduce only the states of the climate system which had been realized earlier. On the other hand, in the case of the analogue technique it is impossible to retrieve all the parameters of interest, since not all of them are supplied with data referring to individual epochs. Initial data that serve as the basis for the learning data sets become more and more scarce further into the past, which also limits the possibilities of this technique. In this case also, the estimates of the effect of new factors (for example, anthropogenic effects) on the climate system are very much limited.

Mathematical climate models are physically-based mathematical descriptions of differing complexity, depending on the number of considered effects and the ways that they can be considered. In principle, a mathematical model permits a detailed simulation of the climate system, but practically it always remains approximate, being confined to a level of physical adequacy and computing possibilities.

Mathematical climate models can be classified into thermodynamic and hydrodynamic. Thermodynamic models give a real-time simulation of temperature variations, but either neglect the field of motion of the atmosphere and the ocean or take it into account in a very simplified form. Hydrodynamic models describe the fields of temperature and motion (as well as other parameters) and can reproduce these fields and the transformations of various kinds of energy in the climatic system. The latter is the principal feature of hydrodynamic models, impossible for other types of models.

Thermodynamic models are subdivided into energy-balance and radiative–convective models. Energy-balance models of zero dimensions are confined to a consideration of the Earth's radiation budget which serves as the basis for calculations of mean equilibrium temperature near the surface. One-dimensional models include a simplified semi-empirical description of the efficient meridional heat transport due to atmospheric and oceanic motions. Radiative–convective models include a consideration of such mechanisms as radiation transfer in the atmosphere, sensible heat transport, dry and wet convection and processes of water vapour transformation in the atmosphere, including cloud formation.

Hydrodynamic atmospheric general circulation models (AGCMs) are based on a four-dimensional simulation of hydro-meteorological variables using finite differences equations or an expression in a finite number of prescribed functions (spectral models). A summary of major prognostic and diagnostic variables considered in the present atmospheric general circulation models is given in Table 5.2.

Present atmospheric general circulation models have from 2 to 10 layers vertically, a horizontal resolution of several hundred kilometres and a time step from 10 to 40 minutes. The spatial and temporal resolutions of the models are interrelated,

**Table 5.2** A summary of major prognostic and diagnostic variables considered in the present AGCMs

| Prognostic variables | Diagnostic variables |
|---|---|
| Surface pressure | Vertical velocity |
| Temperature | Geopotential |
| Horizontal velocity[a] | Air density |
| Water vapour concentration | Cloud amount[b] |
| Soil temperature | Surface albedo[b] |
| Soil moisture | |
| Snow mass | |

[a] In spectral models, instead of horizontal velocity, the vertical vortex component and horizontal divergence are used as prognostic variables, and velocity is diagnosed from these variables.
[b] In some models the cloud amount and surface albedo are prescribed.

resulting in a rapid non-linear increase of required computer time with improved resolution of the model. The computation based on these models with the help of fifth-generation computers (of the type Cray-1 or Cyber-205) for 1 day takes from 30 seconds to several minutes.

A number of physical processes playing an important role in the formation of the climatic regime are not taken into account explicitly by the present atmospheric general circulation models. Such subgrid processes must be parametrized in some way or other, i.e. their effects must be expressed in terms of the variables that are described explicitly for the model grid points. The parametrized subgrid processes are enumerated in Table 5.3. Some atmospheric general circulation model parameters must be either fixed or described as boundary conditions; they are given in Table 5.4.

Ocean circulation models, constituting the second type of interactive models, comprise, in their turn, a hierarchy of models: a swamp model, a model of the mixing layer of a fixed thickness and a model of the mixing layer of varying thickness. In ocean circulation models, temperature, horizontal currents and salinity are prognostic variables, whereas density, pressure and vertical velocity are diagnostic variables.

The system's response to changes in its components can be estimated by prescribed superanomalies, to raise the level of the signal-to-noise ratio of the model itself. Here, to reduce the results to a normal level of anomaly, i.e. to that actually

**Table 5.3** Subgrid processes parametrized in present AGCMs

| |
|---|
| Turbulent heat, moisture and momentum transport between the surface and the atmosphere |
| Turbulent heat, moisture and momentum transport in the atmosphere due to dry and wet convection |
| Water vapour condensation |
| Radiation transfer |
| Formation of cloudiness and its interaction with radiation |
| Formation and melting of snow cover |
| Physics of moisture transport in soil |

**Table 5.4** Fixed parameters and boundary conditions in AGCMs

| |
|---|
| Radius, gravitational acceleration and rate of rotation of the planet |
| Solar constant and orbital parameters of the planet |
| Total mass and composition of the atmosphere |
| Thermodynamic and radiative constants for atmospheric gases and cloudiness |
| Surface albedo |
| The height of the surface above sea level |

expected, the respective transfer functions must be known. This approach serves as the basis for numerical experiments on studies of anthropogenic impacts on climate, when the present level of either substance or thermal energy releases is not enough to obtain statistically significant results.

Atmospheric general circulation models can be used successfully if the following requirements are met: (1) long runs of the models must be accomplished to provide their equilibrium states, (2) the results obtained must be statistically assessed, (3) sufficiently long runs of the models for the results obtained must be statistically significant, or have the possibility to demonstrate the statistical insignificance of the effects obtained, within the limits assumed in a given study, (4) different atmospheric general circulation models must be applied and (5) energy-balance and radiative–convective models must be used as special tools in studies, as well as to reveal physical causes of differences in modelling the same effects with the use of different models.

It may be recognized that the forcing effects of the variability of boundary conditions, determined by prolonged sea surface temperature anomalies and by variations in soil moisture, snow and sea ice cover extent, are major factors which must be taken into account in short-range (seasonal) climate change forecasts. As Miyakoda (1986) notes, of great importance here is a reliable prescription of the forcings under consideration. For example, even maximum negative sea surface temperature anomalies are small (about 3 °C) in comparison with the equator-to-pole temperature contrast (about 28 °C) or the annual sea surface temperature amplitude (5.5 °C). Though a persistent correlation between sea surface temperature and various meteorological parameters may remain for 1–2 seasons, the signal/noise ratio constitutes here only about 20 per cent.

The sea surface temperature anomalies affect the atmosphere mainly through evaporation (with subsequent latent heat release) and sensible heat exchange, which in its turn affects the conditions of water vapour condensation. From data for the North Pacific during the period 1947–1966, the temporal autocorrelation coefficients of sea surface temperature monthly means slowly decrease with an increasing time interval, which testifies to persistent sea surface temperature anomalies. This persistency is clearly seen in the analysis of similar dependences for empirical orthogonal functions. The results obtained show that sea surface temperature anomalies remain substantial even 6 months later and depend on the season. Anomalously warm and cold waters can, for a time, be screened by a thin layer of warm water in the summer and show themselves (on the surface) during the next winter. The spectral analysis of a 28-yr sea surface temperature series in terms of the empirical orthogonal functions revealed the prevailing low frequencies at energy concentrations for frequencies below 1 cycle/year, from which it follows that the external disturbance is characterized mainly by slow variability and therefore can be an important factor of interannual fluctuations in the atmosphere. Slow variability is also typical of the snow and ice cover extent.

Simulation numerical experiments have led to the conclusion that the disturbing forcings of sea surface temperature anomalies and soil moisture variations in the tropical latitudes contribute much to atmospheric variability not only in the tropics but also in middle latitudes, though of course local effects are stronger. The extra-tropical variations in soil moisture and cryospheric parameters tell upon local atmospheric variability (in these cases the scale of forcing varies depending on season and geographical location).

The propagation of the signal from the tropics to mid-latitudes (remote forcing) is determined by zonal (eastern and western) wave fluxes in the upper troposphere (the 200–300 hPa level). In the case of westerly transport the effect propagates at a greater distance, forming global-scale waves. The forcing of boundary conditions is small, which results from weakly manifested respective processes (in this connection, fluxes in the boundary layer and latent heat release in cumulus clouds require thorough analyses).

The atmosphere–ocean interaction in the tropical belt most strongly affects the seasonal and annual variability, since in this case the characteristic time of the prevailing dynamic process in the ocean is comparatively short. If this process is assumed to be connected with the propagation of the Rossby–Blinova baroclinic waves, an increase of phase velocity with decreasing latitude must be taken into account. Whereas at 80° N latitude the time of crossing the Atlantic Ocean by baroclinic waves (7500 km) constitutes 9 yr, at 15° N it decreases to 2 yr and near the equator it is only a year. One more timescale of the atmosphere–ocean interaction is determined by 1-dimensional processes (evaporation, condensation, wind stress and turbulent mixing), constituting about a month in mid-latitudes and somewhat less in the tropics. Hence, from the viewpoint of horizontal heat redistribution, the equatorial zone is a region where ocean dynamics play an important role.

Analysis of data of the sea surface temperature annual change in the equatorial Pacific shows that a westward migration is observed of the warm phase at a phase velocity consistent with that of the Rossby–Blinova wave captured by the thermocline. Results of ocean circulation numerical modelling suggest the same conclusion.

The ocean–atmosphere interaction in the tropics is manifested through an El Niño event, connected with the Southern Oscillation and, probably, with the Indian monsoon. This event consists of an anomalous heating of the upper layer of the Pacific Ocean near its eastern boundary (near the South American coastline), which takes place in some years, with a subsequent westward propagation of warm waters. The period of the sea surface temperature quasi-periodic variability is about 4.5 yr.

The westward propagation of a positive sea surface temperature anomaly is followed by the development of anomalous northern winds across the ITCZ, the latter shifting to the south and intensifying precipitation near the equator. The nature of El Niño is probably connected with a chain of processes starting with an attenuation of the trade wind circulation. The internal Rossby–Blinova waves, stimulated by the Kelvin waves reaching the eastern coastline of the ocean, determine a water temperature increase near the American coastline.

In order to substantiate seasonal forecasts, the following further studies are needed: (1) to draw maps of soil moisture anomalies to improve reliable assessment of the role of this type of forcing, (2) to improve atmospheric general circulation models to assess an exact contribution of the disturbances of the boundary conditions and to raise the signal/noise ratio (in this connection it is important to test the hypothesis of unpredictability of climate changes in the central parts of the continents), (3) to reveal the causes of mysterious time shifting between sea surface temperature anomalies in the tropics and air temperature variations in mid-latitudes, reaching 1–2 seasons, whereas numerical modelling data illustrate a rapid propagation of disturbances and (4) to search for connections among El Niño, the Southern Oscillation and the Indian monsoon (i.e. a more thorough analysis of the nature of El Niño).

Shukla and co-workers (Shukla and Mintz 1982, Shukla and Wallace 1983, Shukla and Fennessy 1988) compared the results of observations with the results of numerical climate modelling in the tropical and mid-latitudes, bearing in mind an assessment of the simulation adequacy for (1) the climatic parameters averaged over different time periods, (2) the spatial and temporal fluctuations within a season, (3) the interannual variability of monthly means (seasonal, annual means) and (4) the interannual variability of the intraseasonal variations. Emphasis has been placed on the low-latitude climate, and the respective numerical modelling results have largely been obtained by using the Goddard Laboratory for Atmospheric Sciences (GLAS) nine-level global climate model.

An analysis of the averaged fields of the basic meteorological parameters (atmospheric pressure, geopotential, temperature, wind, rain rate) calculated for February and July at fixed (from observational data) conditions at the surface level (sea surface temperature, soil moisture, snow and ice cover distribution) has shown that the model gives a realistic pattern of large-scale atmospheric general circulation, both in the winter and in the summer; nevertheless there are substantial differences from the observational data. So, for example, the February sea-level pressure field realistically reflects the existence of the Aleutian and Icelandic minima in the northern hemisphere, but gives an erroneous structure and intensity for the Siberian anticyclone. The most serious differences appear in the southern hemisphere (south of 60° S), where the calculations suggest that the atmospheric general circulation is a large-scale vortex structure, whereas the observations show that the pressure field is zonal.

In July thc high-pressure cells over the North Atlantic and the monsoon minimum are adequately presented, but the orientation of the pressure maximum in the northern Pacific does not correspond to that observed. The location of the February zonal wind maxima in both hemispheres is reliably predicted, but the observed closed maximum near 200 hPa does not manifest itself. The ITCZ is reliably reproduced, as are, in particular, its longitudinal variations. Bearing in mind the strong spatial variability of precipitation and serious observational errors, the calculated and observed fields of precipitation can be considered satisfactory; the zones of extreme precipitation are reliably reproduced. The largest differences are found in the overestimated precipitation in February over the Tibet plateau and in July in the eastern sector of North America.

Since its ability to simulate the atmospheric general circulation variability is the most important criterion of an AGCM's adequacy, Shukla (1986) analysed the results of calculations of the variability of (1) long (low-frequency) planetary waves, (2) the parameter of local temporal structure of disturbances (local mean standard deviation (MSD) of fluctuation periods within 2.5–6 days) connected with frequency, intensity, maturing and disintegration of cyclones and (3) the cases of blocking (with the use of data for GLAS and GFDL models). The calculated values of the parameter of local temporal structure values agree well with those observed in the winter, from the viewpoint of both localization and intensity of the principal regions of cyclonic activity in the northern central Pacific and in the western Atlantic. In the summer, the Atlantic maximum of the parameter of local temporal structure is reliably reproduced, but the Pacific maximum was found to be too weak and shifted to the south. On the whole, the geographical distribution of the variability was simulated fairly reliably, but the amplitudes and vertical structure of the variability differed considerably from those observed.

The ability of atmospheric general circulation models to simulate the intraseasonal variability in mid-latitudes is characterized by reliably described synoptic-scale disturbances, but comparisons of low-frequency (filtered) variability revealed substantial differences. Of particular interest is a description of the cases of blocking, i.e. such situations when the atmospheric general circulation structure in certain regions of mid-latitudes is characterized by a strong stability during time periods longer than the life-times of individual cyclones but shorter than a season. So, for example, the long (7–10 days) anomalies of geopotential (more than 100–300 m) appear in the Pacific (west of the Rocky Mountains), in the Atlantic (west of the Alps and Scandinavian mountains) and west of the Ural Mountains in the former USSR, the local structure of blocking being very similar in all three regions. Shukla (1986) suggested that the cases of blocking could be identified as the situations when, in each of the grid cells in the zone of anomaly of 500 hPa height (determined as a deviation from the seasonal mean change), the amplitude of anomaly larger than 100 m remained stable for 7 days and longer.

Analysis of the numerical modelling data (GLAS model) for 17 winter and 7 summer seasons revealed that the calculated geographical distribution of maximum frequency of occurrence of the blocking cases agrees poorly with that observed (in particular, the calculations underestimate the intensity of the blocking high-pressure ridges). On the whole, the GLAS model cannot reliably simulate the location, intensity and duration of the blocking phenomena, which can be explained by an inadequate spatial resolution and approximately calculated vertical profile of zonal winds (at high altitudes, in particular). Although the role of the slowly varying effect of the boundary conditions (sea surface temperature, snow and ice cover distribution, etc.) in the formation of the blocking situation remains unclear, a prescription in the model of the disturbing effect of boundary conditions from the climatic data may be unacceptable. An important aspect is an inadequate account of the interaction between the zonal flow and orography. Naturally, the inability of the climate model to simulate reliably the blocking cases is a very important problem, since the forecasting of such stable situations is the most promising.

Even with prescribed forcings of the boundary conditions from climatological data, the interannual variability of the monthly mean climatic parameters in mid-latitudes is satisfactorily reproduced, but in the tropical band it is underestimated by a factor of two. Such a result suggests that the internal dynamics of the climate system is the most important factor of the interannual variability of the mid-latitude climate, whereas in the tropics the forcing due to the boundary conditions (sea surface temperature) plays an important role. Similar results were obtained from numerical modelling using the GLAS model. The latter conclusion has been verified experimentally, with varied sea surface temperature field in the band 0–30° N.

As for the interannual variability of the intraseasonal variations, there are still no data from numerical experiments which would make it possible to analyse this variability.

Summing up the numerical experiments on the climatic sensitivity to disturbances in the boundary conditions, Shukla (1986) formulated the available results in the following way.

1. Sea surface temperature:
   (i) A strong effect of the tropical Pacific sea surface temperature anomalies on the intensity of the Walker and Hadley circulation cells in the tropical latitudes and on the quasi-stationary atmospheric general circulation structure in mid-latitudes has been revealed.
   (ii) In the presence of positive (negative) sea surface temperature anomalies in the northern (southern) equatorial Atlantic, droughts appear in north-eastern Brazil.
   (iii) Negative sea surface temperature anomalies in the Arabian Sea and/or positive anomalies in the equatorial Indian Ocean are followed by weakened monsoons in India and adjacent regions.
   (iv) Sea surface temperature anomalies in the northern Pacific affect the quasi-stationary waves in mid-latitudes, but in view of a strong internal mid-latitude atmospheric general circulation variability it is difficult to reveal this effect.
   (v) An effect of sea surface temperature anomalies in the Atlantic Ocean (northern hemisphere) on the atmospheric general circulation in Europe is observed.
2. Soil moisture and albedo:
   (i) Variations in soil moisture affect evaporation and heating of the soil, which lead to changes in local precipitation and circulation.
   (ii) Surface albedo variations (especially near the sub-tropics and desert boundaries), which cause a decrease of precipitation and evaporation, affect the atmospheric general circulation.
   (iii) Large-scale deforestation and irrigation measures can change the local hydrological cycle and atmospheric circulation.
3. Snow cover:
   (i) The vast and stable anomalies of snow cover extent cause local variations in surface air temperature. The snow cover anomalies can also affect the structure of trajectories of long waves and cyclones due to a high snow albedo and strong baroclinicity of the atmosphere near the snow-covered surface.
   (ii) With a large snow cover extent in Eurasia in the winter, a deficit of monsoon precipitation is observed in the zone of the summer Asian monsoon.
   (ii) The growth of the snow cover extent in North America is connected with quasi-stable circulation anomalies.
4. Sea ice:
   (i) Fluctuations of the ice cover extent and ice concentration cause local changes in the ocean–atmosphere heat and moisture exchange.
   (ii) Large anomalies of the sea ice extent cause strong variations in the polar

atmosphere circulation, and sometimes they can affect the atmospheric general circulation in middle and sub-tropical latitudes.

In connection with the principal unpredictability of high-frequency disturbances, the problem appears of frequency filtration and scale-to-scale energy transport. Analysis of the transformation of kinetic and available potential energy in the process of interaction between the components and different zonal wavenumbers, performed by Saltzman (1983), has made it possible to select the prevailing scales of long waves which transmit their kinetic energy to the waves of other scales. The principal feature of this process in the atmosphere is the presence of spectral regions of negative viscosity (the energy transfer from small scales to larger ones), as well as a spectral region of large-scale motions, across which the energy flux is zero.

It appears that the components of atmospheric motions with fluctuation periods much shorter than the duration of the forecast cannot be predicted from initial data with the required accuracy. Therefore, it is extremely important to establish in the multi-frequency system 'atmosphere–ocean–continents' the presence of eigenvectors with periods comparable to, or even exceeding, the time period over which an integration is made.

Such eigenvectors of the system are called the slow variables, in contrast to the high-frequency fast variables. In any problem requiring a simultaneous consideration of both slow and fast variables, an averaging has to be made, which enables one to avoid errors by rejecting the possibility of forecasting the fast variables. An alternative is the filtration of high frequencies by averaging the solution obtained, and then comparing the results with observational data.

One of the non-trivial consequences of this approach has been a detection of natural frequencies, i.e. frequencies of very slow free fluctuations of the atmosphere isolated from the oceans and continents. It transpired that an account of the initial phase of planetary waves is most important with large values of the circulation index, i.e. in cold seasons.

A separate complex of problems is connected with a study of possibilities to apply the theory of non-linear dynamic systems and non-equilibrium thermodynamics to the problems of long-range forecasts. In particular, the calculations point to a possible existence of such regimes of the climatic system when it reaches the stable state after long-period irregular oscillations. It is this paradoxical problem that one faces in long-range forecasts, when nothing definite can be said about the system's behaviour in the very near future, but nevertheless a search for an asymptotic regime might be worthwhile. However, the theory of such phenomena and of the thermodynamics of irreversible processes has been developed only for simple systems; possibilities to generalize the results for such a complicated system as the climatic one have, so far, been no more than the subject of speculative consideration. The situation is much better in the case of simple climate models.

As assessment of the reliability of the empirical prognostic relations is the principal problem of statistical methods for long-range forecasts. At present, independent of the analysis technique applied, it is practically impossible to provide a representative, statistically homogeneous learning set which would guarantee the statistically verified conclusions. A search for correlations based on the limited learning sets, at a given confidence level, may lead to an appearance of false correlations because the statistically significant correlations can be obtained even with a random distribution of the parameters in the sets. The probabilistic character of the statistical forecast produces a problem of an assessment of its significance based in independent criteria: for example, the risk connected with forecasting extreme situations.

The need to search for adequate learning sets for statistical forecasts has given rise to studies of multi-year series of meteorological data, in which the latter are considered as a probabilistic process with a spectrum different from both red and white noise. If it has been assumed earlier that the climatic fluctuations are not that significant and that their account does not improve markedly the quality of long-range physico-statistical forecasts, the point of view that currently prevails is that the long-range forecast must be coordinated with the forecast of the climatic background.

Of great importance is an analysis of the results of comparisons of the functioning of different atmospheric general circulation models. As Gilchrist (1983) noted, such comparisons have usually been concentrated on simulation of mean conditions for a prescribed month – usually for January or July. Little attention has been paid to studies of the day-to-day variability of the model results. This variability is, however, very important in numerous applications, including long-range weather forecasts.

An appearance of the blocking situations is a relatively important aspect of the variability of atmospheric processes. The ability of atmospheric general circulation models to predict the development of such situations, both individually and climatologically, is an important preliminary condition for their application to month-to-season forecasting.

The long-range forecasts have two aspects: (1) a systematic response of the atmosphere to different forcings (boundary conditions and their variations are an example of such forcings) and (2) the internal variability of the atmosphere and its reflection in the properties of the model itself. These aspects are interconnected in the sense that variations in the boundary conditions and their duration can be determined by the internal variability of the atmosphere. It is simpler, however, to model such effects as external with respect to an object under study (the atmosphere).

In this connection, long-range forecasting can be interpreted as an assessment of systematic deviations of atmospheric parameters from the climatological mean, masked with the uncertainty caused by the internal variability (noise) of the atmosphere. As the range of forecasting grows, the uncertainty due to noise and uncertain initial conditions increases, until it is equal to the complete climatological variability; in other words the atmosphere loses its sensitivity to the initial conditions. The respective time interval in general does not coincide with the limit of deterministic predictability, which is determined by the time for saturation of the errors of the energetically active system.

Numerical experiments with an atmospheric general circulation model, aimed at assessing its internal sensitivity, consist of an integration of the model for several years, with the prescribed cycles of the principal boundary conditions (solar radiation, sea surface temperature, etc.). In this case, a comparison of model results obtained for similar calendar months of different years of the run gives a measure of the internal variability. On the whole, the internal variability of the extra-tropical part of the models is approximately on the same level, corresponding to the natural variability of the mid-latitude atmosphere, provided anomalous external forcings are absent. The most important conclusion from this result is that, at the present stage, it is practically impossible to use atmospheric general circulation models to make a forecast for a year, and to solve this problem further studies are needed, including studies of the internal variability of the models themselves.

As an example of individual anomalies, Gilchrist (1983) analysed the ability of the ECMWF community model to simulate the blocking situations in the atmosphere. This ability, along with the respective level of statistical significance of the results, is very important in the case of long-range forecasts. A comparison of the frequency and spatial distributions of the blocking situations in the atmosphere and in the model runs shows the necessity to formulate more clearly the criteria for assessing the blocking phenomenon itself and to clarify what mechanisms (including the forcing boundary conditions) are the key ones for an adequate simulation of the blocking in modelling the general circulation. In a partial case of blocking (for example, the northern mid-latitude droughts), it has been demonstrated that a mechanism exists for self-maintenance of this state at the expense of changing the soil moisture balance and the respective cloud and albedo effects.

Grotch and MacCracken (1988) undertook a comparison of four 3-dimensional climate models (NCAR, GFDL, GISS, OSU) in order to analyse the simulation of the present global mean and regional climates (with respect to that observed) as well as to assess climate changes due to $CO_2$ doubling. The surface air temperature and rain rate have been considered as climatic parameters. Table 5.5 contains the data characterizing the results of temperature calculations for winter (December to February) and summer (June to August). Similar efforts have been further developed later (Cess and Potter 1987, 1988, Cess *et al.* 1989, 1990, 1992). As can be seen, the differences between global mean surface air temperatures do not exceed 2 °C. However, of interest is a substantial (about 50 per cent) overestimation of the amplitude of the surface air temperature annual change in the case of the NCAR model compared with that observed. With the decreasing size of the region of averaging, the difference between calculated and observed surface air temperatures increases, especially in the northern summer, reaching maxima from −4.8 to 6.8 °C; thus, the range of differences constitutes almost 12 °C. Such differences greatly exceed the supposed climate warming due to $CO_2$ doubling.

The capabilities of the models from the viewpoint of simulating the observed precipitation are even more limited. According to two observational series available for the northern hemisphere, the rain rates in the winter average 0.53 and 0.63 mm day$^{-1}$, whereas calculations gave 0.97–1.76 mm day$^{-1}$. In the summer, this range is 1.51–2.34 mm day$^{-1}$ and observations give 1.49 and 1.89 mm day$^{-1}$. As in the case of surface air temperature, the differences increase substantially when the region of

**Table 5.5** Numerical modelling of SAT fields (the summertime SAT values are given in brackets)

| Model | World | North America | USA | Mid-western USA |
|---|---|---|---|---|
| Observed SAT (°C) | 8.5 (13.9) | −5.8 (18.9) | 0.9 (23.0) | −1.5 (23.0) |
| SAT difference (calculated − observed) | | | | |
| NCAR | −1.6 (1.3) | −0.3 (6.0) | −2.1 (6.3) | −0.5 (6.8) |
| GFDL | 1.5 (−0.2) | −1.8 (0.6) | −0.8 (0.1) | −1.3 (3.7) |
| GISS | 0.8 (0.4) | −0.5 (−3.1) | 0.0 (−4.5) | 1.1 (−4.8) |
| OSU | 0.3 (−0.6) | 0.5 (−2.2) | −0.6 (−2.2) | −1.0 (−1.6) |

averaging diminishes. Naturally, understanding the causes and eliminating such differences constitute the most important problem of climate theory; without solving this, reliable climate prediction is impossible. So far, comparisons with observational data show that none of the models can be considered universal, i.e. useful in every condition. Therefore, one can count only on generalized estimates of possible ranges of variability of the climatic characteristics.

Table 5.6 shows the results of calculations of the possible changes in surface air temperature due to $CO_2$ doubling. It follows that the level of the global mean warming varies between 3 and 4.5 °C (although, according to the model of the UK Meteorological Office, it exceeds 5 °C), but as the region of averaging becomes smaller, the range of differences broadens up to 6.5 °C, as in the case of the contiguous US territory (i.e. the USA without Alaska and Hawaii). As for predicted rain rate, although in this case there are some similarities between the results of calculations and observations (for example, the greatest rain rate is observed in the tropics and high latitudes), the spatial distributions of rain rate have little in common. So, for example, for the USA, the annual mean growth of rain rate varies from 2 per cent (in conditions of temperature increase, this corresponds to the drought regime) up to 15 per cent (the regime of weak moistening and possibly an intensification of river run-off). Particularly substantial are the differences between the models in the calculations of climate change characteristics (in particular, the repeatability of extreme events), which are of great practical interest. Sometimes, the differences appear even in the assessments of the sign of variability.

In their analysis of the errors in some current atmospheric general circulation models, Barridge and Sadurny (see Kondratyev and Prokofyev 1983) pointed to the following probable reasons for the appearance of errors: (1) neglecting the hot bands of $CO_2$ in the 15 $\mu$m range, which is equivalent to an additional cooling by about 0.25 deg K day$^{-1}$ at a level of 30 km near the winter pole, (2) an inadequate simulation of the radiative properties of water vapour and carbon dioxide at low temperatures and small optical thicknesses, (3) an assumption of blackness for cirrus clouds, (4) the interaction between radiation and dynamics leads to a positive feedback, with an excess cooling and (5) an appearance of fictitious radiative heat sources at the upper levels of the models (in the stratosphere), connected with an artificial referencing to them of the processes taking place in real conditions at high altitudes.

In addition, the current atmospheric general circulation models are characterized by some potentially weak features connected with the necessity of a more reliable account of the following factors and features of model realizations: (1) the scales of variability of the fields of kinetic energy and geopotential, (2) tropospheric temperature averaged both in time and zonally, (3) the position and structure of the sub-tropical jet streams, (4) the statistics of the trajectories of mid-latitude cyclones,

**Table 5.6** Changes in SAT due to $CO_2$ doubling

| Model | World | North America | USA | Mid-western USA |
|---|---|---|---|---|
| Temperature difference $(2 \times CO_2) - (1 \times CO_2)$ | | | | |
| NCAR | 3.72 (3.30) | 3.74 (2.85) | 3.10 (1.99) | 3.32 (3.36) |
| GFDL | 4.50 (4.06) | 5.65 (5.64) | 4.76 (6.49) | 5.04 (6.65) |
| GISS | 4.64 (4.33) | 5.55 (3.35) | 5.14 (3.7) | 5.57 (3.80) |
| OSU | 3.19 (3.03) | 3.12 (3.33) | 3.35 (3.56) | 3.54 (3.64) |

(5) the 3-dimensional structure of the tropical circulation (including the intensity of the Hadley and Walker cells, as well as their sensitivity to the processes on the surface, in the boundary layer and to the schemes of convection parametrization) and (6) the 3-dimensional structure of the flux of momentum as well as of eddy heat and moisture exchange.

The results discussed above suggest that at present an adequate observational data base has been accumulated which testifies to the existence of climatological anomalies whose amplitude exceeds the noise level due to internal dynamic variability. A deeper understanding has been achieved of the physical mechanisms responsible for such anomalies, positive results have been obtained in their modelling, and the possibility has been demonstrated of long-range forecasts on this basis. All this refers, however, mainly to a time scale of one month. This provides a reliable physical basis for further studies on forecasting the monthly and seasonal mean meteorological parameters. In order to understand the nature of the variability of atmospheric processes and to assess the reliability of numerical climate models, further diagnostic studies are needed based on statistical data on the circulation observed.

The atmospheric general circulation variability manifests itself mainly in the large-scale spatial structure of the fields of meteorological parameters. Studies have shown that the spatial distributions of amplitudes of the long-term fluctuations of meteorological parameters depend weakly on frequency and are characterized by clear local maxima. Such centres of action correspond to the prevailing geographical coordinates of long-lived circulation anomalies. The persistence of large-scale long-term fluctuations in the spatial and temporal spectra determines the possibility of forecasts for periods of a month and season.

In this connection, such characteristic manifestations of anomalies as teleconnections, blocking, cycles of indices and monsoons are of particular concern. The nature of teleconnection phenomena is still unclear, for example, and the statistical evidences and stability of relevant empirical results require further analysis of longer time series of data. The attempts to use the cycle of index for monthly forecasts have proved abortive so far. The blocking is, apparently, one of the manifestations of teleconnections determined by the trigger mechanism of the effect of sea surface temperature anomalies (more specifically, of the heat content of the oceanic mixed layer). At the same time, the statistics of these phenomena are dependent on the criteria used in an objective analysis, which determines the necessity for further improvement of these criteria. The case-to-case variability of the blocking manifestations is very great and this necessitates the development of simple methods for their description and classification based on analysis of long observational series. Systematic empirical studies in these directions are needed which are directly connected with the problems of the Sections programme (see Section 5.3).

Concentrated efforts are needed on studies of the spatial and temporal variability of the sources of diabatic heating, including sensible heat, latent heat and radiation, and also of theoretical analysis with the use, in particular, of 4-dimensional data assimilation.

An analysis of autocorrelations of atmospheric variables, manifested in reproducibility, testifies to the atmospheric memory whose temporal scales increase sharply in the case of averages. This makes it possible, principally, to search for historical analogues for predictions for month-to-season periods. The principal use of this approach requires, however, much longer data series.

The theoretical limit to deterministic predictability is 1–2 weeks for synoptic-scale processes and, probably, up to 4 weeks for planetary-scale events. At the same time, the monthly forecasts are somewhat more successful in cases of anomalous boundary conditions. Some of these conditions provide better predictability, which points to prevailing types of fluctuations. Both observational data and numerical experiments with atmospheric general circulation models show that the principal parameters of boundary conditions on the surface, determining the possible anomalies of atmospheric circulation, are sea surface temperature, soil moisture content, snow cover extent and sea ice edge location.

The sea surface temperature variations should be considered as a decisive factor which determines the variability of the state of the tropical atmosphere. Data of teleconnections between sea surface temperature anomalies and the blocking in mid-latitudes are also well known. The best is the sea surface temperature anomaly reproducibility which reaches 6 months and longer, though sea surface temperatures are characterized by the annual change. In this connection, of particular concern is the concept of the energy-active zones of the ocean that serve the basis for the Sections programme (see Section 5.3).

Considerable experience of using operational (empirical, as a rule) monthly and seasonal forecasts has been gained. The techniques for such forecasts are based on an account of (1) experience of synoptic predictions, (2) stability of relevant predictors, (3) empirically established phase shifts between sea surface temperature anomalies (and other boundary conditions) and atmospheric parameters, (4) analogues, (5) the phases of such phenomena as the quasi-biennial cycle and the Southern Oscillation, (6) teleconnections, (7) solar cyclicity (11-yr and 22-yr cycles, in particular) and (8) the results of retrospective spectral analysis of the time series of the predictant.

The existing techniques for monthly forecasts enable data to be obtained on sea surface temperature monthly means, with variability exceeding the climatological one only by 10 per cent, with a weak reproducibility; whereas, for such parameters as precipitation or sunshine duration, there are almost no substantiated approaches. All the available results are confined to mid-latitudes. Of special importance is a comparison of the results of hydrodynamic modelling with observational data and testing the reliability of statistical interconnections based on independent data series.

The statistical – dynamic approach is of continuing importance, with a variety of applied techniques (numerical methods, qualitative theories of dynamic systems, non-linear mechanics techniques, adjoint hydrodynamic equations, etc.). Major difficulties exist in the parametrization of the sources and sinks of energy, boundary conditions, internal forcings and effects of microturbulence.

The results of numerical experiments on the use of atmospheric general circulation models for monthly forecasts are rather controversial. There are, however, promising indications which point to the feasibility of such forecasts even without account of the boundary conditions. An improved parametrization of subgrid processes is a further step towards advanced models. At the same time, an atmospheric general circulation model can generate a shifted climatology of its own, which is a complicated problem.

From the viewpoint of strategy, the problems of an accurate prescription of the initial state and of the maintenance of the dynamic balance in the model must be solved. Atmospheric general circulation model experiments must be aimed, in particular, at solving the problems of validation through (1) comparing model results with observational series describing the low-frequency atmospheric processes and (2) diagnostic analysis of the results to clarify the comparative significance of different mechanisms determining the atmospheric variability on the chosen time scales.

Despite an adequate simulation by atmospheric general circulation models of the large-scale features of the observed circulation, systematic errors limit their use for climate forecasts 1 to 10 yr ahead. In this connection the reliability of the models must be tested in at least two directions: the reliability of the input parameters, based on comparisons with the observational data in definite regions, and the reliability of the results of model runs. The model results must be widely tested, including an estimation of the reliability of the variables with the use of moments of secondary and higher orders. With potential possibilities of using the models for climate predictions, a new series of intercomparisons of models is needed.

Possibilities now appear to assess the complexity of atmospheric components of the models needed to forecast a year to 10 yr ahead. As for the hydrodynamics of the processes, the present models require a deeper insight into the problem of non-linear relations between the scales resolved and not resolved by a model.

In particular, the applied parametrization of the subgrid turbulence cannot be considered accurate. The information about the sources, sinks and transport of energy and potential entropy in various regimes of flows is inadequate. This information is needed for both understanding the spectral cutting results and an adequate parametrization of the subgrid turbulence. So far, an idealized quasi-geostrophic turbulence theory remains the only approach to these problems.

The necessity of taking into account the orographic effects makes the traditional sigma-system of coordinates less and less suitable for application. With the use of a coordinate system following the profile of the Earth's surface, the errors connected with sharp changes of the profile decrease, on the whole, with increasing horizontal resolution. However, in the vertical the situation can be reversed, and additional studies are needed to exclude steep slopes from the models. In connection with the problem of orographic effects, of the first priority are the Alpine Experiment (ALPEX) data which can serve as the basis for verification of new proposed techniques.

Another problem in analysis of the vertical propagation of energy is an accurate realization of numerical schemes, to avoid an artificial reflection of the planetary-scale Rossby–Blinova waves from the wall at the top of the atmosphere. As shown by the linear analysis of baroclinic instability, a rapid fictitious growth of the subsynoptical-scale waves is also possible at high resolutions.

For a climate forecast more than a year ahead, a strict account of the atmosphere–surface interaction is needed. The total response of the atmosphere to surface conditions is governed by a complicated system of positive and negative feedbacks. Therefore, the parametrization of processes in the boundary layer remains one of the most difficult problems of large-scale atmospheric circulation modelling.

An idealized presentation of snow cover in the first atmospheric general circulation models through run-off and water content of the soil surface layer gave considerable errors in assessing the effects of snow cover in the early and late winter. A detailed description is needed of the thermodynamics of the formation and melting of snow cover, with soil freezing taken into account.

In snow albedo parametrization the surface inhomogeneity (shadowing by vegetation and the dependence on the orientation of slopes) must be considered, as well as the effect of aerosol pollution and changes in the crystalline structure of snow. To substantiate these parametrizations, information is needed on such surface state parameters as snow cover depth, its temperature regime, including vertical temperature profiles, phase composition of soil moisture and meteorological parameters.

In the current models, sea ice is either prescribed or calculated from energy budget relations. As a rule, the ice drift is neglected. Present atmospheric general circulation models adequately describe the annual course of sea ice in the Arctic but only poorly describe it in the Antarctic. Experiments are needed on the sensitivity of climate models to their mode of description of the behaviour of sea ice. The parametrization of the share of the hidden water is most difficult at present. In addition, the values of sea ice albedo (depending on snow covering the ice) and coefficients of the transfer across the ice – air interface must be specified.

In a free atmosphere the emphasis must be placed on the parametrization of processes of condensation and cloud formation as well as the cloud–radiation interaction. Although the condensation and related processes of latent heat release, as well as cloud formation and precipitation, are traditionally considered most important in the climatic system, it is not clear, so far, how realistically the present models predict the distribution of latent heat and precipitation. Model data on global mean precipitation are not enough here, because even if the mean is accurately estimated, the geographical distribution must be known. Model descriptions of the condensation process are inadequate because the model neglects the process of cloud formation (cloudiness is either prescribed from climatological data or derived from simple dependence on relative humidity). Much more adequate observational data on clouds and precipitation are needed to verify new techniques for parametrization. Satellite observations are the most likely source of such data.

The problem consists in developing improved schemes of cloud parametrization in high-resolution models which would explicitly consider the water supply and cloud cover extent as prognostic variables. The required empirical data must largely be supplied by the ISCCP within the WCRP.

In climatic forecasts of a year to 10 yr ahead, a major objective is predicting the regional and interannual variabilities produced by anomalies of basic meteorological parameters. The possibility of these forecasts is ensured by the long memory of oceanic processes in both the tropics and extra-tropical regions as well as by comparatively slow processes connected with soil moisture, the state of vegetation and snow cover, the extent of sea ice, etc. In addition, external forcings may play an important role, being connected for example, with volcanic and anthropogenic aerosols, forest decline, trends in trace gases, variations in the incoming solar radiation, etc. Some of these processes are the sources of noise in the climatic system, whose amplitude is another prognostic aspect of interest.

In modelling these various processes, most important is the simulation of the behaviour of the upper 1 km layer of the ocean, as well as a large-scale atmosphere–sea-ice interaction. Interactive ocean–atmosphere models are the basic way to describe these processes, a simplified description of deep oceanic layers being possible. The simplified climate models are useful in preliminary studies and assessments of the role of individual physical processes. Of great importance are diagnostic studies, whose results serve as the basis for assessing the natural variability of the processes, revealing extreme situations, and developing techniques for empirical and statistical forecasting.

When the timescale of the variability of climatic processes exceeds one year, the atmosphere often becomes a stochastic system, i.e. it completely loses memory of its initial state. Therefore, it would be important to develop a stochastic atmospheric model which would directly provide statistical information, instead of calculating it on the basis of averaged results of continuous integration of a detailed circulation model. In other words, it must be a model of macroturbulence which would give average fields and fluxes, i.e. first and second moments of basic meteorological parameters. The difficulty in developing such models is that the evolution of each of the moments depends on the moments of higher orders. Even the latest turbulent models of large-scale atmospheric dynamics are still unrealistic; they present the atmosphere as spatially homogeneous and almost isotropic. In fact there is, for example, a longitudinal spatial inhomogeneity of climatic characteristics, which is connected with the spatial distribution of the continents. An account of such inhomogeneities in the calculation scheme describing the turbulence unacceptably increases the volumes of calculations which need simplification. One of the possibilities is to close the system of equations by introducing equations for the evolution of second moments, which use numerous tuning parameters. The difficulty is that, in contrast to modelling the microturbulence, with this description of global macroturbulent processes, reverse energetic cascades or zones of negative viscosity appear. Hence, preliminary studies are needed on the realization of this approach.

Simple climate models are still very important. They can most efficiently be used in studies of some feedbacks in the global climate system: assessments of the sensitivity of some climatic characteristics to natural or anthropogenic forcings, diagnostic modelling to analyse the observed or simulated climate changes, and empirical estimates of sensitivity and the development of new concepts. Simple climate models can also be used (1) to describe mechanisms of thermal inertia of the climatic system, (2) to develop parametrizations for the atmospheric general circulation models, (3) to model the responses of the climatic system to natural and anthropogenic forcings (volcanic aerosol, $CO_2$, minor components, etc.) and (4) to model the physical mechanisms of the variability of individual climatic parameters.

## 5.3 The Sections programme

To study the large-scale atmospheric processes responsible for weather and climate changes, we must consider the phenomena occurring in the global atmosphere not only in its interactions with the landmass, but especially in its interactions with the ocean which acts as a gigantic heat reservoir. Attention to the atmosphere–ocean interaction (manifesting itself in heat, water and momentum exchange) is of key importance for the solution of the problems of long-range forecasts of weather and climate changes.

The system of adjoint equations of thermohydrodynamics and the specially developed theory of perturbations, used by Marchuk (1992) and Marchuk *et al.* (1986, 1987, 1989) to study the processes of long-term weather anomalies and climatic changes manifested as air temperature variations over large territories, has shown that short-term and long-term forecasts of temperature anomalies are determined mainly by processes taking place in different regions of the world's oceans, called the energetically active zones of the ocean.

The first theoretical studies by Marchuk *et al.* (1987, 1989) provided a theoretical substantiation of this concept. It has later been developed by Lappo and co-workers (Lappo and Gulev 1985, Lappo and Muzychenko 1989, Lappo *et al.* 1987, 1990) who analysed the ocean–atmosphere energy exchange, and by Marchuk *et al.* (1986, 1987, 1989) who considered the laws of the variability of the Earth's radiation budget components based on satellite data.

It is now understood that long-range changes in the state of the atmosphere are determined by processes of energy, moisture and heat exchange in the ocean–atmosphere system. Based on the concept of energetically active zones of the ocean, a technique has been developed to reveal the atmospheric regions over the ocean and the continent, whose variabilities are interdependent. An approach has appeared to study large-scale climatic processes, without which the development of theoretical bases for large-scale ocean–atmosphere interactions has been difficult.

The technique of adjoint equations has been improved in solving various problems in the field of geophysical sciences. A new aspect has appeared in climate theory, according to which not only the intensity of the oceanic signal is important but also its variability. The measure of variability (heat flux mean standard deviation) is closely connected with the mean standard deviation of the response of the atmosphere over the continent. Thus the use of the adjoint equations formalism has made it possible to select the regions of related variabilities both in the ocean and in the atmosphere and, thereby, to confirm the important information value of energetically active zones of the oceans.

Regular monitoring of the energetically active zones of the oceans revealed a complicated local hydrophysical structure of the waters in these zones as well as their important effect not only on the atmosphere but also on deep layers of the ocean. In its turn, this has necessitated the development of the physical theory of local quasi-stationary thermodynamic structures of the ocean and middle-scale mechanisms for energy transfer in the ocean and in the atmosphere. The spatial inhomogeneity of heat fluxes in the energetically active zones of the oceans and their temporal variability are key factors in establishing a reliable quantitative pattern of heat exchange in the large-scale ocean–atmosphere system.

The highly simplified principal mechanism for the effect of the energetically active zones of the ocean on the formation of climate is that cloud anomalies in energetically active zones through solar radiation screening lead to an anomalous heat content in the oceanic mixed layer. These anomalies, forming largely in the western sectors of the oceans, are then transported by warm currents to the eastern sectors, which causes anomalies in synoptic processes of cyclogenesis which are transported by the westerly zonal flow onto the continents, causing thereby long-term anomalies in weather and climate. These processes are long-term and the climatic connections are out of phase, with a time shift of 2–8 months.

In recent studies based on the analysis of satellite Earth radiation budget data, it has been shown that energetically active zones of the ocean, responsible for the formation of long-term asynchronous connections, are manifested as the regions of abnormally high interannual mean standard deviations of the Earth's radiation budget components (Marchuk *et al.* 1989). Along with this, direct measurements in the near-water atmosphere have shown (Lappo *et al.* 1987) that against the background of the world's oceans, energetically active zones of the ocean are characterized by a higher energy release from the ocean to the atmosphere. Mesoscale numerical modelling (Kondratyev and Khvorostyanov 1986, 1988) has led to the conclusion that a stronger energy release from the ocean to the atmosphere in the form of sensible and latent heat fluxes is determined by horizontal temperature contrasts at the hydrological front between cold and warm currents in energetically active zones of the ocean. This leads also to the formation of clouds and to solar radiation screening, which determine the heat content anomalies in the oceanic mixed layer.

Note a substantial difference between the two mechanisms of energetically active zone formation from the viewpoint of timescales and their determining physical processes. Increased direct energetic activity of the ocean (known as type 1 activity) leads to an intensification of the genesis of fronts and cyclones in energetically active zones, mainly near the western boundaries of the ocean. The cyclones generated here are transported in the zonal flow to the north-east and affect weather and climate, reaching the western coastlines and moving further eastward. The characteristic timescales of these processes are 1–2 weeks.

The second type of energetic activity (type 2), determined by anomalies in cloud cover and solar radiation screening in energetically active zones, is manifested through a slow transport by sea currents of heat content anomalies north-eastward, from the western coastlines of the ocean. When these anomalies reach the western coasts of the continents, the anomalies of baroclinicity and cyclogenesis appear here, which lead to weather and climate variations. The timescale of such energetic activity is several months and differs for various energetically active zones.

Thus, two maxima due to the energetically active zones must exist in the spectrum of weather and climate perturbations: a high-frequency maximum caused by the transport of disturbances through the atmosphere, and a low-frequency maximum determined by the transport of disturbances through oceanic currents (asynchronous interconnection).

A correlation must exist between the amplitudes of disturbances due to the two types of energetic activity. Suppose, for example, that an increased energy release appeared in the Newfoundland energetically active zone. Through energetic activity of the first type, this will affect the weather in Europe 1–2 weeks later (clouds, fronts and cyclones will intensify). At the same time, the heat content of the oceanic mixed layer in the energetically active zone will decrease. If, later, a decrease of energy in the upper quasi-homogeneous layer of the ocean leads to negative anomalies of clouds, to positive anomalies of solar radiation and to the warming in Europe (for example, the process with negative cloud anomaly $\delta_{n-}$ over the Atlantic in the summer of 1975 and positive temperature anomaly $\delta_{T+}$ in the winter of 1976 in Europe), then the respective interconnection exists between energetic activities of types 1 and 2. Another interconnection appears if the primary energetic activity causes an intensified pumping of energy, an increase of cloud amount $\delta_{n+}$ and a cooling in Europe $\delta_{T-}$. For example, the abnormally cold winters of 1966–1967, 1967–1968 and 1968–1969 followed positive anomalies of cloudiness over the Atlantic.

An analysis of conceptual aspects of energetically active zones of the ocean shows that answers to a number of questions are needed: (1) what is the interconnection between direct energetic activity of the ocean in energetically active zones and long-term anomalies of clouds and radiation, i.e. between energetic activities of types 1 and 2, (2) why do the effects of the same ocean basins (energetically active zones) manifest themselves through (a) long-term asynchronous interconnections, (b) direct energetic activity and (c) abnormal mean standard deviations of radiation balance,

(3) why are these water basins permanently located in definite regions and not migrating from year to year, (4) what are the interconnections between energetically active zones and atmospheric as well as oceanic general circulations, centres of action and sea currents, (5) what physical processes determine the existence of energetically active zones and (6) how do energetically active zones affect the formation of climate and its changes?

These and other conceptual questions have been discussed by Marchuk *et al.* (1989). In this monograph it has been shown, in particular, that almost all energetically active zones (about 30) are characterized by an increased leeward cloud amount, i.e. clouds can be considered as a natural atmospheric tracer of the oceanic energetic activity. Clouds as a tracer characterize not only energetically active zones but also the spatial direction of their effect on the atmosphere.

A detailed analysis of the geographical distribution of energetically active zones, together with the characteristics of atmospheric and oceanic general circulations, revealed that almost all energetically active zones are special points of the climatic system: the sources of polar and arctic climatological fronts and points of crossing of these fronts and the ITCZ with the continents, zones of convergence of cold and warm currents, and zones of maximum horizontal sea surface temperature gradients in the regions of cold currents and upwellings.

Thus, first, energetically active zones result from the present structure of the global climate system, and small climatic variations of atmospheric and oceanic general circulations are mainly manifested in key points of the global climate system, in energetically active zones. Secondly, abnormal atmosphere–ocean interactions in energetically active zones (type 1 energetic activity) act, so to speak, as an acupuncture for the whole global climate system and manifest themselves as long-term asynchronous interconnections (type 2 energetic activity). This is how the global climate system energetically active zones feedback is realized.

Climatological analysis of the atmospheric boundary layer where the atmosphere and ocean interact, based on numerous observational data, has shown that the geographical distribution of the most energetically active regions of the atmospheric boundary layer, which have maximum cloud amount, coincides well with the energetically active zones' locations near western and eastern coasts.

In this connection, based on the use of a mesoscale numerical model, the development of low-level cloudiness in the Newfoundland energetically active zone for summer and winter conditions has been modelled, as well as the transformation of longwave and solar radiation by clouds, and sensible and latent heat fluxes, i.e. the indicators of energetic activity have been calculated. The calculations have shown, in particular, that in winter conditions (Kondratyev and Khvorostyanov 1986), at the hydrological front between the Labrador current and the Gulf Stream in the presence of clouds, the ocean loses energy even in daytime when solar heating takes place. Thus, both the winter and summer cloud anomalies are important. These detailed calculations can serve as tests in the parametrization of the anomalies of clouds and radiation in energetically active zones for large-scale models.

A consideration of processes of frontogenesis and cyclogenesis in energetically active zones has led to the conclusion that their development and energetics are substantially affected by an advection of cold air masses from the continent to the ocean and by the energy of phase transitions depending on the intensity of the atmosphere–ocean interaction. These processes affect the cloud amount in energetically active zones and largely determine the anomalies of solar radiation and enthalpy of the quasi-homogeneous layer of the ocean. Analysis of the results of observations of different forms of clouds from weather ships shows that processes of frontogenesis and cyclogenesis in various energetically active zones are different and can affect in different ways the screening of solar radiation; this should be taken into account in the statistical assessment of the role of cloud anomalies over the ocean.

Based on analysis of satellite data and a mesoscale numerical model, Kondratyev *et al.* (1988b) considered the effect of orography on atmospheric processes. Since large mountain regions slow down the westerly zonal flows through orographic friction and often lead to blocking situations, they can be considered as energy-active zones of the land. An adequate account of these zones and a parametrization of orographic friction are particularly important in the models of general circulation and climate, with a good horizontal resolution.

An analysis of the data of sea surface temperature observations in the centres of action in the eighteenth and nineteenth centuries has led to the conclusion that the origin of the Little Ice Age could have been connected largely with temperature anomalies in energetically active zones of the ocean, with respect to the present time. The trend of climatic warming from the mid-nineteenth century has been due to a reduction of these anomalies and probably, to a larger extent than with the greenhouse effect, due to a $CO_2$ increase.

A study of teleconnections of the Earth radiation budget anomalies in key climatic regions of the world (ocean and land) has stimulated further development of studies in the field of application of the time series of satellite observations to uncover the laws of climate formation. In particular, studies of the signals of climate changes, analysis of the predictability of the meteorological fields on the basis of satellite observations, as well as studies of the atmospheric response to anomalous processes in the ocean, etc. have been of great importance (Kondratyev *et al.* 1988a).

Theoretical analysis, verified by global maps of the world's oceans' heat budget based on observational data, and the previously determined empirical relations between the state of the ocean and resultant weather, has confirmed the view that energetically active zones of the ocean are the most important factor determining long-term weather anomalies and climate changes. This concept is particularly important because the available observational means are still inadequate to obtain on a routine basis the necessary meteorological and oceanographic information on a global scale. The Global Weather Experiment has made it possible for the first time to obtain global information for a one-year period. This information has been used to test theoretical models of the atmospheric general circulation and of climate.

Discovering the energetically active zones of the ocean has drastically changed the situation: the concentration of available observational means (mainly ships) for the purpose of monitoring processes taking place in the energetically active zones (for example, their migration and variability) and the use of satellite data permit the development of deeper insights into the processes governing long-term weather anomalies and climate changes and provide long-range weather forecasts with important information. This had led to the development of the Sections programme, the major objective of which is to study the role of the ocean in short-period climate changes. Based on the energetically active zones concept, the Sections programme includes field measurements in different key climatic regions and also numerical modelling experiments using observational data.

The purpose of the Sections programme is to study the effect of long-term variations in the thermodynamic state of the ocean on atmospheric processes (as well as the feedback relations between the atmosphere and processes in the ocean), which results from the ocean–atmosphere interaction concentrated in the energetically active zones of the ocean. For this purpose, consideration of the variability of heat transport by the major systems of oceanic currents (an important factor in the formation of large-scale anomalies in the atmosphere) is of great importance.

Studies of the meridional heat transport have shown, for example, that in low latitudes this transport takes place largely in the ocean. Across the latitudinal circle 25° N, the ocean transports 75 per cent of the heat, whereas only 25 per cent is transported by the atmosphere. In the southern hemisphere, the contribution of the ocean to the poleward heat transport remains substantial even at 60° S. All this means that substantial changes in the heat transport by the principal systems of oceanic currents as well as variations in the location of the major jet streams should seriously affect the development of atmospheric processes. A not lesser, and sometimes more substantial, role in the energetics of the ocean and of the atmosphere is played by mesoscale circulations, which manifest themselves through convection, upwellings, mesoscale vortices and some other phenomena. Their spatial scales are subgrid, which requires the development of a parametrization technique for such processes consisting of a simulation of these processes with the use of meteorological and oceanographic quantities from the equations of thermohydrodynamics (air and water temperatures, wind and current velocities, etc.).

An important aspect of parametrization is the development of the systems which characterize heat, water and momentum exchange between the ocean and the atmosphere as well as radiative processes in the atmosphere and the ocean. A modulation of solar radiation incidence onto the ocean surface by varying cloudiness is a key radiative process.

In the presence of long-term anomalies of cloud amount, strong changes appear in the accumulation of heat by the ocean, which cause sea surface temperature anomalies, which, with the respective time lag, cause weather changes in land areas far from sea surface temperature anomaly centres. Such telecontrol of the weather is observed, for example, in US territory, where the effect of Pacific sea surface temperature anomalies is manifested. The situation described determines the urgency of both field and theoretical studies of the formation of sea surface temperature anomalies and of the role of anomalies as a factor in long-term weather changes.

An analysis of data on the heat budget of the world's oceans has shown, for example, that in January in the northern hemisphere maximum heat fluxes from the ocean to the atmosphere are observed in the western Sargasso Sea, east of Newfoundland, in the Norwegian Sea and east of Japan. Although the maximum heat transfer area of the Newfoundland and Norwegian zones constitutes only 25 per cent of the area of the North Atlantic, their contribution to the total heat flux from the ocean surface to atmosphere north of 40° N reaches 47 per cent.

In the southern hemisphere in July the zones of maximum heat release are located in the region of the Drake Passage, south of Africa, in the region of the Tasman Sea, and south of 60° S, to the Antarctic. The energetically active zones are characterized by the highest variability of ocean surface temperatures. Many of these zones are connected with warm currents flowing from low to high latitudes, including the Gulf Stream and its continuation in the North Atlantic, as well as the Kuroshio Current and its continuation in the North Pacific Ocean.

The existence of the energetically active zone in the Drake Passage is explained by the interaction between the Antarctic Circumpolar Current (encircling the Antarctic) and the Peru and Brasilia Currents. The energetically active zone of the oceans south of Africa is connected with the Agulhas Current, and the energetically active zone in the Tasman Sea is connected with the East Australian Current. Of particular interest is the tropical belt of the world's oceans, where gigantic amounts of solar radiation are absorbed, with subsequent transport of heat to high latitudes. From observational data in the 10–30° N latitudinal belt, the amount of heat released in the upper layer of the ocean, as a result of solar radiation absorption, reaches 350 $W m^{-2}$.

The energetically active zones of the oceans include the zones of migration of the polar ice edge and the regions of powerful monsoon formation in the Indian Ocean and in other regions of the world's oceans as well. The role of the different energetically active zones of the oceans (from the viewpoint of their contribution to the formation of long-term weather anomalies) varies substantially both within a given year and from year to year.

Studies of processes governing the energetically active zones of the oceans' large-scale temporal variability and intensity are the major objective of the Sections programme which has been carried out by a large number of Russian research institutes since the early 1970s. In studies of the effect of the key climatic regions on weather, it is important to determine the factors (predictors) that most affect the persistent large-scale anomalies. As mentioned above, sea surface temperature anomalies constitute, of course, one set of these factors.

The major objectives of the Sections programme are (Marchuk *et al.* 1989): (1) studies of large-scale interaction among the atmosphere, the world's oceans and the continents that form the normal annual course of climate, (2) studies of the temporal variability of major characteristics of the ocean and the atmosphere in energetically active zones of the ocean and (3) studies of relations between heat and dynamics anomalies in energetically active zones of the ocean and of the formation of anomalies of atmospheric circulation and temperature field in the atmosphere over the ocean and continents, by means of quantitative estimation of all the components of the heat budget.

In the course of realization of the Sections programme, the following major problems must be solved: (1) studies of the major climatic factors in the atmosphere and the ocean and the development of models of short-term climate changes for some regions (mid-latitudes) of the northern hemisphere, based on the use of data field measurements in energetically active zones of the ocean, (2) studies of the formation of anomalies of the oceanic parameters in the energetically active zones of the tropical and northern Atlantic and Pacific Oceans as well as in the Norwegian Sea, using a 4-dimensional analysis of observational data, (3) studies of the temporal variability of the major characteristics of the atmosphere and the ocean in energetically active zones of the ocean and the development of a climate model to reproduce the normal annual change of heat supply in the upper oceanic layer and (4) studies of regional-scale interactions between the atmosphere, energetically active zones of the ocean and the continent near the eastern coasts of Asia.

In accomplishing the observational programmes, an assessment of the following characteristics of the atmosphere and the ocean at various latitudes and depths is most important: atmospheric pressure, wind speed and direction, air temperature and humidity, cloud amount and type (from conventional and satellite observations), as well as the direction and velocity of the ocean currents (ship, buoy, satellite and other observations), roughness and salinity. Regular data on these and other characteristics will promote a realization of the sub-programmes of meteorological, aerological, radiation, oceanographic and satellite observations.

Satellites now enable us to determine such characteristics as the amount and type of clouds, sea surface temperature (though sea surface temperature retrievals are still inaccurate), roughness, wind speed and wind stress near the ocean surface and the radiation budget's components for the ocean–atmosphere system and the ocean surface. Thc oceanic fronts, currents, upwellings, mesogyres, internal waves, zones of ice cover and storms are also clearly recognized.

At the test sites chosen within various energetically active zones, specialized observational sub-programmes must be accomplished, with an account of special features of each test site (the ice edge, upwelling, sea surface temperature anomaly, the region of a strong sea current formation, etc.). In these cases specialized equipment becomes very important (for example, to study turbulence in the atmosphere and the ocean, to make thorough temperature measurements in both media, to determine the optical and radiative characteristics of the atmosphere and the ocean, the chemical composition of the air and water, etc.). Flying laboratories operating simultaneously with ships are an important means of ensuring (in addition to ship data) 3-dimensional information. Satellites will provide the background data for enormous regions around the area under study. A variety of observational means attach great importance to the meteorological aspects of measurements and gives the first priority to such problems as maximum unification of instruments, comparability of measurement results and development of techniques for reducing the data to equivalent scales of spatial and temporal averaging.

Naturally, such a large-scale programme as Sections requires much effort in accumulating, processing, archiving and analysing the data. In this connection, the responsibilities must be defined clearly and efforts coordinated between the organizations taking part in the solution of the problems set forth by the programme. Since Sections is an organic component of the WCRP, international scientific cooperation must play an important role in the accomplishment of this programme.

Scientific justification for the Sections programme is based on numerical modelling of the atmospheric general circulation. For the first time in planning such a large-scale experiment, the observational systems and the requirements of data are substantiated on the basis of numerical experiment results, on an account of the specific character of the problem to be solved, and on the need for information to check the central hypothesis which serves as the basis for a theoretical model of long-term forecasts of weather and of climate changes.

## 5.4 The world ocean and climate

The key role of the ocean and of its interaction with the atmosphere in the formation of climate attracts particular attention to processes in the world ocean (World Meteorological Organization 1988b). The ocean covers more than 70 per cent of the globe and is a complicated system in which circulations of various scales constantly appear and disappear, powerful planetary-scale currents and counter-currents are formed, and heat and salts are transported from one region to another. All these processes, mutually adjusted following the laws of hydrothermodynamics, take place under conditions of continuous ocean–atmosphere interaction.

The waters heated in the tropical and subtropical latitudes are formed into currents which transport heat to polar latitudes. Upon the uptake of heat by the atmosphere, the cooled waters flow to the equator to start a new thermodynamic cycle of the atmosphere–ocean system, a gigantic thermal machine. This thermodynamic process is realized via a complicated set of motions governed by physical laws, from small-scale and short-period turbulent pulses and internal gravity waves to planetary-scale oscillations of the ocean surface level.

The typical structure of the world's ocean in the zone of the subtropical and middle latitudes can be described schematically in the following way. First of all, a homogeneous layer forms near the ocean surface, in which a wind-driven intensive mixing takes place. Depending on the extent of roughness, the thickness of this layer may vary from several metres to several tens of metres. This layer is usually called the surface friction layer, or mixed layer; it is a kind of diaphragm for the atmosphere–ocean energy exchange. In the calm state of the ocean the exchange is weak and in stormy conditions it is intensified ten-fold and a hundred-fold. Consequently as a result of storms, the atmosphere annually takes up about 20 per cent of the total heat released by the ocean; this testifies to the importance of this mechanism of energy transfer.

Below the mixed layer there is a baroclinic layer, the most important from the viewpoint of the ocean–atmosphere interaction. It is about 1.5 km thick in the equatorial and mid-latitudinal belt but becomes thinner in the southern and northern latitudes – the zones of a powerful convection and unstable stratification of the waters. In this oceanic layer the water temperature decreases with depth almost linearly; therefore, it is usually called the thermocline layer. This layer results from an unsteady heat flux coming to the world's oceans' surface and from simultaneous effects of the horizontal and vertical turbulent exchange. The heat supplies taken up by the atmosphere can substantially affect atmospheric processes. These heat supplies are accumulated in the thermocline layer and transported by the currents.

Below the thermocline layer the ocean again becomes homogeneous and barotropic. Here the horizontal currents are slow and hydrodynamically stable. Except for the natural friction layer, they do not change with depth and, as a rule, are countercurrents to more intensive but less powerful baroclinic currents.

The formation of currents is strongly affected by centrifugal forces of the rotating Earth which press powerful planetary-scale currents to the eastern coasts of the continents. By the Bernoulli law, this leads to an intensification of coastal currents and the formation of powerful currents – the Gulf Stream in the Atlantic and the Kuroshio Current in the Pacific. These currents are the most important mechanism for heat transport from the equatorial zone to the polar zone. Because of the large water volumes involved in the Gulf Stream, an intensive counter-current is formed below the surface, which also transports cold waters to the southern latitudes. The Kuroshio Current is less intensive, since it is split by the islands of Japan into a number of small streams, but it also plays an important role in the planetary-scale thermodynamic cycle.

Finally, the equatorial east-to-west currents formed by trade winds are particularly important for the dynamics of the atmosphere–ocean system. These currents in the Pacific and Atlantic Oceans are powerful jet streams with the fields of deep counter-currents. The nature of these jet streams differs from that of the Gulf Stream and the Kuroshio Current. These currents, energetically supported by trade winds and driven by them, are formed at the interfaces between gigantic gyres – the northern and southern circulations in the world's oceans. Meeting near the equator, the currents form a kind of boundary layer with intensified eastern currents. Despite the weak character of trade wind circulations in the ocean, they play an important role (as shown by Bjerknes 1969) in the formation of the planetary-scale features of circulation. In particular, it has been established that with decreasing intensity of the Pacific trade winds a strong heating of the surface equatorial waters is observed and there is also an air temperature increase in the tropical region. A large difference occurs between surface air temperatures (hence, pressure in the free atmosphere) at the equator and at the pole. As a result, the planetary-scale atmospheric flows over the mid-latitude Pacific are intensified. In accordance with the law of kinetic energy conservation which approximately holds for the atmosphere, it is followed by simultaneously decreasing intensity of atmospheric flows over the Atlantic.

The scheme for the atmosphere–ocean interaction described above assumes the presence of both constant and continuously appearing factors. Whereas the effect of constant factors is clear (the distribution of the oceans and the continents, ice and vegetation cover, incoming solar radiation, etc.), the effect of such a dynamic factor as the continuously forming and dissipating cloud cover has been poorly studied. Climatic fluctuations take place largely due to non-linear thermodynamic processes, and cloudiness resulting from these processes is one of the major energetic regulators of climate.

A specific aspect of the effect of the oceans on climate is that even small (1–2 °C) sea surface temperature variations, reflecting the variations in the heat content of the oceanic upper layer, can play an important role as a factor which determines the location of powerful convection zones and regions of strong upward flows in the ITCZ. These zones, with a characteristic dense cloudiness, seriously affect the weather-forming processes in the tropics and (to some extent) in the middle latitudes. The rise of cold sub-surface waters due to either tangent wind stress or wave propagation at the thermocline is an oceanic process of much greater importance than the local water transport, stimulating such interannual climatic fluctuations as the Southern Oscillation and the Asiatic monsoon variability.

On time scales of decades and centuries the transport of heat by the sea current systems, which is an important part of the high-latitude heat budget and, hence, determines the formation of the ocean–pole temperature gradient, becomes of fundamental importance. In this connection, the exchange between surface and deep waters and a slow reverse flow to the equator below the thermocline is crucial. The effect of sea surface temperature variations in the tropics on the moisture input to the atmosphere, which affects the cloud distribution and the atmospheric greenhouse effect, is weaker. The river freshwater run-off to the arctic seas can be a strong climatic regulator. It maintains the buoyancy of surface waters and, therefore, more favourable conditions for the formation of sea ice, which, in its turn, intensifies the surface temperature variability, limiting the heat exchange with water masses and intensifying the solar radiation reflection.

On time scales of millenia the oceans control the chemical composition of the atmosphere and, hence, the radiative equilibrium of the climatic system. The concentrations of $CO_2$, $N_2O$ and other minor gaseous constituents are eventually determined by the geochemical equilibrium in the ocean, modulated by interaction with the biosphere.

For many millions of years, radical changes had taken place in the temperature and circulation of the ocean which had reflected (on geological time scales) large-scale climate changes.

One of the key aspects of the effect of the ocean on the formation of climate is connected with heat transport in the ocean. Having analysed this problem in the context of palaeoclimatic changes, Covey and Barron (1988) came to the conclusion that the climatic sensitivity to variations in the heat transport in the ocean can be substantially reduced at the expense of compensating changes in the heat transport in the atmosphere. The possibility of such compensation determines the necessity for further studies of the role of oceanic heat transport in the formation of climate.

The key aspects of the problem of present climate changes are connected with studies of large-scale heat and moisture budgets on the ocean surface as well as wind stress. The amplitude of the annual change of heat supply in the northern hemisphere is $9 \times 10^8\,J\,m^{-2}$, which corresponds to an RMS variability of $\pm 60\,W\,m^{-2}$ in the heat flux across the surface. The Earth's radiation budget in the equatorial band over the ocean constitutes $80\,W\,m^{-2}$ and, from oceanographic observations, the poleward heat flux across 23° N in the Atlantic is $1.1 \times 10^{15}\,W$, which is equivalent to heat release by the ocean of $50\,W\,m^{-2}$. With a 20 per cent accuracy (i.e. $\pm 10\,W\,m^{-2}$), the determination of the enumerated climatic signals (timescales of several weeks and spatial resolution from several hundreds to thousands of kilometres) requires precise measurements which are practically impossible. Thus, an analysis of the energetics of the climatic variability from observational data is still impossible.

A similar conclusion can be drawn from data of numerical modelling of the effect of sea surface temperature anomalies on large-scale atmospheric processes; variations of heat flux across the ocean surface in mid-latitudes, in the tropics, and averaged over the globe constitute 20, 40 and $2\,W\,m^{-2}\,deg\,K^{-1}$, respectively (variations in the ocean surface heat budget at a doubled $CO_2$ concentration are only $3.5\,W\,m^{-2}$). The accuracy $\pm 15\,W\,m^{-2}$ is identical to sea surface temperature variations of $\pm 0.7$ °C in mid-latitudes and $\pm 0.4$ °C in the tropics. The errors in actual measurements are larger. Hencc, thc accuracy of measurements must substantially be increased.

On the other hand, of particular importance is an accumulation of data of continuous observation series for indirect climatic indicators whose analysis can compensate for impossible direct measurements of parameters of climate energetics. In connection with the intensive use of satellite data, their validation by comparing them with the results of control complex sub-satellite observations becomes critically important.

As Manabe (1983) notes, the effect of the ocean on climate is determined by three key factors: the role of the ocean as a source of moisture in the hydrological cycle, the gigantic thermal inertia of the ocean and the horizontal heat transport by the ocean. Two climate models were developed in the Princeton University Geophysical Fluid Dynamics Laboratory (GFDL). One of them considered explicitly the heat transport by the ocean (interactive ocean–atmosphere model); the other used an approximation of a swamp ocean, where the ocean is considered as a wet surface with zero heat capacity, which supplies moisture but does not transport heat.

Based on comparisons of quasi-equilibrium climates calculated with the use of the interactive ocean–atmosphere and swamp ocean models, Manabe (1983) estimated the heat transport in the ocean and its effect on climate with a fixed schematic (sector) geography. Calculations gave lesser quantities of the meridional heat flux transported by the ocean compared with the previous ones obtained from empirical data by Oort and Vonderhaar (1976). The difference between the meridional temperature profiles corresponding to the ocean–atmosphere model and the swamp ocean model characterizes the effect of heat transport by the ocean on the formation of the temperature field. The transport-induced atmospheric heating in high latitudes is confined to the lower troposphere due to the persistent atmospheric stratification, whereas the cooling in low latitudes is propagated through the whole troposphere due to intensive convection. Therefore, the radiative damping of temperature anomalies in the tropics is more rapid than in high latitudes, and it means that an air temperature increase in high latitudes is stronger than the respective decrease in the tropics.

The geographical distribution of sea surface temperature given by the ocean–atmosphere model is characterized by an upwelling-induced local minimum along the equator, whereas the swamp ocean model gives a maximum sea surface temperature in the equatorial band due to maximum insolation. The equatorial upwelling (ocean atmosphere model) is followed by a sharp reduction of precipitation over the tropical ocean. Also, a sharp precipitation along the eastern coasts of the continents in the sub-tropics is intensified due to the northward advection of the warm waters by the sub-tropical gyre. In addition, a current-induced northward shifting of precipitation takes place in mid-latitudes.

The total oceanic circulation substantially affects climate sensitivity to various factors, including an increasing $CO_2$ concentration. Calculations of the meridional profiles of the surface air temperature increase at a four-fold increase of $CO_2$ revealed a substantial difference: in the case of the ocean–atmospheric model, the warming is several times weaker than in the case of the swamp ocean model, in which the ocean is substituted with a static water layer 68 m thick (i.e. no current). This difference is particularly manifested in high latitudes and testifies to a reduced climatic sensitivity to a $CO_2$ increase caused by heat transport in the ocean.

Since a higher air temperature in high latitudes corresponds to the ocean–atmosphere model, which narrows the polar zone of snow and ice cover, the northward retreat of the snow cover and sea ice edge with increasing $CO_2$ takes place in conditions of a weaker insolation and a shorter latitudinal circle. This substantially reduces the effect of the albedo feedback on climate warming. The effect of the ocean not only weakens the sensitivity of climate to external forcings but also delays the reaction by several decades. The latitudinal distribution of the non-stationary reaction to a $CO_2$ increase can considerably differ from variations in the equilibrium climate.

The time-dependent reaction depends critically on the rate of propagation of thermal perturbations from the mixed layer to the deep ocean. The ocean–atmosphere model calculations show, for example, that on year 25 of integration a strong warming of the lower troposphere takes place near the pole, but the temperature of the ice covered water remains almost constant due to the heat-insulating effect of the ice. Near the ice cover edge (75° N) the warming is confined to the surface ice layer due to the stratified halocline. The present interactive ocean atmosphere models are at an initial stage of development. Therefore, their further improvement is an urgent problem. Also urgent is an accumulation of observational data which may be used to test the models' reliability.

As has been mentioned above, the nature of the interannual variability in the tropical Pacific is determined by its connection with the El Niño event consisting of the formation of persistent positive sea surface temperature anomalies along the Equador and Peru coastlines. Variations in sea level in the region of the Galapagos Islands are a representative characteristic of non-seasonal fluctuations near the equator and along the eastern coastline of the ocean. The western Pacific is characterized by a drop in sea level and by a rise of the thermocline during an El Niño event, and the data of observations on Track Island are a source of representative information.

O'Brien and Busalacchi (1983) undertook an analysis of the interannual variability of the tropical Pacific based on the use of a linear model in which the disturbing effect of the wind has been prescribed from monthly mean data of ship observations during five cases of the El Niño event (1963, 1965, 1969, 1972 and 1976). This has shown that the variability of the south-eastern trade winds on both sides of the International Date Line is the reason for the El Niño events. The El Niño event has been caused by the formation of the large-amplitude Kelvin wave front located west of the International Date Line. This front has appeared due to the westward wind stress during the second half of the year, prior to El Niño. Although each of the El Niño events is unique, there are some features in common.

Numerical modelling has led to the conclusion that the interannual variations in zonal wind stress in the central and western Pacific are very important as factors determining the variations in sea level and in the depth of the pycnocline. In the cases considered, the initial phase of El Niño has been connected with the disturbances of the zonal wind field in the western Pacific, and further development has been determined by the wind field features in the central Pacific. The results obtained enable wind field monitoring in the tropical Pacific to be plotted correctly and testify to the need for continuous observations of a detailed structure of the wind field.

As Woods (1983, 1984, 1985, 1991) noted, a concept commonly accepted by that time was that the stratified ocean is weakly turbulent. The estimates based on an account of the properties of internal waves have shown that the vertically integrated rate of energy dissipation varies in the range $1\text{–}10\,\mathrm{W\,m^{-2}}$. At such a low level of dissipation, the turbulence cannot be distributed uniformly, but is concentrated as moving patches within the laminar flow. In fine sunny weather such a regime in the daytime manifests itself at the ocean surface, but at night and in strong winds a turbulent boundary layer forms. The value of the kinetic energy of turbulence in this layer depends on wind shear and on the rate of surface cooling and heating due to solar radiation absorption. The rate of dissipation of the kinetic energy of turbulence changes in the interval $1\,\mu\mathrm{W\,m^{-3}}$ to $1\,\mathrm{mW\,m^{-3}}$. The value of the kinetic energy determines the thickness of the boundary layer (in summer daytime the thickness is about 30 m).

In these conditions, the vertically integrated dissipation of turbulent kinetic energy in the layer 5–20 m constitutes $0.5\text{–}50\,\mathrm{mW\,m^{-1}}$. The turbulence in the boundary layer is continuous and much more intensive, compared with the layer below. Two regimes of turbulence are separated by the thermocline whose depth changes both in time and in space, depending on local astronomical and meteorological factors, as well as on the stratification connected with the diurnal and annual change of the heat content of the oceanic upper layer.

Woods (1983, 1984, 1985, 1991) discussed the significance of the variability of the oceanic boundary layer (mixed layer) for the formation of climate and its changes, the present state of numerical modelling in the interests of climate studies, and relative importance of observations of the mixed layer within the WCRP large-scale oceanographic experiments. To study the climate, simple models of the mixed layer are most useful.

A comparison of the numerical modelling results with observational data has shown that the available models do not meet the requirements of climate studies, including the necessity for reliable simulations of the annual change, which permits the interannual variability to be revealed, as well as such climate trends as that determined by increasing $CO_2$ concentration in the atmosphere.

Woods (1983) considered the dependence of the limited climatic predictability, based on the mixed layer models, on the errors in prescribed input parameters (cloud amount, wind speed, surface cooling, water turbidity). In connection with further improving the models, the following factors claim special attention: (1) the solar heating profile, (2) turbulence below the mixed layer, (3) upwellings and (4) advection. Data on the regional annual change of maximum depth suggest that it is determined by a salinity gradient caused by baroclinic advection below the mixed layer. Such an advection always follows the heat flux divergence due to the need to satisfy the annual mean energy balance.

As Wallace (1985) noted, the present level of understanding of the nature of the interaction between the tropical atmosphere and ocean makes it possible to achieve reliable simulations, on the basis of numerical modelling of (1) the atmospheric response to observed sea surface temperature anomalies and (2) the oceanic response to observed changes in wind stress near the ocean surface in periods of strong disturbances of the atmosphere–ocean system (for example, an abnormal El Niño event in 1982–1983). This made possible long-term weather forecasts with an acceptable reliability for certain limited geographical regions.

The prospects for increasing the reliability of the forecasts and an expansion of the sphere of their application are connected with studies planned within the TOGA programme for a better understanding of the nature of the interseasonal and interannual climatic variability (for example, special effects of individual El Niño events). Since the state of the northern hemisphere tropical ocean surface is stable in the autumn, the data for September can serve as useful information to forecast the January conditions for the next year. This determines the possibility of long-range weather forecasts even in such conditions when a forecast for the interactive climate system based on an account of the atmosphere–ocean interaction is impossible. If the latter problem can be solved, this will make it possible to increase substantially the range of forecasts.

The El Niño-induced teleconnections serve as an empirical basis for long-range forecasts: they manifest themselves, for example, through a mutual correlation of annual mean sea surface temperature values near the equator (6° N to 6° S) in the Pacific east of the International Date Line, precipitation at island stations in the central equatorial Pacific and atmospheric pressure at sea level in Darwin (Australia). The atmospheric response to sea surface temperature anomalies in some regions of the tropical ocean is very complicated; the redistributions of heat and water exchange near the ocean surface change the spatial distribution of the regions of clouds, precipitation and of latent heat release, which affects the atmospheric circulation (including the wind field near the ocean surface), which, in its turn, affects heat and water exchange, etc. The final result of this complicated chain of feedbacks is a redistribution of precipitation over the equatorial Pacific, where intensive precipitation follows positive sea surface temperature anomalies.

Teleconnections appear, shown by the precipitation variations near the equator that are revealed in other equatorial regions and in higher latitudes. A rough analogue to this process is the distribution of waves behind an obstacle; if this obstacle moves, a wave packet propagating along the flow follows it. Similarly, the tropical source or sink of the vortex (in the presence of the homogeneous westerly transport over the rotating sphere) produces linear barotropic Rossby waves whose mean standard deviation is responsible for the teleconnections mechanism. Apparently, this explanation is roughly schematic, since there is no doubt that the interaction between some non-linear dynamic mechanisms plays an important role.

An analysis of the available observational data enables one to characterize an empirical scheme of the effect of El Niño on atmospheric general circulation during the northern hemisphere winter at the level of the jet stream. Each El Niño event is characterized by a saddle formed of two anticyclones, north and south of the zone of intensive precipitation (clouds). North of the anticyclonic gyre, a planetary-scale wave packet propagates, so that an anomalous circulation forms over the northern Pacific and the Gulf of Alaska, while an anticyclone occurs over most of Canada and a cyclone over south-eastern USA. The numerical modelling results revealed the same pattern. On the whole, the same situation was observed in 1982–1983, but some special features took place, too.

The El Niño-induced redistribution of precipitation in the equatorial band shows itself not only at the jet stream level but also near the surface.

An analysis of the interannual variability of climate revealed surprisingly coherent spatial structures of global-scale anomalies of the atmosphere–ocean system, which occur every several years (Rasmusson 1985). The atmospheric general circulation fluctuations induced by the southern oscillation were detected in the early 1920s by Walker. The Southern Oscillation is the strongest in the tropical Pacific, revealed through irregular (2–10 yr) swing oscillations from the phase of low Southern Oscillation index, to which correspond an abnormally high surface atmospheric pressure in the region of Australia and the Indian Ocean and an abnormally low pressure in the south-eastern Pacific, to the counter-phase of a high Southern Oscillation index.

The global-scale oscillations of the pressure field reflect the low-frequency atmospheric general circulation variability in the lower troposphere, which is connected with considerable shiftings of the zones of precipitation in the tropics, causing droughts, much to the detriment of agriculture. The Southern Oscillation signal goes deep into the tropical latitudes of both hemispheres, causing, thereby, considerable changes in atmospheric general circulation. Although the Southern Oscillation event has long been known, its connection with processes in the tropical ocean has become clear only during the last two decades, when the role of the sea surface temperature anomalies induced by El Niño in the tropical Pacific has been established.

It has become clear that both El Niño and the Southern Oscillation are components of one system of global climate change, the El Niño / Southern Oscillation (ENSO), in which sea surface temperature anomalies play a key role. The phases of the 12–24-month evolution of typical ENSO episodes are closely connected with the annual change, manifesting themselves, first of all, through modulations of the annual change. Rasmusson (1985) discussed in detail the specific features of an unusually intensive ENSO event in 1982–1983, which developed, on the whole, similarly to some ENSO events observed earlier. Its special feature is that positive sea surface temperature anomalies have appeared not only in the eastern tropical Pacific but also in the whole global tropical belt. The formation of the global structure of sea surface temperature anomalies and regional anomalies of precipitation located far from the tropical Pacific has rendered it necessary to analyse a complex of events determined by secondary effects and feedbacks, in order to understand whether they are a passive manifestation in the atmosphere of the forcings from the tropical Pacific or whether they are active ENSO components propagating from the sources of anomalies and universally affecting the ENSO. To answer these questions, the system of observations should be enhanced and further numerical experiments are needed.

An analysis by Gill (1985) of the effect on atmospheric general circulation of an abnormal ENSO event in 1982–1983 has led to the conclusion that such forcings have an inhomogeneous global distribution, confirmed by data on precipitation. So, for example, the meteorological conditions in arid zones were characterized by intensified droughts.

The principal feature of tropical atmosphere energetics is latent heat released in the formation of clouds, and, hence, observations of cloud cover can give information about the spatial and temporal variations of heat flux divergence. It is important that the latent heat release is connected with precipitation, though the quantitative aspect of this connection needs further studies. Powerful heat sources in different tropical regions function like a fire heat centre, causing a circulation characterized by two cyclones in the zone of westerly flow in the lower troposphere directed towards the heat source from its western side, and by two anticyclones in the upper troposphere.

An analysis of a series of successive satellite images of cloud cover in late November 1982 has shown that the zone of active convection and intensive precipitation migrated eastward. Usually the warm and lightweight surface waters are blown away westward by trade winds, where a heat centre forms. However, if (as happened in 1982) the zone of intensive precipitation moves towards the ocean, the wind changes its direction and the heat centre can move eastward. If wind changes are rapid, the waves move forward and backward across the ocean basin, causing the respective oscillations.

Special features of equatorial dynamics determine the specific properties of the waves. One of the consequences is a change of sea level in the band along the equator. All this determines the urgency of monitoring the field of winds and currents, as well as the level and temperature of the ocean surface, with the use of various observational means. An approximate scheme of cause-and-effect relations determining the effect of the tropical ocean on the global atmosphere can be characterized in the following way: wind anomalies → changes in the currents and sea level → sea surface temperature anomalies → changes in convection (latent heat release) → wind anomalies.

Of great interest is a comparison of the special features of the processes of atmosphere–ocean interaction in the Pacific and Atlantic Oceans (Hagen, 1989). The Atlantic Ocean, being smaller than the Pacific Ocean by a factor of three, is an excellent laboratory to study processes taking place in the atmosphere–ocean system. A relatively small ocean basin is characterized by a shorter adjustment time than a large basin, since this time is proportional to the time period needed for planetary waves to cross the basin. This means, in particular, that the tropical Atlantic Ocean (in contrast to the Pacific Ocean) has almost no memory. The zonal slope of the thermocline along the equator, the field of density, and the intensity of equatorial currents vary almost in phase with the annual course of the wind field. Even outside the equatorial band, an almost equilibrium response of the tropical ocean to the disturbances is observed. The seasonal conversion of the northern equatorial counter-flow is in phase with the ITCZ shifts in the annual course and associated variations in the wind field.

In view of its relatively small size, the equatorial belt of the Atlantic is characterized by the annual change, dynamically similar to the interannual variability in the Pacific, which improves the conditions for observations of this variability. An analysis of the available observational data and of the results of numerical modelling revealed the surprisingly coherent spatial structures of the response of the tropical Atlantic (in the annual change) to the forcing of the wind field. The dual annual oscillation of the thermocline points to the existence of a seasonal zonal and meridional redistribution of heat, which is an order of magnitude more substantial and is out of phase, compared with heat fluxes across the ocean surface. This means that an account of the dynamic response of the interior of the tropical ocean is most important for understanding and forecasting the role of the effect of the ocean on the climatic system.

As in the Pacific Ocean, the zonal heat transport in the equatorial belt can explain the seasonal and interannual variability of the El Niño events. If, on the scales of seasonal and interannual variability, the Atlantic Ocean affects more weakly the global climate than do the Pacific and Indian Oceans, its effect can prevail on timescales of decades and centuries. The Atlantic Ocean is located along the meridian and functions like a heat pipeline, taking heat from the world's oceans to the North Atlantic. In a narrow band of the tropical Atlantic an intensive interhemispherical meridional heat transport is concentrated, which makes this region the best place on the planet for observations of heat transport in the ocean and its effect on climate.

The interannual variability of the seasonal mean tropical circulation and precipitation is characterized, largely, by changes in location and intensity of the thermally driven Hadley and Walker circulation cells. Although the physical mechanisms responsible for the location and intensity of climatical mean maxima of precipitation (corresponding to the upward branches of these circulation cells) are still poorly studied, there are many observational data and numerical experiment results showing that the interannual fluctuations of the atmospheric general circulation cells are closely connected with variations in the boundary conditions (sea surface temperature, soil moisture) at the surface level.

Shukla and Rasmusson (1985) undertook an overview of observational data and numerical modelling results, which showed that the formation of local interannual atmospheric general circulation variability in different regions of the tropics is largely a regional manifestation of such a global factor as ENSO monsoons, resulting from the interaction of processes in the atmosphere and ocean and on the land surface.

Some of the strongest anomalies of circulation and precipitation contributing most to the interannual variability are closely connected with the annual change; strong droughts and floods in the tropics result from either a modulation of the annual change or shifting of location and time of the maximum amplitude of the annual change. The formation of the atmospheric general circulation interannual variability due to variations in the boundary conditions creates a possibility to forecast the short-term climatic variability far beyond the deterministic predictability. In this connection, the principal features of the atmosphere–ocean interaction in the tropics can be formulated in the following way: (1) sea surface temperature anomalies in the tropics produce a substantial and deterministic effect on the atmospheric general circulation, (2) at given sea surface temperature anomalies these effects can be reliably predicted using atmospheric general circulation models, (3) the tropical sea surface temperature anomalies are mainly determined by wind stress near the ocean surface as well as by heat and water exchange, (4) with prescribed relevant quantities, sea surface temperature anomalies can be predicted using oceanic general circulation models and (5) the predictability of the interactive atmosphere–ocean system can only be higher than that of the atmosphere, since the wind field anomalies in the tropics are determined, first of all, by changes in the heat input, depending on sea surface temperature anomalies.

One of the important circumstances connected with tropical atmosphere predictability consists of the supposition that here (in contrast to mid-latitudes), synoptic-scale instabilities are weak and do not interact with atmospheric general circulation in the tropics, thereby not affecting it substantially. Large-scale heat sources determine the formation of the Hadley and Walker cells as well as monsoons; the location, repeatability and intensity of synoptic-scale disturbances are determined by specific features of the planetary atmospheric general circulation. One of the leading factors determining the variability of large-scale heat sources is that of the sea surface temperature anomalies in the tropics. Since these anomalies are known to be sufficiently stable (for at least several months), even in the case of poor predictability of the atmosphere–ocean system, the range of long-term forecasts can be increased on the basis of an empirical account of the stability of sea surface temperature anomalies. In this connection, the TOGA programme is aimed at studies of only those aspects of predictability that are determined by the world's oceans. However, an account of land surface processes is very important too.

A Report (NOAA 1979) contains the fundamentals of the US strategy on studies of the role of the ocean in changes of global climate, proceeding from a close relation between processes in the atmosphere and ocean and, as a result, global climate variations that manifest themselves through changes in the state of the ocean. The principal features of the ocean as a component of the climatic system ocean–atmosphere–land–cryosphere–biosphere are the following: (1) a larger area of ocean than land (85 per cent in the southern hemisphere and 65 per cent in the northern hemisphere), (2) low temperature (mean temperature 3.5 °C) and large heat capacity, compared with the atmosphere (the heat capacity of the upper 3 m layer of the ocean is equivalent to the total heat capacity of the atmosphere), (3) a stratification of the ocean, at least in tropical and mid-latitudes, where a thin (several hundreds of metres) warm surface layer exists, which determines the stable stratification and isolation of cold waters, lying deeper, from the atmosphere (if processes with a very large characteristic time are not considered), (4) due to high heat capacity and slow currents, changes in the ocean which are much slower than in the atmosphere (the diurnal change of temperature in the open ocean is, as a rule, undetectable, and the annual change is observed only in the upper layer several hundreds of metres thick; in the regions of strong currents, far from the equator, even the annual change can be suppressed by temperature fluctuations due to gyres) and (5) the oceans as major water sources for the atmosphere.

The strong thermal and mechanical inertia of the ocean determines high correlations between the spatial and temporal variabilities of its various characteristics. Understanding the variability of large-scale dynamical processes in the ocean is very important for an assessment of the role of the ocean and the atmosphere in the formation of climate and is potentially substantial in a search for ways of forecasting climate changes. The existence of the oceans on the Earth makes it easier to forecast climate, compared with hypothetical conditions of a planet without oceans: the release of heat accumulated by the ocean to the atmosphere determines the slowly varying impact on the atmosphere. Therefore, an analysis of observational data has revealed correlations between sea surface temperature anomalies and respective climate changes. Probably, the interannual variability of the atmosphere is largely determined by the slowly varying impact of the ocean.

Therefore, in the programme of studies of the role of the ocean in global climate changes, taking into account the fact that the ocean is a major component of the climate system, the emphasis has been placed on studies of processes responsible for sea surface temperature variations and related changes in the atmosphere: (1) studies of processes in the ocean responsible for advection and accumulation of heat as well as the ocean–atmosphere heat exchange as factors of the annual and interannual variability of climate and (2) the development and testing of the observational systems combining conventional and satellite techniques, aimed at accomplishing continuous observations of the ocean as the basis for understanding and forecasting climate changes.

These problems should be solved via both observational and theoretical studies of the following. (1) Processes of large-scale ocean–atmosphere interactions in the tropics, where the interaction is expressed most strongly and can manifest itself during periods of days to weeks, and the seasonal and interannual variability is great. Apparently, the principal mechanism for the formation of large-scale heat anomalies in the equatorial band of the ocean is a redistribution of heat by sea currents. Particularly important are studies of the equatorial Indian Ocean, where the monsoon exemplifies the strongest large-scale and continuous effect of the atmosphere on the ocean, and the Somali Current is one of the strongest in the world; the observational data point to the interconnection of sea surface temperature and precipitation anomalies over India. (2) Processes responsible for heat transport in the ocean, which impact on the global energy balance and related fluctuations of global and regional climate. (Of particular interest are studies of the meridional heat transport both in the ocean and in the atmosphere. According to preliminary estimates, in the northern hemisphere the contribution of the ocean prevails only in low latitudes, but in the southern hemisphere it remains substantial even at 60° latitude). Monitoring should be undertaken of heat transport in the Atlantic Ocean across the latitudinal circle 25° N for at least several years; for this purpose, measurements are needed of the contribution of the Gulf Stream in the Florida Strait as well as of the deep-water and near-surface transport in the remaining part of the ocean. Of great importance are studies of low-frequency variations in the antarctic circumpolar current and of the formation of water masses under the influence of the intensive ocean–atmosphere interaction in the zones of 'roaring forties' and 'furious fifties'; the North Atlantic and the South Pacific are worth being selected as major objects of initial observations. (3) The mechanisms for heat transport in the ocean, including: the transport by average circulation in horizontal gyres (the gyre in the North Atlantic particularly affects the climate) and in the vertical meridional cells of deep circulation, the transport in the near-surface layer of the ocean and the transport by gyres. (In mid-latitudes, as a rule, the heat transport from the ocean to the atmosphere takes place near the eastern coasts of Asia and Africa, the temperature gradients in the zone of cold boundaries of western warm currents create strong horizontal contrasts in the heating of the atmosphere, which substantially affects the atmospheric circulation, causing an intensification of polar jet streams, especially in the northern hemisphere winter.) (4) The transport of heat, water and momentum in the atmospheric boundary layer (especially in stormy conditions in high and mid-latitudes, where the storms occupy about half of the ocean surface) in order to develop techniques for parametrizing processes in the boundary layer. (5) Processes responsible for the variability of the distribution of sea ice and their effect on the seasonal and interannual variations in polar albedo and the ocean–atmosphere interaction. (The variability of sea ice extent and ice concentration, which determines the albedo variations in polar regions, is the strongest indicator of the ocean–atmosphere interaction; in polar regions where poorly studied processes controlling the extent and concentration of ice cover are the leading climate-forming processes, there is a close correlation between the heat flux and sea surface temperature anomalies. Therefore, of particular importance are field experiments near the ice cover edge, one of the major objectives being a study of the meridional heat transport as a factor governing the distribution of sea ice.) (6) mechanisms for vertical and horizontal geochemical exchanges in the ocean, especially the redistribution of $CO_2$ between the ocean and the atmosphere.

Though the fact that the atmosphere–ocean interaction is one of the principal factors of seasonal and interannual variability of climate is indisputable, the role of this interaction in mid-latitudes remains unclear. In this connection, Kushnir and Lau (American Meteorological Society 1991a) performed a numerical modelling of the impact of mid-latitude sea surface temperature anomalies on the atmospheric general circulation with the use of a three-level spectral (with a horizontal resolution corresponding to the consideration of 15 harmonics) atmospheric general circulation model for January conditions with prescribed anomalies of different polarity in the northern hemisphere mid-latitude Pacific Ocean.

The stationary response of atmospheric general circulation was assessed from the data of long-term integration averaged over the last 90 days, and the time-dependent response – from the results of several numerical experiments with prescribed various initial conditions.

The results obtained testify to the complicated nature of the response of atmospheric general circulation on sea surface temperature anomalies in mid-latitudes. A considerable contribution to non-linear mechanisms for the atmosphere–ocean interaction is made by motions of different scales with the presence of both stationary and time-dependent components of interaction as well as with varying spatial distribution of diabatic heat sources and sinks. There is a radical difference in heat balances in the regions of sea surface temperature anomalies for anomalies of different signs. A positive anomaly causes warming of the lower troposphere. Apparently, such a warming is balanced by a meridional heat advection, as predicted by the linear theory. In contrast to this, in conditions of a negative sea surface temperature anomaly such a balance is determined by enhanced convergence of the eddy heat flux over the cold surface of the ocean. In this case, however, the eddy motions and variations in averaged circulation bring forth changes in the distribution of heat sources and sinks, which can affect the character of the atmospheric general circulation response to sea surface temperature anomalies.

With a negative sea surface temperature anomaly, the time-dependent response of atmospheric general circulation differs from an equilibrium one, whereas the positive sea surface temperature anomaly is characterized by transition from initial to equilibrium response. This may testify to the fact that an adjustment of atmospheric flux to a changed sea surface temperature field takes more time than the characteristic time of the planetary waves dispersion. Probably, the reason for such a slow adjustment is an interaction of motions on different scales. Variations in baroclinic activity of the atmosphere over calculated storm-tracks agree with a gradual attenuation of meridional temperature gradient in the Gulf of Alaska, which in its turn corresponds to the sign of sea surface temperature anomaly and gradual intensification of the quasi-stationary positive anomaly of geopotential over Alaska and Canada. The latter is determined, apparently, by enhanced eddy fluxes.

Analysis of observational data in mid-latitudes revealed a sea surface temperature–AGC correlation, provided the atmosphere overtook the ocean – that is, served as a source of perturbations. Though in the opposite case the correlation is weak, nevertheless it deserves attention. Therefore, Miller (American Meteorological Society 1991a) used two versions of the two-layer model of the atmosphere–ocean interactive system to analyse the impact of mid-latitude sea surface temperature anomalies on the atmospheric general circulation, on the supposition that sea surface temperature anomalies were determined by (1) only heat fluxes, (2) heat flux and advection due to sea currents. The results obtained testify to a difficult solution of the problem considered, if only standard techniques of statistical analysis are used, and confirm the previous conclusion about a considerable correlation of atmospheric general circulation anomalies over the North Atlantic and Europe (in some months of the autumn and winter) and sea surface temperature anomalies with a time shift of 1 month.

In the winter of 1987, sea surface temperature anomalies were observed in the Pacific Ocean, typical of El Niño, which continued until the summer (June–July–August), when there was a positive maximum anomaly of 2 °C over the equatorial Pacific and extending from South America to the central Pacific. By the summer of 1988 the El Niño event had transformed into La Niña – that is, a negative sea surface temperature anomaly had formed in the same region. In this connection Sud *et al.* (American Meteorological Society 1991a) undertook a numerical modelling of atmospheric general circulation variations due to sea surface temperature anomalies by calculating for a period of 2 yr (1 January 1987 to 31 December 1988) with the use of an atmospheric general circulation model developed in the Goddard Laboratory for Atmospheric Sciences (GLAS). For comparison purposes, initial sea surface temperature fields were prescribed which corresponded to climatological data and real conditions (other boundary conditions, characterizing the distribution of snow, ice and vegetation cover, were identical).

The results of numerical modelling revealed anomalies of the wind field at the 850 and 200 hPa levels in the tropical western Pacific, which exceeded the anomalies observed. Anomalies of the zonal wind component were overestimated by a factor of two to three, though, on the whole, the agreement of calculated wind field anomalies with those observed was quite satisfactory. The model realistically reproduced a shift of precipitation anomalies following the evolution of warm waters in the period El-Niño–La-Niña for the tropical band of South America. However, the precipitation anomalies observed in north-western India and western tropical Africa were not discovered. For a more reliable simulation of the atmospheric general circulation dynamics, numerical experiments have been started with the use of a new 17-layer model.

To analyse the global structure of interannual variability of annual mean precipitation in July, Guo and Houghton (American Meteorological Society 1991a) performed a numerical modelling of atmospheric general circulation corresponding to a period of a hundred years, both with account of the ocean–atmosphere interaction and without it, with the use of a low-resolution atmospheric general circulation model (a five-layer spectral model with account of 10 harmonics). A special objective of the numerical modelling was to study interrelations between spatial and temporal variability of precipitation and sea surface temperature anomalies in the western tropical Pacific which, as previous studies have shown, affect substantially the processes of the ocean–atmosphere interaction.

Two versions of calculations have been considered: (1) with prescribed annual change of sea surface temperature and ice cover extent and (2) with interactive account of sea surface temperature and ice cover as predictants, pre-calculated on the basis of a thermodynamic model of the ocean mixed layer 50 m thick. Analysis of calculation results showed that calculated global climatological distribution of precipitation (averaged over 100 yr) agreed satisfactorily with that observed. A wavelike structure is one of the most characteristic features of the spatial structure of the precipitation field. Mapping the correlation of precipitation with respect to the reference point (123.75° E; 11.61° N) revealed maxima in the tropics and subtropics of the western Pacific and Atlantic Oceans. The western Pacific–Atlantic structure in the field of precipitation is a prevailing feature of the global precipitation field, closely connected with sea surface temperature anomalies in the region of the Philippines (these anomalies affect appreciably the northern hemisphere atmospheric general circulation). If the effect of the atmosphere–ocean interaction is neglected, then the western Pacific–Atlantic structure changes markedly. As for the mechanism for this structure formation, it remains unclear.

The interannual and interdecadal variability of precipitation in the region of the Sahel is known to correlate with variations in external forcings at the levels of land surface and ocean. In this connection, Towell *et al.* (American Meteorological Society 1991a) performed a numerical modelling of such variability using an 11-level (over the grid 2.5° latitude by 3.7° longitude) climate model with prescribed observed evolution of the sea surface temperature global field for the period 1949–1990 with the data renewed every 5 days. The results obtained show that variations in the global sea surface temperature field affect most the long-term variability of precipitation in the region of the Sahel. Analysis of contributions of various regions of the world's oceans to the formation of precipitation anomalies has shown that the tropical band (30° S to 30° N) plays the principal role; the coefficient of correlation between sea surface temperature variability and precipitation anomalies reaches 0.90 in this case. However, the contribution of sea surface temperature variations in extra-tropical latitudes cannot be ignored; taking account of the whole world's oceans, the value of the correlation coefficient rises to 0.96 (for extra-tropical oceans it is 0.49).

Graham (American Meteorological Society 1991a) analysed a sharp change in atmospheric circulation that took place in the mid-1970s in the wintertime North Pacific and, probably, a related variation in the interactive ocean–atmosphere system in the tropical Pacific. This sharp change was manifested through an expanding Aleutian pressure minimum southward and eastward with an accompanying southward shift of the zone of east–western transport in 1975–1976. In this connection, the laws of atmospheric general circulation have been considered on a decadal scale from the data of observations and numerical modelling with prescribed sea surface temperature field, bearing in mind, in particular, a study of the possible impact of changes in the tropical Pacific in the form of sea surface temperature anomalies on the mid-latitude atmospheric circulation.

Calculations were made with the use of the ECMWF 16-layer spectral (with 21 harmonics, which corresponds to horizontal resolution of 5.625° in latitude and longitude) atmospheric general circulation model. The sea surface temperature fields were prescribed from observational data for the period 1970–1985. Calculated values of 700 hPa level heights in the wintertime northern hemisphere (December–January–February; north of 30° N) were used to determine differences of geopotential averaged over six winters from 1976–1977 to 1981–1982 and, respectively, from 1970–1971 to 1975–1976. To characterize the relations between the data of observation and modelling, a technique of canonical correlation analysis has been used, which is a combination of techniques of empirical orthogonal functions and traditional correlation analysis; this makes it possible to identify the prevailing models of covariances of two fields of meteorological parameters.

The results obtained by Graham (American Meteorological Society 1991a) show that the numerical modelling satisfactorily simulates the observed variability of geopotential, the presence of the signal with a characteristic timescale of 10 yr being of first-priority value for an agreement between observation and calculation data. Similar regularities demonstrate also the prevailing canonical structures, reflecting the effect of sea surface temperature field in the tropics on the formation of atmospheric general circulation in mid-latitudes.

Diagnostic studies, performed during recent years, revealed a strong component of a 2-yr variability of the El Niño / Southern Oscillation event, which suggests the hypothesis of a possible impact of quasi-biennial oscillations in the stratosphere on the ENSO dynamics, from the viewpoint of the appearance of periods of sea surface temperature decrease and increase (Gray *et al.* in American Meteorological Society 1991a, Mason and Tyson 1992). Probably, thc bcginning of thc ENSO warming phase coincides with the eastern phase of the quasi-biennial oscillation attenuation or the transport from the eastern to western phase of the quasi-biennial oscillation in the lower stratosphere in the period January–June.

During the western phase of the quasi-biennial oscillation a natural mechanism appears due to which the penetrating convection will intensify in the band 8–18° S of the region Indonesia to north Australia, which causes an intensification of the Hadley–Walker local circulation cell, which in its turn favours an appearance of strong trade winds in the Pacific and, hence, conditions stimulating the beginning of the ENSO cooling phase. In the case of the opposite situation (eastern phase of the quasi-biennial oscillation), conditions appear that favour the shift of the region of penetrating convection (Indonesia to north Australia) to the equator, with its intensification near the equator (±6° lat.) and suppression in the band 8–18° S. In these conditions the Walker circulation and trade winds cease, and this promotes the beginning and evolution of the ENSO warming phase. The mechanism for modulation of penetrating convection can be explained by the fact that during the western and eastern phases of the quasi-biennial oscillation the contrasting wind stress and statistical stability are formed in the lower stratosphere (the 150–50 hPa layer), which determine this modulation.

## 5.5 Nested climate models

Attempts to use global climate models to reproduce regional climates have not given sufficiently satisfactory results (Klink 1995). Great differences have been found in the prediction of probable regional climate changes caused by $CO_2$ doubling ($2 \times CO_2$). To some extent, such differences are determined by inadequate parametrization of various climate-forming processes. In this connection, Ghan (1992) emphasized the significance of inaccurate treatment of advection and of mixing at progressively shorter spatial scales. Advection is inaccurate at small scales as a result of finite difference or spectral truncation errors. One more factor is climatic noise, or natural (internal) variability of climate. Signal-to-noise ratio for greenhouse warming for individual general circulation model grid cells is much smaller than for global scales. Undoubtedly, finer spatial resolution and climatic noise reduction techniques will improve the reliability of regional scale predictions.

Inadequate results of using the global climate models to simulate regional climates have caused attention to be given to this problem through applying limited area models nested in a general circulation model. The combined limited area / general circulation models can be particularly effective with a complicated topography or coastline, which is typical, for example, of the western USA.

Giorgi *et al.* (1992) discussed the climate numerical modelling results for this region obtained with the use of a general circulation model and a limited area model developed, respectively, at the National Center for Atmospheric Research (NCAR) and Pennsylvania University (mesoscale model MM-4). An important stage of such a numerical modelling is the validation of the results obtained. In this connection, the MM-4 model has been validated on the basis of calculations for the western USA for the periods of two years (1 January 1982 to 31 December 1983) and 16 months (1 January 1988 to 25 April 1989), characterized, respectively, by reduced and increased humidity.

The validation has been concentrated on consideration of two parameters: surface air temperature and precipitation. A comparison with the data of observations has shown that the MM-4 model satisfactorily simulates the surface air temperature annual change (differences do not exceed several degrees). The greatest errors in predictions occur in summer in the Rocky Mountains. The annual change of precipitation is also well simulated. As a rule, the predicted precipitation corresponds better to that observed during the cold half of the year than in summer, and better in coastal regions than in inland areas.

A division of the territory considered into six subregions showed that the precipitation forecast was most reliable in California and the north-western Pacific, where differences in monthly mean and seasonal precipitation do not exceed $\pm(10\text{–}50)$ per cent. Over the Rocky Mountains the calculated precipitation was close to that observed in winter, but rather overestimated in summer. The latter suggested a revision of the mesoscale model by taking into account a less intensive horizontal diffusion along the surface, depending on the gradient of topography. Such a revision provided an almost halved precipitation level in summer over the mountains but almost no affect in winter.

The results obtained make it possible to formulate two principal conclusions, as follows. (1) If the ECMWF data are used to prescribe initial and boundary conditions (fields of wind, temperature, humidity and surface pressure), then the MM-4 simulated regional climate agrees satisfactorily with monthly mean and seasonal values observed at a network of 390 stations. As a rule, the differences between monthly mean and seasonal surface air temperature values do not exceed several degrees, and those of precipitation about 30–40 per cent. It is important that the reliability of the numerical modelling results does not decrease with increasing range of forecast. After a spin-up time of a few days the performance of limited-area models is basically determined by a dynamical equilibrium between the information from the driving large-scale fields and the internal model physics. To reach an equilibrium with the use of models taking into account hydrological processes (including lakes), calculations are needed for a period of months to several years. (2) From the viewpoint of available computer resources, calculations are possible, with the use of a mesoscale model, for a period of many years. In this case, for models with a horizontal resolution of 60 km and a grid of $50 \times 55 \times 14$ points, the Cray Y-MP computer time constitutes about 70 hours. Thus, possibilities of numerical climate modelling using the limited area/general circulation models are quite realistic.

Based on the use of the NCAR model, Bates *et al.* (1992b) undertook a numerical climate modelling in the region of the Great Lakes affecting substantially the climates of adjacent territories. In particular, the thermal inertia of the lakes causes an increase of local surface air temperature and dew point values, of up to 10 deg K. On synoptic scales (about 1000 km) the lakes favour the intensification of centres of low pressure in winter and high pressure in summer. The contribution of the lakes to the formation of snowfalls in winter constitutes 30–50 per cent, and in summer on the contrary a lake-induced (as a lower surface temperature) decrease of cloud amount and precipitation by 20–30 per cent occurs.

Calculations with the use of limited area models have been made with prescribed regularly renewed initial and boundary conditions from the data of objective analysis made at the ECMWF. The lake surface temperature was prescribed from the data of observations. The numerical modelling was made for the 10-day period from 22 December 1985 to 1 January 1986, with varied horizontal resolutions of 30, 60 and 90 km, bearing in mind an analysis of the role of spatial resolution. Naturally, for example, the precipitation field with the 30 km resolution is more inhomogeneous. A comparison with results of conditional calculations in the absence of lakes revealed that the surface air pressure over the lakes was higher by 2.75 hPa than with the lakes present. Preliminary calculations were made with the use of the MM-4 together with a one-dimensional model of the thermal regime, which made it possible to predict the ice-cover onset.

Hostetler *et al.* (1992) analysed possibilities of the combined (in the interactive regime) use of MM-4 and the biospheric model BATS, simulating the heat and moisture exchange in the presence of vegetation cover, as well as a model of the thermal regime of lakes. The MM-4/BATS results of simulating the weather in the western USA have been used as initial data for the model of the thermal regime of lakes, used to calculate the diurnal mean surface temperature and evaporation for Pyramid Lake, Nevada, as well as a catchment-area model to calculate the stream flow discharge for the river Steamboat Creek, Oregon.

The results obtained can be considered positive. The models MM-4/BATS and of the thermal regime of lakes were used (in the interactive regime) to assess the impact of large lakes on regional climate. The MM-4 data were introduced every hour, and also every hour the data of the lake model (surface temperature, evaporation, ice cover) were put in the MM-4. Preliminary results for Pyramid Lake, Nevada, and the Great Lakes show that the interactive model considered ensures quite a realistic simulation of the lake–atmosphere interaction, including the formation of ice cover on the Great Lakes.

Guerrero *et al.* (1992) undertook a numerical modelling of conditions for cyclogenesis in the western North Atlantic for the last glacial maximum, based on the use of a large-scale spectral climate model CCM-1 and nested mesoscale model MM-4 (the data of the large-scale model were used to prescribe initial and boundary conditions for the MM-4). This region is particularly interesting in connection with an intensive cyclogenesis there; this region is characterized by powerful cyclones called bombs (the latter is defined as a cyclone with surface air pressure decreasing by more than 24 hPa per day). Analysis of the CLIMAP data led to the conclusion that such a cyclogenesis had taken place in the period of the last glacial maximum, including the formation of bombs.

Using the CCM-1, calculations were made for the cases corresponding to present conditions (for various versions of CCM-1) and to the last glacial period. A comparison of the numerical modelling results with the data of observations revealed an underestimated (almost halved) intensity of cyclones compared with that observed (the pressure drop per day, expressed in Bergeron – a unit corresponding to a pressure decrease by 24 hPa during 24 hours at the latitude 60° – can be considered a criterion of intensity). The number of cyclones and their intensity predicted with various versions of CCM-1 turned out to be close to observational data.

The intensity of palaeocyclones differs little from the present one (in this case the boundary conditions have been prescribed from the CLIMAP data). Thus, the CCM-1 model could not simulate correctly the intensity of cyclones. Therefore, similar calculations are planned with the use of a mesoscale model. In this case the horizontal resolution of a 16-layer model can be raised to 100 km, whereas the nine-layer CCM-1 model has a resolution of 550 km. An improved parametrization of the processes in the atmospheric boundary layer would be very important.

Bossert *et al.* (1992) used the mesoscale model BATS, developed at Colorado State University, to analyse the laws of formation of the water cycle in a mountainous, arid region of the USA (127.5–97.5° W; 25.5–52.5° N). The 21-layer 3-D model considered has a horizontal resolution of 0.5° in latitude. The calculations cover the period 1–31 January 1988. Initial conditions (fields of wind, temperature, geopotential and relative humidity) have been prescribed from the data of objective analysis at the National Meteorological Center (USA), with boundary conditions renewed every time step.

The simulation of a regional hydrological regime has been made with the use of two approaches: (1) initialization from the data of objective analysis at the National Meteorological Center repeated every 12 hours and (2) a single initialization at the beginning with boundary conditions renewed every 12 hours. A preliminary analysis of the numerical modelling results has shown that after the initial acceleration of the model, the re-initialized climatology resembles in many respects the monthly mean one. Therefore the advantage of the first approach consists in a possibility of simultaneous (parallel) runs during 12-hour intervals, whereas the second approach requires successive calculation, which leads to an increase of computer time.

The results of numerical modelling of regional climate can be considered positive. In particular, the dependence of precipitation on topography is clearly seen (in the presence of a complicated relief) and the model can realistically simulate the fields of liquid and solid precipitation and their accumulation (this testifies to the reliability of parametrization of microphysical processes). The RAMS model is planned to be used to calculate the climate in the basins of some catchment areas in the western USA.

In the period January to May 1983 the atmospheric general circulation in the northern hemisphere turned out to be much more intensive than usual and was followed by numerous strong wintertime storms in the western USA. These storms were connected with a powerful polar jet stream shifted southward (compared with its usual location) by several degrees in latitude. In early winter several storms with precipitation took place at the western coastline, exceeding the norm by 200 per cent. In late winter the storm-tracks continued spreading over the western USA, followed by some wintertime storms which caused intensive snowfalls in the central region of the Rocky Mountains. This precipitation (together with an unusually cold spring followed by a sudden warming in the region of the Rocky Mountains) resulted in local heavy showers in July that caused a maximum high stream flow discharge of the Colorado river followed by floods and damage of constructions.

Based on the use of an advanced MM-4 model, Matthews *et al.* (1992) undertook a numerical modelling of the events mentioned (the model was advanced by using a more reliable parametrization of hydrological processes in soils, the atmospheric boundary layer and radiation processes). Initial and boundary conditions were prescribed from the ECMWF data of objective analysis. The study area of the Colorado river basin is a square with a side of 3000 km centred at 40° N, 116° W. The 13-layer model has a horizontal resolution of 60 km. Analysis of the numerical modelling results has shown that the model provides a good simulation of the situation in question, including its principal feature, a longwave depression with the polar jet stream south-west of it, which is responsible for the formation of intensive precipitation. It was this feature that was responsible for a continuous moist air flow from the Pacific Ocean and frequent storms crossing the regions of the Colorado river. In conditions of a late-spring warming the snow cover melted and a maximum intensity stream flow discharge took place. A numerical modelling reliably simulated details of the mesosynoptic flow across the western USA.

Thus a simulation of mesoscale processes using the atmospheric general circulation model and nested mesoscale model could be considered reliable. The ability of the atmospheric general circulation model to simulate reliably the structure of general circulation and the location of the longwave depression will determine the MM-4 potential from the viewpoint of simulating the mesoscale climatic regularities in the future. Further testing of the CCM-1 with the nested mesoscale unit is planned, bearing in mind an analysis of adequate simulation of climate both at present and in the future, including the doubling of $CO_2$ concentration case.

Medina (1992) discussed the prospects for the analysis of a possible impact of climate changes in the catchment area of the Delaware River (USA) based on the use of a two-dimensional stationary mesoscale model which took into account only physical processes responsible for the formation of orographic clouds. In order to simulate the present regime of precipitation, the model has been realized over a grid with a step of 10 km within the territory 340 km by 440 km, completely including the watershed.

Calculations for the period 1 September 1974 to 31 December 1983 have made it possible to obtain data on daily precipitation and its daily means at locations of 25 rain-gauge stations. A comparison of calculated and observed data for all winter months gave a correlation coefficient of 0.59, whereas in the summer months this coefficient decreased to 0.43, which was determined by the inability of the model to simulate reliably the precipitation during storms with an intensive convection.

The results of numerical modelling for individual days and locations proved unreliable in many cases. Therefore the model can be considered adequate only for climatic assessments on the scale of the whole watershed. It is also useful for assessments of local impacts of global climate changes, provided an initialization is applied with the use of a regional-scale model. The model considered is planned to be used for several watersheds over US territory to assess a possible impact of $CO_2$ doubling on the hydrological regime, initializing from the data of the global climate model.

Since an adequate parametrization of heat and moisture exchange between the surface and the atmosphere is very important for the reliability of global climate models, the problem of taking into account the horizontal inhomogeneity within a spatial grid cell of the model attracts keen attention (Klink 1995). To analyse the ways to solve this problem, Doran *et al.* (1992) undertook measurements of sensible and latent heat fluxes in June 1991 in the region of Boardman (north-eastern Oregon, USA) at some sites of an arid steppe (at 10 locations) as well as on irrigated fields of wheat, maize, potato and soy-beans, followed by aircraft observations along the flight-track crossing the sites under study.

Observations were made at three locations at different altitudes with the use of tethered balloons. Direct (eddy correlation) measurements of $CO_2$ flux were made on the fields of wheat and maize. Two flights of a Convair-580 aircraft were undertaken for multi-spectral air survey of the territory under study. A preliminary analysis of observational results has shown that the differences in heat and moisture exchange in steppe and on agricultural fields are clearly illustrated by the data of both surface and aircraft observations. Maximum values of sensible heat fluxes over the steppe varied within 475–500 $W\,m^{-2}$, whereas the latent heat fluxes turned out to be less by almost an order of magnitude. Over the irrigated areas the fluxes of latent heat reached 600 $W\,m^{-2}$ with the fluxes of sensible heat being between 100 and 200 $W\,m^{-2}$. Within these limits, the heat fluxes on the fields of maize and potato differed most.

In the context of regional climate studies, the consequences of deforestation in the Amazon basin attract special attention; the basin can play an important role in the functioning of the global climate system as one of three quasi-constant centres of intensive convection immersed in the zone of equatorial depression. The existence of the Amazon jungles is partly determined by intensive precipitation and, in their turn, they make an important contribution to the formation of the water cycle. The deforestation during recent decades has led to a significant change of boundary conditions. All this explains a particular interest in assessments of a possible impact of the biospheric dynamics in the Amazon basin on the global climate system, taking account of the relation between natural and anthropogenic impacts.

In this context Marengo and Hastenrath (1992) undertook an analysis of the data from meteorological observations for the Amazon basin which revealed, in particular, a more humid climate in the 1970s but a drier climate during the next decade. The most numerous data for the 1980s revealed two special years, namely 1983, the year of El Niño, which was extremely dry, and 1986 which was much more wet. An examination of observed regularities of the atmospheric general circulation suggested some general conclusions about probable mechanisms for precipitation anomalies.

In the period of wet climate, the centre of convection in the Amazon basin – part of the ITCZ – intensifies and the tropical North Atlantic is characterized by an abnormally cold water surface. In this period a strong Walker circulation cell forms in the Pacific, as well as a zone of downward fluxes over the western coastline of South America, whereas in the Amazon basin the upward motions and convection prevail.

In the dry period the atmospheric circulation is mainly opposite to that observed in the wet period: the ITCZ is shifted northward, and the tropics of the North Atlantic turn out to be relatively warm and the South Atlantic is unusually cold. The described features of atmospheric circulation are characteristic of the period of the season of maximum rains in the northern part of the Amazon basin and are not clearly manifested in other seasons and in the southern part of the Amazon basin.

CHAPTER SIX

# External impacts on the climatic system

The problem of external impacts on the climatic system, and, in particular, anthropogenically induced climate changes, is now of great concern. Here we shall confine ourselves to a discussion of only two of the most important cases of external forcings determined by increased $CO_2$, by other greenhouse gases and by volcanic eruptions.

## 6.1 Carbon dioxide and climate

### 6.1.1 Numerical modelling for $CO_2$ increase

Despite much effort spent on numerical modelling of the impact of $CO_2$ on climate, the problem still remains inadequately studied (Broecker 1995, Tans and Bakwin, 1995). Until recently, the majority of publications had not gone beyond the well-known scheme: an increase of $CO_2$ intensifies the greenhouse effect, which leads to a global climate warming. It has long become clear, however, that both this greatly simplified scheme and the rough assessment that the greenhouse effect is caused by $CO_2$ are far from reality. There are still numerous unsolved problems (particularly concerning a consideration of the effect of the ocean and clouds), connected with various physical aspects of climatic impacts. It must be borne in mind that water vapour contributes most to the formation of the greenhouse effect, and therefore it is critically important to take into account the feedbacks that enable an adequate description of water vapour dynamics in the course of climate change (an account of cloud–radiation interaction is the most difficult aspect of the problem).

Recent studies (Kondratyev 1978b, 1989b, 1990a,b, 1991a, 1992b,d, 1996, Gates 1979, 1988, Houghton 1979, 1984, 1986, Schlesinger and Jiang 1989, Manabe and Wetherald 1980, 1987, Gates *et al.* 1981, Hansen *et al.* 1981, 1988a,b, 1990, 1991, Kellogg and Schware 1981, Manabe *et al.* 1981, 1991, 1992, Mitchell *et al.* 1981, 1987, 1995, Schneider 1981, 1985, 1989, Schneider and Thompson 1981, Clark 1982, 1983, Carbon Dioxide Assessment Committee 1983, Kellogg 1983a,b, 1987, 1991, Kellogg and Bojkov 1983, MacCracken 1983, 1991, Mitchell 1983, 1987, 1989, Manabe 1983, Newell and Hsiung 1984, Gates and MacCracken 1985, Idso 1985, 1987, 1988, 1989, National Climate Program Office of Science 1985, 1988, MacCracken *et al.* 1986, NOAA 1986, 1988, Henderson-Sellers and McGuffie 1987, 1989,

Wigley 1987, 1989a,b, Borisenkov *et al.* 1988, Borisenkov and Kondratyev 1988, Ellsaesser 1988, 1989, 1991, Gates 1988, Gruza and Shevchenko 1988, Hansen 1988, Kellogg and Zhao 1988, Manabe and Stoufer 1988, Schlesinger and Zhao 1988, 1989, US Department of Energy 1988, US Environmental Protection Agency, 1988, Flöhn 1989a,b, Glantz 1989, Henderson-Sellers and Blong 1989, Idso *et al.* 1989, Schlesinger 1989a,b,c, Schlesinger and Zhao 1989, Schlesinger *et al.* 1989, Houghton *et al.* 1990, 1992, 1996, Lindzen 1990, 1991, Washington and Meehl 1993, IPCC 1994, Bouger 1995, Clivar 1995, Keeling *et al.* 1995, Mortier 1995, Kondratyev *et al.* 1996a,b) have shed light on the numerous difficulties connected with obtaining realistic estimates of the $CO_2$ effect on climate. A detailed annotated bibliography has been compiled by Handel and Risby (1992). Here we shall briefly consider the latest results in this field. First of all, they are connected with studies of the multicomponent nature of the greenhouse effect (the role of various minor optically active atmospheric components, the effect of hot bands, temperature dependence, etc.). Also discussed are the poorly studied aspect of the global carbon cycle (see also Chapters 1 and 2), and numerical climate modelling with changes in the atmospheric composition properly taken into account.

The uncertainty of existing assessments is determined by the numerous assumptions on which numerical modelling is based. A particularly serious drawback is a poor knowledge of the role of the ocean in the formation of the global carbon cycle and in the impact of $CO_2$ on climate. There are still no adequate techniques for parametrizing cloud–radiation interaction. The problems of taking into account varying optical properties of cirrus clouds and snow cover remain unsolved. The contribution by the numerous optically active minor gaseous and aerosol components must be estimated more reliably.

The identification of $CO_2$-induced climate changes from observational data is hindered by the masking effect of natural climate variability (this problem has been discussed in detail in Section 2.2). Apparently, not for one or two decades can we expect serious progress in the reliability of numerical modelling of the climate impact of $CO_2$.

The second Report of the National Research Council (1982) published by the US National Academy of Sciences, and documents prepared by the US Department of Energy (MacCracken and Luther 1985) and later by the IPCC (Houghton *et al.* 1990, 1992, 1996) give an accurate enough picture of the development of present-day studies of $CO_2$ climatic implications.

As far back as 1863, J. Tindall clearly formulated the notion of the greenhouse effect. Comparatively transparent to solar radiation, atmospheric water vapour strongly absorbs the thermal emission of the Earth's surface. Some years later, Arrhenius and Chamberlain drew attention to a significant contribution of the $CO_2$ thermal emission to the formation of the heat balance and, consequently, of the temperature field in the atmosphere. Arrhenius' estimates have shown that a doubled $CO_2$ concentration should lead to a global warming by about 6 °C. Callender was the first who suggested the possible impact of anthropogenic releases of $CO_2$ on climate.

At present the anthropogenic growth of $CO_2$ concentrations in the atmosphere is clearly established, and a successful development of climate theory has created the basis for comparatively realistic estimates of the impact of this growth on climate. A group of experts headed by the late J. Charney, formed by the US National Academy of Sciences, prepared and published in 1979 a report in which the results from studies on the problem of $CO_2$ impact on climate were summarized. Some years later, a second similar analytic report was prepared on the status of the problem by 1982, supervised by J. Smagorinsky. In this report, together with a discussion of the problem as a whole, the following aspects were considered in detail: (1) the role of the ocean, (2) the reliability of the conclusions about the impact of $CO_2$ on climate, drawn from analysis of local observational data on the land surface heat balance, (3) the contribution of minor optically active components and aerosols and (4) the perspectives of the identification of the $CO_2$ signal in climate change from observational data (see also Hasselmann 1993).

### 6.1.2 Equilibrium response experiments

In the IPCC Report (Houghton *et al.* 1990) there is a summary (table 3.2) giving the results of model calculations to determine the effect of doubling the present concentration of $CO_2$ in the atmosphere and assuming that equilibrium is re-established. The results presented were for 22 global mixed-layer ocean atmosphere models. The mixed-layer approach falls short of considering full ocean circulation but it is an approximation which is useful for reasons of economy, or necessity, in terms of computing power. In this approach one considers the uppermost layer of the ocean where the oceanic temperature is relatively uniform with depth. It is frequently modelled as a simple slab for which a fixed depth of the mixed layer is prescribed and the oceanic heat storage is calculated; the oceanic heat transport is either neglected, or is treated as being carried only within the mixed layer, or is prescribed from climatology. Variations of mixed-layer depth, oceanic heat flux convergence, and exchanges with the deep ocean, which would entail an additional storage and redistribution of heat, are all neglected as well. Sea ice extents may be determined interactively, but some models already include dynamical effects such as sea ice drift and deformation caused by winds and ocean currents. The simple mixed-layer model has strong limitations for studies of climate change, particularly as it does not allow for the observed lags in heat storage of the upper ocean to be represented, and clearly a coupled ocean–atmosphere model is, in principle, to be preferred. The results already mentioned given in the IPCC Report were values of $\Delta T$, the calculated change in equilibrium surface temperature, and $\Delta P$, the calculated percentage change in precipitation, resulting from a doubling of the concentration of $CO_2$ in the atmosphere. All the models show a significant increase in global mean temperature ranging from 1.9 °C to 5.2 °C, with most results between 3.5 °C and 4.5 °C. The percentage change in precipitation was found to vary between 3 and 15 per cent. However, these are global mean values. The climate models do not just produce estimates for global means; neither does a global mean necessarily convey useful information for everyone. There may well be important regional variations, and for many purposes it is the local rather than the global climate effects which are important. Chapter 5 of the IPCC Report (Houghton *et al.* 1990) describes some of the rather more detailed results of the use of the various models to attempt to determine the consequences of an equilibrium situation with double the present concentration of $CO_2$ in the atmosphere; for details we refer the reader to that chapter of the IPCC Report.

The important thing is, perhaps, not to concentrate too much on the finer spatial details of the results for one time and place with any one particular model, but rather to try to concentrate on what appears to be common to the results of all, or at least most, of the models. The general conclusions are, perhaps, best taken from the Executive Summary of chapter 5 of the IPCC Report (Houghton *et al.* 1990), and they are worth quoting verbatim because they give a very good overall impression of the nature and status of the results.

"1. All models show substantial changes in climate when $CO_2$ concentrations are doubled, even though the changes vary from model to model on a sub-continental scale.

2. The main equilibrium changes in climate due to doubling $CO_2$ deduced from models are given below. The number of *s indicates the degree of confidence determined subjectively from the amount of agreement between models, our understanding of the model results and our confidence in the representation of the relevant process in the model. Five *s indicate virtual certainties, one * indicates low confidence.

■ *Temperature:*

***** the lower atmosphere and Earth's surface warm;

***** the stratosphere cools;

*** near the Earth's surface, the global average warming lies betweeen +1.5 °C and +4.5 °C, with a "best guess" of 2.5 °C;

*** the surface warming at high latitudes is greater than the global average in winter but smaller than in summer (in time dependent simulations with a deep ocean, there is little warming over the high latitude southern ocean);

*** the surface warming and its seasonal variation are least in the tropics.

■ *Precipitation:*

**** the global average increases (as does that of evaporation); the larger the warming, the larger the increase;

*** increases at high latitudes throughout the year;

*** increases globally by 3 to 15 per cent (as does evaporation);

** increases at mid-latitudes in winter;

** the zonal mean value increases in the tropics although there are areas of decrease. Shifts in the main tropical rain bands differ from model to model, so there is little consistency between models in simulated regional changes;

** changes little in subtropical arid areas.

■ *Soil moisture:*

*** increases in high latitudes in winter;

** decreases over northern mid-latitude continents in summer.

■ *Snow and sea-ice:*

**** the area of sea-ice and seasonal snow-cover diminish.

The results from models become less reliable at smaller scales, so predictions for smaller than continental regions should be treated with great caution. The continents warm more than the ocean. Temperature increases in southern Europe and central North America are greater than the global mean and are accompanied by reduced precipitation and soil moisture in summer. The Asian summer monsoon intensifies.

3. Changes in the day-to-day variability of weather are uncertain. However, episodes of high temperature will become more frequent in the future simply due to an increase in the mean temperature. There is some evidence of a general increase in convective precipitation.
4. The direct effect of deforestation on global mean climate is small. The indirect effects (through changes in the $CO_2$ sink) may be more important. However, tropical deforestation may lead to substantial local effects, including a reduction of about 20 per cent in precipitation."

We note, in particular, the comment that predictions for smaller than continental scales should be treated with great caution. Only further improvements of climate models, which would consider a realistic interaction between marine and continental biota, will permit a reliable simulation of possible variabilities of various climatic parameters and, in particular, an assessment of the impact of climate changes on the biosphere.

To obtain these estimates, not only the annual mean or global mean quantities must be numerically modelled but also the annual change and the geographical distribution of calculated climatic parameters, including an estimation of probable errors as well as the transformation of changes by a large-scale model into data for characteristic scales of climate impacts on the biosphere. (One of the possible ways of solving this problem is by the use of empirical statistics.) Moreover, predicted climate changes must be analysed in terms of specific quantities which must be considered in studies of the climatic impact on the biosphere.

Gates (1988) considers the status of climate theory from the viewpoint of the reliability of predicted climate changes due to increasing $CO_2$, as well as methodological aspects of the assessment of climatic impacts on the biosphere. Table 4.46 shows typical sets of input and output parameters of the present three-dimensional atmospheric general circulation models.

Despite different parametrization schemes for subgrid processes, spatial and temporal resolution and calculation algorithms, the present calculations give similar results with respect to the simulation of large-scale climatic features. They may be useful in the analysis of the anthropogenically-induced variability of time series for individual points of the grid. An example for 34° N and 100° W shows that, with a quadrupled $CO_2$, soil moisture decreases in the autumn, but increases in the winter. However, the number of days without precipitation increases in almost every month, particularly in the spring and early summer.

The most important directions that numerical modelling of climate should take are as follows: (1) the development of models to simulate more adequately past and present climates, the calibration and verification of models based on observational data, (2) a comparison of models of different complexity, (3) experiments on the predictability of models useful for climate change forecasts on the scales of months, years and decades as well as assessments of the predictability of natural variability, (4) studies on climate sensitivity aimed at analysis of the models' response to prescribed changes in boundary conditions and various types of forcing and (5) the determination of the models' sensitivity to numerical algorithms and parametrization of physical processes.

Apart from the IPCC Report (Houghton *et al.* 1990, 1992) there is another comparison of atmospheric general circulation models by Randall *et al.* (1992) in which they describe the responses of the surface energy budgets and hydrological cycles of 19 models to a different forcing, namely a uniform increase of 4 deg K in the sea surface temperature. The responses of the simulated surface energy budgets were extremely diverse and were closely linked to the responses of the simulated hydrological cycles. The response of the net surface energy flux is not controlled by cloud effects; instead, it is determined primarily by the response of the latent heat flux. The prescribed warming of the oceans led to major increases in the atmospheric water vapour content and the rates of evaporation and precipitation. The increased water vapour amount drastically increased the downwelling infra-red radiation at the Earth's surface, but the amount of the change varied dramatically from one model to another. Other work on the intercomparison of results obtained from different models is described in Ellingson and Fouquart (1991) and various other papers in a special section of the *Journal of Geophysical Research*, volume 96, 20 May 1991, pages 8921–9157, by Gutowski *et al.* (1991) and by Boer *et al.* (1992).

One can ask, quite reasonably, what is the value of carrying out equilibrium calculations such as those we have just mentioned. Climate is in equilibrium when it is in balance with the radiative forcing. Thus, as long as greenhouse gas concentrations continue to increase, the climate will not reach equilibrium; moreover, even if concentrations of greenhouse gases are eventually stabilized it would be a long time before equilibrium would be established. Thus equilibrium calculations cannot be used directly to provide forecasts of climate changes. The first reason given for carrying out equilibrium simulations (usually with double the concentration of $CO_2$ in the atmosphere) is that they are done with atmospheric–oceanic mixed-layer models which ignore both the deep ocean and changes in the ocean circulation; these require less computer time than time-dependent simulations which include the influence of the deep ocean. Secondly, because these simulations are relatively inexpensive (in terms of computer time and therefore of money) to carry out, they can be used fairly easily for intercomparison of different models and for studying the effects of different parametrizations, e.g. of cloud (Schmitt and Randall 1991, Peterson *et al.* 1992, Potter *et al.* 1992). Thirdly, it appears that apart from areas where the oceanic thermal inertia is high, equilibrium solutions can be scaled and used as approximations to the time-dependent response.

### 6.1.3 Time-dependent response experiments

In the models used in the equilibrium response calculations described above, one supposes that an abrupt change has been made in the atmospheric $CO_2$ concentration and that this leads to a change in the radiative forcing; this is allowed to operate in the model for a period corresponding to a few years until a new equilibrium is established. The atmospheric models used are detailed and highly developed but the treatment of the ocean, as we have already mentioned, is quite primitive. However, in reality, greenhouse gas concentrations are changing continuously, not abruptly. The climate's reaction is not instantaneous, because of the thermal capacity of the oceans. At any given time the response corresponds to only part of the equilibrium change for the corresponding instantaneous value of the forcing. For the remainder, part of the response is delayed by storage in the stably stratified layers of the upper ocean and takes several decades, or even a century, to occur. However, there is also a part of the response which remains invisible for a long time (many centuries) until the deep ocean begins to influence the surface temperature. Apart from the thermal inertia of the oceans there may also be other oceanic effects, e.g. the ocean currents can lead to a spatial redistribution of greenhouse warming. Also, even in the absence of changes in the radiative forcing, the ocean–atmosphere interactions lead to fluctuations which may temporarily mask long-term climate changes.

In order to make reliable predictions of climate change under realistic scenarios of slow changes in the radiative forcing, it is necessary to take the oceanic behaviour properly into account. As we have already mentioned, this means that one has to make use of coupled atmosphere–ocean general circulation models (GCMs). There is discussion of time-dependent model calculations in chapter 6 of the IPCC Report (Houghton *et al.* 1990). The use of a coupled atmosphere–ocean general circulation model means that the computing requirements for a calculation are very high. The experiment which is described involved a 100-yr simulation. Three runs of a model were carried out with (a) no change in atmospheric $CO_2$ concentration, (b) a 1 per cent per annum (compounded) increase in atmospheric $CO_2$ concentration and (c) a 1 per cent per annum (compounded) decrease in atmospheric $CO_2$ concentration. The choice of 1 per cent increase in the $CO_2$ concentration is based on the fact that this has the equivalent effect, in terms of radiative transfer, to the combined effect of the increases in all the various greenhouse gas concentrations at present. By the years 60–80 of the model calculation, the concentration of atmospheric $CO_2$ has built up at a (compound) rate of increase of 1 per cent per annum to approximately double the initial value, but the concentration has been built up slowly and equilibrium has not yet been reached, whereas in the equilibrium calculations described earlier the concentration of $CO_2$ was suddenly doubled and equilibrium was supposed to be re-established rather quickly.

The use of models in time-dependent response calculations is still in its early days. The conclusions from the results that have been obtained so far will be found in the IPCC Report (Houghton *et al.* 1990) and the summary of those conclusions is worth quoting verbatim:

"(a) For a steadily increasing forcing, the global rise in temperature is an approximately constant fraction of the equilibrium rise corresponding to the instantaneous forcing for a time that is earlier by a fixed offset. For an atmospheric model with temperature sensitivity 4 °C for a doubling of $CO_2$, this fraction is approximately 66% with an offset of 11 years. In rough terms, the response is about 60% of the current equilibrium value. Extrapolation using robust scaling principles indicates that for a sensitivity of 1.5 °C the corresponding values are 85%, 6 years and 80% respectively.

(b) The regional patterns of temperature and precipitation change generally resemble those of an equilibrium simulation for an atmospheric model, though uniformly reduced in magnitude. Exceptional regions are around Antarctica and in the northern North Atlantic, where the warming is much less.

(c) These results are generally consistent with our understanding of the present circulation in the ocean, as evidenced by geochemical and other tracers. However, available computer power is still a serious limitation on model capability, and existing observational data are inadequate to resolve basic issues about the relative roles of various mixing processes, thus affecting the confidence level that can be applied to these simulations.

(d) The conclusions about the global mean can be extended, though with some loss of rigour, by using a simple energy-balance climate model with an upwelling – diffusion model of the ocean, similar to that used to simulate $CO_2$ uptake. It is inferred that, if a steadily increasing greenhouse forcing were abruptly stabilized to a constant value thereafter, the global temperature would continue to rise at about the same rate for some 10–20 years, following which it would increase much more slowly, approaching the equilibrium value only over many centuries.

(e) Based on the IPCC Business-as-Usual scenarios, the energy – balance upwelling – diffusion model with best judgement parameters yields estimates of global warming from pre-industrial times (taken to be 1765) to the year 2030 between 1.3 °C and 2.8 °C, with a best estimate of 2.0 °C. This corresponds to a predicted rise from 1990 of 0.7–1.5 °C with a best estimate of 1.1 °C."

Part (e) of the above talks of the IPCC Business-as-Usual scenarios – that is, under the assumption that few or no steps are taken to limit the emissions of greenhouse gases; they are defined precisely in the Report (Houghton *et al.* 1990).

### 6.1.4 Simple models: further studies

An important role has been emphasized of simple climate models (energy-balance, radiative–convective) in the identification of most important fields of the application and interpretation of the results obtained with the use of sophisticated three-dimensional models (MacCracken and Luther 1985). In this connection, of great importance is the coincidence of the estimates of the sensitivity of mean global temperature to increasing $CO_2$ concentrations obtained using simple and sophisticated models (the importance of this coincidence must not be overestimated, however). A very important factor that determines the reliability of the estimates is an account of the effect of $CO_2$ on the whole surface–atmosphere system (but not only radiative heating of the surface), as well as of all the most substantial climate-forming processes and respective feedbacks. (Neglecting the last requirement explains the unreliable estimates of climatic sensitivity from data of local observations of the heat balance.)

Of key importance is an account of the atmosphere–ocean interaction, which slows down the effect of $CO_2$ on climate. Approximate assessments have shown that with doubled $CO_2$ concentrations it takes 3 yr to raise by 20 °C the equilibrium temperature of a 50 m deep mixed layer of the ocean, 30 yr for a 500 m layer and 300 yr for the whole global ocean 5 km deep. (In fact, this phase-shift is much greater, since in the case of climate warming the ocean-to-atmosphere sensible and latent heat fluxes increase, which delays reaching the equilibrium.)

The assessments mentioned above should, however, be considered conditional because, in particular, the main supposition (the basis for assessments), according to which additional heat coming to the ocean behaves like a passive tracer not affecting the processes of exchange responsible for the downward transport of this heat, is generally incorrect. This may be acceptable only for the upper layer of the ocean, where the vertical mixing is mainly determined by winds and seasonal gyres, and where the horizontal circulation is caused by wind stress but not by a surface temperature gradient.

Numerical modelling with an approximate, interactive ocean–atmosphere model, on an assumption of a doubled $CO_2$ concentration and respective $4\,W\,m^{-2}$ increase in energy coming to the ocean surface, has led to the conclusion that a 20 °C rise in sub-polar temperature corresponds to the equilibrium state reached in 1500 yr. About 50 per cent of warming in the tropics falls on the first 5 years, and then the rise in temperature slows down (with 80 per cent of the warming in 100 yr). In high latitudes a 50 per cent warming is only reached in 25 yr.

The effect of polar ice is that its summertime melting prevents the climate from warming, which is clearly manifested in winter. The understanding of the relation between the formation of sea ice and deep waters remains rudimentary, as does that of the climatic factors governing the extent of ice cover in the northern and southern hemispheres. Particular attention must be given to the relative roles of radiation processes and mixing in the ocean in the formation of different annual changes in sea ice extent in the northern and southern hemispheres. While this remains unclear, even a qualitative assessment of the effect of sea ice on the reaction of climate to increasing $CO_2$ concentrations and the formation of deep waters is difficult. In this connection, relevant field observational experiments are very important (Chapter 4).

Two important aspects of cloud–radiation feedback still remain uncertain: (1) the effect of the $CO_2$-induced climate warming on variations in the amount, top height, optical properties and spatial structure of clouds and (2) the reverse effect of these changes on climate. As these problems remain unsolved the supposition that variations in the cloud cover negligibly affect the sensitivity of climate to external perturbations must be regarded as premature and unsubstantiated. Undoubtedly, no climate model can be considered reliable if it is unable to simulate accurately the observed annual change of cloudiness and Earth radiation budget components. The importance of the contribution of various optically active minor atmospheric components (apart from $CO_2$) to the formation of the greenhouse effect is undoubted.

In connection with the problem of the validation of simulation models, it is important to compare the results obtained not only with averaged fields of various parameters but also with statistical characteristics (moments) of higher orders. It is also important to undertake the testing for adequacy of the techniques for a parametrization of various physical processes (the cloudiness–radiation interaction, the formation of snow and ice cover, the heat and moisture exchange, etc.) based on specific field experiments. Since the calibration of present-day climate models may be insufficient, the use of palaeoclimatic data and studies of climate on other planets may be very important.

In simple climate models an account of the ice albedo feedback may be decisive for the prediction of climate warming in high latitudes. Note should be taken, however, that with the appearance of open water, evaporation increases, which can lead to a more intensive formation of overcast stratus clouds, to an increase of the Earth's albedo and, consequently, to a compensation of the $CO_2$-induced warming. Numerical experiments with the use of three-dimensional models have verified the increase in cloud amount by several per cent in the regions where increased $CO_2$ concentrations lead to sea ice melting. A change in the system's albedo in these regions is four times lower than in the surface albedo.

Although one may think that during the last decade the numerical modelling of anthropogenic effects on climate has given consistent results, a more thorough analysis shows that, in fact, there has been a growing uncertainty of empirical and theoretical estimates of the present climate changes (i.e. the period between 1850 and 2100). This is because for such a complicated and multifaceted phenomenon as climate, which is also characterized by numerous feedbacks, the traditional statistically precise estimates of reliability of diagnosis and forecast are very difficult.

As MacCracken (1983) notes, this refers even to calculations of the global mean warming at a doubled $CO_2$ concentration. With the $CO_2$ content and surface air temperature assumed to vary in phase, at $CO_2$ concentrations growing from 280 ppm in 1950 to 345 ppm now, a warming by 0.45, 0.90 or 1.35 °C should have taken place, with a temperature increase due to a $CO_2$ doubling, respectively, by 1.5, 3.0 or 4.5 °C. Bearing in mind that the warming observed during the last century constituted about 0.5 °C and assuming a simple logarithmic extrapolation (depending on $CO_2$), the lower limit of estimates will be most probable.

The uncertainties of existing estimates of anthropogenic impact on climate are determined by four major factors: an inadequate observational data base, imperfect techniques for observational data analyses, the multi-component nature of climate changes, and inadequate numerical models giving controversial estimates of climate sensitivity to various factors (in particular, a consideration of the buffer effect of the ocean is very important).

An inadequate representativity of climatological data on surface air temperature has been demonstrated in a number of publications. For example, Chen (1985) analysed the anomalies of monthly mean and annual mean surface air temperature values (with respect to averages over the whole period under consideration) as well as respective trends for different latitudinal belts. A separate analysis was made of data for the land and the ocean (Table 6.1), based on processing of a more complete data base on surface air temperature for the period 1949–1972 for both hemispheres, averaged over a 1° latitude by 1° longitude grid, though for latitudes north of 75° N and south of 50° S the data are only fragmentary. The data for the northern hemisphere trends agree well with those obtained earlier using quite different processing techniques. However, an analysis of the time series for the whole planet revealed that the surface air temperature decrease has been very small and cannot be considered statistically significant.

Only on regional scales do statistically significant trends show themselves. A statistically reliable temperature decrease took place in the extra-tropical latitudes of the northern hemisphere (80–90° N), but in the same belt of the southern hemisphere a positive trend of surface air temperature was observed, a complete mutual com-

**Table 6.1** Surface air temperature trends (°C/year) between 1949 and 1972

| Latitudinal belt | Ocean and land | Land | Ocean |
|---|---|---|---|
| 90° S to 90° N | −0.002 | −0.006 | −0.003 |
| 0 to 90° N | −0.010[a] | −0.011[a] | −0.005 |
| 0 to 90° S | 0.004 | 0.004 | 0.003 |
| 20 to 90° N | −0.014[a] | −0.014[a] | −0.010[a] |
| 20° S to 20° N | 0.001 | −0.003 | −0.001 |
| 20 to 90° S | 0.09[a] | 0.012[a] | 0.004 |

[a] Data at a 95 per cent statistical significance level.

pensation of surface air temperature changes occurring in both hemispheres. In the equatorial zone, the temperature variations were practically absent. The surface air temperature variations on land and in the ocean are qualitatively similar. Only in the band 20–90° S has a warming over land substantially exceeded a temperature increase over the ocean, but this can result from the effect of urbanization on the surface air temperature observational data. Assessments of the correlation coefficient between surface air temperature over the land and over the ocean gave maximum values for the tropics (0.804 for annual means). The interannual surface air temperature variability is most strongly expressed in polar regions both on land and in the oceans.

Kelly *et al.* (1986) obtained specific data on the secular trend of the annual mean surface air temperature for the northern hemisphere continents for the period 1851–1984 in the form of surface air temperature anomalies with respect to an average for 1946–1960.

The principal difference between new data and those obtained earlier (the limited representativity of all available information) is the conclusion that there was a less strong cooling in the 1880s than had been believed earlier (the difference between the cooling amplitudes constitutes 0.07 °C). The stronger surface air temperature variability in the beginning of the period considered could be explained by inadequate observational data.

The most important feature of the new data is the absence in the late nineteenth century of a long-term surface air temperature trend (the temperature had remained almost constant until about 1920). There is a distinct difference between the temperature levels before and after 1920. During the 5-yr intervals 1915–1919 and 1920–1924, a warming of about 0.3 °C had taken place, which constituted about 75 per cent with respect to total warming during the periods before and after 1920. The warming in the early 1980s has been most intensive, being renewed during the last few years.

The most substantial result of the new analysis of surface air temperature data is the conclusion that there has been a step-wise (discrete) climate warming, which contradicts the interpretation of the secular trend of surface air temperature as determined by either continuously increasing $CO_2$ concentration or by such factors as volcanic eruptions and solar activity.

Nasrallah and Balling (1993) came to the conclusion about the existence of a linear, statistically significant temperature increase of 0.07 °C per decade during the period 1950–1990 over the Middle East region (the land region extending from Morocco to Afghanistan) which is concurrent with the rise in equivalent $CO_2$ concentration from approximately 350 ppm to 430 ppm. Most of this increase has occurred in the spring season (virtually no warming has occurred in the winter months). An analysis of specific features of the spatial distribution of these temperature changes reveals a general cooling effect associated with the increase of atmospheric sulphate concentration levels and a warming effect associated with the degree of human-induced desertification.

Deser and Blackman (1993) have discussed the results of the empirical orthogonal functions (EOF) analysis of the four components of the climatic system – sea surface temperature, surface air temperature, wind and sea level pressure – over the North Atlantic Ocean during winter over the period 1900–1989. The patterns of sea surface temperature and surface air temperature during the periods 1900–1929 and 1939–1968 indicate the warming which was concentrated along the Gulf Stream east of Cape Hatteras and may have been a result of altered ocean currents. Warming also occurred over the Greenland Sea and the eastern sub-tropical Atlantic.

The principal drawback of the radiative–convective climate models widely used to assess the anthropogenic effects on climate is that they lead to overestimated latent heat in the lower atmospheric layers. To eliminate this drawback, Ou and Liou (1985) proposed a parametrization of convection connected with cumulus clouds, to be used together with an improved one-dimensional radiative–convective turbulent model. The use of such a model to assess the climatic sensitivity to a $CO_2$ doubling revealed a reduction of sensitivity, compared with the control case. However, the heating of the upper troposphere increases due to an additional release of latent heat, with convection taken into account. These effects are strongest in the tropics and at a high relative humidity near the surface (90 per cent) in conditions of the mid-latitude standard atmosphere. A substantial effect of convection points to the necessity to use two-dimensional climate models, in which the latitudinal dependence of the latent-heat-induced heating can be taken into account as well as the interaction of convection with an average circulation and horizontal eddy transport of latent and sensible heat.

To estimate the effect on climate of $CO_2$ doubling from 326 to 652 ppm (by volume), Schlesinger *et al.* (1989) used the Oregon State University (OSU) interactive climate model which takes into account a synchronous interaction between the atmosphere and the ocean. The results are characterized by surface air temperature differences at various altitudes between 984 and 396 mbar, and by sea surface temperature differences at depths between 21 and 1150 m, as well as by surface air temperature differences ($2 \times CO_2 - 1 \times CO_2$) calculated for a period of 16 yr. The global annual mean equilibrium surface air temperature increase was 2.82 °C.

The evolution of the global mean surface air temperature reflects a rapid (during the first few months) warming of the atmosphere due to a $CO_2$ doubling, followed by a slower warming of the atmosphere and the ocean. Analysis of the annual mean, zonal mean profile of warming in 16 yr has shown that the surface air temperature increase intensifies from the tropics (about 1 °C) to the sub-tropics (1.5–2.2 °C), is attenuated in mid-latitudes and intensifies in high latitudes up to 2.6 per cent. The warming increases with height in the tropics and sub-tropics but decreases over the rest of the globe.

The surface air temperature increase over the ocean in both hemispheres is intensified from the tropics to mid-latitudes. This trend is similar to a latitudinal variation in excess concentrations of the $^{14}C$ isotope with respect to its quantity before nuclear tests, observed near the ocean surface. During 16 yr the oceanic warming penetrates much deeper in the sub-tropics and mid-latitudes and in the equatorial zone, which is also similar to the variability of the observed depth of excess $^{14}C$ penetration. The warming is reduced with depth in both hemispheres in the 60°–0° latitudinal belt and propagates downward, from a maximum in the near-surface sub-tropical waters towards high latitudes. The maxima observed near 65° N and 60° S at depths of 250–750 m are a very interesting feature of temperature variations with depth. The second of these maxima is not far from the region of shelf ice of the Ross Sea.

Since the 16 yr period of integration is not long enough to reach an equilibrium state, the results obtained were compared with data for the interactive energy balance climate model / box model of the ocean. It was found that the sensitivity of the OSU model to external factors constitutes $0.72\,\mathrm{K\,W^{-1}\,m^2}$, the global mean effective coefficient of the ocean–atmosphere heat exchange is $80\,\mathrm{W\,m^{-2}\,K^{-1}}$, and the effective coefficient of heat diffusion in the ocean $K = 3.2$, 3.8 and $1.5\,\mathrm{cm^2\,s^{-1}}$ (at depths of 50, 250 and 750 m, respectively). The $K$ value averaged by mass ($2.25\,\mathrm{cm^2\,s^{-1}}$) is consistent with the estimates from data on the penetration into the ocean of tritium and $^{14}C$ produced by nuclear tests. A climate model taking account of a box diffusion at $K = 2.25$–$2.50\,\mathrm{cm^2\,s^{-1}}$ simulates well the evolution of respective differences between surface air temperature and the temperature of the oceanic surface layer, obtained in the OSU model. This suggests that the heat transport from the surface deep into the ocean, according to this model, occurs at a rate consistent with the observed intensity of diffusion of tritium and $^{14}C$.

A comparison with results of calculations using an approximate model for 200 yr showed that it takes 75 yr to reach a 63 per cent level of an equilibrium warming, i.e. the ocean greatly delays climate warming because of the downward heat transport and its accumulation by deep layers of the ocean. In conditions of continuous increase of $CO_2$ the time for reaching a 63 per cent level of warming cannot be estimated, but, proceeding from the interactive energy balance model, we may say that the ocean-induced time delay in climatic response $L(t)$ asymptotically approaches $\sqrt{t\tau_e}$, where $t$ is the time and $\tau_e$ is the time to reach a 63 per cent level. Thus, the response of the climatic system progressively slows down with time.

The consequences of such memory of the atmosphere–ocean system can be taken into account from the viewpoint of detecting the effect of increased $CO_2$ on climate from observational data. To identify the $CO_2$ signal it is essential to take into account the existence of other external factors of the thermal impact, both before and after the beginning of the industrial revolution, which have produced the growth in $CO_2$ concentration. Since for the stepwise changes of thermal impacts, $\tau_e = 50$–75 yr, the memory of the climate system is approximately 150–225 yr. From this point of view, the climatic system may never reach the equilibrium state because of the large thermal inertia of the ocean. In this connection, studies of physical processes determining the penetration of the $CO_2$-induced warming deep into the ocean are of fundamental importance. Also urgent is the problem of the stability of the western Antarctic ice sheet under conditions of a considerable warming in deep waters near the shelf ice of the Ross Sea.

Wilson *et al.* (1987) discussed the results from two numerical experiments on climatic sensitivity to a $CO_2$ concentration increase, with an account of the 2 °C increase of the annually changing sea surface temperature for the whole world ocean. In one of the experiments made with the use of an 11-layer model of the UK Meteorological Office (calculations were made for a period of 3 yr) the amount and top height of clouds have been prescribed; in the other (calculations for 2 yr) they have been predictants. In addition, in the second case, a specified scheme of parametrization of longwave radiation transfer has been used, and the geographical variability of albedo has been taken into account. A comparison with the data for the control experiment has given the results shown in Table 6.2.

As can be seen from Table 6.2, the prescribed cloudiness has led to an overestimated warming and underestimated precipitation. However, if sea surface temperature were not prescribed, then the sea surface temperature would go down at a prescribed cloudiness and go up in the case of predicted cloudiness. Bearing in mind that the model climate responds to small sea surface temperature variations approximately linearly, and taking into account an increase of thermal emission by

**Table 6.2** Climate changes due to a $CO_2$ doubling and a 2 deg K SST increase (global means for 2 yr)

| Conditions | Land surface temperature (°C) | Global surface temperature (°C) | Relative humidity (%) | Rainfall (%) | Radiation budget ($W m^{-2}$) |
|---|---|---|---|---|---|
| Prescribed cloudiness | 3.05 | 2.32 | 20 | 5.6 | −1.2 |
| Precalculated cloudiness | 2.96 | 2.29 | 18 | 7.2 | +0.9 |

3.5 $W m^{-2}$ with a doubling of $CO_2$, we can say that the total warming will be zero if the sea surface temperature increase constitutes 1.5 °C and 2.6 °C, respectively, and the global mean surface air temperature will increase by 1.73 °C and 3.08 °C. The latter estimates agree with the data of Hansen *et al.* (1988a), who obtained the values 2.0 °C and 3.2 °C. The calculations of the geographical distribution of precipitation in the period June to August revealed an intensification of precipitation in high latitudes and in most of the ITCZ, but a reduction over most of the continents in northern hemisphere mid-latitudes. In the western Pacific a vast region of strongly reduced precipitation forms, not detected by numerical experiments at a fixed cloudiness.

As Harvey and Schneider (1985) noted, for an adequate numerical modelling of climate change on timescales of an order of $10^0$–$10^4$ yr (glacial–interglacial oscillations, the effect of increasing $CO_2$ on climate), taking place with varying external forcings, of great importance is knowledge of the nature of the time-dependent response to such forcings, determined by the interaction of different components of the climate system. In this connection, an attempt has been made to filter out the contribution of physical processes in the oceans as factors governing the time-dependent response of the ocean, using different versions of energy-balance models. Calculations have been made of the sensitivity of the time-dependent response of sea surface temperature to the forcings, under various assumptions about the processes in the ocean, for which two globally averaged energy balance models, with different parametrizations of the ocean, have been used.

In the first case, a box model of the ocean, subdivided into a few isothermic reservoirs, was used. Calculations made with a prescribed stepwise 2 per cent increase of solar constant (the results obtained can apparently be used in conditions of a comparable increase of $CO_2$ concentration), have shown that it is very important to consider separately the equivalent mixed layer, 30 m thick, and deep layers of the ocean, instead of combining them into one reservoir, as has been done in some interactive models of the system atmosphere–ocean–cryosphere, on timescales of the order of $10^4$ yr. The second model takes into account the interaction of the isothermic mixed layer with the deep layers of the ocean, considering both advection and diffusion. In this case the vertically changing values of the coefficients of heat diffusion $K$ and advection velocity $V$ were prescribed, with an account of various possible feedbacks between $K$ and temperature lapse rate, as well as between $V$ and the equivalent mixed layer heating as a forcing. The temperatures of bottom waters, depending on the response of the mixed layer, were also prescribed.

The time-dependent response of sea surface temperature is relatively insensitive to changes in $K$ and $V$, if the feedbacks are neglected, but it becomes more sensitive with individual feedbacks taken into account. In the case of their combined (complex) account, the response of the sea surface temperature to the forcings may be very complicated and unable to be described with simple models of the ocean (with one characteristic time), despite the use of a complicated model of the overlying atmosphere. With prescribed periodically-varying solar constant, the feedback mechanisms can bring forth both a substantial reduction and an intensification of sea surface temperature amplitude (as a response to disturbances) compared with the conditions without feedbacks. Although the final time of the equivalent mixed layer response is proportional to the thermal inertia of the ocean, two-thirds of this response is determined by the thermal inertia of the equivalent mixed layer.

The results considered testify to the importance of taking into account the feedback mechanisms in the ocean. Numerical experiments are needed with the use of interactive models of larger dimensions, which can, in particular, simulate the ocean–land contrasts, latitudinal variability, and regional and hemispherical cells of the oceanic circulation, with an account of albedo feedbacks.

Harvey and Schneider (1985) obtained estimates of the sensitivity of the time-dependent climatic response to the external forcings prescribed as a 2 per cent increase of the solar constant as well as a doubling (or continuous increase) of $CO_2$ concentrations, with the use of a hemispheric seasonal climate model which takes into account the ocean–land contrasts and the surface–atmosphere interaction. This model is similar to the box model that considers advection and diffusion (the so-called XIII-I model), except that the hemispheric averaging makes possible a separate consideration of the processes in both hemispheres.

A special feature of the new model is a consideration of the ocean–land contrast, the annual change of the mixed layer thickness and of the intermediate layer, as well as a prescription of the annual change of insolation and cloud amount. The values of the coefficients of horizontal heat exchange between the hemispheres have been adjusted so that the observed annual change of climate can be correctly simulated. An account of feedbacks in the ocean was varied in the same way as in the XIII-I model.

Results of calculations revealed the important role of the interaction between the hemispheres and also between the ocean and land in the formation of the time-dependent climatic response to the forcings. A comparison of the data for global mean (XIII-I) and hemispheric models, for the surface and for deep layers of the ocean, made to analyse possibilities of using the XIII-I model, has demonstrated, for example, that the responses of global and hemispheric models (without account of the deep oceanic layers) to a 2 per cent increase of the solar constant are similar during the first half year, but then a difference rapidly occurs. However, an account in the global model of the interaction between mixed layer and deep layers of the ocean through advection and diffusion provides a better agreement, since the interaction with deep layers, which starts to prevail after half a year, is equivalent to a separate account of the thermal inertia of the ocean and land in the hemispheric model.

Calculations have shown that in the case of continuous increase of $CO_2$ concentration, the mixed layer heating only moderately lags behind with respect to the warming of the atmosphere over the land and the ocean, which turns out to be closely interrelated (the air over the land is heated almost as rapidly as over the ocean, due to the land–ocean interaction). The varied account of the feedbacks leads, in this case, to qualitatively the same results as with a stepwise increase of the solar constant, but more weakly influences the time-dependent climatic response, compared with uncertainties in the prescription of the parameters of an equilibrium climate.

During 1 yr with a 2 per cent decrease of the solar constant, prescribed to simulate the effect of a volcanic eruption, the surface air temperature change over the land is about twice as large as the mixed layer temperature change. This testifies to the necessity of distinguishing between the air temperature responses over the land and ocean to disturbances, in comparing the results of model calculations of the volcanic signals with observational data. With the combined effect of short-term and long-term factors (such as the combined effects of eruptions, solar activity and $CO_2$), the shift between variations in air and mixed layer temperatures is insignificant until the long-term effect prevails, although the shifts over the land and ocean, due to short-term disturbances, can be substantial from the viewpoint of the formation of regional climate anomalies caused by such disturbances.

The interrelation of the processes in the atmosphere sets the problem of developing various interactive models. In this connection Owens *et al.* (1985) undertook a numerical modelling of anthropogenic effects on climate and on the ozone layer, using an interactive 1-dimensional radiative–convective model, in which account is taken of the interaction between chemical processes, the minor gases and aerosol components, transport and the temperature field. For a consideration of the radiative heat flux divergence, account is also taken of the absorption of solar radiation by water vapour, $CO_2$, ozone and oxygen. The radiative heat flux divergence due to longwave radiation is calculated with an exact account of such components as water vapour, carbon dioxide and ozone (in these cases, a detailed structure of the absorption spectrum is considered), and the contribution of $CH_4$, $N_2O$, CFC-11 and CFC-12 is estimated with the use of broad-band parametrization of absorption.

Numerical modelling of the stationary (asymptotic) response of the atmosphere has shown that with a doubled $CO_2$ concentration, the surface air temperature rises by 1.7 deg K and total ozone content increases by 2.9 per cent. The supposition of continuing emissions (at the present level) of chlorofluorocarbons (CFCs) has led to a 5.7 per cent decrease of total ozone content, whereas with the combined effect of these emissions and $CO_2$ doubling this decrease drops to 3.5 per cent (in this case the global mean warming reaches 2.2 °C).

The use of specified rates of chemical reactions revealed substantial changes in the vertical profile of ozone concentration due to CFC emissions, compared with those observed earlier; the concentration in the layer 30–40 km decreases, being partially compensated by an increase at altitudes 15–30 km. Estimates of the effect of increasing content of $N_2O$, $CH_4$ and $NO_x$ showed a considerable effect of nitrogen monoxide and methane on the ozone content, if the increase of their emissions continues at the present level (0.2 per cent per year and 1–2 per cent per year, respectively). In particular, a doubled strength of the sources of methane will cause a 4.3 per cent increase of total ozone content and an increase of ozone near the Earth's surface.

Owens *et al.* (1985) obtained estimates of the dynamics of the ozone layer and climate, with two scenarios of changes in $CO_2$, CFCs and $CH_4$ during the century to come. In the case of continuous emissions of methane, the total ozone content increases by 1.5 per cent and the temperature by 2.8 deg K; the ozone becomes substantially redistributed vertically (between the layers 30–40 km and 20–30 km), the ozone concentration strongly increases near the tropopause and the temperature at the level of the stratopause drops (by about 20 deg K).

The results obtained by Owens *et al.* (1985) testify to the complex non-linear interaction between the temperature field and chemical processes; this demonstrates that the consideration of different factors in isolation is inadequate. From the viewpoint of the effect on the ozone layer during the forthcoming century, the role of such minor gaseous components as $CO_2$, CFCs, $N_2O$ and $CH_4$ is similarly important, whereas for the formation of climate changes, the increasing $CO_2$ concentration is decisively important; the $CO_2$ contribution is about twice as large as the total contribution of the other minor gases. The strong sensitivity of the estimates of the effect on ozone to the rate of reactions reveals the necessity for further studies in the field of the kinetics of chemical and photochemical reactions.

As for the reliability of the results of numerical modelling of climate changes due to $CO_2$, the global mean and zonal mean temperatures can be considered reliable. However, the results of calculations of the geographical distribution of such parameters as winds, soil moisture, cloud amount and global radiation cannot still be considered realistic and they should be treated as conditional scenarios.

Bach (1988) analysed possibilities for substantiating the regional scenarios of climate change with the use of available numerical modelling results. Table 6.3 illustrates the estimates of possible regional changes in surface air temperature and precipitation in different regions of Europe (30–70° N; 20° W–20° E) due to a doubling of the concentration of $CO_2$. Since these data are very uncertain in the case of temperature, only qualitative changes are given. Even more uncertain estimates of changes in rain rate enable one to determine only the latitude above (below) which the precipitation is above (below) the norm. Even the estimates of zonal averaging of precipitation are often statistically insignificant.

An analysis of the results of using palaeoanalogues has led to the conclusion that none of the approaches considered makes it possible to analyse the climate in the future. Estimates of changes in atmospheric general circulation in the European region from the data of numerical experiments, with the use of complete 3-

**Table 6.3** Probable scenarios of changes in surface air temperature and rain rate due to a $CO_2$ doubling

| | Winter | Spring | Summer | Autumn |
|---|---|---|---|---|
| *Temperature* | | | | |
| Scandinavia | A | NO | B | NO |
| British Isles | NO | B | NO | B |
| Eastern Europe | A | NO | B | A |
| Central Europe | NO | NO | B | NO |
| Spain | B | NO | A | NO |
| Mediterranean | B | A | A | A |
| North Africa | NO | A | A | A |
| *Rain rate* | 545% | 50% | 65% | 50% |

A, above the norm; B, below the norm; NO, result uncertain.

dimensional climate models and palaeoanalogues, turned out to be similar. In both cases the downward movements in the south of the region intensify and an advection of warm humid air in Scandinavia increases.

The most substantial climate warming is found to be in the Arctic; this is revealed both by palaeodata (the Pleistocene), by the data of present observations and by the numerical modelling results. Numerous numerical experiments reveal a relatively strong increase of winter temperature in the northern European region (especially in Scandinavia) and of summer temperature in the south. Estimates of the annual mean warming range between 2.2 and 4.0 deg K. The estimates of the spatial structure of the fields of temperature and precipitation both in winter and over a year, on average, were statistically significant. On the whole, Bach (1988) states that the present climate models are not very realistic for reliable assessments of regional climate changes under the influence of a $CO_2$ doubling.

A comparison of numerous estimates of the mean global warming with a doubled $CO_2$ content points to the value of $3 \pm 1.5\,^{\circ}C$ as the most probable, which is in accordance with the IPCC assessments (Houghton *et al.* 1990, 1992). Gates *et al.* (1981), Mitchell (1983) and Gates and MacCracken (1985) gave a much lower level of warming (0.2–0.3 °C), but this is probably explained by a prescribed sea surface temperature which strongly limits surface air temperature variations. Stratospheric cooling in the layer 30–45 km reaches 7–11 °C. A doubled $CO_2$ concentration causes a marked intensification of the hydrological cycle. Calculations of zonal mean temperature values give a high-latitudinal warming 2–3 times stronger than in the tropics, where it is limited by the effect of moist convection, smearing the warming through the whole atmosphere. Other effects on the zonal mean fields of climate parameters, which can only be judged qualitatively (quantitative estimates are not quite reliable) are as follows: (1) a marked increase of the annual mean river run-off which takes place in high latitudes, caused by intensification of precipitation due to an intrusion of moist warm air masses north of 60° N, (2) earlier snow melting and later snowfalls due to the high-latitude warming, (3) a reduction in the summer northern hemisphere of soil moisture in middle and high latitudes (north of 35° N), which is explained by an earlier start to the period of relatively strong evaporation (following the phase of snow melting) and by weakening of precipitation in summer and (4) a reduction in the area of polar ice cover.

The observational data available do not permit an assessment of the reliability of the conclusions mentioned above, nor the identification of the signal of the anthropogenically induced climate warming. The absence of convincing evidence for the warming is explained primarily by the following three circumstances: (1) the ocean-caused delay in manifestation of the warming, (2) difficulties in filtering out the $CO_2$ signal from other factors (a change in extra-atmospheric insolation, in aerosol content, etc.) and (3) a limited observation series. In this connection, of great importance is an extensive and long-period programme of climate monitoring as well as the identification of the parameters observed that are most sensitive to the $CO_2$ signal. The latter appear to be, in particular: zonal mean surface air temperature in summer (except for high latitudes, where the signal/noise ratio is lower), zonal mean summertime temperature in the stratosphere and mesosphere where strong anthropogenic cooling should be manifested at a low natural variability, the temperature of surface and deep layers of the ocean and the extent of the polar ice cover.

In connection with a discussion of the causes of global climate change during the last century, Hoffert *et al.* (1985) analysed the results of numerical modelling of the time-dependent reaction of climate to an increasing $CO_2$ concentration in the atmosphere. This analysis suggests that such a reaction on a time scale of about 10–100 yr is probably determined by the simultaneous effects of several external factors (extra-terrestrial insolation, volcanic aerosol, concentration of $CO_2$ and other greenhouse gases), as well as the internal variability of the climatic system, especially the internal dynamics of the ocean.

The development of reliable forecasts of the climatic reaction to external disturbances (with an account of the oceanic thermal inertia that slows down this reaction by several decades) is complicated by lack of a substantiated approach to verifying the results of numerical modelling with the use of observational data.

So far, the test of reliability has mainly been confined to experiments on climatic sensitivity to disturbances and to comparisons of significantly different results obtained with various models. One more complication is connected with the fact that calculations of the time-dependent reaction of climate are very time-consuming and therefore they do not consider the combination of disturbances.

The results of such calculations (based on the use of the known scenarios for fuel consumption dynamics and $CO_2$ emissions) indicate that by 2000 a global warming by 1 °C should have taken place as compared with 1850, which, during the next century, will further increase by 2–5 °C. The reliability of these estimates raises no doubts. It seems, however, that reliable interactive climate models (the atmosphere–ocean system) will not be developed in the very near future, and, in this connection, one of the problems of the first priority is an analysis of the reasons for differences in the equilibrium reaction of climate to disturbances as well as an adequate control of the reliability of the models based on observational data (see also Gaffin *et al.* 1986, Cess *et al.* 1989, 1990, Randall *et al.* 1989, 1992).

In the process of improving the interactive models, it is important to specify their oceanic component from the viewpoint of a detailed consideration of mesoscale and macroscale processes taking place in the world's oceans (mesoscale vortices, internal waves, vertical heat transport due to the small-scale dynamics, etc.). Of particular importance is an understanding of the relative role of mixing both along and across the pycnocline, the mean transport due to the thermocline and tangential-stress components of the oceanic circulation, and the transport and accumulation of heat in the upper oceanic layer.

Further studies are needed of the role of convection and formation of bottom waters, equatorial upwelling, gyres in mid-latitudes and the latitudinal dependence of the oceanic mixed layer as factors of the rate of heat exchange for different parts of the ocean. Models of the oceanic circulation require independent verification from data of oceanographic observations. Special attention must be given to studies of the mechanisms that specifically affect heat and mass transport.

Luther (1983) gave general ideas of the most substantial needs and priorities in climate studies. As was mentioned above, of great importance must be further studies of the role of the oceans and cloud cover (the respective interaction processes) and of the processes in the atmospheric boundary layer and on the land surface as factors of the formation and variability of climate.

As for the oceans, an adequate account is required in climate models of the dynamics and thermodynamic processes taking place in them (currents, upwellings, formation of bottom waters, heat transport from the mixed layer to deeper layers, etc.). Better substantiated estimates are needed of the effect of cloud cover dynamics as a factor of intensification (or smoothing down) of anthropogenically induced climate changes and variations in the properties of clouds when the climate changes. An important problem is an improvement of the parametrization of the dynamics of sea ice and snow cover, as well as its effect on energy exchange in the atmospheric boundary layer (see Chapter 4).

The first priority is an account of the interaction among the climatic system's components with a broad range of time constants typical of them, as well as the interaction among small-scale motions and large-scale motions and vertical transport in the ocean, and the ocean–atmosphere interaction. Forecasting the anthropogenically induced climate change decisively depends on the availability of such models that adequately consider all the important feedback mechanisms and chemical and dynamic processes.

Of course, a reliable forecast is impossible in the absence of substantiated scenarios for variations in the content of the optically active components of the atmosphere. Until the factors of climate variability for the last century are completely understood, the climate forecasts will remain unreliable. In this connection, of critical importance remains the problem of verifying the reliability of numerical climate models by comparing them with observational data (this also refers to palaeoclimates) and by intercomparing the results of models. Special attention is claimed for forecasts of the variability of regional-scale climates and extreme events.

First calculations have recently been made of the effect of increased $CO_2$ concentration on climate, based on the use of new versions of three-dimensional climate models which include a parametrization of the interaction between the atmosphere and oceanic mixed layer. Such models were developed in the Geophysical Fluid Dynamics Laboratory (GFDL), Goddard Institute for Space Studies (GISS), National Center for Atmospheric Research (NCAR), UK Meteorological Office, and the Max Planck Institute for Meteorology.

As Schlesinger and Mitchell (1987) showed, in many respects the results of numerical modelling were qualitatively similar to those obtained earlier with the use of models that treated the ocean as a swamp (in the latter case the heat capacity of the ocean is ignored and insolation is fixed, which provides an equilibrium climate in 300 days, whereas in new models it takes several decades). Both in winter (December–January–February) and in summer (June–July–August) new models reveal (for a doubled $CO_2$ concentration) an increase of the zonal mean temperature near the surface, in the troposphere and lower stratosphere, but reveal a decrease in the stratosphere at altitudes above 20 km (lowering with altitude). Tropospheric warming intensifies with altitude within the band $\pm 30^\circ$ latitude in winter and within the band 40° S to 90° N in summer.

The near-surface warming increases with latitude in both hemispheres in winter but in summer only in the southern hemisphere. The annual change of surface air temperature is negligible in the interval 50° S to 30° N but intensifies in both hemispheres for latitudes greater than 50°. The latest estimates of the global mean surface air temperature increase range between 3.5 and 4.2 °C, which is much narrower than that corresponding to the swamp ocean models (1.3–3.9 °C). However, the results of calculations of the meridional profile of warming differ greatly. So, for example, the surface air temperature increase in the tropics varies from 2 °C (NCAR) to about 4 °C (GISS), while the temperature increase at the level of the tropical tropopause varies from 3 °C (NCAR) to 7 °C (GISS).

There are large differences between the regional effects (geographical distribution and annual change) of $CO_2$ doubling. New calculations gave a global mean increase of rain rate between 7.1 and 11.0 per cent, but the geographical distribution is characterized by regions both of an intensification and of an attenuation of rain, with the largest variability being in the latitudinal band 30° S to 30° N. Outside this band the rain is usually intensified. Large differences take place in the geographical distribution of rain rates. The calculated variations of soil moisture in winter are unexpectedly similar. In summer, there is only a qualitative agreement (an increase in soil moisture over the largest part of the continents).

There is a certain qualitative similarity of cloud variations projected with the use of the swamp and interactive models: an increase of cloudiness in the stratosphere and near the surface in high latitudes in both hemispheres, but a decrease of cloudiness in most of the troposphere and in mid-latitudes. The models that take into account the oceanic mixed layer reveal a strong sensitivity of climate changes to cloud conditions, which manifests itself through (1) a considerable increase of global mean surface air temperature with the projected cloudiness (GFDL) compared with the prescribed frozen cloudiness (GFDL), (2) a negative correlation between variations in cloudiness at the level of the tropical tropopause and surface air temperature revealed by new calculations and (3) estimates of the sensitivity of climate changes to variations in different factors, based on radiative–convective models, with projected changes in climatic parameters, calculated with the use of 3-dimensional models.

Based on the use of the UK Meteorological Office 11-layer interactive climate model, including a simple model of the 50 m mixed layer of the ocean and the thermodynamic model of ice cover (horizontal resolution 5° latitude by 7.5° longitude), Wilson *et al.* (1986, 1987) carried out numerical experiments in order to analyse the effect of $CO_2$ doubling on climate. The data of the control experiments show that the model provides a reliable simulation of the present climate, including the annual change of ice cover extent and sea surface temperature variations (its deviations from observed values do not exceed 2 °C).

The equilibrium state, a doubling of $CO_2$ concentration, is reached in 35 yr with a 5.2 °C increase of the annual mean, global mean temperature and with the year-round decrease of ice cover extent. A 3.5 per cent global mean decrease of cloud amount takes place which is determined by the decreasing amount of middle-level clouds, by the increase of the upper cloud amount and by some extension of low-level clouds in high latitudes. A more intensive climate warming, compared with that obtained in other similar climate models with an interactive account of clouds, can be partially explained by the use of the scheme of parametrization of penetrative convection, instead of the scheme of moist-convective adjustment, and by the contribution of the upper clouds. Variations in the summer precipitation (July–August) are characterized by their intensification in high latitudes and in most of the ITCZ, and by reduction over most of the northern hemisphere continents in mid-latitudes. There is a trend of northward shift of the principal belt of precipitation, with intensified precipitation over India and East Africa and a strong decrease of precipitation over the western Pacific. Apparently, this is caused by intensified warming over the land, due to a decrease in cloud amount.

The necessity for further improvements to climate models determines, in particular, the following directions of perspective studies: (1) analysis of the causes of inadequacy of the models in calculations of regional climate, and their improvement, (2) identification of various climatic characteristics (apart from temperature and precipitation) and a comparison of these with observations and (3) assessment of the representativity of the estimates of the equilibrium climate, from the viewpoint of its comparison with the climate observed. The solution of these and other problems needs combined efforts. It would be worthwhile to organize a centre for archiving the data of climate numerical modelling in order to expand their accessibility.

Schlesinger and Jiang (1988) undertook a simulation modelling for a period of 20 yr to analyse the physical processes which govern the transport of the $CO_2$-induced warming into the ocean. The calculations of global mean values of the ocean heat balance components have shown that, with a $CO_2$ doubling, the rate of the heat input to the ocean is about $3\,W\,m^{-2}$, which corresponds to the warming of the whole world's oceans by about 0.005 °C per year. However, it is important that the heating starts near the surface and gradually penetrates deep into the ocean. By the end of the year 20 of the numerical modelling, the warming (shown in brackets) of different layers of the ocean is: 0.50 m (1.2 °C); 50–250 m (0.9 °C); 250–750 m (0.5 °C); 750–1550 m (0.1 °C) and is practically zero at depths below 1550 m. The heat input to the ocean is largely determined by convection.

A comparison of these results with the results of calculations using the GFDL interactive (but not global) model, for the case of $CO_2$ concentration quadrupling, revealed numerous features in common (this primarily refers to the dominating role of convection as a process responsible for the heat input to the ocean) and even revealed the closeness of quantitative estimates (though the GFDL model calculations proceeded from $CO_2$ concentration quadrupling, while the control integration using the OSU model does not reach an equilibrium climate).

An analysis of the annual mean, zonal mean profile of warming due to $CO_2$ concentration doubling has demonstrated that the heating of the surface layer intensifies from the tropics to mid-latitudes in both hemispheres and gradually penetrates deep into the ocean, reaching deeper waters in the sub-tropics and mid-latitudes compared with the equatorial belt. The meridional profile of warming is very similar to the observed latitudinal distribution of excess $^{14}C$ concentration (compared with its values prior to nuclear tests in the atmosphere). The calculations of zonal mean profiles of variations in the ocean heat balance components showed that the $CO_2$ concentration-doubling-induced heating of the ocean takes place largely due to the transport of heat to deep waters, whereas the meridional heat transport is of secondary importance.

Although over the whole surface of the ocean the heat flux is directed deep into the ocean, it depends considerably on latitude, reaching maximum near 60° N, 75° N, 0°, 40° S and 60° S. The penetration of heat from the surface to deeper layers is at a minimum in the tropics and at a maximum in polar latitudes of both hemispheres (the minimum in the tropics is determined by upwelling of cold waters). Part of the heat produced in the tropics is transported to the poles by the poleward-directed branches of the cells of meridional circulation, which form (in connection with the existence of upwelling) following the law of mass conservation.

The transport of the heat deep into the waters in the sub-tropics is determined by the downward movements existing there. In the sub-tropical regions, gyres are observed which are anticyclonic horizontal circulations in the ocean, driven by westerlies in mid-latitudes of both hemispheres, as well as by the equatorial easterlies. Due to friction in the surface boundary layer of the ocean, the horizontal transport of water occurs in the form of gyres (the Eckman transport process) and, in accordance with the law of mass conservation, downward movements occur (downwelling). Apparently, this gyre-driven downwelling is responsible for the warming going deep into the sub-tropical oceans, where it combines with the poleward heat fluxes from the equatorial belt.

In high latitudes, the suppressed convection serves as an important factor of the heating moving deep into the ocean. Under usual conditions the ocean receives heat from the atmosphere in the tropics and gives it up to the atmosphere in high latitudes. In the case of increasing $CO_2$ concentration, the warming of the ocean surface layer in high latitudes leads to the formation of a stable stratification and reduces the intensity of the convective overturning; this in its turn reduces the heat release from the ocean to the atmosphere in high latitudes and increases the heat input from the atmosphere to the world's oceans. Further studies must be aimed at an analysis of the laws of geographical distribution of the heat input to the ocean, of the contribution of individual oceans and of such zonally asymmetric phenomena and processes as gyres, boundary currents and regional centres of convection.

Weare (1988) tested the ability of the atmospheric general circulation model developed in the University of California (Los Angeles) to simulate the values of the components of the Pacific surface radiation budget, calculated using semi-empirical formulae, from the data of meteorological observations, and the values of components of the radiation budget of the surface–atmosphere system, estimated from the NOAA satellite data (the albedo of the system and the outgoing longwave radiation). The 15-layer model is realized on a 4° latitude by 5° longitude grid.

It is assumed that within a grid cell (below the 100 hPa level) clouds form if: (1) an oversaturation takes place either in the free atmosphere or in the boundary layer and (2) the subgrid convection develops with the formation of cumulus clouds (a parametrization using the Arakawa–Schubert scheme). Clouds formed above the level 400 hPa or at a temperature below −40 °C are considered to consist of ice and their emissivity is taken to be 0.5 (in all other cases the clouds are assumed to emit as a blackbody).

A comparison made using the data for the winter 1 December 1982 to 28 February 1983, after an anomalously intensive El Niño event, revealed considerable differences: the values of the shortwave surface radiation budget and the system's albedo differ by 25 per cent, which should be ascribed to an approximate parametrization of clouds in the climate model. The modelled longwave radiation budget of the surface exceeds systematically (up to 50 per cent) that calculated using a semi-empirical formula (this is probably not connected with an inadequacy of the parametrization of cloud formation). Both observed and calculated outgoing longwave radiation values agree reasonably well (within 5 per cent). Although the numerically-modelled fields of the El Niño-induced anomalies of albedo and outgoing longwave radiation correspond, as a rule, to those observed from the viewpoint of the geographical location, the calculated estimates turned out to be underestimated by a factor of four (possible reasons for such differences have been analysed).

Schlesinger (1988b) analysed the estimates of the sensitivity of global climate warming (a surface air temperature increase) due to a doubled $CO_2$ concentration to a consideration of various feedbacks; these estimates were obtained with the use of energy-balance and radiative–convective models. The criterion of the role of feedbacks is the coefficient

$$f = 1 - (\Delta T_s)_0/\Delta T_s \tag{6.1}$$

where $(\Delta T_s)_0 = G_0 \Delta Q = 1.2\,^\circ C$ is the global mean surface air temperature increase without account of feedback calculated with $\Delta Q = 4\,W\,m^{-2}$ and $G_0 =$

0.3 deg K $W^{-1}$ $m^2$. With $\Delta T_s$ estimated with 17 different radiative–convective models, we obtain $-1.5 \leqslant f \leqslant 0.7$ (Figure 6.1). This broad range of variability is determined by the effect of feedback being considered differently (or ignored) due to (1) an increased moisture content of the atmosphere with the quasi-constant relative humidity, (2) decreased temperature lapse rate, (3) increasing cloud height (provided their temperature remains constant), (4) variations in the cloud cover, (5) variations in the optical thickness of clouds and (6) a decrease of surface albedo because of melting snow and ice cover.

An application of the Oregon State University two-layer radiative–convective model gave the estimates of the coefficient $f$ determined by some feedback mechanisms, as well as in-pair connections with respect to moisture content (Table 6.4). A consideration of in-pair connections shows that the three feedbacks are almost independent, which reflects the additivity of the coefficients (the temperature lapse rate, cloud height and albedo), whereas two others (the amount and optical thickness of clouds) are interdependent (the coefficient $f$ of the in-pair feedback is much less than the sum of individual coefficients). This means that the feedbacks determined by cloud amount and thickness (these happen to be negative when the feedback deter-

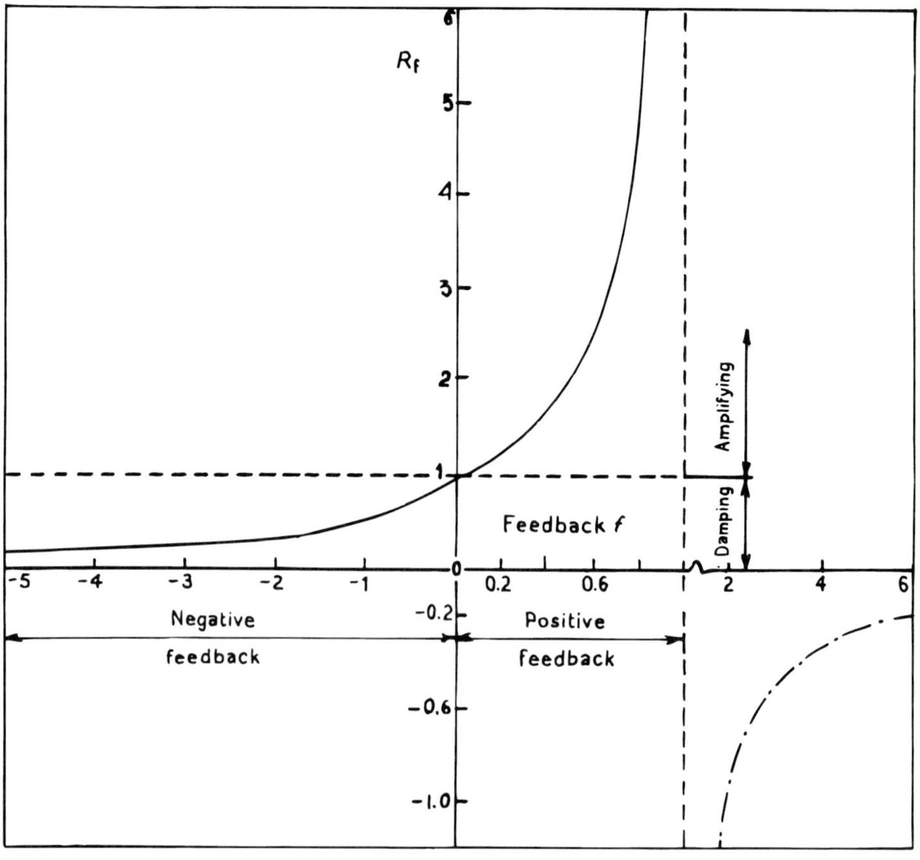


**Figure 6.1** The ratio of the surface temperature change $\Delta T_s$ to the zero feedback value $(\Delta T_s)_0$, $R_f = \Delta T_s/(\Delta T_s)_0$, versus the feedback $f$. After Schlesinger (1988b).

**Table 6.4** The role of various feedbacks

| Feedback factors | Individual feedbacks $T$ (°C) | Individual feedbacks $f$ | In-pair feedbacks $T$ (°C) | In-pair feedbacks $f^a$ |
|---|---|---|---|---|
| Moisture content | 1.94 | 0.340 | | |
| Temperature lapse rate | 0.88 | −0.409 | 1.19 | −0.043 (−0.069) |
| Cloud height | 1.73 | 0.261 | 2.79 | 0.543 (0.601) |
| Cloud amount | 1.38 | 0.074 | 1.81 | 0.291 (0.414) |
| Optical thickness of clouds | 1.39 | 0.079 | 1.70 | 0.248 (0.419) |
| Surface albedo | 1.56 | 0.181 | 2.39 | 0.466 (0.521) |
| Total effect of moisture content, cloud amount and albedo | | | 3.85 | 0.668 (0.782) |

[a] The algebraic sums of in-pair values of $f$ are shown in parentheses.

mined by moisture content is positive) should be considered (within a radiative–convective model) as single in-pair combinations.

As has been mentioned in Chapter 4, cloud feedback deserves special attention. In the analysis of the role of cloud amount variations, $A_c$, the contribution of the feedback, is partially determined by the quantity

$$\delta = -\frac{S_0}{4}\frac{\partial \alpha_p}{\partial A_c} - \frac{\partial R_0}{\partial A_c} \tag{6.2}$$

where $S_0$ is the solar constant, $\alpha_p$ is the planetary albedo and $R_0$ is the Earth's radiation budget. This quantity characterizes the degree of mutual compensation of the effects of cloud amount on albedo and the longwave radiation budget, acting in the opposite direction. According to data of several radiative–convective models, the estimates of $\delta$ are $-100\ \mathrm{W\,m^{-2}}$ (low-level clouds), $-50\ \mathrm{W\,m^{-2}}$ (middle-level clouds) and 50–80 $\mathrm{W\,m^{-2}}$ (upper-level clouds), variations being determined mainly by the variability of cloud emissivity. If $\partial A_c/\partial T > 0$, then the cloudiness of low and middle levels makes a positive contribution to the formation of the coefficient $f$ of the feedback considered, and the low-level clouds make a negative contribution, but at $\partial A_c/\partial T < 0$ the signs of these contributions are reversed.

This complicated situation accounts for the absence of reliable estimates of the coefficient $f$. A similar situation exists with respect to the optical thickness feedback $\tau_c$ (coefficient $f_{OD}$), which is also partially determined by the quantity

$$\phi = \frac{S_0}{4}\frac{\partial \alpha_p}{\partial \tau_c} - \frac{\partial R_0}{\partial \tau_c} \tag{6.3}$$

In the case of radiatively black clouds in the infra-red region, $\partial R_0/\partial \tau_c = 0$ and $\phi = 0$, whereas with non-black clouds $\partial R_0/\partial \tau_c < 0$ and $\phi \gtrless 0$. The estimates of $f_{OD}$ ranged between zero and $-1.32$. As for the $f$ values for other feedbacks, they were as follows: $f_W = 0.3$–$0.4$ (water content), $f_B = 0.15$ (temperature lapse rate for conditions of baroclinic adjustment), $f_M = -0.5$ to $-0.4$ (moist-adiabatic temperature lapse rate), $f_{PC} = -0.65$ (penetrating convection), $f_{SA} = 0.14$–$0.18$ (albedo feedback).

All these values must, however, be considered conditional, bearing in mind the approximate nature of radiative–convective models.

Based on the use of the NCAR community climate model, together with a simple energy-balance swamp model of the ocean, Washington and Meehl (1983, 1984, 1989) performed a numerical modelling of the climate sensitivity to a $CO_2$ concentration doubling for two cases: (1) with an account of the varied albedo feedback, (2) with a prescribed new albedo feedback and solar constant reduced by 2 per cent (AF + SC). The results obtained were compared with the data of numerical experiments carried out earlier, in which another albedo feedback parametrization had been used and the observed solar constant values had been taken.

New numerical experiments started on day 670 of the performed integration (in this way the initial basic state of the climate was prescribed) and covered the subsequent 480 days. In the first case it was assumed that the sea ice albedo was 0.35 (0.70), provided the sea surface temperature was above (below) $-10\,°C$ (the previous parametrization had proceeded from the fixed albedo 0.7 at a temperature below $-1.8\,°C$). The land snow cover albedo was 0.8 for wavelengths between 0 and 0.9 $\mu$m and 0.55 for wavelengths between 0.9 and 4 $\mu$m. Upon complete snow melting, the surface albedo dropped to 0.13.

The results of calculations revealed a strong dependence of the surface air temperature increase due to a $CO_2$ concentration doubling on (1) the specific prescription of albedo feedback (a higher sensitivity of climate warming to $CO_2$ concentration doubling corresponds to a parametrization leading to a more intensive sea ice melting: global mean warming constitutes 1.6 °C instead of 1.3 °C) and (2) special features of the initial basic state of the climate (a more intensive $CO_2$-induced climate warming corresponds to a colder initial climate). So, for example, a decrease of global mean air temperature by several degrees causes a nearly 30 per cent intensification of the response to a $CO_2$ doubling.

An analysis of the meridional cross-sections of warming (in the troposphere) and cooling (in the stratosphere) has shown that in the upper troposphere (10–18 km) the temperature field adjusts itself to varying cloud conditions. In all cases changes in cloudiness were similar.

As a rule, the mid-level cloud amount decreases in almost all latitudes, but the cloud amount at lower and upper levels in middle and high latitudes increases. The decrease of the snow and ice cover extent and mid-level cloud amount which take place in the process of climate warming almost completely compensates for the effect of the growth of lower and upper clouds in extra-tropical latitudes. As a result, changes in the Earth albedo do not exceed 1 per cent.

A three-dimensional numerical climate modelling performed in the USA (GISS, NCAR, GFDL) with the use of models in which cloudiness was interactively considered, had led to the conclusion that with a $CO_2$ concentration doubling the global mean surface air temperature should increase by about 4 °C. The same result was obtained, however, with substantially different contributions of various feedbacks to the surface air temperature formation. So, for example, in the GISS model the cloud feedback plays the leading role, whereas for the NCAR model the ice-albedo feedback is decisively important.

In this connection, Dickinson *et al.* (1987) continued studies of the contribution of albedo feedback using the NCAR model, which showed that the 3.5 °C global warming should drop to 2.2 °C if albedo feedback is not taken into account. It is important that the surface albedo variations in different regions of the globe con-

**Table 6.5** Latitudes and season of maximum albedo variation

| | Latitudinal belt (°) | | | |
|---|---|---|---|---|
| Hemisphere | December–February | March–May | June–August | September–November |
| Northern | 49–87 | 44–62 | 58–76 | 44–62 |
| Southern | 52–71 | 53–71 | 49–67 | 49–67 |

tribute differently to albedo feedback; 80 per cent of this contribution arises from ice and snow cover in the southern hemisphere. Table 6.5 illustrates the location of latitudinal belts where albedo changes in different seasons are at a maximum.

Naturally, the solar radiation absorption intensifies mainly in the spring and summer. If we calculate the feedback parameter $\lambda$ in the relation $\lambda \Delta T_s = \Delta Q$ between surface air temperature variations and variations in heat flux divergence $\Delta Q$ to the surface–troposphere system, it turns out that an account of the snow–ice albedo feedback leads to a decrease in $\lambda$ of about $0.7\,W\,m^{-2}\,K^{-1}$, whereas the total decrease due to the effect of albedo changes because of variations in the snow–ice cover and clouds constitutes $1.1\,W\,m^{-2}\,K^{-1}$ (thus, the contribution of clouds to albedo feedback changes is $0.4\,W\,m^{-2}\,K^{-1}$). Of course, an isolated consideration of albedo feedback is conditional, since the effect of feedbacks is interactive.

Verification of the reliability of climate forecasts is an important but extremely complicated problem. For example, a comparison with data of the present climate observation is useful but limited from the viewpoint of assessing the reliability of the climate, since the climate models are largely tuned to simulate the present climate and can be inadequate to describe the evolution of various feedbacks which are important for climate forecasts. It is also useful to intercompare the climate forecasts based on the use of various models, but the real problem lies in assessing the reliability of the climate forecasts. Testing the models from data on palaeoclimatic warmings is informative to some extent, but such results are uncertain in view of the fragmentary character of data and the impossibility of ascribing the warming to $CO_2$.

As Hall (1988) notes, because of these difficulties, it may be important to complete a theoretical analysis of the uncertainties of climate forecasts by recognition and quantitative estimation of the basic sources of uncertainties and by subsequent numerical experiments of climatic sensitivity to the respective factors which determine the uncertainty of the forecasts (see also Henderson-Sellers 1993). Such an approach makes it possible to obtain more information from the results of model intercomparisons. A great number of sources of uncertainty can be represented mathematically (for example, the uncertainties connected with some introduced assumptions, with parametrizations, with ignoring physical processes and with errors of the input parameters).

Of great concern are palaeovariations of the atmospheric greenhouse effect (Kondratyev and Moskalenko 1984). The estimates have shown that at an early stage of the Earth's evolution, its atmosphere could have contained an abundance of $CO_2$ equivalent to a pressure value of 20 atm. In view of the small area of the continents, a high content of $CO_2$ (of about 10 atm) could have remained during several hundreds of million years. To analyse the possible effect of a high $CO_2$ concentration in the early atmosphere on the climate, Kasting and Ackerman (1986) calculated global mean temperature using a one-dimensional radiative–convective model (with the prescribed moist-adiabatic temperature lapse rate). Because of neglected clouds, a rather high surface albedo was prescribed (0.22).

One of the principal objectives of numerical modelling was to determine the possibility of the formation on the Earth (as it had been on Venus) of the runaway greenhouse effect (such an effect is determined as creating conditions in which water existed only as water vapour or clouds in the atmosphere and where there were neither oceans nor lakes). Consideration was confined to the post-accretion period when the Earth had been heated only at the expense of solar radiation absorption, bearing in mind that a separate detailed analysis of the runaway greenhouse effect would be performed. The second important objective was to study the stability of an atmosphere rich in $CO_2$ with respect to the loss of water due to water vapour photodissociation with subsequent dissipation of hydrogen.

Analysis of the results of calculations has led to the conclusion that the early atmosphere ought to have been stable with respect to the runaway greenhouse effect, resulting in complete evaporation of the ocean, with the $CO_2$ content equivalent at least to a pressure value not less than 100 atm. In conditions of an atmosphere with $CO_2$ pressure of 10–20 atm (the first several hundred million years of the Earth's evolution), the global mean surface temperature had varied within 85–110 °C and the water vapour pressure had been 1.5–10.6 atm.

The early stratosphere ought to have been dry, and hence an oxidizing pre-biotic atmosphere could have formed due to $H_2O$ photodissociation and subsequent dissipation of hydrogen. The present atmosphere is also stable with respect to the formation of the runaway greenhouse effect. With the $CO_2$ content in the present atmosphere assumed to be 100 atm, the surface temperature should reach 233 °C, and the saturating water vapour pressure 29 atm. With atmospheric pressure of 130 atm, the ocean can start boiling only at a temperature approximately 100 °C higher than in the case of 100 atm pressure. The increase of the Earth's albedo due to intensified Rayleigh scattering (with $CO_2$ pressure of 10–20 atm, the albedo increases to 0.4) is the substantial factor of climatic stabilization at a high $CO_2$ content.

No doubt at the initial stage of evolution (about 4.5 billion years ago) the Earth and its atmosphere differed substantially from the present. Their characteristic features included the presence of a weakly expressed continental shell, intensive bombarding with meteors, a strong flux of geothermal heat and, apparently, strong volcanic activity. The early Earth had probably been covered mainly with the oceans, with the volcanoes' tops showing above water, and the speed of its rotation round its axis was 2–4 times faster than that observed today. According to the theory of stellar evolution the solar radiance was 20–35 per cent below that of today, but the level of ultraviolet radiation was higher.

Under such conditions and with the present atmospheric composition, a global glaciation should take place; therefore it is important to explain why this has not happened. In continuation of the studies published, Kiehl and Dickinson (1987) performed new calculations of the surface temperature evolution for the period of 4.5 billion years, with the prescribed dynamics of $CO_2$ concentration and solar constant assumed earlier. These calculations are based on the use of a radiative–convective model in which a more detailed account has been taken of the structure of the infra-red absorption spectrum for the 0–50 km atmospheric layer. The range of wavenumbers 0–2500 $cm^{-1}$ was divided into 500 spectral intervals 5 $cm^{-1}$ wide (in each of them the transmission function is parametrized by the Malkmus model). The continuum absorption by water vapour, both of e-type and p-type, was also taken into account. The reliability of the technique applied was illustrated by a good agreement of calculated radiation fluxes and flux divergences with the data of line-by-line calculations. The delta-Eddington approximation was used to calculate shortwave radiation fluxes (the spectrum was subdivided into 56 intervals). A three-layer cloudiness located at the following levels was prescribed: 250 hPa (0.228); 500 hPa (0.090); 700 hPa (0.313). The thickness of cloud layers was, respectively, 50, 100 and 150 hPa. The surface albedo was 0.1. The Earth's albedo equal to 0.31 corresponds to the model applied.

With an assumption that 4.25 billion years ago the solar constant was 24 per cent below that of today (1360 $W m^{-2}$), and the $CO_2$ mixing ratio was 969 times greater than today (320 ppm), the global mean surface temperature was calculated to have been 3.38 deg K above that of today, whereas the stratospheric temperature was calculated to have been 40–45 deg K below the present value. The global mean surface temperature remains constant with the following relation between the $CO_2$ mixing ratio and the quantity $f_s = 1 - S/S_0$ ($S_0$ is the present value of the solar constant):

$$X_{CO_2} = 7.952 - (20.563 + 130.417 f_s)^{1/2} \tag{6.4}$$

If, for example, $f_s = 0.236$ (as it was 4.25 billion years ago), then $X_{CO_2} = 0.787$, i.e. it ought to have exceeded the present-day estimate by a factor of 428. With an assumption that 4.25 billion years ago the $CH_4$ mixing ratio was $10^{-4}$, to maintain the present surface temperature a $CO_2$ content half as much (in comparison with the present time) would have been enough. The decrease of surface albedo from 0.1 to 0.05 (this corresponds to the Earth covered with ocean, with ice caps at the poles), causing an increase of the solar radiation absorption by 7 $W m^{-2}$ or a decrease of the upper cloud amount by 15 per cent, which would increase the radiation budget by 3.4 $W m^{-2}$, could have been important factors in temperature increase. The combined effect of both these factors could have led to the increase of the absorbed solar radiation by 10–20 $W m^{-2}$.

As has already been pointed out, a new direction of climate change numerical modelling is the use of so-called nested models (see Chapter 5). Based on the use of a mesoscale model MM-4 nested in the NCAR climate model CCM-1, Marinucci *et al.* (1992) performed a numerical modelling of climate changes in western Europe which can occur with a $CO_2$ concentration doubling. The spectral model CCM-1 (taking into account 15 harmonics) was realized over a grid of 4.5° latitude by 7.5° longitude, while the mesoscale model covered the territory 3600 km by 4100 km with a spatial resolution of 70 km. This resolution enables one to consider such relief features as the Alps, the Iberian Plateau, the Pyrenees, the Balkans, the Appenines, the Grampian Mountains (Scotland), the British Isles, the Italian and Hellenic peninsulas, whereas in the case of the large-scale model CCM-1, the British Isles, the Italian and Hellenic peninsulas as well as such large islands as Corsica, Sardinia and Sicily turn out to be smeared.

In the nesting technique initial and boundary conditions for MM-4 are prescribed by interpolation of the data of numerical modelling using the CCM-1 model (this refers to the fields of wind, temperature, humidity and surface pressure). Boundary conditions are renewed every time step, equal to 120 s, by linear interpolation of the CCM-1 data obtained every 12 hours. Also by linear interpolation of the CCM-1 data, information about the fields of sea surface temperature, soil moisture and snow cover needed for mesoscale modelling is prescribed. With the use of the results of numerical modelling of global climate for the cases $1 \times CO_2$ and $2 \times CO_2$, calculations were made for western Europe for 40 months with data averaging over the last 20 months for four seasons (thus, the data for every season were obtained by averaging over 5-month periods).

Analysis of the data of global modelling has shown that they satisfactorily simulate the features of the annual change of large-scale climatology in western Europe, but contradict the observational data from the point of view of regional features, revealing such errors as (1) general weakening and southward drift of the north Atlantic jet stream, (2) underestimated sea surface temperature in the north Atlantic and in the Mediterranean Sea, (3) underestimated temperature and relative humidity in the middle troposphere, (4) serious errors in predicted precipitation for July caused by underestimated soil moisture and overestimated surface temperature and (5) underestimated cloud amount.

Owing to a more reliable account of topography and coastline, the MM-4 provided a more detailed simulation of regional fields of precipitation, air temperature and cloud amount than did the CCM-1. However, the data of MM-4 also suffered from some significant errors: (1) the temperature was underestimated by several degrees over land in January, April and October and overestimated in July, (2) precipitation was underestimated, especially in July, when calculated values turned out to be below those observed by 60–80 per cent (in January, April and October they were underestimated only by 10–35 per cent).

Though the enumerated errors have been determined by imperfect parametrization of some physical processes, a significant contribution is also made by an inadequate global climate model. So, for example, the temperature and precipitation underestimated from the MM-4 data are, apparently, explained by underestimation of temperature and humidity in the troposphere, as well as sea surface temperature, and a relatively weak prescribed jet stream, typical of CCM-1.

Analysis of differences in meteorological parameters for the cases $1 \times CO_2$ and $2 \times CO_2$ (for a possible removal of systematic errors, only climate changes were considered) was confined to a consideration of zonal wind at a level of 300 hPa and precipitation. This analysis has shown that changes in precipitation predicted by global and regional models differ in some cases not only in magnitude but also in sign, though general features of the fields of variability of precipitation and wind turned out to be similar. The characteristic feature was a very substantial annual change of variability of the fields of wind and precipitation. So, for example, in January there was observed a weakened jet stream over the north-eastern Atlantic, central Europe and the eastern Mediterranean, and an accelerated jet stream over the south-western Mediterranean and north-eastern Europe.

Though in the case of $2 \times CO_2$ precipitation should intensify everywhere due to intensified evaporation and increasing water-holding capacity of the warmer troposphere, regional variations in precipitation depend on variations in the atmospheric global circulation. As a rule, an enhancement (reduction) of intensity of jet stream is connected with increase (decrease) of baroclinicity and, hence, increase (decrease) of storm activity. Therefore in January a minimum increase and even reduction of precipitation is observed over the western coastline of Europe (in accordance with the fact that the North Atlantic storm track is located north of the British Isles, where the jet stream is reduced most), but maximum increase over the eastern part of western Europe as well as the central and western Mediterranean, where the intensity of the jet stream is most substantial.

Similar regularities take place in other seasons, too. Specific features of large-scale circulation predicted with the MM-4 model are mainly determined by the boundary conditions prescribed from the CCM-1 data. The effect of subgrid perturbations revealed by the MM4 model is most remarkable near the surface. A combined application of the mesoscale climate model MM-4 and the model BATS, describing hydrological processes in soils, has made it possible to analyse specific features of the hydrological regime. So, for example, in conditions of $2 \times CO_2$ the root zone turned out to be drier over most of continental Europe than at present (except for Scandinavia and the north-western part of the region considered). The drying-up of soils in these conditions is particularly noticeable over the Alps, the Grampian Mountains, the Dinaric Alps (former Yugoslavia) and in the eastern part of continental western Europe. The ($2 \times CO_2$)-induced warming brings forth a decrease of snow accumulation.

The standard approach to the accomplishment of numerical experiments to analyse climatic sensitivity to various factors consists in varying one of the climate-forming factors (most often a doubled $CO_2$ concentration is prescribed in modelling) and subsequent comparison of the results obtained with the results from the control experiment. There is no doubt, however, that the multi-factor climate changes and interactive manifestation of the effects of various factors necessitate a more complicated approach to the solution of the problems of sensitivity, taking account of various interacting factors.

Such an approach has been realized by Henderson-Sellers (1992, 1993) as applied to the BATS scheme, used to parametrize biospheric processes (precisely, heat and moisture exchange in the presence of vegetation cover) in the NCAR global climate model. The atmospheric forcing was determined in this scheme by 12 parameters, including: the top-of-the-atmosphere incident solar radiation, monthly mean surface air temperature, surface air humidity, monthly mean precipitation amount, precipitation distribution through each month, diurnal air temperature range, fractional cloud amount and cloud albedo, precipitable water in the atmospheric column, wind speed, surface air pressure, and ozone transmissivity for solar radiation. The last four parameters were prescribed, while the extra-atmospheric insolation could easily be calculated as a function of latitude, time of day and season. Thus the remaining seven parameters were varied and, hence, the complete factor analysis required $2^7 = 128$ experimental runs.

To save computer time and bearing in mind the probable insignificance of numerous high-order interactions, a decision was made to confine the work to one half fractional factorial experiment (64 runs), which ensured an account of all two-factor and three-factor interactions. Calculations were made for three climate regimes: tropical, temperate and polar (thus the runs totalled 192).

The following limits of variability of the factors considered were prescribed (their letter-notations are given in parentheses): amplitude of the temperature diurnal change (r) = 3.0–6.0 deg K; relative humidity (k) = 0.7–0.8; cloud amount (c) = 0.45–0.7; atmospheric moisture content (w, cm) = 0.8–8.0; precipitation distribution through each month (e) = 5.0–7.30; monthly amount of precipitation (p, mm) = P2 (7, 3, 7, 17, 21, 26, 25, 30, 32, 20, 9, 5 – for individual months), P1 (40, 96, 159, 270, 319, 402, 440, 279, 392, 276, 164, 34); monthly mean surface air temperature (t, K) = T2 (259, 260, 259, 267, 274, 281, 283, 282, 279, 272, 264, 258), T1

(300, 300, 301, 301, 300, 299, 298, 299, 299, 300, 300). The values enumerated have been taken from the data for the following locations characterizing the three climate regimes considered (tropical, temperate, polar): Calabar (5° N, 8° E), Akmolinsk (51° N, 57° E), Nome (64° N, 164° W).

The most important result of the numerical modelling of sensitivity consists in revealing the stability (robustness) of the BATS scheme. Even in extreme and unrealistic cases (for example, snowfall in the zone of tropical forests) a stable computation was preserved, and the results obtained had a physical sense. Table 6.6 characterizes the role of individual factors and their combinations (enumerated in order of priority) with account of variability threshold $\pm 4$ deg K. As can be seen, the surface air temperature is the most important parameter, followed by precipitation, precipitation–temperature interaction, cloud amount; that is, the effect of interactivity manifests itself weakly (it is this that determines certain grounds for analysis of sensitivity to individual isolated factors).

Calculations made with consideration of only the eighth part of all factors (16 runs) gave approximately the same results. The BATS scheme contains 23 parameters characterizing the properties of vegetation cover and soils. The complete factor analysis of sensitivity to variations in these parameters would require $2^{23} = 8\,388\,608$ runs, which is, of course, impracticable. Radically simplified assessments have led to the conclusion that, from the viewpoint of sensitivity of net evapotranspiration to the variability of different parameters, surface roughness and stomatal resistance to photosynthetic radiation are most important for the tropics; for mid-latitudes it is soil porosity, and in polar latitudes it is roughness and soil porosity. Though the calculations considered are inadequate in many respects, they can be very important for planning field experiments from the viewpoint of priority of the parameters measured.

Gordon and Hunt (1994) have studied the intraannual and interannual variability of the greenhouse climate change using a low-resolution four-layer (900, 650, 350 and 100 hPa) spectral (R21) global climatic model. The model included the diurnal and seasonal cycles, smoothed orography, computed sea ice amount and cloud

**Table 6.6** One-, two- and three-factor effects determined as important from extreme values, half-factorial atmospheric forcing experiment

| | | | | | | |
|---|---|---|---|---|---|---|
| Maximum canopy temperature | | | | | | |
| Tropical forests | t | r | c | p | pt | |
| Temperate grassland | t | c | r | | | |
| Polar tundra | t | c | r | | | |
| Total evaporation | | | | | | |
| Tropical forests | p | pt | t | e | ep | |
| Temperate grassland | t | pt | c | p | | |
| Polar tundra | t | pt | c | | | |
| Net carbon gain | | | | | | |
| Tropical forests | t | e | | | | |
| Temperature grassland | t | ct | | | | |
| Polar tundra | t | e | pt | et | p | ept |

*Note*: A threshold of $\pm 4$ deg K was used.
t = surface air temperature; p = precipitation; c = cloud amount; e = precipitation humidity.

**Table 6.7** Annual means and standard deviations (SD) for time temperature series at various points

| | $1 \times CO_2$ | | $2 \times CO_2$ | |
|---|---|---|---|---|
| Point | Mean (K) | SD (deg K) | Mean (K) | SD (deg K) |
| North Pacific | 285.29 | 0.29 | 288.36 | 0.35 |
| Tropical Pacific | 300.98 | 0.61 | 301.71 | 0.41 |
| Indian Ocean | 300.71 | 0.29 | 303.34 | 0.41 |
| Canada | 273.67 | 0.86 | 277.92 | 0.89 |
| China | 299.46 | 0.41 | 301.64 | 0.38 |
| South Africa | 290.06 | 0.31 | 293.79 | 0.44 |

cover, and used implied oceanic heat fluxes to represent transport processes in the oceans. A highly responsive two-layer soil moisture formulation was also incorporated. Greenhouse-induced climate change has been analysed as a difference between 20-yr equilibrated simulations for greenhouse ($2 \times CO_2$) and control ($1 \times CO_2$) conditions. As Gordon and Hunt (1994) pointed out, such a comparison revealed considerable differences from the time-averaged results, with occasions of marked positive and negative temperature deviations. Of particular interest were the periods of negative temperature departures compared with the control run, especially over the northern hemisphere continents. This is a clear demonstration of the complexity of the possible climatic responses under greenhouse conditions.

Table 6.7 illustrates annual means and standard deviations for time series which characterize sea surface or land surface temperatures at certain points. Average figures here show greenhouse warming at all the points, but the analyses of temperature variability for individual points reveals large differences between calculated variability for various points as well as between calculations and observations. For instance, the central Pacific gridpoints are characterized by an enormous range of simulated sea surface temperature variability. In the case of the northern point, rather regular behaviour with interannual variability of about 2 °C in the winter temperature of the control run ($1 \times CO_2$) and the greenhouse simulation represented a simple warming of about 3 °C superimposed on the control. But for the low latitude point, a completely different situation prevailed with very marked interannual variability consisting of warm and cold periods of a few years duration. Considerable variability also took place in the low latitudes of the Indian Ocean. For the land points, an additional source of variability exists in that soil moisture changes can have a noticeable impact on surface temperature, particularly for regions where drought conditions tend to prevail.

The presence of only small differences between the standard deviations of the control and greenhouse ($2 \times CO_2$) runs implies that the same underlying physical processes were operative in the two runs despite the considerable changes in their respective climatologies. The existence (in a number of situations) of very similar variations for both the greenhouse and control runs in the spatial characteristics and amplitude of these deviations suggests that, under greenhouse conditions, occasions will arise when natural variability will override the greenhouse warming. It is important to emphasize in this connection that Gordon and Hunt (1994) could not identify any clear greenhouse fingerprint except changes in the mean values. Further careful analysis is necessary, however, with the application of elaborate statistical methods, as has been done by Hasselmann (1993).

## 6.2 The multifaceted nature of the atmospheric greenhouse effect

The complexity of anthropogenically induced changes in the gas composition of the atmosphere determines the multicomponent nature of the atmospheric greenhouse effect, studied in detail by Kondratyev and Moskalenko (1983, 1984). Their calculations have shown that the mean global atmosphere of the Earth provides a greenhouse effect of $\Delta T = 33.2$ deg K (the difference between planetary temperature and surface temperature, with the following contributions by optically active gaseous components at current concentrations: $H_2O$, 20.6 deg K; $CO_2$, 7.2 deg K; $N_2O$, 1.4 deg K; $CH_4$, 0.8 deg K; $O_3$, 2.4 deg K; and $NH_3$, CFCs, $NO_2$, $CCl_4$, $O_2$ and $N_2$, 0.8 deg K). The data obtained verify that the radiative regime of the atmosphere is governed by the amount of water vapour, carbon dioxide and ozone present in it, with the prevailing influence of water vapour on the greenhouse effect. Having strong, broad rotation–vibration bands, water vapour absorbs most of the thermal emission from the Earth's surface, lowering the outgoing emission. Furthermore, part of the emission from the surface in the 4.8 $\mu$m, 8–13 $\mu$m and 18 $\mu$m transparency windows is lost. The greenhouse effect of water vapour is intensified by the absorption bands of $CO_2$ and ozone.

Consider the effect of variations in the optical characteristics of various atmospheric components on the greenhouse effect. Many processes taking place in the atmosphere exhibit a close correlation (synergism). A rise in tropospheric temperature, for example, leads to an increase in its water vapour content. The spectral transmission functions of the main absorbing components ($CO_2$ and water vapour) strongly depend on temperature, especially in the wings of the bands. The absorption properties of these components increase considerably with temperature. Thus an increase in surface temperature is usually accompanied by an intensified greenhouse effect.

Calculations confirm that increasing $CO_2$ concentration is the most important factor in relation to the anthropogenic impacts on climate. However, increased pollution of the atmosphere as a result of human activities leads to increasing concentrations of other atmospheric components, such as $SO_2$, CO, halocarbons, $HNO_3$ and hydrocarbon compounds. Of great importance is an account of aerosols either directly ejected into the atmosphere or resulting from gas-to-particle transformations. All the enumerated components have absorption bands in the infra-red spectral region which contribute to the greenhouse effect.

To study the effect of the gaseous and aerosol atmospheric components, Kondratyev and Moskalenko (1983, 1984) undertook calculations of radiation fluxes for model atmospheres with different chemical compositions. The calculation scheme involved all the components that affect the radiation absorption in the atmosphere, including water vapour (with account of continuum absorption), carbon dioxide, nitrogen and oxygen (pressure-induced absorption), methane, nitric oxide, sulphur dioxide, nitric acid vapour, ethylene, acetylene, ethane, formaldehyde, chlorofluorocarbons, $CCl_4$, $CF_4$, ammonia and aerosols of different chemical composition and size distribution.

Basic optical components (water vapour and carbon dioxide) govern the radiative regime of the troposphere. Minor components (ozone, methane, nitrous oxide, nitric oxide, carbon monoxide, nitric acid vapour, sulphur dioxide, halocarbons and nitrogen dioxide) affect the radiation transfer markedly only in the atmospheric transparency windows. The spectral structure of the absorption bands of these gases is now being thoroughly studied. The role of minor components increases in the formation of the radiative regime of the stratosphere. When calculating the greenhouse effect, it is most important to consider the radiation absorption by the 6.3 $\mu$m rotational–vibrational band and the rotational bands ($\lambda > 15\,\mu$m) of water vapour; by $CO_2$ 4.3 and 15 $\mu$m bands; $O_3$ 4.74 and 9.6 $\mu$m bands; $N_2O$ 4.5, 7.8 and 8.6 $\mu$m bands; $CH_4$ 3.3 and 7.6 $\mu$m bands; and pressure-induced absorption in $N_2$ 4.3 $\mu$m and $O_2$ 6.5 $\mu$m bands at collisions of molecules $N_2$–$O_2$, $N_2$–$N_2$, and $O_2$–$O_2$. The role of atmospheric rotational vibrational absorption bands 4.67 $\mu$m (CO), 5.3 $\mu$m (NO), 6.3 and 10.6 $\mu$m ($NH_3$), 4.75, 8.8 and 20 $\mu$m ($SO_2$) and 4.9 $\mu$m ($CO_2$) in the troposphere is not very great. In the stratosphere it is important to take into account the radiation transfer in the absorption bands 5.6, 11.2, 20 $\mu$m ($HNO_3$), 5.6, 7.35 and 8.69 $\mu$m ($N_2O$), 9.22 and 11.82 $\mu$m ($CClF_3$), 8.68, 9.13 and 10.93 $\mu$m ($CCl_2F_2$), 7.14, 0.85 and 13.66 $\mu$m ($CFCl_3$), 12.99 $\mu$m ($CCl_4$) and 10 $\mu$m ($CF_4$).

Many of the gaseous components that absorb in the thermal emission spectral region are products of human industrial activities and therefore their concentration can rapidly increase with time. An example based on halocarbons can be considered to be representative. If the mixing ratio of CFCs F-11 and F-12 is now about (1–2) $\times 10^{-10}$, by the end of the century, at the present level of atmospheric pollution, their concentrations may constitute $2 \times 10^{-9}$. The process of chlorofluorocarbon accumulation in the stratosphere is favoured by their long residence time in the atmosphere.

A doubled concentration of nitrous oxide leads to a 0.7 deg K rise of mean temperature, and with double ammonia and methane the surface air temperature is raised by 0.1 deg K and 0.3 deg K, respectively. A 20-fold increase in the chlorofluorocarbon content may lead to a warming of 0.6 deg K to 1 deg K, and the total greenhouse effect resulting from doubled $N_2O$, $CF_4$, $SO_2$ and $HNO_3$ reaches 1.2 deg K.

When considering possible climatic changes due to the anthropogenically induced changes in the chemical composition of the atmosphere, it is necessary to take into account the relations between various climatological factors. For example, an increase in the chlorofluorocarbon content can substantially change the atmospheric ozone concentration field.

Calculations have shown that the greenhouse effect caused by ozone is very sensitive to variations in its vertical profile and is manifested not only through the direct impact of ozone but also, indirectly, through other atmospheric components. The warming can also result from changes in the stratosphere's vertical temperature profile. If a uniform decrease of 25 per cent in the ozone concentration leads to a surface temperature decrease of 0.45 deg K for a mean planetary atmospheric model, a non-uniform decrease in ozone concentrations causes an anti-greenhouse effect of $\Delta T_s = 0.25$ deg K. It is important to consider ozone variations not only in the stratosphere but also in the troposphere, since tropospheric ozone variations often lead to opposite effects from results determined by stratospheric disturbances of the ozone concentration field.

In calculating the greenhouse effect it is important to consider the atmospheric constituents $CF_4$ and $CCl_4$, with strong bands in the region of 10 and 13 $\mu$m. Increased future industrial releases of $CF_4$ and $CCl_4$ can lead to an enhancement of the greenhouse effect by $\Delta T_s = 0.8$ deg K to 1 deg K. Based on the calculations of Kondratyev and Moskalenko (1984), an increase in the $NO_2$ concentration leads to a weak anti-greenhouse effect that depends on the peculiarities of atmospheric vertical temperature profiles and $NO_2$ concentrations.

The effect of aerosols on the atmospheric radiative regime is manifested through the mechanisms of radiation scattering and absorbing by aerosols. Depending on the optical density of aerosols and their size distribution and chemical composition, conditions are created in which aerosols either raise or lower the albedo of the planet. The ambiguity of conclusions can be explained by the variability of aerosol properties that result from the large spatial and temporal variability of the concentration field and size distribution of the aerosols, as well as by the dependence of the chemical composition of aerosols on conditions of their generation.

For small-sized fractions (sub-micrometre) of aerosols, the absorption coefficient always exceeds the scattering coefficient, and in the wavelength region $\lambda > 1\ \mu m$ the attenuation coefficient is determined mainly by absorption of radiation by aerosols. Therefore, sub-micrometre aerosols heat the atmosphere at the expense of solar radiation and emit in the longwave spectral region, absorbing the thermal emission from the surface. Thus the sub-micrometre background aerosol (depending on the type of aerosol) acts like an absorbing gaseous component and intensifies the atmospheric greenhouse effect, raising the surface temperature by 3 deg K for mean global conditions. An increase in the tropospheric aerosol concentration by a factor of 1.5 as a result of human impact will lead to a surface warming of 1.7 deg K.

A doubling of the optical thickness of the stratospheric background aerosol (due to an increase of particles containing free carbon) will raise the temperature $T_s$ by 0.8 deg K.

The foregoing estimates of the greenhouse effect resulting from the atmospheric aerosol sub-micrometre fraction can in part be compensated for by an anti-greenhouse effect of the large-sized aerosol fraction (e.g., sea salt, dust and ice particles). The anti-greenhouse effect is also created by the stratospheric sulphate aerosol layer. However, Kondratyev and Moskalenko (1984) assume that, on average, the planetary greenhouse effect due to the strong absorption of the sub-micrometre aerosol fraction probably prevails over the anti-greenhouse effect of large aerosol particles. Atmospheric pollution due to human industrial activity causes an increase in the concentration of strongly absorbing aerosol particles, which does not lead to substantial variations of the planetary albedo but intensifies the screening of outgoing thermal emissions from the surface.

Clouds are a most important factor determining the radiative regime of the Earth. An increase in cloudiness leads to a higher albedo and decreases the solar radiation flux that reaches the Earth's surface. Yet cloudiness lowers the radiative temperature corresponding to the outgoing emission. Under night-time conditions the greenhouse effect of cloudiness is weak, and in daytime (e.g. at positive surface temperatures) the cooling due to increased reflected solar radiation dominates (as a rule) the greenhouse effect of cloudiness. However, in winter (with snow-covered surfaces) the presence of cloudiness does not raise the albedo substantially. In these conditions the greenhouse effect prevails.

Under the influence of anthropogenic impact, the structural and optical characteristics of the Earth's cloud cover may change substantially. In particular, atmospheric pollution can lead to an increase of the upper cloud cover and to an intensified radiative cooling of the Earth due to the raising of the effective height of the cloud top. At the present time, it is difficult to predict the scale of the greenhouse effect manifestation, due to this factor. It is not excluded that, by its efficiency, it will compete with or even dominate the effect of $CO_2$ warming in the future.

Calculations by Kondratyev and Moskalenko (1984) have shown that an increase in the upper-level cloud cover for a mean planetary model of the atmosphere leads to the anti-greenhouse effect caused by increased planetary albedo, with decreasing shortwave radiation flux coming to the troposphere. The resulting cooling of the surface and troposphere is not compensated for completely by warming due to the varying thermal emission of the planet. Variations in the surface temperature caused by increased upper-level cloud cover (the upper troposphere, lower stratosphere) depend strongly on the choice of the cloud cover model, and temperature variations $\Delta T_s$ therefore lie between 0.7 deg K and 3 deg K. For all cloud models an increase of the upper-level clouds leads to a decrease of the surface temperature.

In a continuation of previous estimates of the contribution by various components to the greenhouse effect, Lacis *et al.* (1981), using a radiative–convective equilibrium model, calculated the mean global surface air temperature increase due to the greenhouse effect caused by anthropogenic emissions into the atmosphere of carbon dioxide, CFCs, ammonia and nitric oxide during the decade of the 1970s as well as with prescribed doubled concentrations of these gaseous components.

Results by Lacis *et al.* (1981) (Table 6.8) show that the total contribution by methane, nitric oxide and CFCs to the climate warming during the period 1970–1980 constituted (taking into account approximate estimates) 50–100 per cent of the $CO_2$ contribution.

The following approximation is possible, with the gas concentration expressed in ppm (except for CFCs, whose concentrations are given in ppb) and at the concentrations $CH_4 < 5$ ppm, $N_2O < 1$ ppm, $CF_2Cl_2$ and $CCl_3F < 2$ ppb, and $\Delta CO_2 < 300$ ppm ($\Delta CO_2$ is an excess concentration of $CO_2$ with respect to its values, equal to 300 ppm):

$$\begin{aligned} T(^\circ C) = {} & 0.57(CH_4)^{0.5} + 2.8(N_2O)^{0.6} - 0.057CH_4 \cdot N_2O \\ & + 0.015CCl_3F + 0.18CCl_2F_2 \\ & + 2.5 \ln[1 + 0.005\Delta(O_2) + 10^{-5}(\Delta CO_2)^2] \end{aligned} \tag{6.5}$$

The third term describes the effect of the overlapping of the bands of methane and nitrous oxide. These estimates were obtained without consideration of the atmosphere–ocean interaction, which slows down the manifestation of the intensified greenhouse effect through a climate warming. The consideration of the

**Table 6.8** Climate warming due to the greenhouse effect of selected gaseous components of the atmosphere

| | Changes during 1970–1980 | | | Arbitrary change | | |
|---|---|---|---|---|---|---|
| Component | $a_0$ (ppb) | $\Delta a$ (ppb) | $\Delta T$ (°C) | $a_0$ (ppb) | $\Delta a$ (ppb) | $\Delta T$ (°C) |
| $CO_2$ | 325 000 | 12 000 | 0.14 | 300 000 | 300 000 | 2.9 |
| $CH_4$ | 1500 | 150 | 0.032 | 1600 | 1600 | 0.26 |
| $N_2O$ | 295 | 6 | 0.016 | 280 | 280 | 0.65 |
| $CCl_3F$ | 0.045 | 0.135 | 0.020 | 0 | 2 | 0.29 |
| $CCl_2F_2$ | 0.125 | 0.190 | 0.034 | 0 | 2 | 0.36 |

$a_0$ = initial concentration of the given component; $\Delta a$ = concentration increase; $\Delta T$ = temperature increase.

atmosphere–ocean interaction, based on the assumption that it is enough to take an oceanic mixed layer 100 m thick, led to the conclusion that in this case the increase in surface air temperature by the year 1980 should have been 50 per cent less, that is about 0.1 °C. Taking into account pre-1970 atmospheric pollution by $CO_2$ and other greenhouse gases, the climate warming for 1970–1980 can be estimated at 0.1–0.2 °C. Since this level of temperature variation is close to the observed mean standard deviations of 10-yr temperature values, it is still impossible to detect an anthropogenically induced climate warming. However, if we proceed from the fact that the total warming during the past 20 years (1970s and 1980s) reached 0.2–0.3 °C, an identification of warming can be realized from observational data by the end of the 1980s. These results testify to the urgency and necessity of the global monitoring of the optically active minor gaseous components of the atmosphere, which contribute markedly to the atmospheric greenhouse effect.

Human activities contribute a variety of optically active minor gaseous components to the atmosphere which add to the formation of a greenhouse effect and can affect the climate to a certain extent. Deepak and Gerber (1983) noted that the related compounds involve gases included in the following groups:

1. Carbon: $CO_2$ and $CH_4$
2. Oxygen: $O_3$
3. Nitrogen: $N_2O$, $NO_2$, $N_2O_5$, $HNO_3$ vapour, $NH_3$
4. Sulphur: $SO_2$, COS, $CS_2$
5. Halogens: F-11, $CFCl_3$; F-12, $CF_2Cl_2$; F-13, $CF_3Cl$; F-21, $CFHCl_2$; F-22, $CF_2HCl$; F-113, $CF_3CCl_3$; F-114, $C_2F_4Cl_2$; F-115, $C_2F_5Cl$; F-116, $C_2F_6$; $CCl_4$; $CH_3Br$; $CH_3Cl$; $CF_4$; $CH_2Cl_2$; $C_2HCl_3$; $CH_3CCl_3$; $C_2Cl_4$
6. Non-methane carbon compounds: $C_2H_2$; $C_2H_4$; $C_2H_6$; $C_3H_8$; $C_6H_6$
7. Peroxyacetylnitrate (PAN).

Deepak and Gerbert (1983) discussed the contribution to the greenhouse effect of the most important of the compounds listed above, as determined by their absorption bands in the 7 $\mu$m to 13 $\mu$m transparency window. Since some of these compounds are optically active both in the longwave and shortwave spectral regions (this pertains mainly to ozone and nitrogen dioxide), their effects can be manifested through both a warming and a cooling of climate, depending on the vertical concentration profile of these components.

Table 6.9 contains estimates of the mean global effect of increased concentrations of different gases on climate. These estimates were obtained using one-dimensional models of the radiative–convective equilibrium (assuming a constant relative humidity of the atmosphere and, as a rule, fixed cloud-top height). In the case of ozone the dependence of its concentration $F(\psi, z)$ on latitude $\psi$ and height $z$ was considered, but a doubled $O_3$ content was applied to the troposphere for all purposes.

With decreased $O_3$ content in the stratosphere, two effects of opposite sign occur: (1) heating of the troposphere and surface due to an increase in the incoming shortwave radiation, but decrease of stratospheric temperature and (2) the latter decreases the longwave radiation flux coming from the stratosphere to the troposphere (hence, cooling). In the case of a homogeneous (vertical) decrease in the $O_3$ concentration in the stratosphere, the effect of longwave cooling prevails. Another situation is observed in the troposphere, where the growth of the $O_3$ content causes

**Table 6.9** Summary of climatic effect caused by optically active minor gaseous components of the atmosphere

| Gas | Centre of the band ($cm^{-1}$) | Background mixing ratio (ppb) | Disturbed mixing ratio (ppb) | Variations in the outgoing emission at the level of the tropopause ($W\,m^{-2}$) | Surface air temperature variation (deg K) |
|---|---|---|---|---|---|
| $CO_2$ | 667 | $330 \times 10^3$ | $660 \times 10^3$ | 4.0 | 3.0 |
| $N_2O$ | 589, 1168, 1285 | 300 | 600 | 0.65 | 0.3–0.4 |
| $CH_4$ | 1306, 1534 | 1500 | 3000 | 0.61 | 0.3 |
| $O_3$ | 1041, 1103 | $F(\psi, z)$ | $2F(\psi, z)$ | 1.24 | 0.9 |
| $CFCl_3$ | 846, 1085, 2144 | 0 | 1 | 0.31 | 0.13 |
| $CF_2Cl_2$ | 915, 1095, 1152 | 0 | 1 | 0.26 | 0.13 |
| $CH_4$ | 632, 1241, 1261, 1283 | 0 | 1 | | 0.07 |
| $CF_2HCl$ | 1117, 1211 | 0 | 1 | | 0.04 |
| $CCl_4$ | 776 | 0 | 1 | | 0.14 |
| $CHCl_3$ | 774, 1220 | 0 | 1 | | 0.1 |
| $CH_2Cl_2$ | 714, 736, 1236 | 0 | 1 | | 0.05 |
| $CH_3Cl$ | 732, 1015, 1400 | 0 | 1 | | 0.013 |
| $CH_3CCl_3$ | 707, 1084 | 0 | 1 | | 0.02 |
| $C_2H_4$ | 949 | 0.2 | 0.4 | | 0.01 |
| $SO_2$ | 518, 1151, 1361 | 2 | 4 | | 0.02 |
| $NH_3$ | 950 | 6 | 12 | | 0.09 |
| $HNO_3$ | 1695, 1333, 850, 459 | $F(\psi, z)$ | $2F(\psi, z)$ | | 0.06 |
| $H_2O$ | 0–2000 | $3 \times 10^3$ | $6 \times 10^3$ | | 0.6 |

a warming due to longwave and shortwave radiation. Therefore, the climate of the troposphere is more sensitive to variations in the $O_3$ content in the troposphere than in the stratosphere. For nitric acid vapour, the dependence of its concentration on height was considered.

Since the contribution of individual minor components to the greenhouse effect is additive, it follows from Table 6.9 that the total contribution is nearly equal to that due to a doubling of $CO_2$ concentration. It is important to note that observational data do not permit a separation between the effects of $CO_2$ and those of minor components. The important role of minor components is determined both by their intensive absorption bands and by linear relations between their contributions to the greenhouse effect and an increase in their concentrations (in the case of $CO_2$ this dependence is logarithmic). New estimates of the minor gaseous constituents' contribution to the greenhouse effect have been obtained by Karol (1988), Wang and Mólnar (1985) and Wang *et al.* (1986, 1991).

To reveal more reliably the effect of increased $CO_2$ concentration on climate, Wang *et al.* (1986) analysed the role of such external factors of climate formation as increased concentrations of greenhouse gases, the effect of tropospheric and stratospheric aerosol, variations in the solar constant, and changes in the surface characteristics and industrial heat releases, the greenhouse gases problem being the main aim of the modelling.

**Table 6.10** The mixing ratio for greenhouse gases

| Component | Chemical composition | Mixing ratio (ppm) | Present rate of growth (% $yr^{-1}$) |
|---|---|---|---|
| Carbon dioxide | $CO_2$ | 340 | 0.4 |
| Stratospheric ozone | $O_3$ | (0.1–10) | 0.03 |
| Tropospheric ozone | $O_3$ | (0.02–0.1) | 0–0.7 |
| Methane | $CH_4$ | 1.7 | 1.2 |
| Carbon monoxide | CO | 0.12 | ? |
| Halocarbon compounds | $CFCl_3$ | $1.9 \times 10^{-4}$ | 5–8 |
| | $CF_2Cl_2$ | $3.5 \times 10^{-4}$ | 5–8 |
| | $CH_3CCl_3$ | $1.3 \times 10^{-4}$ | 5–8 |
| | CFC-113 | $0.3 \times 10^{-4}$ | 15–20 |
| | $CCl_4$, etc. | $1.5 \times 10^{-4}$ | 1–3 |
| Hydrocarbon compounds | $C_2H_4$, etc. | $10^{-3}$ | ? |
| Nitrous oxide | $N_2O$ | 0.3 | 0.2 |
| Nitrogen oxides | $NO_x$ ($NO + NO_2$) | ($3 \times 10^{-5}$–0.015) | ? |
| Sulphur compounds | COS | $5 \times 10^{-4}$ | ? |
| | $CS_2$ | $0.3 \times 10^{-4}$ | ? |
| | $SO_2$ | $1.0 \times 10^{-4}$ | ? |
| Stratospheric water vapour | $H_2O$ | (3–5) | ? |

*Note*: Values in brackets are possible limits to variations.

Table 6.10 characterizes the existing estimates of present mixing ratios for various greenhouse gases and of the respective growth rates. Of great importance is an increase in methane concentration, but the causes still remain unclear. They include intensified anthropogenic emissions due to population growth (rice fields, vegetation burning, etc.) and reduced concentrations of hydroxyl (reactions with OH constitute a major sink for $CH_4$ in the atmosphere). Also poorly understood are factors that govern the growth of $N_2O$ concentration (fossil fuel combustion, denitrification of fertilizers, etc.). The main sources of increasing oxides of nitrogen in the lower troposphere seem to be fossil fuel combustion and denitrification of fertilizers.

The effect of the increase in greenhouse gases' concentration on climate is manifested through the following three mechanisms: (1) direct radiative forcing, (2) climatic–chemical interaction (changes in the atmospheric chemical composition due to greenhouse gases) and (3) dynamically induced feedbacks.

Table 6.11 contains a summary of estimates of the effect of the growth of various greenhouse gases concentrations on the globally averaged surface air temperature and total ozone content obtained with one-dimensional models. Although in numerical climate modelling the effect of tropospheric aerosol is not taken into account as a rule, it is necessary to bear in mind that it reduces by 5 W $m^{-2}$ the amount of solar radiation coming to the Earth's surface and increases the longwave atmospheric emission by 3.5 W $m^{-2}$, which is comparable to the effect of a doubled $CO_2$ concentration.

A substantial uncertainty in the estimates of the anthropogenic impact on climate is connected with the lack of reliable parametrization of cloud formation.

Table 6.12 illustrates a possible increase in surface air temperature due to $CO_2$ and various greenhouse gases. These calculations were made with fixed relative

**Table 6.11** Sensitivity of surface air temperature (SAT) and total ozone content (TOC) to variations in greenhouse gas concentration

| Component | Concentration variations | SAT variations (°C) | TOC variations (%) |
|---|---|---|---|
| $CO_2$ | $330 \rightarrow 660$ ppm | 1.5–4.5 | 3.1 |
| $CH_4$ | $1.7 \rightarrow 3.4$ ppm | 0.2–0.4 | 2.1–3.3 |
| Halocarbon compounds | Continuous emissions at the level in 1977 | 0.5 | (−3.3) − (−5.5) |
| | $0 \rightarrow 1.0$ ppb, $CFCl_3$ | | |
| | $0 \rightarrow 2.7$ ppb, $CF_2Cl_2$ | | |
| | Continuous emissions at a level twice as high as in 1983 | 0.1 | −15.7 |
| | $0 \rightarrow 1.8$ ppb, $CFCl_3$ | | |
| | $0 \rightarrow 5.0$ ppb, $CF_2Cl_2$ | | |
| | $0 \rightarrow 1.5$ ppb, $CH_3CCl_3$ | | |
| $N_2O$ | $0.3 \rightarrow 0.6$ ppm | 0.3–0.4 | (−8.7) − (−10.7) |
| $O_3$ | Doubled concentration in the troposphere | 0.9 | |
| $H_2O$ | 3–6 ppm in the stratosphere | 0.6 | |

humidity, cloud top height and ice cover albedo. A similar study was carried out by Karol (1988, 1991).

Dickinson and Cicerone (1986) summarized studies of the dynamics of the concentration in the atmosphere of some greenhouse gases (carbon dioxide, nitrous oxide, ozone, methane and CFCs) and obtained estimates of possible variations in their concentrations by 2050, as well as the contribution to the formation of the atmospheric greenhouse effect. A consideration of the previous estimates of a doubled $CO_2$-induced warming suggested that a possible range of warming would be 1.0–5.5 °C. Taking account of the oceanic thermal inertia, an estimate was obtained of the anthropogenically-induced warming that had taken place (0.3–1.0 °C). By 2050 the temperature increase could be 1–5 °C or higher. The authors emphasized that a reduction of the uncertainty of the estimates obtained must be an important objective of studies in the future. It will probably take more than a decade to obtain sufficiently reliable estimates.

As Ramanathan *et al.* (1987) show, it is generally accepted that the anomalies of the radiative effects on the atmosphere caused by increasing concentrations of anthropogenic greenhouse gases can constitute a prevailing factor of climate

**Table 6.12** Possible increase in surface air temperature (SAT) due to $CO_2$ and other greenhouse gases

| Component | SAT increase (°C) 1950–1980 | SAT increase (°C) 1980–2010 |
|---|---|---|
| $CO_2$ | 0.25 | 0.39 |
| $CO_2 + NO_x + CH_4 + N_2O$ + Halocarbon compounds + $O_3$ | 0.50 | 0.80 |

changes on a decadal scale and longer, though still there is no convincing cause-and-effect explanation for climate changes for the last 125 years. For example, climate changes on a decadal scale are possible due to the internal variability of the atmosphere–ocean–land–cryosphere system caused by variations in the heat content of the oceans, sea ice and glaciers.

Figure 6.2 illustrates the role of greenhouse minor gaseous constituents (in this case the emphasis is placed on chlorofluorocarbons, CFCs) in the formation of the atmospheric greenhouse effect.

In connection with studies of greenhouse gases, of great importance was the fact that for some multi-atom greenhouse gases their contribution to the atmospheric greenhouse effect, per molecule, is much larger than in the case of $CO_2$. So, for example, in the case of F-11 ($CFCl_3$) and F-12 ($CF_2Cl_2$) this contribution is $10^4$ times larger. This fact draws serious attention to an assessment of the role of numerous secondary greenhouse gases in the formation of the greenhouse effect (Figures 6.3 and 6.4).

Of great importance are the results from studies of the effect of stratospheric ozone disturbances on the climate of the troposphere through the stratosphere–troposphere radiative–dynamic interaction. Recent calculations also show a substantial effect of increased $O_3$ content in the troposphere on the formation of the greenhouse effect. As was mentioned above, numerical modelling results indicate that continuing with the present rate of concentration growth for numerous greenhouse gases during several subsequent decades, an associated surface air temperature increase may exceed the climate warming that has taken place during the last century and will be comparable to the effect of a $CO_2$ concentration increase. The anthropogenically induced climate changes in the stratosphere may also greatly exceed the decadal-scale natural variations. Calculations show, in particular, a temperature sensitivity at the level of the tropical tropopause to variations in the concentrations of ozone and CFCs (Figure 6.5).

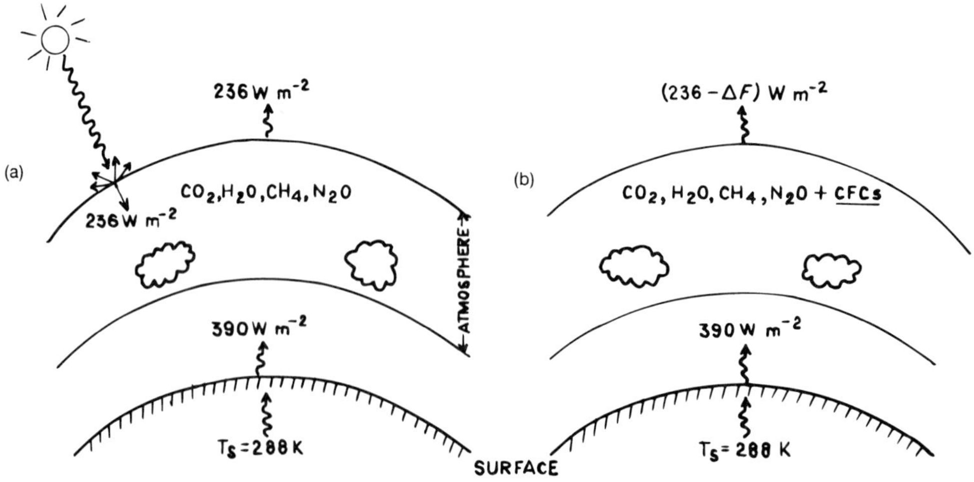


**Figure 6.2** Global energy balance and the greenhouse effect, (a) reference atmosphere and (b) with CFCs. $\Delta F$ = reduction due to increased trapping of infra-red radiation. After Ramanathan *et al.* (1987).

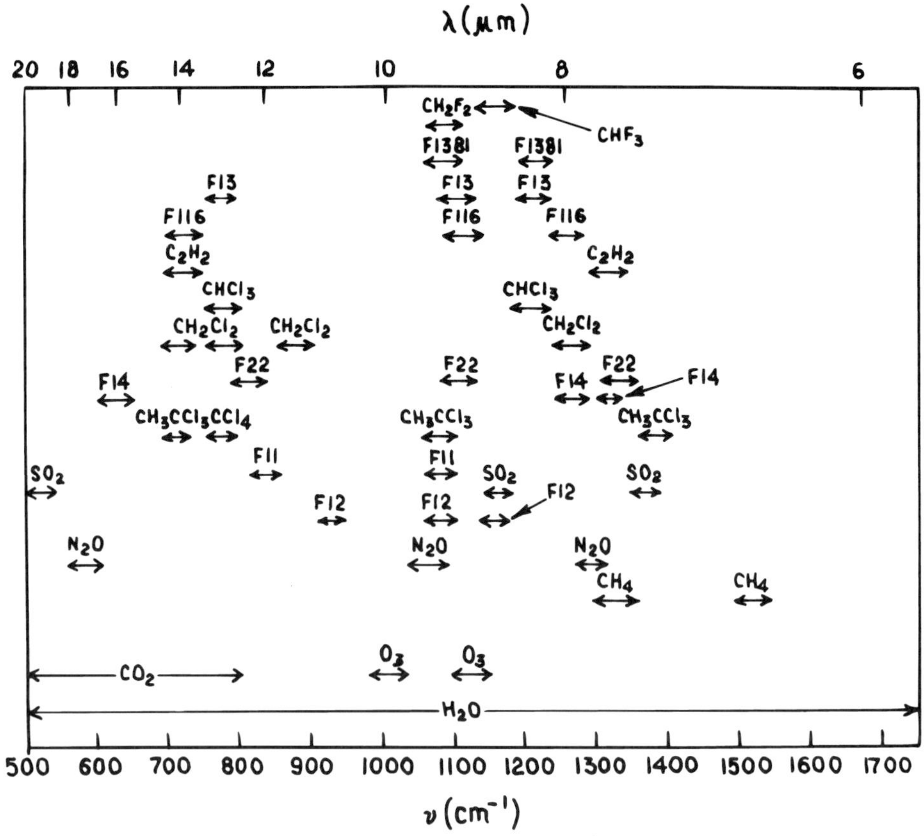


**Figure 6.3** Spectral locations of the absorption features of various trace gases. The spectral region between 7 and 13 μm is referred to as the atmospheric window. The anthropogenic trace gases have the potentials for making it into a 'dirty window'. After Ramanathan *et al.* (1987).

The urgency of the greenhouse gases problem has set the first priority for studies of the interaction among radiative, chemical and dynamic processes which governs the stability of the ozone layer to chemical disturbances. The radiative–dynamic interactions between the troposphere and stratosphere, as well as their greenhouse gases exchange, have become the key aspects of the anthropogenic impact on climate, in view of the importance of considering variations in the content of water vapour and ozone in the stratosphere. There is an urgent need for observations to monitor the long-term trends of the content of not only $CO_2$ but also $O_3$, $H_2O$ and such greenhouse gases as methane, nitrogen monoxide, CFCs, etc. In this connection, various aspects of the problem of the effect of greenhouse gases on climate have been discussed in detail (Ramanathan *et al.* 1987).

An analysis of calculation data on the intensification of the greenhouse effect of greenhouse gases showed that the rate of growth of the effect for the last few decades had increased 3–10 times compared with the period 1850–1960.

The cumulative equilibrium increase of surface air temperature due to $CO_2$ and other greenhouse gases between 1850 and 1980 should have reached 0.7–2 deg K,

(a)

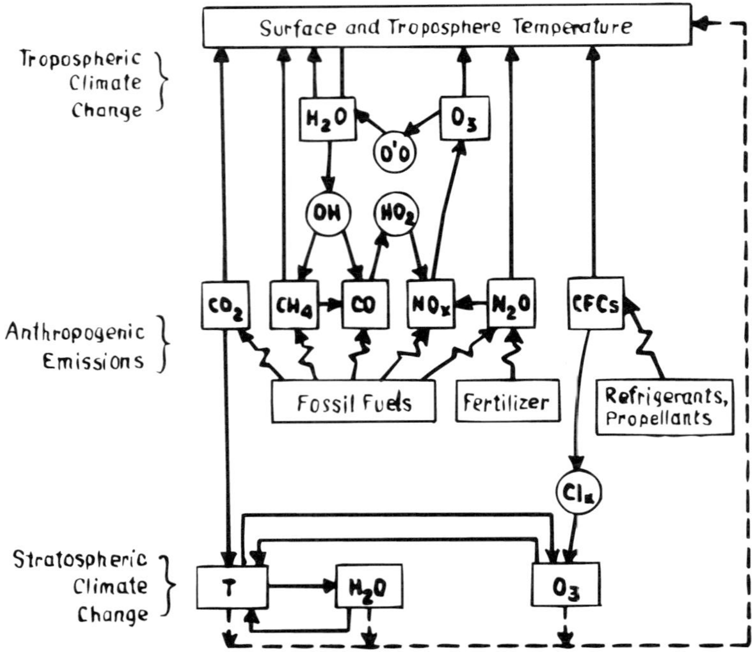


(b)

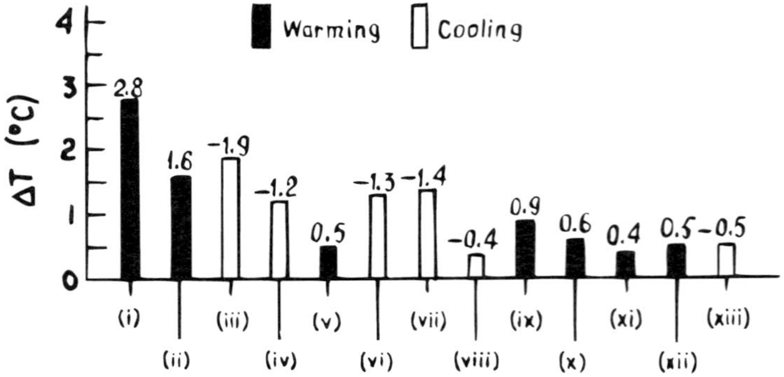


**Figure 6.4** (a) Examples of climate effects due to chemically and radiatively active gases. (b) One-dimensional radiative–convective model estimates of surface temperature effects of various global radiative perturbations. All of the results, except for $CH_4$, and the figure are adapted from Hansen *et al.* (1981). Legend for (b): (i) $CO_2$ (300 ppm → 600 ppm); (ii) solar luminosity (+1%); (iii) stratospheric aerosols, $H_2SO_4$ ($\Delta\tau = +0.02$); (iv) tropospheric aerosols, $H_2SO_4$ ($\Delta\tau = +0.1$); (v) tropospheric aerosols, soot ($\Delta\tau = +0.02$); (vi) land albedo (+0.05); (vii) low cloud (+2% of globe); (viii) middle cloud (+2% of globe); (ix) high cloud (+2% of globe); (x) $N_2O$ (0.28 ppm → 0.56 ppm); (xi) $CH_4$ (1.6 ppm → 3.2 ppm); (xii) $CCl_2F_2$, $CCl_3F$ (0 → 2 ppb each); (xiii) $O_3$ (−25%). After Ramanathan *et al.* (1987).

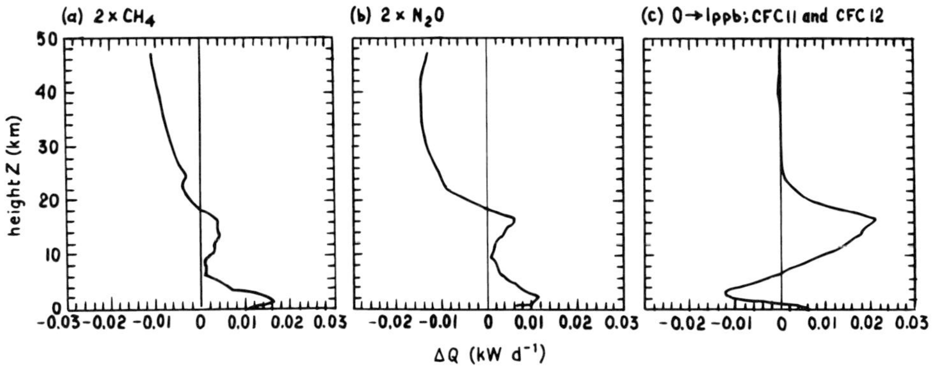


**Figure 6.5** Computed change in radiative heating rates ($\Delta Q$) due to trace gas increase. The calculations are for clear-sky conditions and employ tropical profile. After Ramanathan *et al.* (1987).

with greenhouse gases contributing about 30 per cent (the uncertainty is mainly due to unreliably calculated climate sensitivity to radiative disturbances). Approximate estimates, using a time-dependent one-dimensional model that considered the diffusion in the ocean, showed that the warming between 1850 and 1980 should have reached 0.4–0.9 deg K (with the equilibrium warming ranging between 0.7 and 2 deg K).

Calculations of the warming with different prescribed scenarios of the temporal dynamics of the greenhouse gases content suggested that the possible range of surface air temperature increase between 1980 and 2030 was 0.8–4.1 deg K (Figures 6.6–6.11). Thus for the period of 180 yr (1850–2030) the cumulative equilibrium warming may reach 1.5–6.1 deg K. However, due to the gigantic thermal inertia of the ocean, only 40–50 per cent of warming can be realized by 2030; interesting considerations about the mechanism for delayed warming have been suggested by Wright (1987).

If the present rate of growth of the greenhouse gases concentration continues in forthcoming decades, it means that the non-equilibrium state of the climatic system will intensify. Therefore, it should be borne in mind that although several CFCs (F-13, F-114, F-115, F-14, F-13B1) with a lifetime of about 100–500 yr do not at present contribute noticeably to the formation of the greenhouse effect (which will not happen during the next 50–100 yr), their remaining rate of release may cause substantial changes in the climate of the forthcoming centuries.

Increasing concentrations in the atmosphere of greenhouse gases, such as $CO_2$ (about 0.5 per cent per year), $CH_4$ (0.9 per cent), $N_2O$ (0.25 per cent), $CFCl_3$ (4 per cent), and $CF_2Cl_2$ (4 per cent) necessitate new calculations of the effect of greenhouse gases on climate formation. Previous calculations made with the NCAR CCM-1 coupled climate model (with account of the oceanic mixed layer) have demonstrated that: (1) the contribution of greenhouse gases to the formation of the present climate is substantial and (2) the effect of $CO_2$ and other greenhouse gases is different.

Wang *et al.* (1991) undertook a numerical modelling of climate change (using the CCM-1), with the growth of greenhouse gases (GHGs) concentrations shown in Table 6.13 and corresponding to a scenario where the reduction of greenhouse gases emissions is not foreseen.

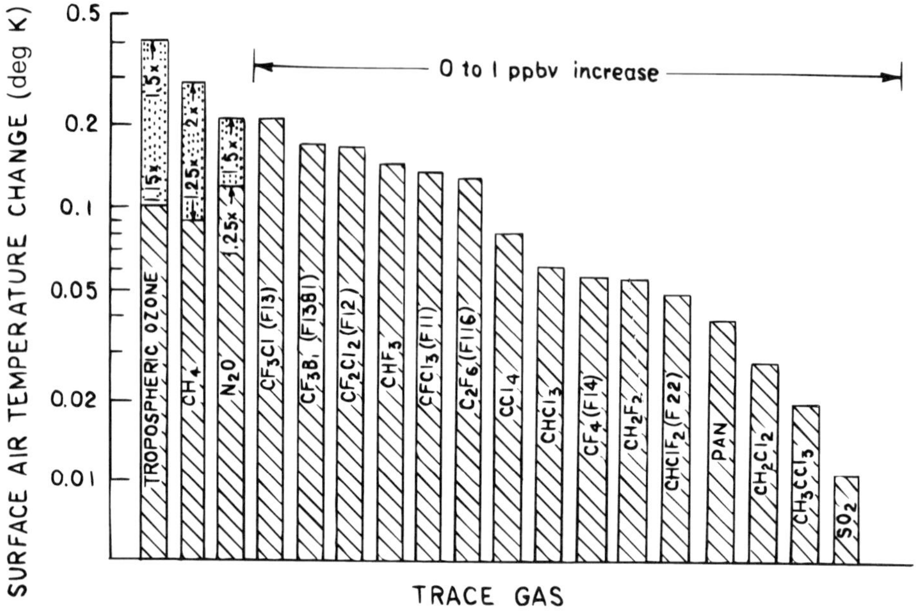


**Figure 6.6** Computed surface air temperature change due to a 0 to 1 ppb increase in trace gas concentrations. Tropospheric $O_3$, $CH_4$ and $N_2O$ increases are also shown for comparison. For these three gases, the factor of increase is indicated in the figure. For example, for tropospheric $O_3$, the bottom part of the shaded region is the temperature change due to a uniform increase in $O_3$ from 0 to 12 km by a factor of 1.16. After Ramanathan *et al.* (1987).

Calculations for the January conditions for a period of 20 yr, with subsequent averaging over the last 10 yr, gave the global mean variation of radiative forcing (Table 6.14) in the stratosphere (STS; 150 hPa), upper troposphere (UTS; 500 hPa), lower troposphere (LTS; 500 hPa) and near the surface (SFC), as well as for the surface–troposphere system (TSS).

As can be seen, the growth of $CO_2$ concentration causes stratospheric cooling, whereas the other greenhouse gases cause warming. Of interest are differences in the vertical distribution of radiative forcing for $CO_2$ and $CH_4$. Two thirds of the total radiative forcing of the system is determined by the contribution of $CO_2$.

Calculations of the temperature field variations have shown that $CO_2$ contributes most to the cooling of the middle stratosphere. The intensification of the high-latitude warming is mainly determined by snow and ice melting and is manifested most in the wintertime northern hemisphere. The cloud amount, as a rule, increases in the stratosphere but decreases in the tropopshere (both promote a warming near the surface which depends strongly, however, on the optical properties of clouds). Calculations of surface air temperature variations revealed four centres of anom-

**Table 6.13** The growth of greenhouse gases concentrations

| Year | $CO_2$ (ppm) | $CH_4$ (ppm) | $N_2O$ (ppm) | $CFCl_3$ (ppb) | $CF_2Cl_2$ (ppb) |
|---|---|---|---|---|---|
| 1990 | 353.9 | 1.717 | 0.309 | 0.280 | 0.484 |
| 2050 | 539.1 | 3.240 | 0.369 | 0.546 | 1.07 |

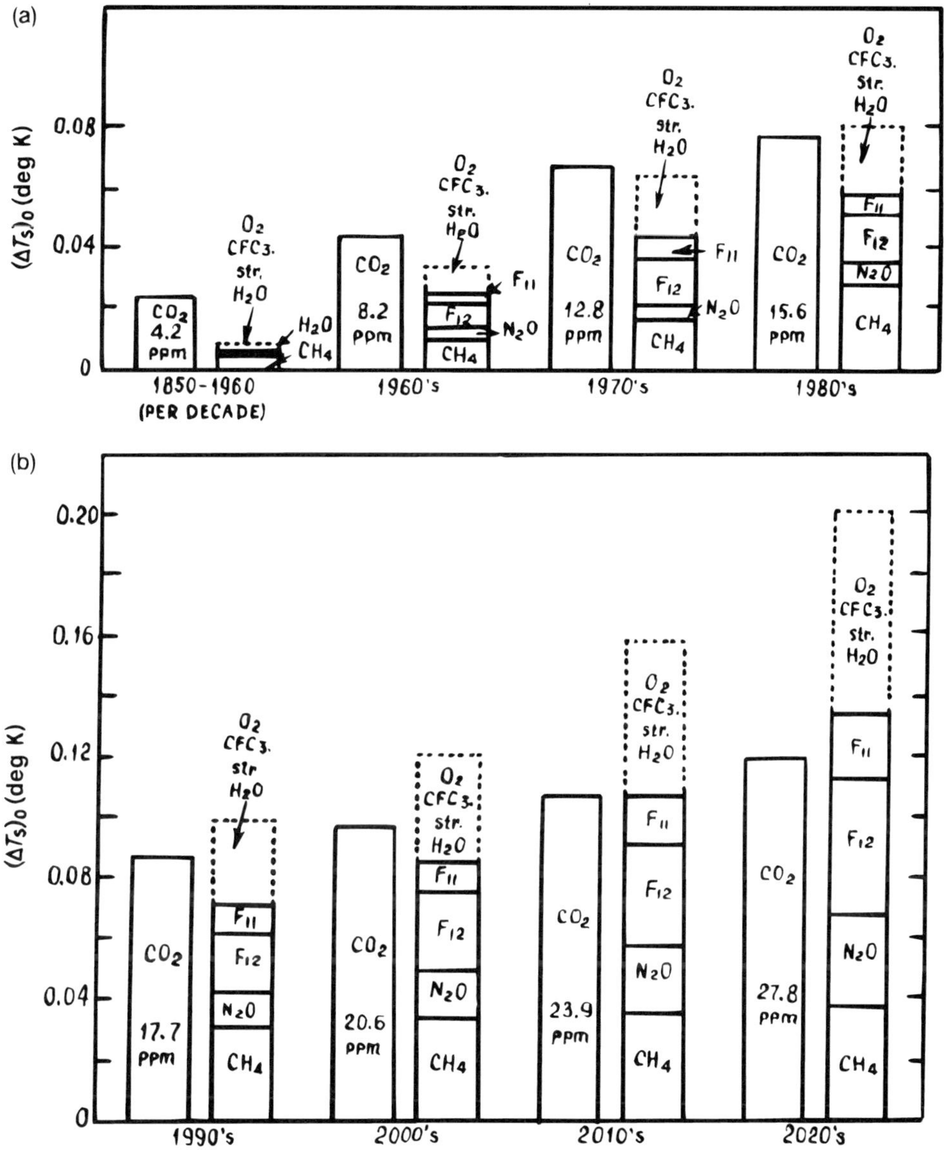


**Figure 6.7** Decadal additions to global mean greenhouse forcing of the climate system. $(\Delta T_s)_0$ is the computed temperature change at equilibrium ($t \to \infty$) for the estimated decadal increases in trace gas abundances, with no climate feedbacks included. Formulas for $(\Delta T_s)_0$ as a function of the trace gas abundances are given by Lacis *et al.* (1981). (a) Past additions. Except for $O_3$ and stratospheric $H_2O$, the estimated trace gas increases are based on measurements. (b) Future additions, 1990–2030. After Ramanathan *et al.* (1987).

alous (by more than 7 °C) warming located near 60° N. The global mean, annual mean surface air temperature increase should constitute 4 °C with simultaneous decrease of total cloudiness, rain rate intensification and water content of the atmosphere.

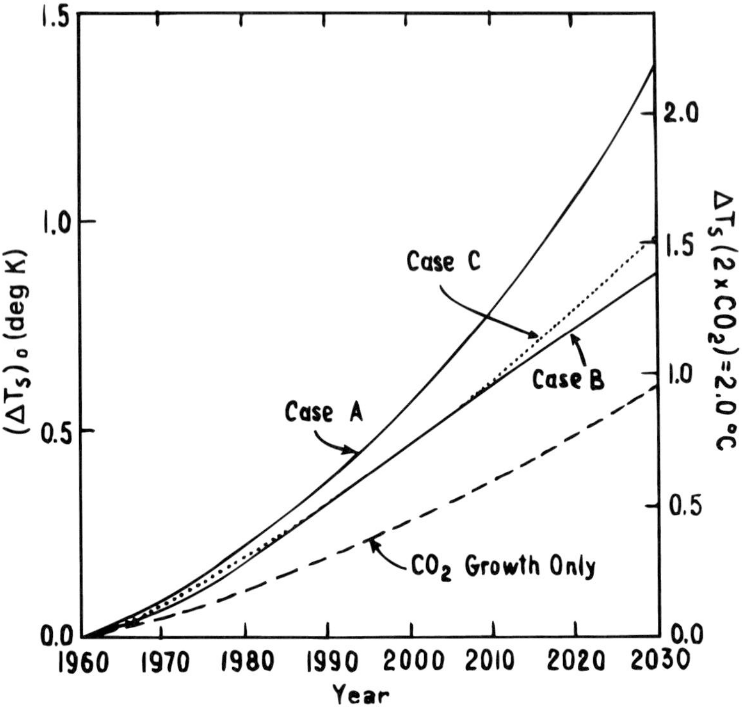


**Figure 6.8** Greenhouse forcing for three trace gas scenarios. Case A uses current growth rates, and Case B assumes substantial reductions in growth. Case C is similar to Case A but with medium growth rate. $(\Delta T_s)_0$ is the equilibrium greenhouse warming for no climate feedbacks. The 'doubled $CO_2$' level of forcing occurs when the $CO_2$ and trace gases added subsequent to 1958 provide a radiative forcing equivalent to doubling of $CO_2$ from 315 ppm to 630 ppm. The '$CO_2$ growth only' curve applies to the Case A scenario. After Ramanathan *et al.* (1987).

Rosen and Gutowski (1991) undertook an analysis of the variability of the fields of zonal-mean component of wind speed ($U$) and momentum ($M$) caused by a $CO_2$ concentration doubling, based on the calculation data and using the GISS, NCAR and GFDL models. Since the dynamics of $M$ is determined largely by the wind field in the tropics, the emphasis has been laid on data for low latitudes, and $M$ has been calculated for the atmospheric layer confined from above to the 200 hPa level

**Table 6.14** Global mean radiative forcing ($W\,m^{-2}$)

| | Radiative forcing | | | | |
|---|---|---|---|---|---|
| Level | $CO_2$ | $CH_4$ | $N_2O$ | $CFCl_3 + CF_2Cl_2$ | Total |
| STS | −0.644 | 0.0496 | 0.0059 | 0.085 | −0.504 |
| UTS | 0.669 | 0.229 | 0.0508 | 0.0735 | 1.03 |
| LTS | 0.867 | 0.171 | 0.067 | −0.011 | 1.01 |
| SFC | 0.551 | 0.220 | 0.055 | 0.188 | 1.11 |
| TSS | 2.11 | 0.619 | 0.173 | 0.251 | 3.14 |

STS, stratosphere; UTS, upper troposphere; LTS, lower troposphere; SFC, near surface; TSS, troposphere–surface system.

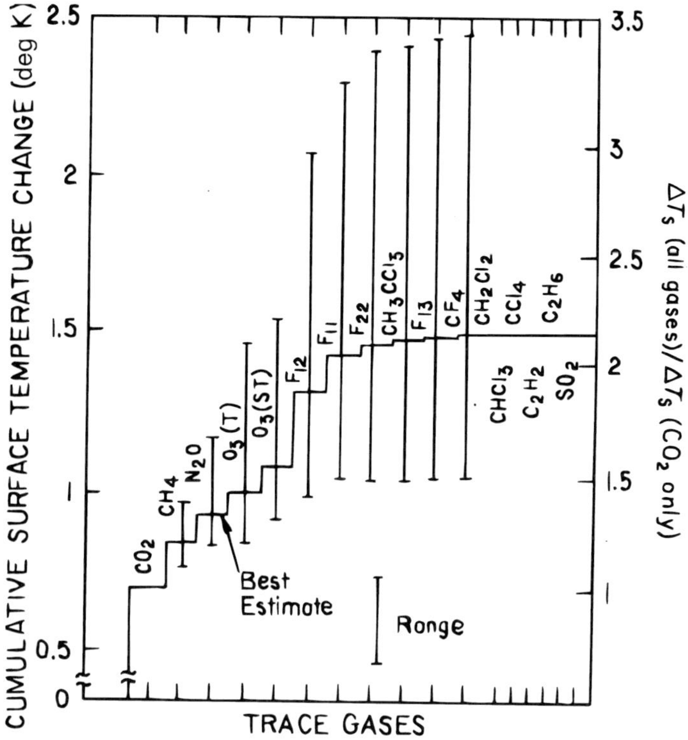


**Figure 6.9** Cumulative equilibrium surface temperature warming due to increases of $CO_2$ and other trace gases over a period of 50 yr from 1980 levels. In this 1-D model calculation, the surface air temperature change is larger than the surface temperature change by about 10 per cent. After Ramanathan *et al.* (1987).

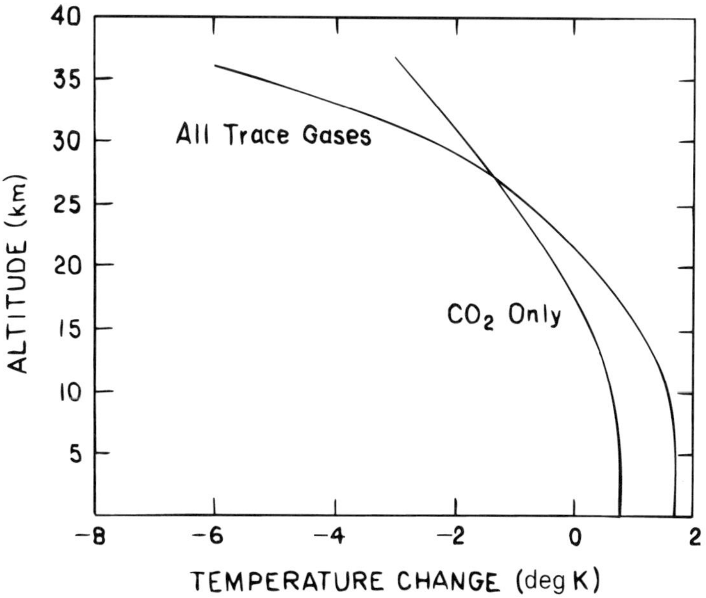


**Figure 6.10** Atmospheric temperature change due to $CO_2$ and other trace gases. After Ramanathan *et al.* (1987).

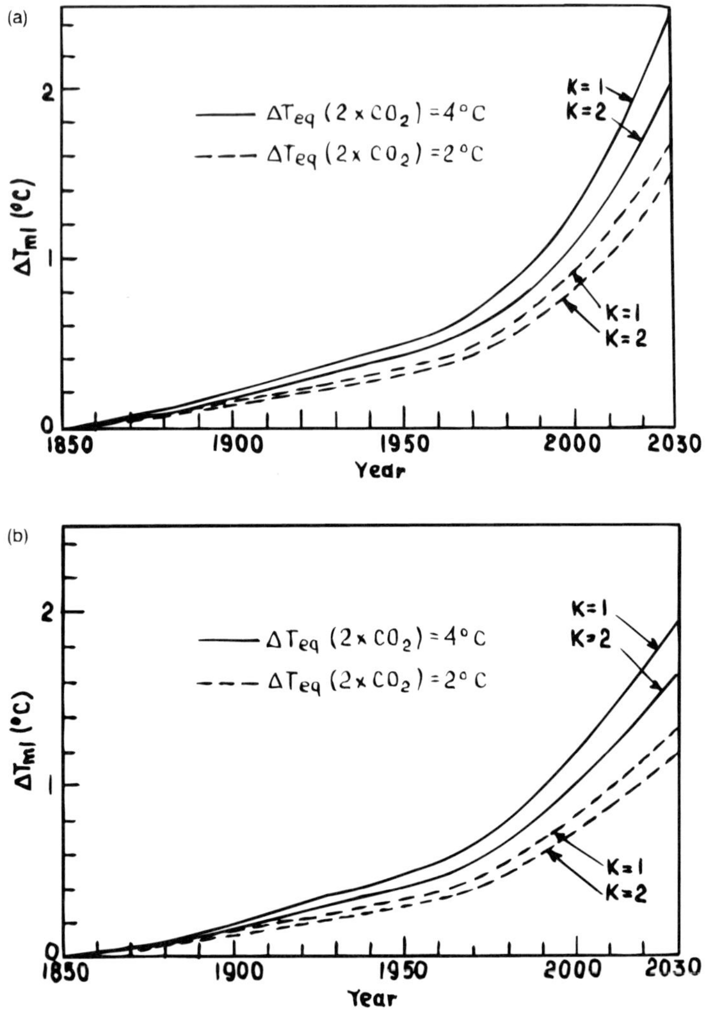


**Figure 6.11** Ocean mixed layer temperature response to the two trace gas scenarios in a 1-D radiative–convective model with a box-diffusion ocean: (a) Case A (current growth rates for $CO_2$, $CH_4$, $N_2O$, $CF_2Cl_2$ and $CFCl_3$ plus an allowance for other trace gases) and (b) Case B (reduced growth rates with only $CO_2$, $CH_4$, $N_2O$, $CF_2Cl_2$ and $CFCl_3$ included). Results are shown for two values of the equilibrium climate sensitivity and two values of the vertical diffusion coefficient ($cm^2\,s^{-1}$). After Ramanathan *et al.* (1987).

($M$(200)). The results considered refer to two seasons: December–February and June–August.

The results of control integration show that the models satisfactorily simulate the observed fields of $U$ and $M$(200), in particular the annual change, though the GISS model gave unacceptable results for winter, which transformed into unreliable data on the annual change. According to these three models, in the ($2 \times CO_2$) conditions $M$(200) substantially decreases, which can be detected in the process of further observations as the $CO_2$ signal. For the GISS and NCAR models, the $M$(200) sensitivity is mainly determined by atmospheric dynamics in the tropics. Since the calcu-

**Table 6.15** Scenarios of the concentrations of greenhouse gases

| Gas | A | B | C | D | E |
|---|---|---|---|---|---|
| $CO_2$ (ppm) | 333.0 | 660.0 | 330.0 | 353.9 | 539.1 |
| $CH_4$ (ppm) | 0 | 0 | 1.700 | 1.717 | 3.247 |
| $N_2O$ (ppm) | 0 | 0 | 0.300 | 0.309 | 0.369 |
| $CFCl_3$ (ppb) | 0 | 0 | 1.000 | 0.280 | 0.546 |
| $CF_2Cl_2$ (ppb) | 0 | 0 | 1.000 | 0.484 | 1.070 |

lated changes in $U$ are comparable to interannual variations (noise), this suggests that the prospects are poor for finding the $CO_2$ signal in the data of zonal wind observations in low latitudes.

The interannual variability of the general atmospheric circulation in low latitudes is known to be largely determined by low-frequency oscillations, including a well-known El Niño / Southern Oscillation event and less well-known Northern Oscillation event, which manifests itself as a negative correlation between atmospheric pressure at the sea level at the Ship N location (30° N, 140° W) and Manila (14.31° N, 212° E). It was also found that such low-frequency oscillations in the tropical ocean–atmosphere system can substantially affect weather and climate not only in the tropics but also in mid-latitudes. So, for example, the ENSO can affect precipitation in semi-arid regions of northern China due to the processes of interaction between sea surface temperature in the eastern equatorial Pacific and maximum pressure in the western sub-tropical Pacific. The Northern Oscillation can markedly affect the surface air temperature monthly mean in the USA (for example, with a negative Northern Oscillation a decrease of surface air temperature is observed in the south-west, with an increase along the western coastline).

So far, the connection between the ENSO and the Northern Oscillation remains unclear. These circumstances have prompted Liang *et al.* (1996) to undertake a numerical modelling of the effect of the greenhouse gases concentration increase in the atmosphere and, respectively, intensification of the atmospheric greenhouse effect on low-frequency oscillations, using the NCAR CCM-1 climate model. Table 6.15 characterizes the greenhouse gases' mixing ratios for five scenarios, A–E. Table 6.16 contains the results of calculations of the annual mean, global mean values of the basic parameters of climate ($T_s$ is the sea surface temperature, $R$ is the rain rate, $C$ is the cloud amount, $Q$ is the total water content of the atmosphere) averaged over the last 10 yr of the multi-year integration.

A comparative analysis of the global fields of surface pressure for cases A and E revealed marked differences between the regional variabilities of climate. There is a correlation between pressure at sea level and the Southern Oscillation index. One of

**Table 6.16** Scenarios of the climatic impact of the intensification of the greenhouse effect

| Parameter | A | B | C | D | E |
|---|---|---|---|---|---|
| $T_s$ (K) | 283.1 | 287.0 | 287.9 | 288.8 | 292.8 |
| $Q$ (mm) | 17.7 | 22.5 | 23.8 | 25.2 | 32.3 |
| $R$ (mm d$^{-1}$) | 3.03 | 3.26 | 3.33 | 3.38 | 3.64 |
| $C$ (%) | 48.3 | 47.1 | 46.2 | 46.0 | 44.0 |

the substantial variations caused by the greenhouse effect consists of the disappearance of a positive correlation between pressure and the Northern Oscillation index in the Russian territory. On the whole, the results under consideration point to a considerable sensitivity of low-frequency oscillations, including both the ENSO and the Northern Oscillation, to the greenhouse effect enhancement.

Based on the use of a previously suggested algorithm to recognize tropical cyclones from data of everyday calculations using the GFDL climate model (calculations for a period of many years have been made with different horizontal resolution and varied parametrization of cloud dynamics), Broccoli and Manabe (1991) analysed the effect of the ($2 \times CO_2$)-induced global warming on the frequency of tropical cyclones. All the models reproduced a climatological distribution of the frequency of tropical cyclones which was qualitatively similar to that observed. In the case of a $CO_2$ doubling the results depended strongly on the approach to cloud parametrization. With 'frozen' cloudiness (i.e. with cloud feedback neglected), the tropical cyclone intensified, whereas in conditions of predicted cloudiness (with the cloud feedback taken into account) the effect was the opposite.

The coupled climate models suggested earlier for the atmosphere–ocean system have been based on taking into account a mixed layer 50 m thick (as a rule) and parametrization of heat transport under the mixed layer as the process of diffusion. A serious shortcoming of such models is that they cannot simulate the effect of climate changes on the oceanic circulation and, in particular, on the heat transport in the ocean and heat supply by deep layers of the ocean. Meanwhile, such processes strongly affect the equilibrium climate and the response of the non-equilibrium climate to an external forcing. No doubt, a sufficiently reliable numerical modelling of the climatic impact of the greenhouse effect enhancement is possible only on the basis of a coupled model of the ocean–atmosphere system.

In this connection, there is no doubt that the usually considered spasmodic doubling of $CO_2$ concentration is of no more than academic interest, and this has prompted Cubasch *et al.* (1991, 1992) to undertake new calculations of the effect of continuously increasing $CO_2$, with the use of a coupled climate model. The atmospheric component of this model (ECHAM) is the ECMWF spectral (T21) modified numerical model of weather forecast, and the oceanic 11-level component (LSG), over a grid of 5.6° latitude by 5.6° longitude, and is based on a system of primitive equations which describe large-scale geostrophic motions; it contains also an interactive unit simulating the dynamics of sea ice. The atmosphere–ocean interaction is determined by the processes of energy-exchange and mass-exchange. Both components have been integrated synchronously but with different time steps. All calculations have been made for a period of hundred years (1985–2084) with three prescribed scenarios of $CO_2$ growth and one control scenario with a constant $CO_2$ concentration, corresponding to 1985 (the correspondence of the data for the control case with those of observations is, on the whole, satisfactory).

The surface air temperature measured according to the control scenario turned out to be quasi-stationary, which should be explained as a manifestation of the internal variability. With the supposed spasmodic doubling of $CO_2$, first, the surface air temperature rapidly increases (approximately 1 deg K over 5 yr) and then the phase of slow linear increase sets in up to (finally) 1.7 deg K over 100 yr, without any sign of saturation, which agrees with a great difference between the time constants for the upper layer of the ocean (about 10 yr) and the deep waters (from several hundreds to one thousand years).

According to scenario A of the continuous rapid increase of $CO_2$, the surface air temperature increases only by 0.5 deg K for the first 40 yr, but then every 10 yr the surface air temperature grows by 0.35 deg K, with total increase by 2.6 deg K during 100 yr. According to scenario D (slow growth of $CO_2$), the surface air temperature increase over 100 yr is less than 0.6 deg K. The mapping of the surface air temperature increase (scenario A) for the last 10 yr (2075–2084) demonstrated a much lower warming over the ocean than over the land (in some regions of the oceans cooling even takes place). A maximum of warming, by about 12 deg K, takes place where the ice cover retreats (especially in the Arctic).

From the viewpoint of further studies (in connection with the contribution by greenhouse gases), special attention is claimed for the problem of the interaction among chemical, radiative and dynamic processes (Figure 6.12), as well as an adequate account of the role of the ocean and clouds as climate-forming factors (so far, even the sign of the cloud feedback has not been determined reliably, and the estimates of the delaying effect of the ocean vary from several decades to a century). Since greenhouse gases are characterized by strong spatial and temporal inhomogeneity, and the estimates of the effect of greenhouse gases on climate have been obtained, as a rule, using one-dimensional models, these estimates should be considered only as illustrative, bearing in mind the necessity of more reliable calculations based on the use of two-dimensional and three-dimensional models.

Based on the use of a two-dimensional radiative–convective climate model, Lal and Jain (1988) performed calculations of a possible effect on climate of changing concentrations of some optically active minor gaseous components of the atmosphere observed in the Antarctic. Table 6.17 shows the estimates of the contributions of the springtime ozone decrease and of the trends of increasing concentrations of $CH_4$ and $CO_2$.

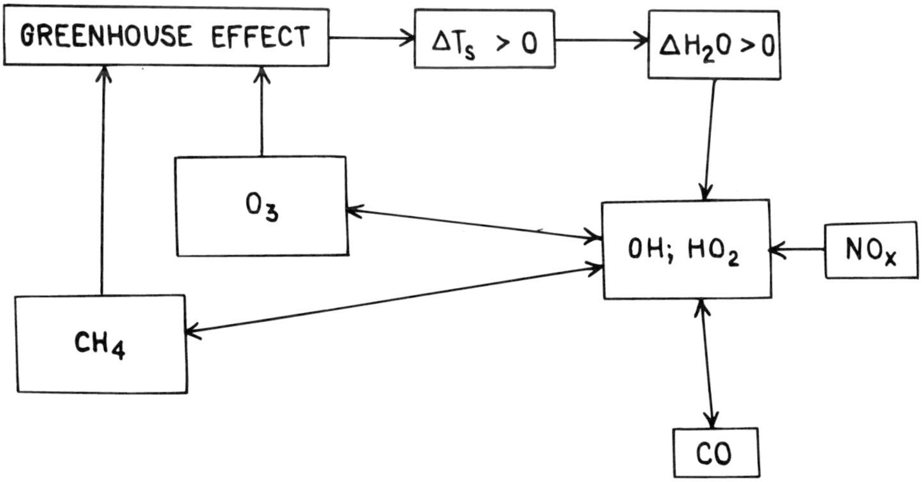


**Figure 6.12** Schematic of the climate–chemistry interactions involving OH and $HO_2$ chemistry in the troposphere. Surface temperature increase ($\Delta T_s > 0$) due to the greenhouse effect of trace gases (including $CO_2$) leads to an increase in $H_2O$ which contributes to an increase in OH. The addition of $NO_x$, CO and $CH_4$ reacts with OH and $HO_2$ to alter $CH_4$ and tropospheric $O_3$, which in turn modulates the greenhouse effect. After Ramanathan *et al.* (1987).

**Table 6.17** Possible springtime variations in surface air temperature (SAT) (September–November) in the Antarctic (60–90° S)

| Case | SAT (K) |
|---|---|
| Equilibrium value of SAT | 232.19 |
| (a) $\Delta O_3/O_3 = -30\%$ ($10 \leqslant z \leqslant 30$ km) | −0.47 |
| (b) $\Delta O_3/O_3 = -30\%$ ($10 \leqslant z \leqslant 30$ km) $\Delta CH_4 = 1.94\%$ | −0.09 |
| (c) $\Delta O_3/O_3 = -30\%$ ($10 \leqslant z \leqslant 30$ km) $\Delta CH_4 = 1.94\% + \Delta CO_2 = 3.64\%$ | 0.31 |

$z$ = Height of ozone layer.

As can be seen, the springtime decrease of ozone causes a climate cooling (due to the prevailing contribution of varying radiative heat exchange), but the prevailing opposite greenhouse effect of $CH_4$ and $CO_2$ causes a warming.

Detailed calculations of the mutual effects of changes in the ozone concentrations and temperature field in the southern hemisphere based on a 2-dimensional model have been made by Vupputuri (1988a,b). The results of calculations shown in Figures 6.13 to 6.16 illustrate the interaction between chemical and dynamic mechanisms for the formation of the springtime ozone minimum (the 'ozone hole') in the Antarctic (Figure 6.16).

Estimates of the climatic effect of variations in the content of the optically active components are hindered by the necessity to consider the radiative–chemical and dynamic interactions among them, the results of which can be both additive and mutually balanced radiative effects. For example, CO oxidation (in the presence of NO) in the troposphere is a source of tropospheric ozone, and hence variations in

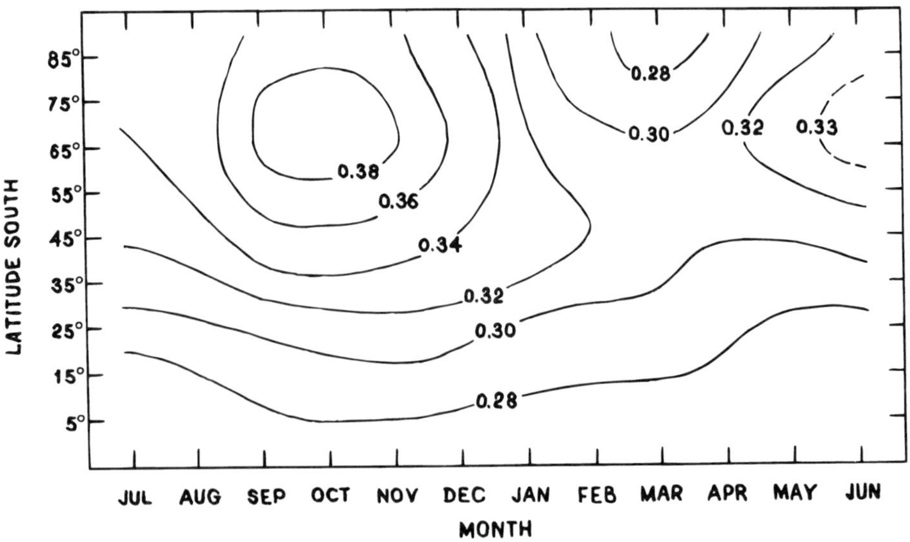


**Figure 6.13** The computed distribution of zonal mean total ozone (cm STP) as a function of latitude and month. After Vupputuri (1988a).

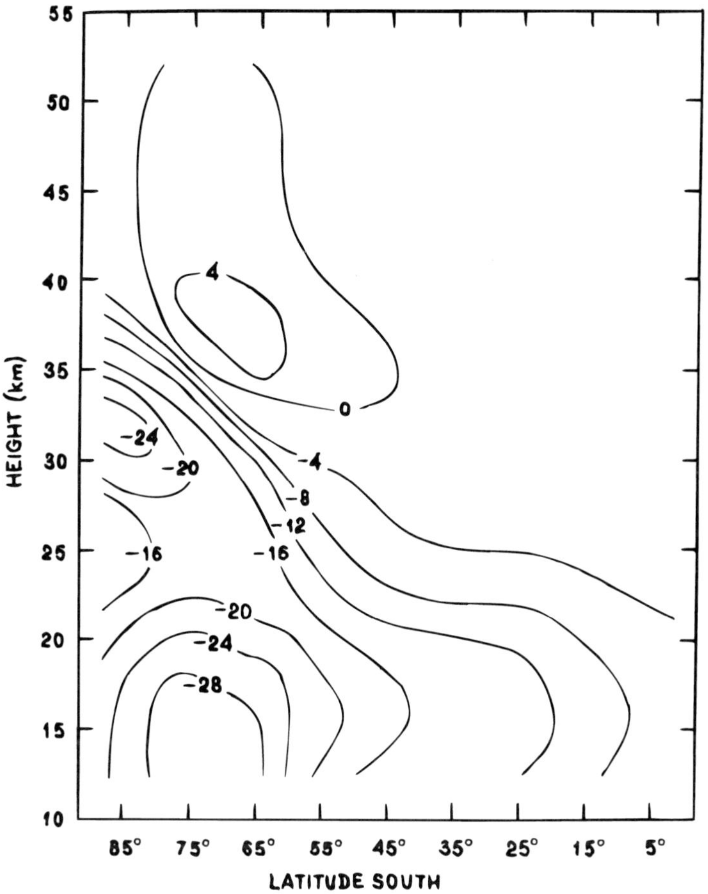


**Figure 6.14** The computed latitude–height cross-section of ozone mixing ratio changes (%) for October. After Vupputuri (1988a).

the concentrations of these gases can indirectly affect climate, although CO and NO do not have a direct radiative effect. A similar situation is observed with respect to CFCs which affect stratospheric ozone.

The radiative–dynamic interaction between the troposphere and stratosphere can play an important role. The effect of such pollutants as CO, NO, $CH_4$, among others, is determined by their influence on the chemically active hydroxyl, whose content, in turn, depends on the amount of water vapour in the troposphere, which increases with the warming caused by the intensified greenhouse effect.

Such a climate–chemical interaction is also substantial in the stratosphere: $CO_2$-induced cooling of the stratosphere can lead to a considerable increase in the $O_3$ content of the upper stratosphere at the expense of the temperature dependence of the rates of chemical reactions determining the formation and destruction of ozone. Naturally, complicated relations exist between the processes of the anthropogenic impact on the stratosphere, the radiative–dynamic interaction between the troposphere and stratosphere, and the climatic effect of the anthropogenically induced changes in the composition of the troposphere and stratosphere.

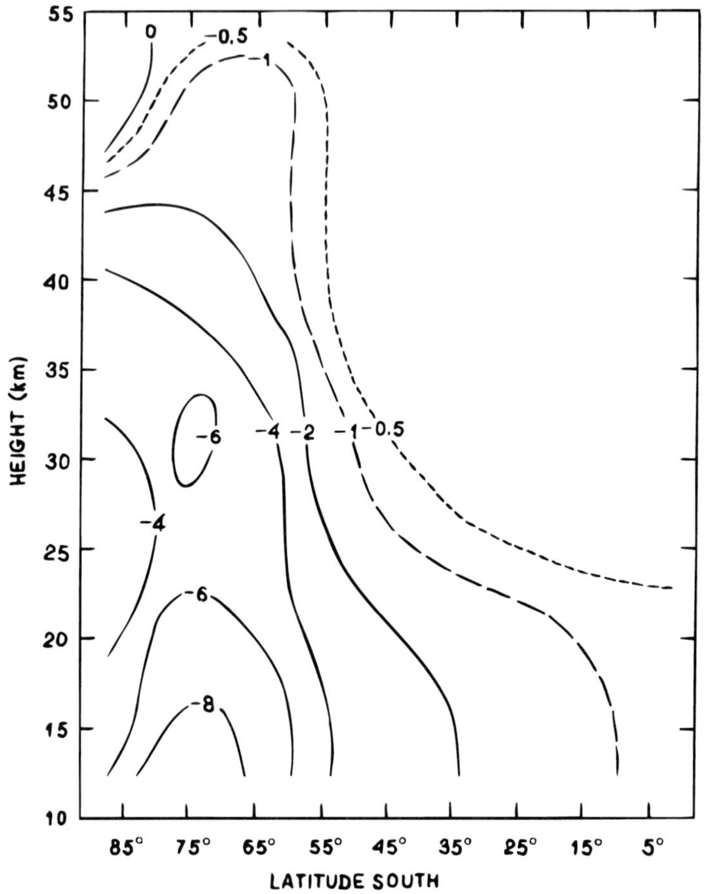


**Figure 6.15** The computed latitude–height cross-section of temperature changes (°C) for October. After Vupputuri (1988a).

To resolve these difficult problems reliably, the following steps are necessary: (1) to continue global monitoring of the optically active minor gaseous components, bearing in mind an identification of trends on the order of several tenths of 1 per cent per year, (2) to perform independent monitoring of changes in the ozone content of the troposphere and stratosphere and to reveal transformations of the vertical profile of $O_3$ concentrations, (3) to develop one-dimensional and two-dimensional climate models with an interactive account of chemical and climate-forming processes, (4) to apply three-dimensional general circulation models to assess the climate sensitivity to atmospheric ozone variations, (5) to develop general circulation models capable of reproducing the dynamics of water vapour in the stratosphere, taking into account chemical sources and sinks, (6) to enlarge substantially a set of the spectroscopic parameters of minor gaseous components (there are still no data on the intensity of bands for some of them) and to make direct measurements of transmission functions (at a spectral resolution not less than $5\,cm^{-1}$) under both laboratory and field conditions and (7) to obtain more reliable information about variations in the extra-atmospheric insolation, especially in the spectral intervals 200–210 nm and 300–320 nm.

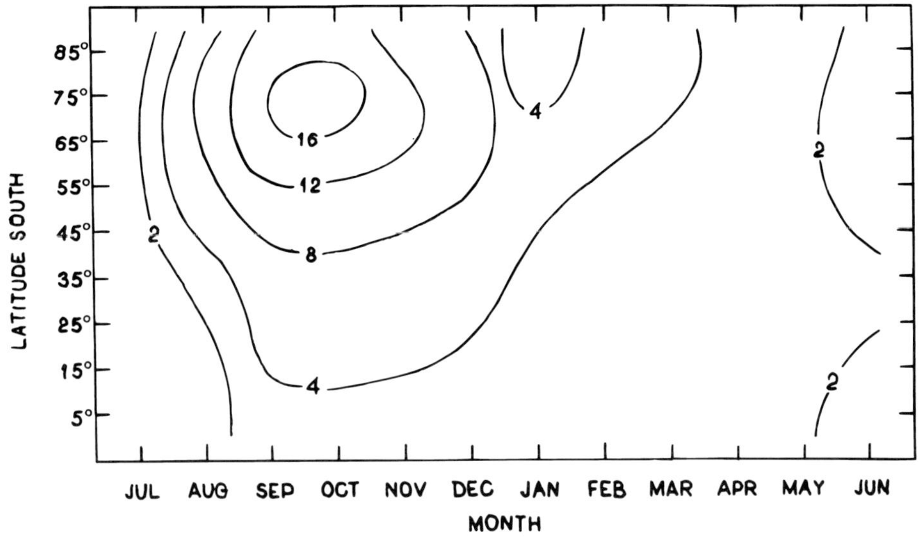


**Figure 6.16** The computed distribution of zonal mean total ozone reduction (%) as a function of latitude and month. After Vupputuri (1988a).

Bojkov (1984) reviewed studies on tropospheric ozone, stimulated by the discovery of anthropogenic effects that lead to an increase of the ozone content in the troposphere. His review revealed considerable effects of ozone on the atmospheric greenhouse effect and on climate. There is an urgent need for studies of tropospheric chemical processes on the processes in the stratosphere. Such impacts are caused, in particular, by the following factors: (1) the role of the troposphere as a source of numerous gaseous components for the stratosphere ($NO_x$, OH, $CH_4$, CO, $CH_3Cl$, $CH_3CCl_3$, etc.), one of the most important sinks of which is their reaction with hydroxyl, (2) changes in the content of such components as $NO_x$, $CH_4$ and CO which affect the tropospheric ozone content and hence the thermal regime of the troposphere and (3) the contribution by increased tropospheric ozone of several per cent to variations in the total content of atmospheric ozone.

The absence of specific means for observations of tropospheric ozone precludes adequate confidence in the reliability of available information and of preliminary conclusions based on any analysis of this information. Although the surface ozone concentration was measured at hundreds of locations, the data obtained are tainted by the effects of local sources of pollution and cannot be used as a reliable source of information about ozone in the free atmosphere (except for background and alpine stations). Data that were more representative were obtained using the Umkehr technique and ozone-sondes. However, a sufficiently long series of regular observations made with similar ozone-sondes is available for only two locations: Höhenpeissenberg, Germany, and Talville and Paiern, Switzerland.

An analysis of available data has made it possible to uncover basic features of the spatial and temporal variability of the troposphere's ozone content which can be characterized by an increase of the mixing ratio with height. A substantial annual change and meridional variability of tropospheric ozone with maxima near 33° N and 55° N as well as interhemispheric asymmetry (with the tropospheric ozone content less in the southern hemisphere than in the northern hemisphere) have been observed.

From the Höhenpeissenberg observations, the annual changes of the surface air ozone concentration are characterized by a minimum in December–January, a subsequent rapid increase in February–March, a maximum in May–June and a rapid decrease beginning in September. During the period 1971–1972 to 1981–1982, the total tropospheric ozone content increased by about 30 per cent. This increase was confirmed by data from other stations.

A consideration of all the available data has led to the conclusion that during the last two decades of ozone-sonde observations, a trend was detected for a total tropospheric ozone content increase at a rate of 0.7–3 per cent per year (at the time, the total tropospheric ozone content fluctuated only slightly, by $\pm 2$ per cent, from which it follows that total tropospheric ozone content variations do not reflect variations in the total ozone content). This positive total tropospheric ozone content trend should be considered to have been anthropogenically induced.

In this regard Bojkov (1984) briefly reviewed the factors involved in the formation of tropospheric ozone from the viewpoint of the income of stratospheric ozone and of photochemical processes of ozone destruction. The tropospheric ozone balance is presented in Table 6.18.

Calculations made using a 1-dimensional radiative–convective model have shown that an increase in total tropospheric ozone content in Höhenpeissenberg from 1967 should have led to a 0.3 °C warming of the troposphere, an increase comparable to a $CO_2$-induced warming for the same period.

Stolarski *et al.* (1992) summarized the results of analysis of the data of conventional and satellite observations that characterized the global dynamics of total ozone content. The data for 1989 showed that in that year the ozone hole in the Antarctic was deep and long-standing (hence, the conditions in 1987, when the total ozone content minimum was particularly substantial, cannot be considered unusual). In 1990 (as in 1987 and 1989) the ozone hole was also deep and remained so till December – the latest moment for ozone hole disappearance – compared with that observed earlier. The ozone hole recorded in 1991 turned out to be similar to that observed in 1989.

Thus, 4 of the 5 yr prior to 1992 are characterized by the formation of deep and long-standing ozone holes. The quasi-biennial modulation of the ozone-hole intensity observed earlier has not manifested itself during the last 3 yr. During the 4 yr 1987, 1989, 1990 and 1991 the area covered by the ozone hole remained practically constant. By the present time, a statistically significant decreasing trend for total

**Table 6.18** Preliminary assessments of the tropospheric ozone balance (Mt $yr^{-1}$)

| Sources and sinks | Northern hemisphere | Southern hemisphere | World |
|---|---|---|---|
| Income from the stratosphere | 400 | 250 | 650 |
| Destruction at surface level | 450 | 180 | 630 |
| Photochemical destruction | 1300 | 570 | 1870 |
| Photochemical formation | 1350 | 500 | 1850 |

ozone content during all seasons has been observed in middle and high latitudes of both the northern and southern hemispheres.

From the total ozone content in the southern hemisphere mid-latitudes a total ozone content decrease was statistically significant during the whole year everywhere south of 25° S. This conclusion agrees with the results of long-term observations at several Dobson stations. The total ozone content decrease observed earlier in mid-latitudes of the northern hemisphere during the spring and the summer continued and intensified. Almost all stations recorded a statistically significant total ozone content decreasing summertime trend, starting from 1970, which was confirmed by the satellite TOMS and SAGE data.

Analysis of the results obtained with the TOMS instrumentation revealed a latitudinal structure of total ozone content trends in the northern hemisphere mid-latitudes, which shows itself also in the Dobson data, but less vividly. Within the tropical band (25° S to 25° N) there were no statistically significant total ozone content trends. The data of SAGE and ozone-sondes show that a decrease of ozone concentration in the northern hemisphere mid-latitudes is concentrated mainly in the lower stratosphere. At altitudes below 25 km a decrease for 10 yr constitutes about 10 per cent, which agrees with the total ozone content data. The observed trends of ozone content (from SAGE data and by conversion technique) at different altitudes in the upper stratosphere agree qualitatively (from the viewpoint of the vertical ozone profile) with results of numerical modelling. However, the level of measured decrease turned out to be lower. Of the last 20 yr, the tropospheric ozone content increased by about 10 per cent (and more) during 10 yr in the northern hemisphere mid-latitudes.

It was assumed that before 1979 tropospheric ozone had been chemically inert and that its major source has been the flux of ozone from the stratosphere, but subsequent investigations revealed an important role of photochemical processes in the formation of the tropospheric field of ozone content. So, for example, in a pure atmosphere the processes of oxidation of methane and non-methane hydrocarbons can serve as the sources of ozone, and in conditions of the urban atmosphere the sources are carbon oxides. In this case the formation of ozone through reactions of oxidation depends critically on the level of the nitrogen oxides concentration: if the mixing ratio for $NO_x$ is less than approximately 10 ppt (by volume), then ozone does not form but gets destroyed. The observed higher content of ozone in the northern hemisphere troposphere, compared with the southern hemisphere, suggested the supposition about anthropogenic origin of this contrast.

Analysis of observations in the Arctic has led to the conclusion about the existence of other possible mechanisms for the formation (or destruction) of tropospheric ozone. One of them is the springtime depletion of ozone in the atmospheric boundary layer under the influence of bromoform released to the atmosphere by decaying algae; another is the pollution of the atmosphere by nitrogen oxides.

In this connection, Hofmann *et al.* (1992) discussed the results of observations of the vertical ozone profile in the Arctic (near Kiruna, Sweden: 68° N, 21° E) in January to early February 1990 using a digital electrochemical ozone-sonde, which revealed episodes of almost complete absence of ozone in the layer under temperature inversion, whose thickness was several kilometres. A negative correlation was found between the total $NO_2$ content in the atmosphere and the ozone mixing ratio at an altitude of $(1 \pm 0.5)$ km during most of the period of observations. They showed that the discussed destruction of ozone could be explained as a result of its reactions with pollutants, for example $NO_x$, with subsequent heterogeneous processes with participation of $NO_3$ in conditions of low-level insolation (in a given continental location the effect of bromoform could not, apparently, be significant).

With a low level of insolation and increased content of $NO_x$, reactions of ozone depletion predominante over the usual smog reactions of ozone formation:

$$O_3 + NO \rightarrow O_2 + NO_2$$

$$O_3 + NO_2 \rightarrow NO_3 + O_2$$

$$NO_2 + NO_3 + M \rightarrow N_2O_5 + M \tag{6.6}$$

with subsequent heterogeneous removal of $N_2O_5$ and $NO_3$ through the formation of $HNO_3$. Estimates have shown that a decrease of the ozone mixing ratio from 25 ppb to zero can occur within 2 days with the NO flux equal to $3.2 \times 10^{11}\,cm^{-2}\,s^{-1}$. However, an impact of other reactions is also possible. The results obtained show that the tropospheric ozone depletion in the Arctic connected with the arctic haze can be more typical and substantial than it has been thought before.

Recent observations in the Arctic have revealed episodic decreases of ozone concentration in the atmospheric boundary layer from 30–40 ppb (by volume) to a level below the detection threshold during a period of less than one day in conditions of very stable atmospheric boundary layer stratification. The ozone depletion began at sunrise (with the beginning of the polar day) and continued during March–April. A high correlation was observed between the ozone concentration decrease and the concentration of gaseous bromine, which turned out to be much higher than during the remaining part of the year. To explain the respective cause-and-effect connection, it is necessary to answer two questions: (1) what is the source of filtered bromine (f-Br is the total concentration of bromine, both gaseous and in particles, which can be measured with a filter sample) and (2) what are the ways of transformation of gaseous HBr and Br-containing organic compounds into more active forms of bromine.

McConnell *et al.* (1992) supposed that in the period of long polar night the concentration of Br (released by sea salts) in the snow cover grows, with subsequent release of Br to the atmosphere in the form of $Br_2$ at sunset (the presence of sea salts on snow surface can be explained by the permanent existence of polynyas and leads in the Arctic). In general, the gas-to-particle conversion of $Br_2$ into HBr or Br-containing organic compounds should not cause substantial ozone depletion in the atmospheric boundary layer. One may think, however, that HBr and Br-containing organic compounds are washed out from the atmosphere by aerosol particles and ice crystals, and during this process photostimulated heterogeneous reactions of repeated formation and release of $Br_2$ to the atmosphere take place.

Such a cyclic process of aerosol–atmosphere bromine exchange can maintain the level of concentration of Br atoms and BrO radicals, sufficient to provide an agreement of the ozone depletion intensity with the observational data. The atoms of Br appearing with $Br_2$ photolysis, can cause a fast (within 1–2 days) depletion of ozone in the process of subsequent catalytic reactions:

$$Br + O_3 \rightarrow BrO + O_2$$

$$BrO + BrO \rightarrow 2Br + O_2 \tag{6.7}$$

HCHO and $CH_3CHO$ can act as accelerators of the process of ozone depletion. To verify the hypothesis discussed, further laboratory experiments and field observations are needed (in particular, to study the vertical distribution of Br in snow cover).

To substantiate countermeasures and policy with respect to undesirable consequences of climate changes caused by increases in the concentration of greenhouse gases, Jain and Bach (1992) developed a one-dimensional energy-balance climate model, with a combined account of the energy-economic and biochemical climatic aspects of the problem. Calculations for 240 yr (1860–2100) were made with a time step of 1 yr. For the period 1860–1985, actual data of greenhouse gases trends were taken, and for the period 1985–2100 different scenarios of greenhouse gases concentration changes were prescribed.

A case of extrapolation of the present greenhouse gases trends (Business as Usual) was used as the basis for comparison. The greenhouse gases considered included: $CO_2$, $CH_4$, $N_2O$, CFC-11, CFC-12, CFC-113, CFC-114, CFC-115, HCFC-22, halon-1301, $CCl_4$, $CH_3CCl_3$, tropospheric ozone and stratospheric water vapour. Maximum permissible values of global climate warming of 0.1 °C per 10 yr or 2 °C for the period 1860–2100 were prescribed as criteria for countermeasures to reduce the greenhouse gases emissions.

The climate model was supplemented with (1) models of biogeochemical $CO_2$ cycle (three reservoirs: atmosphere, mixed layer and deep layers of the ocean with account of interaction with the biosphere), (2) a simple balance model to predict concentrations of CFC, halon and $N_2O$ and (3) a model of chemical reactions determining the concentration of methane. The enumerated models ensure the input data of greenhouse gases concentration for the climate model. One more model estimates sea level taking account of (1) thermal expansion, (2) mountain glaciers thawing and (3) ablation of the Greenland ice sheet.

Testing the reliability of this climate model by comparing calculated and observed secular trends of surface air temperature and sea level gave, on the whole, positive results. An acceptable agreement was obtained from comparison with data for a three-dimensional global model developed at the Max Planck Institute (Hamburg). Thus, the model under discussion can be considered to be useful to substantiate countermeasures to reduce greenhouse gases emissions and the respective policy.

According to the results obtained, further emissions will lead to a fast increase of global mean surface air temperature, which can be prevented only with countermeasures to reduce greenhouse gases emissions. A reduction of emissions of CFCs and halons is most costly and requires much effort. Of great importance are the preservation of tropical forests, and reforestation, which will enable the reduction of net $CO_2$ emissions to the atmosphere. The most efficient measure would be to reduce industrial $CO_2$ emissions; this is to be expected, first of all, from industrial countries.

An interesting plan to reduce greenhouse gases emissions was suggested by the Ankette Commission (Germany), which foresaw an individual approach to various countries, depending on the level of their development (see Table 6.19). As is seen, a reduction of emissions by developing countries has not been foreseen, but finally (by 2100) a 50 per cent reduction of global emission would be ensured. This means that, with respect to the Business as Usual scenario of continuation of emissions, the strategy determined by the considered scenario of emission reduction, ensures a

**Table 6.19** Proposals on a differentiated reduction of greenhouse gases emissions (%) by various countries

| Year | Industrial countries with various levels of economy: High | Moderate* | Weak | Moderate* | Other countries | Developing countries | Global means |
|---|---|---|---|---|---|---|---|
| 1990 | +5 | +5 | +5 | +5 | +5 | +11 | +6 |
| 1995 | +5 | +7 | +8 | +6 | +10 | +24 | +10 |
| 2000 | −10 | −4 | +5 | −4 | +15 | +37 | +4 |
| 2005 | −30 | −15 | −5 | −20 | +25 | +50 | −5 |
| 2010 | −37 | −21 | −11 | −27 | +15 | +53 | −10 |
| 2020 | −50 | −35 | −25 | −40 | −5 | +60 | −20 |
| 2030 | −60 | −50 | −43 | −54 | −22 | +63 | −30 |
| 2050 | −80 | −80 | −80 | −80 | −55 | +70 | −50 |
| 2100 | −80 | −80 | −80 | −80 | −55 | +70 | −50 |

* Two different versions of moderate.

decrease of the level of the global mean surface air temperature within 0.9–1.7 °C (depending on the assumed sensitivity of the climate system to a $CO_2$ doubling). An accomplishment of the countermeasures discussed will also prevent a dangerous rise of sea level.

Of course, the reliability of the results considered should be analysed in the light of assessing aerosol climatic effects which are very significant and lead to a substantial decrease (if not annihilation) of greenhouse warming. The mitigation measures have been discussed in detail in the IPCC Reports (Houghton *et al.* 1990, 1992, 1996).

## 6.3 Volcanoes and climate

Studies of volcanic eruptions are interesting from the point of view of understanding both the laws of palaeoclimate and present-day climate changes (Kondratyev 1985b, 1992b, Kondratyev and Galindo 1997). Volcanic eruptions have contributed to the formation of the atmosphere and oceans. The products of explosive volcanic eruptions reach the stratosphere and remain there for a year or longer, changing the chemical composition of the stratosphere and thereby affecting the Earth's radiation budget. It is these variations in the Earth's radiation budget caused by volcanic aerosol that may be responsible for post-eruption climate changes. We shall discuss here an impact of explosive volcanic eruptions on global climate, in addition to what has been written in Chapter 4.

Volcanic eruptions are well known to be a most powerful natural experiment on the environmental impact on climate. From the viewpoint of tropospheric and stratospheric dynamics, the monitoring of the global-scale transport of eruption products is of great interest.

Benjamin Franklin proposed that an eruption of the Hekla volcano in Iceland could have been a reason for the severe winter of 1783–1784 in Europe. Since then, numerous studies have been undertaken in a search for correlations between individual volcanic eruptions and adverse weather conditions or between serial eruptions and climate anomalies. These studies have revealed that volcanic eruptions may be one of the important factors of climate changes.

Of all the possible causes of climate changes, volcanic eruptions are among the most adequately documented and understood. For example, a trend towards a stratospheric temperature increase was observed after the 1963 Agung eruption in Indonesia. This trend was superimposed on a stratospheric temperature variation that was quasi-biennial and increasing in amplitude, and could be explained by the Agung eruption. This increase was clearly pronounced in the southern hemisphere, where the variability of the amplitude of quasi-biennial oscillations is small, whereas in the northern hemisphere (with a small input from the eruptive cloud) there was no stratospheric temperature increase. These observational data agree with theoretical estimates, from which it follows that a stratospheric warming is mainly caused by absorption of the upward thermal emission flux and, to a smaller degree, by the solar radiation absorption. In the troposphere and at the surface there was a global-scale temperature decrease (larger in the northern hemisphere, 0.6 °C). This decrease could not, however, have been totally caused by the eruption.

Following the 1815 Tambora volcanic eruption in Indonesia 'a year without sun' was observed, and a summertime temperature decrease in New England and western Europe (with respect to the climatic mean) reached 1–2.5 °C.

Mass and Schneider (1977) analysed the consequences of numerous eruptions causing an increase of several tenths in the aerosol optical thickness in the stratosphere in the visible wavelength range. They found that a statistically reliable temperature decrease a year after the eruption would constitute 0.3 °C and after 2 yr 0.1 °C and more.

Based on processing the aerosol remote-sounding data from the SAGE-1 satellite for March 1979 to February 1980, a period without strong volcanic eruptions, McCormick and Wang (1987) proposed a base model of the background stratospheric aerosol in the form of zonally averaged (monthly means and zonal means) vertical profiles of the coefficient of aerosol extinction at a wavelength of 1 $\mu$m in the tropics, middle and high latitudes of both hemispheres (the corresponding data have been tabulated and graphed). The similarity of the seasonal mean vertical profiles of the coefficient of aerosol extinction $\beta$ in the latitudinal bands (75° S to 40° S; 40° S to 20° S; 20° S to 20° N; 20° N to 40° N; 40° N to 75° N) has made it possible to suggest the following polynomial approximation:

$$\log \beta = a + bz + cz^2 + dz^3 \qquad (6.8)$$

where $z$ is the height with respect to the troposphere in km. The values of the coefficients for different seasons are given in Table 6.20.

Although volcanoes have long been considered a cause of glaciations, this hypothesis remains unsubstantiated. Analysis of the aerosol particle content in the Greenland ice cores has not revealed any increase in the aerosol content during the early stages of the Wisconsin glaciation cycle.

Lough and Fritts (1987) performed an analysis of surface air temperature variability for the period 1602–1900 from dendroclimatic data at 77 locations over the US territory and south-west Canada, to reveal the effect of volcanic eruptions on climate. The results obtained show that although, over most of the US territory, post-eruption temperature decreases had taken place, in some states in the western

**Table 6.20** Values of coefficients of polynomial approximation of the vertical aerosol profile

| Season | *a* | *b* | *c* | *d* |
|---|---|---|---|---|
| March–May | –3.60 | $-3.59 \times 10^{-2}$ | $6.30 \times 10^{-3}$ | $-3.17 \times 10^{-4}$ |
| June–August | –3.78 | $-1.79 \times 10^{-2}$ | $-5.66 \times 10^{-4}$ | $-1.27 \times 10^{-4}$ |
| September–November | –3.67 | $-3.26 \times 10^{-2}$ | $-2.99 \times 10^{-4}$ | $-1.20 \times 10^{-4}$ |
| December–February | –3.50 | $-5.42 \times 10^{-2}$ | $4.01 \times 10^{-4}$ | $-1.21 \times 10^{-4}$ |
| Mean global | –3.64 | $-4.77 \times 10^{-2}$ | $-1.46 \times 10^{-3}$ | $-1.71 \times 10^{-4}$ |

*Note*: See Equation (6.8) for the polynomial approximation.

USA a warming was observed. This reflected the inhomogeneity of the spatial distribution of the climatic system's response to volcanic eruptions, with a clearly expressed annual course of coolings and warming in the western USA that is most substantial in the summer. The effect of eruptions was strongest in low latitudes. These conclusions have also been confirmed by independent dendroclimatic data for other regions elsewhere.

Rampino *et al.* (1988) reviewed the data on the effect of volcanic eruptions on climate and concluded that there was a small decrease of surface air temperature after strong explosive volcanic eruptions and much more substantial heating of the stratosphere (up to 3–4 °C). The results of analysis of the acidity of the ice cores are a reliable indicator of the chronology and scale of eruptions in the past. The results illustrate changes in weather and climate taking place after the strongest eruptions. All these eruptions were, however, much weaker than those revealed from palaeodata, which could have led to an onset of a volcanic winter.

The role of volcanic eruptions in the present temperature trend in the northern hemisphere is not clear, since the post-1940 cooling was not preceded by an enhanced volcanic activity. Theoretical estimates revealed a strong effect of the composition and size distribution of volcanic aerosol on climate changes. For example, even with a small amount of an absorbing silicate aerosol, a considerable increase of solar radiation absorption may take place, and the presence of silicate particles larger than 0.5 $\mu$m can cause a warming at the surface as a result of the thermal emission of such particles. Calculations of temperature variations at the surface and in the stratosphere made by Hansen *et al.* (1978) after the Agung eruption revealed good agreement with observations.

Bryson and Goodman (1982) analysed the modulation of the incoming direct solar radiation under the influence of volcanically-induced changes in atmospheric transparency, using data of the longest series of actinometric observations. To estimate the temporal dynamics of the volcanic aerosol content in the stratosphere, it is assumed that the characteristic reaction time for the formation of the $H_2SO_4$ aerosol is 115 days and that the time for deposition of the particles is about 400 days. In view of a rapid gravitational settling of ash particles, the mean annual content of the sub-micrometre $H_2SO_4$ aerosol is assumed to be twice as large as that of ashes directly ejected into the stratosphere. With these assumptions, one can describe the temporal dynamics of the ash and $H_2SO_4$ components of eruptive aerosols, typical of an explosive eruption.

Until recently, observations of atmospheric transparency that were used to estimate the climatic impact of eruptions have been a major source of information on their atmospheric effect. Clear-sky solar radiation data for 1880–1980, used to identify the secular trend of atmospheric optical thickness, revealed a distinct correlation between transparency decrease and large-scale volcanic eruptions. These include the following eruptions among others: Pelée and Soufrière in 1902, Katmai in 1912, Bezymiannaya in 1955 and Agung in 1963. A comparison of the secular trend of the optical thickness and a synthesized chronology of powerful volcanic eruptions, taking into account the moment of eruption and the subsequent decrease of the concentration of erupted aerosols for low, middle and sub-polar latitudes, yielded a correlation coefficient of +0.87. An additional consideration of moderate eruptions raised the correlation coefficient to +0.95. This leads to the belief that variations in atmospheric transparency are almost completely governed by volcanic eruptions.

The residual variability of the optical thickness (about 5–10 per cent) reveals an increasing trend that may be ascribed to an anthropogenic impact. Attempts to find 11-yr or 22-yr cycles in the residual variability have not been successful. This attests to the absence of the impact of solar activity on incoming solar radiation. An analysis of data on clear-sky global radiation and mean annual surface air temperature after 1920 yielded a correlation coefficient of 0.91. Although this correlation should not be considered as proof of the respective cause-and-effect connection, a volcanically-induced modulation of the mean hemispherical temperature secular trend can nevertheless be supposed to take place.

Examination of this assumption, by calculating the temperature variations in the twentieth century with the use of an energy-balance climate model (the mean hemispheric cloud cover is assumed to be fixed) showed that with a prescribed observed trend of the optical thickness, temperature variations close to those observed are obtained. Using this as a basis, Bryson and Goodman (1982) drew the conclusion that the observed variability of the mean surface air temperature in the northern hemisphere is determined primarily by volcanic eruptions. On the other hand, it follows that with the volcanically induced variability excluded, it should be possible to filter out the trend due to an increase in the atmospheric $CO_2$ concentration. This attempt, however, failed; the residual variability shows no trend of temperature increase. Even the forecast taking into account a $CO_2$ concentration increase until 2020 (with the optical thickness of the atmosphere fixed at the level of 1975) revealed a future warming during the next 50 yr that was less than the warming observed in recent years. In the opinion of Bryson and Goodman, in forecasting possible future climate changes the problem of possible prediction of volcanic eruptions rather than the runaway $CO_2$-induced greenhouse effect must be of primary importance. It must be emphasized, however, that this conclusion is of questionable validity (Kondratyev 1985b).

Note in this connection that, based on the use of the technique of superimposed epochs, a correlation has been found between low-latitude volcanic eruptions and a positive sea surface temperature anomaly in the eastern Pacific (90–180° W, 0–10° S), whereas the extra-tropical eruptions were followed by a negative sea surface temperature anomaly. In particular, for three seasons after an eruption in the tropics, a sea surface temperature increase was statistically substantial at the 90 per cent confidence level, i.e. the sea surface temperature exceeded the norm by more than $2.81\,\sigma/\sqrt{N}$ (where $N$ is the number of eruptions considered equal to 12, and $\sigma$ is the sea surface temperature mean standard deviation during a season).

To test the reliability of these results (with the use of the technique of superimposed epochs), Parker (1988) processed another sea surface temperature data base for the latitudinal band 20° S to 20° N, east of 170° W. Although the new results confirm the conclusions mentioned above, an inadequate volume of the initial data base and random successions of extreme sea surface temperature anomalies suggest that the new results are not sufficiently reliable. In this connection the development of realistic numerical models of the situation under discussion is of great importance.

Using the technique of superimposed epochs, Nicholls (1988) analysed the anomalies of atmospheric pressure in Darwin (12° S, 131° E) as an indicator of the El Niño / Southern Oscillation event before and after 10 volcanic eruptions in the tropical latitudes. This analysis was made from the data of mean monthly pressures for the period 1898–1984 and revealed clear correlations. Against the clearly expressed trend of increased pressure during the whole period, it was found that from 9 months to the moment of eruption was a time of negative pressure anomalies, and that in the post-eruption period a positive anomaly formed, reaching a maximum 4 months after the eruption (see also Portman and Gutzler 1995, Robock and Mao 1995).

The results obtained reflect the possibility of a relation between the ENSO and low-latitude eruptions. However, a limited number of samples precludes a conclusion being drawn about the statistical reliability of correlation. There is no doubt that the supposition about the ENSO being caused by volcanic eruptions cannot be considered substantiated, since the pressure anomaly and the trend of pressure mentioned above appear long before the moment of eruption. On the other hand, if a maximum positive anomaly of pressure 4 months after the eruption is the result of the pressure increasing trend appearing before the eruption, an opposite assumption with respect to eruptions being caused by anomalies in the atmosphere and ocean, forming in the course of the ENSO, can be considered more likely. The results under discussion should rather be considered as a manifestation of random fluctuations.

The continuing numerical modelling to detect a volcanic signal from surface air temperature observations has not, so far, given reliable enough results; information about the level and statistical significance, the geographical distribution and duration of a volcanic signal remains contradictory, which is partly explained by inadequate data on both explosive eruptions and temperature. In this connection, Bradley (1988) undertook an analysis of data of some catalogues of eruptions to assess the reliability of their information content, as well as of a new homogeneous data series on surface air temperature for the northern hemisphere conditions, bearing in mind that powerful explosive eruptions (to which a volcanic eruption index of about 5–6 corresponds) simultaneously affect the surface air temperature, with maximum response 2–3 months after the eruption and with a subsequent gradual attenuation for 2–3 yr. The volcanic signal is strongest in the summer and in the autumn, but is almost zero in the winter. On the whole, the surface air temperature decrease over the northern hemisphere continent reaches about 0.4 °C. It appears that there are also secondary maxima of the surface air temperature anomaly a year after the primary maximum and a third, weaker, maximum 1 yr later.

The latitude at which an eruption takes place is an important factor which determines the volcanic signal. If the high-latitude eruptions are strongest in high- and mid-latitudes, then the effect of low-latitude eruptions is confined, largely, to the tropical and middle latitudes. Thus it seems that middle latitudes are much more subject to the effect of both low-latitude and high-latitude explosive eruptions. The eruptions with a volcanic eruption index of about 4 somehow affect the surface air temperature field only during the first post-eruptive month.

Depending on the season, the temperature decrease varies between 0.05 and 0.1 °C. The effect of large explosive eruptions on the decrease of surface air temperature over the land for the last 100 yr (except for the strongest five eruptions) has been characterized by fast formation and short life-time of the volcanic signal.

For substantial manifestation of the volcanic signal in simultaneous variability of temperature, explosive eruptions of the intensity considered should repeat regularly almost every 3 yr, but during the last 100 yr there was no such repeatability. The conclusion can be drawn that assumptions with respect to the leading role of eruptions in the formation of the temperature trend during the last century are not justified.

Assessment of post-eruption variations in the Earth's radiation budget is a key aspect of the climatic impact of volcanic eruptions. Using the two-stream approximation and a three-layer model of the surface–troposphere–stratosphere system, Lenoble *et al.* (1982) and Lenoble (1986) estimated the effect of a stratospheric aerosol layer constituting a 75 per cent water solution of sulphuric acid or background aerosol particles (Table 6.21) on the system's radiation budget and its components, as well as on the surface temperature (the surface is considered to be an isotropic reflector). The distribution of the particles number density by radius, $n(r)$, is prescribed by a modified gamma function:

$$n(r) = A_r^{\alpha} \exp(-br^{\gamma}) \quad (6.9)$$

where $A_r$, $\alpha$, $b$, and $\gamma$ are parameters. A comparison of the applied two-stream approximation to an accurate solution has shown that the errors of an approximate calculation of the diffuse transmission do not exceed 10 per cent with the single scattering albedo $\omega \leqslant 0.6$, optical thickness $\tau \leqslant 0.5$ and normally incident solar rays. Here the system's albedo is calculated accurately, and the error in calculations of albedo variations, due to changed properties of the stratospheric aerosol layer, does not exceed 10 per cent.

Calculations of the sensitivity of mean monthly albedos of the system to the choice of aerosol model almost always revealed in both cases an increase of albedo by several tenths of 1 per cent. Table 6.22 illustrates that the system's radiation budget decreases since the contribution by backscattering prevails.

An estimation of the mean global surface temperature decrease (at $\tau_s = 0.03$) gave $-0.45$ °C (volcanic aerosol) and $-1.1$ °C (background aerosol). An account of the

**Table 6.21** The size distribution and optical properties of stratospheric aerosols

| | Parameters | | | | | |
|---|---|---|---|---|---|---|
| Aerosol model | $\alpha$ | $b$ | $\gamma$ | $\omega_s^{eff}$ | $b_s^{eff}$ | $c_s = \tau_{IR}^{eff}/\tau_s^{eff}$ |
| Background aerosol | 1.0 | 18 | 1.0 | 0.998 | 0.189 | 0.0383 |
| Volcanic aerosol | 1.0 | 16 | 0.5 | 0.994 | 0.169 | 0.0217 |

$\omega_s^{eff}$, $b_s^{eff}$ are effective values of the single scattering albedo and backscattering; $\tau_{IR}^{eff}$, $\tau_s^{eff}$ are effective values of the optical thicknesses for the shortwave and longwave radiation, respectively; $\tau_s = 0.03$. See Equation (6.9).

**Table 6.22** Variations in the surface–troposphere–stratosphere system's radiation budget $\Delta B$ ($W\,m^{-2}$)

| | Source of radiation budget variations | | | |
|---|---|---|---|---|
| Aerosol model | Backscattering | Shortwave radiation absorption | Greenhouse effect | Total |
| Background aerosol | −28.5 | 0.0923 | 1.56 | −26.8 |
| Volcanic aerosol | −26.4 | 2.28 | 0.887 | −23.2 |

albedo feedback resulted in an intensified cooling of −0.8 °C and −2.15 °C, respectively. A weaker effect of volcanic aerosols on climate would exist if there were a lower single-scattering albedo (e.g. stronger absorption).

When discussing the results of observations of volcanic aerosols, it is important to assess possibilities for quantitative characteristics of the eruption intensity. In this regard, Kelly and Sear (1982) analysed the adequacy of Lamb's dust veil index (DVI), which was introduced to characterize quantitatively the impact of volcanic aerosols on the Earth's radiation budget and climate during the several years following an eruption. An estimate of the DVI, taking into account observational data and empirical and theoretical studies of the possible climatic implications of the dust veil, is based on the nature of such implications as understood in 1960. The DVI is calculated by averaging over the maximum possible number of DVI estimates determined with the use of various techniques.

Five available techniques were mutually calibrated (normalized) to give the DVI value 1000 for the 1883 Krakatoa eruption. Techniques B and C have been favoured. According to technique B, the mean global DVI is

$$\mathrm{DVI} = 0.97 R_{\max} E_{\max} t_{\mathrm{mo}} \tag{6.10}$$

where $R_{\max}$ is the post-eruption maximum extinction of direct solar radiation determined as a monthly mean for the mid-latitudes of the given hemisphere (maximum extinction can sometimes be reached only 2 yr after the eruption), $E_{\max}$ is the geographical extent of the dust veil in conditional units as a function of the volcano's latitude ($E_{\max} = 1.0$ for the band 20° N to 20° S and 0.3 for latitudes 40°), and $t_{\mathrm{mo}}$ is the lifetime of the dust veil (in months).

Technique C gives

$$\mathrm{DVI} = 52.5 \mathrm{T}_{\mathrm{D}\,\max} E_{\max} t_{\mathrm{mo}} \tag{6.11}$$

where $\mathrm{T}_{\mathrm{D}\,\max}$ is the mean temperature decrease (°C) in the mid-latitudes of the given hemisphere during the year when the effect of an eruption is at a maximum. Approximate estimates of the DVI, which may be considered as showing only an order of magnitude, are explained by different available observational data and hence by different techniques chosen for the estimation.

Recently specified data on the chronology of eruptions and a more adequate understanding of the nature of their climatic impact (e.g. the role of the secondary $H_2SO_4$ aerosol) necessitate a revision of the understanding of the DVI.

The climatic implications of volcanic eruptions have recently been of great concern. For example, an extensive complex of observations of volcanic emissions and their propagation in the atmosphere has been carried out with the use of conventional means (e.g. ground-based, aircraft, balloon) as well as satellites. Substantial progress has been reached in the development of the techniques for numerical modelling of climate and large-scale transport. An important effort to find an optional linear filter (fingerprint) for the detection of a given time-dependent, multivariate climate change signal in the presence of natural climate variability noise has been made by Hasselmann (1993).

Of principal interest are attempts to reveal a volcanic signal in the observed climate changes. The difficulty of the solution to this problem is that the interannual variability of surface air temperature is determined by various processes which can be classified as connected with the external forcings, internal regular oscillations and random processes, which appear to be largely due to the dynamic instability of the atmosphere. The most probable external factors are large-scale volcanic eruptions and increased $CO_2$ concentrations in the atmosphere (to date, there is no convincing evidence for the effect of solar activity on climate). Regular changes in the atmospheric general circulation in the North American region have been related to the El Niño / Southern Oscillation. A considerable contribution to atmospheric variability is made by its internal instability. Because of the varying intensity and location of cyclones and jet streams, a random variability of heat fluxes occurs in the atmosphere, leading to a random variability in air temperature.

To understand and forecast long-term climate variability, it is important to analyse the role of each of the atmospheric general circulation mechanisms and to detect their signals in surface air temperature. With this in mind, Robock (1984a) undertook an analysis of air temperature monthly means for 1891–1981 in the northern hemisphere on a $5^\circ \times 10^\circ$ grid. The parameter volcanically-weighted temperature anomaly was taken as a criterion of the effect of volcanoes on climate:

$$\mathrm{TW} = [\textstyle\sum T \cdot \mathrm{DVI}(t - L)]/\textstyle\sum \mathrm{DVI}(t - L) - (1/91) \textstyle\sum T \tag{6.12}$$

where TW is the volcanically weighted anomaly (from data for the whole period), DVI is the dust veil index characterizing the power of eruption (after Mitchell), $t$ is the time (in years) and $L$ is the time shift (from 2 to 4 yr). The last term characterizes the mean anomaly of temperature for 1881–1975 and is very small for all the months. An analysis of calculations of the annual change of zonally averaged values of TW for negative $L$ did not uncover any regularities, but at $L = 0$ negative anomalies prevail, being most substantial in polar regions during winter.

Maximum response of the atmosphere to a volcanically-induced disturbance is observed about 2–3 yr after an eruption and is manifested by cooling everywhere except in the Arctic in summer. These results agree well with energy-balance calculations and can be explained as being a result of the effect of sea ice thermal inertia (ice thermal inertia feedback) and, to a lesser extent, of the snow-albedo feedback. The reaction of the temperature field to the ENSO (pressure anomalies in Darwin, Australia, are its indicator) is manifested either synchronously or with a moderate time shift of about 6 months. Although in this case the zonally averaged reaction is close to zero, it results from mutual compensation of strong changes of the opposite sign at different longitudes.

Thus if the signal of external effects is most clearly detected in mean zonal temperature values, the spatial structure of the temperature field can serve as an indicator of internal variability. If, by analogy to the last equation, we calculate the mean weighted values of temperature using its instantaneous mean hemisperical values as weighting coefficients, trends are revealed that can be considered to be caused by increased $CO_2$ concentrations, which are, as a rule, positive, except for north-eastern and north-western North America and Asia, where the trends are of the opposite sign. This interpretation must, however, be specified with an account of the effect of long-term ocean–atmosphere interaction.

Gilliland and Schneider (1984) developed a new seasonal energy-balance climate model in which 10 units are considered that characterize average conditions for the hemispheres: the atmosphere over the land, the atmosphere over the ocean, the land surface, the mixed layer and deep layers of the ocean, the surface air temperature being a predictant. All principal input parameters (the depth of the mixed layer, cloud amount, albedo, insolation, hemispherical energy exchange) vary during a year and are prescribed from observational data (when possible) or assessed from physical considerations. The values of unreliably estimated parameters (for example, the coefficients characterizing the interaction between the units) have been chosen under conditions of minimum mean standard deviation from observed seasonal mean surface air temperature. The time step is 1 day and the interval of averaging 1 month. The spatially-weighted hemispherical mean surface air temperature over the land in the northern hemisphere (5–85° N) and in the southern hemisphere (0–45° S) were taken as the principal characteristics of the observed climate, which provides for the most reliable comparison with observational data (though in this case, too, the inadequate representativity of the observational data should be borne in mind).

Based on the use of this model, Gilliland and Schneider (1984) obtained estimates of the contribution to climate change of the three most widely discussed external factors of climate, namely volcanic eruptions, increasing $CO_2$ concentrations and solar constant variations. An analysis of the numerical modelling results has made it possible to reveal the existence of independent volcanic impacts on climate in both hemispheres, whereas the manifestations of the effects of $CO_2$ and extra-atmospheric insolation (the solar constant sinusoidal cycle was taken with a period of 76 yr) are on a global scale. An estimation of the mean global surface air temperature increase due to a $CO_2$ doubling gave $(1.6 \pm 0.3)\,°C$; its statistical significance remains, however, uncertain because of the use of numerous adjustable parameters. With an account of this estimate, the increase of $CO_2$ from 314 ppm in 1959 to more than 340 ppm in 1983 (i.e. by 8 per cent) should have caused an equilibrium warming of $(0.33 \pm 0.17)\,°C$.

With the time period for the content of anthropogenic sulphate aerosol to double in the atmosphere assumed to be 10 yr (this agrees with some observational data), then in 3 to 4 decades the content of the background stratospheric aerosol will reach the level observed after strong volcanic eruptions (such as of Agung and El Chichón). Naturally, in these conditions the $CO_2$-induced warming will be substantially compensated. This testifies to the extreme urgency of monitoring the dynamics of stratospheric aerosol. One of the best ways of detecting the $CO_2$ signal is to improve the reliability of the estimates of other factors of climate change, whose contribution must be filtered out for the $CO_2$ signal to be detected.

It seems that any climatic signal (including that caused by either volcanic eruptions or increasing $CO_2$ concentration) can be detected only with observational data averaged over vast regions. The most rational will be a weighted average, taking into account the results of numerical modelling from the viewpoint of revealing the geographical regions and periods of the year with the most substantial climate changes. Belle and Abdullah (1985) considered an example of such averaging of the surface air temperature observational data for 1958–1981 as the background data and for a later period as a source of information about the signal of the El Chichón eruption, with an account of numerical modelling results, according to which a maximum post-eruption surface air temperature decrease should have been observed in 1983, reaching 0.3–0.5 °C in the tropics and 0.7–1.2 °C in polar regions. The processing of data for 1983 revealed, however, a warming (exceeding $2\sigma$), which could apparently be explained by the effect of the almost simultaneously occurring and unusually intensive ENSO event.

From the data of the global network of 63 aerological stations for the period winter 1957–1958 to summer 1983, Angell (1986a,b, 1987, 1988a,b) and Angell and Korshover (1987) analysed the variability of tropospheric temperature for seven climatic zones in both hemispheres (the representativity of data for the southern hemisphere needs further studies) and for the world as a whole (after the 1963 Agung and 1982 El Chichón eruptions). For extra-tropical latitudes, information has been obtained on surface air temperature and the temperature of the 850–300 hPa layer (1.5–9 km), and for the tropics calculations have been made of the temperature of the 300–100 hPa layer (9–16 km).

The results obtained by Angell and Korshover (1987) show that during the first year after the eruption of Agung, a decrease of average surface air temperature for the northern hemisphere by 0.34 °C was observed and the El Chichón eruption was followed by a temperature increase of 0.37 °C (at the 0.5 per cent level of statistical significance according to the Student criterion $t$). These changes in the hemispherical mean temperature are largely caused by the contribution of the mid-latitude band, where the respective variations reached −0.36 °C (Agung) and +1.27 °C (El Chichón). Apparently, the trend of temperature increase is connected with the effect of an abnormally strong positive sea surface temperature anomaly in the eastern equatorial Pacific (ENSO) observed in 1982–1983. The connection between changes of the sea surface temperature anomaly and surface air temperature in the tropics is direct and in mid-latitudes is indirect.

Analysis of meteorological observations made after the 1982 El Chichón eruption showed a warming of the lower stratosphere or near the surface, as it had been after the 1963 Agung eruption (8° S). The troposphere in both hemispheres was abnormally warm in 1983 (a year after the eruption). The absence of the volcanic signal in the field of tropospheric temperature in 1983 can be explained by the ENSO event in 1982–1983, unusual in its variability, and by a high positive correlation between the amplitude of sea surface temperature variations in the tropics (reflecting the ENSO intensity) and the tropospheric temperature (Angell 1988a).

So, for example, in the period 1960–1986 the coefficient of correlation between sea surface temperature in the region 0–10° S, 80–180° W and zonally averaged temperature of the 300–250 hPa layer in the equatorial zone constituted 0.66 (at the 1 per cent significance level), with a time shift (towards a delay in the tropospheric temperature response) of one season. In the sub-tropics of the northern and southern hemispheres, maximum values of the correlation coefficient were equal to, respectively, 0.48 and 0.52 (at the 5 per cent significance level) with a time shift of 2 seasons. In moderate and high latitudes a maximum correlation took place, with a time shift of 2–3 seasons. There is evidence that in the period of ENSO a thermal disturbance forms in the troposphere near the equator and then propagates to higher latitudes, attenuating as it propagates.

If we use the results obtained to filter out the ENSO signal, it turns out that after the El Chichón eruption, the same (by amplitude) cooling of the atmosphere

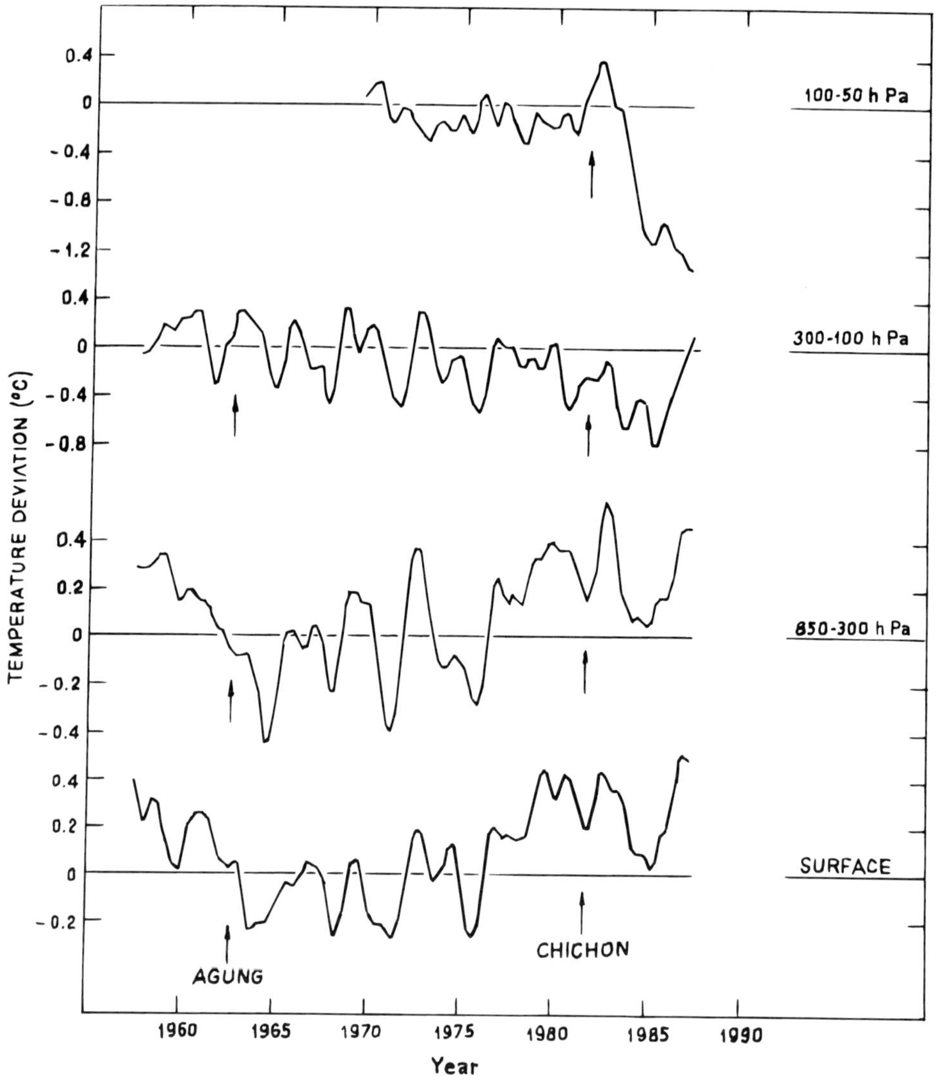


**Figure 6.17** Variation in global temperature at the Earth's surface and in 850–300 hPa, 300–100 hPa and 100–50 hPa layers. A 1-2-1 weighting has been applied twice to successive seasonal deviations from long-term seasonal means, with the first two and last two deviations twice averaged. Abscissa tick marks are in the northern summer (JJA), and note the change in ordinate scale. After Angell (1988b).

(about 0.5 °C) took place as after the Agung eruption but it was more short-lived. The use of the same correction procedure as applied to observational data after six strong volcanic eruptions in the tropics for the last century (Krakatoa, Soufrière, Pelée, Santa Maria, El Chichón and Agung) has led to the conclusion that there was a decrease of the mean surface temperature of the northern hemisphere continents of about 0.3 °C, whereas after the Katmai eruption in Alaska it was only 0.1 °C. As for the cooling of the whole Earth's surface, in this case the data on the volcanic signal in the past are uncertain and controversial, being strongly dependent on the volcanically-induced sea surface temperature increase in the eastern equatorial Pacific.

Angell (1988b) undertook a new attempt to analyse the available observational data to detect the volcanic signal (Figures 6.17 to 6.21).

So far, the most convincing evidence for the post-El Chichón tropospheric cooling, in accordance with theoretical estimates, is a decrease of zonal mean surface

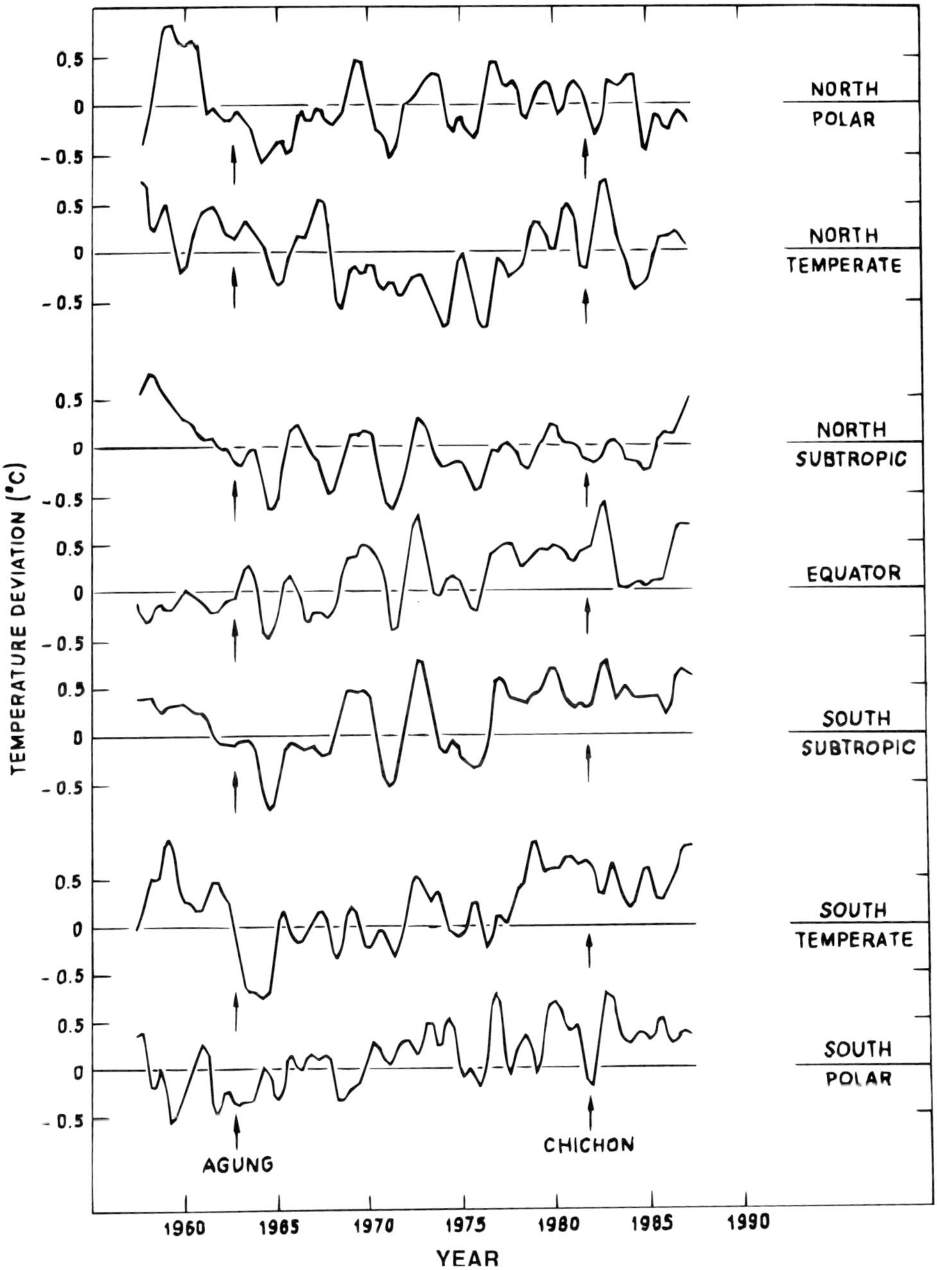


**Figure 6.18** Variation of the 850–300 hPa temperature in three tropical and four extra-tropical climatic zones. Otherwise see caption to Figure 6.17. After Angell (1988b).

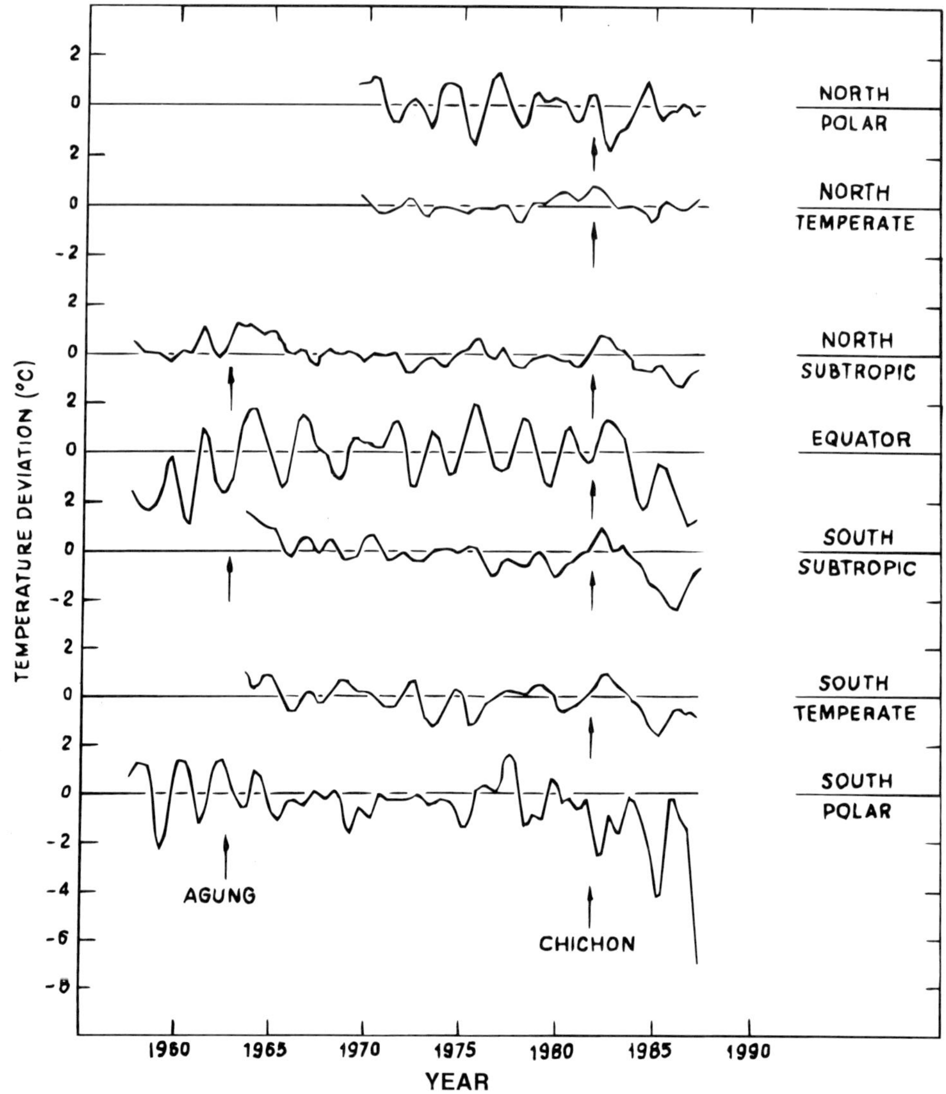


**Figure 6.19** Variation of the 100–50 hPa temperature in three tropical and four extra-tropical climatic zones. Otherwise see caption to Figure 6.17. After Angell (1988b).

air temperature in the northern hemisphere sub-tropics (10–30° N). In contrast to this, a sharp warming took place in the region 0–10° S, 180–80° W. Since the sea surface temperature had returned to the norm by late 1983, one could expect a manifestation of the El-Chichón-induced temperature decrease in 1984. The results considered suggest that it is impossible to detect the volcanic signal without other factors of climate formation being taken into account (see also Chapter 2).

As has been mentioned earlier, strong volcanic eruptions are an excellent natural model of the response of the global climatic system to radiative disturbances caused by the formation of post-eruption aerosol in the stratosphere. The strong 1963 Agung eruption on the island of Bali (8° S, 115° E) was, apparently, second in power to the Krakatoa eruption of 1883 and was adequately documented by data from

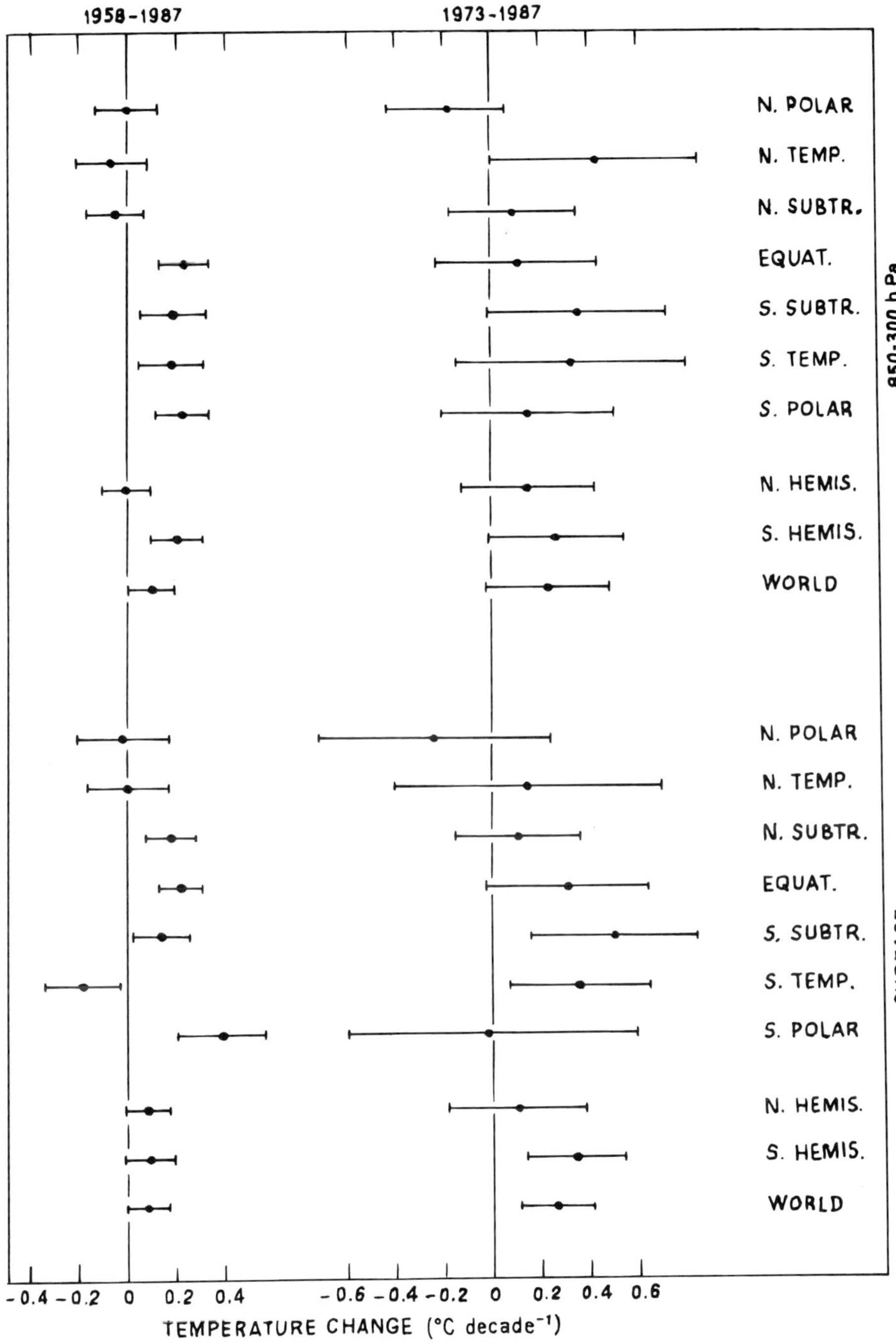


**Figure 6.20** Surface and 850–300 hPa temperature changes (in °C per decade) during 1958–1987 (left) and 1973–1987 (right) in climatic zones, hemispheres and world, based on linear regression applied to year-average temperatures. The horizontal bars are 95% confidence limits, estimated from the standard deviation of annual temperatures about the line of regression. After Angell (1988b).

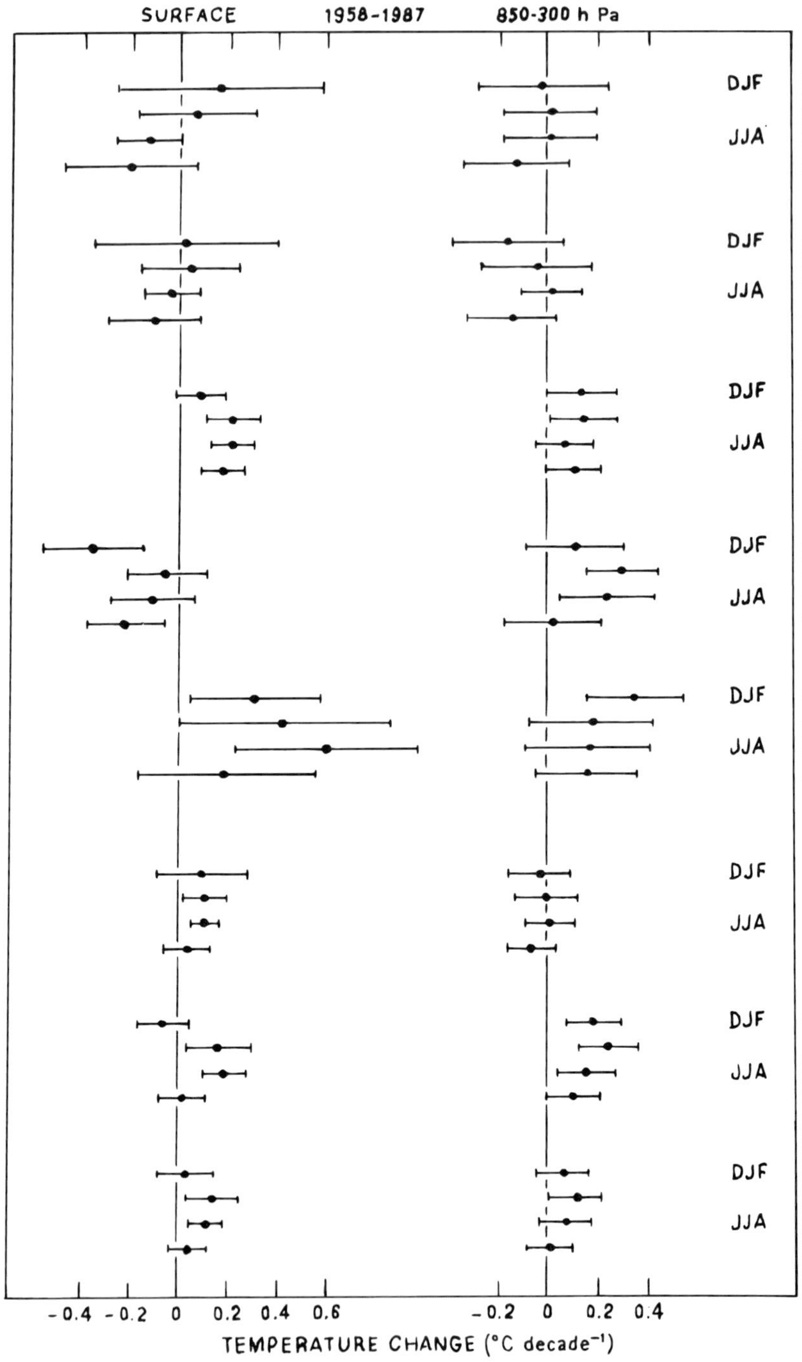


**Figure 6.21** Surface and 850–300 hPa temperature changes by season (DJF is December–January–February, etc.) during 1958–1987 in climatic zones, hemispheres and world. From top to bottom the trends are for DJF, MAM, JJA, and SON, but for simplicity only the trends for DJF and JJA are labelled. Otherwise see caption to Figure 6.17. After Angell (1988b).

numerous observations. The post-eruption global distribution of aerosol can be identified through observations of the aerosol optical effects.

For an approximate assessment of the effect of increasing concentration of stratospheric aerosol on the thermal regime of the atmosphere in low latitudes, where the radiative forcing is at a maximum, Hansen *et al.* (1978, 1992) applied a one-dimensional radiative convective model. The case of the background atmosphere with a stratospheric aerosol optical thickness of 0.005 was considered as the control. The post-eruption dynamics of aerosol were superimposed on this background condition, on the assumption that 20, 60 and 20 per cent of aerosol were concentrated, respectively, in the layers at altitudes of 22–26, 19–22 and 16–19 km. It was assumed that the aerosol is of sulphuric acid and that the heat capacity of the surface is equivalent to the existence of a 70 m layer of oceanic water.

A comparison of calculated and observed (in Australia) temperature changes at the 60 and 100 hPa levels revealed a satisfactory agreement, bearing in mind that part of the observed temperature increase could have been determined by quasi-biennial variability. The calculated change of the tropospheric mean temperature in the band 30° N to 30° S, from the moment of eruption to the beginning of 1966, agreed with the data of observations: namely, a temperature decrease of about several tenths of a degree Celsius, with a time constant of about 1 yr.

Table 6.23 summarizes the results of calculations for the cases of sulphate aerosol, which weakly absorbs solar radiation, and silicate aerosol, which is a strong absorber of solar radiation. A considerable difference between the results illustrates the importance of an adequate account of the aerosol optical properties. It seems that the properties of real aerosols are intermediate between the characteristics of sulphates and silicates. The results obtained, which show that the quantity, sign and phase shift of temperature variations in the troposphere and stratosphere, calculated with a one-dimensional climate model, agree with those observed after the Agung eruption, testify to the possibility of applying this approximate model to assess the climatic effects caused by changes in atmospheric composition.

Kondratyev and Moskalenko (1984) considered the special features of radiative heat exchange in the atmosphere disturbed by an evolving post-eruption gas–dust cloud. An atmosphere disturbed by products of volcanic eruption is characterized by increased concentrations of $CO_2$, $SO_2$ and $CH_4$, as well as by considerable turbidity due to volcanic aerosol. There is probably also an increased concentration

**Table 6.23** Variations in stratospheric and tropospheric temperature due to sulphate and silicate aerosols

| | Time (days) | | | | | | | |
|---|---|---|---|---|---|---|---|---|
| | 30 | 60 | 120 | 180 | 360 | 540 | 720 | 1000 |
| Sulphates | | | | | | | | |
| $T_{55\ \mathrm{hPa}}$ | 0.63 | 2.15 | 4.77 | 5.34 | 3.75 | 1.90 | 0.84 | 0.02 |
| $T_{\mathrm{tr}}$ | −0.01 | −0.03 | −0.12 | −0.23 | −0.48 | −0.54 | −0.51 | −0.44 |
| Silicates | | | | | | | | |
| $T_{55\ \mathrm{hPa}}$ | 3.8 | 8.1 | 12.8 | 13.1 | 10.5 | 7.4 | 5.2 | 2.8 |
| $T_{\mathrm{tr}}$ | −0.01 | 0.03 | 0.08 | −0.13 | −0.22 | −0.23 | −0.19 | −0.13 |

$T_{55\ \mathrm{hPa}}$ = stratospheric temperature; $T_{\mathrm{tr}}$ = tropospheric mean temperature

of nitrogen oxides, whose concentration in the products of volcanic emissions has not been reliably confirmed.

A thick gas–dust cloud of volcanic origins, under the influence of diffusion and circulation, propagates rapidly, covering a vast band of latitudes. Simultaneously, a gas-to-particle conversion leads to a strong increase of the sulphate aerosol concentration in the stratosphere.

As has been mentioned above, the observations revealed the bimodal particle size distribution of stratospheric aerosol, characterized by the modes of the newly formed and old stratospheric aerosol. At the initial stage, the newly formed aerosol is characterized by the radius $r = 0.003$–$0.01\ \mu m$. As a result of the process of coagulation, the size of particles increases, and, upon reaching a critical size, due to sedimentation, they fall out from the aerosol cloud. The old aerosol evolves independently. The time of sinking for the particles of stratospheric aerosol cloud is in the range 0.8–1.5 yr (depending on the environmental conditions and the power of volcanic eruption). The long life-time of the stratospheric volcanic cloud determines its significant effect on the planetary-scale radiative heat exchange and climate.

Models of the optical characteristics of the stratospheric aerosol cloud, with an account of the evolution of its chemical composition and size distribution, have been constructed on the basis of data from size distribution measurements and calculated data on the coefficients of absorption, scattering and phase function for different types of stratospheric aerosol. The characteristics of molecular absorption for $CO_2$, $H_2O$, $N_2O$, CO, $O_3$ and $CH_4$ have been determined via numerical modelling. The radiation absorption has been taken into account from experimental data. Spectral transmission functions have been obtained taking account of the temperature effect. The climatic implication of the greenhouse effect has been assessed with a 1-dimensional model described by Kondratyev and Moskalenko (1984).

The volcanic aerosol caused the most substantial changes in the radiative regime of the stratosphere in the altitude range 20–30 km. The results of numerical modelling have shown that the albedo $A_s$ of the surface–aerosol-cloud system depends on the surface albedo, the optical characteristics of the aerosol, and the particle size distribution in the aerosol cloud. If the surface albedo $A_0 > 0.39$ then the stratospheric aerosol causes a decrease of the system's albedo, $A_s$. If $A_0 < 0.39$ then the stratospheric aerosol raises the system's albedo. With the optical thickness of the eruptive cloud $\tau_a = 0.3$ (for $\lambda = 0.55\ \mu m$) the global radiation decreases by almost 10 per cent and the planetary albedo increases by 7 per cent, the stratospheric temperature rising by 6 deg K (due to the absorption of thermal emission from the surface and the troposphere). The effect of the planetary albedo increase due to the reflection of shortwave radiation is partly compensated by the absorption of ultraviolet radiation by $SO_2$, which is ejected by volcanic eruption.

Vupputuri and Blanchet (1984) undertook a numerical modelling of the effect of the El Chichón eruption on the temperature field and chemical composition of the atmosphere, using a 1-dimensional, interactive radiative–convective climate model, with an account of chemical reactions (photochemical sources and sinks) at a prescribed relative humidity of the atmosphere, and one-layer cloudiness with a constant top height of 6.5 km and cloud amount 0.446 (surface albedo 0.2). The calculations of radiative flux divergence were made with the use of the delta-Eddington approximation. In the photochemical model, the reactions of oxygen, hydrogen and nitrogen cycles were taken into account (concentrations of $H_2O$, $H_2$ and $CH_4$ were taken from observations). The interaction of volcanic aerosol and photochemical processes is determined by the change in the rate of photodissociation of $O_3$ and $NO_2$. Numerical modelling was undertaken for a mean global atmospheric model, with a constant optical thickness of the eruptive aerosol of about 0.1 (model 1), as well as with an account of varying optical thickness (model 2, Figure 6.22).

Figure 6.23 shows results of calculations of the vertical temperature profile with and without account of the interaction of volcanic aerosol with ozone (model 1). As can be seen, the interactive calculation provides good agreement with data for the standard atmosphere, except for the tropospheric layer (which can be explained by lack of account of the horizontal and vertical heat transport). The data of Figure 6.24 illustrate the results of calculations of the vertical profile of the ozone mixing ratio, which confirm the qualitative similarity of the profiles but a systematic underestimation of calculated quantities. Figure 6.25 shows results of calculations of the vertical profile of temperature changes, which characterize the effect of the optical properties of aerosol and the role of the radiative flux divergence due to longwave radiation. These data reveal a strong effect of the optical properties on stratospheric heating. An account of the contribution of the thermal emission provides a considerable warming of the stratosphere (due to absorption of the upward radiation fluxes in the transparency window) and an attenuation of the tropospheric cooling (due to the thermal emission of the aerosol cloud). If the stratospheric aerosol is of sulphuric acid, then the stratospheric heating is totally determined by the absorption of the upward longwave radiation flux.

Figure 6.26 shows the effect of volcanic aerosol on the vertical profile of the ozone mixing ratio, and Figure 6.27 illustrates the role of the height of maximum aerosol concentration in the formation of the vertical temperature profile. Lowering the level of maximum concentration brings forth a considerable reduction of stratospheric heating and tropospheric cooling.

Figures 6.28 to 6.30 show the results of numerical modelling of the evolution of the post-eruption temperature field (with the aerosol content dynamics from data of

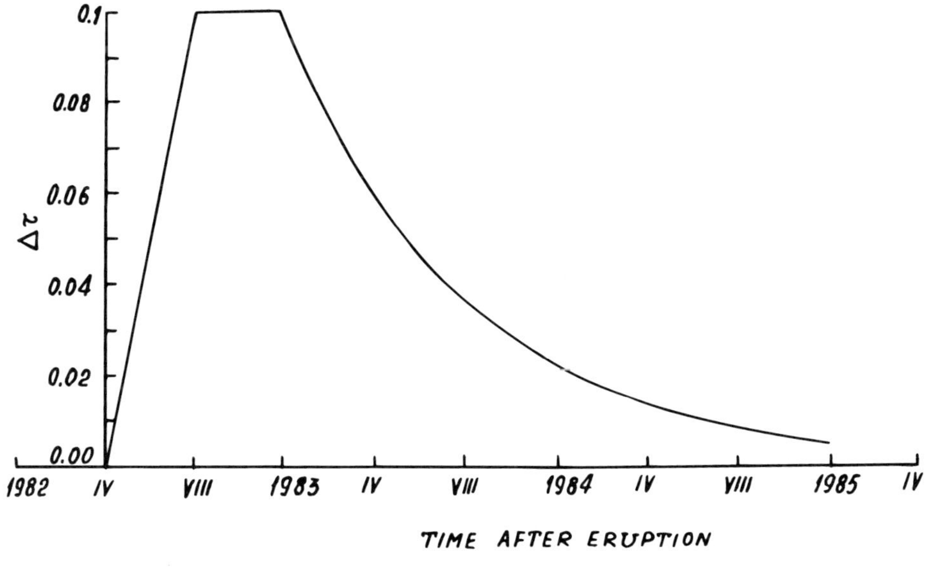


**Figure 6.22** Supposed variation in aerosol optical thickness due to El Chichón eruption.

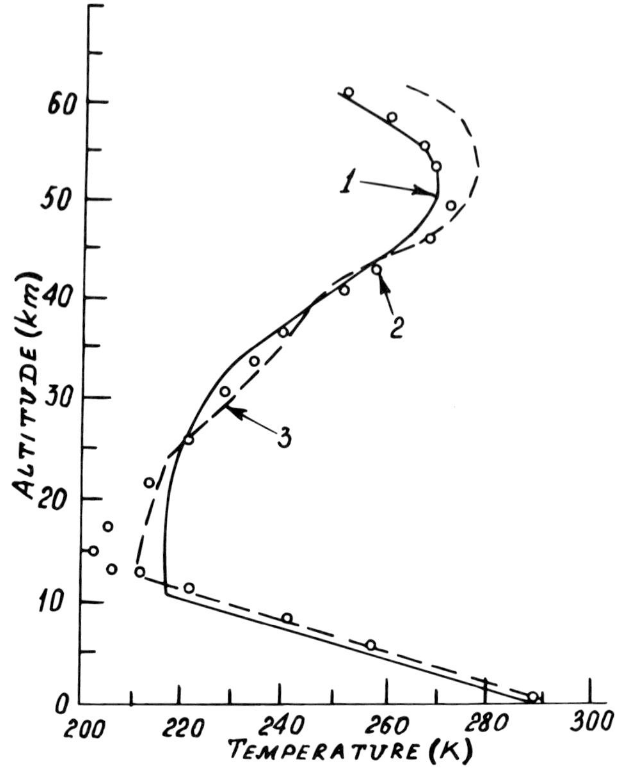


**Figure 6.23** Results of calculations of the vertical temperature profile with and without account of the interaction between stratospheric aerosol and ozone: 1, the 1962 US standard atmosphere; 2, with the interaction taken into account; 3, with the interaction neglected (with ozone content constant).

Figure 6.22), of the radiative regime and water vapour concentration, and Figure 6.31 shows the decrease of the ozone content. As follows from Figure 6.28, the response of the stratosphere to volcanic disturbance is much faster than that of the troposphere. At the level of maximum aerosol concentration (25 km), the stratospheric warming 5 months after the disturbance reaches about 4 °C, whereas maximum cooling near the surface (about 0.9 °C) occurs in about 10 months. Changes in the water vapour concentration (Figure 6.30) follow the temperature variations. The same (but to a lesser extent) occurs in the ozone content dynamics (Figure 6.31); a maximum of ozone content lags behind with respect to the maximum of stratospheric heating by about a month. From the data of satellite observations, the post-El-Chichón drop of the ozone content constituted 5 per cent.

MacCracken and Luther (1983) made calculations of the effect of the post-El-Chichón stratospheric aerosol on the radiative regime of the atmosphere, the results of which were then used in numerical modelling of the effect of the eruption on climate for a 2-dimensional, zonal, dynamic-statistical climate model, with a prescribed latitude-dependent annual change of cloudiness. Table 6.24 shows the results of calculations of variations in the fluxes of direct solar (S) and scattered (H) as well as global (G) radiation at the level of the Earth's surface, with the given aerosol

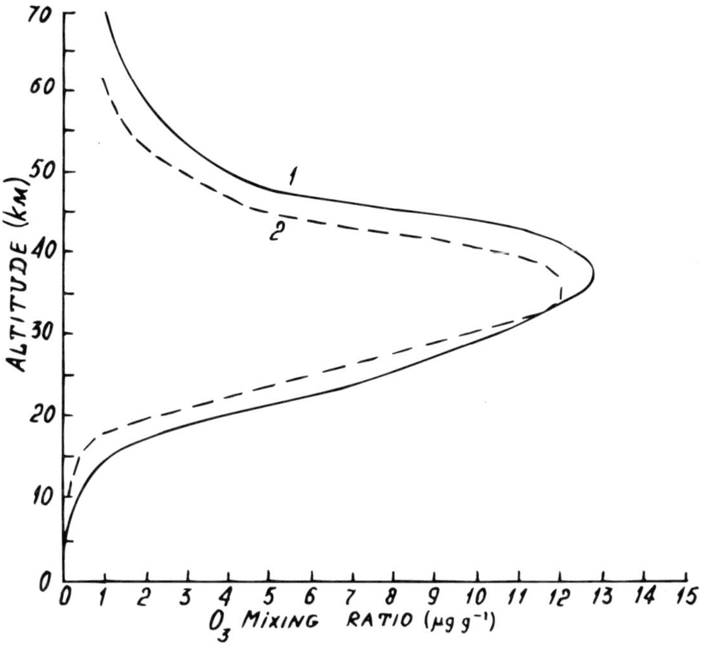


**Figure 6.24** Results of calculations of the vertical profile of the ozone mixing ratio with account of the interaction between stratospheric aerosol and ozone: 1, the 1962 US standard atmosphere; 2, calculation results.

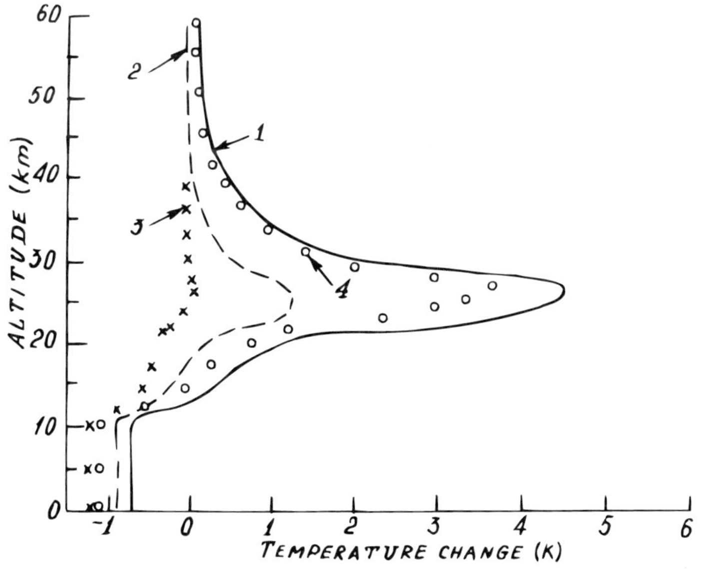


**Figure 6.25** The impact of the optical properties of volcanic aerosol on the vertical profile of temperature variations due to eruption: 1, optical characteristics corresponding to ash (absorption index is 0.002); 2, $H_2SO_4$ (75% solution) aerosol (absorption index is zero); 3, same as 2 but without account of radiative heat flux divergence; 4, same as 1 but without account of radiative heat flux divergence.

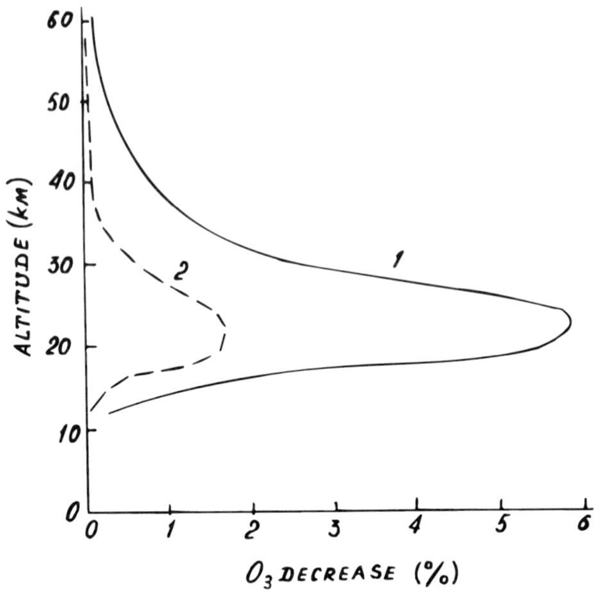


**Figure 6.26** A decrease in ozone content due to volcanic aerosol: 1, ash; 2, $H_2SO_4$ aerosol.

optical thickness of 0.255 at the wavelength of 0.55 $\mu$m (this corresponds to conditions for July 1982 in Hawaii). These results agree well with the observational data. In assessing the equilibrium effect of climate, it is assumed that the aerosol optical thickness of the infra-red region constitutes 5.9 per cent with respect to the visible. The main results of numerical modelling of the equilibrium state are that there has

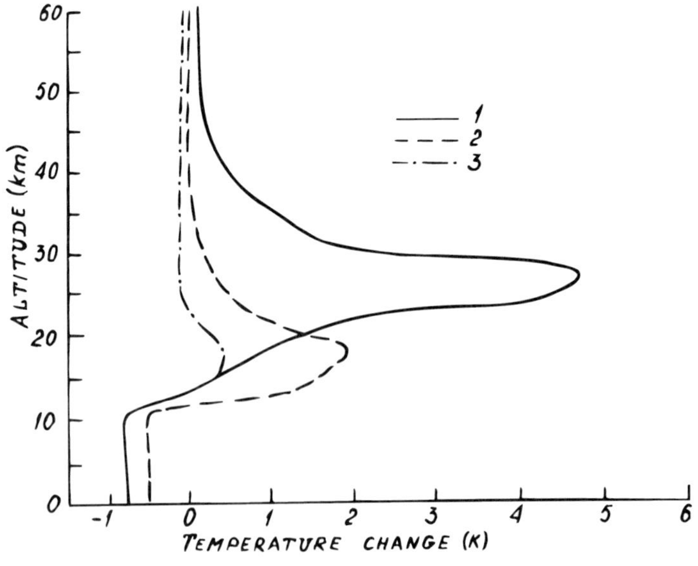


**Figure 6.27** The impact of the level of a maximum of volcanic aerosol concentration on the vertical temperature profile: 1, ash (the height of the maximum is 25 km); 2, ash (17 km); 3, $H_2SO_4$ aerosol (17 km).

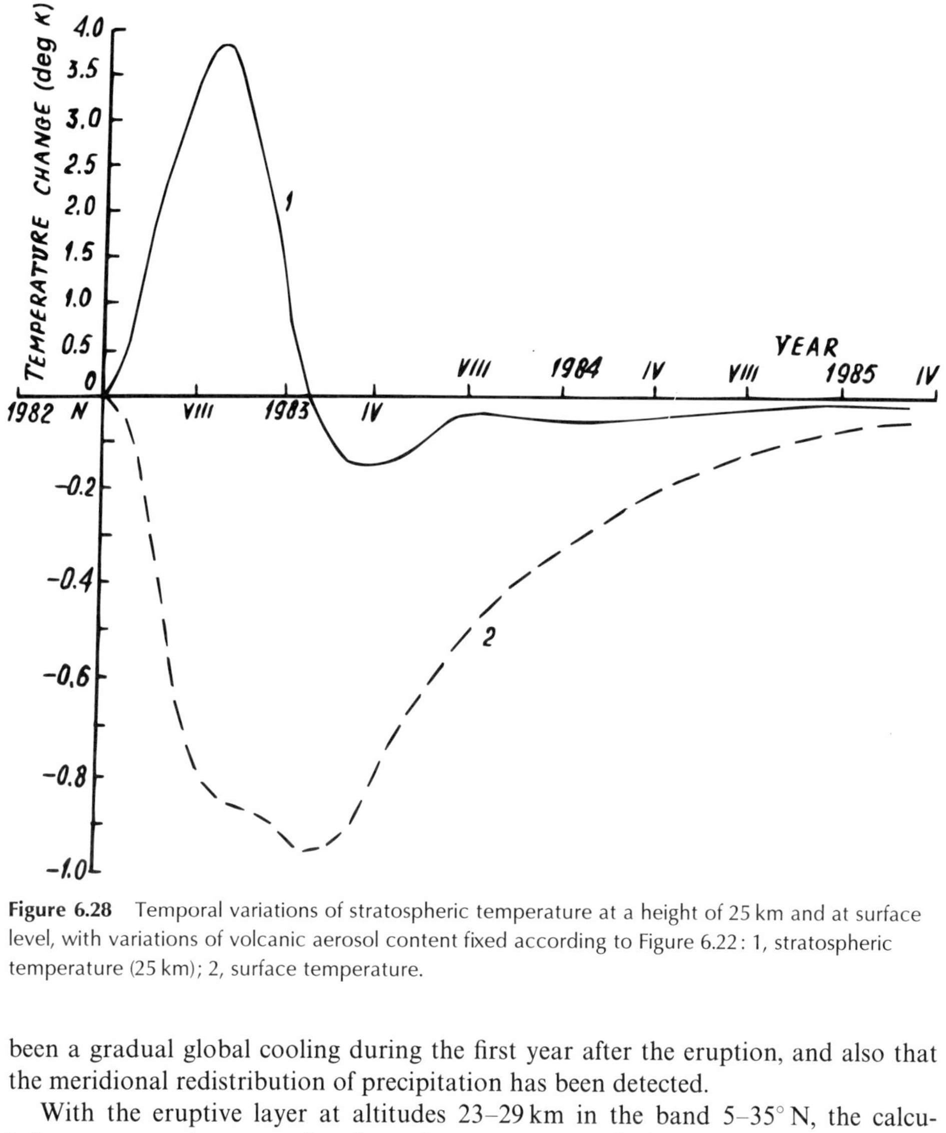


**Figure 6.28** Temporal variations of stratospheric temperature at a height of 25 km and at surface level, with variations of volcanic aerosol content fixed according to Figure 6.22: 1, stratospheric temperature (25 km); 2, surface temperature.

been a gradual global cooling during the first year after the eruption, and also that the meridional redistribution of precipitation has been detected.

With the eruptive layer at altitudes 23–29 km in the band 5–35° N, the calculations gave a decrease of surface air temperature of about 1.0 deg K at latitude

**Table 6.24** Volcanically-induced changes in the shortwave radiation fluxes

| | Sun elevation | |
|---|---|---|
| Parameter | 60° | 30° |
| ΔS | −19% | −30% |
| H/S | | |
| without aerosol | 0.04 | 0.07 |
| with aerosol | 0.28 | 0.46 |
| ΔG | −1.2% | −4.9% |

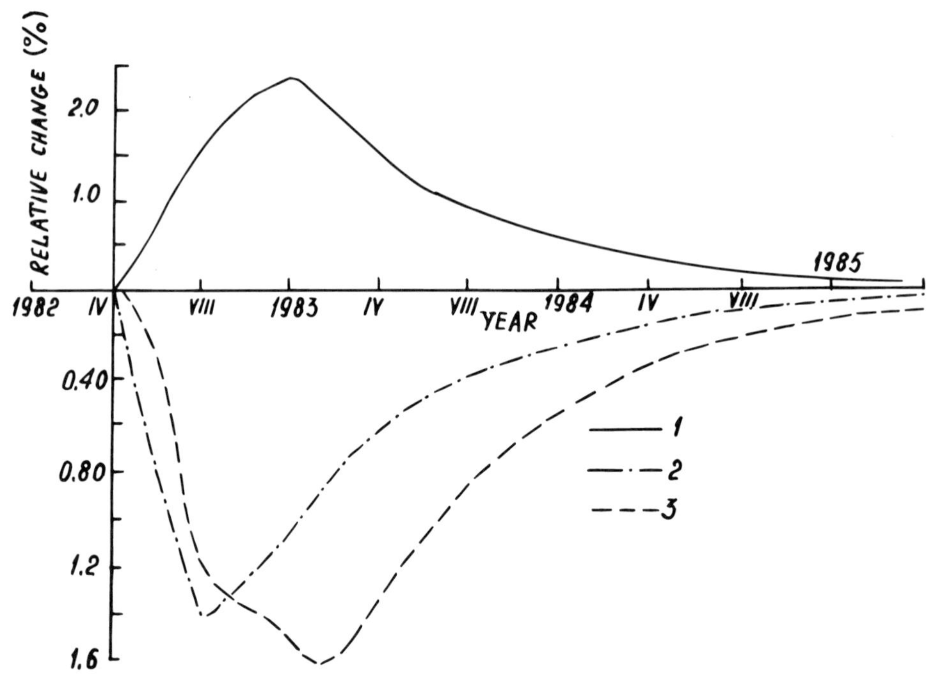


**Figure 6.29** Post-eruption changes in the radiative regime: 1, surface–atmosphere system's albedo; 2, shortwave balance of the surface; 3, atmospheric thermal emission.

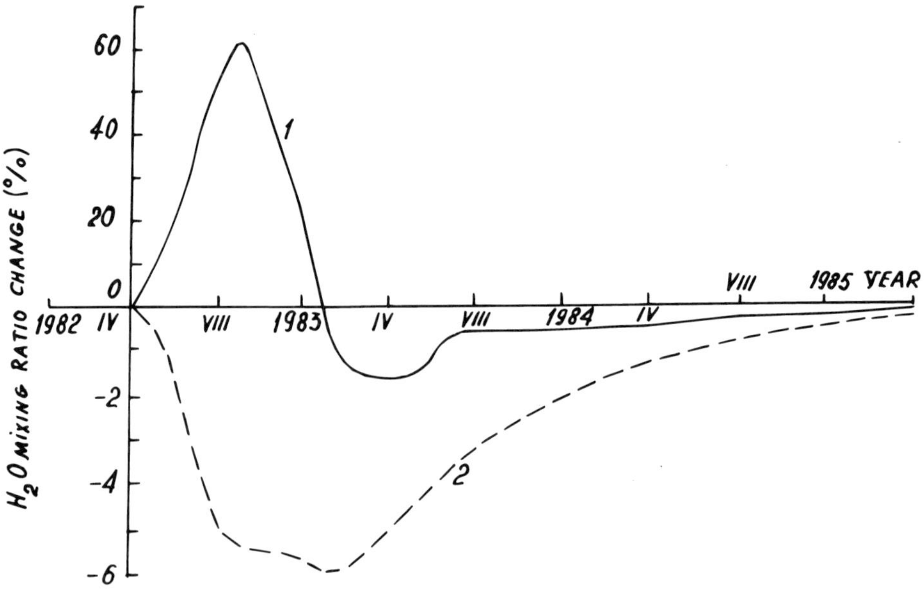


**Figure 6.30** Post-eruption variations in the water vapour mixing ratio: 1, at a height of 25 km; 2, at the surface level.

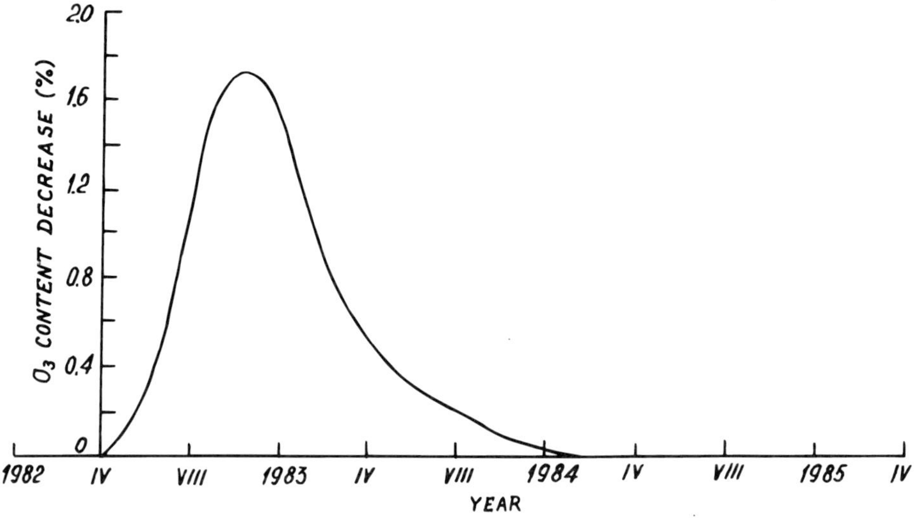


**Figure 6.31** The post-eruption decrease of total ozone content, from April 1982 onwards.

20° N, rising to 1.4 deg K near 70° N. Averaged values were 0.9 deg K (northern hemisphere) and 0.3 deg K (southern hemisphere). The calculations revealed a southward shift of the ITCZ and an intensification of the Hadley cell in the northern hemisphere, which favours the transport of heat from the southern hemisphere oceans, determining thereby the indirect effect of the eruption on the southern hemisphere climate. The shift of the ITCZ causes an intensification of precipitation in the southern hemisphere sub-tropics. The intensity of the global moisture cycle decreases, and this is manifested through a decrease of precipitation and evaporation.

Since the assessments of the equilibrium climate have led to the conclusion that it is possible to detect the consequences of the El Chichón eruption for several years afterwards, MacCracken and Luther (1983) obtained estimates of the time-dependent regime using a seasonal model of a 70 m layer of the ocean. In this case, a time-dependent meridional profile of the aerosol optical thickness was prescribed, with a maximum of 0.3 in July. The calculations for the period 1 April 1982 to 1 July 1983 revealed a gradual global temperature decrease of about 0.2 deg K for the last 6 months of integration. The results obtained do not permit the assessment of a possible duration of cooling and its subsequent intensification. Numerical modelling has led to the conclusion that precipitation decreased from 5 per cent to more than 10 per cent north of the ITCZ during the first post-eruption year. MacCracken and Luther (1983) emphasized the necessity of continuing numerical experiments in order to test the reliability of climate models applied to assess the climatic sensitivity to increasing $CO_2$ concentrations. Such experiments will improve the understanding of the causes of climate change during the century to come.

Charlock (1983) calculated the vertical profile of the infra-red radiative heating in the stratosphere (at altitudes 12–28 km) at the equator in July for conditions simulating the El-Chichón-disturbed stratosphere. It is assumed that the eruptive aerosol layer 4 km thick (optical thickness 0.25 at wavelength 0.69 $\mu$m) is located at altitudes 18 or 25 km. In both cases (with a cloudless atmosphere) there is approximately equal relative heating of the aerosol layer due to the absorption of the thermal radiation from the surface in the transparency window 700–1240 $cm^{-1}$, which is only partially compensated by cooling due to the emission from the aerosol layer.

The absolute value of radiative flux divergence in the lower aerosol layer is about twice as much as in the upper layer. The aerosol-induced heating is determined only by the contribution of the transparency window, since for the remaining spectral interval the radiative flux divergence is negative. The total radiative temperature change in the lower layer is positive and in the upper layer it is negative.

An increase of the total downward thermal radiation flux at the surface level, determined by the volcanic stratospheric layer, constitutes only 0.1–0.3 $W m^{-2}$ and at a height of 10 km it varies within 0.7–1.2 $W m^{-2}$ (these calculations were made for latitudes 0° N and 35° N). Therefore the effect of aerosol on the tropospheric climate is determined by its influence on shortwave radiation transport. With clouds in the troposphere, not only the quantity but also the sign of the aerosol effect on the radiative regime of the stratosphere can change, since in this case the upward flux of the thermal emission absorbed by aerosol decreases. The radiative flux divergence change is approximately proportional to the height of cloud top, with its values exceeding 4 km.

Studies considered above with the aim of assessing the meteorological consequences of eruptions have been preliminary, since they have been based on the prescribed aerosol content in the stratosphere and a simplified atmospheric general circulation modelling. In this connection Hunt (1977) performed a numerical modelling of the effect of volcanic aerosol on climate, with an account of realistically reproduced diffusion of eruption products from their source in tropical latitudes (the 1983 Krakatoa eruption has been considered, with the geographical coordinates 6° S, 105° E).

Numerical modelling was undertaken with the use of a three-dimensional atmospheric general circulation model, which made it possible to assess both the direct effect of eruption products on the radiative and thermal regimes and the indirect effect on the temperature field through the atmospheric dynamics. The 18-level hemispherical atmospheric general circulation model (the finite-difference analogue to the system of primitive equations in stereographic projection) used in calculations is a further modification of the GFDL and NOAA atmospheric general circulation models.

An integration has been made for annual mean conditions without account of orography and land/ocean contrasts, but with an account of the hydrological cycle. For comparison purposes, the results of the previous numerical modelling have been used, referring to the case of the undisturbed atmosphere. Neglect of the thermal inertia of the ocean leads to an increased sensitivity of the model to the forcings, compared with real atmospheric conditions.

A consideration of radiative flux divergences due to shortwave and longwave radiation is based on prescribed climatological distributions of water vapour, $CO_2$, $O_3$, clouds and surface albedo. The initial source of aerosol is fixed as a ring located at a height of 23 km and 3° latitude wide, encircling the equator. Since the atmospheric general circulation model contains a wall at the equator, the aerosol diffusion is confined to the southern hemisphere.

It is assumed that, on reaching the lower level (0.85 km), the aerosol is immediately washed out from the atmosphere (this sink for aerosol is a single one). In accordance with the aerosol model suggested by Deirmendjian for the first phase upon the Krakatoa eruption, it was assumed that the aerosol particle size distribution is characterized by prevailing large (2 $\mu$m) particles. The initial total number of particles, determining their mixing ratio, is $8 \times 10^{23}$ particles.

The particles are assumed to be of silicate (purely scattering) with refractive index 1.56, independent of wavelength within the shortwave radiation spectrum. This corresponds to the attenuation coefficient 2.9 km at a wavelength of 0.45 $\mu$m, with the particle concentration $10^8\,m^{-3}$. The effect of stratospheric aerosol on the longwave radiation transfer is not taken into account, while in the case of shortwave radiation it is determined only by backscattering.

The special feature of the radiation parametrization of the model is that it does not permit an exact account to be taken of the effect of backscattering, since part of the radiation scattered back is manifested as local heating (up to 10 °C at the upper level). However, this defect of the model can be considered as a certain merit, since observations point to a strong stratospheric heating after powerful eruptions.

Calculations of atmospheric general circulation were made with a time step of 10 minutes, but the effect of variations in the radiative flux divergence is taken into account every 24 hours. The volcanic experiment was started on day 254 of the control integration and continued for 150 days, when volcanic aerosol was propagated over the globe. For 130 days the control and volcanic experiments were performed simultaneously.

The results of the effects of eruptions were assessed by comparing the 10-day running means, starting from the first day of the volcanic experiment. Also, additional calculations were made for 14 days, with a diffusion of aerosol considered as a passive tracer (the interaction with the radiative factors was neglected).

A comparison of calculated values of the variability of hemispherical mean kinetic energy of the atmosphere has shown that substantial differences, compared with control calculations, occurred only during the second and third weeks, when the concentration of aerosol in tropical latitudes was high. If the effect of radiative factors on the propagation of aerosol is not taken into account, then in this case the values of kinetic energy are close to those of the control. The effect of volcanic aerosol on the wind field is weak and consists, as a rule, of a small attenuation of mean zonal winds and of some changes in the synoptic wind fields, which would be difficult to detect in conditions of the real atmosphere. The temperature field changes substantially (though these changes have apparently been overestimated because of neglect of the thermal inertia of the ocean). There is a decrease of the mean hemispherical surface temperature in the tropics of 0.7 deg K. In middle and high latitudes the effect of eruptions is masked by variations in local weather conditions. Such a limited response of the mid-latitude and high-latitude atmosphere is determined by the hemispheric (but not global) nature of the model and by the comparatively short duration of the experiment (a consideration of global disturbances made in other publications revealed a maximum of response to a high-latitude eruption). The temperature changes obtained outside the tropical band are largely caused by variations of meridional energy transport (especially of latent heat of condensation), but not by the direct effect of stratospheric dust loading on the solar radiation attenuation. The eruption does not affect the hydrological cycle to any great extent. This conclusion appears to be incorrect, if the development of the processes is considered for a longer time period.

The principal conclusion from the numerical experiment carried out by Hunt (1977) is that the effect of individual large eruptions on climate cannot be of long duration. More serious consequences must take place in the case of a series of prolonged successive eruptions. Numerical experiments should be continued to assess the effect of volcanic eruptions on climate, with the use of a more realistic atmospheric general circulation model which takes into account the oceanic circulation and the annual change.

The radiation parametrization of the model must also be more precise, especially from the viewpoint of considering the effect of aerosol on the longwave radiation transport. Consideration of the hydrological cycle and of the surface characteristics must be improved. It is also important to consider the effects of different types of volcanoes located at various latitudes. Unfortunately, the observational data characterizing the products of volcanic eruptions, their propagation and transformation, are scarce.

Based on the use of the energy-balance climate model suggested by Sellers, Robock (1984a,b, 1985) obtained estimates of the climatic consequences of eruptions, with the following assumptions: (1) the characteristic time of gas-to-particle conversion in the formation of sulphate stratospheric aerosol is 60 days, (2) the exponential decay of the eruptive cloud begins 500 days after the eruption, (3) the spatial and temporal distribution of the optical thickness of cloud is retrieved with an account of maximum optical thicknesses of 0.30–0.36 (this corresponds to the data of observations on Hawaii) and (4) the disturbance of the global surface air temperature field is determined only by variations in the radiative regime of the surface–atmosphere system under the influence of the eruptive cloud. The numerical modelling carried out separately for the land and for the world's oceans revealed that the initial response (cooling) of the climate to an eruption was faster over the land (which is, of course, explained by the large thermal inertia of the ocean), but that two years later these differences were smoothed. Naturally, the mean zonal values of cooling $\Delta T$ are largely determined by the ocean's contribution, in view of the large area of the surface of the ocean. The most substantial climate cooling ought to take place not in the region of eruption, but in the spring and autumn of 1984 in the region of the north pole, which can be explained by the influence of non-linear feedbacks determined by high albedos of snow and ice cover (albedo feedback), as well as by the thermal inertia of the ocean. The formation of maxima of $\Delta T$ in the spring and autumn is determined by the superposition of a higher sensitivity of polar regions to the effects on climate.

Numerical experiments performed by Robock (1984a), without account of feedback, revealed a linear response of climate to radiative disturbances and a maximum variability in the region of the greatest density in the eruptive cloud. Both intensification and phase shift of $\Delta T$ were absent at the poles. The climate of polar regions, which with its maximum variability is most sensitive to eruptions, is important from the viewpoint of detecting climate changes due to increasing concentrations of $CH_4$ and $CO_2$ for the following reasons: (1) the effect of eruptions is characterized by similar spatial distributions, but of an opposite sign, masking the effect of $CO_2$ and (2) the maximum internal climatic variability in high latitudes can partly be the result of previous eruptions. This means that, when comparing the estimates of $CO_2$-induced climatic variability (or due to eruptions) and considering the signal/noise ratio as a criterion of this variability, we are in fact dealing with the ratio signal/(signal + noise). It follows that a detection of the $CO_2$ signal is possible only after filtering out the volcanic signal. Calculations by Robock (1984a) of mean hemispherical and mean global values of climate cooling showed that the northern hemisphere response is faster and stronger than that of the southern hemisphere. When the undisturbed state returns, the effect of thermal inertia is manifested: the northern hemisphere is heated more rapidly than the southern hemisphere, and therefore by 1991 (in 10 yr) the climate anomalies in both hemispheres should equalize. Numerical modelling led to the conclusion that the process of returning to the undisturbed state would start in 1985–1987, with considerable anomalies preserved in the northern hemisphere polar regions which would vanish only in 1990. The difference between theory and observations mentioned in Chapter 2 illustrates the multifaceted nature of climate changes and the conditional character of the estimates obtained with only one of the factors taken into account.

After the 1982 El Chichón eruption in México a sufficiently complete satellite monitoring of the eruptive aerosol propagation was accomplished for the first time. However, on the whole, the results of observations turned out to be insufficient. The situation changed in 1991, when the eruption of Mount Pinatubo occurred in the Philippines. By this time the scientific community had deployed a variety of satellite, aircraft, balloon and surface instruments which could be used to monitor the consequences of eruptions not only from the viewpoint of global propagation of erupted products but also a diverse monitoring of the evolution of volcanic aerosol properties and its impact on the atmospheric radiation regime and climate.

It should be emphasized that of principal importance was a combination of conventional and satellite observational means, the conventional means having been used not only for the remote sounding of the eruptive cloud but also for direct measurements of the characteristics of volcanic aerosol. No doubt, an experience of the complex monitoring of the Pinatubo eruption can be considered as a convincing and, at the present time, most complete illustration of the fruitful combination of both conventional and satellite observational means to study the environmental variability.

An explosive eruption of the andesite island volcano Mount Pinatubo (15.14° N; 120.35° E) on 15 June 1991 in the Philippines (the island Luson) has affected most of the stratosphere, compared with the previous data of satellite observations (the previous eruption of Mount Pinatubo took place about 635 yr ago). Gigantic ejections of $SO_2$ and other products of the eruption led to the formation of a stratospheric aerosol layer at altitudes of 30 km and higher, reliably recorded by satellite observations (McCormick 1992). At altitudes above 20 km the aerosol cloud moved rapidly westward, and its front made a complete revolution round the Earth in 3 weeks.

Three weeks after the eruption (7 July 1991), aircraft measurements were begun of the characteristics of the eruptive cloud, covering (during six flights from 7 to 14 July) a region 4.5° S to 37° N; 80° W to 45° E. The complex of the airborne instruments included: zenith-looking lidar and correlation spectrometer, a side-looking Fourier spectrometer and a multi-channel radiometer to measure upward and downward fluxes of total, scattered and direct solar radiation.

A data base was obtained from synchronous satellite observations using the TOMS spectrometer to map the total ozone content (these data were processed to retrieve the content of $SO_2$), the AVHRR and the occultation sounder SAGE. A preliminary analysis of the observational data showed that the mass of ejected products was about 20 Mt, which exceeded three times the power of the 1982 El Chichón eruption. During the first several months the basic aerosol mass remained within the tropical band 30° N to 20° S.

From the data of optical measurements, the total mass of eruptive $H_2SO_4$ aerosol constituted 20–30 Mt (these estimates agree with the data on $SO_2$). The aerosol optical thickness at the wavelength 0.5 $\mu$m varied between 0.2 to 0.4. An additional absorption of solar radiation by eruptive aerosol has caused a considerable increase of stratospheric temperature. So, for example, in September the temperature rise at the 30 hPa level in the band 0–30° N reached 3.5 °C. A decrease of $NO_2$ content was observed over New Zealand, which agreed with the hypothesis on stimulating the process of ozone depletion under the influence of heterogeneous reactions on the surface of aerosol particles. In view of the great significance of the Mount Pinatubo eruption studies, we shall discuss the relevant results in some detail.

### 6.3.1 Satellite observations

The available satellite remote-sensing sources that can be used to monitor the consequences of volcanic eruptions are rather diverse, which allows the acquisition of rich information. Not only are digital images in the visible and infra-red spectral intervals transmitted from meteorological satellites available, but also there are data on the content in the stratosphere of minor gaseous and aerosol components (the instrumentation SAGE and TOMS), aerosol optical thickness of the atmosphere (AVHRR), etc.

The processing of the Nimbus-7 TOMS data (this instrumentation had been successfully operating since the launch of Nimbus-7 in 1978) made it possible to retrieve the total $SO_2$ content in the stratosphere with an error of about $\pm 30$ per cent. Initial data on the $SO_2$ content refer to several days (11–14 June 1991) prior to the cataclysm, when several weak eruption-precursors had taken place.

The first eruptive cloud about 100 $km^2$ was observed on 12 June and contained about 25 kt of $SO_2$. By 13 June it had moved westward at a distance of 1100 km (appearing over Vietnam), and its mass had reached 110 kt. On 14 June this cloud was hardly seen at all. On 13 June a second similar cloud (100 $km^2$, 15 kt) was observed over the western edge of the island of Luson, but it was observed only for 24 hours. On 15 June a third cloud was recorded (7500 $km^2$, about 450 kt) stretching for 1600 km west of the volcano Pinatubo to south Vietnam.

The 15 June cataclysmic eruption began before noon and continued until the morning of 16 June. It was discovered from the TOMS data only on 16 June (at which time the satellite had been over the zone of eruption). The mass of eruptive $SO_2$ at that moment was estimated at 15.5 Mt (the value could be underestimated in view of some malfunction of the instruments). By 23 June the eruptive cloud had covered a region of about 15 million $km^2$ stretching a distance of 10 000 km from Indonesia to central Africa, being rather uniformly distributed. Two weeks after the eruption (30 June) the length of the cloud, located in the latitudinal band 10° S to 20° N, grew to 16 000 km, and the mass of $SO_2$ reached 12 Mt. Thus two weeks after the eruption the eruptive cloud contained about 60 per cent of the initially ejected $SO_2$. This gaseous cloud can be observed during its motion round the globe for 22 days.

Table 6.25 contains comparative data characterizing various eruptions. Bearing in mind the large power of the Pinatubo eruption, one can suppose a possibility of serious climatic consequences of this eruption.

**Table 6.25** Comparative characteristics of various eruptions from the TOMS data

| Characteristics of eruptions | Nevarupta/Katmai 1912 | El Chichón 1982 | Nevado del Ruiz 1985 | Pinatubo 1991 |
|---|---|---|---|---|
| $SO_2$ amount (kt) | $(5.2–20) \times 10^3$ | $7 \times 10^3$ | 750 | $2 \times 10^4$ |
| Ash ejections (g) | $3.4 \times 10^{16}$ | $3.0 \times 10^{15}$ | $4.8 \times 10^{13}$ | $1.0 \times 10^{16}$ |
| Gas/ash ratio | $(2–6) \times 10^{-4}$ | $2.3 \times 10^{-3}$ | 0.015 | $1.9 \times 10^{-3}$ |
| VEI | 6 | 4–5 | 3 | 5–6 |

VEI = volcanic eruption intensity.

Continuous functioning of the occultation photometer SAGE-II carried by the ERBS satellite has provided a long observation series on the content of some minor gaseous components and aerosol in the atmosphere, beginning in 1984. From the data on aerosol extinction at the wavelength 1020 nm for June–August 1991, McCormick and Veiga (1992) analysed the evolution of spatial distribution of eruptive aerosol (both its $H_2SO_4$ and solid components).

Approximately 1 month after the eruption these data referred largely to middle and tropical latitudes. In this period and later on, the aerosol was observed at altitudes up to 29 km, being mainly concentrated in the layer 20–25 km. At first, the aerosol optical thickness reached a maximum in the band 10° S to 30° N, but by the end of July, high values of aerosol optical thickness had been observed at least to 70° N, with maxima north of 30° N in the layer below 20 km.

In both hemispheres there was observed a correlation of anticyclonic systems of high pressure with an advection of the volcanic material of the sub-tropics, located at the level 21 km (16 km) to the zones of mid-latitude jet streams in the southern (northern) hemisphere. By August, the aerosol optical thickness in the southern hemisphere atmosphere had increased by an order of magnitude, compared with early July, at the expense of increasing aerosol content above 20 km. The mass of eruptive aerosol was estimated at 20–30 Mt, which considerably exceeded the estimates (about 12 Mt) obtained for the El Chichón eruption.

A successful development of the technique to retrieve the aerosol optical thickness of the atmosphere at the wavelength 0.5 $\mu$m from the NOAA AVHRR data (the retrieval error being 0.03–0.05) has made it possible, starting from July 1987, to perform a regular global mapping of aerosol optical thickness over the world's oceans. The availability of such information has also opened up possibilities to monitor the dynamics of such phenomena as aerosol outbreaks from the Sahara desert and Arabia.

To analyse the variability of the optical properties of the atmosphere after the eruption of Pinatubo, Stowe *et al.* (1992) undertook a combined daily mapping of aerosol optical thickness for 12 weeks after the eruption. The maps have shown that the outbreaks of desert aerosol mentioned above are overshadowed by the scales of eruption consequences. The aerosol optical thickness retrieval technique is based on comparing measured outgoing shortwave radiation values with results of calculations for various optical models of the atmosphere for pixels 1 km by 4 km in size with the use of a multi-spectral (channels 0.63 and 3.7 $\mu$m) algorithm to recognize the clear-sky cases, similar to the sea surface temperature retrieval algorithm (the aerosol optical thickness is mapped over the grid 1° latitude by 1° longitude).

Analysis of a series of global maps of aerosol optical thickness has shown that the eruptive aerosol revolves round the globe in 21 hours. The distribution of aerosol is characterized by horizontal inhomogeneity even 2 months after the eruption (in August the aerosol optical thickness maxima were observed near Kamchatka and over the Bering Sea, as a result of forest fires in Siberia). If the criterion for recognition of the eruptive aerosol layer is an aerosol optical thickness value exceeding 0.1, it turns out that 2 months after the eruption it covered 42 per cent of the Earth's surface, which is twice as much as after the El Chichón eruption.

By the end of the first 2 months the aerosol layer had been confined to the band 20° S to 30° N, with individual clusters in higher latitudes. A maximum of aerosol optical thickness was observed on 23 August and was 0.31. The global mass of emitted $SO_2$ equal to 13.6 Mt corresponds to observed aerosol optical thickness values (the mass of $H_2SO_4$ aerosol is larger than that of $SO_2$ by a factor of 1.8).

A decrease of the Earth's radiation budget due to the Earth's albedo increase at the expense of scattering on eruptive aerosols (by about 1.3 per cent) reached 2.5 $W\,m^{-2}$, which was equivalent to a global mean surface air temperature decrease of not less than 0.5 °C, provided aerosols were uniformly distributed over the globe during the next 2–4 yr. The value of the Earth's radiation budget decrease by 2.5 $W\,m^{-2}$ is obtained from the outgoing shortwave radiation increase by 4.3 $W\,m^{-2}$ and outgoing longwave radiation decrease by 1.8 $W\,m^{-2}$. Of course, such global cooling may not manifest itself, in view of multifactor surface air temperature variability (for example, the ENSO). A consequence of the growth of stratospheric aerosol content can also be its destructive impact on the ozone layer due to the ozone-depleting heterogeneous chemical reactions and a weak oxidation of precipitation (with the $H_2SO_4$ aerosol particles depositing onto the troposphere).

### 6.3.2 Aircraft observations

Three weeks after the Mount Pinatubo eruption, NASA scientists undertook an extensive programme of aircraft observations of the characteristics of the eruptive products plume (Valero and Pilewskie 1992). Six flights in the period 7–14 July 1991 covered the region 34° N to 4.5° S, 76° W to 46° E. The flying laboratory 'Electra' has a flight-ceiling of 5.5–7.0 km, which in some cases generated serious difficulties in the remote sounding of the stratosphere through tropospheric layers.

The airborne complex included, in particular, the instruments to measure the fluxes of total, direct solar and scattered radiation: a multi-channel radiometer with an oscillating shadow ring to measure direct solar radiation flux by the Sun-shadow method (TDDR) and a multi-channel flux radiometer, MFR. These devices, with identical spectral channels (380, 412, 500, 675, 785, 865 and 1064 nm) were zenith-looking and nadir-looking, respectively. The data on direct solar radiation enabled one to find the aerosol optical thickness, with the Rayleigh component being calculated from measured atmospheric pressure and values of extra-atmospheric fluxes of solar radiation found by extrapolation (using the Langley method) of the data of surface observations at Mauna Loa observatory, with an error of below 0.5 per cent. The errors of radiation flux measurements did not exceed 1–2 per cent.

Analysis of observational data revealed an aerosol optical thickness increase near to the volcano (15.15° N), but aerosol optical thickness maxima were recorded near 10° N, which reflected specific motion of the eruptive cloud. A maximum of aerosol optical thickness was recorded on 13 July at the wavelength 412 nm and reached 0.49.

The available data on the spectral dependence of aerosol optical thickness, characterized by a rapid decrease of aerosol optical thickness with increasing wavelength in the near infra-red, allowed the retrieval of the effective size of particles as a parameter of the aerosol size distribution. Here it was necessary to classify the aerosol optical thickness spectra into two groups, corresponding to the data for 7, 8, 12 and 13 July, when the aerosol was mainly monomodal, and for 10 and 14 July, when the aerosol was bimodal.

From the data for 8 and 13 July, the values of the effective radius were 0.18 and 0.35 $\mu$m, and the masses of aerosol in the air column were 35 and 80 mg $m^{-2}$, respectively. The bimodal size distribution observed on 10 July was characterized by particle radii 0.1 and 0.8 $\mu$m with the mass of aerosol 40 mg $m^{-2}$. Apparently, the bimodal size distribution reflects the process of gas-to-particle conversion of the fine-disperse fraction of $H_2SO_4$ droplets in the presence of larger particles of volcanic origin.

Calculations of the increase of the surface–atmosphere system albedo as a function of Sun elevation, made in various suppositions, gave maximum values up to 0.12, for surface albedo = 0.05, Sun elevation = 10° and aerosol optical thickness = 0.5.

Lidar sounding from 'Electra' in July 1991 (Nd:YAG lidar, wavelength 532 nm) revealed a multi-layer structure and horizontal inhomogeneity of the aerosol plume in the stratosphere 3 to 4 weeks after the Pinatubo eruption (Winker and Osborn 1992a). The scattering ratio was determined as the ratio of the sum of the coefficients of aerosol and Rayleigh backscattering to the coefficient of Rayleigh backscattering, with the use of normalization from the data for the level near the tropopause at an altitude of 16 km.

The thickness of aerosol layers between the altitudes 17 and 26 km varied from less than 1 km to several kilometres. The layers north of 20° N were relatively weak (scattering ratio 3 and less), but south of 15° N the scattering ratio values exceeded 10, as a rule. A maximum of scattering ratio recorded on 12–13 July at an altitude of about 25 km reached 80 and more, but north of 11° N this thick aerosol layer suddenly disappeared.

Assessments of the mass of aerosol in the vertical atmospheric column, made in supposition that the aerosol is pure $H_2SO_4$, have given 95 mg $m^{-2}$, which (with a conditional recalculation) is equivalent to a global mass of 8 Mt. Apparently, by the moment of observation, only half the eruptive $SO_2$ had been converted into $H_2SO_4$ aerosol; that is, finally, the mass of sulphate aerosol could reach 16 Mt (the El Chichón eruption had given about 12 Mt of aerosol).

The polarization lidar carried by 'Electra' generated a linearly polarized signal at the wavelength 532 nm. The signal of backscattering on stratospheric volcanic aerosol was recorded for two orthogonal polarizations with subsequent calculation of both the scattering ratio and depolarization ratio determined as a ratio of signals for perpendicular and parallel components of polarization (aircraft observations of depolarization ratio for the volcanic plume were made for the first time). The observation data, with vertical resolution 15 m, were averaged to the resolution 150 m.

Analysis of depolarization ratio data revealed aerosol layers with different depolarization signatures, which is determined by special chemical composition, physical state and shape of particles (Winker and Osborn 1992b). The depolarization of backscattered light is known to reflect the presence of scattering on anisotropic particles. In this case, possible reasons of depolarization can be the presence of crystallized $H_2SO_4$ droplets, aggregates of frozen droplets and silicate particles of an irregular shape. The data for 12–13 July testify to the existence of a layer with an intensive depolarization at an altitude of 22 km and a layer with a low depolarization near the 25 km level. Apparently, in the latter case there was almost pure $H_2SO_4$ aerosol with a small amount of ashes or other non-spherical particles, and in the first case there was a mixed aerosol. For a reliable explanation of the causes of depolarization, data are needed of direct measurements of the aerosol size distribution, the shape and chemical composition of particles.

Since the global mass of $SO_2$ ejected by Pinatubo was estimated from the TOMS data at 20 Mt, it follows that the transformation of all $SO_2$ into $H_2SO_4$ aerosol should result in 41 Mt of aerosol. Hoff (1992) performed aircraft observations with the use of the zenith-looking ultraviolet correlation spectrometer COSPEC-V (measurements of scattered radiation intensity were made at seven wavelengths in the interval 297–315 nm) to estimate the $SO_2$ content in the air column.

The results of observations for different atmospheric masses gave values of $SO_2$ content from 8.9 to 64.2 $\mu$m atm m. Maximum values, corresponding to minimum masses, were observed near 14° N. Apparently, the most reliable value should be between the value averaged over all data (24.6 $\mu$m atm m) and data of near-noon observations (40.9 $\mu$m atm m).

The $SO_2$ content increased between 12 and 14 July, which could be explained by the input of the most powerful plume of eruption products at an altitude of 26 km, revealed from the lidar sounding data. The $SO_2$ global mass was estimated at 8.8 Mt (with a global mean content in air column of 25 $\mu$m atm m) to 14.4 Mt (41 $\mu$m atm m). By the moment of the observations discussed, about 25 Mt of eruptive aerosol should have formed.

Using the airborne infra-red Fourier spectrometer, Mankin *et al.* (1992) measured the Sun spectra at Sun elevations from 0° to 15° after the Pinatubo eruption to retrieve (by comparing measured absorption spectra with those calculated) the total content of $SO_2$, HCl and $O_3$ in the stratosphere from the data of wavenumbers 1364–1368, 2926 and 1141 $cm^{-1}$, respectively; the resolution of the spectrometer constituted 0.06 $cm^{-1}$. The errors in retrieving the total content of $SO_2$, HCl and $O_3$ were, respectively, 25 per cent, 20 per cent and 16 per cent.

Analysis of the results obtained has led to the conclusion that measured values of the total ozone content did not differ considerably from the normal ones and, in contrast to the data of observations after the El Chichón eruption of 1982, the HCl content did not increase markedly. Thus, the Pinatubo eruption has not led to any increase of chlorine in the stratosphere. The estimates of the $SO_2$ content are within those obtained from the data of the correlation spectrometer. A maximum of $SO_2$ was recorded near 15° N, with considerable amounts of $SO_2$ in the band 15–30° N.

With $SO_2$ supposed to be uniformly distributed over the layer 20–30 km, its average mixing ratio turns out to be 32 ppbv. Proceeding from the fact that the $SO_2$ plume covers the latitudinal band 20° S to 20° N, the mass of gas in the plume 4 weeks after the eruption should be 5.5 Mt. Such estimates should be considered very approximate. It follows from them, however, that 3 to 4 weeks after the eruption, a large quantity of eruptive $SO_2$ remains in the stratosphere. Aircraft observations at mid-tropospheric altitudes did not permit one to obtain data on volcanically-induced changes in the stratospheric water vapour content.

### 6.3.3 Balloon observations

Of great importance for analysis of the eruption consequences were balloon observations. In the period 16 July to 29 August, seven launches of the aerosol radio-sonde were made at Laramie (41° N, Wyoming, USA) to monitor the dynamics of stratospheric aerosol before and after the Pinatubo aerosol products had got to the region of Laramie. Apart from the sensors for atmospheric pressure, temperature and ozone, the complex of aerosol radio-sonde instruments included a condensation nuclei counter (the radii of particles exceeding 0.01 $\mu$m), a fast-operating eight-channel aerosol particle counter (the radii of particles exceeding 0.15, 0.25, 0.50, 1.0, 2.0, 3.0, 10.0 $\mu$m) and a special two-channel ($\geqslant$0.15 and 0.25 $\mu$m) particle counter to assess the volatility of particles. In some launchings the aerosol radio-sonde carried also a three-cascade inpactor to sample the aerosol for its subsequent electron-microscopic analysis. The aerosol radio-sonde ceiling varied within 35.5–38.8 km.

Analysis of the vertical profiles of condensation nuclei from the data of all launches, performed by Deschler *et al.* (1992), showed that first the eruptive cloud could be observed on 16 July. (An increase of condensation nuclei concentrations near the tropopause due to, probably, oilfield fires in Kuwait, took place on 17 June.) By 26 July, the aerosol number concentration below the level of the east–west transport in the upper stratosphere had increased to a record value of about 50 $cm^{-3}$ at a height of 17 km. This persistent lower aerosol layer, determined by a rapid northward transport of particles, was also revealed from the data of all other launches of aerosol radio-sondes.

The first signs of the layer formed above 20 km (directly in the zone of east–west transport) were noticed in the form of a thin layer at a level of 23 km (this layer was also recorded from the data of lidar sounding in Boulder, Colorado on 29 July to 2 August). By 6 August it had ceased to exist, but was again observed over Laramie on 29 August with a decreased aerosol concentration after the aerosol particles had circled the Earth. The vertically stretched layer of eruptive aerosol remained until the end of August at altitudes from the level of the tropopause to 22 km, and in mid-September the aerosol was observed at altitudes up to 30 km.

An increase of condensation nuclei concentrations at altitudes 15–20 km, due to processes of sulphate aerosol gas-to-particles formation, exceeded the saturation level for the condensation nuclei counter. This means that the condensation nuclei concentration was more than 500 $cm^{-3}$; that is, it exceeded by two orders of magnitude the background value for the altitudes 18–20 km.

Experiments with heating the aerosol to 150 °C showed that 95–98 per cent of it was volatile, which permitted one to consider it as droplets of concentrated $H_2SO_4$ water solution, and the characteristic time for $SO_2$ conversion into $H_2SO_4$ to be equal to 38 days (previous observations of smoke aerosol from Kuwait had given about 50 per cent of volatile aerosol).

The aerosol size distribution was characterized by large particles above the level of reversal of wind direction in the stratosphere (in the layer 22–23 km), compared with the layer 16–17 km. The altitudes below 20 km are characterized by particles with a modal radius of about 0.07 $\mu$m, whereas above 20 km the radius increases to 0.35 $\mu$m (in the case of a monomodal log-normal size distribution).

In the lower aerosol layer and on the boundaries of the upper layer the aerosol size distribution was first (from the data for 2 August) bimodal. The high concentration of small droplets of $H_2SO_4$ aerosol identified from the relevant observation data suggested the conclusion that the droplets had formed either due to homogeneous nucleation or through condensation on small ions.

During three launches of the aerosol radio-sonde in Laramie on 26 and 30 July and on 14 August 1991 the volcanic aerosol was sampled at altitudes between 15.5 and 37.5 km using a three-cascade impactor ensuring a separation of particles with a diameter of more than 4, 1, and 0.25 $\mu$m (Sheridan *et al.* 1992). These samples fixed on a thin film substrate were analysed with the use of an electron microscope, which showed that almost all (>99 per cent) small size aerosol particles were submicrometre $H_2SO_4$ droplets which later (under the influence of ammonia) transformed into $(NH_4)_2SO_4$ particles. The remaining particles – eruptive aerosol – were much larger sulphite and sulphate ones with an addition of the Earth crust components up to 10 $\mu$m in size.

Data from the aerosol radio-sonde optical counter showed that a maximum of the July size distribution of particles larger than 0.15 $\mu$m radius in the layers of stratospheric aerosol exceeded that in August. This fact was verified; a layer of sub-micrometre aerosol deposited onto the film in July was 20–30 per cent thicker (heavier) than in August. A detailed analysis of the small particles of sulphate aerosol suggested the conclusion that almost no investigated particle revealed the presence of either solid or dissolved condensation nuclei. From 98 per cent to more than 99 per cent of particles turned out to contain $H_2SO_4$. On this basis one can suppose that the $H_2SO_4$ aerosol resulted from homogeneous nucleation. This supposition agrees with experimental assessments of the volatility of particles, which testify that 95–98 per cent of volcanic aerosol consists of $H_2SO_4$ water solution droplets. The 30 July samples revealed a quantity of large (super-micrometre) particles in the form of soil particles covered with a layer of $H_2SO_4$, and the 26 July data showed that the small sulphate aerosol particles were, as a rule, surrounded by one or numerous rings of smaller sulphur-containing particles (probably, these particles had resulted from small droplets deposited onto the film).

### 6.3.4 Surface observations

So far, only part of the results of surface observations of the eruptive cloud has been published. However, these preliminary results are an important addition to the data discussed above. So, for example, the surface lidar soundings performed at Mauna Loa observatory (in Hawaii: 19.53° N, 155.58° E; 3400 m above sea level) detected on 1 July 1991 the Pinatubo-ejected volcanic plume. In this connection, Goldman *et al.* (1992) processed the Sun spectra recorded with the Fourier spectrometer, to retrieve the $SO_2$ content in the air column (Table 6.26).

The retrieval was made by comparing measured and calculated absorption spectra for the $\nu_3$ band of $SO_2$ centred at wavenumber 1362.06 $cm^{-1}$ (a stronger $\nu_1$ band at 1157.71 $cm^{-1}$ could not be used because of considerable overlapping with water vapour and $CH_4$ absorption bands).

As can be seen from Table 6.26, by September the $SO_2$ content had dropped to the threshold of the spectrometer sensitivity (ca. $10^{13}$ mol $cm^{-2}$); that is, by this time, practically all eruptive $SO_2$ had transformed into $H_2SO_4$ aerosol.

Naturally, lidar sounding was the most intensive means of surface observations. Since there had not been large-scale volcanic eruptions since 1986, at the latest, which could have led to an increase of the aerosol content in the stratosphere, one

**Table 6.26** Summarized results of $SO_2$ content assessment

| Date 1991 | Spectral resolution ($cm^{-1}$) | Content ($10^{16}$ molecules $cm^{-2}$) |
|---|---|---|
| 11 May | 0.004 | $<0.9$ |
| 9 July | 0.010 | $(5.1 \pm 0.5)$ |
| 10–24 September | 0.004 | $<0.9$ |

may consider that by June 1991, when Mount Pinatubo erupted, the aerosol content was close to the background level. Lidar sounding (ruby-laser) performed at Mauna Loa (De Foor *et al.* 1992) showed that the post-eruption stratospheric aerosol plume had reached Hawaii on 1 June and had been concentrated in the layer 21.5–22.8 km, with maximum scattering ratio 22.5 (in June the scattering ratio was below 1.5). At that time the tropopause was at an altitude of 14.2 km.

Analysis of the data of subsequent soundings in July–August showed that the plume was, as a rule, below 26–27 km (the height of the tropopause varied within 14–16 km), whereas after the El Chichón eruption the eruptive aerosol had reached the 30 km level. Both a general increase of the aerosol backscattering coefficient and its cyclic variability, reflecting apparently the movement of the plume round the globe with a zonal mean velocity of 70–80 km $h^{-1}$, have been observed during the evolution of the volcanic plume.

An increase of the total (for the layer 15.8–33 km) normalized backscattering coefficient NRBS in July–August can be approximated by the following dependence on the day of the year $D$, counted from 1 July, when $D = 180$:

$$\text{NRBS} = (0.374 \pm 0.87)(D - 180) + (1.16 \pm 0.76) \times 10^{-3} \qquad (6.13)$$

The correlation coefficient in this case is 0.67. The volcanically-induced decrease of the solar radiation transmission by the atmosphere constituted 13 per cent and approximately corresponded to that observed after the El Chichón eruption.

Analysis of the results of lidar soundings (Nd:YAG laser) performed in Garmisch-Partenkirchen (Germany: 47.6° N, 11.1° E) revealed a first arrival of the stratospheric plume of Pinatubo-erupted products on 1 July 1991, when the aerosol backscattering coefficient exceeded the background level, but on the following day there was no increase, and during the next months rather variable vertical profiles of backscattering coefficient were recorded (Jäger 1992).

The vertical stratification of aerosol can be explained by special features of atmospheric circulation. The aerosol, concentrated in the layer 15–17 km and observed in early July, moved to the observation location by the shortest trajectory. The two-layer structure of the volcanic plume recorded in August agrees with typical features of the transport in the summertime mid-latitudes: the west-to-east transport at altitudes below 20 km and the transport in the opposite direction above 20 km. Upon the transition to the characteristic wintertime circulation regime (beginning in September), with the prevailing west-to-east transport through the thickness of the eruptive aerosol layer, one thick layer formed at altitudes from the tropopause to about 28 km and the northward transport of erupted products intensified. All these regularities are similar to those observed after the El Chichón eruption.

Lidar aerosol soundings of the troposphere and stratosphere performed by Post *et al.* (1992) at three locations near Boulder (Colorado, USA) with the use of three lidars at wavelengths 10.591 $\mu$m ($CO_2$ laser), 0.694 $\mu$m (ruby-laser) and 0.574 $\mu$m (Nd:YAG laser) detected an arrival of Pinatubo-ejected aerosols, beginning on 27 July 1991 (probably the persistent stratified cloudiness taking place during 5 days prior to this date had prevented an earlier detection of eruptive aerosols).

In 1982 the El Chichón eruptive plume was first observed over Boulder only 4 months after the eruption, whereas the Pinatubo eruptive cloud reached this place after only 1.4 months, though Pinatubo is located 2° lat. south of El Chichón. No doubt, these differences are determined by special circulation features in the stratosphere and (or) peculiar ejections of volcanic matter. The difference is that after the Pinatubo eruption the backscattering in the stratosphere immediately intensified, whereas in the case of El Chichón it only happened 8 months after the eruption.

The eruptive cloud from Pinatubo was characterized by strong variability, which necessitated frequent observations during the initial phase of the cloud evolution. The calculations of the ratio of total backscattering (for the layer from the tropopause to 30 km) for $CO_2$ lidars and in the visible revealed a decrease of this ratio with time reflecting a decrease of the effective size of aerosol particles.

In December 1980 Johnston *et al.* (1992) started daily observations in Lauder (New Zealand: 45° S, 170° E) of the twilight sky brightness in zenith at sunrise and sunset, to retrieve the total content of $NO_2$ in the atmosphere from the data at wavelengths of about 450 nm, where the $NO_2$ absorption band is located. An analysis of the total content of $NO_2$ observation series for 11 yr revealed, apart from diurnal and annual changes, a long-term variability determined, apparently, by the El Chichón eruption (1982) and solar activity (during an 11-yr cycle). On the whole, the total content of $NO_2$ was less in the mid-1980s than at the beginning and the end of that decade.

During the whole period of observations, except for 1991, a minimum of total content of $NO_2$ in the annual change took place in June. However, in the period July–September 1991 a gradual decrease of monthly mean morning quantities of total content of $NO_2$ took place, which had not been observed earlier.

Since part of the maximum decrease of total content of $NO_2$ in October, amounting to 34–45 per cent (sunrises) and 30–40 per cent (sunsets), compared with the data for 1981, 1989 and 1990 (with a similar level of solar activity), could be determined by the contribution of an additional scattering on volcanic aerosol, a conclusion can be made that the real total content of $NO_2$ decrease constituted more than 20 per cent and was probably determined by heterogeneous chemical reactions with participation of aerosol. It is known that the reaction

$$N_2O_5 + H_2O \rightarrow 2HNO_3$$

on the surface of aerosol particles must lead to a decrease of the $NO_2$ and $O_3$ content and to an increase of $HNO_3$. Since there was no observed decrease of ozone after the Pinatubo eruption, the possibility of $HNO_3$ increase must be checked.

### 6.3.5 Climatic consequences of the Mount Pinatubo eruption

The stratospheric heating is known to be the most clear-cut manifestation of climatic consequences of volcanic eruptions (Kondratyev 1985b, Kondratyev and Galindo 1997). In this connection, based on the data of aerological sounding, Labitzke and McCormick (1992) undertook a comparison of stratospheric temperatures at the 30 and 50 hPa levels in the northern hemisphere, beginning from

June 1991, with their monthly means over 20 yr (1965–1984) and 26 yr (1964–1989). This comparison revealed a considerable increase of temperature in July, August, September and October in the latitudinal band 0–30° N.

In September and October there was an increase of temperature at the 30 hPa level, reaching in some locations 3.5 °C at the level of statistical significance from $2\sigma$ to $3\sigma$ (in June the temperature was below and in July at the level of the multi-annual mean). Monthly mean, zonal mean temperatures at the 30 hPa level near 20° turned out in September and October to be approximately, by 2.5 °C, above averages over 26 yr, with diurnal mean temperatures increasing almost by 3 °C. Still greater increase of stratospheric temperature occurred south of 20° N.

It is natural to suppose that all these changes are determined by enhanced solar radiation absorption in the stratosphere by volcanic aerosol. One should also expect a simultaneous warming of the southern hemisphere stratosphere, especially in the band 0–20° S, taking into account data of satellite and lidar soundings on the spatial distribution of eruptive aerosols, which show that maximum heating of the stratosphere should take place in the layer 23–26 km in the band 35° N to 20° S.

An important peculiar feature of the Mount Pinatubo eruption is the fact that it took place in a year when, in the course of quasi-biennial oscillations as a major manifestation of the interannual variability of stratospheric temperature, a transition took place in the lower stratosphere, from west-to-east to east-to-west transport (cold phase of the quasi-biennial oscillation), whereas during the eruptions of Agung in 1963 and El Chichón in 1982 an opposite situation took place. It is also important that if during those two eruptions the Sun moved from the equator to the sub-tropics, during the Pinatubo eruption it was located farther north and moved approximately southward. Apparently the transport of eruptive aerosol to high latitudes of both hemispheres, which started in the autumn of 1981, ought to have reduced the stratospheric heating in the tropical band and intensified the global warming (see also Robock and Mao 1995).

The 15 June 1991 Pinatubo eruption was followed, apparently, by ejections of gases and dust to altitudes above 25 km, the most powerful in this century, and it can be expected to produce significant climatic consequences. Hansen *et al.* (1992) obtained preliminary estimates of such consequences with the use of the GISS climate model.

With this aim in view, three scenarios of volcanically-induced changes of the stratospheric aerosol optical thickness have been prescribed: (1) E1: aerosol properties correspond to those prescribed for the conditions of the El Chichón eruption (aerosols are droplets of a 75 per cent $H_2SO_4$ water solution; aerosol optical thickness decreases from 1 immediately after the eruption to 0 six months later; eruptive aerosols, initially uniformly distributed in the band 0–30° N, then propagate over the globe, with aerosol optical thickness in the band 30–90° N being doubled, compared with the band 30° N to 90° S); (2) 2 × E1: prescribed doubled aerosol optical thickness, compared with E1; (3) P: the same temporal change of aerosol optical thickness as in the models E1 and 2 × E1 (exponential decrease 10 months after the eruption with a time constant of 1 yr), but aerosol optical thickness is greater by a factor of 1.7 than for E1 and there is a more realistic global distribution of aerosol optical thickness than before (hemispherically uniform distribution of aerosol had been reached by January 1992).

Numerical modelling has shown that in the case of scenario P the Pinatubo-induced global mean change of the Earth's radiation budget in early 1992 could reach about 4 W $m^{-2}$. This means that the effect of the volcanically-induced cooling considerably exceeds the greenhouse warming due to an accumulation of all greenhouse gases during the period from the beginning of the industrial revolution. A decrease of global mean surface air temperature, which by the end of 1992 should have constituted about 0.5 °C, exceeds the global mean surface air temperature mean standard deviation by a factor of approximately three.

Calculations of surface air temperature changes for scenarios E1 and 2 × E1 have demonstrated that the eruption-induced cooling is too weak to change the situation of 1991 as one of the warmest years in the current century, in view of the strong inertia of the climate system and weak initial perturbing effect. However, for scenario 2 × E1 (as in the case of P) the decrease of surface air temperature reaches 0.5 °C, and, if this estimate is realistic, the predicted cooling will considerably exceed the warming by about 0.2 °C, characteristic of the El Niño effect.

Though regional estimates of surface air temperature changes are unreliable, they have shown that if, without account of the effect of eruptions, the probability of unusually cold winters in Moscow ought to have decreased from 33 per cent in the 1950s to 15–20 per cent at present, then, bearing in mind the contribution of the Pinatubo eruption, the probability of severe winters in the period 1992–1994 should increase to 30–50 per cent. The reliability of the numerical modelling results is still substantially limited by imperfection of the climate model applied, which does not take into account, for example, the interaction with deep layers of the ocean and thermal inertia of the climate system. The properties of clouds and aerosol require a more adequate consideration.

The key directions of prospective studies on volcanoes and climate may be formulated as follows:

1. The complex global monitoring of gas and aerosol products of eruptions, with the use of all available conventional and satellite techniques, bearing in mind detailed studies of the chemical composition and physical properties of eruptive products near the eruption and in the whole global atmosphere (sulphur and nitrogen compounds claim special attention).
2. Further studies of gas-to-particle conversions responsible for the formation of stratospheric aerosol.
3. Numerical modelling of regional and global impacts on climate (first of all, air temperature and precipitation).
4. A search for the most sensitive indicators of the effect of eruptions on climate.
5. Palaeoclimatic reconstructions that characterize the effect of eruptions on climate.

A reliable assessment of the effect of volcanic eruptions on climate is only possible based on numerical modelling with the use of complete climate models based on adequate empirical information.

So far, we can state that none of the strong volcanic eruptions has been adequately documented. Therefore, there is a need for a specialized international programme which would combine studies on numerical modelling and observational subprogrammes, aimed at (1) regular investigations of the active volcanoes and their effect on the atmosphere and (2) alert observations with the use of regular networks and specialized observational means after strong volcanic eruptions.

CHAPTER SEVEN

# Conclusion

In conditions of intensive industrial and technological progress, human activities depend more and more on weather and climate, in particular for agriculture, energy production and transport. An important fact is that the impact of human activities on the environment and climate is increasing, for example, CFCs affecting the ozone layer and the climatic consequences of increased $CO_2$ and other minor gaseous constituents. This is why the Second UN Conference on Environment and Development which was held in Brazil in June 1992 paid so much attention to problems of global climate change and global ozone dynamics which have been discussed in detail in the Report prepared by the Intergovernmental Panel on Climate Change (IPCC). Special attention should be given to the atmospheric sciences (Committee on Atmospheric Sciences 1980), the following three aspects being of the first priority: (1) weather predictions (including the development of techniques for local now-casting, as well as synoptic forecasts, with emphasis on precipitation and dangerous weather conditions), (2) the climate (including the possibility of forecasting climatic anomalies with a characteristic time scale of a month and longer and the climatic response to regional variations in surface properties, for example as a result of urbanization and deforestation) and (3) atmospheric chemistry (including global biogeochemical cycles, especially those that can affect the regional and global climate, problems of regional chemistry and atmospheric pollution).

The current observational means have substantially improved the weather forecasts 1 to 12 hours ahead. In this connection further effort is needed for a better understanding of the laws of the various mesoscale processes. During the last two decades the reliability of 48-hour forecasts of the fields of temperature and winds has been greatly improved. The 24-hour forecasts of precipitation zones have become more reliable, mainly due to the development and application of high-resolution models for limited geographical regions. However, predictions of amounts of precipitation and of the start and end times are less successful. Further theoretical studies and meso-meteorological field experiments are needed here, aimed at improving techniques for precipitation forecasts 24 to 72 hours ahead. The emphasis must be placed on the development of the forecasting technique and the ways of distributing information about the formation and evolution of local dangerous meteorological conditions with a characteristic time scale of less than 24 hours.

The most serious errors in the forecasts of large-scale motions appear to be connected with an inadequate simulation of long (planetary-scale) waves, whose variability determines the location of anomalies in temperature and the jet streams that stimulate the evolution of cyclones. The errors are determined by three factors: (1) an inadequate description of the atmospheric state from data of the global network, (2) imperfect techniques for an assimilation of observational data in numerical models and (3) drawbacks in the physical and mathematical constructions of the models. The use of various data bases will make it possible to determine whether a denser network plays the key role in improving the reliability and enlarging the range of the forecasts.

There are now empirical data that illustrate the potential predictability of climate changes a month and more ahead. This particularly refers to such persistent large-scale weather-forming processes as blocking or stable trajectories of the cyclones. Therefore it may be possible in the near future to develop techniques to forecast the monthly and seasonal climatic anomalies. A major strategy should consist of combined empirical and theoretical studies aimed at the recognition of such non-random and potentially predictable quantities as soil moisture supply or sea surface temperature. It must be borne in mind here that the climatic signals are most easily detected in specific regions of the globe. This makes the Sections programme particularly important.

An analysis of the response of global and regional climates to regional-scale changes in the properties of the surface is very important. These changes take place under the influence of both natural factors (ice and snow cover extent, desert surface) and anthropogenic factors (urbanization, deforestation, heat and moisture releases by power stations, etc.). The local-scale consequences of these changes have been well studied, but they require more reliable quantitative estimates. The effects on regional and global climates remain inadequately studied, though a rapid development of human activities determines the growing urgency of these problems. In this connection, it is most important to intensify fundamental studies of the processes of the atmosphere–ocean interaction and cloud dynamics as well as to continue developing regional and global climate models. To accomplish this, the programme of satellite observations must be enhanced (such as an experiment on studies of the Earth's radiation budget) and more efficient computing facilities must be provided.

The strategy of climate studies should foresee the development in the following interdependent directions: the climatic data base and methods for data analysis, studies of the key climatic features (for example, monsoons), analysis of the climate-forming processes (especially cloud–radiation interaction), a statistical analysis of the climatic data, the development of climate models, analysis of climate sensitivity to various factors and estimations of the predictability of the climatic averages and anomalies.

With respect to enhancing the volume and diversity of climate data, further development of satellite observations is of primary importance. Studies of the water cycle and of latent heat as factors of atmospheric energetics require more reliable data on evaporation and precipitation for every region of the world. To understand the atmosphere–land-surface interaction, data on soil moisture, albedo, snow and vegetation cover, and evapotranspiration are needed. A consideration of the ocean–atmosphere interaction should be based on data on sea surface temperature and the temperature of the upper mixed layer. Information on sea ice is also very important.

To determine a possible Sun-induced climatic variability, continuous measurements are needed of the solar constant, ultraviolet radiation and high-energy particles, using satellites, balloons and aircraft. Further success of the WCRP and its components (TOGA, WOCE, GEWEX, ISCCP, ISLSCP, GCOS), as well as a realization of the IGBP, will open up new horizons in this respect.

Planning of an observational system in the interests of the climate problem is impossible without more-reliably-substantiated requirements for observational data, based on further development of climate models. Of course, the process of planning will go forward stage-by-stage, as the observational system improves and the process of numerical climate modelling develops. The first stage is based on the use of data from operational meteorological satellites and on a combination of conventional means of observations.

Investigations have revealed the particular urgency to study some climatic phenomena which are relatively independent components of the global climate system. These include (1) the mode of circulation in the equatorial Pacific as a separate dynamic system in which interactions are revealed between large-scale fields of wind and precipitation, (2) the atmospheric and ocean temperatures at the interface, and between the sea currents, (3) the monsoon circulation (which is probably connected with the equatorial mode or quasi-biennial oscillations in the tropical stratosphere), which is apparently the most important climatic subsystem and can seriously affect the climate of the mid-latitude northern hemisphere, (4) the quasi-biennial oscillations and (5) the blocking (the anticyclonic systems in the middle latitudes which can be persistent for many weeks and sometimes cause catastrophic droughts) of long (planetary-scale) waves.

From the viewpoint of understanding the climatic laws, the following physical processes are of the first priority: (1) the cloud–radiation interaction as well as the effect of natural and anthropogenic aerosols on the radiative regime of the atmosphere and cloud formation, (2) the atmosphere–ocean interaction, the role of the ocean in the accumulation and transport of heat (contributions from the oceanic gyres, the wind-driven roughening of surface water and small-scale horizontal diffusion), (3) the land-surface processes (the heat and moisture budgets at the surface level, a consideration of the soil moisture, albedo, snow and ice cover characteristics and evapotranspiration), (4) the effect of $CO_2$ and other minor gaseous constituents on the atmospheric radiative regime and climate and (5) solar–climatic interrelations.

In view of the rudimentary state of development in atmospheric chemistry, studies of the minor gaseous constituents which either enter the atmosphere or leave it due to natural and anthropogenic processes, and the role of the minor gaseous constituents in the physical processes that govern weather and climate, have become very important. Priority must be given here to the aspects of atmospheric chemistry that are directly relevant to the climate problem. This concerns, in particular, studies of the cycles of the optically active minor components (primarily water vapour, carbon dioxide, ozone, sulphur and nitrogen compounds) that govern the atmospheric radiative regime, as well as the components affecting the optically active compounds (the hydroxyl radical, the oxides of nitrogen, hydrocarbons and CFCs). Considerable progress in atmospheric chemistry can be achieved only with an enhancement of the programme of global measurements of the atmospheric chemical composition using aircraft, balloons and satellites (field studies, in both polar regions, of the global ozone layer dynamics is a good example). Laboratory studies are urgently needed of the rates of chemical reactions, as are developments of complex chemical and dynamic atmospheric models to study biogeochemical cycles, wash-out processes and interactions between chemical and meteorological processes.

Special attention should be paid to gas-to-particle reactions that produce aerosol particles from gaseous compounds (in particular, the chain of processes of $SO_2$ oxidation into $H_2SO_4$ and the formation of sulphate aerosol, including dimethyl sulphide transformation). About half the mass and 90 per cent of the surface of atmospheric aerosol particles (both liquid and solid) are products of gas-to-particle conversions. The fact that the surface area of all the particles exceeds that of the globe, illustrates the possibilities of reactions on the surface of particles with gas components.

Studies in the field of regional atmospheric chemistry and pollution, which have recently been initiated, are becoming important. Studies include the harmful effect of acid rain, caused by industrial releases of acid compounds, on forests, gardens, lakes and crops, and, to a lesser extent, on constructions, and also the worsening visibility due to pollution. To estimate the extent of regional pollution, extensive investigations are needed of the atmospheric chemical composition, and of key processes governing the atmospheric composition and the long-range transport of pollutants, as well as changes which they exhibit prior to scavenging. Such studies must be based on both laboratory and field measurements (surface and aircraft) and on numerical modelling (arctic haze and tropospheric ozone problems are relevant illustrations).

Bearing in mind the importance and urgency of the various problems, the following order of their priority may be established: (1) a better understanding of the weather-forming processes, and weather forecasts, with emphasis on precipitation in cyclones, (2) further studies of the climatic system and its variability on a time scale of seasons to decades and (3) studies of biogeochemical cycles and balances as well as their connection with atmospheric processes.

Problems of solar–climatic relations and the impact on weather also deserve serious attention. The results from studies obtained during the last two decades have given new empirical evidence for numerous feedbacks between the various manifestations of solar activity, weather and climate. The planned observations from satellites will help to accomplish a more adequate, reliable and continuous monitoring of the processes on the Sun, which opens prospects for intensive searches for physical mechanisms of the Sun–weather–climate relations that are not clear so far. In this connection, studies of the following aspects claim serious attention: variations in the chemical composition of the upper atmospheric layers caused by variations in the electromagnetic and corpuscular emissions of the Sun, the electrical interaction between different atmospheric layers due to solar activity, the response of stratospheric and mesospheric dynamics to a heating of the ozone layer caused by varying ultraviolet radiation, and the dynamic stratosphere–troposphere interaction in order to determine a possible impact of the upper atmosphere warming on tropospheric dynamics.

We conclude by mentioning that, while this book has been concerned with the scientific aspects of observing global change, the question of anthropogenic influences on the climate is of great concern to many thinking people on Earth at present and is associated with many political questions (Kondratyev *et al.* 1994). The problem is, as should have emerged from this book, that it is very difficult to separate the anthropogenic influences from the natural causes of climate change. We mentioned in Section 6.1 the conclusion of the 1990 IPCC Report (Houghton *et al.* 1990) that a doubling of the atmospheric concentration of $CO_2$ would lead to global warming near the Earth's surface of 'between +1.5 °C and +4.5 °C, with a "best guess" of 2.5 °C'. This is not, however, the same as observing a warming and being able to say categorically that that warming is due to anthropogenic production of $CO_2$ ; it could arise from natural causes.

The interpretation given to the data has changed somewhat over the last five years. In the 1990 IPCC Report (Houghton *et al.* 1990) it was noted that the mean surface air temperature has increased by between 0.3 °C and 0.6 °C over the last 100 years. The report then said 'The size of this warming is broadly consistent with predictions of climate models, but it is also of the same magnitude as natural climate variability. Thus the observed increase could be largely due to this natural variability; alternatively this variability and other human factors could have offset a still larger human-induced greenhouse warming. The unequivocal detection of the enhanced greenhouse effect from observations is not likely for a decade or more.' By the time of the IPCC Report *Climate Change 1995* (Houghton *et al.* 1996) the balance of opinion had shifted; this report said 'Since the 1990 IPCC Report, considerable progress has been made in attempts to distinguish between natural and anthropogenic influences on climate' and it gave various details, all under a general heading 'The balance of evidence suggests a discernible human influence on global climate'. Those readers who are interested will find it instructive to read the various IPCC Reports (Houghton *et al.* 1990, 1992, 1996), while those who are interested in the political aspects will find some of these discussed by Kondratyev *et al.* (1994) and Kondratyev (1997).

# References


ACKERMAN, S. A. and CHANG, H., 1992, Radiative effects of airborne dust on regional energy budgets at the top of the atmosphere. *Journal of Applied Meteorology*, **31**, 223–233.

ACKERMAN, S. A., FREY, R. A. and SMITH, W. L., 1992, Radiation budget studies using collocated observations from Advanced Very High Resolution Radiometer, High-Resolution Infrared Sounder/2, and Earth Radiation Budget Experiment instruments. *Journal of Geophysical Research*, **97**, 11513–11525.

ADAMENKO, V. N., KONDRATYEV, K. YA and SINIAKOV, S. A., 1991, Deposition from the atmosphere of heavy metals at the North Pole as compared to background regions of north-western Europe. *Izvestiya All-Union Geographical Society*, **123**, 316–323.

ADAMENKO, V. N., MASANOVA, M. D. and TCHETVERIKOV, A. F., 1982, *Climate Change Indication: Analysis and Interpretation Techniques* (Leningrad: Gidrometeoizdat).

ADEM, J., 1988, Possible causes and numerical simulation of the northern hemispheric climate during the last deglaciation. *Atmosfera*, **1**, 17–38.

ADEM, J., 1990, Review of the development and applications of the Adem thermodynamic climate model. *Climate Dynamics*, **5**, 145–160.

ADEM, J. and MENDOZA, V. M., 1987, Sensitivity experiments on ocean temperature predictions with a thermodynamic climate model. *Geofisica Internacional*, **26**, 525–544.

ADEM, J. and MENDOZA, V. M., 1988, Recent numerical-thermodynamic experiments on sea surface temperature prediction. *Geofisica Internacional*, **27**, 309–326.

AKIMOTO, H. and NARITA, H., 1994, Distribution of $SO_2$, $NO_x$ and $CO_2$ emissions from fuel combustion and industrial activities in Asia with $1° \times 1°$ resolution. *Atmospheric Environment*, **28**, 213–226.

AMERICAN METEOROLOGICAL SOCIETY, 1986, *Extended Summaries of the Third Conference on Climate Variations and Symposium on Contemporary Climate: 1850–2100, held in Los Angeles on 8–11 January 1985* (Boston: American Meteorological Society).

AMERICAN METEOROLOGICAL SOCIETY, 1991a, *Proceedings of the Fifth Conference on Climate Variations, 1991* (Boston: American Meteorological Society).

AMERICAN METEOROLOGICAL SOCIETY, 1991b, Policy Statement of the American Meteorological Society on Global Climate Change. *Bulletin of the American Meteorological Society*, **72**, 57–61.

AMERICAN METEOROLOGICAL SOCIETY, 1991c, *Preprints. Second Symposium on Global Change Studies, held in New Orleans on 14–18 January 1991* (Boston: American Meteorological Society).

AMERICAN METEOROLOGICAL SOCIETY, 1991d, *Second Symposium on Global Change Studies, New Orleans, 14–18 January 1991* (Boston: American Meteorological Society).

AMERICAN METEOROLOGICAL SOCIETY, 1992, *Preprints. Third Symposium on Global Change Studies held in Atlanta, Georgia, on 5–10 January 1992* (Boston: American Meteorological Society).

AMERICAN METEOROLOGICAL SOCIETY, 1995a, *Fourth Conference on Polar Meteorology and Oceanography, 15–20 January 1995* (Dallas, Texas: American Meteorological Society), J1–J12.

AMERICAN METEOROLOGICAL SOCIETY, 1995b, *Sixth Symposium on Global Change Studies, 15–20 January 1995* (Dallas, Texas: American Meteorological Society), J1–J9.

ANGELL, J. K., 1981, Comparison of variations in atmospheric quantities with sea surface temperature variations in the equatorial eastern Pacific. *Monthly Weather Review*, **109**, 230–243.

ANGELL, J. K., 1986a, Annual and seasonal global temperature changes in the troposphere and low stratosphere, 1960–1985. *Monthly Weather Review*, **114**, 1922–1930.

ANGELL, J. K., 1986b, Environmental impact of volcanism. *Proceedings of Norman D. Watson Symposium* (Rhode Island: University of Rhode Island), pp. 9–17.

ANGELL, J. K., 1987, Rocketsonde evidence for a stratospheric temperature decrease in the Western Hemisphere during 1973–1985. *Monthly Weather Review*, **115**, 2569–2577.

ANGELL, J. K., 1988a, Impact of El Niño on the delineation of tropospheric cooling due to volcanic eruptions. *Journal of Geophysical Research*, **93**, 3697–3706.

ANGELL, J. K., 1988b, Variations and trends in tropospheric and stratospheric global temperatures, 1958–1987. *Journal of Climate*, **1**, 1296–1313.

ANGELL, J. K., 1989, On the relation between atmospheric ozone and sunspot number. *Journal of Climate*, **2**, 1404–1416.

ANGELL, J. K. and KORSHOVER, J., 1983, Global temperature variations in the troposphere and stratosphere, 1958–1982. *Monthly Weather Review*, **111**, 901–921.

ANGELL, J. K. and KORSHOVER, J., 1987, Variability in United States cloudiness and its relation to El Niño. *Journal of Climate and Applied Meteorology*, **26**, 580–584.

ANTHES, R. A., BRETHERTON, F. P. and SCHIMEL, D. C., 1992, Climate system modelling program. *Proceedings of the Third Symposium on Global Change Studies, held in New Orleans on 14–18 January 1991* (Boston: American Meteorological Society), pp. 1–16.

ANTHES, R. A., KUO, Y.-H., BAUMHERNER, D. P., ERRICO, R. M. and BETTGE, T. W., 1985, Predictability of mesoscale atmospheric motions. *Advances in Geophysics*, **28B**, 152–202.

ARDANUY, P. E., KYLE, K. E. and HOYT, D., 1992, Global interrelationships among the Earth's radiation budget, cloudiness, volcanic aerosols, and surface temperature. *Journal of Climate*, **5**, 1120–1135.

ARDANUY, P. E., STOWE, L. L., GRÜBER, A. and WEISS, M., 1991, Shortwave, longwave and net cloud-radiative forcing as determined from Nimbus-7 observations. *Journal of Geophysical Research*, **96**, 18537–18549.

ARDANUY, P. E., STOWE, L. L., GRÜBER, A., WEISS, M. and LONG, C. S., 1989, Longwave cloud radiative forcing as determined from Nimbus-7 observations. *Journal of Climate*, **1**, 766–799.

AUER, L. H. and KAO, C. Y. J., 1991, Further analysis of the global outgoing longwave radiative flux observed by Nimbus. *Journal of Geophysical Research*, **96**, 17367–17370.

AUSBEL, J. H., GRÜBER, A. and NAKICENOVIC, N., 1988, Carbon dioxide emissions in a methane economy. *Climate Change*, **12**, 245–263.

BACH, W., 1986, Nuclear war: the effects of smoke and dust on weather and climate. *Progress in Physical Geography*, **10**, 315–363.

BACH, W., 1987, Kann die Menschheit einen Atomkrieg überlegen? In *Verantwortung für den Frieden Naturwissenschaftler – Initiative*. Köln.

BACH, W., 1988, *Modelling the Climatic Effects of Trace Gases: Reduction Strategy and Options for a Low Risk Policy*. Preprint (Munster: University of Munster).

BACH, W., CRANE, A. J., BERGER, A. L. and LONGHETTO, A. (editors), 1983, *Carbon Dioxide: Current Views and Developments in Energy/Climate Research* (Dordrecht: Reidel).

BADER, D. C., MACCRACKEN, M. C. and MALONE, R. C., 1992, The Computer hardware, Advanced Mathematics and Model Physics (CIIAMMP) climate modeling program. *Proceedings of the Third Symposium on Global Change Studies, held in New Orleans on 14–18 January 1991* (Boston: American Meteorological Society), pp. 140–142.

BAKAN, S., CHLOND, A., CUBASCH, U., FEICHTER, J., GRAF, H., GRASSL, H., HASSELMANN, K., KIRCHNER, I., LATIF, M., ROECKNER, E., SAUSEN, R., SCHLESE, U., SCHRIEVER, D., SCHULT, I., SCHUMANN, U., SIELMANN, F. and WELME, W., 1991, Climate response to smoke from the burning oil wells in Kuwait. *Nature*, **351**, 367–371.

BALACHANDRAN, N. K. and RIND, D., 1995, Modelling the effects of UV variability and the QBO on the troposphere–troposphere system. Part I: The middle atmosphere. *Journal of Climate*, **8**, 2058–2079.

BALLING, R. C., 1992, *The Heated Debate: Greenhouse Predictions versus Climate Reality*. Pacific Research Institute, 195 pp.

BALLING, R. C. and IDSO, S. B., 1989, Historical temperature trends in the United States and the effect of urban population growth. *Journal of Geophysical Research*, **94**, 3359–3364.

BARNETT, T. P., 1986, Detection of changes in the global troposphere temperature field induced by greenhouse gases. *Journal of Geophysical Research*, **91**, 6659–6667.

BARNETT, T. P., DÜMENIL, L., SCHLESE, U., ROECKNER, E. and LATIF, M., 1989, The effect of Eurasian snow cover on regional and global climate variations. *Journal of Atmospheric Science*, **46**, 661–685.

BARNETT, T. P. and SCHLESINGER, M. E., 1987, Detecting changes in global climate induced by greenhouse gases. *Journal of Geophysical Research*, **92**, 14772–14780.

BARNSTON, A. G., LIVEZEY, R. E. and HALPERT, M. S., 1991, Modulation of Southern Oscillation–northern hemisphere mid-winter climate relationships by the QBO. *Journal of Climate*, **4**, 203–217.

BARRIE, L. A., 1986, Arctic air pollution: an overview of current knowledge. *Atmospheric Environment*, **20**, 643–664.

BARRIE, L. A., 1991, Chemical changes in the Arctic troposphere at polar sunrise. *International Conference on the Role of the Polar Regions in Global Change. Proceedings of a Conference held June 11–15, 1990, at the University of Alaska Fairbanks*, edited by G. WELLER, C. L. WILSON and B. A. B. SEVERIN (Fairbanks: Geophysical Institute and Center for Global Change and Arctic Systems Research), p. 673.

BARRIE, R. G. and CHORLEY, R. J., 1988, *Atmosphere, Weather and Climate* (New York: Routledge).

BARRON, E. J., 1991, EOS: climate and hydrology. *Proceedings of the Second Symposium on Global Change Studies, held in New Orleans on 14–18 January 1991 (Preprints)* (Boston: American Meteorological Society), pp. 2.8–2.10.

BASSETT, G. W. and LIN, Z., 1993, Breaking global temperature records after Mt. Pinatubo. *Climatic Change*, **25**, 179–184.

BATES, G. T., GIORGI, F. and HOSTETLER, S. W., 1992b, Regional climate modeling of the Great Lakes basin. *Proceedings of the Third Symposium on Global Change Studies, held in New Orleans on 14–18 January 1991* (Boston: American Meteorological Society), pp. 18–19.

BATES, G. T. and MEEHL, G. A., 1986, The effect of $CO_2$ concentration on the frequency of blocking in a general circulation model coupled to a simple mixed layer ocean model. *Monthly Weather Review*, **114**, 687–701.

Bates, T. S., Calhoun, J. A. and Quinn, P. K., 1992a, Variations in the methanolsulfonate to sulfate molar ratio in submicrometer marine aerosol particles over the South Pacific Ocean. *Journal of Geophysical Research*, **97**, 9859–9866.

Bauer, P. and Schluessel, P., 1993, Rainfall, total water, ice water and water vapor over sea from polarized microwave simulations and Special Sensor Microwave/Imager data. *Journal of Geophysical Research*, **98**, 20737–20760.

Belle, T. L. and Abdullah, A., 1985, Detecting global climate change predicted by climate models. *Third Conference on Climate Variation and Symposium on Contemporary Climate: 1850–2100, 9–10 January 1985, Los Angeles* (Boston: American Meteorological Society), pp. 89–90.

Bengtsson, L., 1985, Medium-range forecasting – the experience of ECMWF. *Bulletin of the American Meteorological Society*, **66**, 1133–1146.

Bengtsson, L., 1991, Advances and prospects in numerical weather prediction. *Quarterly Journal of the Royal Meteorological Society*, **117**, 855–902.

Bengtsson, L. and Shukla, J., 1988, Integration of space and in-situ observations to study global climate change. *Bulletin of the American Meteorological Society*, **69**, 1130–1143.

Berger, A. (editor), 1980, *Climatic Variations and Variability: Facts and Theory.* (Dordrecht: Reidel).

Berger, A., 1984, Hiver nucléaire ou les conséquences climatiques d'un conflit nucléaire généralisé. *Revue des Questions Scientifiques*, **155**, 461–493.

Berger, A., Gallee, H., Fichocet, T., Marsiat, T. and Tricot, C., 1990, Testing the astronomical theory with a coupled climate–ice sheet model. *Global and Planetary Change*, **3**, 125–141.

Berger, A., Loutre, M. F. and Tricot, C., 1993, Insolation and Earth's orbital periods. *Journal of Geophysical Research*, **98**, 10341–10362.

Berger, A. and Tricot, C., 1986, Global climatic changes and astronomical theory of paleoclimates. In *Earth Rotation: Solved and Unsolved problems*, edited by A. Cazanova (Dordrecht: Reidel), pp. 111–129.

Berger, W. H. and Labeyrie, L. D. (editors), 1987, *Abrupt Climatic Change: Evidence and Implications.* (Dordrecht: Reidel).

Bernard, H. W., 1993, *Global Warming Unchecked: Signs to Watch For* (Indiana University Press).

Bernhardt, K. and Mäder, C., 1987, A statistical evaluation of reports on significant weather events since the year 1000 A.D. *Zeitschrift für Meteorologie*, **37**, 120–130.

Bernhardt, K. H., 1987, Aufgaben der Klimadiagnostik in der Klimaforschung. *Gerlands Beiträge zur Geophysik*, **96**, 113–126.

Bess, T. D., Smith, G. L. and Charlock, T. P., 1989, A ten-year monthly data set of outgoing longwave radiation from Nimbus-6 and Nimbus-7 satellites. *Bulletin of the American Meteorological Society*, **70**, 480–491.

Betts, A. K. and Harshvardhan, 1987, Thermodynamic constraints on the cloud liquid water feedback in climate models. *Journal of Geophysical Research*, **92**, 8493–8489.

Bjerknes, J., 1969, Atmospheric teleconnections from the Equatorial Pacific, *Monthly Weather Review*, **97**, 163–172.

Blake, D. and Rowland, F., 1986, World-wide increase in tropospheric methane, 1978–1983. *Journal of Atmospheric Chemistry*, **4**, 43–62.

Blanchet, J. P., 1989, The effects of arctic aerosols on a general circulation model simulation: a sensitivity study. In *Aerosols and Climate*, edited by P. V. Hobbs and M. P. McCormick (Hampton, VA: Deepak), pp. 373–383.

Blanchet, J. P., 1991, The influence of Arctic haze and radiatively active trace gases on the Arctic climate. *International Conference on the Role of the Polar Regions in Global*

*Change. Proceedings of a Conference held June 11–15, 1990, at the University of Alaska Fairbanks*, edited by G. WELLER, C. L. WILSON and B. A. B. SEVERIN (Fairbanks: Geophysical Institute and Center for Global Change and Arctic Systems Research), p. 693.

BLANCHET, J. P. and LIST, R., 1987, On radiative effects of anthropogenic aerosol components in Arctic haze and snow. *Tellus*, **39B**, 293–317.

BLOOMFIELD, P., 1992, Trends in global temperature. *Climatic Change*, **21**, 1–16.

BLOOMFIELD, P. and NYCHKA, D., 1992, Climate spectra and detecting climate change. *Climatic Change*, **21**, 275–287.

BOARD ON ATMOSPHERIC SCIENCES AND CLIMATE (editor), 1989, *Ozone Depletion, Greenhouse Gases, and Climate Change. Proceedings of a Joint Symposium by the Board on Atmospheric Sciences and Climate Sciences, Mathematics and Resources, National Research Council* (Washington DC: National Academy Press).

BOER, G. J., ARPE, K., BLACKBURN, M., DÉQUÉ, M., GATES, W. L., HART, T. L., LE TREUT, H., ROECKNER, E., SHEININ, D. A., SIMMONDS, I., SMITH, R. N. B., TOKIOKA, T., WETHERALD, R. T. and WILLIAMSON, D., 1991, An intercomparison of the climate studies simulated by 14 atmospheric general circulation models. CAS/JSC Working Group on Numerical Experimentation, Report No. 15 *WMO/TD No. 425* (Geneva: World Meteorological Organization).

BOER, G. J., ARPE, K., BLACKBURN, M., DÉQUÉ, M., GATES, W. L., HART, T. L., LE TREUT, H., ROECKNER, E., SHEININ, D. A., SIMMONDS, I., SMITH, R. N. B., TOKIOKA, T., WETHERALD, R. T. and WILLIAMSON, D., 1992, Some results from an intercomparison of the climates simulated by 14 atmospheric general circulation models. *Journal of Geophysical Research*, **97**, 12771–12786.

BÖHME, W., 1989, Programme und Kooperationen in der Klimaforschung-Mittel zum Verständnis des Klimasystems als eine Voraussetzung für die Bewältung der Klimaproblems. *Zeitschrift für Meteorologie*, **39**, 185–192.

BOJKOV, R. D., 1984, Tropospheric ozone, its changes and possible radiative effects. WMO Special Environment Report No. 86 (Geneva: World Meteorological Organization).

BOLIN, B., DÖÖS, B., JAGER, J. and WARRICK, R. (editors), 1986, *The Greenhouse Effect, Climate Change, and Ecosystems* (Chichester: Wiley).

BOLLE, H. J., 1982, Aerosol research within the World Climate Research Programme. In *Light Absorption by Aerosol Particles* (Berlin: Spectrum), pp. 1–51.

BOLLE, H. J. and RASOOL, S. I. (editors), 1985, *Development of the Implementation Plan for the International Satellite Land-Surface Climatology Project (ISLSCP). World Climate Paper WMO/TD No. 46* (Geneva: World Meteorological Organization).

BOLLE, N., SEILER, W. and BOLIN, B., 1986, Other greenhouse gases and aerosols. In *The Greenhouse Effect, Climate Change, and Ecosystems*, edited by B. BOLIN, B. DÖÖS, J. JAGER and R. WARRICK (Chichester: Wiley), pp. 157–203.

BONAN, G. B., 1991, The sensitivity of ecosystem $CO_2$ flux in the boreal forests of interior Alaska to climatic parameters. *International Conference on the Role of the Polar Regions in Global Change. Proceedings of a Conference held June 11–15, 1990, at the University of Alaska Fairbanks*, edited by G. WELLER, C. L. WILSON and B. A. B. SEVERIN (Fairbanks: Geophysical Institute and Center for Global Change and Arctic Systems Research), pp. 391–395.

BONAN, G. B., POLLARD, D. and THOMPSON, S. L., 1993, Influence of subgrid-scale heterogeneity in leaf area index, stomatal resistance, and soil moisture on grid-scale land–atmosphere interactions. *Journal of Climate*, **6**, 1882–1897.

BONY, S., LE TREUT, H., DUVEL, J. P. and KANDEL, R. S., 1992, Satellite validation of GCM-simulated annual cycle of the earth radiation budget and cloud forcing. *Journal of Geophysical Research*, **97**, 18061–18081.

BORGER, G. R. and VERNEKAR, A. D., 1988, Validation of outgoing longwave radiation in a general circulation model. *Monthly Weather Review*, **116**, 106–119.

BORISENKOV, E. P. (editor), 1988, *Climate Oscillations during the Last Thousand Years* (Leningrad: Gidrometeoizdat).

BORISENKOV, E. P., KAROL, I. L. and MELESHKO, V. P., 1988, Results of MHO studies in the field of climate theory and atmospheric general circulation. In *Main Geophysical Observatory by the 70th Anniversary of the Soviet Power* (Leningrad: Gidrometeoizdat), pp. 18–37.

BORISENKOV, E. P. and KONDRATYEV, K. YA., 1988, *Carbon Cycle and Climate* (Leningrad: Gidrometeoizdat).

BORUCKI, W. J., POLACH, J. B., TUON, O. B., WOODWARD, H. T. and WIEDMAN, E. R., 1980, The influence of solar UV radiation on climate. *Proceedings of Conference on Ancient Sun, Fossil Fuel records, Earth, Moon and meteorites*, Boulder, CO, 513–522.

BORZENKOVA, I. I., VINNIKOV, K. YA., SPIRINA, L. P. and STEKHNOVSKII, D. I., 1976, Variation of northern hemisphere air temperature from 1881 to 1975. *Meteorology and Hydrology* (*Meteorologia i Gidrologia*), No. 7, 27–35.

BOSSERT, J. E., KAO, C. Y. J., ROADS, J. O., CHEN, S. C. and HEGOSHI, K., 1992, Regional-scale simulations of the western U.S. climate. *Proceedings of the Third Symposium on Global Change Studies, held in New Orleans on 14–18 January 1991* (Boston: American Meteorological Society), pp. 24–28.

BOUGER, B., 1995, Rôle de la végétation terrestre dans le cycle global du $CO_2$. *Comptes rendus des Séances de l'Académie d'Agriculture de France*, **81**, No. 5, 109–123.

BOVILLE, B., 1986, Detecting $CO_2$-induced temperature changes in the stratosphere. *Advances in Space Research*, **6**, 45–50.

BOWLING, S. A., 1991, Problems with the use of climatological data to detect climatic change at high latitudes. *International Conference on the Role of the Polar Regions in Global Change. Proceedings of a Conference held June 11–15, 1990, at the University of Alaska Fairbanks*, edited by G. WELLER, C. L. WILSON and B. A. B. SEVERIN (Fairbanks: Geophysical Institute and Center for Global Change and Arctic Systems Research), pp. 206–209.

BOZÓ, L. and PÁVLÖLGYI, T., 1988, A global model on the climatic effects of sulfur compounds and $CO_2$. *Idöjárás*, **92**, 210–221.

BRADLEY, R. S., 1988, The explosive volcanic eruption signal in northern hemisphere continental temperature records. *Climatic Change*, **12**, 221–243.

BRADLEY, R. S., DIAZ, H. F., EISCHEID, J. K., JONES, P. D., KELLY, P. D. and GOODESS, C. M., 1987, Precipitation fluctuations over northern hemisphere land areas since the mid-19th century. *Science*, **237**, 171–175.

BRADLEY, R. S. and KLEMIG, F. T., 1993, Recent changes in the North American arctic boundary layer in winter. *Journal of Geophysical Research*, **98**, 8851–8858.

BRASSEUR, G., 1993, The response of the middle atmosphere to long-term and short-term solar variability: a two-dimensional model. *Journal of Geophysical Research*, **98**, 23079–23090.

BRASSEUR, G. and DERUDDER, A., 1987, The potential impact on atmospheric ozone and temperature of increasing trace gas concentrations. *Journal of Geophysical Research*, **92**, 10902–10920.

BRASSEUR, G., DERUDDER, A., LEATING, G. M. and PITTS, M. C., 1987, Response of middle atmosphere to short-term solar ultraviolet variations. 2: Theory. *Journal of Geophysical Research*, **92**, 903–914.

BREITENBERGER, E. and WENDLER, G., 1991, A statistical study of atmosphere–sea ice interactions in the Southern Ocean. *International Conference on the Role of the Polar Regions in Global Change. Proceedings of a Conference held June 11–15, 1990, at the University of Alaska Fairbanks*, edited by G. WELLER, C. L. WILSON and B. A. B. SEVERIN (Fairbanks: Geophysical Institute and Center for Global Change and Arctic Systems Research), p. 320.

BRIEGLEB, B. P., 1992, Longwave band model for thermal radiation in climate studies.

*Journal of Geophysical Research*, **97**, 11475–11485.

Brinkmann, W. A. R., 1976, Surface temperature trend for the northern hemisphere – updated. *Quaternary Research*, **6**, 355–358.

Broccoli, A. J. and Manabe, S., 1991, Will global warming increase the frequency of tropical cyclones? *Proceedings of the Fifth Conference on Climate Variations* (Boston: American Meteorological Society), p. 46.

Broecker, W. S., 1995, Chaotic climate., *Scientific American*, **275**, No. 5, 44–50.

Bromwich, D. H. and Robasky, F. M., 1993, Recent precipitation trends over the polar ice sheets. *Meteorology and Atmospheric Physics*, **51**, 259–273.

Browning, K. A., 1990, Rain, rainclouds and climate, *Quarterly Journal of the Royal Meteorological Society*, **116**, 1025–1052.

Brühl, C. and Crutzen, P., 1987, Scenarios of possible changes in atmospheric temperatures and ozone concentrations due to man's activities, estimated with a 1-D coupled photochemical climate model. *Climate Dynamics*, **2**, 173–203.

Bryan, K., Manabe, S. and Spelman, M. J., 1988, Interhemispheric asymmetry in the transient response of a coupled ocean–atmosphere model to a $CO_2$ forcing. *Journal of Physical Oceanography*, **18**, 851–967.

Bryan, K. and Spelman, M. J., 1985, The ocean's response to a $CO_2$-induced warming. *Journal of Geophysical Research*, **90**, 11679–11688.

Bryson, R. A. and Goodman, B. M., 1982, The climate effect of explosive volcanic activity: analysis of the historical data. *Proceedings of the Symposium on Atmospheric Effects and Potential Climate Impact of the 1980 Eruptions of Mount St. Helens, NASA Conference Publications*, **2240**, 191–202.

Budd, W. F. and Simmonds, I., 1991, The impact of global warming on the Antarctic mass balance and global sea level. *International Conference on the Role of the Polar Regions in Global Change. Proceedings of a Conference held June 11–15, 1990, at the University of Alaska Fairbanks*, edited by G. Weller, C. L. Wilson and B. A. B. Severin (Fairbanks: Geophysical Institute and Center for Global Change and Arctic Systems Research), pp. 489–494.

Budyko, M. I., 1985, The climate of the future. In *Marx–Lenin Concept of the Current Global Problems* (Moscow: Mysl), pp. 265–273.

Budyko, M. I., 1989a, Ways of anthropogenic climate change. *Meteorology and Hydrology* (*Meteorologia i Gidrologia*), No. 9, 11–21.

Budyko, M. I., 1989b, An empirical assessment of the forthcoming climate change. *Meteorology and Hydrology* (*Meteorologia i Gidrologia*), No. 10, 5–14.

Budyko, M. I. and Groisman, P. Ya., 1989, The warming in the 1980s. *Meteorology and Hydrology* (*Meteorologia i Gidrologia*), No. 3, 5–10.

Budyko, M. I. and Groisman, P. Ya., 1991, Expected climate changes by the year 2000. *Meteorology and Hydrology* (*Meteorologia i Gidrologia*), No. 4, 84–94.

Budyko, M. I., Israel, Yu. A. and Yanshin, A. L., 1991, Global warming and its consequences. *Meteorology and Hydrology* (*Meteorologia i Gidrologia*), No. 12, 5–10.

Buriez, J. C., Bonnel, B., Fouquart, Y., Geleyn, J. F. and Morcrette, J. J., 1988, Comparison of model-generated and satellite-derived cloud cover and radiation budget. *Journal of Geophysical Research*, **93**, 3705–3719.

Callendar, G. S., 1961, Temperature fluctuations and trends over the Earth. *Quarterly Journal of the Royal Meteorological Society*, **87**, 1–12.

Campbell, I. M., 1986, *Energy and Atmosphere: A Physico-Chemical Approach.* (Chichester: Wiley).

Carbon Dioxide Assessment Committee, 1983, *Changing Climate. Report of the Carbon Dioxide Assessment Committee* (Washington DC: National Academic Press).

CAS, 1977, *Report of the Informal Meeting of Experts on the Role of Sea Ice in the Climate System, 24–28 October 1977* (Geneva: WMO).

Cess, R. D., 1985, Nuclear war: illustrative effects of atmospheric smoke and dust upon solar radiation. *Climatic Change*, **7**, 237–252.

Cess, R. D., Nemesure, S., Dutton, E. G., DeLuisi, J. J., Potter, G. L. and Morcrette, J. J., 1993, The impact of clouds on the shortwave radiation budget of the surface atmosphere system: interfacing measurements and models. *Journal of Climate*, **6**, 308–316.

Cess, R. D. and Potter, G. L., 1987, Exploratory studies of cloud radiative forcing with a general circulation model. *Tellus*, **39A**, 460–473.

Cess, R. D. and Potter, G. L., 1988, A methodology for understanding and intercomparing atmospheric climate feedback processes in general circulation models. *Journal of Geophysical Research*, **93**, 8305–8314.

Cess, R. D., Potter, G. L., Blanchet, J. P., Boer, G. J., Ghan, D. J., Kiehl, J. T., Le Treut, H., Li, Z. X., Liang, X. Z., Mitchell, J. F. B., Morcrette, J. J., Randall, D. A., Riches, M. R., Roeckner, E., Schlese, U., Slingo, A., Taylor, K. E., Washington, W. M., Wetherald, R. T. and Yagai, I., 1989, Interpretation of cloud–climate feedbacks as produced by 14 atmospheric general circulation models. *Science*, **245**, 513–515.

Cess, R. D., Potter, G. L., Blanchet, J. P., Boer, G. J., Delvenio, A. D., Déqué, M., Dymnikov, V., Galin, V., Gates, W. L., Chan, S. J., Kiehl, J. T., Lacis, A. A., Le Treut, H., Li, Z. X., Liang, X. Z., McAvaney, B. J., Meleshko, V. P., Mitchell, J. F. B., Morcrette, J. J., Randall, D. A., Rikus, L., Roeckner, E., Royer, J. F., Schlese, U., Sheinin, D. A., Slingo, A., Sokolov, A. P., Taylor, K. E., Washington, W. M., Wetherald, R. T., Yagai, I. and Zhang, M. H., 1990, Intercomparison and interpretation of climate feedback processes in 19 atmospheric general circulation models. *Journal of Geophysical Research*, **95**, 16601–16616.

Cess, R. D., Potter, G. L., Gates, W. L., Morcrette, J. J. and Corsetti, L., 1992, Comparison of General Circulation Models to Earth Radiation Budget Experimental data: computation of clear-sky fluxes, *Journal of Geophysical Research*, **97**, 20421–20426.

Chahine, M. T., 1992, Global Energy and Water Cycle Experiment (GEWEX). *Proceedings of the Third Symposium on Global Change Studies, held in New Orleans on 14–18 January 1991* (Boston: American Meteorological Society), pp. 7–9.

Chamberlain, J. W. and Hunten, D. M., 1987, *Theory of Planetary Atmospheres. An Introduction to Their Physics and Chemistry*. Second Edition (San Diego: Academic Press).

Charles, C. D., Rind, D., Jouzel, J., Koster, R. D. and Fairbanks, R. G., 1994, Glacial–interglacial changes in moisture sources for Greenland: influences on the ice record on climate. *Science*, **263**, 508–511.

Charlock, T. P., 1983, The effects of volcanic aerosols on the thermal infrared budget of the lower stratosphere. *Proceedings of the Fifth Conference on Atmospheric Radiation* (Boston: American Meteorological Society), pp. 350–353.

Charlock, T. P., Cattani-Carnes, K. M. and Rose, F., 1988, Fluctuation statistics of outgoing longwave radiation in a general circulation model and in satellite data. *Monthly Weather Review*, **116**, 1540–1554.

Charlock, T. P. and Ramanathan, V., 1985, The albedo field and cloud radiative forcing produced by a general circulation model with internally generated cloud optics. *Journal of Atmospheric Science*, **42**, 1408–1429.

Charlock, T. P., Rose, F. G. and Cattani-Carnes, K. M., 1989, Cross correlations between the radiation and atmospheric variables in a general circulation model and in satellite data. *Monthly Weather Review*, **117**, 212–220.

Charlson, R. J., Langner, J., Rodhe, H., Leovy, C. B. and Warren, S. G., 1991, Perturbation of the northern hemisphere radiative balance by backscattering from anthropogenic sulfate aerosols. *Tellus*, **43A–B**, 152–163.

CHARLSON, R. J., LOVELOCK, J., ANDREAE, M. and WARREN, S., 1987, Oceanic phytoplankton, atmospheric sulfur, cloud albedo, and climate. *Nature*, **326**, 655–661.

CHARLSON, R. J., SCHWARTZ, S. E., HALES, J. M., CESS, R. D., COAKLEY, J. A., HANSEN, J. A. and HOFMANN, D. E., 1992, Climate forcing by anthropogenic aerosols. *Nature*, **255**, 423–430.

CHARNEY, J. G., QUIRK, W. J., CHOW, S. M. and KORNFIELD, J., 1977, A comparative study of the effects of albedo change on drought in semi-arid regions. *Journal of Atmospheric Science*, **34**, 1366–1385.

CHEN, R. S., 1982, *Combined Land/Sea Surface Air Temperature Trends, 1959–1972.* Report no. COO-2195-42, Department of Meteorology and Physical Oceanography, Massachussetts Institute of Technology, Cambridge.

CHEN, R. S., 1985, Surface air temperature over land and ocean, 1949–1972. *Extended Summaries of the Third Conference on Climate Variations and Symposium on Contemporary Climate: 1850–2100* (Boston: American Meteorological Society), pp. 39–40.

CHERTOCK, B., FROUIN, R. and SOMERVILLE, C. J., 1991, Global monitoring of net solar irradiance at the ocean surface: climatological variability and the 1982–1983 El Niño. *Journal of Climate*, **4**, 639–650.

CHERTOCK, B. and SUD, Y. C., 1993, Comparison of ocean surface solar irradiance in the GLA General Circulation Model and satellite-based calculations. *Journal of Climate*, **6**, 560–567.

CHERUY, G., KANDEL, R. S. and DUVEL, J. P., 1991a, Outgoing longwave radiation and its diurnal variation from combined ERBE and Meteosat observations. 1: Estimating OLR from Meteosat data. *Journal of Geophysical Research*, **96**, 22611–22622.

CHERUY, G., KANDEL, R. S. and DUVEL, J. P., 1991b, Outgoing longwave radiation and its diurnal variation from combined ERBE and Meteosat observations. 2: Using Meteosat data to determine the longwave diurnal cycle. *Journal of Geophysical Research*, **96**, 22623–22630.

CHRISTY, J. R., 1993, Impacts on global temperatures due to Mt. Pinatubo and the 91–92 ENSO. *Proceedings of the Seventeenth Annual Climate Diagnostic Workshop.* (Washington DC: NOAA), pp. 13–17.

CHRISTY, J. R., MCNIDER, R. T. and ROBERTSON, F. R., 1991, Comparison of MSU and CCM1 tropospheric temperatures for 1979–1986. *Proceedings of the Fifth Conference on Climate Variations* (Boston: American Meteorological Society), pp. 28–31.

CHRISTY, J. R. and TRENBERTH, K. E., 1991, Monitoring global monthly mean surface temperatures. *Proceedings of the Fifth Conference on Climate Variations* (Boston: American Meteorological Society), pp. 127–129.

CHYLEK, P. and RAMASWAMY, V., 1984, Effect of graphitic carbon on the albedo of clouds. *Journal of Atmospheric Science*, **41**, 3076–3094.

CHYLEK, P., RAMASWAMY, V. and SRIVASTAVA, V., 1983, Albedo of soot-contaminated snow. *Journal of Geophysical Research*, **88**, 10837–10843.

CHYLEK, P., SRIVASTAVA, V., CAHENZLI, L., PINNICK, R. G., DOD, R. L., NOVAKOV, T., COOK, T. L. and HINDS, B. D., 1987, Aerosol and graphitic carbon content of snow. *Journal of Geophysical Research*, **92**, 9801–9810.

CICERONE, R. J. and OREMLAND, R. S., 1988, Biogeochemical aspects of atmospheric methane. *Global Biogeochemical Cycles*, **2**, 299–327.

CLARK, W. C. (editor), 1982, *Carbon Dioxide Review: 1982* (New York: Oxford University Press).

CLARK, W. C., 1983, The carbon dioxide review – guest editorial. *Climatic Change*, **5**, 3–6.

CLARK, W. C., 1985, Scales of climate impacts. *Climatic Change*, **7**, 5–27.

CLARK, W. C. and MUNN, R. E. (editors), 1986, *Sustainable Development of the Biosphere.* (Cambridge: Cambridge University Press).

CLAUSSEN, M. and ESCH, M., 1994, Biomes computed from simulated climatologies. *Climate Dynamics*, **9**, 235–244.

CLEBAUX, C., COLIN, R., SIMON, P. C. and GRANIER, C., 1993, Infrared cross sections and global warming potential of 10 alternative hydrohalocarbons. *Journal of Geophysical Research*, **98**, 10451–10497.

CLIVAR, 1995, A study of climate variability and predictability. *Science Plan. World Climate Paper WMO/TD No. 690* (Geneva: World Meteorological Organization).

COAKLEY, J. A. and CESS, R. D., 1985, Response of the NCAR community climate model to the radiative forcing by the naturally occurring tropospheric aerosol. *Journal of Atmospheric Science*, **42**, 1677–1692.

COHEN, J. and RIND, D., 1991, The effect of snow cover on the climate. *Journal of Climate*, **4**, 689–706.

COLMAN, R. A., MCAVENEY, B. J., FRASER, J. R., RIKUS, L. S. and DAHNI, R. R., 1993, Snow and cloud feedbacks modelled by an atmospheric general circulation model. *Climate Dynamics*, **9**, 253–265.

COMMITTEE ON ATMOSPHERIC SCIENCES (CAS), 1977, *Report of the Informal Meeting of Experts on the Role of Sea Ice in the Climate System* (Geneva: World Meteorological Organization).

COMMITTEE ON ATMOSPHERIC SCIENCES (CAS), 1980, *The Atmospheric Sciences: National Objectives for the 1980s* (Washington DC: National Academy of Sciences).

COSPAR, 1978, Status of satellite observing possibilities for studies of climate physical processes. Report of the COSPAR WG-6, Boulder, CO, 113 pp.

COVEY, C., 1987, Environmental studies of nuclear war: a recent synthesis and future prospects – an editorial/review essay. *Climate Change*, **10**, 1–10.

COVEY, C., 1988, Atmospheric and oceanic heat transport: simulations versus observations. *Climatic Change*, **13**, 149–160.

COVEY, C. and BARRON, E., 1988, The role of ocean heat transport in climatic change. *Earth Science Reviews*, **24**, 429–445.

COVEY, C., TAYLOR, K. E. and DICKINSON, R. E., 1991, Upper limit for sea ice albedo feedback contribution to global warming. *Journal of Geophysical Research*, **96**, 9169–9174.

CRACKNELL, A. P., 1994a, Climate change – the background. In *Remote Sensing and Global Climate Change*, edited by R. A. VAUGHAN and A. P. CRACKNELL (Berlin: Springer), pp. 1–33.

CRACKNELL, A. P., 1944b, Basis and structure of climate models. In *Remote Sensing and Global Climate Change*, edited by R. A. VAUGHAN and A. P. CRACKNELL (Berlin: Springer), pp. 135–176.

CRACKNELL, A. P. and HAYES, L. W. B., 1991, *Introduction to Remote Sensing* (London: Taylor and Francis).

CRANE, R. G., 1986, Remote sensing and polar climate. *Earth and Mineralogical Science*, **55**, 38–41.

CROLL, J., 1867, On the change in the obliquity of the ecliptic, its influence on the climate of the polar regions and on the level of the sea. *Philosophical Magazine*, **33**, 426–445.

CROMMELYNCK, D., 1994, Status of our knowledge of the solar constant after the "Atlas-1 and -2" and EURERA missions. *Studies of the Earth from Space* (*Issledovania Zemli iz Kosmosa*), No. 5, 18–21.

CROWLEY, T. J., 1990, Are there any satisfactory analogs for a future greenhouse warming? *Journal of Climate*, **3**, 1282–1292.

CROWLEY, T. J., 1993, Geological assessment of the greenhouse effect. *Bulletin of the American Meteorological Society*, **74**, 2363–2374.

CROWLEY, T. J. and BAUM, S. K., 1994, General circulation model study of late carboniferous interglacial climates. *Paleoclimates Data and Modelling*, **1**, 3–22.

CROWLEY, T. J., CRISTE, T. A. and SMITH, N. R., 1993, Reassessment of Crête (Greenland) ice core acidity/volcanism link to climate change. *Geophysics Research Letters*, **20**, 209–212.

Crowley, T. J. and North, G. R., 1988, Abrupt climate change and extinction events in Earth history. *Science*, **240**, 996–1002.

Crowley, T. J. and North, G. R., 1991, *Paleoclimatology* (New York: Oxford University Press).

Crowley, T. J. and Parkinson, C. L., 1988, Late Pleistocene variations in Antarctic sea ice. I: Effect of orbital isolation changes. *Climate Dynamics*, **3**, 85–92.

Crutzen, P. J., 1985, The global environment after nuclear war. *Environment*, **27**, 6–11, 34–37.

Crutzen, P. J. and Birks, J. W., 1982, The atmosphere after a nuclcar war: twilight at noon. *Ambio*, **11**, 114–125.

Crutzen, P. and Hahn, J., 1986, *Schwarzer Himmel, Auswirkungen einer Atomkrieges auf Klima und Globale Umwelt* (Frankfurt: Fischer Verlag).

Cubasch, K., Hasselmann, K., Höck, H., Maier-Reimer, E., Mikolajewicz, U., Santer, B. D. and Sausen, R., 1991, Climate change prediction with a coupled ocean–atmosphere model. *Proceedings of the Fifth Conference on Climate Variations* (Boston: American Meteorological Society), pp. 525–528.

Cubasch, V., Hasselmann, K., Höck, H., Maier-Reimer, E., Mikolajewicz, U., Santer, B. D. and Sausen, R., 1992, Time-dependent greenhouse warming computations with a coupled ocean–atmosphere model. *Climate Dynamics*, **8**, 55–69.

Cullen, M. J. P., 1993, The unified forecast/climate model. *Meteorological Magazine*, **122**, 81–94.

Curran, R. J., 1987, NASA's plans to observe the Earth's atmosphere with lidar. *IGARSS '87: International Geoscience and Remote Sensing Symposium held in Ann Arbor, MI, on 18–21 May, 1987* (Washington D.C.: I.E.E.E.), pp. 379–380.

Curry, J. A., 1986, Interactions among turbulence, radiation, and microphysics in Arctic stratus clouds. *Journal of Atmospheric Science*, **43**, 90–106.

Curry, J. A. and Ebert, E. E., 1989, Sensitivity of the thickness of Arctic sea ice to the optical properties of clouds. *Annals of Glaciology*, **14**, 43–46.

Curry, J. A., Herman, G. F. and Ebert, E. E., 1988, Mean and turbulent structure of the summertime Arctic cloudy boundary layer. *Quarterly Journal of the Royal Meteorological Society*, **114**, 715–746.

Curry, J. A., Schramm, J. L. and Ebert, E. E., 1993, Impact of clouds on the surface radiation balance of the Arctic ocean. *Meteorology and Atmospheric Physics*, **51**, 197–218.

d'Almeida, G. A., Koepke, P. and Shettle, E., 1991, *Atmospheric Aerosols: Global Climatology and Radiative Characteristics* (Hampton, VA: Deepak).

Damon, P. E. and Kunen, S. M., 1976, Global cooling? *Science*, **193**, 447–453.

Darnell, W. L., Staylor, W. F., Gupta, S. K., Ritchey, N. A. and Wilber, A. C., 1992, Seasonal variation of surface radiation budget derived from International Satellite Cloud Climatology Project C1 data. *Journal of Geophysical Research*, **97**, 15741–15760.

D'Arrigo, R. D. and Jacoby, G. C., 1993, Secular trends in high Northern latitude temperature reconstructions based on tree rings. *Climatic Change*, **25**, 163–168.

Deepak, A. (editor), 1992, *Atmospheric Aerosols: Their formation, Optical Properties and Effects* (Hampton, VA: Spectrum).

Deepak, A. and Gerber, H. F. (editors), 1983, Report of the experts' meeting on aerosols and their climatic effects, Williamsburg, Va, 23–30 March 1983, World Climate Paper No. 55 (Geneva, World Meteorological Organisation).

De Foor, T. E., Robinson, E. and Ryan, S., 1992, Early lidar observations of June 1991 Pinatubo eruption plume at Mauna Loa Observatory, Hawaii. *Geophysical Research Letters*, **19**, 187–190.

Delworth, T., Manabe, S. and Stouffer, R. J., 1993, Interdecadal variations of the thermohaline circulation in a coupled ocean–atmosphere model. *Journal of Climate*, **6**,

1993–2011.

DESCHLER, T., HOFMANN, D. J., JOHNSON, B. J. and ROZIER, W. R., 1992, Balloon-borne measurements of the Pinatubo aerosol size distribution and volatility at Laramie, Wyoming during the summer of 1991. *Geophysical Research Letters*, **19**, 199–202.

DESER, C. and BLACKMAN, M. L., 1993, Surface climate variations over the North Atlantic Ocean during winter: 1900–1989. *Journal of Climate*, **6**, 1743–1753.

DESER, C. and WALLACE, J. M., 1987, El Niño events and their relation to the Southern Oscillation: 1925–1986. *Journal of Geophysical Research*, **92**, 14189–14196.

DETWILLER, R. P. and HALL, C. A., 1988, Tropical forests and the global carbon cycle. *Science*, **239**, 42–47.

DEWEY, K. F., 1987, Satellite maps of snow cover frequency for the northern hemisphere. *Journal of Climate and Applied Meteorology*, **26**, 1210–1219.

DEWEY, K. F. and HEIM, R. JR., 1982, A digital archive of northern hemisphere snow cover, November 1966 through December 1980. *Bulletin of the American Meteorological Society*, **63**, 1132–1141.

DIAZ, H. F., 1991, Changes in regional-scale precipitation for global land areas during the past century. *Proceedings of the Fifth Conference on Climate Variations* (Boston: American Meteorological Society), pp. 35–37.

DIAZ, H. F., BRADLEY, R. S. and EISCHEID, J. K., 1989, Precipitation fluctuations over global land areas since the late 1800s. *Journal of Geophysical Research*, **94**, 1195–1210.

DICKINSON, R. E., 1982, Modelling climate changes due to carbon dioxide increases. In *Carbon Dioxide Review* (New York: Oxford University Press), pp. 101–133.

DICKINSON, R. E., 1984, Modelling evapotranspiration for three-dimensional global climate models. In *Climate Processes and Climate Sensitivity. Geophysical Monograph, Volume* 5, edited by American Geophysical Union, pp. 58–72.

DICKINSON, R. E., 1985, Status of the formulation of the land-surface hydrological processes in GCMs. *World Climate Papers*, No. 96, Appendix B.

DICKINSON, R. E., 1986a, Global climate and its connections to the biosphere. In *Climate–Vegetation Interactions*, edited by C. ROSENSWEIG and R. DICKINSON, Report OIES-2 (Boulder, CO: University Cooperation of Atmospheric Research), pp. 5–8.

DICKINSON, R. E., 1986b, Impact of human activities on climate – a framework. In *Sustainable Development of the Biosphere* (Cambridge: Cambridge University Press), pp. 253–291.

DICKINSON, R. E., 1986c, The climate system and modelling of future climate. In *The Greenhouse Effect, Climatic Change and Ecosystems*, edited by B. BOLIN, B. DÖÖS, J. JAGER and R. WARRICK (Chichester: Wiley), pp. 207–270.

DICKINSON, R. E., 1987, Evapotranspiration in global climate model. *Advances in Space Research*, **7**, 17–26.

DICKINSON, R. E., 1989, Uncertainties of estimates of climatic change: a review. *Climatic Change*, **15**, 5–14.

DICKINSON, R. E. and CICERONE, R. J., 1986, Future global warming from atmospheric trace gases. *Nature*, **319**, 109–115.

DICKINSON, R. E. and HANSON, B., 1984, Vegetation-albedo feedbacks. In *Climate Processes and Climate Sensitivity*, edited by J. HANSEN and T. TAKAHASHI, Geophysical Monograph 29, Maurice Ewing Volume 5 (Washington: American Geophysical Union), pp. 180–186.

DICKINSON, R. E., MEEHL, G. A. and WASHINGTON, W. M., 1987, Ice-albedo feedback in a $CO_2$-doubling simulation. *Climatic Change*, **10**, 241–248.

DIXON, R. K., BROWN, S., HOUGHTON, R. A., SOLOMON, A. M., TREXLER, M. C. and WISNEWSKI, H., 1994, Carbon pools and flux of global forest ecosystem. *Science*, **263**, 185–190.

DOE, 1995, *Cloud–Radiation Feedbacks: The Current State of the Science. Elements of Change*. US Department of Energy DOE/ER-066 OT, UC-402. Washington, DC, 122 pp.

DORAN, J. C., BARNES, F. J., COULTER, R. L. and CRAWFORD, T. L., 1992, A field study of the effects of inhomogeneities of surface sensible and latent heat fluxes. *Proceedings of the Third Symposium on Global Change Studies* (Boston: American Meteorological Society), pp. 73–78.

DOZIER, J., 1992, Editor's corner. *The Earth Observer*, **4**, 1–7.

DRUYAN, L. M. and HASTENRATH, S., 1991, Modelling the differential impact of 1984 and 1950 sea-surface temperatures on Sahel rainfall. *International Journal of Climatology*, **11**, 367–380.

DUTTON, E. G., 1990, Comments on 'Major volcanic eruptions and climate: a critical evaluation'. *Journal of Climate*, **3**, 587–588 (see Mass and Portman 1989); reply, MASS, C. F. and PORTMAN, D. A. *ibid*, **3**, 589–590.

DUVEL, J. P. and BRÉON, F. M., 1991, The clear-sky greenhouse effect sensitivity to a sea surface temperature change. *Journal of Climate*, **4**, 1162–1169.

DUVEL, J. P. and KANDEL, R. S., 1988, Tropospheric response to tropical diurnal forcing observed by Meteosat. *Geophysical Research Letters*, **15**, 788–791.

EBERT, E. E. and CURRY, J. A., 1992, A parametrization of ice cloud optical properties for climate models. *Journal of Geophysical Research*, **97**, 3831–3836.

EDDY, J. A., 1977, Climate and the changing sun, *Climatic Change*, **1**, 173–190.

EHHALT, D. H., 1986, Atmospheric trace gases: interaction with the biosphere. In *Climate–Vegetation Interactions*, edited by C. RUSENZWEIG and R. DICKINSON. Report OIES-2 (Boulder, CO: UCAR).

ELLINGSON, R. G. and FOUQUART, Y., 1991, The intercomparison of radiation codes in climate models: An overview. *Journal of Geophysical Research*, **96**, 8925–8927.

ELLIOTT, W. P. and ANGELL, J. K., 1987, The relation between Indian monsoon rainfall, the southern oscillation, and hemispheric air and sea temperature: 1884–1984. *Journal of Climate and Applied Meteorology*, **26**, 943–948.

ELLIS, J. S. and VONDERHAAR, T. H., 1976, *Zonal Average Earth Radiation Budget Measurements from Satellites for Climate Studies.* Atmospheric Science Report No. 240, Department of Atmospheric Science, Colorado State University, Fort Collins.

ELLSAESSER, H. W., 1985, Do the recorded data of the last century indicate a $CO_2$ warming? *Extended Abstracts of the 3rd Conference on Climatic Variation and Symposium on Contemporary Climate: 1850–2100, held in Los Angeles on 8–11 January 1985* (Boston: American Meteorological Society), pp. 87–88.

ELLSAESSER, H. W., 1987a, Interpretation of our present terrestrial climatic record. UCRL Preprint 97354, LLNL, Livermore, CA.

ELLSAESSER, H. W., 1987b, The Sahelian drought – first evidence of a $CO_2$-induced climatic change? *Proceedings of the 11th Annual Climate Diagnostics Workshop* (Washington DC: Climate Analysis Center, US Dept. of Commerce), pp. 201–204.

ELLSAESSER, H. W., 1988, Table for Mike Riches. 9 February 1988. Private communication, 12 pp.

ELLSAESSER, H. W., 1989, A different view of the climatic effect of $CO_2$ – updated. Preprint UCRL-87940, August.

ELLSAESSER, H. W., 1991, Comment on 'The global warming debate heats up: an analysis and perspective' by S. H. Schneider. *Bulletin of the American Meteorological Society*, **72**, 1009.

ELLSAESSER, H. W., MACCRACKEN, M. C., WALTON, J. J. and GROTCH, S. L., 1986, Global climatic trends as revealed by the recorded data. *Review of Geophysics*, **24**, 745–792.

ENGARDT, M. and RODHE, H., 1993, A comparison between patterns of temperature trends and sulfate aerosol pollution. *Geophysical Research Letters*, **20**, 117–120.

ERICKSON, D. J. and EATON, R. E., 1993, Global biodynamical cycling estimates with CZCS satellite data and general circulation models. *Geophysical Research Letters*, **20**, 683–686.

Esser, G., 1987, Sensitivity of global carbon pools and fluxes to human and potential climatic impacts. *Tellus*, **39B**, 245–260.

Fegley, B. and Prinn, R. G., 1988, Oxygen production by large impacts in the Earth's early atmosphere. *Proceedings of the Conference on Origins of the Earth held in Berkeley, California, on 1–3 December 1988* (Houston: Lunar and Planetary Institute, NASA), pp. 18–19.

Finlayson-Pitts, B. J. and Pitts, J. N., 1986, *Atmospheric Chemistry: Fundamentals and Experimental Techniques, Volume* 4 (New York: Wiley).

Fischer, G. (editor), 1989, *Climate Simulations with the ECMWF T21 Model in Hamburg. Part III: Diagnosis of Response Experiments.* Model Report No. 7, Meteorological Institute of Large Scale Atmospheres, Stuttgart.

Fletcher, J. O., 1991, How climate changes. *International Conference on the Role of the Polar Regions in Global Change. Proceedings of a Conference held June 11–15, 1990, at the University of Alaska Fairbanks*, edited by G. Weller, C. L. Wilson and B. A. B. Severin (Fairbanks: Geophysical Institute and Center for Global Change and Arctic Systems Research), p. 149.

Flöhn, H., 1985, *Das Problem der Klimaänderungen in Vergangenheit und Zukunft.* (Darmstadt: Wissenschaftliche Buchgesellschaft).

Flöhn, H., 1986, Singular events and catastrophes now and in climatic history. *Naturwissenschaften*, **73**, 123–149.

Flöhn, H., 1987, Air–sea interaction processes as models for abrupt climatic changes. In *Abrupt Climatic Change*, edited by W. H. Berger and L. D. Labeyrie (Dordrecht: Reidel), pp. 23–30.

Flöhn, H., 1988, Treibhauseffekt und Klimaentwicklung. In *Atel-Mitteilungen*, pp. 4–10.

Flöhn, H., 1989a, Ändert sich unser Klima? *Mannheimer Forum 88/89* (Mannheim: von Boehringer), pp. 135–189.

Flöhn, H., 1989b, Wo bleibt das Erwärmungsignal? Das $CO_2$ Klima-problem in globaler Sicht. *Die Geowissenschaften*, **7**, 31–37.

Flöhn, H., 1993a, Climate evolution during the last millenium: what can we learn from it? In *Global Change in the Perspective of the Past*, edited by J. A. Eddy and H. Oeschger (Chichester: Wiley), pp. 295–316.

Flöhn, H., 1993b, Physical 3D-climatology from Hann to the satellite era. *Geophysical Monographs No. 75, IUGG Volume 15, American Geophysical Union*, 9–20.

Flöhn, H. and Kapala, A., 1989, Changes of tropical sea air interaction processes over a 30-yr period. *Nature*, **338**, 244–246.

Flöhn, H., Kapala, A., Knoche, H. R. and Mächel, H., 1992, Water vapour as an amplifier of the greenhouse effect: new aspects. *Meteorologische Zeitschrift*, **1**, 1–13.

Foley, J. A., Taylor, K. E. and Ghan, S. J., 1991, Planktonic dimethylsulfide and cloud albedo: An estimate of the feedback response. *Climatic Change*, **18**, 1–15.

Folland, C. K. and Kates, F., 1984, Changes in decadally averaged sea surface temperature over the world, 1861–1980. In *Milankovitch and Climate*, edited by A. Berger, J. Imbrie, J. Hays, G. Kukla and B. Saltzman, part II (Hingham, Mass.: Reidel), pp 721–727.

Folland, C. K. and Parker, D. E., 1989, Comparison of corrected sea surface and air temperature for the globe and the hemispheres, 1956–1988. *Proceedings of the 13th Annual Climate Diagnostics Workshop* (Washington DC: US Department of Commerce), pp. 160–165.

Folland, C. K., Parker, D. E. and Kates, F. E., 1984, Worldwide marine temperature fluctuations 1856–1981. *Nature*, **310**, 670–673.

Foster, J. L. and Rango, A., 1982, Snow cover conditions in the northern hemisphere during the winter of 1981. *Journal of Climatology*, **2**, 171–183.

Frankignoul, C. and Molin, A., 1988, Response of the GISS general circulation model to a midlatitude sea surface temperature anomaly in the North Pacific. *Journal of Atmospheric Science*, **45**, 95–108.

FRIED, A., HERY, B. and REGAZZI, R. A., 1992, Measurements of carbonyl sulfide in automotive emissions and an assessment of its importance to the global sulfur cycle. *Journal of Geophysical Research*, **97**, 14621–14634.

FRITTS, H. C. and LOUGH, J. M., 1985, An estimate of average annual temperature variations for N. America. 1602 to 1961. *Climatic Change*, **7**, 203–224.

FRÖHLICH, C., 1982, Variability of solar 'constant' on time scales from minutes to years. *Journal of Geophysical Research*, **92**, 792–800.

FRÖHLICH, C., 1994, Review of space observations of total solar irradiance. In *The Sun as a Variable Star; Solar and Stellar Irradiance Variations*, edited by J. M. PAPP, C. FRÖHLICH, H. S. HUDSON and S. K. SOLANKI (Cambridge: Cambridge University Press), pp. 38–51.

GAFFIN, S. R., HOFFERT, M. I. and VOLK, T., 1986, Nonlinear coupling between surface temperature and ocean upwelling as an agent in historical climate variations. *Journal of Geophysical Research*, **91**, 3944–3950.

GALINDO, I., 1965, Turbidiometric estimations in Mexico City using the Volz sun photometer. *Pure and Applied Geophysics*, **60**, 189–196.

GALINDO, I., 1975, Physikalische und mathematische Untersuchungen zur atmosphaerischen Trübung. *Archives for Meteorology, Geophysics and Bioclimatology, Series B*, **23**, 225–254.

GALINDO, I., 1978, On the presence of saharian aerosol at the western part of the Atlantic Ocean. *Zeitschrift für Meteorologie*, **28**, 352–360.

GALINDO, I., 1984, Anthropogenic aerosols and their regional scale climatic factors. In *Aerosols and their Climatic effects* (Hampton, VA: Deepak), pp. 245–259.

GALINDO, I., 1992a, Cambios climático regionales como componentes del cambio climático global. *Ciencia*, **45**, 21–27.

GALINDO, I., 1992b, Extinction of short-wave solar radiation due to El Chichón stratospheric aerosol. *Atmosfera*, **5**, 253–268.

GALLIMORE, R. G. and KUTZBACH, J. E., 1989, Effects of soil moisture on the sensitivity of a climate model to Earth orbital forcing at 9000 YR BP. *Climatic Change*, **14**, 175–206.

GARRAT, J. R., 1993, Sensitivity of climate simulation to land-surface and atmospheric boundary-layer treatment, a review. *Journal of Climate*, **6**, 419–449.

GATES, W. L., 1979, The effect of the ocean on the atmospheric general circulation. Climate Research Institute Report, Oregon State University, Corvallis.

GATES, W. L., 1981, Paleoclimatic modelling – a review of problems and prospects for the pre-Pleistocene. Climatic Research Institute Report, Oregon State University, Corvallis.

GATES, W. L., 1988, Status of global model simulations of climatic effects of increased atmospheric carbon dioxide. *Report of the 4th Session of the CAS/JSC W.G. on Numerical Experimentation held in Downsview, Ontario, Canada, on 26–30 September*, WMO/TD-N278, Appendix B.

GATES, W. L., COOK, K. H. and SCHLESINGER, M. E., 1981, Preliminary analysis of experiments on the climatic effects of increased $CO_2$ with an atmospheric general circulation model and a climatological ocean. *Journal of Geophysical Research*, **86**, 6385–6393.

GATES, W. L. and MACCRACKEN, M. C., 1985, The challenge of detecting climate changes induced by increasing carbon dioxide. In *Detecting the Climatic Effects of Increasing Carbon Dioxide* (Washington DC: US Department of Energy), pp. 1–12.

GAUTIER, D., 1987, Comparative evolution of the atmospheres of giant planets. *Europhysics News*, **18**, 116–119.

GELEYN, J. F. and TANRÉ, D., 1984, Short-term climatic response to the atmospheric aerosol radiative forcing. *Proceedings of International Radiation Symposium, held in Perugia, Italy, 21–28 August 1984* (Hampton, VA: Deepak), pp. 243–246.

GELLER, M. A., 1992, Solar-activity induced stratospheric change and tropospheric climate. *Symposium on Solar Constant and Spectral Distribution of Solar Irradiation, Hamburg, 17–28 August 1981* (Boulder, CO: IUGG), pp. 124–132.

GERASIMOV, K. P. and VELICHKO, A. A. (editors), 1987, *Palaeogeography of Europe for the last 100 years* (Moscow: Nauka).

GERBER, H. E. and DEEPAK, A. (editors), 1984, *Aerosols and their Climatic Effects* (Hampton, VA: Deepak).

GHAN, S. J., 1992, The GCM credibility gap. Editorial, *Climatic Change*, **21**, 345–346.

GILCHRIST, A., 1983, Increased carbon dioxide concentrations and climate: the equilibrium response. In *Carbon Dioxide: Current Views and Developments in the Energy/Climate Research* (Dordrecht: Reidel), pp. 219–258.

GILL, A. E., 1985, An overview of the dynamics of the tropical oceans and global atmosphere. *International Conference on the TOGA Science Program, held in Paris on 17–21 September 1984*, WCRP Publication Series No. 4 (Geneva: WMO), pp, I.1–I.12.

GILLE, J. C., SMYTHE, C. M. and HEATH, D. F., 1984, Observed ozone response to variations in solar ultraviolet radiation. *Science*, **225**, 315–317.

GILLILAND, R. L. and SCHNEIDER, S. H., 1984, Volcanic $CO_2$ and solar forcing of northern and southern hemisphere surface air temperatures. *Nature*, **310**, 38–41.

GINZBURG, A. S., 1988, *The Planet Earth in 'Afternuclear' Epoch* (Moscow: Nauka).

GIORGI, F., BATES, G. T. and NILMAN, S. J., 1992, A regional climate model for the Western United States. 1: Model validation. *Proceedings of the Third Symposium on Global Change Studies* (Boston: American Meteorological Society), pp. 10–17.

GIORGI, F. and MARINUCCI, M. R., 1992, Validation of a regional atmospheric model over Europe: Sensitivity of wintertime and summertime simulations to selected physics parametrizations and lower boundary conditions. *Quarterly Journal of the Royal Meteorological Society*, **117**, 1171–1207.

GLANTZ, M. A., 1989, The cases scenario approach to climate related impact analysis. *Idöjárás*, **93**, 10–22.

GLEICK, P. H., 1989, The implications of global climatic changes for international security. *Climatic Change*, **15**, 309–326.

GLOERSEN, P. and CAMPBELL, W. J., 1988, Variations in the Arctic, Antarctic, and global sea ice covers during 1978–1987 as observed with the Nimbus-7 scanning multichannel microwave radiometer. *Journal of Geophysical Research*, **93**, 10666–10674.

GLOWIENSKA-HENSE, R. and HENSE, A., 1992, The effct of an Arctic polynya on the northern hemisphere mean circulation and eddy regime: a numerical experiment. *Climate Dynamics*, **7**, 155–163.

GOLDMAN, A., MURCRAY, F. J., RINSLAND, C. P., BLATHERWICK, R. D., DAVID, S. J., MURCRAY, F. H. and MURCRAY, D. G., 1992, Mount Pinatubo $SO_2$ column measurements from Mauna Loa. *Geophysical Research Letters*, **19**, 183–186.

GOLOMBEK, A. and PRINN, R. G., 1986, A global three-dimensional model of the circulation and chemistry of $CFCl_3$, $CF_2Cl_2$, $CH_3CCl_3$, $CCl_4$ and $N_2O$. *Journal of Geophysical Research*, **91**, 3985–4002.

GOODESS, C., PALUTIKOF, P. and DAVIES, T. D., 1992, *The Nature and Causes of Climate Changes.* Lewis Publ. 248 pp.

GOODY, R. M., HASKINS, R., ABDOU, W. and CHEN, I., 1995, Detection of climate forcing using emission spectra. *Studies of the Earth from Space*, No. 5, 22–23.

GOODY, R. M. and WALKER, J. C. G., 1972, *Atmospheres* (New York: Prentice-Hall).

GORDON, A., 1991, The Southern Ocean: its involvement in global change. *International Conference on the Role of the Polar Regions in Global Change. Proceedings of a Conference held June 11–15, 1990, at the University of Alaska Fairbanks*, edited by G. WELLER, C. L. WILSON and B. A. B. SEVERIN (Fairbanks: Geophysical Institute and Center for Global Change and Arctic Systems Research), pp. 249–255.

GORDON, A. H., 1992, Interhemispheric contrasts of mean global temperature anomalies.

*International Journal of Climatology*, **12**, 1–10.
GORDON, H. B. and HUNT, B. G., 1987, Interannual variability of the simulated hydrology in a climate model – implications for drought. *Climate Dynamics*, **1**, 113–130.
GORDON, H. B. and HUNT, B. G., 1991, Droughts, floods, and sea-surface temperature anomalies: a modelling approach. *International Journal of Climatology*, **11**, 347–366.
GORDON, H. B. and HUNT, B. G., 1994, Climatic variability within an equilibrium greenhouse simulation. *Climate Dynamics*, **9**, 195–212.
GORSHKOV, V. G., 1984, Possible role of biota of the ocean in global carbon cycle. *Oceanology (Okeanologia)*, **24**, 453–459.
GORSHKOV, V. G., 1986, The role of the ocean in the absorption of anthropogenic $CO_2$ releases from the atmosphere. *Izvestia All-Union Geographical Society*, **118**, 386–395.
GORSHKOV, V. G., 1987a, Anthropogenic disturbances of global carbon cycle. *Izvestia All-Union Geographical Society*, **119**, 106–118.
GORSHKOV, V. G., 1987b, Variations in the global content of organic carbon in the oceanic and land biota. *Il Nuovo Cimento*, **10C**, 365–380.
GORSHKOV, V. G., 1995, *Physical and Biological Bases of Life Stability. Man, Biota, Environment* (Berlin: Springer-Verlag).
GORSHKOV, V. G. and KONDRATYEV, K. YA., 1988, Conceptual aspects of ecological studies: the role of energy and mass exchange. *Vestnik of the USSR Academy of Science*, No. 10, 62–70.
GORSHKOV, V. G., KONDRATYEV, K. YA. and LOSOV, K. S., 1996, Future of the contemporary civilisation. *Vestnik Russian Academy of Sciences*, No. 2, 15–24.
GORSHKOV, V. G., KONDRATYEV, K. YA. and SHERMAN, S. G., 1989, The global carbon cycle change: the Le Chatelier principle in the biotic response to changing $CO_2$ concentrations in the atmosphere (Contribution I). *Izvestia All-Union Geographical Society*, **121**, 284–293.
GÖTZ, G., MÉSZAROS, E. and VALI, G., 1991, *Atmospheric Particles and Nuclei* (Budapest: Akadémiai Kiado).
GRAF, J. E., 1987, Earth observing systems: the earth research system of the 1990s. *AIAA Paper*, No. 320, 1–13.
GRAHAM, N. E., BARNETT, T. P., CHERVIN, R. M., SCHLESINGER, M. E. and SCHLESE, U., 1989, Comparisons of GCM and observed surface wind fields over the tropical Indian and Pacific oceans. *Journal of Atmospheric Science*, **46**, 760–788.
GRASSL, H., 1973, Aerosol influence on radiative cooling. *Tellus*, **25**, 386–395.
GRASSL, H., 1974a, Erwärmung und Abkühlung durch atmosphärisches Aerosol. *Annalen der Meteorologie, Neue Folge*, **9**, 55–59.
GRASSL, H., 1974b, Atmospheric absorbers in the window region and their influence on radiative heating and cooling. *Proceedings of the International Conference on Structure, Composition and General Circulation of the Upper and Lower Atmospheres and Possible Anthropogenic Perturbations, 14–25 January 1974, Toronto*, **2**, 1129–1142.
GRASSL, H., 1975, Albedo reduction and radiative heating of clouds by absorbing aerosol particles. *Beiträge zur Physik der Atmosphäre*, **48**, 199–210.
GRASSL, H., 1988, What are the radiative and climatic consequences of the changing concentration of atmospheric aerosol particles? In *Changing Atmosphere: Report Dahlem Workshop, held in Berlin on 1–6 November 1987* (Chichester: Wiley), pp. 187–199.
GRASSL, H., 1992, The influence of aerosol particles on radiation parameters of clouds. *Időjárás*, **86**, 60–75.
GRAVES, C. E., LEE, W. H. and NORTH, G. R., 1993, New parametrizations and sensitivities for simple climate models. *Journal of Geophysical Research*, **98**, 5025–5036.
GRAY, B. M., 1974, Early Japanese winter temperatures. *Weather*, **29**, 103–107.
GRIBBIN, J., 1983, *Future Weather: Carbon Dioxide, Climate, and the Greenhouse Effect* (New York: Penguin).
GROTCH, S. L., 1987, Some considerations relevant to computing average hemispheric tem-

perature anomalies. *Monthly Weather Review*, **115**, 1305–1317.
GROTCH, S. L., 1988, Regional intercomparisons of general circulation model predictions and historical climate data. Carbon Dioxide Research Division Report TRO41 (DOE/NBB-0084), Washington DC.
GROTCH, S. L. and MACCRACKEN, M. C., 1988, Comparison of model simulations for four general circulation models. *CDIAC Communications, Oak Ridge, TN*, Summer, pp. 6–8.
GROTCH, S. L. and MACCRACKEN, M. C., 1991, The use of general circulation models to predict regional climatic change. *Journal of Climate*, **4**, 286–303.
GROVEMAN, B. S. and LANDSBERG, H. E., 1979a, *Reconstruction of Northern Hemisphere Temperature: 1579–1880*. Publication 79–181 Meteorological Program, University of Maryland.
GROVEMAN, B. S. and LANDSBERG, H. E., 1979b, *Data Appendices to Publication No. 79–181. Appendices A, B, C*. Publication 79–182, Meteorological Program, University of Maryland.
GRUBB, M., KOCH, M., THOMSON, K., MUNSON, A. and SULLIVAN, F., 1993, *The Earth Summit Agreements: A Guide and Assessments* (New York: Earthscan).
GRUZA, G. V., RAN'KOVA, E. YA. and ROCHEVA, E. V., 1989, An analysis of global data on the surface air temperature variations during instrumental observations. *Meteorology and Hydrology* (*Meteorologia i Gidrologia*), No. 1, 22–31.
GRUZA, G. V. and SHEVCHENKO, N. N., 1988, Climatic signal, climatic noise, and extreme possibilities of long-term forecasts. *Meteorology and Hydrology* (*Meteorologia i Gidrologia*), No. 6, 5–14.
GUERRERO, T. A. L., BARRON, E. J. and WARNER, T. T., 1992, Linking a general circulation model and a mesoscale model to examine cyclogenesis in the western north Atlantic for the last glacial maximum. *Proceedings of the Third Symposium on Global Change Studies* (Boston: American Meteorological Society), pp. 41–47.
GULEV, S. K., LAPPO, S. S. and TERESHCHENKOV, V. P., 1991, Monitoring and climate changes of the World Ocean. *Izvestia Akademii Nauk SSSR, Fizika Atmosfery i Okeana*, **27**, 1043–1053.
GUNARATHAM, M. and YOUNG, H., 1995, Enhancing global sinks of carbon dioxide. *Search*, **26**, No. 8, 236–238.
GUPTA, S. K., STAYLOR, W. F., DARNELL, W. O., WILBER, A. C. and RITCHEY, N. A., 1993, Seasonal variation of surface and atmospheric cloud radiative forcing over the globe derived from satellite data. *Journal of Geophysical Research*, **98**, 20761–20778.
GUTOWSKI, W. J., GUTZLER, D. S., PORTMAN, D. and WANG, W. C., 1988, Surface energy balance of three general circulation models: current climate and response to increasing atmospheric $CO_2$. Technical Report 042, DOE/ER-60422-H1, Washington DC.
GUTOWSKI, W. J., GUTZLER, D. S. and WANG, W. C., 1991, Surface energy balances of three general circulation models: implications for simulating regional climate change, *Journal of Climate*, **4**, 121–134.
HAGEN, E., 1989, ENSO-events, northern hemisphere air temperatures, and sea-surface temperatures of the northern Atlantic. *Tropical Ocean–Atmosphere Newsletter*, No. 48, 8–11.
HALL, D. K., 1988, Assessment of polar climate change using satellite technology. *Review of Geophysics*, **26**, 26–39.
HAMEED, S. and CESS, R., 1983, Impact of a global warming on biospheric sources of methane and its climatic consequences. *Tellus*, **35B**, 1–7.
HAMEED, S. and CURRIE, R. G., 1986, Cyclic variations in Canadian and United States drought. *Cycles*, **37**, 59–64.
HAMEED, S., SPERBER, K. R. and CESS, R. D., 1989, Genesis of the southern oscillation within the atmosphere: illustration with global general circulation models. *Geophysical Research Letters*, **16**, 691–694.

HANDEL, M. D. and RISBY, J. S., 1992, An annotated bibliography on the greenhouse effect and climate change. *Climatic Change*, **21**, 97–225.

HANSEN, J. E., 1988, GCM climate change predictions: implications for global measurement needs. *Report of the NASA Workshop on the Detection of the Greenhouse Effect on Climate from Space Observations, 29 September–10 October 1987*, edited by R. A. SCHIFFER (Washington DC: NASA), Appendix C.

HANSEN, J., FUNG, I., LACIS, A., LEBEDEFF, S., RIND, D., RUEDY, R., RUSSELL, G. and STONE, P., 1988a, Global climate changes as forecast by the GISS 3-D model. *Journal of Geophysical Research*, **93**, 9341–9364.

HANSEN, J., JOHNSON, D., LACIS, A., LEBEDEFF, S., LEE, P., RIND, D. and RUSSELL, G., 1981, Climate impact of increasing atmospheric carbon dioxide. *Science*, **213**, 957–966.

HANSEN, J., LACIS, A., RIND, D., RUSSELL, G., STONE, P., FUNG, I., RUEDY, R. and LERNER, J., 1984, Climate sensitivity: analysis of feedback mechanisms. In *Climate Processes and Climate Sensitivity*, edited by J. HANSEN and T. TAKAHASHI. Geophysical Monograph 29, Maurice Ewing Volume 5 (Washington: American Geophysical Union), pp. 130–163.

HANSEN, J., LACIS, A., RUEDY, R. and SATO, M., 1992, Potential climate impact of Mount Pinatubo eruption, *Geophysical Research Letters*, **19**, 215–218.

HANSEN, J. and LEBEDEFF, S., 1987, Global trends of measured surface air temperature. *Journal of Geophysical Research*, **92**, 13345–13372.

HANSEN, J. and LEBEDEFF, S., 1988, Global surface air temperatures: update through 1987, *Geophysical Research Letters*, **15**, 323–326.

HANSEN, J., RIND, D., DELGENIO, A., LACIS, A., LEBEDEFF, S., PRATHER, M. and RUEDY, R., 1988b, Regional greenhouse climate effects. In *Coping with Climate Change: Proceedings of the Second North American Conference on Preparing for Climate Change*, edited by J. C. TOPPING (Washington DC: Climate Institute).

HANSEN, J., RIND, D., DELGENIO, A., LACIS, A., LEBEDEFF, S., PRATHER, M., RUEDY, R. and KARL, T., 1991, Regional green-house, climate effects. In *Greenhouse-Gas-Induced Climatic Change*, edited by M. E. SCHLESINGER (New York: Elsevier), pp. 211–229.

HANSEN, J., ROSSOW, W. and FUNG, I., 1990, The missing data on global climate change. *Issues in Science and Technology*, **7**, 62–69.

HANSEN, J. E. and TAKAHASHI, T. (editors), 1984, *Climate Processes and Climate Sensitivity* (Washington DC: American Geophysical Union).

HANSEN, J. E., WANG, W. C. and LACIS, A. A., 1978, Mount Agung eruption provides test of a global climatic perturbation. *Science*, **199**, 1065–1068.

HANSEN, J. E. and WILSON, H., 1993, Commentary on the significance of global temperature records. *Climatic Change*, **25**, 185–192.

HANSON, K. MAUL, G. A. and KARL, T. R., 1989, Are atmospheric 'greenhouse' effects apparent in the climatic record of the contiguous US (1985–1987) *Geophysical Research Letters*, **16**, 49–52.

HAO, W. M. and WARD, D. E., 1993, Methane production from global biomass burning. *Journal of Geophysical Research*, **98**, 20657–20662.

HARRISON, E. F., MINNIS, P., BARKSTROM, B. R., RAMANATHAN, V., CESS, R. D. and GIBSON, G. G., 1990, Seasonal variation of cloud radiative forcing derived from the Earth Radiation Budget Experiment. *Journal of Geophysical Research*, **95**, 18687–18703.

HARSHVARDHAN, RANDALL, D. A., CORSETTI, T. G. and DAZLICH, D. A., 1989, Earth radiation budget and cloudiness simulations with a general circulation model. *Journal of Atmospheric Science*, **46**, 1922–1942.

HARTMANN, D. L. and DOELLING, D., 1991, On the net radiative effectiveness of clouds. *Journal of Geophysical Research*, **96**, 869–891.

HARTMANN, D. L., KOWALEWSKY, K. and MICHELSON, M. L., 1991, Diurnal variations of outgoing longwave radiation and albedo from ERBE scanner data *Journal of Climate*, **4**, 598–617.

HARTMANN, D. L., OCKERT-BELL, M. E. and MICHELSON, M. L., 1992, The effect of cloud type on Earth's energy balance: global analysis. *Journal of Climate*, **5**, 1281–1304.

HARVEY, L. D. D., 1988, On the role of high latitude ice, snow, and vegetation feedbacks in the climatic response to external forcing changes. *Climatic Change*, **13**, 191–224.

HARVEY, L. D. D., 1989, Transient climatic response to an increase of greenhouse gases. *Climatic Change*, **15**, 15–30.

HARVEY, L. D. D. and SCHNEIDER, S. H., 1985, Transient climate response to external forcing on 100–104 year time scales. 2: Sensitivity experiments with a seasonal, hemispherically averaged, coupled atmosphere, land, and ocean energy balance model. *Journal of Geophysical Research*, **90**, 2207–2222.

HARWELL, M. A., 1984, *Nuclear Winter* (Berlin: Springer).

HASSELMANN, K., 1993, Optimal fingerprint for the detection of time-dependent climate change. *Journal of Climate*, **6**, 1957–1971.

HEATH, D. F. and SCHLESINGER, B. M., 1984, Temporal variability of UV solar spectral irradiance from 160–400 nm over periods of the evolution and rotation regions from maximum to minimum phases of the sunspot cycle. *Proceedings of the International Radiation Symposium, Perugia, Italy, 21–28 August 1984* (Hampton, VA: Deepak), pp. 315–319.

HEIMANN, M., 1995, Dynamics of the carbon cycle. *Nature*, **375**, 629–630.

HENDERSON-SELLERS, A., 1983, *The Origin and Evolution of Planetary Atmospheres* (Bristol: Adam Hilger).

HENDERSON-SELLERS, A., 1986, Increasing cloud in a warming world. *Climatic Change*, **9**, 267–310.

HENDERSON-SELLERS, A., 1987a, Climate is a cloudy issue. *New Scientist*, **115**, 37–39.

HENDERSON-SELLERS, A., 1992, Critical factors for modelling surface hydrology. *Proceedings of the Third Symposium on Global Change Studies* (Boston: American Meteorological Society), pp. 79–85.

HENDERSON-SELLERS, A., 1993, A factorial assessment of the sensitivity of BATS land-surface parametrization scheme. *Journal of Climate*, **6**, 227–247.

HENDERSON-SELLERS, A. and BLONG, R., 1989, *Greenhouse Effect: Living in a Warmer Australia* (Kensington: New South Wales University Press).

HENDERSON-SELLERS, A., DICKINSON, R. E., DURBRIDGE, T. B., KENNEDY, P. J., MCGUFFIE, K. and PITMAN, A. J., 1993, Tropical deforestation: modeling local to regional scale climate change. *Journal of Geophysical Research*, **98**, 7289–7316.

HENDERSON-SELLERS, A., DICKINSON, R. E. and WILSON, M. F., 1988, Tropical deforestation: important processes for climate models. *Climatic Change*, **13**, 43–68.

HENDERSON-SELLERS, A. and MCGUFFIE, K., 1984, The cryosphere – credible or confusing climate monitor? In *Proceedings of International Radiation Symposium held in Perugia, Italy, on 21–28 August 1984* (Hampton, VA: Deepak), pp. 235–240.

HENDERSON-SELLERS, A. and MCGUFFIE, K., 1987, *A Climate Modelling Primer* (Chichester: Wiley).

HENDERSON-SELLERS, A. and MCGUFFIE, K., 1989, Diagnosis of cloud amount increase from an analogue model of a warming world. *Atmosfera*, **2**, 67–102.

HENDERSON-SELLERS, A. and ROBINSON, P. J., 1986, *Contemporary Climatology* (London: Longman).

HENDERSON-SELLERS, A. and WILSON, M. F., 1983, Albedo observations of the Earth's surface for climate research. *Philosophical Transactions of the Royal Society of London*, **309A**, 285–294.

HENDERSON-SELLERS, B., 1987b, Modelling sea surface temperature rise resulting from increasing atmospheric carbon dioxide concentrations. *Climatic Change*, **11**, 349–360.

HOBBS, P. V. (editor), 1993, *Aerosol–cloud–climate interactions* (New York: Academic Press).

HOBBS, P. V. and McCORMICK, M. P. (editors), 1988, *Aerosols and Climate* (Hampton, VA: Deepak).

HOFF, R. M., 1992, Differential $SO_2$ column measurements of the Mt. Pinatubo volcanic plume. *Geophysical Research Letters*, **19**, 125–128.

HOFFERT, M. I., FREI, A. and NARAYANAN, V. K., 1988, Application of solar max ACRIM data to analysis of solar-driven climatic variability on Earth. *Climatic Change*, **13**, 267–285.

HOFFERT, M. I., GAFFIN, S., WANG, Z. Y., HSIEH, C. T. and VOLK, T., 1985, Interannual oscillations in past and future temperature records: models and data analysis. *Extended Abstracts of the 3rd Conference on Climatic Variation and Symposium on Contemporary Climate: 1850–2100, held in Los Angeles on 8–11 January 1985* (Boston: American Meteorological Society), pp. 118–119.

HOFMANN, D. J., FERGUSON, E. E., JOHNSTON, P. V. and METTHEWS, W. A., 1992, Tropospheric ozone variations in the Arctic during January 1990. *Planetary and Space Science*, **40**, 203–210.

HOLLOWAY, J. L. and MANABE, S., 1971, Simulation of Climate by a global circulation model. *Monthly Weather Review*, **99**, 335–370.

HOLZAPFEL-PSCHORN, A. and SEILER, W., 1986, Methane emission during a cultivation period from an Italian rice paddy. *Journal of Geophysical Research*, **91**, 11803–11814.

HOOD, L. L., JIRIKOWIC, J. L. and McCORMACK, J. P., 1993, Quasi-decadal variability of the stratosphere: influence of long-term solar ultraviolet radiation. *Journal of Atmospheric Science*, **50**, 3941–3958.

HOOPER, N. J. and SHERMAN, J. W., 1986, Temporal and spatial analysis of civil marine satellite requirements. NOAA Technical Report NESDIS 16, Washington DC.

HORVÁTH, F., 1988, Global sea-level changes. *Időjárás*, **92**, 69–74.

HOSKINS, B. J. and PIERCE, R. P. (editors), 1983, *Large Scale Dynamical Processes in the Atmosphere* (New York: Academic Press).

HOSKINS, B. J. and VALDES, P. J., 1990, On the existence of storm-tracks. *Journal of Atmospheric Science*, **47**, 1854–1864.

HOSTETLER, S. W., BATES, G. T. and GIORGI, F., 1992, Nesting hydrological models within a regional climate model. *Proceedings of the Third Symposium on Global Change Studies* (Boston: American Meteorological Society), Abstract p. 86.

HOUGHTON, J. T., 1979, Greenhouse effects of some atmospheric constituents. *Philosophical Transactions of the Royal Society of London*, **A290**, 515–521.

HOUGHTON, J. T. (editor), 1984, *The Global Climate* (Cambridge: Cambridge University Press).

HOUGHTON, J. T., 1986, *The Physics of Atmospheres*. Second edition (Cambridge: Cambridge University Press).

HOUGHTON, J. T., CALLANDER, B. A. and VARNEY, S. K. (editors), 1992, *Climate Change 1992: The Supplementary Report to the IPCC Scientific Assessment* (Cambridge: Cambridge University Press).

HOUGHTON, J. T., JENKINS, G. J. and EPHRAUMS, J. J. (editors), 1990, *Climate Change: the IPCC Scientific Assessment*. Volume 1 (Cambridge: Cambridge University Press).

HOUGHTON, J. T., MEIRA FILHO, L. G., CALLANDER, B. A., HARRIS, N., KATTENBERG, A. and MASKELL, K. (editors), 1996, *Climate Change 1995: The Science of Climate Change* (Cambridge: Cambridge University Press).

HOUZE, R. A., 1993, *Cloud Dynamics* (London: Academic Press).

HOYT, D. V. and EDDY, J. A., 1983, Solar irradiance modulation by active regions from 1969 through 1981. *Geophysical Research Letters*, **10**, 509–512.

HOYT, D. V. and SCHATTEN, K. H., 1993, A discussion of plausible solar irradiance variations, 1700–1992. *Journal of Geophysical Research*, **98**, 18895–18906.

HSUING, J., NEWELL, R. E. and HOUGHTBY, T., 1989, The annual cycle of oceanic heat storage and meridional heat transport. *Quarterly Journal of the Royal Meteorological Society*, **115**, 1–28.

Hu, Y. X. and Stamnes, K., 1993, An accurate parametrization of the radiative properties of water clouds suitable for use in climate models. *Journal of Climate*, **6**, 728–742.

Huang, T. Y. W. and Brasseur, G. P., 1993, Effect of long-term solar variability in a two-dimensional interactive model of the middle atmosphere. *Journal of Geophysical Research*, **98**, 20413–20427.

Huggett, R. J., 1991, *Climate, Earth Processes and Earth History* (Heidelberg: Springer).

Hulme, M., 1992, A 1951–80 global land precipitation climatology for the evaluation of general circulation models. *Climate Dynamics*, **7**, 57–72.

Humphreys, W. J., 1940, *Physics of the Air* (New York: McGraw-Hill).

Hunt, B. G., 1977, A simulation of the possible consequences of a volcanic eruption on the general circulation of the atmosphere. *Monthly Weather Review*, **105**, 247–260.

Hunt, B. G., 1981, An examination of some feedback mechanisms in the carbon dioxide climate problem. *Tellus*, **33**, 78–88.

Hunt, B. G., 1988, Nonlinear influences – a key to short-term climatic perturbations. *Journal of Atmospheric Science*, **45**, 387–395.

Hunt, B. G. and Gordon, H. B., 1988, The problem of 'naturally' occurring drought. *Climate Dynamics*, **3**, 19–34.

Hunt, B. G. and Gordon, H. B., 1989, Diurnally varying regional climatic simulations. *International Journal of Climatology*, **9**, 331–356.

Hunt, G. E. (editor), 1985, *Recent Advances in Planetary Meteorology* (Cambridge: Cambridge University Press).

Hunter, D. E., Schwartz, S. E., Wagener, R. and Benkoutz, C. M., 1993, Seasonal, latitudinal and secular variations in temperature trend: evidence for influence of anthropogenic sulfate. *Geophysical Research Letters*, **20**, 2455–2458.

Husar, R. B., Stowe, L. L. and Deepak, A., 1991, *Reports of the Expert Meeting on Global Aerosol Data System (GADS)* (Hampton, VA: Deepak).

IASC, 1996, *Arctic Systems: Natural Environments, Human Actions, Nonlinear Processes. Executive Summary* (Oslo: International Arctic; Report No. 3, 21 pp.

ICSU, International Council of Scientific Unions, 1988, *Solar–Terrestrial Energy Program (STEP) 1990–1995: A study of Energy Transfer Mechanisms in the Solar–Terrestrial System* (London, Paris: International Council of Scientific Unions).

Idso, S. B., 1985, An upper limit to global surface air temperature. *Archives for Meteorology, Geophysics and Bioclimatology*, **34**, 141–144.

Idso, S. B., 1987, A clarification of my position on the $CO_2$–climate connection. *Climatic Change*, **10**, 81–86.

Idso, S. B., 1988, Me and the modellers: perhaps not so different at all. *Climatic Change*, **12**, 93–94.

Idso, S. B., 1989, *Carbon Dioxide and Global Change: Earth in Transition* (Tempe, AZ: IBS Press).

Idso, S. B., 1991a, The aerial fertilization effect of $CO_2$ and its implications for global carbon cycling and maximum greenhouse warming. *Bulletin of the American Meteorological Society*, **72**, 962–966.

Idso, S. B., 1991b, Carbon dioxide and the fate of the Earth. *Global Environmental Change*, **1**, 178–182.

Idso, S. B., 1991c, Comment on 'Modelling the seasonal contribution of a $CO_2$ fertilization effect on the terrestrial vegetation to the amplitude increase in atmospheric $CO_2$ at Mauna Loa Observatory' by G. H. Kohlmaier *et al. Tellus*, **43B**, 338–341.

Idso, S. B., 1991d, Reply. *Bulletin of the American Meteorological Society*, **72**, 1910–1913.

Idso, S. B., 1992, The DMS–cloud albedo feedback effect: greatly underestimated? *Climatic Change*, **21**, 427–434.

Idso, S. B., Kimball, B. A., Anderson, M. G. and Mauney, J. R., 1989, Greenhouse warming could magnify positive effects of $CO_2$ enrichment on plant growth. *CDIAC Communications, Oak Ridge National Laboratory*, Winter, pp. 8–9.

INGRAM, W. J., 1989, Modelling cloud feedbacks on climate change. *Weather*, **44**, 303–310.

INGRAM, W. J., WILSON, C. A. and MITCHELL, J. F. B., 1989, Modelling climate change: an assessment of sea ice and surface albedo feedbacks. *Journal of Geophysical Research*, **94**, 8609–8622.

INSTITUTE OF THE ARCTIC AND ANTARCTIC (IAA), 1989, *Programme to Study the Climate of Polar Regions, 1989* (Leningrad: Gidrometeoizdat).

INTERAGENCY ARCTIC RESEARCH POLICY COMMITTEE, 1985, *Federal Arctic Research. Detailed listing of existing US programs. Integral Compilation, September 1985*. DOE/ER-0251.

INTERNATIONAL ASSOCIATION OF METEOROLOGY AND ATMOSPHERIC PHYSICS (IAMAP), 1989, *Brief review papers and abstracts, Volumes I, II. Fifth Scientific Assembly of the International Association of Meteorology and Atmospheric Physics, held in University of Reading, UK, on 31 July–12 August 1989.*

IPCC, 1994, *Radiative Forcing of Climate Change. The 1994 Report of the Scientific Assessment Working Group of IPCC. Summary for Policymakers, 1994*, WMO/UNEP Geneva, pp. 1–28.

ISY, 1988, *Report of the ISY Mission to Planet Earth Conference. Durham, NH, April 29–May 1, 1988*, 38 pp.

IVLEV, L. S., 1982, *Chemistry and Structure of Atmospheric Aerosols* (Leningrad: LGU Publishers).

IVLEV, L. S. and ANDREEV, S. D., 1986, *Optical Properties of Atmospheric Aerosols* (Leningrad: LGU Publishers).

JACOBOWITZ, H., SMITH, W. L., HOWELL, H. B., NOGLE, F. W. and HICKEY, J. R., 1979, The first 18 months of planetary radiation budget measurements from the Nimbus-6 ERB experiment. *Journal of Atmospheric Science*, **36**, 501–507.

JACOBY, G. C. and COOK, E. R., 1981, Past temperature variations inferred from a 400-year tree-ring chronology from Yukon Territory, Canada. *Arctic and Alpine Research*, **13**, 409–418.

JACOBY, G. C. and D'ARRIGO, R., 1989, Reconstructed northern hemisphere annual temperature since 1671 based on high-latitude tree-ring data from North America. *Climatic Change*, **14**, 39–60.

JÄGER, H., 1992, The Pinatubo eruption cloud observed by lidar at Garmisch-Partenkirchen. *Geophysical Research Letters*, **19**, 191–194.

JAIN, A. K. and BACH, W., 1992, A climate-policy model: the effectiveness of countermeasures to reduce the man-made greenhouse effect. *Proceedings of the Third Symposium on Global Change Studies* (Boston: American Meteorological Society), pp. 102–109.

JAWOROWSKI, Z., SEGALSTAD, T. V. and HISDSDAL, V., 1992, *Atmospheric $CO_2$ and Global Warming: A Critical Review*. Second revised edition (Oslo: Norsk Polarinstitutt).

JENKINS, G. S., 1993, A general circulation model study of the effects of faster rotation rate, enhanced $CO_2$ concentration, and reduced solar energy forcing: implications for the faint young sun paradox. *Journal of Geophysical Research*, **98**, 20803–20812.

JENNINGS, H. (editor), 1993, *Atmospheric Aerosols* (Tucson, AZ: Arizona Press).

JENTSCH, V., 1991a, An energy balance climate model with hydrological cycle. 1: Model description and sensitivity to internal parameters. *Journal of Geophysical Research*, **96**, 17169–17179.

JENTSCH, V., 1991b, An energy balance climate model with hydrological cycle. 2: Stability and sensitivity to external forcing. *Journal of Geophysical Research*, **96**, 17181–17193.

JET PROPULSION LABORATORY, 1985, *Earth Systems Science Committee Working Group on Imaging and Tropospheric Sounding, Final Report*. D-2415 (Maryland: NOAA).

JOHNSTON, P. V., MCKENZIE, R. L., KEYS, J. G. and MATTHEWS, W. A., 1992, Observations of depleted stratospheric $NO_2$ following Pinatubo volcanic eruption. *Geo-*

*physical Research Letters*, **19**, 211–214.

JONES, P. D., WIGLEY, T. M. L. and KELLY, P. M., 1982, Variations in surface air temperature, I, Northern hemisphere, 1881–1980. *Monthly Weather Review*, **110**, 59–70.

JONES, P. D., RAPER, S. C. B., BRADLEY, R. S., DIAZ, H. F., KELLY, P. M. and WIGLEY, T. M. L., 1986, Northern hemisphere surface air temperature variations: 1851–1984. *Journal of Climate and Applied Meteorology*, **25**, 161–179.

JONES, P. D., WIGLEY, T. M. L. and FARMER, G., 1990, Marine and land temperature data sets: a comparison and a look at recent trends. In *Greenhouse-Gas-Induced Climatic Change: A Critical Appraisal of Simulations and Observations*, edited by M. E. SCHLESINGER (New York: Elsevier).

JSC/CCCO, 1987, Report of the Fourth Session of the JSC/CCCO Working Group on satellite observing systems for climate research, 1987, Tokyo, Japan, 20–24 October 1986, World Climate Paper No. 124.

KAHL, J. D. W., SERREZE, M. C., STONE, R. S., SHIOTANI, S., KISLEY, M. and SCHNELL, R. C., 1993, Tropospheric temperature trends in the Arctic: 1958–1986. *Journal of Geophysical Research*, **98**, 12825–12838.

KANDEL, R. S., 1980, *Earth and Cosmos* (Oxford: Pergamon).

KANDEL, R. S., 1993, Simple climate models and the greenhouse effect. In *Carbon Dioxide: Current Views and Developments in the Energy/Climate Research* (Dordrecht: Reidel), pp. 179–218.

KARL, T. R., 1988, Multi-year fluctuations of temperature and precipitation: the gray area of climate change. *Climatic Change*, **12**, 179–198.

KARL, T. R. and JONES, P. D., 1989, Urban bias in area-averaged surface air temperature trends. *Bulletin of the American Meteorological Society*, **70**, 265–270.

KARL, T. R., KUKLA, G. and GAVIN, J., 1986, Relationship between decreased temperature range and precipitation trends in the United States and Canada, 1941–1980. *Journal of Climate and Applied Meteorology*, **25**, 1878–1886.

KARL, T. R., KUKLA, G. and GAVIN, G., 1987, Recent temperature changes during overcast and clear skies in the United States. *Journal of Climate and Applied Meteorology*, **26**, 698–711.

KARL, T. R. and QUAYLE, R. G., 1988, Climate change in fact and in theory: are we collecting the facts? *Climatic Change*, **13**, 5–18.

KARLSTROM, T. N. V., 1961, The glacial history of Alaska: its bearing on paleoclimatic theory. *Annals of the New York Academy of Sciences*, **95**, 290–340.

KAROL, I. L., 1988, *An Introduction to the Earth's Climate Dynamics* (Leningrad: Gidrometeoizdat).

KAROL, I. L., 1991, Assessments of the potential greenhouse effect of halogenhydrocarbons. *Meteorology and Hydrology* (*Meteorologia i Gidrologia*), No. 7, 57–63.

KAROLY, D. J., 1987, Southern hemisphere temperature trends: a possible greenhouse gas effect? *Geophysical Research Letters*, **14**, 1139–1141.

KAROLY, D. J., 1989, Northern hemisphere temperature trends: a possible greenhouse gas effect? *Geophysical Research Letters*, **16**, 465–468.

KAROLY, D. J., 1991, On fingerprint detection of greenhouse climate change. *Proceedings of the Fifth Conference on Climate Variations* (Boston: American Meteorological Society), pp. 1–4.

KASAHARA, A. and DA SILVA DIAS, P. L., 1986, Response of planetary waves to stationary tropical heating in a global atmosphere with meridional and vertical shear. *Journal of Atmospheric Science*, **43**, 1893–1911.

KASTING, J. F., 1989, Long-term stability of the Earth's climate. *Global and Planetary Change (special issue)*, **1**, 83–96.

KASTING, J. F. and ACKERMAN, T. P., 1986, Climatic consequences of very high carbon dioxide levels in the earth's early atmosphere. *Science*, **234**, 1383–1385.

KASTING, J. F. and POLLACK, J. B., 1984, Effects of high $CO_2$ levels on surface temperature and atmospheric oxidation state of the early earth. *Journal of Atmospheric Chemistry*, **1**, 403–428.

KASTING, J. F., TOON, O. B. and POLLACK, J. B., 1988, How climate evolved on the terrestrial planets. *Scientific American*, **258**, 46–53.

KAUFMAN, Y. J., FRASER, R. S. and MAHONEY, R. L., 1991, Fossil fuel and biomass burning effect on climate – heating or cooling? *Journal of Climate*, **4**, 578–588.

KEELING, C. D., WHORF, T. P., WAHLEN, M. and VAN DER PLICHT, 1995, Interannual extremes in the rise of atmospheric carbon dioxide since 1980. *Nature*, **375**, 666–670.

KEELING, C. D., WHORF, T. P., WONG, C. S. and BELLAGAY, R. D., 1985, The concentration of atmospheric carbon dioxide at ocean weather station P from 1969 to 1981. *Journal of Geophysical Research*, **90**, 10511–10528.

KELLOGG, W. W., 1983a, Impacts of a $CO_2$-induced climate change. In *Carbon Dioxide: Current Views and Developments in the Energy/Climate Research* (Dordrecht: Reidel), pp. 379–414.

KELLOGG, W. W., 1983b, Feedback mechanisms in the climate system affecting future levels of carbon dioxide. *Journal of Geophysical Research*, **88**, 1263–1269.

KELLOGG, W. W., 1987, Mankind's impact on climate: the evolution of an awareness. *Climatic Change*, **10**, 113–136.

KELLOGG, W. W., 1991, Response to sceptics of global warming. *Bulletin of the American Meteorological Society*, **72**, 499–511.

KELLOGG, W. W. and BOJKOV, R. D., 1983, *Report of the JSC/CAS Meeting of Experts on Detection of Possible Climate Change, Moscow, 3–6 October 1982* (Geneva: WMO).

KELLOGG, W. W. and SCHWARE, R., 1981, *Climate Change and Society: Consequences of Increasing Atmospheric Carbon Dioxide* (Boulder, CO: Westview Press).

KELLOGG, W. W. and ZHAO, Z. C., 1988, Sensitivity of soil moisture to doubling of carbon dioxide in climate model experiments. Part I: North America. *Journal of Climate*, **1**, 348–366.

KELLY, P. M., GOODESS, C. M., JONES, P. D. and WIGLEY, T. M. L., 1986, Recent changes in surface air temperature over the northern land masses. *Proceedings of the First WMO Workshop on the Diagnosis and Prediction of Monthly and Seasonal Atmospheric Variations over the Globe*, WMO Technical Document No. 87, pp. 59–63.

KELLY, P. M., JONES, P. D., SEAR, C. B., CHERRY, B. S. G. and TAVAKOL, R. K., 1982, Variations in surface air temperatures, 2, Arctic regions, 1881–1980. *Monthly Weather Review*, **110**, 71–83.

KELLY, P. M. and SEAR, C. B., 1982, The formulation of Lamb's dust index. *Proceedings of the Symposium on Atmospheric Effects and Potential Climatic Impact of the 1980 Eruptions of Mount St Helens*, NASA Conference Publication, No. 2240, pp. 293–298.

KEPPENNE, C. L. and GHIL, M., 1992, Adaptive filtering and prediction of the Southern Index Oscillation. *Journal of Geophysical Research*, **97**, 20449–20454.

KERR, R. A., 1989b, Hansen vs. the world on the greenhouse threat. Greenhouse models vs. reality. *Science*, **244**, 1041–1043.

KERR, R. A., 1989c, Volcanoes can muddle the greenhouse. *Science*, **245**, 127–128.

KHESHGI, H. S. and WHITE, B. S., 1993, Does recent global warming suggest an enhanced greenhouse effect? *Climatic Change*, **23**, 121–139.

KHMELEVTSOV, S. S., 1988, *Studies of Climate Using Energy-balance Models* (Leningrad: Gidrometeoizdat).

KIEHL, J. T., 1985, Searching for the radiative signal of increasing carbon dioxide and other trace gases. Detecting the climatic effects of increasing carbon dioxide DOE-ER-0232 (Washington DC: US Department of Energy), pp. 13–28.

KIEHL, J. T., BOVILLE, B. A. and BRIEGLEB, P., 1988, Response of a general circulation model to changes in ozone. *Journal of Atmospheric Science*, **45**, 1798–1817.

KIEHL, J. T. and DICKINSON, R. E., 1987, A study of the radiative effects on enhanced atmospheric $CO_2$ and $CH_4$ on early Earth surface temperatures. *Journal of Geophysical Research*, **92**, 2991–2998.

KIM, Y. and CESS, R. D., 1993, Effect of anthropogenic sulfate aerosols on low-level cloud albedo over oceans. *Journal of Geophysical Research*, **98**, 14883–14886.

KLIMONT, Z., AMANN, M., COFALA, J., GYÁRLAS, F., KLAASSEN, G. and SCHÖPP, W., 1994, An emission inventory for the Central European Initiative, **28**, 235–246.

KLINK, K., 1995, Surface aggregation and subgrid-scale climate. *International Journal of Climatology*, **15**, 1215–1240.

KO, M. K. W., SCHNEIDER, H. R., SHIA, R. L., WEINSTEIN, D. K. and SZE, N. D., 1993, A two-dimensional model with coupled dynamics, radiation, and photochemistry. 1: Simulation of the middle atmosphere. *Journal of Geophysical Research*, **98**, 20429–20440.

KOBAYASHI, T., 1993, Effects due to cloud geometry on biases in the albedo derived from radiance measurements. *Journal of Climate*, **6**, 120–128.

KOCH, L., 1945, The east Greenland ice. *Medd. Groenl.*, **130**, 1–373.

KODAMA, K., KOBAYASHI, D., ISHIKAWA, N. and WAKAHAMA, G., 1991, On the effect of global warming on the snowmelt in an Arctic permafrost area. *International Conference on the Role of the Polar Regions in Global Change. Proceedings of a Conference held June 11–15, 1990, at the University of Alaska Fairbanks*, edited by G. WELLER, C. L. WILSON and B. A. B. SEVERIN (Fairbanks: Geophysical Institute and Center for Global Change and Arctic Systems Research), p. 199.

KOENIG, G., LIOU, K.-N. and GRIFFIN, M., 1987, An investigation of cloud/radiation interactions using three-dimensional nephanalysis and Earth radiation budget data bases. *Journal of Geophysical Research*, **92**, 5540–5554.

KOLOMEEV, M. P., MALYSHEV, S. L. and NIKONOV, S. A., 1991, Tropospheric aerosol and climate: the results of numerical experiments with a 3-D seasonal energy-balance model. *Izvestiya Akademii Nauk SSSR, Fizika Atmosfery i Okeana*, **27**, 369–375.

KOMHYR, W. D., GAMMON, R. H., HARRIS, T. B., WATERMAN, L. S., CONWAY, T. J., TAYLOR, W. R. and THONING, K. W., 1985, Global atmospheric $CO_2$ distribution and variations from 1968–1982 NOAA/GMCC $CO_2$ flask sample data. *Journal of Geophysical Research*, **90**, 5567–5596.

KOMHYR, W. D., HARRIS, T. B., WATERMAN, L. S., CHIN, J. F. S., and THONING, K. W., 1989, Atmospheric carbon dioxide at Mauna Loa observatory. 1: NOAA global monitoring for climatic change measurements with a nondispersive infrared analyzer, 1974–1985. *Journal of Geophysical Research*, **94**, 8533–8548.

KONDRATYEV, K. YA., 1972, Complex Atmospheric Energetics Experiment. WMO Bulletin No. 12.

KONDRATYEV, K. YA. (editor), 1973, *Effects of Aerosols on Radiation Transfer: Possible Climatic Consequences* (Leningrad: LGU Publishers).

KONDRATYEV, K. YA., 1976, Aerosol and Climate. Trudy GGO, Issue 381, pp. 3–66.

KONDRATYEV, K. YA. (editor), 1978a, *Atmospheric Aerosols and their Effect on Radiation Transfer* (Leningrad: Gidrometeoizdat).

KONDRATYEV, K. YA., 1978b, Once more on a possible impact of $CO_2$ on climate. *Zeitschrift für Meteorologie*, **37**, 299–308.

KONDRATYEV, K. YA., 1980, *Radiative Factors of the Present Global Climate Change* (Leningrad: Gidrometeoizdat).

KONDRATYEV, K. YA., 1982a, World Climate Research Programme: The present state, perspectives, and the role of satellite observations. *Progress in Science and Technology. Meteorology and Climatology*, Volume 8 (*Itogi Nauki i Tekhniki. Meteorologia i Klimatologia*) (Moscow: VINITI).

KONDRATYEV, K. YA., 1982b, An experience gained from earth observation from space and planning of the optimal observation systems. *Journal of the Meteorological Society of Japan, Series II*, **60**, 593–602.

KONDRATYEV, K. YA., 1983a, *Satellite Climatology* (Leningrad: Gidrometeoizdat).

KONDRATYEV, K. YA., 1983b, The Earth's radiation budget, aerosol and clouds. *Progress in Science and Technology. Meteorology and Climatology*, Volume 10 (*Itogi Nauki i Tekhniki, Meteorologia i Klimatologia*) (Moscow: VINITI).

KONDRATYEV, K. YA., 1985a, Climate as observed from space. In *Biosphere and Climate from Space Studies. The First All-Union Conference Plenary Reports, Baku*, pp. 13–28.

KONDRATYEV, K. YA., 1985b, Volcanoes and climate. *Progress in Science and Technology. Meteorology and Climatology*, Volume 13 (*Itogi Nauki i Tekhniki. Meteorologia i Klimatologia*) (Moscow: VINITI).

KONDRATYEV, K. YA., 1985c, The atmospheric greenhouse effect and climate. *Izvestiya All-Union Geographical Society*, **117**, No. 3, 301–311.

KONDRATYEV, K. YA., 1986a, Anthropogenic impacts on the Arctic atmosphere. *Izvestiya All-Union Geographical Society*, **117**, No. 3, 196–202.

KONDRATYEV, K. YA., 1986b, *Changes in Global Climate* (Rotterdam: Balkema).

KONDRATYEV, K. YA., 1986c, Natural and anthropogenic climate changes. *Progress in Science and Technology. Meteorology and Climatology*, Volume 16 (*Itogi Nauki i Tekhniki. Meteorologia i Klimatologia*) (Moscow: VINITI).

KONDRATYEV, K. YA., 1987a, The problems of climate and the role of radiation studies. In *Progress in Hydrometeorology and Environmental Control* (Leningrad: Gidrometeoizdat), pp. 57–97.

KONDRATYEV, K. YA., 1987b, Global climate. *Progress in Science and Technology. Meteorology and Climatology*, Volume 17 (*Itogi Nauki i Tekhniki. Meteorologia i Klimatologia*) (Moscow: VINITI).

KONDRATYEV, K. YA., 1988a, Comparative meteorology of planets. *Progress in Science and Technology. Astronomy*, Volume 37 (*Itogi Nauki i Tekhniki. Astronomia*) (Moscow: VINITI).

KONDRATYEV, K. YA., 1988b, *Climate Shocks: Natural and Anthropogenic* (New York: Wiley).

KONDRATYEV, K. YA., 1989a, Global dynamics of ozone. *Progress in Science and Technology. Meteorology and Climatology*, Volume 19 (*Itogi Nauki i Tekhniki. Meteorologia i Klimatologia*) (Moscow: VINITI).

KONDRATYEV, K. YA., 1989b, The International Geosphere–Biosphere Program (IGBP): the role of observations from space. In *Ecological Sustainability of Regional Development*, edited by L. KAIRIUSTIS, A. BURACAS and A. STRASZAK (Warszawa: IIASA, Polish Academy of Science), pp. 49–63.

KONDRATYEV, K. YA., 1989c, The International Geosphere–Biosphere Program: the role and place of Sun–atmosphere interrelationships. *Geofisica Internacional*, **28**, 453–466.

KONDRATYEV, K. YA., 1990a, International Space Year: priorities and perspectives. *Studies of the Earth from Space*, (*Issledovania Zemli iz Kosmosa*), No. 1, 3–15.

KONDRATYEV, K. YA., 1990b, International Space Year: priorities and perspectives. *Il Nuovo Cimento*, **13C**, 835–846.

KONDRATYEV, K. YA., 1990c, *Planet Mars* (Leningrad: Gidrometeoizdat).

KONDRATYEV, K. YA., 1991a, New assessments of global climate change. *Atmosfera*, **4**, 177–188.

KONDRATYEV, K. YA., 1991b, *Atmospheric Aerosols* (Leningrad: Gidrometeoizdat).

KONDRATYEV, K. YA., 1992a, Global climate monitoring. *Studies of the Earth from Space*, (*Issledovania Zemli iz Kosmosa*), No. 1, 96–106.

KONDRATYEV, K. YA., 1992b, *Global Climate* (Leningrad: Nauka Press).

KONDRATYEV, K. YA., 1992c, The Second World Conference on Climate. II: Assessments of global climate changes and measures to prevent global changes. *Izvestiya All-Union Geographical Society*, **123**, 473–480.

KONDRATYEV, K. YA., 1992d, On the Framework Convention on Climate Change. *Il Nuovo Cimento*, **156C**, 87–98.

KONDRATYEV, K. YA., 1992e, Correction of the scientific programme and instrument complex for Earth Observation System (EOS) satellites. *Studies of the Earth from Space, (Issledovania Zemli iz Kosmosa)* No. 6, 120–123.

KONDRATYEV, K. YA., 1993, Global Climate Observing System. *Studies of the Earth from Space, (Issledovania Zemli iz Kosmosa)* No. 6, 104–115.

KONDRATYEV, K. YA., 1994, Perspectives of the programme Mission to Planet Earth and the Earth Observing System. *Studies of the Earth from Space, (Issledovania Zemli iz Kosmosa)* No. 5, 105–108.

KONDRATYEV, K. YA., 1996, Greenhouse warming versus aerosol cooling in the context of global climate change. *Energy Conservation and Management*, **37**, 763–768.

KONDRATYEV, K. YA, 1997, Comments on "Open Letter to Ben Santer." *Bulletin of the American Meteorological Society*, **78**, 689–691.

KONDRATYEV, K. YA., ADAMENKO, V. N., HENDERSON-SELLERS, B., FILATOV, N. N. and POZDNYAKOV, D. V., 1990b, Using large lakes as analogues for oceanographic studies. In *Modelling Marine Ecosystems*, Volume 2, edited by A. M. DAVIES (Boca Raton: CRC Press), pp. 299–344.

KONDRATYEV, K. YA. and BINENKO, V. I., 1984, *Impact of Cloudiness on Radiation and Climate* (Leningrad: Gidrometeoizdat).

KONDRATYEV, K. YA., BINENKO, V. I. and MELNIKOVA, I. N., 1996b, Solar radiation absorption by cloud and cloudless atmosphere. *Meteorology and Hydrology. (Meteorologia i Gidrologia)*, No. 2, 14–23.

KONDRATYEV, K. YA., BUZNIKOV, A. A. and POKROVSKY, O. M., 1992, Global ecology: remote sensing. *Itogi Nauki i Tekhniki. Meteorologia i Klimatologia*, Volume 12 (Moscow: VINITI).

KONDRATYEV, K. YA., CRACKNELL, A. P. and VAUGHAN, R. A., 1994, Politics and climate change. In *Remote Sensing and Global Climate Change*, edited by R. A. VAUGHAN and A. P. CRACKNELL (Berlin: Springer), pp. 465–473.

KONDRATYEV, K. YA., DYACHENKO, L. I. and KOZODEROV, V. V., 1988a, *The Earth's Radiation Budget* (Leningrad: Gidrometeoizdat).

KONDRATYEV, K. YA. and GALINDO, I., 1997, *Volcanic Activity and Climate* (Williamsburg, VA: Deepak).

KONDRATYEV, K. YA., GRIGORYEV, A. A. and POKROVSKY, O. M., 1983a, The effect of aerosols on climate and aerosol climatology on the basis of observations from space. Space observations of aerosols and ozone. *Advances in Space Research*, **2**, 3–10.

KONDRATYEV, K. YA. and HUNT, G. E., 1982, *Weather and Climate on Planets* (Oxford: Pergamon).

KONDRATYEV, K. YA., IVANOV, V. A., POZDNIAKOV, D. V. and PROKOFYEV, M. A., 1985, Natural and anthropogenic aerosols: a comparative analysis. *Pontificia Academia Scientiarum. Scripta Vararia*, No. 56, 281–303.

KONDRATYEV, K. YA., JOHANNESSEN, O. M. and MELENTYEV, V. V., 1996a, *High Latitude Climate and Remote Sensing* (Chichester: Wiley/Praxis).

KONDRATYEV, K. YA. and KHVOROSTYANOV, V. I., 1986, A 3-D numerical model of mesoscale processes in the oceanic energy-active zones, with cloud–radiation interaction. *Izvestiya Akademii Nauk SSSR, Fizika Atmosfery i Okeana*, **22**, 1142–1153.

KONDRATYEV, K. YA. and KHVOROSTYANOV, V. I., 1988, The role of clouds and radiation in the formation of the energy-active zones of the ocean (a numerical experiment). *Izvestiya Akademii Nauk SSSR, Fizika Atmosfery i Okeana*, **24**, 899–912.

KONDRATYEV, K. YA. and KOTLIAKOV, V. M., 1991, Polar regions and their contribution to the formation of global ecological changes. *Izvestiya Akademii Nauk SSSR, Fizika Atmosfery i Okeana*, Series of Geography, No. 1, 36–46.

KONDRATYEV, K. YA. and KOZODEROV, V. V., 1984, Anomalies of the Earth's radiation budget and ocean mixed layer dynamics as indicators of energy-active zones. In *Itogi Nauki i Tekhniki*, Volume 4 (Moscow: VINITI), 284 pp.

KONDRATYEV, K. YA., KRUPENIO, N. N. and SELIVANOV, A. S., 1987, *Planet Venus.* (Leningrad: Gidrometeoizdat).

KONDRATYEV, K. YA., LOGINOV, V. F. and KRAVCHUK, E. G., 1990a, The solar radiation trends as an indicator of anthropogenic aerosol change. *Doklady Akademii Nauk SSSR*, **315**, 341–344.

KONDRATYEV, K. YA. and MELENTYEV, V. V., 1986, *Remote Sensing of Clouds and Atmospheric Water Content From Space* (Leningrad: Gidrometeoizdat).

KONDRATYEV, K. YA. and MOSKALENKO, N. I., 1983, Key problems of studies of the planets of the Solar system (greenhouse effect of the planctary atmospheres). *Progress in Science and Technology. Studies of Space*, Volume 19 (*Itogi Nauki i Tekhniki. Iseledovania Kosmicheskogo Prostranstva*) (Moscow: VINITI).

KONDRATYEV, K. YA. and MOSKALENKO, N. I., 1984, Greenhouse effect of the atmosphere and climate. *Progress in Science and Technology. Meteorology and Climatology*, Volume 12 (*Itogi Nauki i Tekhniki. Meteorologia i Klimatologia*) (Moscow: VINITI).

KONDRATYEV, K. YA., MOSKALENKO, N. I. and POZDNIAKOV, D. V., 1983b, *Atmospheric Aerosol* (Leningrad: Gidrometeoizdat).

KONDRATYEV, K. YA., MOSKALENKO, N. I., SAFEROVA, T. M., SKVORTSOVA, S. Y. and PARZHIN, S. N., 1991, Simulation of atmosphere optical characteristics using ES line computer. *Doklady Akademii Nauk SSSR*, **318**, 569–571.

KONDRATYEV, K. YA. and NIKOLSKY, G. A., 1978, Solar activity and climate. *Doklady Akademik Nauk SSSR*, **243**, 607–610.

KONDRATYEV, K. YA. and NIKOLSKY, G. A., 1995a, Solar activity and climate. 1: Observation data ozone and condensation hypothesis. *Studies of the Earth from Space* (*Issledovania Zemli iz Kosmosa*), No. 5, 3–17.

KONDRATYEV, K. YA. and NIKOLSKY, G. A., 1995b, Solar activity and climate. 2: Direct impact of extraatmospheric spectral distribution of solar radiation. *Studies of the Earth from Space* (*Issledovania Zemli iz Kosmosa*), No. 6, 3–17.

KONDRATYEV, K. YA. and PROKOFYEV, M. A., 1983, Physical bases of climate forecasts for a period from one month to several decades. *Progress in Science and Technology. Meteorology and Climatology*, Volume 11 (*Itogi Nauki i Tekhniki. Meteorologia i Klimatologia*) (Moscow: VINITI).

KONDRATYEV, K. YA. and TER MARKARYANTS, N. E. (editors), 1976, *Complete Radiation Experiment* (Leningrad: Gidrometeoizdat).

KONDRATYEV, K. YA., TOROYAN, G. R. and KHVOROSTYANOV, V. I., 1988b, An interpretation of satellite observations of orographic cloudiness on the basis of a mesoscale numerical model. *Studies of the Earth from Space* (*Issledovania Zemli iz Kosmosa*), No. 1, 3–15.

KONDRATYEV, K. YA., VASILYEV, O. B., IVLEV, L. S., NIKOLSKY, G. A. and SMOKTY, O., 1973, *The Impact of Aerosol on Radiation Transfer: Possible Climatic Implications* (Leningrad: LGUP).

KÖPPEN, W., 1873, Concerning multiyear periods of weather, especially the 11-year period of temperature (in German). *Zeitshrift Österreichisches Gesellschaft für Meteorologie*, **8**, 241–248, 257–267.

KRAPIVIN, V. F. and VILKOVA, L. P., 1987, Model experiments on the assessment of the distribution of excess $CO_2$ among the biospheric reservoirs. *Meteorology and Hydrology* (*Meteorologia i Gidrologia*), No. 7, 51–58.

KUKLA, G. and KARL, T. R., 1992, Recent rise of the nighttime temperatures in the northern hemisphere. DOE Research Summary, No. 14.

KUTZBACH, J. E. and GUETTER, P. J., 1986, The influence of changing orbital parameters and surface boundary conditions on climate simulations for the past 18 000 years. *Journal of Atmospheric Science*, **43**, 1726–1759.

KVENVOLDEN, K., 1988, Methane hydrates and global climate. *Global Biogeochemical Cycles*, **2**, 221–229.

LABITZKE, K., 1987, Sunspots, the QBO, and the stratospheric temperature in the north polar region. *Geophysical Research Letters*, **14**, 535–537.

LABITZKE, K. and MCCORMICK, M. P., 1992, Stratospheric temperature increases due to Pinatubo aerosols. *Geophysical Research Letters*, **19**, 207–210.

LABITZKE, K., NAUJOKAT, B. and ANGELL, J. K., 1986, Long-term temperature trends in the middle stratosphere of the northern hemisphere. *Advances in Space Research*, **6**, 7–16.

LABITZKE, K. and VAN LOON, H., 1988, Associations between the 11-year solar cycle, the QBO and the atmosphere. Part 1: The troposphere and stratosphere in the northern hemisphere in winter. *Journal of Atmospheric and Terrestrial Physics*, **50**, 197–206.

LABITZKE, K. and VAN LOON, H., 1989, Recent work correlating the 11-year solar cycle with atmospheric elements grouped according to the phase of the quasi-biennial oscillation. *Space Science Reviews*, **49**, 239–258.

LABITZKE, K. and VAN LOON, H., 1992, Association between the 11-year solar cycle and the atmosphere. Part V: Summer. *Journal of Climate*, **5**, 240–251.

LACIS, A., HANSEN, J., LEE, P., MITCHELL, T. and LEBEDEFF, S., 1981, Greenhouse effect of trace gases, 1970–1980. *Geophysical Research Letters*, **8**, 1035–1038.

LACIS, A., HANSEN, J. and SATO, M., 1992, Climate forcing by stratospheric aerosols. *Geophysical Research Letters*, **19**, 1607–1610.

LACIS, A. A., WUEBBLES, D. J. and LOGAN, J. A., 1989, Radiative forcing of global climate by changes in the vertical distribution of ozone. *Journal of Geophysical Research*, **95**, 9971–9981.

LAL, M. and JAIN, A. K., 1988, Climatic effect of observed changes in atmospheric trace gases at Antarctica. *Atmospheric Environment*, **22**, 2047–2048.

LAMARCHE, V. C. and STOCKTON, C. W., 1974, Chronologies from temperature-sensitive bristlecone pines at upper treeline in western United States. *Tree Ring Bulletin*, **34**, 21–45.

LAMB, H. H., 1977, *Climate Present, Past and Future*, Volume 2, *Climate History and the Future* (London: Methuen).

LAMB, H. H., 1981, Climatic changes and food production: observations and outlook in the modern world. *GeoJournal*, **5.2**, 101–112.

LANGE, M. A., 1991, Antartic sea ice: its development and basic properties. *International Conference on the Role of the Polar Regions in Global Change. Proceedings of a Conference held June 11–15, 1990, at the University of Alaska Fairbanks*, edited by G. WELLER, C. L. WILSON and B. A. B. SEVERIN (Fairbanks: Geophysical Institute and Center for Global Change and Arctic Systems Research), pp. 275–283.

LANGNER, J. and RODHE, M., 1991, A global three-dimensional model of the troposphere sulfur cycle. *Journal of Atmospheric Chemistry*, **13**, 225–263.

LANGNER, J., RODHE, M., CRUTZEN, P. J. and ZIMMERMAN, P., 1992, Anthropogenic influence on the distribution of tropospheric sulphate aerosol. *Nature*, **358**, 712–715.

LAPPO, S. S. and GULEV, S. K., 1985, Phase differences in annual changes in thermal characteristics of the oceans, continents, atmosphere, and ice. *Doklady Akademii Nauk SSSR*, **284**, 1471–1476.

LAPPO, S. S., GULEV, S. K. and ROZHDESTVENSKY, A. E., 1990, *Large-scale Thermal Interaction in the System 'Ocean–Atmosphere' and Energy-active Zone of the World Ocean* (Leningrad: Gidrometeoizdat).

LAPPO, S. S. and MUZYCHENKO, A. G., 1989, The oceanic climate regions in the North Pacific. In *Large-scale Atmosphere–Ocean Interaction and the Formation of the Hydrophysical Fields* (Moscow: Gidrometeoizdat), pp. 42–47.

LAPPO, S. S., SOBCHENKO, E. A. and TERZIEV, F. S., 1987, The laws of the formation of the Newfoundland EAZO of the North Atlantic ocean. In *Progress in Science and Technology. Atmosphere, Ocean, Cosmos – the programme 'Sections'*, Volume 8 (Moscow: VINITI), pp. 227–232.

LASHOF, D. A., 1989, The dynamic greenhouse: feedback processes that may influence future concentrations of atmospheric trace gases and climatic change. *Climatic Change*, **14**, 213–242.

LASZLO, I. and PINKER, R., 1993, Shortwave cloud radiative forcing at the top of the atmosphere and at the surface and the atmosphere column as determined from ISCCP C1 data. *Journal of Geophysical Research*, **98**, 2703–2713.

LATIF, M., STERL, A., MAIER-REIMER, E. and JUNGE, M. M., 1993, Climate variability in a coupled GCM. Part I: The tropical Pacific. *Journal of Climate*, **6**, 5–21.

LAU, K. M., 1985a, Elements of a stochastic-dynamical theory of the long-term variability of the El Niño/southern oscillation. *Journal of Atmospheric Science*, **42**, 1552–1558.

LAU, K. M. and CHAN, P. H., 1983, Climate variability and atmospheric teleconnection from satellite observed cloud fluctuations. *Proceedings of the Fifth Conference on Atmospheric Radiation, held in Baltimore, MD, on 31 October–4 November 1983* (Boston: American Meteorological Society), pp. 214–216.

LAU, N. C., 1985b, Modelling the seasonal dependence of the atmospheric response to observed El Niños in 1962–1976. *Monthly Weather Review*, **113**, 1970–1996.

LAU, N. C., 1986, Diagnosis of intra-seasonal oscillations appearing in GCM experiments conducted at GFDL. World Climate Paper No. 121 (WMO/TD-N138), pp. 63–68.

LAURMANN, J. A., 1985, The global greenhouse warming problem: setting the priorities. A guest editorial. *Climatic Change*, **7**, 261–266.

LAUSCHER, F., 1981, *Secular Variations in 10-year Mean and Extreme Annual Mean Temperatures over the Whole Globe* (in German), Publication 252, Zentralanstalt, für Meteorologie und Geodynamik, Wien, Austria.

LAW, K. S. and PYLE, J. A., 1993, Modeling trace gases in the troposphere. Ozone and odd nitrogen. *Journal of Geophysical Research*, **98**, 18377–18400.

LAWRENCE, M. G., 1993, An empirical analysis of the strength of the phytoplankton–dimethylsulfide–cloud–climate feedback cycle. *Journal of Geophysical Research*, **98**, 20663–20674.

LEAN, J., 1987, Solar ultraviolet irradiance variations: a review. *Journal of Geophysical Research*, **92**, 839–868.

LEDLEY, T., 1991a, The climatic response to meridional sea-ice transport. *Journal of Climate*, **4**, 147–163.

LEDLEY, T., 1991b, Snow on sea ice: competing effects in climate shaping. *Journal of Geophysical Research*, **96**, 17195–17208.

LEDLEY, T., 1993, Variations of snow on sea ice: a mechanism for producing climate variations. *Journal of Geophysical Research*, **98**, 10401–10410.

LEDLEY, T. S. and CHU, S., 1994, Global warming and the growth of ice sheet. *Climate Dynamics*, **9**, 213–220.

LENNERTZ, D., 1982, The European Space Agency's Earth observation satellite programmes in the 1980s. *Satellite Remote Sensing and Developing Countries. Proceedings of the EARSeL-ESA Symposium. Igls, 2–21 April, 1982* (Paris: European Space Agency), pp. 21–23.

LENOBLE, J., 1986, Detection of stratospheric aerosol characteristics in relation with climate impact. *Advances in Space Research*, **6**, 67–72.

LENOBLE, J., TANRÉ, D., DESCHAMPS, P. Y. and HERMAN, M., 1982, A simple method to compute the change in earth atmosphere radiation balance due to a stratospheric aerosol layer. *Journal of Atmospheric Science*, **39**, 2565–2582.

LEWIS, J. S. and PRINN, R. G., 1984, *Planets and Their Atmospheres. Origin and Evolution* (Orlando, FL: Academic Press).

LI, S. M. and BARRIE, L. A., 1993, Biogenic sulfur aerosol in the Arctic troposphere. 1: Contributions of total sulfate. *Journal of Geophysical Research*, **98**, 20613–20622.

LI, S. M., BARRIE, L. A. and SIROS, A., 1993, Biogenic sulfur aerosol in the Arctic troposphere. 2: Trends and seasonal variations. *Journal of Geophysical Research*, **98**, 20623–20632.

LI, Z. Q. and LEIGHTON, H. G., 1991, Scene identification and its effect on cloud radiative forcing in the Arctic. *Journal of Geophysical Research*, **96**, 9175–9188.

LI, Z. Q. and LEIGHTON, H. G., 1993, Global climatologies of solar radiation budgets at the surface and in the atmosphere from 5 years of ERBE data. *Journal of Geophysical Research*, **98**, 4919–4930.

LIANG, X.-Z., WANG, W.-C. and DUDEK, M. P., 1996, Northern hemispheric interannual teleconnection patterns and their changes due to the greenhouse effect. *Journal of Climate*, **9**, 465–479.

LIN, X. and COAKLEY, J. A., 1993, Retrieval of properties for semitransparent clouds from multispectral infrared imagery data. *Journal of Geophysical Research*, **98**, 18501–18514.

LINDER, K. and INGLIS, M., 1989, *The Potential Impacts of Climate Change on Electric Utilities: Regional and National Estimates.* Report of ICF Inc. to the US Environmental Protection Agency.

LINDZEN, R. S., 1990, Some coolness concerning global warming. *Bulletin of the American Meteorological Society*, **71**, 288–299.

LINDZEN, R. S., 1991, Review of Climate Change. The IPCC Scientific Assessment. *Quarterly Journal of the Royal Meteorological Society*, **117**, 651–652.

LINGLE, C. S., BRENNER, A. C., ZWALLY, H. J. and DIMARZIO, J. P., 1991, Multiyear elevation changes near the west margin of the Greenland Ice Sheet from satellite radar altimetry. *International Conference on the Role of the Polar Regions in Global Change. Proceedings of a Conference held June 11–15, 1990, at the University of Alaska Fairbanks*, edited by G. WELLER, C. L. WILSON and B. A. B. SEVERIN (Fairbanks: Geophysical Institute and Center for Global Change and Arctic Systems Research), pp. 35–42.

LIOU, K. N., 1992, *Radiation and Cloud Processes in the Atmosphere* (Oxford: Oxford University Press).

LIOU, K. N. and OU, S. C., 1989, The role of cloud microphysical processes in climate: An assessment from a one-dimensional perspective. *Journal of Geophysical Research*, **94**, 8599–8608.

LIOU, K. N. and ZHENG, Q., 1989, An investigation of the role of cloud/radiation in a general circulation model. *IRS '88: Current Problems in Atmospheric Radiation*, edited by J. LENOBLE and F. GELEYN (Hampton, VA: Deepak), pp. 119–122.

LONDON, J. and WHITE, G. F. (editors), 1984, *The Environmental Effects of Nuclear War* (Boulder, CO: Westview Press).

LORENZ, E. N., 1990, Can chaos and intransitivity lead to interannual variability? *Tellus*, **42A**, 378–389.

LOTH, B., GRAF, H. F. and OBERHUBER, J. M., 1993, Snow cover model for global climate simulations. *Journal of Geophysical Research*, **98**, 10451–10464.

LOUGH, J. M. and FRITTS, H. C., 1987, An assessment of the possible effects of volcanic eruptions on North American climate using tree-ring data, 1602 to 1900 A.D. *Climatic Change*, **10**, 219–240.

LOVINS, A. B., LOVINS, L. H., KRAUSE, F. and BACH, W., 1981, *Least Cost Energy. Solving the $CO_2$ Problem* (Andover, MA: Brick House).

LURIA, M., VAN VALIN, C. C., GALLOWAY, J. N., KEENE, W. C., WELLMAN, D. L., SIEVERING, H. and BOATMAN, J. F., 1989, The relationship between dimethylsulfide and particulate sulfate in the mid-Atlantic ocean atmosphere. *Atmospheric Environment*, **23**, 139–148.

LUTHER, F. M., 1983, Detecting the radiative effects of carbon dioxide: Developing a strategy. *Proceedings of the Fifth Conference on Atmospheric Radiation, 31 October–4 November 1983* (Boston: American Meteorological Society), pp. 166–169.

MCBEAN, G. A., 1989, Global energy and water fluxes. *Weather*, **44**, 285–290.

MCCONNELL, J. C., HENDERSON, G. S., BARRIE, L., BOTTENHEIM, J., NIKI, H., LANGFORD, C. H. and TEMPLETON, E. M. J., 1992, Photochemical bromine production implicated in Arctic boundary-layer ozone depletion. *Nature*, **355**, 150–152.

McCormick, M. P., 1992, Initial assessment of the stratospheric and climatic impact of the 1991 Mount Pinatubo eruption: Prologue. *Geophysical Research Letters*, **19**, 149–150.

McCormick, M. P. and Veiga, R. E., 1992, SAGE-II measurements of early Pinatubo aerosols. *Geophysical Research Letters*, **19**, 155–158.

McCormick, M. P. and Wang, P. H., 1987, Background stratospheric aerosol reference model. *Advances in Space Research*, **7**, 73–80.

MacCracken, M. C., 1983, Is there climatic evidence now for carbon dioxide effects?, ARPA Paper N83-62.3, Lawrence Livermore National Laboratory, Livermore, CA.

MacCracken, M. C., 1991, Ten key questions indicating the level of current uncertainty in forecasting climatic change, UCRL-1D-106243, Livermore, CA.

MacCracken, M. C., Cess, R. D. and Potter, G. L., 1986, Climatic effects of anthropogenic Arctic aerosols: an illustration of climate feedback mechanisms with one- and two-dimensional climate models. *Journal of Geophysical Research*, **91**, 14445–14450.

MacCracken, M. C. and Ellsaesser, H. W., 1983, Climatic data bases for detecting $CO_2$-induced climate change. Preprint. UCRL-Draft, Lawrence Livermore Laboratory, Livermore, CA.

MacCracken, M. C. and Kukla, G. J., 1985, Detecting the climatic effects of carbon dioxide: volume summary. In *Detecting the Climate Effects of Increasing Carbon Dioxide*, DOE/ER-0235, US Department of Energy, Washington DC, pp. 1663–1676.

MacCracken, M. C. and Luther, F. M., 1983, Radiative and climatic effects of the El Chichón eruption. *Proceedings of the Fifth Conference on Atmospheric Radiation* (Boston: American Meteorological Society), pp. 346–349.

MacCracken, M. C. and Luther, F. M. (editors), 1985, *The Potential Climatic Effects of Increasing Carbon Dioxide* (Washington DC: US Department of Energy).

MacCracken, M. C. and Modes, H., 1982, The first detection of carbon dioxide effects: Workshop summary 8–10 June 1981, Harpers Ferry, West Virginia. *Bulletin of the American Meteorological Society*, **63**, 1164–1178.

McElroy, M. B. and Salawitch, R. J., 1989, Changing composition of the global stratosphere. *Science*, **243**, 763–770.

McLaren, A. S., Bourke, R. H. and Weaver, R. L. S., 1991, Contour mapping of Arctic Basin ice roughness parameters. *International Conference on the Role of the Polar Regions in Global Change. Proceedings of a Conference held June 11–15, 1990, at the University of Alaska Fairbanks*, edited by G. Weller, C. L. Wilson and B. A. B. Severin (Fairbanks: Geophysical Institute and Center for Global Change and Arctic Systems Research), p. 79.

Madden, R. A., 1991, The effect of changing spatial coverage on estimates of the global mean surface air temperature. *Proceedings of the Fifth Conference on Climate Variations* (Boston: American Meteorological Society), pp. 8–11.

Makarova, E. A., Kharitonov, A. V. and Kazachevskaya, T. V., 1991, *Solar Radiation Flux* (Moscow: Nauka Press).

Malone, R. C., Auer, L. H., Glatzmaier, G. A., Wood, M. C. and Toon, O. B., 1986, Nuclear winter: three-dimensional simulations including interactive transport, scavenging and solar heating of smoke. *Journal of Geophysical Research*, **91**, 1039–1054.

Manabe, S., 1983, Carbon dioxide and climatic change. *Advances in Geophysics*, **25**, 39–82.

Manabe, S. and Bryan, K., 1985, $CO_2$-induced change in a coupled ocean–atmosphere model and its paleoclimatic implications. *Journal of Geophysical Research*, **90**, 11689–11708.

Manabe, S., Spelman, M. J. and Stouffer, R. J., 1992, Transient responses of a coupled ocean–atmosphere model to gradual changes of atmospheric $CO_2$. Part II: Seasonal response. *Journal of Climate*, **5**, 105–126.

Manabe, S. and Stouffer, R. J., 1980, Sensitivity of global climate model to an increase of $CO_2$ concentration in the atmosphere. *Journal of Geophysical Research* **85**, 5529–5554.

MANABE, S. and STOUFFER, R. J., 1988, Two stable equilibria of a coupled ocean–atmosphere model. *Journal of Climate*, **1**, 841–866.

MANABE, S., STOUFFER, R. J., SPELMAN, M. J. and BRYAN, K., 1991, Transient responses of a coupled ocean–atmosphere model to gradual changes of atmospheric $CO_2$, Part I: Annual mean response. *Journal of Climate*, **4**, 785–818.

MANABE, S. and WETHERALD, R. T., 1980, On the distribution of climate resulting from an increase in $CO_2$ content of the atmosphere. *Journal of Atmospheric Science*, **37**, 99–118.

MANABE, S. and WETHERALD, R. T., 1987, Large-scale changes of soil wetness induced by an increase in atmospheric carbon dioxide. *Journal of Atmospheric Science*, **44**, 1211–1235.

MANABE, S., WETHERALD, R. T. and STOUFFER, R. J., 1981, Summer dryness due to an increase of atmospheric $CO_2$ concentration. *Climatic Change*, **3**, 347–386.

MANKIN, W. G., COFFEY, M. T. and GOLDMAN, A., 1992, Airborne observations of $SO_2$, HCl, and $O_3$ in the stratospheric plume of the Pinatubo volcano in July 1991. *Geophysical Research Letters*, **19**, 179–182.

MANLEY, G., 1974, Central England temperatures: Monthly means 1659 to 1973. *Quarterly Journal of the Royal Meteorological Society*, **100**, 389–405.

MARCHETTI, C., 1988, How to solve the $CO_2$ problem without tears. *Proceedings of the World Hydrogen Conference, held in Moscow on 25–29 September* (private communication).

MARCHUK, G. I., 1992, *Conjugated Equations and Analysis of Complex Systems* (Moscow: Nauka).

MARCHUK, G. I., DYMNIKOV, V. P. and ZALESNY, V. B., 1987, *Mathematical Models in Geophysical Hydrodynamics and Numerical Methods for Their Realization* (Leningrad: Gidrometeoizdat).

MARCHUK, G. I., KONDRATYEV, K. YA. and KOZODEROV, V. V., 1989, Energy-active zones: conceptual bases. Part 1. *Progress in Science and Technology. Atmosphere, Ocean, Cosmos – the Programme 'Sections'*, Volume 10 (*Itogi Nauki i Tekhniki. Atmosfera, Okean, Kosmos – Programma 'Raxrezy'*) (Moscow: VINITI).

MARCHUK, G. I., KONDRATYEV, K. Y.A, KOZODEROV, V. V. and KHVOROSTYANOV, V. I., 1986, *Clouds and Climate* (Leningrad: Gidrometeoizdat).

MARENGO, J. A. and HASTENRATH, S., 1992, Case studies of extreme climate events in the Amazon basin. *Proceedings of the Third Symposium on Global Change Studies* (Boston: American Meteorological Society), pp. 95–97.

MARINUCCI, M. R., GIORGI, F. and VISCONTI, G., 1992, A $2 \times CO_2$ climate change scenario over Europe generated with a limited area model nested in a general circulation model. *Proceedings of the Third Symposium on Global Change Studies* (Boston: American Meteorological Society), pp. 33–40.

MARLAND, G., ROTTY, R. M. and TREAT, N. L., 1985, $CO_2$ from fossil fuel burning: global distribution of emissions. *Tellus*, **37B**, 243–258.

MASON, S. J. and TYSON, P. D., 1992, The modulation of sea surface temperature and rainfall associations over southern Africa with solar activity and the quasi-biennial oscillation. *Journal of Geophysical Research*, **97**, 5847–5856.

MASS, C. and SCHNEIDER, S. H., 1977, Statistical evidence on the influence of sunspots and volcanic dust on long-term temperature records. *Journal of Atmospheric Science*, **34**, 1995–2004.

MASS, C. F. and PORTMAN, D. A., 1989, Major volcanic eruptions and climate: a critical evaluation. *Journal of Climate*, **2**, 566–573; see also *ibid.* **3**, 587–588, 589–590.

MATISHOV, G. P., 1987, *The World Ocean and the Glaciation of the Earth* (Moscow: Mysl).

MATSON, M. and WIESNET, D. R., 1981, New data base for climate studies. *Nature*, **289**, 451–456.

MATTENCCI, G., 1989, Orbital forcing in a stochastic resonance model of the late-Pleistocene climatic variations. *Climate Dynamics*, **3**, 179–190.

MATTHEWS, D., GIORGI, F. and BATES, G., 1992, Regional model simulation of an extreme precipitation event in the Colorado river basin. *Proceedings of the Third Symposium on Global Change Studies* (Boston: American Meteorological Society), pp. 20–21.

MATTHEWS, E., 1983, Global vegetation and land use: new high-resolution data bases for climate studies. *Journal of Climate and Applied Meteorology*, **22**, 474–487.

MATTHEWS, E., 1984, Vegetation, land-use and seasonal albedo data sets: documentation of archived data type. NASA Technical Memorandum 86107.

MATTHEWS, E., 1986, Data needs and data bases for climate studies. *Proceedings ISLSCP Conference, held in Rome, 2–6 December 1985* (Paris: European Space Agency), pp. 191–203.

MECHERIKUNNEL, A. T., LEE, R. B., KYLE, H. L. and MAJOR, E. R., 1988, Intercomparison of solar total irradiance data from recent spacecraft measurements. *Journal of Geophysical Research*, **93**, 9503–9509.

MEDINA, J. G., 1992, Range of applicability of a simple local-scale precipitation model to the Delaware river basin. *Proceedings of the Third Symposium on Global Change Studies* (Boston: American Meteorological Society), pp. 29–32.

MEEHL, G. A., 1990, Development of global couple atmosphere general circulation model. *Climate Dynamics*, **5**, 19–34.

MEEHL, G. A., 1991, A Reexamination of the mechanism of the semi-annual oscillation in the southern hemisphere. *Journal of Climate*, **4**, 911–926.

MEEHL, G. A., 1992a, Effect of tropical topography on global climate. *Annual Review of Earth and Planetary Sciences*, **20**, 85–112.

MEEHL, G. A., 1992b, Global coupled models: atmosphere, ocean, sea ice. In *Climate System Modeling*, edited by K. E. TRENBERTH (Cambridge: Cambridge University Press), pp. 555–581.

MEEHL, G. A., 1993, A coupled air–sea biannual mechanism in the tropical Indian and Pacific regions: role of the oceans. *Journal of Climate*, **6**, 31–41.

MEEHL, G. A. and ALBRECHT, B. A., 1988, Tropospheric temperatures and southern hemisphere circulation. *Monthly Weather Review*, **116**, 953–960.

MEEHL, G. A., BRANSTATOR, G. W. and WASHINGTON, W. M., 1993a, Tropical Pacific interannual variability and $CO_2$ climate change. *Journal of Climate*, **6**, 42–63.

MEEHL, G. A. and WASHINGTON, W. M., 1986, Tropical response to increased $CO_2$ in a GCM with a simple mixed layer ocean: similarities to an observed Pacific warm event. *Monthly Weather Review*, **114**, 667–674.

MEEHL, G. A. and WASHINGTON, W. M., 1988, A comparison of soil moisture sensitivity in two global climate models. *Journal of Atmospheric Science*, **45**, 1476–1492.

MEEHL, G. A., WASHINGTON, W. M. and KARL, T. R., 1993b, Low frequency variability and $CO_2$ transient climate change. Part I: Time averaged differences. *Climate Dynamics*, **8**, 117–133.

MEESON, B. W., CORPREW, F. E., MCMANUS, J. M. P., MYERS, D. M., CLOSS, J. W., SUN, K.-J., SUNDAY, D. J. and SELLERS, P. J., 1996, Global data sets for land–atmosphere models are now available. *The Earth Observer*, **7**, No. 5, 22–25.

MELESHKO, V. P., 1991, Calculations of anthropogenic climate changes with coupled global ocean–atmosphere models: problems and ways of further advancement of the models (overview). *Izvestiya Akademii Nauk SSSR, Fizika Atmosfery i Okeana*, **27**, 691–723.

MELESHKO, V. P., SOKOLOV, A. P. and SHEININ, D. A., 1991, A model of the atmosphere and upper oceanic layer for climate studies and long-range weather forecasts. *Meteorology and Hydrology* (*Meteorologia i Gidrologia*), No. 5, 5–14.

MÉSZÁROS, E., 1981, *Atmospheric Chemistry: Fundamental Aspects* (Budapest: Akad. Kaido).

MÉSZÁROS, E., 1985, On the climatic effects of radiatively active trace gases and aerosol particles. *Idöjárás*, **89**, 70–76.

MÉSZÁROS, E., 1987, Past climatic changes as revealed by recent research in Earth sciences. *Idöjárás*, **91**, 302–308.

MÉSZÁROS, E., 1988, Contemporary variations in atmospheric composition. *Idöjárás*, **92**, 101–109.

MILANKOVITCH, M. M., 1920, Théorie mathématique des phénomènes thermiques produits par la radiation solaire. *Académie Yougoslave des Sciences et des Arts de Zagreb* (Paris: Gauthier-Villars).

MILLER, G. H. and DE VERNAL, A., 1992, Will greenhouse warming lead to Northern Hemisphere ice-sheet growth? *Nature*, **355**, 244–246.

MINNIS, P., HARRISON, E. F., STOWE, L. L., GIBSON, G. G., DENN, F. M., DORLLING, D. R. and SMITH, W. L., 1993, Radiative climate forcing by the Mount Pinatubo eruption. *Science*, **259**, 1411–1415.

MINNIS, P. and WIELICKI, B. A., 1988, Comparison of cloud amounts derived using GOES and LANDSAT data. *Journal of Geophysical Research*, **93**, 9385–9403.

MINNIS, P., YOUNG, D. F. and HARRISON, E. F., 1991, Examination of the relationship between outgoing infrared window and total longwave fluxes using satellite data. *Journal of Climate*, **4**, 1114–1133.

MINTZ, Y., SELLERS, P. S. and WILLMET, C. J., 1983, On the design of an interactive biosphere for the GLAS general circulation model. NASA Technical Memorandum No. 84975. Goddard Space Flight Center, Greenbelt, MD.

MITCHELL, C. S., POTTER, G. L., ELLSAESSER, H. W. and WALTON, J. J., 1981, Case study of feedbacks and synergism in a double $CO_2$ experiment. *Journal of Atmospheric Science*, **38**, 1906–1910.

MITCHELL, J. F. B., 1983, The seasonal response of a general circulation model to changes in $CO_2$ and sea temperatures. *Quarterly Journal of the Royal Meteorological Society*, **109**, 113–152.

MITCHELL, J. F. B., 1987, Simulation of climate change due to increased atmospheric carbon dioxide. *The Meteorological Magazine*, **116**, 361–376.

MITCHELL, J. F. B., 1989, The greenhouse effect and climate change. *Review of Geophysics*, **27**, 115–140.

MITCHELL, J. F. B. and HILLS, T. S., 1986, Sea-ice and the antarctic winter circulation: a numcrical experiment. *Quarterly Journal of the Royal Meteorological Society*, **112**, 953–970.

MITCHELL, J. F. B., JOHNS, T. C., GREGORY, J. M. and TETT, S. F. B., 1995, Climate response to increasing levels of greenhouse gases and sulphate aerosols, *Nature*, **376**, 501–504.

MITCHELL, J. F. B. and SENIOR, C. A., 1989, The antarctic winter: simulations with climatological and reduced sea-ice extents. *Quarterly Journal of the Royal Meteorological Society*, **115**, 225–246.

MITCHELL, J. F. B. and WARRILOW, D. A., 1987, Summer dryness in northern mid-latitudes due to increased $CO_2$. *Nature*, **330**, 238–240.

MITCHELL, J. F. B., WILSON, C. A. and CUNNINGTON, W. M., 1987, On $CO_2$ climate sensitivity and model dependence of results. *Quarterly Journal of the Royal Meteorological Society*, **113**, 293–322.

MITCHELL, J. M., 1961, Recent secular changes of global temperature. *Annals of the New York Academy of Sciences*, **95**, 235–250.

MIYAKODA, K., 1986, Status of experimental monthly predictions and proposals for specific initiatives. *Monthly Weather Review*, N110, Appendix B, 13 pp.

MIZEX group, 1989 MIZEX East, 1987, *EOS. Transactions, American Geophysical Union*, **70**, 545–555.

MOKHOV, I. I., 1989, The diagnosis of the features of the annual change of the atmospheric temperature regime in a GCM. *Physics of the Atmosphere and Ocean* (*Fizika Atmosfery i Okeana*), **25**, 143–150.

MOKHOV, I. I., 1993, *Diagnostics of the Structure of the Climatic System* (St Petersburg: Gidrometeoizdat).

MOKHOV, I. I. and SCHLESINGER, M. E., 1993, Analysis of global cloudiness. Comparison of Meteor, Nimbus-7 and ISCCP satellite data. *Journal of Geophysical Research*, **98**, 12849–12868.

MÓLNAR, G. and MIKA, J., 1992, Combined physical–statistical estimate of global changes due to solar constant variations based on model results. *Proceedings of the Symposium on the Solar Constant and Spectral Distribution of Solar Irradiance, Hamburg, 17–18 August 1981* (Boulder, CO:IUGG), pp. 146–154.

MÓLNAR, G. and WANG, W. C., 1993, Effects of cloud optical property feedbacks on the greenhouse warming. *Journal of Climate*, **6**, 814–821.

MOLNIA, B. F., 1991, One glacier's retreat, a global warming does not make: an argument for systematic monitoring of North American glaciers. *International Conference on the Role of the Polar Regions in Global Change. Proceedings of a Conference held June 11–15, 1990, at the University of Alaska Fairbanks*, edited by G. WELLER, C. L. WILSON and B. A. B. SEVERIN (Fairbanks: Geophysical Institute and Center for Global Change and Arctic Systems Research), p. 89.

MONASTERSKY, R., 1995, World climate panel charts path for action. *Science News*, **148**, No. 19, 293.

MORCRETTE, J. J., 1991, Radiation and cloud radiative properties in the European centre for medium range weather forecasts forecasting system. *Journal of Geophysical Research*, **96**, 9121–9132.

MORTIER, F., 1995, Dynamique du $CO_2$ dans la biosphère et conséquences sur la photosynthèse, les arbres et la forët. Technical Publication, National Bureau of Forests, No. 29, 1–159.

MURRAY, R., 1992, Some notable features of Manloy's Central England temperatures with special reference to very warm years. *Weather*, **47**, 98–103.

MURTHY, J. M., 1995, Transient response of the Hadley Centre coupled ocean–atmosphere model to increasing carbon dioxide. III: Analysis of global-mean response using single models. *Journal of Climate*, **8**, 456–514.

MYSAK, L. A., MANAK, D. K. and MARSDEN, R. F., 1991, Greenland sea ice anomalies during 1901–1984 and their relation to an interdecadal Arctic climate cycle. *International Conference on the Role of the Polar Regions in Global Change. Proceedings of a Conference held June 11–15, 1990, at the University of Alaska Fairbanks*, edited by G. WELLER, C. L. WILSON and B. A. B. SEVERIN (Fairbanks: Geophysical Institute and Center for Global Change and Arctic Systems Research), pp. 284–289.

NAGURNY, A. P., 1986, Modelling the continental and marine ice in climate models. *Progress in Science and Technology. Meteorology and Climatology*, Volume 13 (*Itogi Nauki i Tekhniki* (*Meteorologia i Gidrologia*) (Moscow: VINITI).

NAKAMURA, N. and OORT, A. H., 1988, Atmospheric heat budgets of the polar regions. *Journal of Geophysical Research*, **93**, 9510–9524.

NAMAIS, J., 1980, Some concomitant regional anomalies associated with hemispherically averaged temperature variations. *Journal of Geophysical Research*, **85**, 1585–1590.

NAMAIS, J., 1983, Short period climatic variations. *Collected Works, 1934 through 1982*. University of California, San Diego.

NASA, 1977, Proposed NASA Contribution to the Climate Program. NASA, Goddard Space Flight Center, Greenbelt, MD.

NASA, 1984, Earth Observing System. Science and Mission Requirements Working Group Report. Volume 1 (Parts 1 and 2). NASA Technical Memorandum No. 86129, Goddard Space Flight Center, Greenbelt, MD.

NASA, 1987, Space opportunities for tropospheric chemistry research. *Proceedings of the*

*Workshop held in New York City, NY, on 9–13 September 1985.* NASA Conference Publications, No. 2450.

NASA, 1991, *Reference Handbook EOS*, Goddard Space Flight Center, Greenbelt, MD.

NASA EARTH SCIENCE APPLICATIONS DIVISION, 1987, *The Program and Plans for FY 1987–1988–1989* (Washington DC: NASA).

NASRALLAH, H. A. and BALLING, R. C., 1993, Spatial and temporal analysis of middle eastern temperature changes. *Climatic Change*, **25**, 153–161.

NATIONAL CLIMATE PROGRAM OFFICE, 1988, *National Climate Program. Five-Year Plan, 1989–1993* (Rockville, MD: US Department of Commerce).

NATIONAL CLIMATE PROGRAM OFFICE OF SCIENCE, 1985, *Interagency Research Report for Assessing Climatic Effects of Nuclear War. Report to the Office of Science, Technology and Policy* (Washington DC: NOAA).

NATIONAL RESEARCH COUNCIL (US), 1982, *Carbon Dioxide and Climate: A Second Assessment* (Washington DC: $CO_2$ Climate Review Panel).

NESME-RIBES, E., FERREIRA, E. N., SADOURNY, R., LE TREUT, H., LI, Z. X., 1993, Solar dynamics and its impact on solar irradiance and terrestrial climate. *Journal of Geophysical Research*, **98**, 18923–18936.

NEWELL, R. E. and DOPPLICK, T. G., 1979, Questions concerning the possible influence of anthropogenic $CO_2$ on atmospheric temperature. *Journal of Applied Meteorology*, **18**, 822–825.

NEWELL, R. E. and HSIUNG, J., 1984, Sea surface temperature, atmospheric $CO_2$ and the global energy budget: some comparisons between the past and present. In *Climatic Changes on a Yearly to Millenial Basis* (Dordrecht: Reidel), pp. 533–561.

NICHOLLS, N., 1988, Low latitude volcanic eruptions and the El Niño–southern oscillation. *Journal of Climatology*, **8**, 91–95.

NICOLIS, C., 1987, Long-term climatic variability and chaotic dynamics. *Tellus*, **39A**, 1–9.

NICOLIS, C., 1988, Transient climatic response to increasing $CO_2$ concentration: some dynamical scenarios. *Tellus*, **40A**, 50–60.

NJAN, E. C., 1989, A generalized theory of sun–climate/weather link and climate change. *Il Nuovo Cimento*, **12C**, 597–612.

NOAA, 1979, *An Ocean Climate Research Plan* (Boulder, CO: NOAA Environmental Research Laboratory), 46 pp.

NOAA, 1986, *Annual Report 1986. National Climate Program* (Rockville, MD: National Climate Program Office)

NOAA, 1988, *National Climate Program Annual Report 1987* (Rockville, MD: NOAA).

NOAA, 1991, *Product Development Plans for Operational Satellite products.* NOAA Climate and Global Change Program. SPE Special Reports No. 5, University Corporation for Atmospheric Research (UCAR), Boulder, CO.

NORWEGIAN RESEARCH COUNCIL, 1992, Nansen Centennial Arctic Programme Plan for Scientific Expedition to the Arctic Ocean 1994–1996 (Oslo: Norwegian Research Council), 54 pp.

NSF-UCAR LONG-RANGE PLANNING COMMITTEE, 1987, *The Atmospheric Sciences: A vision for 1989–1994. Report of the NSF-UCAR Long-Range Planning Committee* (Washington DC: National Academic Press).

O'BRIEN, J. J. and BUSALACCHI, A. J., 1983, Pacific ocean response to El Niño conditions. In *Large-scale Oceanographic Experiments in the WCRP. Papers presented at the JSC/CCCO Study Conference in Tokyo, 10–21 May 1983.* WCRP Publications Series, 2, No. 1, 111–122.

OCKERT-BELL, M. E. and HARTMANN, D. L., 1992, The effect of cloud type on Earth's energy balance: results for selected regions *Journal of Climate*, **5**, 1157–1171.

OECHEL, W. C., 1991, Effects of global change on net ecosystem carbon flux of Arctic tussock tundra. *International Conference on the Role of the Polar Regions in Global Change. Proceedings of a Conference held June 11–15, 1990, at the University of Alaska*

*Fairbanks*, edited by G. Weller, C. L. Wilson and B. A. B. Severin (Fairbanks: Geophysical Institute and Center for Global Change and Arctic Systems Research), p. 369.

Oerlemans, J., 1989, A projection of future sea level. *Climatic Change*, **15**, 151–174.

Oglesby, R. J., Marshall, S. and Saltzman, B., 1993, A comparison of GCM sensitivity to $CO_2$ and solar luminosity changes. AGU 1993 Annual Meeting, EOS Suppl., p. 122.

Oliver, H. R., 1992, Studies of surface energy balance of sloping terrain. *International Journal of Climatology*, **12**, 55–68.

Oort, A. H., 1983, *Global Atmospheric Circulation Statistics*. NOAA Professional Paper No. 14 (Washington DC: Government Printing Office), pp. 1958–1973.

Oort, A. H., Pan, Y. H., Reynolds, R. W. and Ropelewski, C. F., 1987, Historical trends in the surface temperature over the oceans based on the COADS. *Climate Dynamics*, **2**, 29–38.

Oort, A. H. and Vonderhaar, T. H., 1976, On the observed annual cycles in the ocean–atmosphere heat balance over the northern hemisphere. *Journal of Physical Oceanography*, **6**, 781–800.

Osborn, T. J. and Wigley, T. M. L., 1994, A simple model for estimating methane concentration. *Climate Dynamics*, **9**, 181–193.

Ou, S. C. S. and Liou, K. N., 1985, Cumulus convection and climatic temperature perturbations. *Journal of Geophysical Research*, **90**, 2223–2332.

Owens, A. J., Hales, C. H., Filkin, D. L., Miller, C., Steed, J. M. and Jesson, J., (1985), A coupled one-dimensional radiative–convective chemistry-transport model of the atmosphere. 1: Model structure and steady state perturbation calculations. *Journal of Geophysical Research*, **90**, 2283–2312.

Paci, G. and De Leffe, A., 1983, Status and future plans for the first European remote sensing satellite ERS-1. In *Remote Sensing Applications in Marine Science and Technology*, edited by A. P. Cracknell (Dordrecht: Reidel), pp. 67–81.

Palmer, T. N., 1993, A nonlinear dynamical perspective on climate change. *Weather*, **48**, 314–325.

Palmer, T. N. and Branković, C., 1989, The 1988 US drought limited to anomalous sea surface temperature. *Nature*, **338**, 54–57.

Paltridge, G. W., 1989, Meteosat data as an index of fluctuations in Earth radiation budget on monthly time scales. *Atmosfera*, **2**, 125–129.

Paltridge, G. W., 1991, Rainfall-albedo feedback to climate. *Quarterly Journal of the Royal Meteorological Society*, **117**, 647–650.

Paltridge, G. and Woodruff, S., 1981, Changes in global surface temperature from 1880 to 1977 derived from historical records of sea surface temperature. *Monthly Weather Review*, **109**, 2427–2434.

Parish, T. R., 1991, On the relationship between Antarctic katabatic winds and large-scale tropospheric circulation. *International Conference on the Role of the Polar Regions in Global Change. Proceedings of a Conference held June 11–15, 1990, at the University of Alaska Fairbanks*, edited by G. Weller, C. L. Wilson and B. A. B. Severin (Fairbanks: Geophysical Institute and Center for Global Change and Arctic Systems Research), p. 191.

Parker, D. E., 1985, On the detection of temperature changes induced by increasing atmospheric carbon dioxide. *Quarterly Journal of the Royal Meteorological Society*, **3**, 587–601.

Parker, D. E., 1988, Stratospheric aerosols and sea-surface temperatures. *Journal of Climatology*, **8**, 87–90.

Parkinson, C. L., 1989, Dangers of multiyear averaging in analyses of long-term climate trends. *Climate Dynamics*, **4**, 39–44.

Parkinson, C. L., 1991, Strengths and weaknesses of sea ice as a potential early indicator of climate change. *International Conference on the Role of the Polar Regions in Global Change. Proceedings of a Conference held June 11–15, 1990, at the University of Alaska Fairbanks*, edited by G. Weller, C. L. Wilson and B. A. B. Severin (Fairbanks: Geophysical Institute and Center for Global Change and Arctic Systems Research), pp. 17–21.

Parungo, F. and Hicks, B., 1993, Sulfate aerosol distributions and cloud variations during El Niño anomalies. *Journal of Geophysical Research*, **98**, 2662–2675.

Parungo, F., Kopcewicz, B. and Nagamoto, C., 1992, Aerosol particles in the Kuwait oil fire plumes: Their morphology, size distribution, chemical composition, transport, and potential effect on climate. *Journal of Geophysical Research*, **97**, 15867–15882.

Pearman, G. I. (editor), 1988, *Greenhouse: Planning for Climate Change* (Melbourne: Scientific and Industrial Research Organization, Commonwealth of Australia).

Pearman, G. I., 1991, Changes in atmospheric chemistry and the greenhouse effect: A southern hemisphere perspective. *Climatic Change*, **18**, 131–146.

Pearman, G. I. (editor), 1992, *Limiting Greenhouse Effects. Controlling Carbon Dioxide Emissions* (Chichester: Wiley).

Peng, L., Chou, M. D. and Arking, A., 1987, Climate warming due to increasing atmospheric $CO_2$: simulations with a multilayer coupled atmosphere–ocean seasonal energy balance model. *Journal of Geophysical Research*, **92**, 5505–5521.

Penner, J. B., Charlson, R. J., Hales, J. M., *et al.*, 1993, Quantifying and minimizing uncertainty of climate forcing by anthropogenic aerosols. DOE/WBB-0092T. UC-402. Under contract NW-7405-Eng.-48. US Department of Energy, Washington DC.

Penner, J. E., 1986, Uncertainties in the smoke source term for 'nuclear winter' studies. *Nature*, **324**, 922–926.

Pepin, R. O., 1988, On the loss of Earth's primordial atmosphere. *Proceedings of Conference on the Origin of the Earth, held at Berkeley, California, on 1–3 December 1988* (Boston: American Meteorological Society), pp. 68–69.

Peterson, T. C., Barnett, T. P., Roeckner, E. and Vonderhaar, T. H., 1992, An analysis of the relationship between cloud anomalies and sea surface temperature anomalies in a global circulation model. *Journal of Geophysical Research*, **97**, 20497–20506.

Philips, O. L. and Gentry, A. H., 1994, Increasing turnover through time in tropical forests. *Science*, **263**, 954–958.

Philips, T. J., Gates, W. L. and Arpe, K., 1991, Temporal sampling considerations in global climate modeling. *Proceedings of the Fifth Conference on Climate Variations* (Boston: American Meteorological Society), pp. 20–23.

Physikalisch-Meteorologisches Observatorium und Weltstrahlungszentrum, 1992, *Jahresbericht 1991*, Publication No. 659, Davos.

Pitari, G., Rizi, V., Ricciardulli, L. and Visconti, G., 1993, High-speed civil transport impact; role of sulfate, nitric acid trihydrate and ice aerosols studied with a two-dimensional model including aerosol physics. *Journal of Geophysical Research*, **98**, 23141–23164.

Pitt, D. and Nilsson, S., 1993, *Protecting the Atmosphere. The Climate Change Convention and its Context* (London: Earthscan).

Pittock, A. B., Walsh, K. and Frederiksen, J. S., 1989, General circulation model simulation of mild nuclear winter effects. *Climatic Dynamics*, **3**, 191–206.

Pollack, J. B., 1979, Climate change on the terrestrial planets. *Icarus*, **37**, 479–553.

Pollack, J. B. and McKay, P., 1985, The impact of polar stratospheric clouds on the heating rates of the winter polar stratosphere. *Journal of Atmospheric Science*, **42**, 245–262.

Pollack, J. B., Rind, D., Lacis, A., Hansen, J. E., Sato, M. and Ruedy, R.,

1993, GCM simulation of volcanic aerosol forcing, Part I: Climate changes induced by steady-state perturbations. *Journal of Climate*, **6**, 1719–1742.

Portman, D. A. and Gutzler, D. S., 1995, Explosive volcanic eruptions, the El Niño – Southern Oscillation, and US climate variability. *Journal of Climate*, **9**, 17–33.

Posey, J. W. and Clapp, P. F., 1964, Global distribution of normal surface albedo. *Geofisica Internacional*, **4**, 33–48.

Post, M. J., Grund, C. J., Langford, A. O. and Proffitt, M. H., 1992, Observations of Pinatubo ejecta over Boulder, Colorado, by lidars of three different wavelengths. *Geophysical Research Letters*, **19**, 195–198.

Potter, G. L., Kiehl, J. T. and Cess, R. D., 1987, A clarification of certain issues related to the $CO_2$–climate problem. *Climatic Change*, **10**, 87–96.

Potter, G. L., Slingo, J. M., Morcrette, J. J. and Corsetti, 1992, A modeling perspective on cloud radiative forcing. *Journal of Geophysical Research*, **97**, 20507–20518.

Práger, T. and Kovács, E., 1988, Investigation of the climate modifying effects of atmospheric trace gases and aerosol particles by a radiative–convective model. *Idöjárás*, **92**, 153–162.

Prather, M., 1988, European sources of halocarbons and nitrous oxide: update 1986. *Journal of Atmospheric Chemistry*, **6**, 375–406.

Prentice, I. E., Sykes, M. T. and Cramer, W., 1993, A simulation model for the transient effects of climate change on forest landscapes. *Ecological Modelling*, **65**, 51–70.

Prinn, R., Cunnold, D., Rasmussen, R., Simmonds, P., Atreya, F., Crawford, A., Fraser, P. and Rosen, R., 1987, Atmospheric trends in methylchloroform and the global average for the hydroxyl radical. *Science*, **238**, 945–950.

Putaud, J. P., Belviso, S., Nguyen, B. C. and Mihalopoulos, N., 1993, Dimethylsulfide, aerosols and condensation nuclei over the tropical northeastern Atlantic ocean. *Journal of Geophysical Research*, **98**, 14863–14871.

Quinn, P. K., Covert, D. S., Bates, T. S., Kapustin, V. N., Ramsey-Bell, D. C. and McInnes, L. M., 1993, Dimethylsulfide/cloud condensation nuclei/climate system: relevant size-resolved measurements of the chemical and physical properties of atmospheric aerosol particles. *Journal of Geophysical Research*, **98**, 10411–10428.

Radok, U. and Wendler, G., 1991, Plateau weather. *International Conference on the Role of the Polar Regions in Global Change. Proceedings of a Conference held June 11–15, 1990, at the University of Alaska Fairbanks*, edited by G. Weller, C. L. Wilson and B. A. B. Severin (Fairbanks: Geophysical Institute and Center for Global Change and Arctic Systems Research), pp. 192–198.

Ramanathan, V., 1981, The role of ocean–atmosphere interactions in the $CO_2$ climate problem. *Journal of Atmospheric Science*, **38**, 918–930.

Ramanathan, V., 1982, The role of ocean–atmosphere interactions in the $CO_2$ problem. *American Institute of Physics Conference Proceedings*, No. 82, pp. 275–291.

Ramanathan, V., 1987, The role of earth radiation budget studies in climate and general circulation research. *Journal of Geophysical Research*, **92**, 4075–4095.

Ramanathan, V., 1988a, The greenhouse theory of climate change: a test by an inadvertent global experiment. *Science*, **240**, 293–299.

Ramanathan, V., 1988b, The radiative and climatic consequences of the changing atmospheric composition of trace gases. In *Changing Atmosphere: Report of Dahlem Workshop, Berlin, 1–6 November 1987* (Chichester: Wiley), pp. 159–186.

Ramanathan, V., Barkstrom, B. R. and Harrison, E. F., 1989a, Climate and the Earth's radiation budget. *Physics Today*, May, 22–32.

Ramanathan, V., Callis, L. B. and Boughner, R. E., 1976, Sensitivity of surface temperature and atmospheric temperature to perturbations in the stratospheric concentration of ozone and nitrogen dioxide. *Journal of Atmospheric Science*, **33**, 1092–1112.

Ramanathan, V., Callis, L., Cess, R., Hansen, J., Isaksen, I., Kuhn, W., Lacis, A., Luther, F., Mahlman, J., Rach, R. and Schlesinger, M., 1987,

Climate–chemical interactions and effects of changing atmospheric trace gases. *Review of Geophysics*, **25**, 1441–1482.

RAMANATHAN, V., CESS, R. D., HARRISON, E. F., MINNIS, P., BARKSTROM, B. R., AHMAD, E. and HARTMANN, D., 1989b, Cloud-radiative forcing and climate: results from the Earth radiation budget experiment. *Science*, **243**, 57–63.

RAMANATHAN, V. and COLLINS, W., 1991, Thermodynamic regulation of ocean warming by cirrus clouds deduced from observations of the 1987 El Niño. *Nature*, **351**, 27–32.

RAMASWAMY, V., 1988, Aerosol radiation forcing and model responses. In *Aerosol and Climate*, edited by P. HOBBS and M. P. MCCORMICK (Hampton, VA: Deepak).

RAMASWAMY, V. and CHEN, C. T., 1993, An investigation of the global solar radiative forcing due to changes in cloud liquid water path. *Journal of Geophysical Research*, **98**, 16703–16712.

RAMASWAMY, V. and FREIDENREICH, S. M., 1992, A study of broadband parameterizations of the solar radiative interactions with water vapour and water drops. *Journal of Geophysical Research*, **97**, 11487–11512.

RAMASWAMY, V. and KIEHL, J. T., 1985, Sensitivities of the radiative forcing due to large loadings of smoke and dust aerosols. *Journal of Geophysical Research*, **90**, 5597–5614.

RAMPINO, M. R., SANDERS, J. E., NEWMAN, W. S. and KÖNIGSSON, L. K. (editors), 1987, *Climate: History, Periodicity, and Predictability* (New York: Van Nostrand Reinhold).

RAMPINO, M. R., SELF, S. and STOTHERS, R. B., 1988, Volcanic winters. *Annual Review of Earth and Planetary Science*, **16**, 73–99.

RAMPINO, M. R. and VOLK, T., 1988, Mass extinctions, atmospheric sulphur and climate warming at the K/T boundary. *Nature*, **332**, 63–65.

RANDALL, D. A., CESS, R. D., BLANCHET, J. P., BOER, G. J., DAZLICH, D. A., DELGENIO, A. D., DEQUE, M., DYMNIKOV, V., GALIN, V., GHAN, S. J., LACIS, A. A., LETREUT, H., LI, Z. D., LIANG, X. Z., MCAVANEY, B. J., MELESHKO, V. P., MITCHELL, J. F. B., MORCRETTE, J. J., POTTER, G. L., RIKUS, L., ROECKNER, E., ROYER, J. F., SCHLESE, U., SHEININ, D. A., SLINGO, J., SOKOLOV, A. P., TAYLOR, K. E., WASHINGTON, W. M., WETHERALD, R. T., YAGAI, I. and ZHANG, M. H., 1992, Intercomparison and interpretation of surface energy fluxes in atmospheric general circulation models. *Journal of Geophysical Research*, **97**, 3711–3724.

RANDALL, D. A., HARSHVARDHAN, DAZLICH, D. A. and CORSETTI, T. G., 1989, Interactions among radiation, convection, and large-scale dynamics in a general circulation model. *Journal of Atmospheric Science*, **46**, 1943–1970.

RANGO, A., 1984, *Global Hydrological Cycle*. NASA Technical Memorandum No. 86129, Goddard Space Flight Center, Greenbelt, MD, Part 2, pp. 1–4.

RANGO, A. and MARTINEC, J., 1987, Large scale effects of seasonal snow cover and temperature increase on runoff. *IAHS Publications*, No. 166, 121–127.

RAO, P. K., HOLMES, S. J., ANDERSON, R. K., WINSTON, J. S. and LEHR, P. E., 1990, *Weather Satellites: Systems, Data and Environmental Applications* (Boston: American Meteorological Society).

RASCH, P. J., BOVILLE, B. A. and BRASSEUR, G. P., 1995, A three-dimensional general circulation model with coupled chemistry for the middle atmosphere. *Journal of Geophysical Research*, **100**, 9041–9071.

RASMUSSON, E. M., 1985, Climate variability and southern oscillation. *International Conference on the TOGA Science Program, WCRP Publication Series, No. 4* (Geneva: WMO), pp. III.11–III.21.

RASMUSSON, E. M., ARKIN, P. A., KRUEGER, A. F., QUIROZ, R. S. and REYNOLDS, R. W., 1983, The equatorial Pacific atmospheric climate during 1982–1983. *Tropical Ocean–Atmosphere Newsletter*, No. 21, 2–3.

RATAN, D., SMITH, G. L. and BESS, T. D., 1991, A twelve year data set of albedo and absorbed solar radiation from Nimbus-6 and Nimbus-7 satellites. *Proceedings of the Fifth Conference on Climate Variations* (Boston: American Meteorological Society), pp. 380–383.

RAVAL, A. and RAMANATHAN, V., 1985, Observational determination of the greenhouse effect. *Nature*, **342**, 758–761.

RECK, R. A., 1983, The role of aerosols in the climate systems: results of numerical experiments in climate models. Space observations of aerosols and ozone. *Advances in Space Research*, **2**, 11–18.

RECK, R. A. and HUMMEL, J. R., 1981, Influence of aerosol optical properties on surface temperatures computed with a radiative convective model. *Atmospheric Environment*, **15**, 1727–1771.

REEBURGH, W. S. and WHALEN, S. C., 1991, The role of tundra and taiga systems in the global methane budget. *International Conference on the Role of the Polar Regions in Global Change. Proceedings of a Conference held June 11–15, 1990, at the University of Alaska Fairbanks*, edited by G. WELLER, C. L. WILSON and B. A. B. SEVERIN (Fairbanks: Geophysical Institute and Center for Global Change and Arctic Systems Research), p. 370.

REID, G. C., 1991, Solar total irradiance variations and the global sea surface temperature record. *Journal of Geophysical Research*, **96**, 2835–2844.

REITAN, C. H., 1974, A climatic model of solar radiation and temperature change. *Quaternary Research*, **4**, 25–38.

REVELLE, R., 1983, Methane hydrates in continental slope sediments and increasing atmospheric carbon dioxide. In *National Research Council, Changing Climate* (Washington DC: National Academy Press), pp. 252–261.

RIND, D., 1986, The dynamics of warm and cold climates. *Journal of Atmospheric Science*, **43**, 3–24.

RIND, D., 1987, The doubled $CO_2$ climate: impact of the sea surface temperature gradient. *Journal of Atmospheric Science*, **44**, 3235–3268.

RIND, D., 1988a, The doubled $CO_2$ climate and the sensitivity of the modeled hydrologic cycle. *Journal of Geophysical Research*, **93**, 5385–5412.

RIND, D., 1988b, Dependence of warm and cold climate depiction on climate model resolution. *Journal of Climate*, **1**, 965–997.

RIND, D., 1995, Drying out the tropics. *New Scientist*, **146**, No. 1976, 36–40.

RIND, D. and CHANDLER, M., 1991, Increased ocean heat transport and warmer climate. *Journal of Geophysical Research*, **96**, 7437–7461.

RIND, D., GOLDBERG, R. and RUEDY, R., 1989, Change in climatic variability in the 21st century. *Climatic Change*, **14**, 5–38.

RIND, D., HEALY, R., PARKINSON, C. and MARTINSON, D., 1995, The role of sea ice in $2 \times CO_2$ climate model sensitivity, Part I: The total influence of sea ice thickness and extent. *Journal of Climate*, **8**, 449–463.

RIND, D., SUOZZO, R. and BALACHANDRAN, N. K., 1988a, The GISS global climate/middle atmosphere model. Part 2: Model variability due to interactions between planetary wave drag. *Journal of Atmospheric Science*, **45**, 371–386.

RIND, D., SUOZZO, R., BALACHANDRAN, N. K., LACIS, A. and RUSSELL, G., 1988b, The GISS global climate/middle atmosphere model. Part 1: Model structure and climatology. *Journal of Atmospheric Science*, **45**, 329–370.

ROADS, J. O., CHEN, S. C. and KAO, J., 1991, Global aspects of the hydrological cycle. *Proceedings of the Fifth Conference on Climate Variations* (Boston: American Meteorological Society), pp. 54–55.

ROBINSON, D. A., DEWEY, K. F. and KEIM, R. R., 1993, Global snow cover monitoring: an update. *Bulletin of the American Meteorological Society*, **74**, 1689–1698.

ROBOCK, A., 1984a, Climate model simulations of the effects of the El Chichón eruption. *Geofisica Internacional*, **23**, 403–414.

ROBOCK, A., 1984b, Snow and ice feedbacks prolong effects of nuclear winter. *Nature*, **310**, 667–670.

ROBOCK, A., 1985, An updated climate feedback diagram. *Bulletin of the American Meteorological Society*, **66**, 786–787.

ROBOCK, A. and MAO, J., 1995, The volcanic signal in surface temperature observations, *Journal of Climate*, **8**, 1086–1103.

RODHE, H. and LANGNER, J., 1993, Atmospheric concentration of DMS and its oxidation products estimated in a global 3-D model. In *Dimethylsulfide: Oceans, Atmosphere and Climate* edited by G. RESTELLY and G. ANGELETTI (Berlin: Springer), pp 333–343.

ROECKNER, E., 1988, Roeckner replies. *Nature*, **235**, 304.

ROECKNER, E., SCHLESE, U., BIERCAMP, J. and LUEWE, P., 1987, Cloud optical depth feedbacks and climate modelling. *Nature*, **329**, 138–140.

ROEKERER, J. G., 1994, Solar variability effects on climate. *Paleoclimate Research* (in press).

ROMANOV, V. F., 1988, On a parametrization of the effects of synoptic vortex processes in climatic models of atmospheric general circulation. *Izvestiya Akademii Nauk SSSR, Fizika Atmosfery i Okeana*, **24**, 367–380.

ROMANOV, V. F., ARISKINA, N. V., VASILYEV, V. F. and LAGUN, V. E., 1987, *Atmospheric Energetics in Polar Regions* (Leningrad: Gidrometeoizdat).

ROOTS, E. F., 1989, Climate Change: high-latitude regions. *Climatic Change*, **15**, 223–254.

ROPELEWSKI, C. F., HALPERT, M. S., CHELLISH, M., *et al.*, 1992, Surface temperature and precipitation patterns during the 1980s: climate variability or global change? *Proceedings of the Third Symposium on Global Change Studies* (Boston: American Meteorological Society), pp. 110–117.

ROSEN, R. D. and GUTOWSKI, W. J., 1991, Sensitivity of zonal winds and atmospheric angular momentum to doubling of $CO_2$. *Proceedings of the Fifth Conference on Climate Variations* (Boston: American Meteorological Society), pp. 32–34.

ROSENBERG, N. J., 1993, A methodology called 'MINK' for study of climate change impacts and responses on the regional scale. *Climatic Change*, **24**, 1–6.

ROSENBERG, N. J., EASTERLING, W. E., CROSSON, P. R. and DARMSTADTER, J. (editors), 1989, *Greenhouse Warming: Abatement and Adaption. Proceedings from Resources for the Future* (Washington DC: Resources for the Future).

ROTMANS, J., 1990, *IMAGE: An Integrated Model to Assess the Greenhouse Effect* (Dordrecht: Kluwer).

ROWE, C. M., 1993a, Global land-surface albedo modelling. *International Journal of Climatology*, **13**, 473–495.

ROWE, C. M., 1993b, Incorporating landscape heterogeneity in land surface albedo models. *Journal of Geophysical Research*, **98**, 5037–5043.

ROWLAND, F. S. and ISAKSEN, I. S. A. (editors), 1988, *The Changing Atmosphere* (Chichester: Wiley).

ROWNTREE, P. R. and SANGSTER, A. B., 1986, Modelling the impact of land surface changes on Sahel rainfall. In *Research Activities in Atmospheric and Ocean Model, Report No. 9* (*WMO*/TD-*N141*), pp. 7.27–7.30.

ROYER, T. C., 1991, High latitude climate forcing by 18.6-yr lunar tidal fluctuations. *International Conference on the Role of the Polar Regions in Global Change. Proceedings of a Conference held June 11–15, 1990, at the University of Alaska Fairbanks*, edited by G. WELLER, C. L. WILSON and B. A. B. SEVERIN (Fairbanks: Geophysical Institute and Center for Global Change and Arctic Systems Research), p. 150.

RUSSELL, J. M., GORDLEY, L. L., PARK, J. H., DRAYSON, S. R., HASKETH, W. D., CICERONE, R. J., TUCK, A. F., FREDERICK, J. E., HARRIES, J. E. and CRUTSEN, P., 1993, The halogen occultation experiment. *Journal of Geophysical Research*, **98**, 10777–10798.

SAGAN, C., 1984, Nuclear war and climatic change – guest editorial. *Climatic Change*, **6**, 1–4.

SALBY, M. L. and SHEA, D. J., 1991, Correlation between solar activity and the atmosphere: an unphysical explanation. *Journal of Geophysical Research*, **96**, 22579–22597.

SALTZMAN, B., 1983, *Theory of Climate. Advances in Geophysics*, Volume 25 (New York: Academic Press).

SANKAR-RAO, M., 1986, An overview of climate models. *Proceedings of the Indian Academy of Science, Earth and Planetary Science*, **95**, 5447–5484.

SARMIENTO, J. and TOGGWEILER, J., 1984, A new model for the role of the oceans in determining atmospheric $pCO_2$. *Nature*, **308**, 621–624.

SATO, N., SELLERS, P. J., RANDALL, D. A., SCHNEIDER, E. K., SHUKLA, J., KINTER, J. L., HOU, Y. T. and ALBERTAZZI, E., 1989, Effects of implementing the simple biosphere model in a general circulation model. *Journal of Atmospheric Science*, **46**, 2757–2782.

SAUSEN, R., BARTHEL, K. and HASSELMANN, K., 1987, Coupled ocean–atmosphere models with flux correction. *Climate Dynamics*, **2**, 145–164.

SAVCHENKO, V. G. and NAGURNY, A. P., 1987, *The Impact of Heat Fluxes from the Ocean on the High-Latitude Climate Oscillations* (Leningrad: Gidrometeoizdat).

SAVCHENKO, V. G., NAGURNY, A. P. and MAKSHTAS, A. P., 1989, The sea ice response to atmospheric aerosol pollution. *Meteorology and Hydrology* (*Meteorologia i Gidrologia*), No. 4, 102–108.

SCHATTEN, K. H., 1988, A model for solar constant secular changes. *Geophysical Research Letters*, **15**, 121–124.

SCHIFFER, R. A. (editor), 1987, *Report of the NASA Workshop on the Detection of the Greenhouse Effect on Climate from Space Observations, 29 September–10 October 1987* (Washington DC: NASA).

SCHLESINGER, M. E., 1986, Equilibrium and transient climatic warming induced by increased atmospheric $CO_2$. *Climate Dynamics*, **1**, 35–52.

SCHLESINGER, M. E., 1988a, Negative or positive cloud optical depth feedback? *Nature*, **335**, 303–304.

SCHLESINGER, M. E., 1988b, Quantitative analysis of feedbacks in climate model simulations of $CO_2$-induced warming. In *Physically-Based Modelling and Simulation of Climate and Climatic Change*. Volume II, edited by M. E. SCHLESINGER (Dordrecht: Reidel), pp. 653–736.

SCHLESINGER, M. E., 1988c, How to make models for behaviour of clouds? *Nature*, **336**, 315–316.

SCHLESINGER, M. E., 1988d, *Physically-Based Modelling and Simulation of Climate and Climate Change*. Part 2 (Dordrecht: Kluwer).

SCHLESINGER, M. E., 1989a, Model projections of the climatic changes induced by increased atmospheric $CO_2$. In *Climate and the Geo-Sciences: A Challenge for Science and Society in the 21st Century*, edited by A. BERGER and S. H. SCHNEIDER (Dordrecht: Kluwer), pp. 375–415.

SCHLESINGER, M. E., 1989b, Quantitative analysis of feedbacks in climate model simulation. In *Understanding Climate Change*, edited by A. BERGER, R. E. DICKINSON, J. W. KIDSON, Geophysical Monograph 52, IUGG, Volume 7 (Washington DC: AGUM), pp. 177–187.

SCHLESINGER, M. E., 1989c, *Research on $CO_2$-induced Climatic Change and Development of a Physically-Based CGM Cloud Parametrization*. NSF Foundation Grants ATM-8712033 and ATM-8719459. Oregon State University, Corvallis, OR.

SCHLESINGER, M. E. (editor), 1991, *Greenhouse-Gas-Induced Climatic Change* (New York: Elsevier).

SCHLESINGER, M. E. and JIANG, X., 1988, The transfer of $CO_2$-induced warming into the ocean: an analysis of simulations by the OSU coupled atmosphere–ocean general circulation model. *Climate Dynamics*, **3**, 1–18.

SCHLESINGER, M. E. and JIANG, X., 1989, The timescale of climate change induced by

increased carbon dioxide. Preprint (Corvallis, OR: Oregon State University).
SCHLESINGER, M. E., JIANG, K. and CHARLSON, R. J., 1993, Implications of anthropogenic atmospheric sulphate for the sensitivity of the climate system. In *Climate Change and Energy Policy*, edited by L. ROSEN and R. GLASSER (New York: American Institute of Physics), pp. 75–108.
SCHLESINGER, M. E. and MITCHELL, J. F. B., 1987, Climate model simulations of the equilibrium climatic response to increased carbon dioxide. *Review of Geophysics*, **25**, 760–798.
SCHLESINGER, M. E., OH, J. H. and ROSENFELD, D., 1988, A parametrization of the evaporation of rainfall. *Monthly Weather Review*, **116**, 1887–1895.
SCHLESINGER, M. E. and ZHAO, Z. C., 1989, Seasonal climatic changes induced by doubled $CO_2$ as simulated by the OSU atmospheric GCM/mixed-layer ocean model. *Journal of Climate*, **2**, 459–495.
SCHLESINGER, M. E., ZHAO, Z. C. and VICKERS, D., 1989, Design and critical appraisal of an accelerated integration procedure for atmospheric GCM/mixed-layer ocean models. *Journal of Climate*, **2**, 641–655.
SCHMITT, C. and RANDALL, D. A., 1991, Effects of surface temperature and clouds on $CO_2$ forcing. *Journal of Geophysical Research*, **96**, 9159–9168.
SCHNEIDER, H. R., KO, M. K. W., SHIA, R. L. and SZE, N. D., 1993, A two-dimensional model with coupled dynamics, radiative transfer and photochemistry. 2: Assessment of the response of stratospheric ozone to increased levels of $CO_2$, $N_2O$, $CH_4$ and CFC. *Journal of Geophysical Research*, **98**, 20441–20449.
SCHNEIDER, S. H., 1981, Climatic impact assessment in the $CO_2$ context – an editorial. *Climatic Change*, **3**, 345–346.
SCHNEIDER, S. H., 1985, Large-scale three-dimensional modelling aspects of 'nuclear winter'. *Extended Abstracts of the 3rd Conference on Climate Variation and Symposium on Contemporary Climate: 1850–2100, held in Los Angeles, CA, on 8–11 January 1985* (Boston: American Meteorological Society), pp. 8–9.
SCHNEIDER, S. H., 1989, The greenhouse effect: science and policy. *Science*, **243**, 771–780.
SCHNEIDER, S. H., 1994, Detecting climatic change signals: are there any 'fingerprints'? *Science*, **263**, 341–348.
SCHNEIDER, S. H. and THOMPSON, S. L., 1981, Atmospheric $CO_2$ and climate: importance of the transient response. *Journal of Geophysical Research*, **86**, 3135–3147.
SCHNELL, R. C., SHERIDAN, P. J. and KAHL, J. D., 1991, AGASP-III, polar lows and CEAREX Norwegian Arctic flight program, Spring 1989. *International Conference on the Role of the Polar Regions in Global Change. Proceedings of a Conference held June 11–15, 1990, at the University of Alaska Fairbanks*, edited by G. WELLER, C. L. WILSON and B. A. B. SEVERIN (Fairbanks: Geophysical Institute and Center for Global Change and Arctic Systems Research), p. 694.
SCHOEBERL, N. R., BHARTIA, P. K., HOLSENRATH, E. and TORRES, E., 1993, Tropical ozone loss following the eruption of Mt. Pinatubo. *Geophysical Research Letters*, **20**, 29–32.
SCHÖNWIESE, C. D. and RUNGE, K., 1991, Some updated statistical assessments of the surface temperature response to increased greenhouse gases. *International Journal of Climatology*, **11**, 237–250.
SCHRYER, D. R. (editor), 1982, *Heterogeneous Atmospheric Chemistry* (Washington, DC: American Geophysical Union).
SCHWARTZ, S. E., 1993, Comparing the global heating and cooling from fossil fuel combustion. DOE Research Summary, No. 28.
SCHWARZKOPF, M. D. and RAMASWAMY, V., 1993, Radiative forcing due to ozone in the 1980s: dependence on altitude of ozone change. *Geophysical Research Letters*, **20**, 205–208.
SCUDERI, L. A., 1993, A 2000-year tree ring record of annual temperatures in the Sierra

Nevada mountains. *Science*, **259**, 1433–1436.

SELF, S., RAMPINO, M. R. and BARBERA, J. J., 1981, The possible effects of large 19th and 20th century volcanic eruptions on zonal and hemispheric surface temperatures. *Journal of Volcanology and Geothermal Research*, **11**, 41–60.

SELLERS, P. and DORMAN, J. L., 1987, Testing the Simple Biosphere Model (SiB) using point micrometeorological and biophysical data. *Journal of Climate and Applied Meteorology*, **26**, 622–651.

SELLERS, P. J. and HALL, F. G. (editors), 1987, First ISLSCP Field Experiment (FIFE). Konza Prairie, 1987–1989. Experiment Plan, Appendices.

SELLERS, P. J., HALL, F. G., ASRAR, G., STREBEL, D. E. and MURPHY, R. E., 1988, The First ISLSCP Field Experiment (FIFE). *Bulletin of the American Meteorological Society*, **69**, 22–27.

SELLERS, P. J., HALL, F. G., ASRAR, G., STREBEL, D. E. and MURPHY, R. E., 1992, An overview of the First International Satellite Land Surface Climatology Project (ISLSCP) Field Experiment (FIFE). *Journal of Geophysical Research*, **97**, 18345–18371.

SELLERS, W. D., 1982, The response of a climate model to variations in the solar constant. *Symposium on the Solar Constant and Spectral Distribution of Solar Irradiance, Hamburg, 17–28 August 1981*. Coll. ext. abstr. (Boulder, CO:IUGG), pp. 134–138.

SENIOR, C. A. and MITCHELL, J. F. B., 1993, Carbon dioxide and climate: the impact of cloud parametrization. *Journal of Climate*, **6**, 393–418.

SERGIN, V. YA. and ORESHKO, A. P., 1987, A time-averaged global atmospheric circulation model. *Meteorology and Hydrology* (*Meteorologia i Gidrologia*), No. 7, 13–22.

SHAW, G. E., 1983, Bio-controlled thermostasis involving the sulfur cycle. *Climatic Change*, **5**, 297–303.

SHAW, G. E., 1987, Aerosols as climate regulators: a climate–biosphere linkage? *Atmospheric Environment*, **21**, 985–986.

SHAW, G. E., 1988, Arctic aerosols: a review. *Reviews of Geophysics*, **26**, 89–112.

SHAW, G. E., 1989a, Aerosol transport from sources to ice sheets. In *The Environment Record in Glacier Ice Sheets*, edited by H. OESCHGER and C. C. LANGWAY (Chichester: Wiley), pp. 13–27.

SHAW, G. E., 1989b, A new proposal for the operating machinery in the global cloud albedo–marine phytoplankton feedback loop. Preprint (Geophysical Institute, University of Alaska).

SHAW, G. E., 1995, Contamination of the Arctic. *Izvestiya Russian Geographical Society*, **127**, No. 1, 3–14.

SHAW, G. E., STAMNES, K. and HU, Y. X., 1993, Arctic haze: perturbation to the radiation field. *Meteorology and Atmospheric Physics*, **51**, 227–238.

SHERIDAN, P. J., SCHNELL, R. C., HOFFMANN, D. J. and DESCHLER, T., 1992, Electron microscope studies of Mt. Pinatubo aerosol layers over Laramie, Wyoming, during summer 1991. *Geophysical Research Letters*, **19**, 203–206.

SHERMAN, J. W., 1985, NOAA N-ROSS/ERS-1 environmental data development (NNFEDD) activity. NOAA Technical Report NESDIS-14, Washington DC.

SHERSTIUKOV, B. G. and LOGINOV, V. F., 1986, *Short-period Cyclic Variations in the Lower Atmosphere and the Heliogeophysical Processes* (Moscow: Gidrometeoizdat).

SHUKLA, J., 1986, SST anomalies and blocking. *Advances in Geophysics*, **29**, 443–452.

SHUKLA, J. and FENNESSY, M. J., 1988, Prediction of time-mean atmospheric circulation and rainfall: influence of Pacific sea surface temperature anomaly. *Journal of Atmospheric Science*, **45**, 9–28.

SHUKLA, J. and MINTZ, Y., 1982, Influence of land-surface evapotranspiration on the Earth's climate. *Science*, **215**, 1498–1501.

SHUKLA, J. and RASMUSSON, E. M., 1985, Variability of the tropical circulation. *International Conference on the TOGA Science Program, held in Paris on 17–21 September 1984*. WCRP Publication Series, No. 4 (Geneva: WMO), pp. II.24–II.39.

SHUKLA, J. and WALLACE, J. M., 1983, Numerical simulation of the atmospheric response to equatorial Pacific sea surface temperature anomalies. *Journal of Atmospheric Science*, **40**, 1613–1630.

SIEGENTHALER, U. and SARMIENTO, J. L., 1993, Atmospheric carbon dioxide and the ocean. *Nature*, **365**, 119–125.

SINGER, S. F., 1988, Re-analysis of the nuclear winter phenomenon. *Meterological and Atmospheric Physics*, **38**, 228–239.

SIRÉN, G., 1961, Pine at the timberline as an indicator of climatic fluctuations in northern Fennoscandia in historic times (in Swedish). *Communicationes. Instituti Forestalis Fenniae*, **54** (2), 1–66.

SKLYAROV, YU. A., 1994, Solar constant (state of research). *Studies of the Earth from Space, (Issledovania Zemli iz Kosmosa)*, No. 4, 96–110.

SLINGO, A. and SLINGO, J. M., 1988, The response of a general circulation model to cloud longwave radiative forcing. I: Introduction and initial experiments. *Quarterly Journal of the Royal Meteorological Society*, **114**, 1027–1062.

SLINGO, A. and SLINGO, J. M., 1991, Response of the National Center for Atmospheric Research Community Climate Model to improvements in the Representation of clouds. *Journal of Geophysical Research*, **96**, 15341–15357.

SLINGO, A., WILDERSPIN, R. C. and SMITH, R. N. B., 1989, Effect of improved physical parametrizations on simulations of cloudiness and the Earth's radiation budget. *Journal of Geophysical Research*, **94**, 2281–2302.

SLINGO, J. M., 1987, The development and verification of a cloud prediction scheme for the ECMMF model. *Quarterly Journal of the Royal Meteorological Society*, **113**, 899–928.

SMAGORINSKY, J., 1984, The problem of climate and climate variations. World Climate Paper, No. 872.

SMITH, E. A., VONDERHAAR, T. H., HICKEY, J. B. and MASCHHOFF, R., 1985, The nature of the short period fluctuations in solar irradiance received by the Earth. *Climatic Change*, **5**, 211–236.

SMITH, G. L., RATAM, D. and BESS, T. D., 1991, Variability of absorbed solar radiation. *Proceedings of the Fifth Conference on Climate Variations* (Boston: American Meteorological Society), pp. 384–387.

SMITH, L. D. and VONDERHAAR, T. H., 1991, Clouds–radiation interactions in a general circulation model: impact upon the planetary radiation balance. *Journal of Geophysical Research*, **96**, 893–914.

SMITH, R. M. B. and PEARSON, D. W., 1986, An experiment with a prognostic cloud water variable in a general circulation model. In *Research Activities in Atmosphere and Ocean Model*, Report No. 9 (WMO/TD-N141), pp. 4.27–4.30.

SMITH, S. D., MUENCH, R. D. and PEASE, C. H., 1990, Polynyas and leads: an overview of physical processes and environment. *Journal of Geophysical Research*, **95**, 9461–9480.

SMITH, T. M., LEEMANS, R. and SHUGART, H. H., 1992, Sensitivity of terrestrial carbon storage to $CO_2$-induced climate change: Comparison of four scenarios based on general circulation models. *Climatic Change*, **21**, 367–384.

SODEN, B. J., 1992, Validation of cloud forcing simulated by the National Center for Atmospheric Research Community Climate Model using observations from the Earth Radiation Budget Experiment. *Journal of Geophysical Research*, **97**, 18137–18159.

SOLOMON, A., 1986, Transient response of forests to $CO_2$-induced climate change: simulation modelling experiments in eastern North America. *Oecologia*, **68**, 567–579.

SOMERVILLE, R. C. J., 1985, Climate stabilization by cloud optical thickness feedbacks. *Extended Abstracts of the 3rd Conference on Climate Variation and Symposium on Contemporary Climate: 1850–2100, held in Los Angeles, CA, on 8–11 January 1985* (Boston: American Meteorological Society), pp. 101–102.

SOMERVILLE, R. C. J. and REWER, L. A., 1984, Cloud optical thickness feedbacks in the $CO_2$ climate problem. *Journal of Geophysical Research*, **89**, 9668–9672.

SORNETTE, A., DUBOIS, J., CHEMINÉE, J. L. and SORNETTE, D., 1991, Are sequences of volcanic eruptions deterministically chaotic? *Journal of Geophysical Research*, **96**, 11931–11946.

SPÄNKUCH, D., 1984, Die Wechselwirkung von Strahlung und Bewölkung als ein wesentlicher Klimabildender Process. *Arbeit von Plenum und Klassen A&W der DDR, Berlin*, **9**, 37–54.

SPENCER, R. W. and CHRISTY, J. R., 1991, A physical interpretation of brightness temperatures observed by the microwave sounding units based upon Raobs. *Proceedings of the Fifth Conference on Climate Variations* (Boston: American Meteorological Society).

SPERBER, K. P., HAMEED, S. GATES, W. L. and POTTER, G. L., 1987, Southern oscillation simulated in a global climate model. *Nature*, **329**, 140–142.

STANHILL, G. and MORESHET, S., 1992, Global radiation climate changes: the world network. *Climate Change*, **21**, 57–75.

STEFFEN, K., ABDALATI, W. and STROEVE, J., 1993, Climate sensitivity studies of the Greenland ice sheet using satellite AVHRR, SMMR, SSM/I and in situ data. *Meteorology and Atmospheric Physics*, **51**, 239–258.

STEPHENS, G. L. and GREENWALD, T. J., 1991a, The Earth's radiation budget and its relation to atmospheric hydrology. 1: Observations of the clear-sky greenhouse effect. *Journal of Geophysical Research*, **96**, 15311–15324.

STEPHENS, G. L. and GREENWALD, T. J., 1991b, The Earth's radiation budget and its relation to atmospheric hydrology. 2: Observations of cloud effects. *Journal of Geophysical Research*, **96**, 15325–15340.

STEPHENS, G. L., RANDALL, D. A., WITTMEYER, I. L. and DAZLICH, D. A., 1993, The Earth's Radiation Budget and its relation to atmospheric hydrology. 3: Comparison of observations over the oceans with a GCM. *Journal of Geophysical Research*, **98**, 4931–4950.

STOCKER, T. F. and MYSAK, L. A., 1992, Climatic fluctuations on the century time scale: a review of high-resolution proxy data and possible mechanisms. *Climatic Change*, **20**, 227–250.

STOLARSKI, R., BOJKOV, R., BISHOP, L., ZEREFOS, C., STAEHELIN, J. and ZAWODNY, J., 1992, Measurement trends in stratospheric ozone. *Science*, **256**, 342–349.

STONE, P. H. and YAO, M. S., 1987, Development of a two-dimensional zonally averaged statistical-dynamical model. Part II: The role of eddy momentum fluxes in the general circulation and their parametrization. *Journal of Atmospheric Science*, **44**, 3769–3786.

STONE, R. S. and KAHL, J. D., 1991, Cloud radiative effects and associated changes in tropospheric temperatures and winds at the South Pole. *International Conference on the Role of the Polar Regions in Global Change. Proceedings of a Conference held June 11–15, 1990, at the University of Alaska Fairbanks*, edited by G. WELLER, C. L. WILSON and B. A. B. SEVERIN (Fairbanks: Geophysical Institute and Center for Global Change and Arctic Systems Research), pp. 184–189.

STOWE, L. L. (editor), 1988, *Report of the Earth Radiation Budget and Requirements Review, Rosslyn, Virginia, 30 March–3 April 1987* (Washington DC: NOAA).

STOWE, L. L., CAREY, R. M. and PELLAGRINO, P. P., 1992, Monitoring the Mt. Pinatubo aerosol layer with NOAA/11 AVHRR data. *Geophysical Research Letters*, **19**, 159–162.

SUD, Y. C., WALKER, G. K. and SMITH, W. E., 1991, Analysis of a general circulation model simulation of the atmospheric response to the observed Sea Surface Temperature Anomalies of January and February 1983, *Journal of Climate*, **4**, 107–115.

TANS, P. P. and BAKWIN, P. S., 1995, Climate change and carbon dioxide forever. *AMBIO*, **24**, 376–378.

TARKO, A. M., BOGATYREV, B. G. and KIRILENKO, A. P., 1992, The role of countries and regions in the formation of the global atmospheric carbon dioxide budget. *Doklady Akademii Nauk SSSR*, **322**, 610–613.

THOMAS, G. and HENDERSON-SELLERS, A., 1987, Evaluation of satellite derived land cover characteristics for global climate modelling. *Climatic Change*, **11**, 313–348.

THOMAS, G. E., OLIVERO, J. J., JENSEN, E. J., SCHROEDER, W. and TOON, O. B., 1989, Relation between increasing methane and the presence of ice clouds at the mesopause. *Nature*, **338**, 490–492.

THOMPSON, S. L. and POLLARD, D., 1995, A global climate model (GENESIS) with a land-surface transfer scheme (LSX). Part II: $CO_2$ sensitivity. *Journal of Climate*, **8**, 1104–1121.

THOMPSON, S. L. and SCHNEIDER, S. H., 1986, Nuclear winter reappraised. *Foreign Affairs*, Summer, 981–1005.

TICKELL, C., 1977, *Climate Change and World Affairs* (Harvard: Center for International Affairs).

TINSLEY, B. A. and HEELIS, R. A., 1993, Correlations of atmospheric dynamics with solar activity evidence for a correction via the solar wind, atmospheric electricity and cloud microphysics. *Journal of Geophysical Research*, **98**, 10375–10384.

TRABALKA, J. R. (editor), 1985, *Atmospheric Carbon Dioxide and the Global Carbon Cycle*. DOE/ER-0239 (Washington DC: US Dept of Energy).

TRABALKA, J. R. and REICHLE, D. E. (editors), 1986, *The Changing Carbon Cycle: A Global Analysis* (New York: Springer).

TRENBERTH, K. E. (editor), 1992, *Climate System Modelling* (Cambridge: Cambridge University Press).

TRENBERTH, K. E., BRANSTATOR, G. W. and ARKIN, P. A., 1988, Origins of the 1988 North American drought. *Science*, **242**, 1640–1645.

TSELONIDIS, G., LACIS, A. A., ROSSOW, W. B. and RIND, D., 1991, Significance of cloud optical thickness variations in climate simulations. *Proceedings of the Fifth Conference on Climate Variations* (Boston: American Meteorological Society), pp. 262–265.

TSONIS, A. A. and ELSNER, J. B., 1989, Testing the global warming hypothesis. *Geophysical Research Letters*, **16**, 795–798.

TUCKER, C. J., FUNG, I. Y., KEELING, C. D. and GAMMON, R. H., 1986, Relationship between atmospheric $CO_2$ variations and a satellite-derived vegetation index. *Nature*, **319**, 195–199.

TUCKER, G. B., 1991, Confidence in modelling future climate: A southern hemisphere perspective. *Climatic Change*, **18**, 195–204.

TURCO, R. P., TOON, O. B., ACKERMAN, T., POLLACK, J. B. and SAGAN, C., 1983, Nuclear winter: global consequences of multiple nuclear explosions. *Science*, **222**, 1283–1292.

TURNER, J., 1991, Detection of high latitude atmospheric circulation changes using satellite data. *International Conference on the Role of the Polar Regions in Global Change. Proceedings of a Conference held June 11–15, 1990, at the University of Alaska Fairbanks*, edited by G. WELLER, C. L. WILSON and B. A. B. SEVERIN (Fairbanks: Geophysical Institute and Center for Global Change and Arctic Systems Research), pp. 14–16.

TWOMEY, S. A., PIEPGRASS, M. and WOLFE, T. L., 1984, An assessment of the impact of pollution on global cloud albedo. *Tellus*, **36B**, 356–366.

UNIVERSITY CORPORATION FOR ATMOSPHERIC RESEARCH (UCAR), 1978, *Status of Satellite Observing Possibilities for Studies of Climate Physical Processes*. Report of the COSPAR WG-6. (Boulder, CO: UCAR).

US ARCTIC RESEARCH COMMISSION, 1989, *Arctic Research for an Arctic Nation* (Washington DC: Arctic Research Commission).

US DEPARTMENT OF ENERGY, 1982, *Proceedings of the Workshop on First Detection of Carbon Dioxide Effects, held in Harpers Ferry, West Virginia. US Department of Energy Conference 8106214* (Washington DC: DOE).

US DEPARTMENT OF ENERGY, 1984, *Glaciers, Ice Sheets, and Sea Level: Effect of a $CO_2$-induced Climatic Change*. Report of a Workshop, held in Seattle, Washington, on 13–15 September 1984. US Department of Energy, DOE/ER.602351 (Washington DC: DOE).

US Department of Energy, 1988, *Carbon Dioxide and Climate: Summaries of Research FY 1988*. DOE/ER-0385 (Washington DC: Carbon Dioxide Research Division, DOE).

US Environmental Protection Agency (EPA), 1988, Report to Congress on the Effects of Global Climate Change.

US National Research Council, 1985, *The Effects on the Atmosphere of a Major Nuclear Exchange. US National Research Council Report* (Washington DC: National Academic Press).

Valero, F. P. J. and Pilewskie, P., 1992, Latitudinal survey of spectral optical depths of the Pinatubo volcanic cloud-derived particle sizes, columnar mass loadings, and effects on planetary albedo. *Geophysical Research Letters*, **19**, 163–166.

van den Dool, H. M., Krijnen, H. J. and Schuurmans, C. J. E., 1978, Average winter temperature at De Bilt (The Netherlands): 1634–1977. *Climatic Change*, **1**, 310–330.

van Loon, H., 1991, Temperature trends and circulation changes in the northern hemisphere in winter. *Proceedings of the Fifth Conference on Climate Variations* (Boston: American Meteorological Society), pp. 12–15.

van Loon, H. and Labitzke, K., 1993, Review of the decadal oscillations in the stratosphere of the Northern hemisphere. *Journal of Geophysical Research*, **98A**, 18919–18922.

Varatxox, C., 1988, New results of the strato-mesospheric cooling of the northern hemisphere (1969–1978), *Earth, Moon and Planets*, **41**, 191–196.

Vasilyev, V. F. and Romanov, V. F., 1989, Synoptic vortex processes and the seasonal peculiarities of the atmospheric climatic regime. *Meteorology and Hydrology* (*Meteorologia i Gidrologia*), No. 4, 21–33.

Vaughan, R. A., 1987, *Remote Sensing Applications in Meteorology and Climatology* (Dordrecht: Reidel).

Vaughan, R. A. and Cracknell, A. P., 1994, *Remote Sensing and Global Climate Change* (Berlin: Springer).

Velichko, A. A. (editor), 1987, *Climates of the Earth in the Geological Past* (Moscow: Nauka).

Velikhov, E. E. (editor), 1987, *Climate and Biological Consequences of a Nuclear War* (Moscow: Nauka).

Viner, D. and Hulme, M., 1993a, What carbon dioxide concentrations should be used with climate change scenarios? UK Department of the Environment Technical Note No. 3, Norwich.

Viner, D. and Hulme, M., 1993b, The UK Met Office high resolution GCM transient experiment. UK Department of the Environment Technical Note No. 4, Norwich.

Vinnikov, K. Ya., 1986, *The Sensitivity of Climate* (Leningrad: Gidrometeoizdat).

Vinnikov, K. Ya., Groisman, P. Ya., Luguina, K. M. and Golubev, V. A., 1987, Variations in average northern hemisphere air temperature between 1841 and 1985. *Meteorology and Hydrology* (*Meteorologia i Gidrologia*), No. 1, 45–55.

Vinnikov, K. Ya., Gruza, G. V., Zakharov, V. F., Kirillov, A. A., Kovyneva, N. P. and Ran'kova, E. Ya., 1980. Current climatic changes in the northern hemisphere (in Russian). *Meteorologia i Gidrologia*, No. 6, 5–17.

Vinnikov, K. Ya. and Lemeshko, N. A., 1987, Soil moisture content and a sink for the USSR territory at a global warming. *Meteorology and Hydrology* (*Meteorologia i Gidrologia*), No. 12, 96–103.

Vinnikov, K. Ya. and Lugina, K. M., 1982, Some problems of monitoring the global thermal regime of the northern hemisphere (in Russian). *Meteorologia i Gidrologia*, No. 11, 5–14.

von Storch, H. and van Loon, H., 1986, Atmospheric response to SST anomalies in the SPCZ area. A pilot study. In *Research Activities in Atmosphere and Ocean Model*. Report No. 9, Geneva WMO/TD-N141, pp. 7.21–7.24.

Vupputuri, R. K. R., 1985, The effect of ozone photochemistry on atmospheric and surface temperature changes due to increased $CO_2$, $N_2O$, $CH_4$, and volcanic aerosols in the atmosphere. *Atmosphere–Ocean*, **23**, 359–374.

VUPPUTURI, R. K. R., 1988a, Potential effects of anthropogenic trace gas emissions on atmospheric ozone and temperature structure and surface climate. *Atmospheric Environment*, **22**, 2809–2818.

VUPPUTURI, R. K. R., 1988b, The interactive effects of large injections of smoke, dust and $NO_x$ on atmospheric temperature and ozone structure and surface climate. In *Natural and Man-Made Hazards*, edited by M. I. EL-SABH and T. S. MURTHY (Dordrecht: Reidel), pp. 643–668.

VUPPUTURI, R. K. R., 1988c, The possible effects of Tambora eruption in 1815 on atmospheric thermal and chemical structure and surface climate. Preprint (Toronto: Canadian Climate Centre).

VUPPUTURI, R. K. R. and BLANCHET, J. P., 1984, The possible effects of El Chichón eruption on atmospheric thermal and chemical structure and surface climate. *Geofisica Internacional*, **23**, 433–447.

WADHAMS, P., 1994, Remote sensing of snow and ice and its relevance to climate change processes. In *Remote Sensing and Global Climate Change*, edited by R. A. VAUGHAN and A. P. CRACKNELL (Berlin: Springer), pp. 303–339.

WALISER, D. E. and GAUTIER, C., 1993, A satellite-derived climatology of the ITCZ. *Journal of Climate*, **6**, 2162–2174.

WALLACE, J. M., 1985, Atmospheric response to equatorial sea-surface temperature anomalies. *International Conference on the TOGA Science Program, held in Paris on 17–21 September 1984. WCRP Publication Series No. 4* (Geneva: WMO), pp. II.1–II.12.

WALSH, J. E. and CRANE, R. G., 1992, A comparison of GCM simulations of Arctic climate. *Geophysical Research Letters*, **19**, 29–32.

WALSH, J. E. and INCEK, D. R., 1981, On the use of snow cover as a short-term climatic predictor. *Proceedings of the 6th Annual Climate Diagnostic Workshop held in Columbia University, New York, on 14–16 October 1981* (Washington DC: NOAA), pp. 170–177.

WALSH, J. E., ZWALLY, H. J. and WEATHERLY, J. W., 1991, Antarctic sea ice and temperature variations. *International Conference on the Role of the Polar Regions in Global Change. Proceedings of a Conference held June 11–15, 1990, at the University of Alaska Fairbanks*, edited by G. WELLER, C. L. WILSON and B. A. B. SEVERIN (Fairbanks: Geophysical Institute and Center for Global Change and Arctic Systems Research), pp. 263–268.

WANG, C. and CHANG, J. S., 1993a, A three-dimensional numerical model of cloud dynamics, microphysics, and chemistry. 1: Concepts and formulation. *Journal of Geophysical Research*, **98**, 14827–14844.

WANG, C. and CHANG, J. S., 1993b, A three-dimensional numerical model of cloud dynamics, microphysics, and chemistry. 3: Redistribution of pollutants. *Journal of Geophysical Research*, **98**, 16787–16798.

WANG, C. and CHANG, J. S., 1993c, A three-dimensional numerical model of cloud dynamics, microphysics, and chemistry. 4: Cloud chemistry and precipitation chemistry. *Journal of Geophysical Research*, **98**, 16799–16808.

WANG, W. C., DUDEK, M. P., LIANG, X. J. and KIEJHL, J. T., 1991, A general circulation model study of the climatic effect of atmospheric trace gases. *Proceedings of the Fifth Conference on Climate Variations* (Boston: American Meteorological Society), pp. 5–7.

WANG, W. C. and MÓLNAR, G., 1985, A model study of the greenhouse effects due to increasing atmospheric $CH_4$, $N_2O$, $CF_2Cl_2$, and $CFCl_3$. *Journal of Geophysical Research*, **90**, 12971–12980.

WANG, W. C., WUEBBLES, D. J., WASHINGTON, W. M., ISAACS, R. G. and MÓLNAR, G., 1986, Trace gases and other potential perturbations to global climate. *Review of Geophysics*, **24**, 110–140.

WANG, W. C., ZHUANG, Y. C. and BOJKOV, R., 1993, Climate implications of observed changes in ozone vertical distributions in middle and high latitudes of the Northern Hemisphere. *Geophysical Research Letters*, **20**, 1567–1570.

WARNECKE, G., HUCH, M. and GERMANN, K., 1991, *Tätort 'Erde'. Menschliche Eingriffe in Naturraum und Klima* (Berlin: Springer).

WARRICK, R. A., BARROW, E. M. and WIGLEY, T. M. L. (editors), 1993, *Climate and Sea Level Change: Observations, Projections and Implications* (Cambridge: Cambridge University Press).

WASHINGTON, W. M. and MEEHL, G. A., 1983, General circulation model experiments on the climatic effects due to a doubling and quadrupling of carbon dioxide concentration. *Journal of Geophysical Research*, **88**, 6600–6610.

WASHINGTON, W. M. and MEEHL, G. A., 1984, Seasonal cycle experiment on the climate sensitivity due to a doubling of $CO_2$ with an atmospheric general circulation model coupled to a simple mixed-layer ocean model. *Journal of Geophysical Research*, **89**, 9475–9503.

WASHINGTON, W. M. and MEEHL, G. A., 1989, Climate sensitivity due to increased $CO_2$: experiments with a coupled atmosphere and ocean general circulation model. *Climate Dynamics*, **4**, 1–38.

WASHINGTON, W. M. and MEEHL, G. A., 1993, Greenhouse sensitivity experiments with penetrative cumulus convection and tropical cirrus albedo effects. *Climate Dynamics*, **8**, 211–223.

WASHINGTON, W. M. and PARKINSON, C. L., 1986, *An Introduction to Three-Dimensional Climate Modeling* (Mill Valley, CA: University Science Books).

WATSON, R. T., MERRA FILHO, G. G., SANHUEZA, E. and JANETUS, A., 1992, Sources and sinks. In Houghton *et al.* (1992), pp. 25–46.

WEARE, B. C., 1988, A comparison of radiation variables calculated in the UCLA general model to observations. *Journal of Climate*, **1**, 485–499.

WEARE, B. C., MOKHOV, I. I. and PROJECT MEMBERS, 1995, Evaluation of total cloudiness and its variability in the Atmospheric Model Intercomparison Project. *Journal of Climate*, **3**, 2224–2238.

WEAVER, A., LOEWENSTEIN, M., PODOLSKE, J. R., STRAHAN, S. E., PROFITT, M. H., AITKING, K., MARGITAN, J. J. and TOON, O. B., 1993, Effect of Pinatubo aerosol on stratospheric ozone at mid-latitudes. *Geophysical Research Letters*, **20**, 2515–2518.

WEBB, M. J., SLINGO, A. and STEPHENS, G. L., 1993, Seasonal variations of the clear-sky greenhouse effect: the role of change in atmospheric temperatures and humidities. *Climate Dynamics*, **9**, 117–130.

WEBB, T. and WIGLEY, T. M. L., 1985, What past climates can indicate about the warmer world. In *The Potential Climatic Impact of Carbon Dioxide*, edited by M. C. MACCRACKEN and F. M. LUTHER. DOE/ER-0237 (Washington DC: US Department of Energy), pp. 237–257.

WEBSTER, F., 1984, *An Ocean Climate Research Strategy* (Washington DC: National Academic Press).

WEISENSTEIN, D. K., KO, M. K. W., RODRIGUEZ, S. M. and SZE, N. D., 1993, Effects on stratospheric ozone from high-speed civil transport: sensitivity to stratospheric aerosol loading. *Journal of Geophysical Research*, **98**, 23133–23140.

WELLER, G., 1993, Polar meteorology. *Meteorology and Atmospheric Physics*, **51**, 141–146.

WELLER, G. and MEITNER, R., 1993, Satellite derived outgoing longwave radiation, surface temperature and sea ice concentration off the coast of Adélie-land, Eastern Antarctica. *Meteorology and Atmospheric Physics*, **51**, 219–226.

WELLER, G., WILSON, C. L. and SEVERIN, B. A. B. (editors), 1991, *International Conference on the Role of the Polar Regions in Global Change. Proceedings of a Conference held June 11–15, 1990, at the University of Alaska Fairbanks* (Fairbanks: Geophysical Institute and Center for Global Change and Arctic Systems Research).

WETHERALD, R. T. and MANABE, S., 1981, Influence of seasonal variation upon the sensitivity of a model climate. *Journal of Geophysical Research*, **86**, 1194–1204.

WETHERALD, R. T. and MANABE, S., 1988, Cloud feedback processes in a general circulation model. *Journal of Atmospheric Science*, **45**, 1397–1415.

WHITE, R. M., 1989, Greenhouse policy and climate uncertainty. *Bulletin of the American Meteorological Society*, **70**, 1123–1127.

WIELICKI, B. A. and PARKER, L., 1992, On the determination of cloud cover from satellite sensors: the effect of sensor spatial resolution. *Journal of Geophysical Research*, **97**, 12799–12823.

WIGLEY, T. M. L., 1983, The preindustrial carbon dioxide level. *Climatic Change*, **5**, 315–320.

WIGLEY, T. M. L., 1987, The effect of model structure on projections of greenhouse-gas-induced climatic change. *Geophysical Research Letters*, **14**, 1135–1138.

WIGLEY, T. M. L., 1989a, The greenhouse effect: scientific assessment of climate change. The Prime Minister's Seminar on Global Climate Change, 26 April 1989 (London), pp. 1–16.

WIGLEY, T. M. L., 1989b, Possible climate change due to $SO_2$-derived cloud condensation nuclei. *Nature*, **339**, 365–367.

WIGLEY, T. M. L., 1993, Balancing the carbon budget. Implications for projections of future carbon dioxide concentration change. *Tellus*, **45B**, 409–425.

WIGLEY, T. M. L., ANGELL, J. K. and JONES, P. D., 1985, *Analysis of the Temperature Record/Detecting the Climatic Effects of Increasing Carbon Dioxide.* DOE/EOR-0235 (Washington DC: US Department of Energy), pp. 55–90.

WIGLEY, T. M. L., INGRAM, M. J. and FARMER, G., 1981, *Climate and History. Studies in Past Climates and Their Impact on Man* (Cambridge: Cambridge University Press).

WIGLEY, T. M. L., JONES, P. D., KELLY, P. M. and RAPER, S. C. B., 1989, Statistical significance of global warming. *Proceedings of the 13th Annual Climate Diagnostics Workshop, held in Cambridge, MA* (Boston: American Meteorological Society), pp. A-1–A-8.

WIGLEY, T. M. L. and RAPER, S. C. B., 1987, Thermal expansion of sea water associated with global warming. *Nature*, **330**, 127–131.

WIGLEY, T. M. L. and RAPER, S. C. B., 1990, Future changes in global-mean temperature and thermal-expansion-related sea level rise. In *Climate and Sea Level Change: Observations, Projections and Implications*, edited by R. A. WARRICH and T. M. L. WIGLEY (Cambridge: Cambridge University Press), pp. 111–133.

WIGLEY, T. M. L. and SCHLESINGER, M. E., 1985, Transient response of the $CO_2$-induced climate change. *Extended Abstracts of the 3rd Conference on Climate Variation and Symposium on Contemporary Climate: 1850–2100, held in Los Angeles on 8–11 January 1985* (Boston: American Meteorological Society), pp. 138–139.

WILLETT, H. C., 1950, Temperature trends of the past century. *Proceedings of the Royal Meteorological Society, Centenary Special Volume*, 195–206.

WILLIAMS, R. S. and FERRIGNO, J. G., 1991, Satellite monitoring of areal changes in the glacier component of the Earth's cryosphere. *International Conference on the Role of the Polar Regions in Global Change. Proceedings of a Conference held June 11–15, 1990, at the University of Alaska Fairbanks*, edited by G. WELLER, C. L. WILSON and B. A. B. SEVERIN (Fairbanks: Geophysical Institute and Center for Global Change and Arctic Systems Research), p. 44.

WILLSON, R. C., 1984, Measurements of solar total irradiance and its variability. *Space Science Review*, **38**, 203–242.

WILLSON, R. C. and HUDSON, H. S., 1988, Solar luminosity variations in solar cycle 21. *Nature*, **322**, 810–814.

WILSON, C. A. and MITCHELL, J. F. B., 1986, Diurnal variation and cloud in a general circulation model. *Quarterly Journal of the Royal Meteorological Society*, **112**, 347–369.

WILSON, C. A. and MITCHELL, J. F. B., 1987, A doubled $CO_2$ climate sensitivity experiment with a global climate model including a simple ocean. *Journal of Geophysical Research*, **92**, 13315–13343.

WILSON, C. A., MITCHELL, J. F. B., INGRAM, W. J. and CUNNINGTON, W. M., 1986, A 2 × $CO_2$ climate sensitivity experiment with a global climate model including a simple ocean. In *Research Activities in Atmospheric and Ocean Model*, Report No. 9 (WMO/TD-N141), Geneva, pp. 7.34–7.46.

WILSON, M. F., HENDERSON-SELLERS, A., DICKINSON, R. E. and KENNEDY, P. J., 1987, Sensitivity of the biosphere–atmosphere transfer scheme (BATS) to the inclusion of variable soil characteristics. *Journal of Climate and Applied Meteorology*, **26**, 341–362.

WINKER, D. M. and OSBORN, M. T., 1992a, Airborne lidar observations of the Pinatubo volcanic plume. *Geophysical Research Letters*, **19**, 167–170.

WINKER, D. M. and OSBORN, M. T., 1992b, Preliminary analysis of observations of the Pinatubo volcanic plume with a polarization-sensitive lidar. *Geophysical Research Letters*, **19**, 171–174.

WOOD, E. F., 1991, *Land Surface–Atmosphere Interactions for Climate Modelling* (Dordrecht: Kluwer).

WOODBURY, G. E. and MCCORMICK, M. P., 1986, Zonal and geographical distribution of cirrus clouds determined from SAGE data. *Journal of Geophysical Research*, **91**, 2775–2785.

WOODS, J. D., 1983, Climatology of the upper boundary layer of the ocean. In *Large-scale oceanographic experiments in the WCRP. Papers presented at the JSC/CCCO Study Conference, held in Tokyo on 10–21 May 1983*, WCRP Publication Series, Volume 2, No. 1, pp. 147–179.

WOODS, J. D., 1984, The upper ocean and air–sea interaction in global climate. In *The Global Climate*, edited by J. T. HOUGHTON (Cambridge: Cambridge University Press), pp. 141–187.

WOODS, J. D., 1985, The World Ocean Circulation Experiment. *Nature*, **314**, 501–511.

WOODS, J. D., 1991, Global ocean observing and climate forecasting. *Science in Parliament*, **48**, 4–10.

WOODWARD, W. A. and GRAY, H. L., 1993, Global warming and the problem of testing for trend in time series data. *Journal of Climate*, **6**, 953–962.

WOODWELL, C., 1986, Global warming: and what we can do about it. *Amicus Journal*, **8**, Fall, 8–12.

WOODWELL, G. M., 1989, The warming of the industrialized middle latitudes, 1985–2050: causes and consequences. *Climate Change*, **15**, 31–50.

WORLD CLIMATE RESEARCH PROGRAMME, 1985, *First Implementation Plan for the World Climate Research Programme*. WCRP Publication Series, No. 5.

WORLD METEOROLOGICAL ORGANIZATION (WMO), 1983, *Report of the JSC/CAS Meeting of Experts on Detection of Possible Climate Change, Moscow, 3–6 October 1982*, edited by W. W. KELLOGG and R. D. BUJKOV (Geneva: WMO).

WORLD METEOROLOGICAL ORGANIZATION (WMO), 1985a, Report of the First Session of the JSC Scientific Steering Group on Land Surface Processes and Climate (Geneva, 21–25 January 1985). *World Climate Papers*, No. 96 (WMO/TD-N43), ICSU-WMO, Geneva.

WORLD METEOROLOGICAL ORGANIZATION (WMO), 1985b, Report of the Third Session of the JSC/CCCO Working Group on Satellite Observing Systems for Climate Research, Madison, USA, 29 April to 2 May 1985. *World Climate Papers*, No. 105 (WMO/TD-N71).

WORLD METEOROLOGICAL ORGANIZATION (WMO), 1986, Report of the International Conference on the Assessment of the Role of Carbon Dioxide and of Other Greenhouse Gases in Climate Variations and Associated Impacts, Villach, Austria, 9–15 October 1985. WMO Technical Note, No. 661.

World Meteorological Organization (WMO), 1987a, Report of the Fourth Session of the JSC/CCCO Working Group on Satellite Observing Systems for Climate Research, held in Tokyo, Japan, on 20–24 October 1986. *World Climate Papers*, No. 124, Appendices.

World Meteorological Organization (WMO), 1987b, Report of the Workshop on Space Systems Possibilities for a Global Energy and Water Cycle Experiment. In *World Climate Papers*, No. 137, pp. 1–38.

World Meteorological Organization (WMO), 1988a, Concept of the Global Energy and Water Cycle Experiment. Report of the JSC Group on GEWEX (Montreal, Canada 8–12 June 1987 and Pasadena, USA, 5–9 January 1988). WCRP-5 (WMO/TD-N215), Geneva, Annexes.

World Meteorological Organization (WMO), 1988b, Modelling the sensitivity and variations of the ocean–atmosphere system. Report of a Workshop at the ECMWF (11–13 May 1988). WCRP-15 (WMO/TD-N254), Geneva.

World Meteorological Organization (WMO), 1988c, Report of the Fourth Session of the CAS/JSC Working Group on Numerical Experimentation, Downsview, Ontario, Canada, 26–30 September 1988 (WMO/TD-N278).

World Meteorological Organization (WMO), 1988d, WMO/IOC Inter-Governmental TOGA Board Report of the First Session (Geneva, 2–6 November 1987). WMO Special Report (WMO/TD-N216), Geneva, Appendices.

World Meteorological Organization (WMO), 1988e, The global description. WOCE Core Project 1 Planning Meeting, Washington DC, 10–14 November 1986. WCRP-2 (WMO/TD-N205), WMO, Geneva.

World Meteorological Organization (WMO), 1988f, Sea ice and climate. WCRP-18 (WMO/TD-N272), WMO, Geneva, Annexes.

World Meteorological Organization (WMO), 1988g, Water resources and climatic change; sensitivity of water-resource systems to climate change and variability. WCAP-4 (WMO/TD-N247), Geneva.

World Meteorological Organization (WMO), 1989a, Report of the Third Session of the Working Group on Sea Ice and Climate (Oslo, Norway, 31 May–3 June 1988). WCRP-18 (WMO/TD-N272), Geneva, Annexes.

World Meteorological Organization (WMO), 1989b, Land surface processes and climate. Third Session of the JSC Working Group, Manhatten, Kansas, 29 June–3 July 1987. WCRP-8 (WMO/TD-N232), Geneva, Appendices.

World Meteorological Organization (WMO), 1989c, Proceedings of International WOCE Scientific Conference, held in Paris on 28 November–2 December 1988, WCRP-21 (WMO/TD-N295).

World Meteorological Organization (WMO), 1991a, Global Energy and Water Cycle Experiment (GEWEX). Report of the First Session of the JSC Scientific Steering Group for GEWEX held in Pasadena, USA on 7–10 February 1989. WCRP-25 (WMO/TD-N321), ICSU-WMO, Geneva.

World Meteorological Organization (WMO), 1991b, Global climate modelling. Report of Second Session of WCRP Steering Group and Global Climate Modelling held in Bristol, UK on 18–20 November 1991. WCRP-7 (WMO/TD-N482), Geneva.

World Meteorological Organization (WMO), 1991c, Global Energy and Water Cycle Experiment (GEWEX), 1991, World Climate Research Paper No. 57 (Geneva: WMO).

World Meteorological Organization (WMO), 1995, Global climate in 1994. World Meteorological Organization Publication No. 826 (Geneva: WMO).

World Ocean Circulation Experiment (WOCE), 1986, Status Report on US WOCE Planning. US WOCE Planning Report No. 3. Texas A&M University, College Station, Texas.

World Ocean Circulation Experiment (WOCE), 1989, The US Contribution to

WOCE Numerical Modelling, US WOCE Planning Report No. 14. Texas A&M University, College Station, Texas.

WRIGHT, P. B., 1987, Variations in tropical Atlantic sea surface temperatures and their global relationships. Max-Planck Institut für Meteorologie. Report No. 12, Hamburg.

WU, M. L. C. and CHANG, L. H. A., 1991, Differences in global data sets of atmospheric and surface parameters and their impact on outgoing longwave radiation and surface downward flux parameters. *Journal of Geophysical Research*, **96**, 9227–9262.

WU, M. L. C. and CHANG, L.-P., 1992, Longwave radiation budget parameters computed from ISLCP and HIRS2/MSU products. *Journal of Geophysical Research*, **97**, 10083–10102.

WUEBBLES, D. J., 1988, Global atmospheric chemistry and its role in climate change. Division of Environmental Chemistry, American Chemical Society Abstract (Los Angeles).

WUEBBLES, D. J., 1991, *Primer on Greenhouse Gases* (Boca Raton, FL: Lewis).

WUEBBLES, D. J. and EDMONDS, J., 1988, *A Primer on Greenhouse Gases*. TRO40 (Washington DC: US Department of Energy, Carbon Dioxide Division), 100 pp.

WUEBBLES, D. J., GRANT, K. E., CANNEL, P. S. and PENNER, J. E., 1989, The role of atmospheric chemistry in climate change. *APCA Journal*, **39**, 22–28.

WUEBBLES, D. J., KINNISON, D. E., GRANT, K. E. and LEAN, J., 1991, The effect of solar flux variations and trace gas emissions on recent trends in stratospheric ozone and temperature. *Journal of Geomagnetism and Geoelectricity*, **43**, 709–718.

WUEBBLES, D. J., MACCRACKEN, M. C. and LUTHER, E. M., 1984, A proposed reference set of scenarios for radiatively active atmospheric constituents. UCRL Preprint-90605. Lawrence Livermore National Laboratory, Livermore, CA.

WYLIE, D. P. and MENZEL, W. P., 1989, Two years of cloud cover statistics using VAS. *Journal of Climate*, **2**, 380–391.

YENTSCH, V., 1985, A coupled atmosphere–ocean climate model with ice and cloud feedbacks, Research Activities in Atmosphere and Ocean Modelling, Report No. 8 (WMO/TD-N63), Geneva, pp. 6.1–6.3.

ZANDER, R., RINSLAND, G. P., FARMER, C. B. and NORTON, R. H., 1987, Infrared spectroscopic measurements of halogenated source gases in the stratosphere with the ATMOS instruments. *Journal of Geophysical Research*, **92**, 9836–9850.

ZHAO, Z. C. and KELLOG, W. W., 1988, Sensitivity of soil moisture to doubling of carbon dioxide in climate model experiments. Part II: the Asian monsoon region. *Journal of Climate*, **1**, 367–378.

ZHONG, W., HAIGH, J. D. and PYLE, J. A., 1993, Greenhouse gases in the stratosphere. *Journal of Geophysical Research*, **98**, 2995–3004.

ZUBAKOV, V. A., 1988, Paleoclimatic stages and changes: content and causes. *Trudy Gosudarstvennogo Gidrologicheskogo Instituta*, No. 330, 69–104.

ZUEV, V. E. and KABANOV, M. V., 1987, *Optics of Atmospheric Aerosol* (Leningrad: Gidrometeoizdat).

ZWALLY, H. J., COMISO, J. C. and WALSH, J. E., 1991, Variability of Antarctic sea ice. *International Conference on the Role of the Polar Regions in Global Change. Proceedings of a Conference held June 11–15, 1990, at the University of Alaska Fairbanks*, edited by G. WELLER, C. L. WILSON and B. A. B. SEVERIN (Fairbanks: Geophysical Institute and Center for Global Change and Arctic Systems Research), p. 22.

ZWALLY, H. J., PARKINSON, C. L. and COMISO, J. C., 1983, Variability of Antarctic sea ice and $CO_2$ change. *Science*, **220**, 1005–1012.

ZWIERS, F. W. and BOER, G. J., 1987, A comparison of climates simulated by a general circulation model when run in the annual cycle and perpetual modes. *Monthly Weather Review*, **115**, 2626–2644.

# Index